marginal productivity of money, 582–83, 584, 599
operating expenses, 52
phoning customers on hold, 514–15
point elasticity function, 274
price-demand function, 105, 107, 193–94, 207–8, 229, 254, 273, 290, 328, 348, 363, 515
 for all-silk power ties, 259
 for collectible doll, 231–32
 for glazed presidential inauguration mugs, 259
 for mini picture frames, 259
 for new book, 219
 for new bookshelf, 219
 for number of DVD players sold, 261
 for oranges, 260–61
 for pizza, 448
 for single-occupant tent, 259
 for spark plug, 232
price function, 207, 218, 219
price of mountain bikes, 166–67
price setting, 270
pricing, demand-based, 345
producers' surplus, 478, 518
production
 of auto antitheft devices, 231
 average value of, 591–92, 601
 daily cost schedule for, 106
 maximizing, 577–80, 582–84, 599, 600
 optimal increase in, 579–80, 584, 600
production function, 559
 Cobb-Douglas, 532, 533, 559–60
production runs, 343
profit
 calculation of, 463
 daily, 328, 342, 346, 347, 600
 marginal, 304, 385, 395–96, 434, 489
 maximizing, 325–27, 329, 331, 342, 570–71, 599, 600
 monthly, 207, 330–31, 599
 rate of change of, 499–500
 over time, 517
 total, 363
 from two products, 597
 weekly, 597, 599
profit function, 149, 159–60, 207, 289, 303, 518, 532, 559
 annual, 213
 average, 315–16
 marginal, 316
 recovering, 360–61, 363

from two types of p
related rates in, 256–57
research and development expenditures, 108
revenue, 54, 290, 330, 599
 daily, 254–55, 342, 347
 as function of demand, 345
 marginal, 285–86, 384, 417, 552–53, 559
 maximizing, 566–67, 570–71, 599
 maximum, 330, 331, 342
 monthly, 260, 273, 330–31
 percentage error, 216
 price adjustment to increase, 268
 price setting to maximize, 270
 rate of change of, 499, 500, 519
 in solid waste management industry, 240
 from two products, 597
 weekly, 315, 331, 597, 599
 yearly, 559
revenue function, 107–8, 208, 220, 416, 532
 recovering, 363
 from two types of product, 532
sales
 annual retail, 221
 calculation of, 52
 customer response to, 78, 108
 of import cars, 80
 monthly, 171–72, 187
 percentage error, 220
 rate of, 405, 435
 rate of change of, 358–60, 499
 of sunglasses, 39
sales after advertising, 299, 304, 462–63, 516
 in two media, 531, 558, 597
sales projection percentage error, 216
supply function, 468–69, 477, 518
U.S. exports, 149
U.S. imports, 137, 149

Chemistry
radioactive decay, 428

Communication
cellular phone service subscribers, 272
commercial FM radio stations, 272
international phone service billing rate of change, 435
phoning customers on hold, 514–15
rumor propagation, 78–79

Computers
capital value of leased game, 513
clock speed of CPU, 187
learning new system, 429
web site growth rate, 401–2

Construction
of box
 maximizing volume of, 334–35, 341, 349
 minimizing material for, 600
of enclosure, 348
 for pet dog, 322–24, 329, 572
of fencing, 341
 minimizing cost of, 583
of garden, 329
home building rate, 396
of parking lot, 583
of playground, 329
of pool and walkway, 348
volume of pyramid, 597
of window, 349

Crime and criminology
aggravated assaults, 40
arson arrests, 52
conviction rate, 52, 53
drug abuse arrest rate, 363
drug arrests, 52–53
prison population
 rate of change function for, 407
prison term length, 514

Demographics. *See also* Actuarial; Population
births, 300–301, 305
deaths, 304
 infant, rate of change in, 406
 postneonatal, 329
people living alone, 198
poverty threshold, 216

Distance. See Height

Economics. *See also* Business
average sale price for home, 518
average wage, 461–62
civilians employed by executive branch, 311–13
defense budget, 348
demand functions, 271
federal debt, 305
food expenditures by household, 519

(continued on back cover)

Brief Calculus

Brief Calculus

The Study of Rates of Change
Updated Edition

Bill Armstrong
Lakeland Community College

Don Davis
Lakeland Community College

PRENTICE HALL
Upper Saddle River, New Jersey 07458

Editor in Chief: Sally Yagan
Executive Project Manager: Ann Heath
Assistant Vice President of Production and Manufacturing: David W. Riccardi
Executive Managing Editor: Kathleen Schiaparelli
Senior Managing Editor: Linda Mihatov Behrens
Production Editor: Bob Walters
Manufacturing Buyer: Alan Fischer
Manufacturing Manager: Trudy Pisciotti
Marketing Manager: Patrice Lumumba Jones
Marketing Assistant: Vince Jansen
Director of Marketing: John Tweeddale
Development Editor: Tony Palermino
Editor in Chief, Development: Carol Trueheart
Associate Editor, Mathematics/Statistics Media: Audra J. Walsh
Editorial Assistant/Supplements Editor: Joanne Wendelken
Art Director: Maureen Eide
Assistant to the Art Director: John Christiana
Interior Designer: Jill Little
Cover Designer: Daniel Conte
Art Editor: Grace Hazeldine
Art Manager: Gus Vibal
Director of Creative Services: Paul Belfanti
Photo Editor: Beth Boyd
Cover Photo: Pete Turner / The Image Bank
Art Studio: Academy Artworks

© 2003 by Prentice-Hall, Inc.
Upper Saddle River, New Jersey 07458

Printed in the United States of America
10 9 8 7 6 5 4 3 2 1

ISBN: 0-13-101987-2

Prentice-Hall International (UK) Limited, *London*
Prentice-Hall of Australia Pty. Limited, *Sydney*
Prentice-Hall of Canada Inc., *Toronto*
Prentice-Hall Hispanoamericana, S.A., *Mexico*
Prentice-Hall of India Private Limited, *New Delhi*
Prentice-Hall of Japan, Inc., *Tokyo*
Pearson Education Asia Pte. Ltd.
Editora Prentice-Hall do Brasil, Ltda., *Rio de Janeiro*

Contents

Preface

Audience and Prerequisites

Brief Calculus: The Study of Rates of Change is intended for a one or two-term calculus course for students majoring in business, economics, social or life sciences. Students who have completed an appropriate college algebra or precalculus course are prepared to study the topics presented in this textbook. The text is designed for both two-year and four-year schools.

Style and Approach

In contrast to many brief and applied calculus textbooks on the market, our book is written for the **student.** Our conversational writing style explains the mathematics clearly and simply and employs real data from the Internet to motivate and underscore the relevance of calculus in a wide range of applications. The book draws from the best ideas of reform mathematics, yet is built on a tradition of solid mathematics. As reviewer Rene Barrientos from Miami-Dade Community College put it, "This is the first post-reform calculus text that emphasizes applications without giving up the symbolic manipulations and coverage required of a rigorous calculus course."

The core concepts of calculus are introduced in applied settings using the concept commonly known as the Rule of Three (numerical, graphical, and algebraic). The Rule of Three is one tool we use to aid students' understanding of new ideas. The exploration of concepts and conjectures using graphing calculator technology has been seamlessly incorporated into examples and exercises where appropriate. We recognize that the graphing calculator is only a tool to aid in the understanding of mathematics. Nowhere does this technology overshadow the mathematics; it simply augments the mathematics and allows the power and relevance of calculus to shine through.

Content Highlights

Applications

The applications in *Brief Calculus: The Study of Rates of Change* are the heart of our textbook. Reviewers have lauded the quantity, quality, and variety of applications in our text calling them superior to that of competitors. Application examples and exercises usually end with the phrase "...and interpret" so that students are asked not only to determine the numerical solution to an application, but also to write the meaning of the solution in the context of the application itself.

Many application models such as U.S. Imports from China, the number of cases of tuberculosis in the U.S., and the percentage of 3 to 5 year olds enrolled in preschool are derived from real data gathered from sources such as the U.S. Census Bureau web site and are indicated by the "On the Web" icon. We incorporate real life applications taken from business and economics, the biological and life sciences, and the social and physical sciences throughout the entire text.

Rate of Change Theme

In Chapter 1, the basic algebraic and transcendental functions are reviewed in a concise, yet comprehensive manner. These functions are used to introduce the core concept of Chapter 1, that the average rate of change of a function over an interval is equivalent to the slope of a secant line over an interval. Appropriate units and proper interpretation of solutions are stressed. For example, a solution to an application may read, "This means during the period from 1987 to 1991, U.S. imports from China increased at a rate of 3.18 billion dollars per year." By introducing this idea in Chapter 1, we create a smooth transition to computing the instantaneous rate of change and the derivative in Chapter 2. The remaining chapters continue to emphasize the rate of change theme, along with the importance of correct units and a reasonable interpretation of the solution. If any student who uses *Brief Calculus: The Study of Rates of Change* is asked, "What is calculus?", the student will inevitably proclaim "It is the study of rates of change!"

To supplement the rate of change theme, the differential is introduced in Chapter 3 and is used as a mathematical tool to introduce new topics in later sections. After a simple introduction and explanation, the differential is used to introduce linear approximations, marginal analysis, and for measuring rates and errors. In Chapter 4, the differential is used to derive the elasticity of demand formula.

Rule of Three (plus one)

Concepts such as functions, rates of change, limits, derivatives, marginal analysis, optimization, integrals, differential equations, and partial derivatives are analyzed numerically, graphically, and algebraically. In addition, we stress the verbal approach through an emphasis on interpretation of solutions. The Rule of Three (plus one) is particularly powerful in Chapter 2 when limits and the derivative are introduced.

Graphs and Art

Our experience in the classroom indicates that many students at the brief calculus level are visual learners. Because of this observation, our text contains an abundance of graphs, pictures, charts, and tables. This visual approach acts to reinforce the mathematical concepts of calculus and to show that many real life applications begin with numerical data. Moreover, the charts and tables visually demonstrate to students that they are working frequently with real data taken from genuine sources.

Exercises

For many instructors, the exercise sets in a textbook are one of the most important components. We have taken great care to provide exercises that meet user demand in terms of quantity and quality. With greater than 3500 exercises from which to choose, we believe instructors and students will have no problem finding the types of problems they need ranging from routine skill and practice type problems to multi-step skill and concept based applications. We use a combination of real data-based problems, with their inherent complexities, and realistic problems that offer beneficial, but tidier solutions.

The exercises have been carefully written in matching odd-even pairs and are graded by level of difficulty. Care was given to write the exercises and solutions in a style and approach that is consistent with the text. Each exercise has been scrutinized in every draft of the manuscript to insure accuracy and appropriateness.

Chapter Features

Chapter Openers

The first page of each chapter includes a photo and a pair of graphs that foreshadow the fundamental ideas to be presented in the chapter. "What We Know" reiterates what information has been learned in previous chapters and "Where Do We Go" tells what new concepts will be studied. Taken together, these features help create a road map to guide students through the book and underscore the connections between topics.

Flashbacks

Selected functions and examples used earlier in the textbook are revisited to introduce new concepts and are denoted by the Flashback icon, . The Flashback concisely reviews an example from a previous section and then extends the problem by considering other questions that may be asked. In this manner, new topics are motivated in a natural way. Moreover, the Flashback often reviews skills and concepts from previous chapters that are needed. We believe that his pedagogical technique of using functions and applications previously introduced allows students to concentrate on new concepts using familiar applications.

On the Web

Many of the rich and varied applications in the textbook have been researched on the Internet. The "On the Web" icon, , denotes applications which use models based on data gathered from the Internet. This feature impresses on students the fact that calculus can be applied to real world problems. We have performed extensive research to insure that we have included applications from a variety of disciplines including problems taken from business and economics, the social sciences, the biological and life sciences, and the physical sciences.

Interactive Activities

Extensions to worked out examples are denoted by the "Interactive Activity" icon, . Many Interactive Activities are used to examine the problem from another perspective using the Rule of Three, while others explore additional properties of recently introduced topics. Others ask the student to do an exploration and make a conjecture to a completed example. Interactive Activities may serve many purposes: instructors may assign them as critical thinking exercises, they may be used as a springboard for classroom discussion, or may provide a vehicle for a collaborative activity. Solutions to selected Interactive Activities are given on the textbook's companion web site (www.prenhall.com/armstrong).

Checkpoints

At strategic points in each section, examples are followed by an exercise denoted with the "Checkpoint" icon, ✓. Each checkpoint asks students to work a particular, odd-numbered problem in the exercise set and helps to insure that a recently introduced skill or concept is understood. We have carefully chosen a parallel problem that requires a similar solution process to encourage students to check their grasp of the concept or skill. This pedagogical tool promotes better interaction between the text and student and encourages students to develop good study habits. Students who make use of the checkpoints will learn to take ownership of the course material.

Technology Notes

Additional tips and instructions for using graphing calculators are indicated with the "Technology Notes" icon, . These notes do not give keystroke commands, instead they offer tips based on common questions that students may have. Some Technology Notes refer to the **online calculator manuals** found at the textbook's companion web site (www.prenhall.com/armstrong).

From Your Toolbox

Whenever a previously introduced key definition, theorem, or property is needed, it is quickly reviewed and denoted by the "From Your Toolbox" icon, . This feature allows students to stay on task with the topic at hand without having to interrupt their reading to flip back to review previous material.

Notes

Immediately following many definitions, theorems, or properties, brief "Notes" are included to clarify a mathematical idea verbally, and to provide additional insights to help students understand the material.

Section Projects

At the end of each section, a *Section Project* presents a series of questions that ask students to explore the idea presented. We designed these questions to challenge, rather than discourage, the student. Many of the projects provide real data collected from the Internet and ask students to use the regression capabilities of their calculator to produce a model for the data, then to apply the recently introduced calculus concepts to the model. Instructors may use these section projects as a standard hand-in assignment or for a collaborative activity.

Supplements

Student's Solution Manual (0-13-085882-X)

Written by Matthew Hudock, Saint Philips College, San Antonio TX. This booklet contains complete, worked out solutions to all of the odd numbered exercises and review problems in the text.

Companion Web Site

This site (www.prenhall.com/armstrong) is designed to complement the text by offering a variety of teaching and learning resources including: a list of chapter objectives, a readiness quiz for each chapter to help students assess their preparedness for the chapter contents, solutions to many of the Interactive Activities, a set of destinations with links to other course related sites, the online graphing calculator manuals referenced in the Technology Notes, and a bulletin board for submitting and answering questions.

Instructor's Solution Manual (0-13-085885-4)

Written by Matthew Hudock, Saint Philips College, San Antonio TX. This booklet contains complete, worked out solutions to all even-numbered exercises and review problems in the text.

Test Item File (0-13-085881-1)

Written by Laurel Technical Service, Inc. This volume contains hardcopy of the test items available in PH Custom Test.

PH Custom Test: Windows (0-13-040295-8) Macintosh (0-13-040297-4)

PH Custom Test is a menu-driven random test generator available on either a Windows or Macintosh platform. The system incorporates a unique editing function that allows the instructor to enter additional problems, or alter existing problems in the test bank using a full set of mathematical notation. The test system offers free-response, multiple-choice, and mixed exams. An almost unlimited number of quizzes, review exercises, and chapter tests may be generated quickly and easily. The system will also save time by producing answer keys, student worksheets, and a gradebook for the instructor, if desired

ACKNOWLEDGMENTS

We owe a debt of gratitude to many individuals who helped us shape and refine *Brief Calculus: The Study of Rates of change.* Our first draft reviewers included: Martin Bonsangue, California State University at Fullerton, Fred Bakenhus, St. Philips College, Biswa Datta, Northern Illinois University, Matthew Hudock, St. Philips College, Anthony Macula, SUNY Geneseo, and Thomas Ordayne, University of South Carolina at Spartanburg. After thoughtful revision, we sought guidance from Rene Barrientos, Miami-Dade Community College, Mark Burtch, Arizona State University, Karabi Datta, Northern Illinois University, Adrienne Goldstein, Miami-Dade Community College, John Grima, Glendale

Community College, Michael Kirby, Tidewater Community College, Zhuangyi Lui, University of Minnesota at Duluth, and Martha Pratt, Mississippi State University. Our final panel of reviewers included Mark Burtch and Peter Casazza, Univ. of Missouri at Columbia.

We owe special thanks to Tony Palermino. Tony served not only as our Developmental Editor, but also our mentor, coach, advisor, and advocate. Tony's persistent and top quality professional work was essential in preparation of the manuscript. Without Tony's guidance, our text would not be where it is today. Matthew Hudock deserves thanks for his analysis and meticulous reviews and for preparation of exercise set solutions.

We thank Rollie Santos, Ph.D., from the Department of Economics at Lakeland Community College, who acted as advisor in many of the applications in business and economics; and Sue Hill, Ph.D., of the Department of Biology at Lakeland Community College, who provided materials and insight into many biological and life science applications. Teaching assistants Jamie Smolko and Denise Kerr were very helpful in the development of solutions to exercises in early drafts of the manuscript. The Lakeland Community College Applied Calculus students piloted the text in the manuscript and page proof phases and provided important student feedback.

Many people involved in the management and production of the text deserve recognition. We are grateful to Sally Yagan and Paige Akins for believing in us, Ann Heath for her scheduling and crisis management skills, Bob Walters for his formidable production abilities, Kathy Boothby Sestak for her can-do attitude, and to Patrice Jones for his innovative marketing techniques. We also thank the Prentice Hall sales staff for their sales efforts and enthusiasm. We appreciate the quality services of Laurel Technical Services who did a first-rate job in writing review and test bank exercises, and of Academy Artworks for their preparation of the graphs, tables, and art in the text, and Sara Beth Newell and Joanne Wendelken for copying the manuscript and routing our numerous telephone calls.

Finally, we owe personal thanks to our families. Our wives, Lisa and Melissa, were supportive and patient during the entire process. To our children Austin and Dylan; Randy, Rusty, and Ronnie who understood why sometimes Dad could not come out and play in the backyard.

Bill Armstrong
Don Davis

OVERVIEW

CHAPTER OPENERS

Each chapter begins with an **outline** and a **pictorial essay** that provides a visual orientation to the chapter contents.

CHAPTER 2

Limits, Instantaneous Rate of Change, and the Derivative

The release of CFC's and other pollutants can cause smog.

The release of CFC's into the atmosphere by the U.S. from 1989–1995 can be modeled by $f(x) = -1.2x^3 + 13.37x^2 - 69.65x + 328.71$.

In 1994, the amount of CFC's released by the U.S. into the atmosphere was decreasing at a rate of $38.81 \frac{\text{thousand metric tons}}{\text{year}}$.

What We Know

In Chapter 1, we reviewed algebra and also learned about an important rate of change called an *average rate of change*. We saw how an average rate of change is equivalent to the slope of a secant line over an interval.

Where Do We Go

In this chapter, we will see how the limit concept is used to introduce a new type of rate of change called an *instantaneous rate of change*. We will see how an instantaneous rate of change is equivalent to the slope of a tangent line at a specific point.

111

What We Know reviews material learned in the previous chapter.

Where Do We Go introduces the major ideas that will be covered in this chapter. Taken together, they provide a roadmap to connect the topics and ideas presented in the book.

---

228   Chapter 4 • Additional Differentiation Techniques

 CHECKPOINT 3   Now work Exercise 33.

Applications

In our first application, we apply the Generalized Power Rule to a rational exponent function.

EXAMPLE 4   **Applying the Generalized Power Rule to Rational Exponent Functions**

The death rate caused by heart disease in the United States can be modeled by

$$f(x) = 336.18(x + 1)^{-0.06}, \qquad 0 \le x \le 15$$

where x represents the number of years since 1980 and $f(x)$ represents the death rate (measured in deaths per 100,000 people) caused by heart disease. Determine $f'(x)$. Evaluate $f'(2)$ and interpret.

SOLUTION

Using the Generalized Power Rule, we get

$$f'(x) = \frac{d}{dx}[336.18(x + 1)^{-0.06}]$$
$$= (336.18) \cdot (-0.06)(x + 1)^{-0.06-1} \cdot (1)$$
$$= -20.1708(x + 1)^{-1.06}$$

Notice that $x = 2$ corresponds to the year 1982. Continuing, we have

$$f'(2) = -20.1708(3)^{-1.06} \approx -6.29$$

So in 1982 the death rate caused by heart disease was decreasing at a rate of about

CHECKPOINTS

At strategic points in each section, examples are followed by a **checkpoint** icon. Each checkpoint tells the student to work an odd-numbered problem in the exercise set that reinforces the example just presented. Working the checkpoints is a good study habit that will result in better understanding of the material.

EXAMPLE 5 — Interpreting the Secant Line Slope

The U.S. imports from China for the years 1987 to 1996 can be modeled by

$$f(x) = 0.32x^2 + 1.64x + 3.98, \qquad 1 \le x \le 10$$

where x represents the number of years since 1986 and $f(x)$ represents the dollar value, in billions, of goods imported.

(a) Make a table of function values for $x = 1, 2, 3, \ldots, 10$. Use these values when calculating parts (b) and (c).

(b) Determine the average rate of change in U.S. imports from China from 1987 to 1991 and interpret. Include appropriate units.

(c) Determine the average rate of change in U.S. imports from China from 1991 to 1996 and interpret. Include appropriate units.

(d) Compare the results from parts (b) and (c).

SOLUTION

(a) A numerical table of values for the model is shown in Table 1.3.1.

TABLE 1.3.1

x	1	2	3	4	5	6	7	8	9	10
$f(x)$	5.94	8.54	11.78	15.66	20.18	25.34	31.14	37.58	44.66	52.38

(b) First we see that the year 1987 corresponds to $x = 1$ and the year 1991 corresponds to $x = 1991 - 1986 = 5$. This means that the increment in x is $\Delta x = 5 - 1 = 4$. So we need to compute the difference quotient

$$m_{\text{sec}} = \frac{f(x + \Delta x) - f(x)}{\Delta x} = \frac{f(1 + 4) - f(1)}{4} = \frac{f(5) - f(1)}{4}$$

The values from Table 1.3.1 give us

$$m_{\text{sec}} = \frac{20.18 - 5.94}{4} = \frac{14.24}{4} = 3.56 \quad \text{or} \quad 3.56 \, \frac{\text{billion dollars}}{\text{year}}$$

This means that during the period from 1987 to 1991, U.S. imports from China increased at an average rate of 3.56 $\frac{\text{billion dollars}}{\text{year}}$. Notice that the amount of imports increased, since the secant line slope is

---

EXAMPLE—ON THE WEB

Every example in the book is titled and features a fully worked-out solution. Many of the applications in the textbook have been researched on the Internet. The **On the Web** icon denotes applications which use models based on real data gathered from the Internet.

FLASHBACK

Selected functions and examples used earlier in the textbook are revisited to introduce new concepts and are set off as a **Flashback.** Revisiting previously introduced material allows students to concentrate on the new concept and to see how more can be learned about a problem by investigating it with a different technique.

Flashback — U.S. IMPORTS FROM CHINA REVISITED

In Section 1.3, the U.S. imports from China for the years 1987 to 1996 were modeled by

$$f(x) = 0.32x^2 + 1.64x + 3.98,^* \qquad 1 \le x \le 10$$

where x represents the number of years since 1986 and $f(x)$ represents the dollar value, in billions, of goods imported. The graph of the model is shown in Figure 2.3.1.

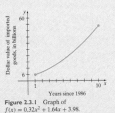

Figure 2.3.1 Graph of $f(x) = 0.32x^2 + 1.64x + 3.98$.

Determine the average rate of change for U.S. imports from China from 1990 to 1993 and interpret.

Flashback Solution

Recall that the average rate of change is simply equivalent to the *slope of the secant line* through two points. Since 1990 corresponds to an x-value of 4 and 1993 corresponds to an x-value of 7, we need to determine the corresponding y-values for these x-values to get the two points that the secant line passes through. We determine the y-value when $x = 4$ by simply substituting $x = 4$ into our model as follows:

$$y = f(x) = 0.32x^2 + 1.64x + 3.98$$
$$= 0.32(4)^2 + 1.64(4) + 3.98 = 15.66$$

We now know one point that the secant line passes through is (4, 15.66). Similarly, when $x = 7$, we get $y = 31.14$. Hence, the second point that the secant line passes through is (7, 31.14). See Figure 2.3.2. Now the secant line slope through these two points is simply

$$m_{\text{sec}} = \frac{y_2 - y_1}{x_2 - x_1} = \frac{31.14 - 15.66}{7 - 4} = 5.16$$

So our answer is that the average rate of change for U.S. imports from China from 1990 to 1993 was 5.16 $\frac{\text{billion dollars}}{\text{year}}$. Recall from Chapter 1 that this means that over the period 1990 to 1993 U.S. imports from China increased 5.16 $\frac{\text{billion dollars}}{\text{year}}$ on average.

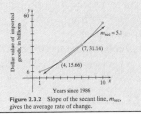

Figure 2.3.2 Slope of the secant line, m_{sec}, gives the average rate of change.

---

*The data and model given in exercises and examples preceded by the "On the Web" icon are based on information gathered at the U.S. Census Bureau web site. These exercises and examples appear throughout this textbook.

There are a couple of important items in the Flashback that we must review:

- Since we found a *rate*, we must include *appropriate units*.
- The secant line has a positive slope indicating that from 1990 to 1993 U.S.

Extensions to worked-out examples appear as **Interactive Activities.** Many ask the student to complete the Rule of Three (plus one) or explore additional properties of recently introduced topics. Solutions to selected Interactive Activities appear on the companion Web site at *www.prenhall.com/armstrong.*

✓ CHECKPOINT 3

Now work Exercise 39.

EXAMPLE 5 | **Analyzing a Limit Involving $\frac{0}{0}$**

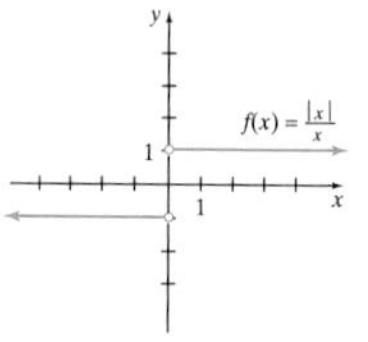

Determine $\lim\limits_{x \to 0} \frac{|x|}{x}$.

SOLUTION

If we try substituting, we get

$$\lim\limits_{x \to 0} \frac{|x|}{x} = \frac{0}{0}$$

By letting $f(x) = \frac{|x|}{x}$, we can use a table and a graph to determine this limit. See Table 2.1.4 and Figure 2.1.12.

$f(x) = \frac{|x|}{x}$

Figure 2.1.12

Interactive Activity

Recall that $|x|$ is defined to be

$$|x| = \begin{cases} x, & \text{if } x \geq 0 \\ -x, & \text{if } x < 0 \end{cases}.$$

Use this definition to algebraically verify that $\lim\limits_{x \to 0} \frac{|x|}{x}$ does not exist.

TABLE 2.1.4

	$x \to 0^-$					$x \to 0^+$			
x	-1	-0.1	-0.01	-0.001	0	0.001	0.01	0.1	1
$f(x)$	-1	-1	-1	-1		1	1	1	1

As we can see numerically and graphically, since the left-hand limit does not equal the right-hand limit, we conclude that $\lim\limits_{x \to 0} \frac{|x|}{x}$ does not exist. ✦

For Example 6, we need to recall Limit Theorem 1, which states that the limit of a constant is that constant.

EXAMPLE 6 | **Analyzing Limits Involving Two Variables**

Determine the following limits:

(a) $\lim\limits_{h \to 0}(3x + 2h)$ (b) $\lim\limits_{h \to 0} \dfrac{5xh + 2h^2}{h}$

(b) From the table and the graphs in Figures 1.6.1a and b, we see that any real number can be substituted for x, so the domain of the functions $f(x) = 2^x$ and $g(x) = \left(\frac{1}{2}\right)^x$ is $(-\infty, \infty)$.

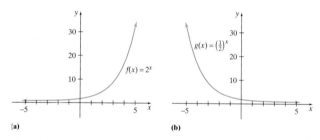

(a) (b)

Figure 1.6.1 ◼

Others ask the student to explore additional properties of recently introduced topics. Interactive Activities may be assigned as homework or may be used as a springboard for classroom discussion or group work. Some solutions appear at *www.prenhall.com/armstrong.*

Interactive Activity

For Table 1.6.1, make another column of the table that calculates the **successive ratios** of the function values by dividing each value by the preceding value. For example, for the function $g(x) = \left(\frac{1}{2}\right)^x$, the ratios would begin $\frac{16}{32}, \frac{8}{16}, \dots$. What pattern do you notice?

Notice that $f(x) = 2^x$ is increasing on its domain and as $x \to \infty, f(x) \to \infty$. Consequently, we call this kind of function an **exponential growth function.** Whereas, notice that $g(x) = \left(\frac{1}{2}\right)^x$ is decreasing on its domain and, as $x \to \infty$, $g(x) \to 0$. We call this kind of function an **exponential decay function.** (See Properties of General Exponential Functions at the top of p. 71.)

For the function $f(x) = a \cdot b^x$, the value of a can affect the end behavior. First, we see that with any value of a, $f(0) = a \cdot b^0 = a \cdot 1 = a$. This means that the y-intercept of the graph of $f(x) = a \cdot b^x$ is at $(0, a)$. To see the effect of a, compare the graphs of $f(x) = 2 \cdot 3^x$ and $g(x) = -2 \cdot 3^x$ in Figure 1.6.2.

FROM YOUR TOOLBOX

Whenever a previously introduced key definition, theorem, or property is needed, it is reviewed in **From Your Toolbox.** This feature allows the student to stay on task without having to flip back to review previous material.

From Your Toolbox

1. The *price–demand function* p gives us the price $p(x)$ at which people buy exactly x units of product.
2. The cost of producing x units of product with variable costs m and fixed costs b is given by the *cost function:*

$$C(x) = mx + b = \begin{pmatrix} \text{variable} \\ \text{costs} \end{pmatrix} x + \begin{pmatrix} \text{fixed} \\ \text{cost} \end{pmatrix}$$

   Note that since variable costs are often expressed as a function, $C(x)$ may be a higher-order polynomial function.

3. The total revenue R generated by producing and selling x units of product at price $p(x)$ is given by the *revenue function:*

$$R(x) = x \cdot [p(x)] = \begin{pmatrix} \text{quantity} \\ \text{sold} \end{pmatrix} \cdot \begin{pmatrix} \text{unit} \\ \text{price} \end{pmatrix}$$

4. The profit P generated after producing and selling x units of a product is given by the *profit function:*

$$P(x) = R(x) - C(x) = \text{revenue} - \text{cost}$$

TECHNOLOGY NOTES

The graphing calculator is incorporated into the text at appropriate junctures. Important tips and instructions for using graphing calculators are given in marginal **Technology Notes.** Many of the notes refer to the online calculator manual found on the companion Web site *www.prenhall.com/armstrong*.

TECHNOLOGY NOTE
Due to the limitations of graphing calculators, you may not see a hole in the graph at $x = 2$. Using ZDECIMAL or selecting the x-axis window so that $x = 2$ is the midpoint of the graphing interval should show the hole. See Figure 2.1.2.

Figure 2.1.2

It appears from Table 2.1.1 that if we start to the left of $x = 2$ or to the right of $x = 2$, as we allow x to approach 2, our functional values are approaching 4. We can say that,

"The limit of $\dfrac{x^2 - 4}{x - 2}$, as x approaches 2, is 4."

Using an arrow for the word *approaches* and *lim* as shorthand for the word limit, the mathematical notation for this English sentence is

$$\lim_{x \to 2} \frac{x^2 - 4}{x - 2} = 4$$

You should interpret this limit notation to mean that as x gets closer and closer to 2, from both sides of 2, $\dfrac{x^2 - 4}{x - 2}$ gets closer and closer to 4.

For a graphical perspective, let's graph $y = \dfrac{x^2 - 4}{x - 2}$. Figures 2.1.2 and 2.1.3 show the graph of $y = \dfrac{x^2 - 4}{x - 2}$, where Figure 2.1.3 is the result of utilizing the ZOOM IN command.

Notice in Figures 2.1.3b and c that we have used the TRACE command to get as close to 2 as possible from the left of 2 and from the right of 2, respectively. We now have graphical support for our numerical work. That is, numerically and graphically we believe that $\lim\limits_{x \to 2} \dfrac{x^2 - 4}{x - 2} = 4$.

$$\lim_{x \to 1} \frac{x^2 - 1}{x - 1} = 2$$

TECHNOLOGY NOTE
Many graphing calculators have a TABLE feature. Consult the on-line calculator manual at www.prenhall.com/armstrong

TABLE 2.1.2

x	0	0.9	0.99	0.999	1	1.001	1.01	1.1	2
		$x \to 1$ FROM LEFT				$x \to 1$ FROM RIGHT			
$f(x)$	1	1.9	1.99	1.999		2.001	2.01	2.1	3

Figure 2.1.4 shows the graph of $f(x) = \dfrac{x^2 - 1}{x - 1}$. In Figure 2.1.5 we have used the ZOOM IN command to capture that part of the graph where $x = 1$, and Figure 2.1.6 is the result of our ZOOM IN.

We now use the TRACE command to get as close to $x = 1$ as possible, from both sides of $x = 1$, and observe the corresponding y-values. See Figures 2.1.7a and b.

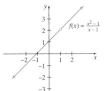
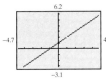
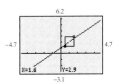

Figure 2.1.5

SECTION SUMMARY

Each section ends with a brief **Summary** that reviews the key ideas and formulas presented. A bulleted list of important functions with their formulas facilitates a quick review.

SECTION EXERCISES

Each section concludes with a comprehensive set of exercises that begins with basic skills and moves on to more conceptually challenging applications. The **Applications** feature both realistic and real data-based problems as noted by the On the Web symbol.

SUMMARY

In this section, we revisited the business functions and from them derived the marginal business functions. These functions are found by differentiation, and they determine the cost, revenue, or profit for producing one more item of a product. Then we discussed the average business functions, which were found by taking the business function and dividing by the independent variable. Finally, we discussed the marginal average business functions, which were the derivatives of the average business functions.

Important Functions
- **Marginal cost function:** $MC(x) = C'(x)$
- **Marginal revenue function:** $MR(x) = R'(x)$
- **Marginal profit function:** $MP(x) = P'(x)$

- **Average cost function:** $AC(x) = \dfrac{C(x)}{x}$

- **Average profit function:** $AP(x) = \dfrac{P(x)}{x}$

- **Marginal average cost function:** $MAC(x) = \dfrac{d}{dx}\left(\dfrac{C(x)}{x}\right)$

- **Marginal average profit function:** $MAP(x) = \dfrac{d}{dx}\left(\dfrac{P(x)}{x}\right)$

SECTION 3.2 EXERCISES

For Exercises 1–6, assume $C(x)$ is in dollars and complete the following:

(a) Determine the marginal cost function MC.
(b) For the given production level x, evaluate $MC(x)$ and interpret.

(c) Evaluate the actual change in cost by evaluating $C(x + 1) - C(x)$ and compare with the answer to part (b).

1. $C(x) = 23x + 5200$; $x = 10$
2. $C(x) = 14x + 870$; $x = 12$

SECTION 4.2 EXERCISES

In Exercises 1–10, determine the derivative for the following functions.

1. $f(x) = 5\ln x$
2. $f(x) = -8\ln x$
3. $f(x) = \ln x^6$
4. $f(x) = \ln x^4$
5. $f(x) = 4x^3 \cdot \ln x$
6. $f(x) = 12x^3 \cdot \ln x$
7. $f(x) = \dfrac{3x^5}{\ln x}$
8. $f(x) = \dfrac{12}{\ln x}$
9. $f(x) = 10 - 12\ln x$
10. $f(x) = -2 + 8\ln x$

In Exercises 11–24, determine the derivative for the following functions.

11. $g(x) = \ln(x + 7)$
12. $g(x) = \ln(2 - x)$
13. $g(x) = \ln(2x - 5)$
14. $g(x) = \ln(3x + 4)$
15. $g(x) = \ln(x^2 + 3)$
16. $g(x) = \ln(3x^3 - 11)$
17. $g(x) = \ln\left(\sqrt{2x + 5}\right)$
18. $g(x) = \ln\left(\sqrt[3]{4x + 2}\right)$
19. $g(x) = (\ln x)^6$
20. $g(x) = (\ln x)^4$
21. $g(x) = \sqrt{x} \cdot \ln\left(\sqrt{x}\right)$
22. $g(x) = 4x^5 \cdot \ln(3x^3)$
23. $g(x) = \dfrac{x^2 + 2x + 3}{\ln(x + 5)}$
24. $g(x) = \dfrac{4x^3 - x + 2}{\ln(x + 7)}$

For Exercises 25–34, determine the derivative for the following functions.

25. $f(x) = \log_{10} x$
26. $f(x) = \log_5 x$
27. $f(x) = 6\log_3 x$
28. $f(x) = 11\log_4 x$
29. $f(x) = x^2 \log_9 x$
30. $f(x) = 2x^5 \log_8 x$
31. $f(x) = \log_2(5x + 3)$
32. $f(x) = \log_5(3x + 9)$
33. $f(x) = \log_{10}\left(\dfrac{x + 3}{x^2 + 1}\right)$
34. $f(x) = \log_2\left(\dfrac{x^3}{x^2 - 1}\right)$

For Exercises 35–42, determine an equation for the line tangent to the graph of the function at the given point.

35. $f(x) = \ln x$; $(2, \ln 2)$
36. $f(x) = \ln x$; $(1, 0)$
37. $f(x) = \ln\sqrt{2x - 1}$; $(1, 0)$
38. $f(x) = \ln(3x)$; $(2, \ln 6)$

APPLICATIONS

43. A research assistant in biology finds in an experiment that at low temperatures the growth of a certain bacteria culture can be modeled by

$$f(t) = 750 + 12\ln t, \quad t \geq 1$$

where t represents the number of hours since the start of the experiment and $f(t)$ represents the number of bacteria present.

(a) Determine $f'(t)$.
(b) Evaluate and interpret $f(12)$ and $f'(12)$.

44. The city of Plantersville has enacted new zoning laws in order to curb the growth of the city's population. They find that the population can be modeled by

$$P(x) = 10,000 + 100\ln x, \quad x \geq 1$$

where x represents the number of years since the laws were adopted and $P(x)$ represents the city's population.

(a) Determine $P'(x)$.
(b) Evaluate and interpret $P(20)$ and $P'(20)$.

45. Prescription drug companies have found that the popularity of the new drug Vectrum has dwindled and can be modeled by

$$f(x) = 150 + 5\log_2 x, \quad x \geq 1$$

where x represents the number of years that the drug has been on the market and $f(x)$ represents the number of prescriptions written for the drug annually in thousands.

(a) Determine $f'(x)$.
(b) Evaluate $f'(2)$ and $f'(10)$ and interpret each.

46. The urban school district of Molisburg has started a new educational campaign in an attempt to reduce the increase in lice found in the elementary school student population. The number of children who contracted lice can be modeled by

$$g(t) = 200 + 8\log_3 t, \quad t \geq 1$$

where t represents the number of years since the new educational campaign has been enacted and $g(t)$ represents the number of students who are diagnosed with lice annually.

(a) Determine $g'(t)$.
(b) Evaluate $g'(2)$ and $g'(7)$ and interpret each.

47. The life expectancy for African-American females in the United States can be modeled by

(c) Write the equation of the tangent line at $x = 3$, and determine y on the tangent line when $x = 15$. Interpret and compare to $f(15)$.

48. The life expectancy for white females in the United States can be modeled by

$$f(x) = 75.32 + 1.29 \ln x, \qquad 1 \le x \le 26$$

where x represents the birth year since 1969 and $f(x)$ represents the life expectancy in years.
(a) Determine $f'(x)$.
(b) Evaluate and interpret $f(3)$ and $f'(3)$. Compare to part (b) in Exercise 47.
(c) Write the equation of the tangent line at $x = 3$, and determine y on the tangent line when $x = 15$. Interpret and compare to $f(15)$. Compare to part (c) in Exercise 47.

49. The annual per capita consumption of light and skim milk in the United States can be modeled by

$$f(x) = 10.12 + 2 \ln x, \qquad 1 \le x \le 16$$

where x represents the number of years since 1979 and $f(x)$ represents the annual per capita consumption of light and skim milk in gallons.
(a) Graph f in the viewing window [1, 16] by [10, 17].
(b) Determine $f'(x)$.
(c) Evaluate and interpret $f'(5)$ and compare to $f'(10)$.

50. The average expenditure for a new domestic car in the United States can be modeled by

$$f(x) = 15{,}302.93 + 1685.66 \ln x, \qquad 1 \le x \le 7$$

where x represents the number of years since 1980 and $f(x)$ represents the average expenditure for a new domestic car in dollars.
(a) Graph f in the viewing window [1, 7] by [15,000, 19,000].

SECTION PROJECT

The revenue, in billions of dollars, generated in the solid waste management industry in the United States from 1980 to 1996 are displayed in Table 4.2.2.

TABLE 4.2.2

YEAR	x	REVENUE (IN $BILLIONS)
1980	1	52.0
1990	11	146.4
1994	15	172.5
1995	16	180.0
1996	17	184.3

Source: U.S. Census Bureau web site.

(a) Use your calculator to determine a logarithmic regression model of the form

$$f(x) = a + b \ln x, \qquad 1 \le x \le 17$$

where x represents the number of years since 1979 and $f(x)$ represents the revenue generated in the solid waste management industry, in billions of dollars.

(b) Determine $f'(x)$.

SECTION PROJECT

At the end of each section, a **Section Project** presents the student with a series of questions that challenge the student to explore the ideas presented. Many of the projects provide real data collected from the Internet and ask the students to use their calculator to produce a model for the data.

CHAPTER REVIEW EXERCISES

Rounding out each chapter is a group of more than 100 review exercises that test students' understanding of all of the topics covered in the chapter. The **Section and Review Exercises** provide the user with greater than 3500 exercises in total.

CHAPTER REVIEW EXERCISES

1. For $f(x) = \dfrac{1 - \sqrt{1 - 2x - x^2}}{x}$, complete the table to numerically estimate the following.
(a) $\lim\limits_{x \to 0^-} f(x)$ (b) $\lim\limits_{x \to 0^+} f(x)$ (c) $\lim\limits_{x \to 0} f(x)$

x	−0.1	−0.01	−0.001	0	0.001	0.01	0.1
$f(x)$				?			

2. For $f(x) = (\sqrt{4 - x})^5$, complete the table to numerically estimate the following.
(a) $\lim\limits_{x \to 4^-} f(x)$ (b) $\lim\limits_{x \to 4^+} f(x)$ (c) $\lim\limits_{x \to 4} f(x)$

x	3	3.9	3.99	4	4.01	4.1	5
$f(x)$				?			

For Exercises 3 and 4, use your calculator to graph the given function. Use the ZOOM IN and TRACE commands to graphically estimate the indicated limits. Verify your estimate numerically.

3. $f(x) = x^3 - 2x$; $\lim\limits_{x \to 3.1} f(x)$

4. $f(x) = \sqrt{x} + \dfrac{1}{|x|}$; $\lim\limits_{x \to 2.5} f(x)$

For Exercises 5–10, determine the indicated limit algebraically.

5. $\lim\limits_{x \to 2} (7x^3 - 10x)$

6. $\lim\limits_{x \to -1} \dfrac{x^2 - 1}{x + 1}$

7. $\lim\limits_{x \to -3} |x - 5|$

8. $\lim\limits_{x \to 0} \sqrt{36 - 8x}$

9. $\lim\limits_{x \to 10} \dfrac{x + 10}{x^2 - 100}$

10. $\lim\limits_{x \to 2.2} (x + 9)^3$

11. Use the graph of f to find the following:

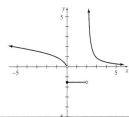

(a) $\lim\limits_{x \to -4} f(x)$ (b) $\lim\limits_{x \to 0^-} f(x)$ (c) $f(0)$
(d) $\lim\limits_{x \to 2} f(x)$ (e) $\lim\limits_{x \to 3} f(x)$ (f) $f(3)$

Determine the indicated limit in Exercises 12–14.

12. $\lim\limits_{h \to 0} (x + 2h)^2$ 13. $\lim\limits_{h \to 0} \dfrac{2x^2 h - 9h}{h}$ 14. $\lim\limits_{h \to 0} \dfrac{6x^3 h^2 + h}{h}$

15. For $f(x) = 3x^2$, find
(a) $f(2 + h)$ (b) $\lim\limits_{h \to 0} \dfrac{f(2 + h) - f(2)}{h}$

16. For $f(x) = \dfrac{9}{x}$, find
(a) $f(4 + h)$ (b) $\lim\limits_{h \to 0} \dfrac{f(4 + h) - f(4)}{h}$

17. Acme Stuffed Animals, Inc, is introducing a new line of teddy bears. The total cost of producing Scare Bear (with glowing eyes) is projected to be $C(x) = 36{,}000 + \sqrt{10{,}000x}$, where x is the number of units made and $C(x)$ is the cost in dollars.
(a) Find and interpret $C(400)$.
(b) Find and interpret $\lim\limits_{x \to 100} C(x)$.
(c) Find and interpret $AC(x) = \dfrac{C(x)}{x}$ for $x = 25$.

18. A toy speedboat moves away from a dock, and its distance from the dock is given by

$$d(t) = \begin{cases} 2t^2, & 0 \le t \le 3 \\ 12t - 18, & 3 < t \end{cases}$$

where $d(t)$ is in feet and t is in seconds.
(a) Graph d for $0 \le t \le 6$.
(b) Find $\lim\limits_{t \to 3^-} d(t)$.
(c) Find $\lim\limits_{t \to 3^+} d(t)$.
(d) Find $\lim\limits_{t \to 3} d(t)$.
(e) Describe the behavior of the toy speedboat.

19. A pastry chef in a commercial test kitchen is fine-tuning a pie-filling recipe. Colleagues acting as tasters have rated various recipes on a scale of 1 to 10. The average rating $R(s)$ appears to be a function of the sugar content s.

$$R(s) = 8 - \dfrac{s^2 - 48s + 512}{50}$$

where s is the number of tablespoons of sugar.
(a) Graph R in the viewing window [0, 40] by [0, 10].
(b) Find $R(20)$, $R(25)$, and $R(30)$.
(c) Find $\lim\limits_{h \to 0} \dfrac{R(24 + h) - R(24)}{h}$.

CHAPTER

Functions, Models, and Average Rate of Change

The Mall of America outside of Minneapolis, Minnesota, offers U.S. and Canadian consumers the ultimate shopping experience.

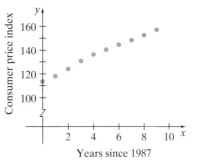

Scatterplot of the U.S. consumer price index from 1987 to 1997.

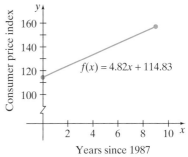

Function modeling the U.S. consumer price index from 1987 to 1997.

What We Know

We begin our study of calculus having learned the basics of algebra. These include equation solving, factoring, and the graphing of functions.

Where Do We Go

In this chapter, we will review the function concept, the properties of various types of functions, and the **average rate of change** of functions over an interval. We will see how data can be used to form or **model** functions.

SECTION 1.1   THE COORDINATE SYSTEM AND FUNCTIONS

Plotting Points

We start our study of functions by examining how they appear visually. The graph on which functions are plotted is called the **Cartesian plane.**

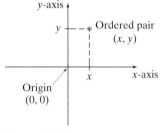

y-axis

y — Ordered pair
(*x, y*)

x   *x*-axis

Origin
(0, 0)

Figure 1.1.1

The Cartesian plane can be thought of as two number lines that are perpendicular to one another, as shown in Figure 1.1.1. The point at which the lines cross is called the **origin.** The horizontal number line, or horizontal axis, locates the values of the **independent variable,** which is usually denoted by *x*. The vertical axis locates the values of the **dependent variable,** which is usually denoted by *y*. These are commonly called the *x* and *y* axes, respectively. Points are plotted as **ordered pairs** in the form (independent variable, dependent variable), which, most of the time, are of the form (*x, y*). Before we continue, let's plot some points on a Cartesian plane. Figure 1.1.2 shows the graph of some ordered pairs. To plot the ordered pair (−3, 1), we start at the origin and move to the left 3 units and then up 1 unit. In a similar fashion, we plotted the remainder of the points shown in Figure 1.1.2.

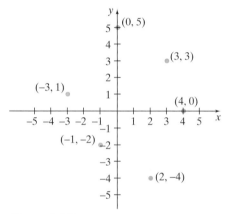

Figure 1.1.2

In the study of applied calculus, we often use tabular data as a basis for forming a mathematical model. When we plot data, we are producing what is called a **scatterplot.** Frequently, the values of the independent variable are a measure of time, primarily in years. The dependent variable then represents some phenomenon in business, life, or social science that is related to time.

EXAMPLE 1

Creating a Scatterplot

The data in Table 1.1.1 represent the U.S. imports of products from France for the years 1990 to 1994. The value of the imports is in billions of U.S. dollars. Let *x* represent the number of years since 1990, and let *y* represent the value in billions of dollars of U.S. imports from France. Construct ordered pairs to represent the data, and plot the data.

SOLUTION

To avoid dealing with large values of the independent variable *x*, it is convenient to let *x* represent the number of years since 1990 (see Table 1.1.2). The year 1990 corresponds with *x* = 0, 1991 corresponds to *x* = 1, and so on. Notice that we get

TABLE 1.1.1

YEAR	IMPORTS
1990	10.7
1991	13.3
1992	14.8
1993	15.3
1994	16.8

Source: U.S. Census Bureau web site.

TABLE 1.1.2

YEARS SINCE 1990	IMPORTS
0	10.7
1	13.3
2	14.8
3	15.3
4	16.8

these values by taking the year and subtracting 1990. This process is called **standardizing the values.** So constructing these ordered pairs gives (0, 10.7), (1, 13.3), (2, 14.8), (3, 15.3), and (4, 16.8). Using these ordered pairs, we get the scatterplot shown in Figure 1.1.3.

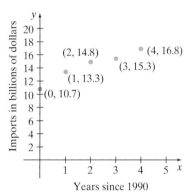

Figure 1.1.3

✓ CHECKPOINT 1

Now work Exercise 11.

Notice in Example 1 that the possible values of x came from the set {0, 1, 2, 3, 4} and the possible values of y came from the set {10.7, 13.3, 14.8, 15.3, 16.8}. These sets of numbers have special names in mathematics. The set of all possible values of the independent variable, in this case x, is called the **domain**; the set of all possible values of the dependent variable, in this case y, is called the **range.** These two sets of numbers are critical in the study of functions.

Function Notation and Evaluating Functions

Many times, a dependent relationship exists between two phenomena. The price of a concert ticket may **depend** on the popularity of the band, the cost of manufacturing may depend on the quantity manufactured, and the life expectancy of a person may depend on the year in which she or he was born. These examples exhibit a direct relationship, or correspondence, between independent variable values (price, quantity, year) and the dependent variable values (popularity, cost, life length). We write these relationships between the independent and dependent variables using **functions.**

A function f can be thought of as a process in which an independent value x in the domain is mapped to a dependent value y in the range. This is illustrated in Figure 1.1.4.

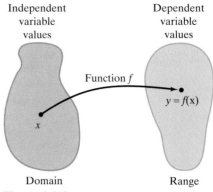

Figure 1.1.4

Function, Domain, and Range

A **function** is a rule, or a set of ordered pairs, that assigns to each element in the domain one and only one element in the range. The **domain** is the set of all possible independent variable values. The **range** is the set of all possible dependent variable values.

Functions are usually expressed by first naming the function with a letter like f or g, and then writing the independent variable letter in parentheses. For example, in the expression $f(x)$ (read "f of x"), f represents the function and $f(x)$ represents the value of f at x, where x is the independent variable. It is often convenient to represent $f(x)$ with the dependent variable y. This means that on our Cartesian plane we can now label the vertical axis with $f(x)$, $g(x)$, or any other function name that we wish to use.

Functions are not only expressed as tables and graphs, but also as mathematical expressions. For example, the function $f(x) = \frac{x+2}{5}$ means: "Take the independent value and add 2; then divide that sum by 5."

The process of "plugging" in various x-values into a function is called **evaluating** the function. To evaluate the function $f(x) = \frac{x+2}{5}$ at $x = 13$, denoted by writing $f(13)$, we substitute 13 for every x. This gives

$$f(x) = \frac{x+2}{5}$$

$$f(13) = \frac{13+2}{5} = \frac{15}{5} = 3$$

EXAMPLE 2 | **Evaluating Functions**

For the function $g(x) = -2x + 6$, evaluate $g(-2)$, $g(0)$, and $g(3)$, and write the results as ordered pairs.

SOLUTION

Evaluating the function, we get $g(-2) = -2(-2) + 6 = 4 + 6 = 10$, so we get the ordered pair $(-2, 10)$.

Now $g(0) = -2(0) + 6 = 0 + 6 = 6$. This gives us the ordered pair $(0, 6)$.

And $g(3) = -2(3) + 6 = -6 + 6 = 0$, which produces the third ordered pair $(3, 0)$.

Now let's take a look at a relationship that is not a function. Figure 1.1.5 shows the graph of relationship R. Since $(11, 3)$ and $(11, -3)$ are on the graph, R assigns both 3 and -3 to the value 11. Applying our definition of function, we see that R is not a function. A visual way to tell whether a graph represents a function is to use the **vertical line test.**

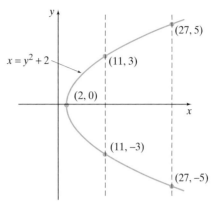

Figure 1.1.5

Vertical Line Test

If every vertical line drawn through a graph intersects the graph at only one point, then the graph represents a function.

EXAMPLE 3 | **Determining if a Graph Represents a Function**

Use the vertical line test to determine which of the graphs in Figures 1.1.6a, b, and c do not represent functions.

SOLUTION

We see in Figure 1.1.6c that vertical lines will intersect the graph at only one place, but in Figures 1.1.6a and b they will intersect at more than one point. So the first two graphs do not represent functions, while the third graph is a function.

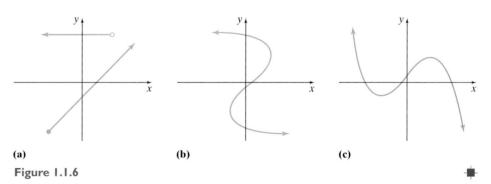

(a)   (b)   (c)

Figure 1.1.6

✓ CHECKPOINT **2**

Now work Exercise 21.

Many times, we need to express the domain and range of a function as some portion of the real number line. To write this in a convenient form, we use **interval**

notation. Let's say that we have the inequality notation $2 \le x < 5$; that is, all the x-values from 2 to 5, including 2. We can conveniently write this section of the real number line using interval notation as $[2, 5)$. Intervals with two endpoints are called **finite intervals.** If the endpoints are included, the interval is **closed**. If the endpoints are not included, the interval is **open**.

	Finite Intervals		
Number Line	**Interval Notation**	**Inequality Notation**	**Interval Type**
a b	$[a, b]$	$a \le x \le b$	Closed
a b	(a, b)	$a < x < b$	Open
a b	$(a, b]$	$a < x \le b$	Half-open
a b	$[a, b)$	$a \le x < b$	Half-open

Intervals that extend indefinitely in at least one direction are called **infinite intervals.** We use the **infinity** symbol ∞ to indicate that the numbers extend positively (or to the right on the number line) without bound. We use $-\infty$ to represent **negative infinity** to show that the numbers extend negatively (or to the left on the number line) without bound. Since ∞ and $-\infty$ are only **concepts** and not real numbers themselves, we always write these endpoints as open, using parentheses.

	Infinite Intervals		
Number Line	**Interval Notation**	**Inequality Notation**	**Interval Type**
a	$[a, \infty)$	$x \ge a$	Unbounded closed
a	(a, ∞)	$x > a$	Unbounded open
b	$(-\infty, b]$	$x \le b$	Unbounded closed
b	$(-\infty, b)$	$x < b$	Unbounded open
a	$(-\infty, \infty)$	All x in $\mathbb{R}$	Number line

Now that we have reviewed interval notation, let's return to the analysis of functions, domains, and ranges. We may think of the domain as the set of all possible values that can be *plugged in* for the independent variable of the function. The range is the set of numbers that *come out* of the function. An illustration of this *plugged-in* and *come out* process is given in Figure 1.1.7.

To find numbers that cannot be in the domain, think of the number properties learned in algebra. We cannot divide a number by zero, for example, $\frac{3}{0}$ is not defined. And we cannot take the square root of a negative number (that is, $\sqrt{-9}$ does not have a *real* number solution). (You might have heard of complex numbers, but in applied calculus we only consider real numbers.) So, generally, when we look for domain values of functions, we exclude values that can make the denominator of a fraction zero or make the expression under a square root (or any even index root) function negative. (Another restriction will arise in Section 1.7, when we study logarithmic functions.)

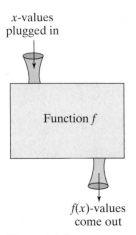

x-values
plugged in

Function f

$f(x)$-values
come out

Figure 1.1.7

EXAMPLE 4 | **Finding the Domain of a Function Algebraically**

Determine the domain of the following functions. Show the domain on a real number line and by writing in interval notation.

(a) $f(x) = \sqrt{5x - 2}$ 　　(b) $g(x) = \dfrac{\sqrt{x-2}}{x^2 - 4x}$

SOLUTION

(a) To find the domain for f, we need the x-values so that the radicand (the expression under the square root symbol) is greater than or equal to zero. As an inequality, this means that we need

$$5x - 2 \geq 0$$
$$5x \geq 2$$
$$x \geq \frac{2}{5}$$

This domain is shown on the number line in Figure 1.1.8.

In interval notation, the domain is $\left[\dfrac{2}{5}, \infty\right)$.

$\frac{2}{5}$ ← → x

Figure 1.1.8

(b) This function requires us to examine the restrictions on both the numerator and denominator. In the numerator, we see that we need values of x such that

$$x - 2 \geq 0$$
$$x \geq 2$$

For the denominator, we can factor x from each term to get

$$x^2 - 4x = x(x - 4)$$

If we set the factored form equal to zero and solve, we get

$$x(x - 4) = 0$$
$$x = 0 \quad \text{or} \quad x = 4$$

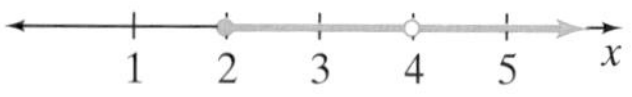

So we must also exclude the values of 0 and 4 from the domain, since they make the denominator of g zero. The resulting domain is shown on the number line in Figure 1.1.9. Since the domain of g is the x-values for which both the numerator and denominator are defined, the domain of g is $[2, 4) \cup (4, \infty)$. ∎

1　2　3　4　5 ← → x

Figure 1.1.9

✓ CHECKPOINT **3**

Now work Exercise 43.

Before continuing, we summarize how to determine the domain of a function algebraically.

Determining the Domain of a Function
To determine the domain of a function, we exclude independent variable values that
1. Make the denominator of a fraction zero.
2. Produce a negative result under an even-indexed radical.

　　Using purely algebraic techniques, finding the range of a function can be difficult. Many functions require techniques learned in calculus to determine the range, but the range is relatively easy to determine when we see the graph of the function. We illustrate how in Example 5.

EXAMPLE 5

Determining Domains and Ranges From a Graph

Determine visually the domain and range for the function in Figure 1.1.10, and write your answer using interval notation.

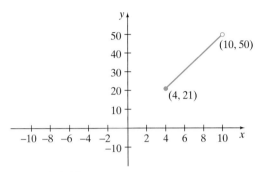

Figure 1.1.10

SOLUTION

First, notice that the left endpoint of the function is a solid dot and the right endpoint is a hollow dot. This means that the domain is the half-open interval [4, 10). Notice that the lowest point on the graph is at the ordered pair (4, 21) and the highest point is near (10, 50). So we see that the range of the function is [21, 50). ∎

Now let's introduce a function that we will see frequently in this text, the **price–demand function.**

Price–Demand Function

The **price–demand function** p gives us the price at which people buy exactly x units of product.

Generally, as the price of a product decreases, more and more people will buy the product. So for the price–demand function as the x-values increase, the p-values decrease. This is shown in Figure 1.1.11.

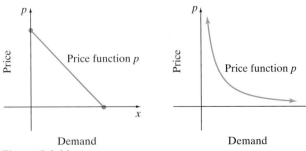

Figure 1.1.11

EXAMPLE 6 | **Determining the Domain and Range of a Price–Demand Function**

The price–demand function for electronic organs sold at Red River Mall is given by

$$p(x) = 16,000 - 320x$$

The graph of p is given in Figure 1.1.12. Here, x represents the number of electronic organs that people buy at Red River Mall weekly and $p(x)$ represents the price per organ in dollars.

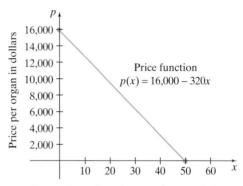

Figure 1.1.12

(a) Determine the domain and the range of the function.
(b) Evaluate $p(15)$ and interpret.

SOLUTION

(a) Notice that the function applies only for x-values up to 50 units. For other values, $p(x)$ is not defined. We may express this restriction on x using interval notation by writing

$$p(x) = 16,000 - 320x, \qquad [0, 50]$$

We can see that the possible values for x are in the interval $[0, 50]$, which gives us our domain. Since the range is the set of all possible function values, we can see from the graph that the range is $[0, 16,000]$.

(b) Evaluating the price–demand function at $x = 15$, we get

$$p(15) = 16,000 - 320(15) = 16,000 - 4800 = 11,200$$

This means that when the weekly demand for organs is 15, the price of an organ is $11,200.

SUMMARY

In this section we have reviewed the fundamental properties of functions.

• A **function** is a rule that assigns to each element in the domain one and only one element in the range.
• The **domain** of a function is the set of all possible independent variable values.
• The **range** of a function is the set of all possible dependent variable values.
• The **price–demand function** p gives us the price $p(x)$ at which people buy exactly x units of product.

SECTION 1.1 EXERCISES

For Exercises 1–4, use the table to construct ordered pairs and plot them in the Cartesian plane.

1.

x	$f(x)$
−2	3
1	5
4	4

2.

x	$g(x)$
1	2.1
2	2.3
3	2.9

3.

x	−4	−3.5	−2	2
$f(x)$	−5	−6	−8	−11

4.

x	0	−1	−2
$f(x)$	0	1.1	1.5

For Exercises 5–10, make a table and write the ordered pairs based on the scatterplots.

5.

6.

7.

8.

9.

10.

For Exercises 11–14, standardize the values of the independent variable based on its definition and then make a scatterplot of the data. (Data obtained from the U.S. Census Bureau web site, except for Exercise 13.)

✓ 11. Let x represent the number of years since 1980.

YEAR	PUBLIC EXPENDITURES AT PRIVATE UNIVERSITIES AND COLLEGES (IN BILLIONS OF DOLLARS)
1980	41
1983	44
1984	46
1985	49
1988	51

12. Let x represent the number of years since 1980.

YEAR	NUMBER OF VISITS TO PUBLIC LAND FOR RECREATION (IN MILLIONS)
1987	56
1988	57
1989	61
1990	71
1991	72

13. Let x represent the elapsed time in seconds.

Time	Speed (in miles per hour)
0	0
1	40
2	120
3	160
4	180

14. Let x represent the number of years since 1960.

Year	Per Capita Water Consumption (in gallons)
1960	339
1965	403
1970	427
1975	451
1980	440
1985	380

For Exercises 15–18, determine which of the tables represent functions.

15.

Domain	Range
0	1
2	1
3	2
4	5

16.

Domain	Range
−5	−2
−3	1
−2	5
−1	7
0	10
1	6

17.

Domain	Range
−3	1
−2	0
−2	3
−1	4
0	5
1	6

18.

Domain	Range
2	1
−2	3
0	0
−1	2
−3	7
2	4

For Exercises 19–24, determine if the graph represents a function.

19.

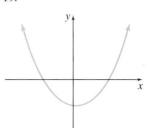

20.

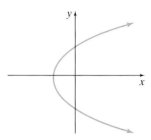

✓ 21.

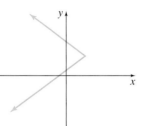

22.

23.

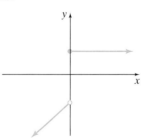

24.

For Exercises 25–35, let $f(x) = 2x - 1$ and $g(x) = x^2$. Evaluate each expression and write the solution as an ordered pair.

25. $f(2)$ 26. $f(-1)$ 27. $f(1)$ 28. $f(-2)$

29. $g(3)$ 30. $g(1)$ 31. $g(0)$ 32. $f\left(\dfrac{1}{2}\right)$

33. $g\left(\dfrac{1}{2}\right)$ 34. $f(-0.5)$ 35. $g(-0.25)$

36. Use the graph of the function f to answer parts (a) through (f).

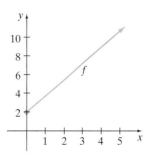

(a) What is the independent variable?
(b) What is the dependent variable?

(c) What is the value of $f(x)$ when $x = 0$?

(d) What is $f(3)$?

(e) What is the domain of f?

(f) What is the range of f?

37. Use the graph of the function f to answer parts (a) through (f).

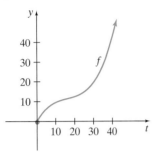

(a) What is the independent variable?

(b) What is the dependent variable?

(c) What is the value of $f(t)$ when $t = 10$?

(d) What is $f(30)$?

(e) What is the domain of f?

(f) What is the range of f?

38. Use the graph of the function g to answer parts (a) through (g).

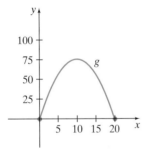

(a) What is the independent variable?

(b) What is the dependent variable?

(c) What is the value of $g(x)$ when $x = 20$?

(d) What is $g(10)$?

(e) What are the values of x when $g(x) = 50$?

(f) What is the domain of g?

(g) What is the range of g?

39. Copy and fill in the following table.

Interval Notation	Inequality Notation	Number Line
$[-\frac{1}{2}, 5)$		$\xrightarrow{\hspace{2cm}} x$
	$x \geq -2$	$\xrightarrow{\hspace{2cm}} x$
		$\xleftarrow{\hspace{1cm}} \overset{\diamond}{-3} \quad x$
	$-1 < x < 10$	$\xrightarrow{\hspace{2cm}} x$
$(-\infty, 3) \cup (3, \infty)$		$\xrightarrow{\hspace{2cm}} x$

For Exercises 40–51, determine the domain of the function algebraically, and write the domain using interval notation.

40. $f(x) = 2x$

41. $g(x) = -10x$

42. $f(x) = \dfrac{x^2 - 36}{x - 6}$

✓ 43. $g(x) = 3x^2 - 10$

44. $f(x) = x^3 - 6x^2$

45. $y = \dfrac{x - 10}{5 - x}$

46. $g(x) = \sqrt{x - 6}$

47. $f(x) = \sqrt{6 - x}$

48. $f(x) = \dfrac{\sqrt{x - 1}}{x - 4}$

49. $y = \sqrt{x^2 + 3}$

50. $y = \dfrac{10x}{x^2 - 25}$

51. $g(x) = \dfrac{x - 1}{2x^2 - 4x}$

In Exercises 52–55, graph the function on your calculator and use the graph to determine the domain and range of the function. Write your solutions using interval notation.

52. $g(x) = x^2 - 4$

53. $f(x) = \dfrac{x^2 + 5}{10}$

54. $f(x) = \sqrt{9 - x}$

55. $y = \dfrac{x - 2}{x^2 + 6}$

56. The price–demand function for the new Hanford hand-held game unit is given by

$$p(x) = 120 - 0.1x, \qquad 0 \leq x \leq 80$$

Here x represents the number of units sold per day and $p(x)$ represents the price of the unit.

(a) Complete the following table of values. (In other words, determine the values for the price when $x = 0, \ 10, \ 20, \ 30, \dots, 80$.)

x	0	10	20	30	40	50	60	70	80
$p(x)$									

(b) Make a graph of p on the interval $[0, 80]$.

(c) If $[0, 80]$ is the domain, what is the range of the function?

(d) Evaluate $p(45)$ and interpret the answer.

57. The price–demand function for the new Teddy Bear line of designer lingerie is given by

$$p(x) = \dfrac{165 - x}{4}, \qquad 0 \leq x \leq 35$$

Here x represents the number of units sold per day, and $p(x)$ represents the price of the unit.

(a) Complete the following table of values.

x	0	5	10	15	20	25	30	35
$p(x)$								

(b) Make a graph of p on the interval $[0, 35]$.

(c) If $[0, 35]$ is the domain, what is the range of the function?

(d) Evaluate $p(22)$ and interpret the answer.

58. For the price–demand function in Exercise 56, determine the desired price for each unit if the number sold is 65 units per day.

59. For the price–demand function in Exercise 57, determine the desired price for each unit if the number sold is 18 units per day.

For Exercises 60–65, write the domain and range of the function using interval notation.

60.

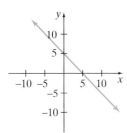

61.

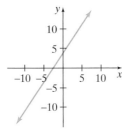

62.

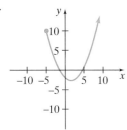

63.

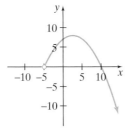

64.

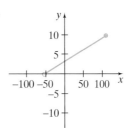

65.

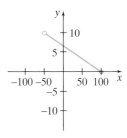

SECTION PROJECT

Determine if the following can be represented by functions.

(a) A person's astrological sign is a function of his or her birth date.

(b) A person's income is a function of how long he or she works each day.

(c) The voter turnout for an election is a function of the weather.

(d) The distance a delivery person is from a pizzeria is a function of the number of deliveries made.

(e) The number of U.S. Congressmen a state has is a function of its population.

SECTION 1.2   LINEAR FUNCTIONS AND AVERAGE RATE OF CHANGE

The first type of function that we will study is called a **linear function.** When we say things like "The cost of a phone call is a function of the length of the call" or "The price of a cab ride is a function of the distance to the destination," we are referring to linear functions.

Another critical concept that we address in this section is **average rate of change.** This kind of rate is a common component of linear functions. If we know the average rate of change of a linear function (also known as the **slope** of the line), it is usually a short process to write the function itself. Statements such as "On average, the price of used cars is going up $100 per year" or "My computer equipment is depreciating at about $80 each month" are using the average rate of change concept.

Linear Functions

We begin our investigation of linear functions with an example of how it is used in an everyday application. Suppose that The Fashion Mystique wants to print promotional flyers for its annual sidewalk sale. The store manager goes to the local copy center and finds that there is a $5 setup charge and a 10 cent copying fee for each flyer produced. The manager is handed the pricing schedule shown in Table 1.2.1.

There must be a functional relationship between the values in the two columns since the cost of the copies is a **function** of the number of copies made. Since the cost of the job **depends** on the number of copies produced, we believe that the number of copies is the independent variable, and the cost is the dependent variable.

TABLE 1.2.1

NUMBER OF COPIES	COST
100	$15
500	$55
1000	$105

This function consists of two parts. The first is the setup charge, which is a flat fee that is assessed no matter how many copies are made. It is what economists refer to as a **fixed cost.** The other part is the copying fee, which changes as the number of copies changes. Economists call this type of cost **variable costs.** We can use this combination of fixed and variable costs to verify the cost of making 500 flyers. For example, the $5 setup fee plus 10 cents times 500 for the copying fee gives

$$\text{Cost for } 500 = \left(\begin{array}{c} \text{price per} \\ \text{copy} \end{array} \right) \cdot \left(\begin{array}{c} \text{number} \\ \text{of copies} \end{array} \right) + \left(\begin{array}{c} \text{setup} \\ \text{fee} \end{array} \right)$$

$$= (0.10) \cdot 500 + 5$$

$$= 50 + 5 = 55 \text{ or } \$55$$

This is an exact numerical representation of what a linear cost model looks like. We can now define this type of cost function.

Cost Function

The cost $C(x)$ of producing x units of a product is given by the **cost function**

$$C(x) = \left(\begin{array}{c} \text{variable} \\ \text{costs} \end{array} \right) \cdot \left(\begin{array}{c} \text{units} \\ \text{produced} \end{array} \right) + \left(\begin{array}{c} \text{fixed} \\ \text{costs} \end{array} \right)$$

NOTE: Some cost functions are not linear because the variable cost is a function itself. However, every cost function is made up of a variable cost expression and a fixed cost.

EXAMPLE 1 | **Determining a Cost Function**

Determine a cost function, C, for the printing job just described.

SOLUTION

The variable costs are 10 cents per copy and the fixed costs are 5 dollars, which means that the cost function for the printing job can be written as

$$C(x) = 0.10x + 5$$

where x represents the number of copies made and $C(x)$ represents the cost, in dollars, of making the copies. Figure 1.2.1 shows a graph of the cost function and the points from the pricing schedule given in Table 1.2.1.

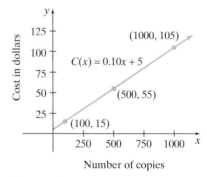

Figure 1.2.1

The cost function in Example 1 is one example of a linear function. Notice that the three points of the graph all *line up.* This is precisely why we call functions of this type **linear.** A linear function is made up of two key parts: a fixed number called a constant and a variable part that is the product of a number and a variable.

Linear Function

A **linear function** has the form

$$f(x) = mx + b$$

where m is the **average rate of change** or **slope** of the line and b is a **constant**, where m and b are real numbers.

The number m has a very important meaning. Let's return to the printing job described in Example 1 and use the graph to compute the cost per copy, denoted $\frac{\text{cost}}{\text{copy}}$. To do this, we calculate the difference between two of the points $(500, 55)$ and $(1000, 105)$ as follows:

$$\frac{\text{cost}}{\text{copy}} = \frac{\text{difference in cost}}{\text{difference in copies}}$$

$$= \frac{105 - 55}{1000 - 500}$$

$$= \frac{50}{500} = \frac{1}{10} = 0.1$$

We say that the cost per copy is increasing at an average rate of 0.10. If we think of the cost as the variable y and the copies as x, we have the following definition.

Average Rate of Change

The **average rate of change** of y with respect to x is

$$\frac{\text{difference in } y}{\text{difference in } x} = \frac{y_2 - y_1}{x_2 - x_1}$$

(Notice that the average rate of change in the cost of the copies is 0.10, which is exactly the same as what we called the variable costs.) When referring to a linear function, the average rate of change is also called the **slope** of the line. See Figure 1.2.2.

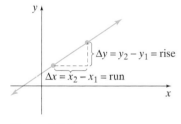

Figure 1.2.2

Slope

The **slope** of a line, denoted by m, is a measurement of the steepness of the line. Given two points on a line, (x_1, y_1) and (x_2, y_2), the slope of the line is computed by

$$m = \frac{y_2 - y_1}{x_2 - x_1}$$

The slope of the line also gives the average rate of change of y with respect to x.

NOTE: All of the following are ways to interpret slope:

$$\text{Average rate of change} = m = \frac{y_2 - y_1}{x_2 - x_1}$$

$$= \frac{\text{rise}}{\text{run}} = \frac{\text{difference in } y}{\text{difference in } x}$$

$$= \frac{\text{change in } y}{\text{change in } x} = \frac{\Delta y}{\Delta x}$$

EXAMPLE 2 | **Computing Average Rate of Change**

For the points $P = (2, -3)$ and $Q = (4, 2)$:

(a) Compute the average rate of change between the points.

(b) If the independent variable value increases by 1 unit, how will this affect the dependent value?

SOLUTION

(a) Here we will let P be the first point and Q be the second point. This means that (x_1, y_1) is the ordered pair $(2, -3)$ and (x_2, y_2) is the ordered pair $(4, 2)$. Using the definition of average rate of change we get

$$m = \frac{y_2 - y_1}{x_2 - x_1} = \frac{2 - (-3)}{4 - 2} = \frac{5}{2} = 2.5 \, \frac{\text{dependent units}}{\text{independent unit}}$$

(b) Since the average rate of change is $m = 2.5$, we see that if x, the independent value, increases by 1 unit then the dependent value will increase by 2.5 or $\frac{5}{2}$ units. This change is illustrated in Figure 1.2.3.

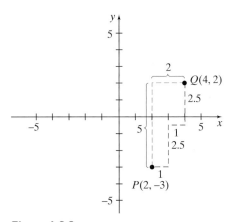

Figure 1.2.3

Interactive Activity

 Rework Example 2a, except let P be the second point and Q be the first point.

✓ CHECKPOINT 1 _____ Now work Exercise 9.

The Interactive Activity following Example 2 illustrates that the average rate of change between two points has only one numerical value. In other words, the **slope of a line is unique.**

Equations of Lines

There are several ways to write equations of lines. The type in the definition $f(x) = mx + b$ is called the **slope–intercept form.** Since linear equations are critical in our study of graphing and calculus, let's explore other ways to write a linear function.

We said at the beginning of this section that if the average rate of change of a line is known, it is a short process to write its linear function. To show how easy it is, let's say that we know that the slope of a line is $m = 3$ and a point on the line is $(4, 2)$. Now we call any other point on the line (x, y). See Figure 1.2.4.

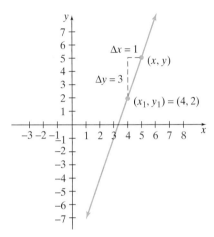

Figure 1.2.4

Using the definition of average rate of change, $m = \frac{y_2 - y_1}{x_2 - x_1}$, gives us

$$3 = \frac{y - 2}{x - 4}$$

where $(x_1, y_1) = (4, 2)$ and $(x_2, y_2) = (x, y)$. Multiplying each side of the equation by $x - 4$ yields

$$y - 2 = 3(x - 4)$$

We now have an equation for the line with slope $m = 3$ through the point $(4, 2)$. Since this equation uses a known slope and a known point, it is called the **point–slope form.**

> #### Point–Slope Form
> The equation of a line with known slope m and known point (x_1, y_1) is given by
> $$y - y_1 = m(x - x_1)$$

Many times, we want to use our calculator to graph a line with a known point and a known slope. We can do this by solving the point–slope form equation for y.

$$y - y_1 = m(x - x_1) \qquad \text{Point–slope form}$$
$$y = m(x - x_1) + y_1 \qquad \text{Solve for } y.$$

We call this form the **calculator friendly** or simply **CF form** of a line.

Calculator Friendly or CF Form
A linear function with a known slope m and known point (x_1, y_1) is given by
$$y = m(x - x_1) + y_1$$

EXAMPLE 3 | **Writing a Linear Function in Point–Slope and CF Form**

For the points $(2, -3)$ and $(4, 2)$:

(a) Write an equation in point–slope form using the slope found in Example 2.

(b) Write a function in the CF form and graph the function.

(c) Write the function in the slope–intercept form $f(x) = mx + b$.

SOLUTION

(a) In Example 2, we found the slope between these two points to be $m = 2.5$. Replacing (x_1, y_1) with $(2, -3)$, we get

$$y - (-3) = 2.5(x - 2)$$

or simply

$$y + 3 = 2.5(x - 2)$$

(b) Figure 1.2.5 has the graph. If we solve the solution to part (a) for y, we get

$$y = 2.5(x - 2) - 3$$

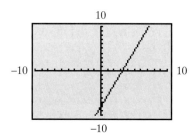

Figure 1.2.5

TECHNOLOGY NOTE
Throughout the textbook, the viewing window for calculator-generated graphs will appear on the perimeter of the graph. In text, we denote the viewing window by writing [Xmin, Xmax] by [Ymin, Ymax].

(c) To get the slope–intercept form $f(x) = mx + b$, we just need to simplify the function that is given in the CF form.

$$
\begin{aligned}
y &= 2.5(x - 2) - 3 \\
&= 2.5x - 2(2.5) - 3 \\
&= 2.5x - 5 - 3 \\
&= 2.5x - 8
\end{aligned}
$$

Since y is just another name for $f(x)$, we see that the slope–intercept form is $f(x) = 2.5x - 8$. ■

Interactive Activity
Rework Example 3, except use the point $(4, 2)$ when using the point–slope form.

Intercepts

Two important points on the graph of a linear function are its **intercepts.** These are the points where the graph crosses the x (independent) and y (dependent) axes, as shown in Figure 1.2.6.

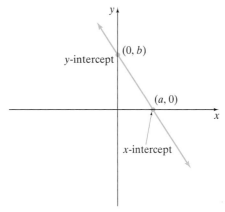

Figure 1.2.6

The figure also suggests how to find the intercepts. To get the y-intercept, we can replace x with zero and solve the resulting equation for y. To get the x-intercept, we replace y with zero and solve the resulting equation for x.

EXAMPLE 4 | **Finding the Intercepts of a Function**

Determine the x- and y-intercepts of the graph of the linear function

$$y = \frac{2}{3}(x+1) + 3.$$

SOLUTION

To find the y-intercept, we replace x with zero and solve for y. This gives

$$y = \frac{2}{3}(0+1) + 3 = \frac{2}{3} + 3 = \frac{11}{3}$$

So the y-intercept is $\left(0, \frac{11}{3}\right)$.

To get the x-intercept, we replace y with zero and solve for x to get

$$0 = \frac{2}{3}(x+1) + 3$$

$$-3 = \frac{2}{3}(x+1)$$

$$\frac{-9}{2} = x + 1$$

$$\frac{-11}{2} = x$$

Thus, the x-intercept is $\left(\frac{-11}{2}, 0\right)$. These points are highlighted on the graph in Figure 1.2.7.

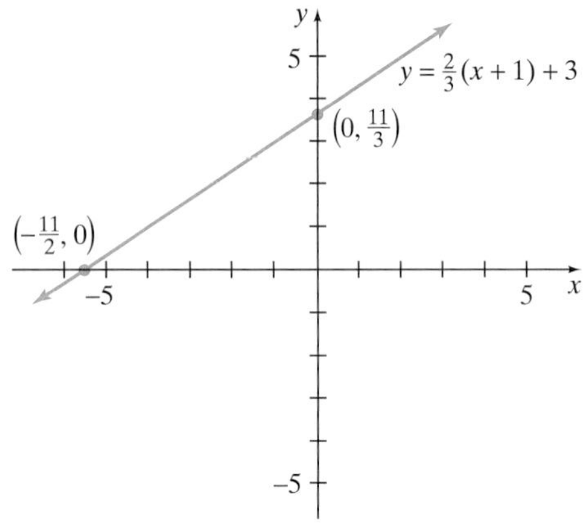

Figure 1.2.7

Notice that the y-intercept can be written in the form $(0, b)$, and if we evaluate the linear function $f(x) = mx + b$ when $x = 0$, we get $f(0) = m(0) + b = b$. This means that the y-intercept of the graph of a linear function is the same as the constant b. This is why the form $f(x) = mx + b$ is called the **slope–intercept form** of a line. Now let's look at how intercepts can be used in an application.

EXAMPLE 5 | **Writing a Depreciation Model and Interpreting Intercepts**

The value of many products is said to **depreciate**, meaning that they decrease in value, as time goes on. The amount that the value of the product loses each year is called the **depreciation rate.** Many businesses incorporate depreciation models into their fiscal plans and for scheduling upgrades of equipment. For example, the Clark Clipboard Company recently purchased a metal press for \$15,000. The press is estimated to depreciate at a rate of $2500 \, \frac{\text{dollars}}{\text{year}}$.

(a) Write a linear function in the form $f(x) = mx + b$ for the value of the press after x years.

(b) Determine the intercepts and interpret each.

SOLUTION

(a) We know that when the press is new, meaning that its age is $x = 0$, its value is \$15,000. Thus, $f(0) = 15,000$, which means that the point $(0, 15,000)$ is on the graph. Notice that this point is the y-intercept. Now, since the rate at which the value of the press **decreases** is $2500 \, \frac{\text{dollars}}{\text{year}}$, the average rate of change in the press's value is $-2500 \, \frac{\text{dollars}}{\text{year}}$. Since the average rate of change is the same as the slope of a line, we have $m = -2500$. So the value of the press after x years is $f(x) = -2500x + 15,000$.

(b) From part (a) we know that the y-intercept is $(0, 15,000)$. To find the x-intercept, we set $f(x)$ equal to zero and solve the equation for x.

$$0 = -2500x + 15,000$$
$$-15,000 = -2500x$$
$$\frac{-15,000}{-2500} = x$$
$$6 = x$$

The y-intercept, at the point $(0, 15,000)$, means that when the press has just been purchased its value is $15,000. The x-intercept of $(6, 0)$ means that after 6 years of use the press has no value. Notice that the reasonable domain of $f(x)$ is $[0, 6]$. A graph of this depreciation function is shown in Figure 1.2.8.

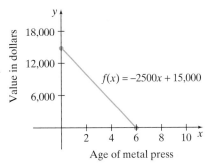

Figure 1.2.8

✓ **CHECKPOINT 2**

Now work Exercise 41.

We found in Example 5 that, as time went on, the value of the metal press decreased. On the other hand, as the x-values increased in the function in Example 4, so did the function values. These examples serve as motivation for the following important behaviors of functions.

Increasing and Decreasing Functions

1. A function is said to be **increasing** if the function values $y = f(x)$ increase as the x-values increase.
2. A function is said to be **decreasing** if the function values $y = f(x)$ decrease as the x-values increase.

NOTE: In simple terms, this means that a line is increasing if it runs uphill when viewed from left to right, and is decreasing if it runs downhill when viewed from left to right.

EXAMPLE 6 | **Identifying Increasing and Decreasing Functions**

Graph each line and classify the function as increasing, decreasing, or neither. State the slope of the line.

(a) $h(x) = 2x - 1$ (b) $g(x) = -3(x + 2) + 1$ (c) $f(x) = 4$

SOLUTION

(a)

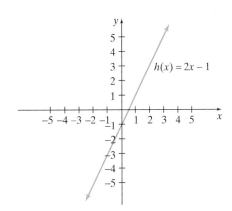

Figure 1.2.9

The graph "runs uphill" when viewed from left to right (Figure 1.2.9), so the function values increase as the x-values increase. This means that the function is increasing. The function has the slope–intercept form $h(x) = mx + b$, so we can directly read that the slope of the line is $m = 2$.

(b)

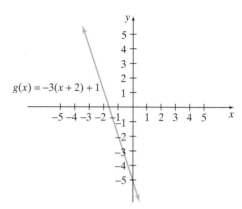

Figure 1.2.10

The graph "runs downhill" when viewed from left to right (Figure 1.2.10), so the function values decrease as the x-values increase. This means that the function is decreasing. The function has the CF form $g(x) = m(x - x_1) + y_1$, so the slope of the line is $m = -3$.

(c)

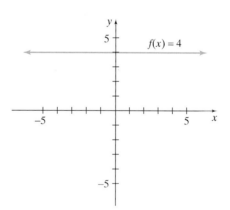

Figure 1.2.11

For this function, the function values stay the same as the x-values increase (Figure 1.2.11). We call this function a **constant function**, since the function is made up of only a constant. Notice that if we write the function in the form $f(x) = mx + b$ it would look like $f(x) = 0 \cdot x + b$. This means that the slope of the line is $m = 0$. ▬

A natural question to ask is "What is the slope of a vertical line?" A line of this type has the form $x = a$, where a is a constant. Since the change in x is zero, the formula for slope, $m = \frac{y_2 - y_1}{x_2 - x_1}$, would have zero in the denominator. Since division by zero is not defined, we can say that **vertical lines have undefined slope.** We summarize lines and their graphs in Table 1.2.2.

TABLE 1.2.2

Type of Line	Value of Slope m	Diagram
Increasing	Positive	
Decreasing	Negative	
Horizontal	Zero	
Vertical	Undefined	

Marginal Cost

Often in economics it is desirable to know the cost of producing the $(x + 1)^{\text{st}}$ item of a product. This cost is commonly called the **marginal cost.**

At the beginning of the section, we found that the cost function of copying promotional flyers for The Fashion Mystique was given by $C(x) = 0.10x + 5$, where x represents the number of copies produced and $C(x)$ represents the cost of producing x flyers. The cost of producing 110 flyers is given by $C(110)$. This cost is

$$C(110) = 0.10(110) + 5 = 11 + 5 = 16$$

So the cost of producing 110 flyers is $16. The cost of producing 111 flyers is found by evaluating $C(111)$. This is

$$C(111) = 0.10(111) + 5 = 11.1 + 5 = 16.1$$

The value of $C(111) - C(110)$ is then $16.1 - 16 = 0.1$, which is the difference between the cost of producing 110 and 111 flyers. In other words, we have found that the cost of producing the 111th flyer is $0.10, or 10 cents.

Recall, on page 15, that we discovered the cost is increasing at an average rate of $0.10 per copy. Thus, the marginal cost for the 111th copy is the same as the average rate of change for the cost of the copies. The fact that the marginal cost and the slope (given by the variable costs) are the same is no coincidence.

For linear cost functions, variable costs and marginal costs are equal.

We will study marginal functions in more depth in Chapter 3.

EXAMPLE 7

Determining Marginal Costs

The ProAudio Company manufactures DVD disks. It determines that the weekly fixed costs are $14,000 and the variable costs are $2.60 per disk.

(a) Determine the linear cost function C and interpret $C(1500)$.

(b) Identify the marginal cost. At a 1500 per week production level, what is the cost of manufacturing the 1501st disk?

SOLUTION

(a) The manufacturing process has fixed costs of $14,000 and variable costs of $2.60, so the linear cost function is $C(x) = 2.60x + 14,000$. Evaluating at $x = 1500$ gives

$$C(1500) = 2.60\,(1500) + 14,000 = 3900 + 14,000 = 17,900$$

Thus, the weekly cost of producing 1500 DVD disks is $17,900.

(b) Since the marginal cost and the variable cost are the same for a linear function, the marginal cost for the manufacturing process is $2.60. Thus, at a 1500 disk per week production level, the cost of manufacturing the 1501st disk is $2.60. ∎

Piecewise-defined Functions

Often one simple function alone cannot represent everyday applications. For example, a taxi may charge a flat fee of $3 up to the first mile driven and an additional $0.25 for 1 mile and for each mile afterward. To express a function for the cost of the taxi, we must write the function in two parts. Up to the first mile, the x-values go from 0 to 1, and the cost is given by the constant function, which we will call $y_1 = 3$. For x-values greater than or equal to 1, we have to add 25 cents for each mile ridden to the $3. So for the x-values greater than 1, the cost is given by a second piece that we will call $y_2 = 0.25x + 3$. So we can join the two pieces y_1 and y_2 to get the cost of riding x miles in the taxi as

$$f(x) = \begin{cases} 3, & 0 < x < 1 \\ 0.25x + 3, & x \geq 1 \end{cases}$$

A graph of f is given in Figure 1.2.12. These kind of functions which are defined for specific intervals of the domain are called **piecewise-defined functions.** Let's try some examples of graphing this new type of function.

TECHNOLOGY NOTE
To see how to graph piecewise-defined functions, see the online calculator manual at www.prenhall.com/armstrong

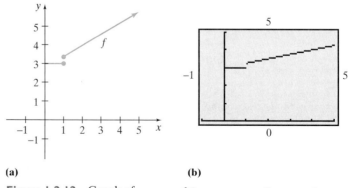

(a)   (b)

Figure 1.2.12 Graph of $f(x) = \begin{cases} 3, & 0 < x < 1 \\ 0.25x + 3, & x \geq 1 \end{cases}$

EXAMPLE 8 | **Graphing a Piecewise-defined Function**

For the piecewise-defined function $f(x) = \begin{cases} x+2, & x < 1 \\ 3-x, & x \geq 1 \end{cases}$

(a) Evaluate $f(0)$, $f(1)$, and $f(3)$.

(b) Make an accurate graph of the function.

SOLUTION

(a) Since the x-value 0 is in the interval $(-\infty, 1)$, we evaluate using the "top" piece to get $f(0) = 0 + 2 = 2$. To get $f(1)$, we see that $x = 1$ is the left end point of our interval $[1, \infty)$, so we use the "bottom" piece to evaluate and get $f(1) = 3 - 1 = 2$. Finally, $f(3)$ is in the interval $[1, \infty)$, so we use the bottom piece again to get $f(3) = 3 - 3 = 0$.

(b) On the interval of x-values $(-\infty, 1)$, we plot the linear equation $y = x + 2$ shown in Figure 1.2.13a. Then, on the interval $[1, \infty)$, we plot the equation $y = 3 - x$ as in Figure 1.2.13b. Finally, the graph of f is found by plotting the two graphs on the same Cartesian plane, as shown in Figure 1.2.13c.

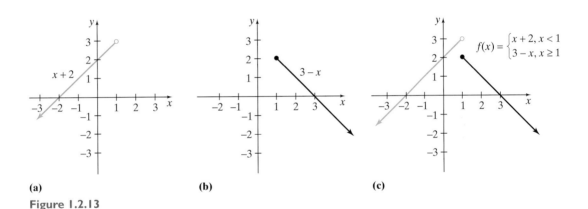

(a)   **(b)**   **(c)**

Figure 1.2.13

We use a solid dot to show which end point is included when $x = 1$ and a hollow dot to indicate that the endpoint is not included when $x = 1$.

Absolute Value Function

A piecewise-defined function that is often used is the **absolute value function.** Recall from algebra that the absolute value of any real number x, denoted by $|x|$, is defined by

$$|x| = \begin{cases} -x, & x < 0 \\ x, & x \geq 0 \end{cases}$$

In words, when taking the absolute value of any number, just write that number if the value is positive or zero, but if the number is negative, take its opposite. For example, $|8| = 8$, since $8 > 0$; and $|-5| = -(-5) = 5$, since $-5 < 0$. Notice how the definition of absolute value looks like a piecewise-defined function. It seems natural to define the absolute value function as a piecewise-defined function.

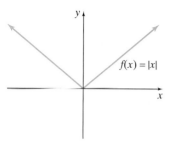

Figure 1.2.14

Absolute Value Function

The **absolute value function** $f(x) = |x|$ is defined by

$$f(x) = |x| = \begin{cases} -x, & x < 0 \\ x, & x \geq 0 \end{cases}$$

A graph of $f(x) = |x|$ is shown in Figure 1.2.14. Notice that the domain of the function is $(-\infty, \infty)$, the set of real numbers. Since an absolute value cannot be negative, the range of the absolute value function is $[0, \infty)$.

There are two key points to recognize about this type of function. First, the graph of the function $f(x) = |x|$ has a sharp "corner" at the origin. Second, the variable x inside the absolute value bars can be replaced by **any** algebraic expression.

EXAMPLE 9

Rewriting and Graphing Piecewise-defined Functions

For the function $f(x) = |x - 5|$, graph the function and rewrite $f(x) = |x - 5|$ as a piecewise-defined function.

SOLUTION

A graph of the function $f(x) = |x - 5|$ is given in Figure 1.2.15.

TECHNOLOGY NOTE
Most graphing calculators use abs(to denote the absolute value operator.

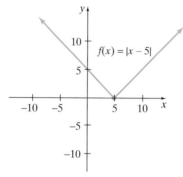
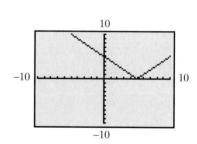

Figure 1.2.15

Interactive Activity

 Use your calculator to verify the solution of Example 9 by comparing the graphs of the absolute value function to the piecewise-defined function.

We see from the graph that the corner has shifted 5 units to the right to the point $(5, 0)$. Also from the graph, the piecewise-defined function will have one piece for x-values less than 5 and another for x-values greater than or equal to 5. For $x \geq 5$, we see that $|x - 5| = (x - 5) = x - 5$. For the x-values less than 5, we have to take the opposite of $x - 5$ to get $|x - 5| = -(x - 5) = -x + 5$. This application of the definition of absolute value allows us to write the function in piecewise-defined form as

$$f(x) = |x - 5| = \begin{cases} -x + 5, & x < 5 \\ x - 5, & x \geq 5 \end{cases}$$

SUMMARY

In this section, we learned that the slope of a line is also its **average rate of change** and is computed by

$$m = \frac{y_2 - y_1}{x_2 - x_1}$$

We also found that the value of the slope indicated whether the linear function was increasing or decreasing:

- If $m > 0$, the function is increasing.
- If $m < 0$, the function is decreasing.
- If $m = 0$, the function is the constant function.
- If m is undefined, the line is vertical.

The slope, along with a point on the line, can be used to write **linear equations** three ways:

- **Slope–intercept** form: $f(x) = mx + b$
- **Point–slope** form: $y - y_1 = m(x - x_1)$
- **Calculator friendly (CF)** form: $y = m(x - x_1) + y_1$

The linear cost function has the form $f(x) = mx + b$, where m is the variable costs and b represents the fixed costs. This same model, $f(x) = mx + b$, can be used as a depreciation model, where b represents the cost of a new item and m is the depreciation rate.

Functions that have different rules based on the x-values are called **piecewise-defined functions,** and a special type of piecewise-defined function is the **absolute value function.**

SECTION 1.2 EXERCISES

For Exercises 1–8, find the average rate of change between the two points.

1. $(4, 8)$ and $(5, 3)$

2. $(4, 8)$ and $(3, 5)$

3. $(2, 3)$ and $(-4, 8)$

4. $(2, 2)$ and $(-4, 4)$

5. $(5, 6.1)$ and $(7, 8.3)$

6. $(0, 1.25)$ and $(4, 5)$

7. $(a - 1, b - 1)$ and (a, b) where a and b are constants

8. $(1 - a, 1 - b)$ and $(a - 1, b)$ where a and b are constants

For Exercises 9–14, determine the average rate of change for the given graphs.

 9.

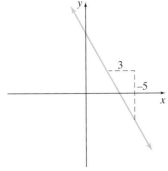

10.

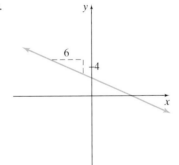

11.

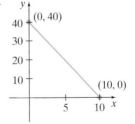

12.

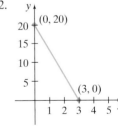

13.

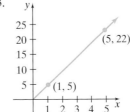

14.

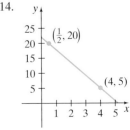

For Exercises 15–22, write an equation of the line in point–slope form if possible and graph.

15. The line contains the points $(2, 6)$ and $(5, -3)$.

16. The line contains the points $(3, 2)$ and $(-1, -2)$.

17. The line is given by the values in Table 1.2.3.

TABLE 1.2.3

x	y
3	5
5	13

18. The line is given by the values in Table 1.2.4.

TABLE 1.2.4

x	y
2	4
0	4

19. The line has a y-intercept at -3 and passes through the origin.

20. The line has an x-intercept at 2 and a y-intercept at 4.

21. The line is vertical and passes through the point $(-2, 9)$.

22. The line is horizontal and passes through the point $(-3, 1)$.

For Exercises 23–33, write an equation of the line in CF form.

23. A line has a slope of 3 and passes through the point $(8, 4)$.

24. A line has a slope $-\dfrac{5}{8}$ and $f(0) = 3$.

25. A line has a slope of 5 and a y-intercept at -13.

26. A line has an average rate of change of $-\dfrac{2}{3}$ and an x-intercept at 3.

27. A line has an average rate of change of 0.3 and an x-intercept at -4.

28. The graph that contains the data points $(1, 3)$ and $(-10, 1)$.

29. The graph that contains the data points in Table 1.2.5.

TABLE 1.2.5

x	$f(x)$
-1	2
3	1

30. The graph that contains the data points in Table 1.2.6.

TABLE 1.2.6

x	$f(x)$
0	2
-3	4

31. A line has an x-intercept at $(-12, 0)$ and a y-intercept of $\left(0, \frac{3}{2}\right)$.

32. A linear function f has function values of $f(5) = 1$ and $f(-7) = -1$.

33. A linear function g has function values of $g(0) = 6$ and $g(4) = 1.3$.

For Exercises 34–40, determine the x- and y-intercepts for each graph.

34.

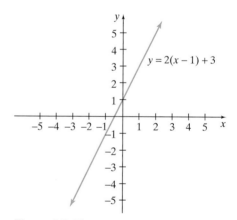

Figure 1.2.16

35.

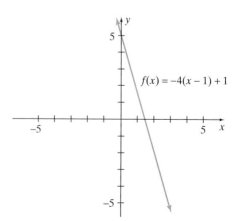

Figure 1.2.17

36. The graph of the function $f(x) = \dfrac{x}{2} - 4$.

37. The graph of the function $g(x) = 120 - 0.3x$.

38. The graph of the function $f(x) = \dfrac{750 - 35x}{10}$.

39. The graph of the function $g(x) = 0.2(x - 5.1) + 10$.

40. The graph of the linear equation $y - 3 = -2(x + 4)$.

Answer Exercises 41–44 by using the depreciation model.

✓ 41. A new photocopy machine initially costs $25,000 and depreciates at a rate of $1250 per year.
(a) Determine an equation for the depreciation function.
(b) In how many years will the machine be worth $10,000?

42. A new multimedia computer costs $800 and depreciates at a rate of $150 \dfrac{\text{dollars}}{\text{year}}$.
(a) Determine an equation for the depreciation function.
(b) When will the computer depreciate to the point that it has no worth?

43. A factory invests $90,000 in a new machine press that depreciates at a rate of $6000 per year.
(a) Determine an equation for the depreciation function.
(b) How long will it take for the press to have half its original value?

44. A XQ-383 sports car costs $40,000 and depreciates $3000 per year.
(a) Determine an equation for the depreciation function.
(b) How much will the car be worth in five years?

For Exercises 45–53, classify each of the linear functions as increasing or decreasing, and identify the slope of the line.

45. The function $f(x) = -4.5(x - 2)$

46. The function $g(x) = \dfrac{100 - 5x}{2}$

47. The function $f(x) = \dfrac{500 + 60x}{15}$

48. The function $g(x) = 4(x - 6) + 8$

49. The function $f(x) = 5 + 4x$

50. The graph of the function containing the points in Table 1.2.7.

51. The linear function containing the points in Table 1.2.8.

TABLE **1.2.7**	
x	$f(x)$
1	3
7	5

TABLE **1.2.8**	
x	y
10	5
1	2

52.

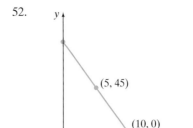

(5, 45)

(10, 0)

53.

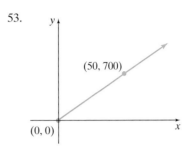

(50, 700)

(0, 0)

In Exercises 54–59, write the linear cost function for the given information in the slope–intercept form $C(x) = mx + b$.

54. The variable costs are $70\dfrac{\text{dollars}}{\text{unit}}$, and the fixed costs are $10,000.

55. The variable costs are $147.75 per unit, and the fixed costs are $2700.

56.

NUMBER OF UNITS	COST
5	$1250
23	$5580

57.

QUANTITY	COST
10	$7000
50	$30,000

58. The fixed costs are $14,500, and it costs $100,100 to produce 75 units of product.

59. The variable costs are $80\dfrac{\text{dollars}}{\text{unit}}$, and the cost of producing 40 units is $3850.

60. The Dirt Rocket Company produces spokes for off road motorcycles. It determines that the daily cost in dollars of producing x quick-replace spokes is $C(x) = 1250 + 3x$.
(a) What are the fixed and variable costs for producing the spokes?
(b) Evaluate $C(400)$ and interpret.
(c) How many spokes can be produced for $1820?
(d) What is the marginal cost for the spokes?

(e) If the current daily production level is 100 spokes, what is the cost in dollars of producing the 101st spoke?

61. The Fontana Vitamin Company determines that the cost of producing x bottles of a new supplement is given by the linear cost function $C(x) = 1.25x + 550$.

(a) What are the fixed and variable costs for producing the supplement?

(b) Evaluate $C(70)$ and interpret.

(c) How many bottles of the supplement can be produced for $693.75?

(d) What is the marginal cost for the supplement?

(e) If the current production level is 70 bottles, what is the cost of producing the 71st bottle of the supplement?

62. The Get-It-Now florist determines that the cost in dollars of arranging and delivering x get-well bouquets is given by the cost function $C(x) = 10 + 32x$.

(a) What are the fixed and variable costs for producing the bouquets?

(b) Evaluate $C(15)$ and interpret.

(c) How many bouquets can be produced and delivered for $8010?

(d) What is the marginal cost for the bouquets?

(e) If the current production level is 40 bouquets, what is the cost of producing the 41st bouquet?

For Exercises 63–68, make an accurate graph of the piecewise-defined functions.

63. $f(x) = \begin{cases} 4 - 0.2x, & x < 5 \\ 5, & x \geq 5 \end{cases}$

64. $g(x) = \begin{cases} 2x, & x \leq 0 \\ -3x, & x > 0 \end{cases}$ 65. $y = \begin{cases} -3, & x < 4 \\ 3, & x > 4 \end{cases}$

66. $f(x) = \begin{cases} \dfrac{x+6}{2}, & x \leq 1 \\ \dfrac{7}{2}, & x > 1 \end{cases}$ 67. $g(x) = \begin{cases} -4, & x < 5 \\ x - 2, & x \geq 5 \end{cases}$

68. $f(x) = \begin{cases} -2, & x < 0 \\ 2x + 4, & 0 \leq x \leq 5 \\ -x, & x > 5 \end{cases}$

For Exercises 69–74, graph the absolute value function then rewrite the function in piecewise form.

69. $f(x) = |x + 1|$ 70. $y = |x - 3|$

71. $g(x) = |6 - 2x|$ 72. $f(x) = |10 - x|$

73. $g(x) = |3x - 15|$ 74. $f(x) = |4x + 12|$

75. The Lesky Truck Rental Company charges a flat fee of $20 plus 15 cents per mile driven for renting a moving truck.

(a) What are the fixed costs? What are the variable costs?

(b) Write a linear function C, where x represents the number of miles driven and $C(x)$ represents the rental costs.

(c) What would be a realistic domain for this function?

(d) Evaluate $C(120)$ and interpret.

(e) What is the cost of the 121st mile driven? What is this cost called?

76. Consider the daily cost schedule (Table 1.2.9) for the cost of producing x Never Die light bulbs.

TABLE 1.2.9

BULBS PRODUCED	COST (IN DOLLARS)
0	1050
100	1250
215	1480

(a) What are the fixed costs? Explain how you determined this from Table 1.2.9.

(b) How much does the cost increase for each additional bulb produced? How do you classify this kind of cost?

(c) Write a linear cost function C, where x represents the number of bulbs produced and $C(x)$ represents the cost.

(d) What would the cost be if 260 bulbs were produced?

77. Consider the price schedule for the repair cost of a Ripen Microwave oven (Table 1.2.10).

TABLE 1.2.10

HOURS LABOR	REPAIR COST (IN DOLLARS)
0	50
1	65
2	80

(a) What are the fixed costs? Explain how you determined this from Table 1.2.10.

(b) How much does the repair cost increase for each additional hour of labor? How do you classify this kind of cost?

(c) Write a linear repair cost function C, where x represents hours of labor and $C(x)$ represents the repair cost.

(d) What would the repair cost be if the microwave oven took 4.5 hours to fix?

🌐 78.* The annual total expenditures on pollution abatement in the United States can be modeled by

$$f(x) = 2x + 48.17, \qquad 0 \le x \le 20$$

where x represents the number of years since 1972 and $f(x)$ represents the expenditures in billions of constant 1987 dollars.

(a) Is the amount of money spent on pollution abatement generally increasing or decreasing? At what rate?

(b) Use your calculator to graph the function on the interval $[0, 20]$. What years does this interval correspond to?

(c) Evaluate $f(12)$ and interpret.

(d) In what year did the annual expenditures approach 80 billion dollars?

🌐 79. The amount spent annually in college bookstores in the United States can be modeled by

$$f(x) = 0.19x + 1.67, \qquad 0 \le x \le 15$$

where x represents the number of years since 1982 and $f(x)$ represents the amount spent in billions of dollars.

(a) How much does the amount of spending increase in college bookstores each year?

(b) According to the model, how much was spent in college bookstores in 1985?

(c) According to the model, how much was spent in college bookstores in 1990?

(d) How much more was spent in college bookstores in 1990 compared to 1985?

_____

The data and model given in exercises and examples preceded by the web site icon are based on information gathered at the U.S. Census Bureau web site. These exercises and examples appear throughout this textbook.

SECTION PROJECT 🌐

The number of disabled Korean War veterans receiving compensation is shown in Table 1.2.11.

(a) If x represents the number of years since 1980 and $f(x)$ represents the number of disabled Korean War veterans (in thousands), compute the average rate of change between the points $(0, 241)$ and $(5, 226)$. Write the average rate of change using proper units.

(b) Is the number of disabled Korean War veterans receiving compensation increasing or decreasing?

TABLE 1.2.11

YEAR	NUMBER OF DISABLED KOREAN WAR VETERANS (IN THOUSANDS)
1980	241
1985	226

Source: U.S. Census Bureau web site.

(c) Write a linear mathematical model for the number of disabled Korean War veterans receiving compensation since 1980.

(d) Evaluate $f(15)$ and interpret.

(e) Compute the x-intercept of the graph of the function. Interpret its meaning using a brief sentence.

(f) Given the fact that the Korean War ended in 1953, does it make sense for there to be an x-intercept for this model? Explain why or why not using a brief sentence.

SECTION 1.3   QUADRATIC FUNCTIONS AND AVERAGE RATE OF CHANGE ON AN INTERVAL

In this section, we turn our attention to quadratic functions. This family of functions has many applications in the physical laws of motion, economics, and the social sciences. We then extend our knowledge of the average rate of change by discussing the secant line slope of a function on an interval.

Properties of Quadratic Functions

After the linear function, the next most common type of function we study is the **quadratic function.** The quadratic function may be called a **second-degree function** since the largest exponent for the independent variable is 2.

Quadratic Function
A function of the form $f(x) = ax^2 + bx + c$ is called a **quadratic function,** where a, b, and c are real numbers and $a \ne 0$.

Since there are no x-values that can make a denominator zero or give us the square root of a negative number, the domain of a quadratic function is the set of real numbers $(-\infty, \infty)$. The shape of a quadratic function is called a **parabola.**

EXAMPLE 1

Graphing and Evaluating a Quadratic Function

The percentage of people taking the Scholastic Aptitude Test (SAT) who intend to study business and commerce can be modeled by

$$f(x) = -0.09x^2 + 2.01x + 9.63, \qquad 1 \le x \le 22$$

where $f(x)$ is the percentage and x is the number of years since 1974. Evaluate $f(10)$ and interpret the result.

SOLUTION

Evaluating the model at $x = 10$ gives

$$f(10) = -0.09(10)^2 + 2.01(10) + 9.63 = -9 + 20.1 + 9.63 = 20.73$$

Thus, in 1984 about 20.73% of all the people taking the SAT intended to study business and commerce. See Figure 1.3.1.

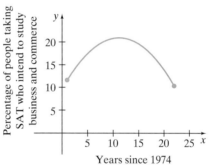

Figure 1.3.1   Graph of
$f(x) = -0.09x^2 + 2.01x + 9.63$ on $[1, 22]$.

Figure 1.3.1 shows that the graph appears to have a peak or maximum value when x has the value of about 11. This peak is called the **vertex** of the graph. To the left of the vertex, the graph appears to be increasing. To the right, the graph appears to be decreasing.

The vertex of a parabola is one of its most important features, since it gives us the maximum or minimum value of the function. Usually, finding a maximum or minimum value requires tools from calculus, but in the case of quadratic functions, we can use a simple formula to find the x-coordinate of the vertex. (You will derive this property in Chapter 5.)

Vertex of a Parabola

The x-coordinate of the vertex of a parabola given by $f(x) = ax^2 + bx + c$ is

$$x = \frac{-b}{2a}$$

The y-coordinate can be found by evaluating $f\left(\dfrac{-b}{2a}\right)$.

The way that the parabola $f(x) = ax^2 + bx + c$ opens is determined by the value of a, which is called the **leading coefficient.** If a is a positive number, the graph opens up, and if a is negative, the graph opens downward. If the graph of a parabola opens up, the vertex is a **minimum,** and if the graph opens down, the vertex is a **maximum.** Consequently, if we know the y-coordinate of the vertex and the direction in which the graph opens, we can readily determine the range of any quadratic function. For example, consider the function $f(x) = -x^2 + x + 6$. Since the leading coefficient of the quadratic function is $-1 < 0$, we know that the graph opens down. This means that the vertex is a maximum point. The x-coordinate of the vertex is given by $x = \frac{-b}{2a} = \frac{-1}{2(-1)} = \frac{1}{2}$. The y-coordinate of the vertex is

$$f\left(\frac{1}{2}\right) = -\left(\frac{1}{2}\right)^2 + \left(\frac{1}{2}\right) + 6$$

$$= -\frac{1}{4} + \frac{1}{2} + 6 = \frac{25}{4}$$

Since the maximum function value is $y = \frac{25}{4}$, we see that the range of the function is $\left(-\infty, \frac{25}{4}\right]$.

Determining Zeros of Quadratic Functions

The real **zeros** of a function f are given by the x-intercepts of its graph. They can also be called the **roots** of the function, since they are the solutions to the equation $f(x) = 0$.

Many times, the zeros, or roots, of a function can be found algebraically by the process of **factoring.** We just set $f(x)$ equal to zero and factor, as shown in Example 2.

EXAMPLE 2 | **Determining the Zeros of a Quadratic Function by Factoring**

Determine the zeros of the function $f(x) = 4x^2 + 6x - 4$. Check the result numerically.

SOLUTION

To find the zeros for this function, we need to solve the quadratic equation $4x^2 + 6x - 4 = 0$. So, by factoring, we get

$$
\begin{array}{ll}
4x^2 + 6x - 4 = 0 & \text{Given equation} \\
2(2x^2 + 3x - 2) = 0 & \text{Factoring out the constant 2} \\
2x^2 + 3x - 2 = 0 & \text{Dividing both sides by the constant 2} \\
(x + 2)(2x - 1) = 0 & \text{Factoring the expression } 2x^2 + 3x - 2 \\
(x + 2) = 0 \quad \text{or} \quad (2x - 1) = 0 & \text{Setting each factor equal to zero} \\
x = -2 \quad \text{or} \quad x = \dfrac{1}{2} & \text{Solving for } x
\end{array}
$$

We can check our solutions by evaluating $f(x) = 4x^2 + 6x - 4$ at both solutions to see if the function value is zero. Checking at $x = -2$, we get

$$f(-2) = 4(-2)^2 + 6(-2) - 4 = 16 - 12 - 4 = 0$$

At the second zero $x = \frac{1}{2}$, we get

$$f\left(\frac{1}{2}\right) = 4\left(\frac{1}{2}\right)^2 + 6\left(\frac{1}{2}\right) - 4 = 1 + 3 - 4 = 0$$

So the zeros of the function $f(x) = 4x^2 + 6x - 4$ are $x = -2$ and $x = \frac{1}{2}$. ◼

Many times, the quadratic that we wish to solve is not easily factorable. To algebraically solve these types, we use the **quadratic formula.** The derivation of this formula appears in Appendix C.

Quadratic Formula

The zeros of the quadratic function $f(x) = ax^2 + bx + c$ are given by the **quadratic formula**

$$x = \frac{-b \pm \sqrt{b^2 - 4ac}}{2a}$$

$b^2 - 4ac$ is called the **discriminant.**

If $b^2 - 4ac > 0$, the graph looks like

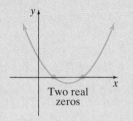

or

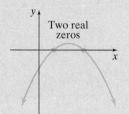

If $b^2 - 4ac = 0$, the graph looks like

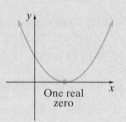

or

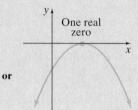

If $b^2 - 4ac < 0$, the graph looks like

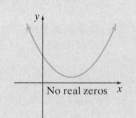

or

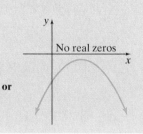

EXAMPLE 3

Determining Zeros with the Quadratic Formula

Determine the zeros of the quadratic function $f(x) = 2x^2 - 6x + 3$.

SOLUTION

To find the zeros of the function, we need to solve the quadratic equation $2x^2 - 6x + 3 = 0$. Since this equation is not easily factorable, we determine the roots by evaluating the quadratic formula using $a = 2$, $b = -6$, and $c = 3$. This gives us

$$x = \frac{-(-6) \pm \sqrt{(-6)^2 - 4(2)(3)}}{2(2)} = \frac{6 \pm \sqrt{36 - 24}}{4} = \frac{6 \pm \sqrt{12}}{4}$$

$$= \frac{6 \pm 2\sqrt{3}}{4} = \frac{2(3 \pm \sqrt{3})}{4} = \frac{3 \pm \sqrt{3}}{2}$$

So the two real zeros of the function $f(x) = 2x^2 - 6x + 3$ are $x = \frac{3 - \sqrt{3}}{2}$ and $x = \frac{3 + \sqrt{3}}{2}$.

✓ CHECKPOINT 1

Now work Exercise 19.

Sometimes even the quadratic formula can be unwieldy when locating zeros of quadratic functions. In some situations, we use a calculator to approximate the solutions to quadratic equations. To approximate the zeros of $f(x) = 2x^2 - 6x + 3$, we can use the ZERO or ROOT command on our calculator. Figure 1.3.2a shows the approximation of the zero of the leftmost x-intercept, and the rightmost is shown in Figure 1.3.2b. So the approximations for the zeros of the function are $x = 0.63$ and $x = 2.37$, rounded to the nearest hundredth.

TECHNOLOGY NOTE
To see how to use the ZERO or ROOT command on your calculator, consult the online graphing calculator manual at www.prenhall.com/armstrong

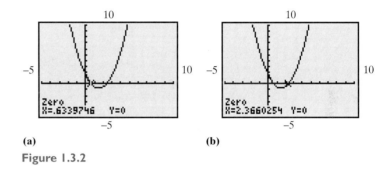

(a)   (b)

Figure 1.3.2

Secant Line Slope and the Difference Quotient

In Section 1.2, we introduced the average rate of change of a line. This definition is reviewed in the From Your Toolbox to the left.

At times, we want to know what the average rate of change would be for a **nonlinear function** over a specified interval. We can find this rate by computing the slope of a **secant line,** which is a line that passes through two points on a curve. To determine the slope of the secant line, let's suppose that a curve defined by a nonlinear function f has two distinct points, $P(x_1, y_1)$ and $Q(x_2, y_2)$. The average rate of change, or slope, of a line passing through P and Q is

$$m = \frac{y_2 - y_1}{x_2 - x_1} = \frac{f(x_2) - f(x_1)}{x_2 - x_1}$$

We call the change from P to Q in the x-coordinate $\triangle x$ (read "delta x"). This means that $\triangle x = x_2 - x_1$. If we let $x_1 = x$, then $x_2 = x + \triangle x$. See Figure 1.3.3.

From Your Toolbox

The **slope** of a line, denoted by m, is a measurement of the steepness of the line. Given two points on a line, (x_1, y_1) and (x_2, y_2) the slope is computed by

$$m = \frac{y_2 - y_1}{x_2 - x_1} = \frac{\text{change in } y}{\text{change in } x}$$
$$= \frac{\Delta y}{\Delta x}$$

The slope of the line also gives the **average rate of change** of y with respect to x.

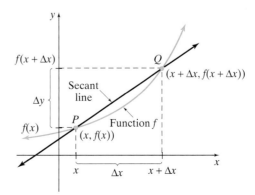

Figure 1.3.3

We can see from the graph that the coordinates of these two points are $P(x, f(x))$ and $Q(x + \Delta x, f(x + \Delta x))$. Now, determining the $\frac{\text{change in } y}{\text{change in } x}$, we get

$$\text{Secant line slope} = \frac{\text{change in } y}{\text{change in } x}$$

$$= \frac{f(x + \Delta x) - f(x)}{(x + \Delta x) - x}$$

$$= \frac{f(x + \Delta x) - f(x)}{\Delta x}$$

This fraction represents the **average rate of change** between the points P and Q. The fraction $\frac{f(x + \Delta x) - f(x)}{\Delta x}$ is also called the **difference quotient.**

Difference Quotient

The average rate of change of a function f on an interval $[x, x + \Delta x]$ is equivalent to the slope of the secant line through the points $(x, f(x))$ and $(x + \Delta x, f(x + \Delta x))$, denoted by m_{sec}, and is given by the **difference quotient**

$$m_{\text{sec}} = \frac{f(x + \Delta x) - f(x)}{\Delta x}$$

where $\Delta x \neq 0$. The units of this average rate of change are $\frac{\text{units of } f}{\text{unit of } x}$.

NOTE: The denominator of the difference quotient, Δx, is also called the **increment in x.** The numerator $f(x + \Delta x) - f(x)$ is the same as Δy and is called the **increment in y.**

EXAMPLE 4 | **Determining the Secant Line Slope**

Find the average rate of change of $f(x) = 2x^2 + 1$ on the interval $[2, 5]$.

SOLUTION

The first x-value is $x = 2$, and the increment in x is $\Delta x = 5 - 2 = 3$. Using the difference quotient gives us

$$m_{\text{sec}} = \frac{f(x + \Delta x) - f(x)}{\Delta x} = \frac{f(2 + 3) - f(2)}{3} = \frac{f(5) - f(2)}{3}$$

$$= \frac{(2(5)^2 + 1) - (2(2)^2 + 1)}{3} = \frac{(51) - (9)}{3} = \frac{42}{3} = 14$$

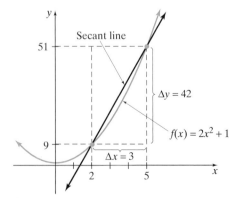

Figure 1.3.4

Since the slope of the secant line on $[2, 5]$ is 14, we conclude that the average rate of change of f on $[2, 5]$ is 14 or $14 \frac{\text{units of } f}{\text{unit of } x}$. The graph of f and this secant line is shown in Figure 1.3.4.

In applications, the secant line slope can be an indicator of the general trend in the rate of change of some phenomena on some interval. This is illustrated in Example 5.

EXAMPLE 5

Interpreting the Secant Line Slope

The U.S. imports from China for the years 1987 to 1996 can be modeled by

$$f(x) = 0.32x^2 + 1.64x + 3.98, \qquad 1 \leq x \leq 10$$

where x represents the number of years since 1986 and $f(x)$ represents the dollar value, in billions, of goods imported.

(a) Make a table of function values for $x = 1, 2, 3, \ldots, 10$. Use these values when calculating parts (b) and (c).

(b) Determine the average rate of change in U.S. imports from China from 1987 to 1991 and interpret. Include appropriate units.

(c) Determine the average rate of change in U.S. imports from China from 1991 to 1996 and interpret. Include appropriate units.

(d) Compare the results from parts (b) and (c).

SOLUTION

(a) A numerical table of values for the model is shown in Table 1.3.1.

TABLE 1.3.1

x	1	2	3	4	5	6	7	8	9	10
$f(x)$	5.94	8.54	11.78	15.66	20.18	25.34	31.14	37.58	44.66	52.38

(b) First we see that the year 1987 corresponds to $x = 1$ and the year 1991 corresponds to $x = 1991 - 1986 = 5$. This means that the increment in x is $\Delta x = 5 - 1 = 4$. So we need to compute the difference quotient

$$m_{\text{sec}} = \frac{f(x + \Delta x) - f(x)}{\Delta x} = \frac{f(1 + 4) - f(1)}{4} = \frac{f(5) - f(1)}{4}$$

The values from Table 1.3.1 give us

$$m_{\text{sec}} = \frac{20.18 - 5.94}{4} = \frac{14.24}{4} = 3.56 \quad \text{or} \quad 3.56 \, \frac{\text{billion dollars}}{\text{year}}$$

This means that during the period from 1987 to 1991, U.S. imports from China increased at an average rate of 3.56 $\frac{\text{billion dollars}}{\text{year}}$. Notice that the amount of imports increased, since the secant line slope is positive.

(c) For the years from 1991 to 1996, the corresponding x-values are $x = 5$ and $x = 10$, respectively. Thus, $\Delta x = 10 - 5 = 5$. Now we compute the difference quotient as

$$m_{\text{sec}} = \frac{f(5+5) - f(5)}{5} = \frac{f(10) - f(5)}{5}$$

$$= \frac{52.38 - 20.18}{5} = \frac{32.2}{5}$$

$$= 6.44 \quad \text{or} \quad 6.44 \; \frac{\text{billion dollars}}{\text{year}}$$

This means that during the period from 1991 to 1996, U.S. imports from China increased at an average rate of $6.44 \; \frac{\text{billion dollars}}{\text{year}}$.

(d) By comparing the rates in part (b) to those in part (c), we can see that total U.S. imports from China, on average, were increasing much faster between 1991 to 1996 than between 1987 to 1991. This comparison is shown graphically in Figure 1.3.5.

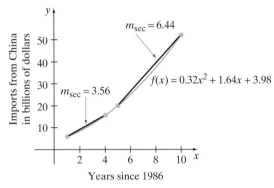

Figure 1.3.5

The difference quotient is a fundamental component of our understanding of differential calculus, so we will continue to compute the secant line slope throughout the remainder of Chapter 1 and into Chapter 2.

SUMMARY

This section featured the introduction of the **quadratic function**, $f(x) = ax^2 + bx + c$, which is a second-degree polynomial function. We found that every graph of a quadratic function is a parabola, and the **vertex of the parabola** was found at the coordinates $\left(\frac{-b}{2a}, f\left(\frac{-b}{2a} \right) \right)$. Roots, also known as zeros of the quadratic function, or x-intercepts of the graph, can be found in three ways:

- Algebraic methods like factoring
- The **quadratic formula,** $x = \frac{-b \pm \sqrt{b^2 - 4ac}}{2a}$
- The capabilities of the calculator

Finally, we introduced the **slope of the secant line** and found that this type of average rate of change could be computed by the difference quotient

$$m_{\text{sec}} = \frac{f(x + \Delta x) - f(x)}{\Delta x}$$

We found that the **difference quotient** could determine the average rate of change over an interval for a nonlinear function. As you will see in upcoming sections, the difference quotient will be used for more than just quadratic functions.

SECTION 1.3 EXERCISES

For Exercises 1–10:

(a) Determine the coordinates of the vertex algebraically.

(b) Use the result from part (a) to determine the intervals where the function is increasing and where it is decreasing.

1. $f(x) = x^2 + 6x + 5$ 2. $f(x) = x^2 - 4x + 2$

3. $g(x) = -x^2 + 6x + 6$ 4. $g(x) = -2x^2 - 4x + 5$

5. $g(x) = 5x^2 + 6x - 3$ 6. $g(x) = 5x^2 - 7x + 2$

7. $f(x) = 0.14x^2 + 0.5x - 0.3$

8. $f(x) = 0.81x^2 + 3.24x - 0.4$

9. $f(x) = -0.9x^2 - 1.8x + 0.5$

10. $f(x) = -0.35x^2 + 2.8x - 0.3$

For Exercise 11–18, find the zeros of the quadratic function by factoring.

11. $f(x) = x^2 - 16$ 12. $f(x) = x^2 - 49$

13. $f(x) = -2x^2 - 4x$ 14. $f(x) = -13x^2 - 39x$

15. $f(x) = x^2 - 5x + 6$ 16. $f(x) = x^2 + 2x - 8$

17. $g(x) = 6x^2 - 5x - 50$ 18. $g(x) = 8x^2 + 14x + 3$

In Exercises 19–24, find the real number roots of the quadratic function by the quadratic formula or write *no real roots.*

✓ 19. $f(x) = x^2 - x - 1$ 20. $f(x) = x^2 - 3x - 2$

21. $f(x) = 11x^2 - 7x + 1$ 22. $f(x) = 4x^2 - 12x + 11$

23. $g(x) = 2x^2 + 2x + 1$ 24. $f(x) = 9x^2 - 12x + 8$

In Exercises 25–30, use the ZERO or ROOT capabilities of your calculator to approximate the zeros of the quadratic function. Round to the nearest hundredth. If the graph of the function has no *x*-intercepts, write *no real zeros.*

25. $f(x) = x^2 - 2x - 24$ 26. $f(x) = x^2 - 2x - 15$

27. $f(x) = 4x^2 - 12x + 9$ 28. $g(x) = 9x^2 + 24x + 16$

29. $g(x) = x^2 - 5x + 8$ 30. $f(x) = 1.53x^2 - 3x + 2.67$

In Exercises 31–36, use the difference quotient to determine the slope of the secant line for each function over the indicated intervals.

31. $f(x) = -5x + 1$
(a) *x* changes from $x = 0$ to $x = 6$.
(b) *x* changes from $x = 0$ to $x = 3$.
(c) *x* changes from $x = 3$ to $x = 4$.

32. $f(x) = 3x - 5$
(a) *x* changes from $x = 0$ to $x = 8$.
(b) *x* changes from $x = 0$ to $x = 4$.
(c) *x* changes from $x = 4$ to $x = 8$.

33. $f(x) = x^2 + x$
(a) *x* changes from $x = 0$ to $x = 4$.
(b) *x* changes from $x = 0$ to $x = 2$.
(c) *x* changes from $x = 2$ to $x = 4$.

34. $f(x) = x^2 - 2x$
(a) *x* changes from $x = 0$ to $x = 6$.
(b) *x* changes from $x = 0$ to $x = 3$.
(c) *x* changes from $x = 3$ to $x = 6$.

35. $f(x) = x^2 - x + 3$
(a) *x* changes from $x = 0$ to $x = 2$.
(b) *x* changes from $x = 0$ to $x = 1$.
(c) *x* changes from $x = 1$ to $x = 2$.

36. $f(x) = 2x^2 - 3$
(a) If $x = 2$ and $\Delta x = 3$.
(b) If $x = 2$ and $\Delta x = 1$.
(c) If $x = 2$ and $\Delta x = 0.5$.

APPLICATIONS

37. The number of parking citations that the Sampsonburg Police Department gave over a five-year period can be modeled by

$$f(x) = \frac{1}{5}x^2 + 100x + 30, \qquad 0 \le x \le 5$$

where *x* represents the number of years since record taking of the citations began and $f(x)$ represents the number of citations given. Determine $\dfrac{f(5) - f(0)}{5}$ and interpret.

38. The number of unbreakable sunglasses sold at the U-C-Me specialty store during a long-term sales promotion can be modeled by

$$g(x) = \frac{1}{10}x^2 + 50x + 10, \qquad 0 \le x \le 10$$

where *x* represents the number of months since the promotion began and $g(x)$ represents the number of sunglasses sold. Determine $\dfrac{g(10) - g(0)}{10}$ and interpret.

39. If a rock is thrown from the ground with an initial velocity of 80 feet per second, then its height can be modeled by

$$s(t) = -16t^2 + 80t, \qquad 0 \le t \le 5$$

where t represents the number of seconds since the rock was thrown and $s(t)$ represents the rock's height in feet.

(a) Evaluate $s(3)$ and interpret.

(b) Determine the average rate of change in the rock's height for $t = 1$ to $t = 3$ and interpret.

40. If a model rocket is launched from a 3-foot platform with an initial velocity of 150 feet per second, then its height can be modeled by

$$s(t) = -16t^2 + 150t + 3$$

where t represents the number of seconds since launch and $s(t)$ represents the height in feet.

(a) Evaluate $s(8)$ and interpret.

(b) Determine the average rate of change in the rocket's height for $t = 5$ to $t = 8$ and interpret.

41. The number of bacteria in a colony after t hours is given by

$$g(t) = t^2 + 8t + 2000, \qquad 0 \le t \le 24$$

where t is the number of hours since the colony was established and $g(t)$ represents the number of bacteria.

(a) Evaluate $g(3)$ and interpret.

(b) Determine the average rate of change in the increase of the colony's population for $t = 3$ to $t = 6$ and interpret.

42. The number of seeds dispersed by a plot of dandelions each day in April can be modeled by

$$f(x) = 40x - x^2, \qquad 1 \le x \le 30$$

where x represents the number of days in April and $f(x)$ represents the number of seeds dispersed.

(a) Evaluate $f(7)$ and interpret. Does the graph of the model open up or down? How do you know this?

(b) On which day(s) of the month is the dispersion of seeds the greatest?

(c) Determine the average rate of change in the dispersion of the seeds between $x = 24$ and $x = 29$ and interpret.

43. The annual number of aggravated assaults in the United States can be modeled by

$$f(x) = -3.08x^2 + 40.35x + 305.89, \qquad 0 \le x \le 10$$

where x represents the number of years since 1986 and $f(x)$ represents the number of aggravated assaults per 100,000 people.

(a) Evaluate $f(2)$ and interpret. Does the graph of the model open up or down? How do you know this?

(b) Use your calculator to make a graph of the annual aggravated assault rate for $0 \le x \le 10$. During what year was the number of aggravated assaults per 100,000 people the greatest from 1986 to 1996?

(c) Determine the average rate of change in the number of aggravated assaults per 100,000 people between $x = 4$ and $x = 9$.

44. Courthouse records show that the average property tax on a three-bedroom home in Independence, during the 1990s can be modeled by

$$f(x) = 22x^2 + 35x + 500, \qquad 0 \le x \le 9$$

where x represents the number of years since 1990 and $f(x)$ is the average property tax, measured in dollars.

(a) Does the graph of the model open up or down? How do you know this?

(b) The domain of the model consists of x-values in the interval $[0, 9]$. What is the corresponding range?

(c) Determine the average rate of change in the property tax from 1991 to 1995.

45. The number of calories consumed each day per person in the United States can be modeled by

$$f(x) = 0.89x^2 - 1.93x + 3306.27, \qquad 0 \le x \le 15$$

where x represents the number of years since 1974 and $f(x)$ represents the number of calories consumed each day per person.

(a) Evaluate $f(6)$ and interpret.

(b) Make a graph of the model in the window $[0, 15]$ by $[3200, 3500]$.

(c) During what year did the calorie consumption exceed 3400 calories each day?

(d) Determine the average rate of change in the number of calories consumed each day per person on the interval $[1, 11]$ and interpret.

46. The number of lawyers in private practice in the United States can be modeled by

$$f(x) = 0.37x^2 - 1.43x + 190.34, \qquad 0 \le x \le 30$$

where x represents the number of years since 1960 and $f(x)$ represents the total number of lawyers in private practice, measured in thousands.

(a) Evaluate $f(22)$ and interpret.

(b) Make a table of the function $f(x)$ for the values $x = 0, 5, 10, 15, 25,$ and 30.

(c) Determine the average rate of change in the number of lawyers in private practice from 1970 to 1980, and interpret using appropriate units.

(d) Determine the average rate of change in the number of lawyers in private practice from 1980 to 1990, and compare the result to part (c).

Section Project

The number of bowling alleys in the United States can be modeled by

$$f(x) = -4.2x^2 + 2.2x + 8622.97, \qquad 0 \le x \le 20$$

Here x represents the number of years since 1975, and $f(x)$ represents the number of bowling alleys open for business in the United States that year.

(a) According to the model, how many bowling alleys were open for business in the United States in 1978?

(b) Does the graph of the function open up or down? How do you know this from inspecting the model?

(c) Make a graph of the model in the viewing window $[0, 20]$ by $[7000, 8700]$. Is the graph consistent with your answer to part (b)?

(d) Determine the average rate of change in the number of bowling alleys in the United States from 1975 to 1985 and interpret.

(e) Determine the average rate of change in the number of bowling alleys in the United States from 1985 to 1995 and interpret.

(f) Does the average rate of change during the time periods in parts (d) and (e) indicate a steady decline or a sharp decline in bowling establishments?

SECTION 1.4   OPERATIONS ON FUNCTIONS

In this section we extend our discussion to functions of higher degree. In general, we can call this family of functions **polynomial functions.** While continuing to build up our tools to prepare for the study of calculus, we will also discuss **operations on functions,** which are commonly used to determine the **business functions.** Finally, we will revisit the difference quotient and examine more applications of the secant line.

Operations on Functions

The basics of arithmetic with real numbers—addition, subtraction, multiplication, and division—are called mathematical **operations.** We can perform these operations on functions.

Operations on Functions

Let f and g be functions. Then, for all x-values for which both f and g exist, we have

1. The **sum** of f and g is $(f + g)(x) = f(x) + g(x)$.
2. The **difference** of f and g is $(f - g)(x) = f(x) - g(x)$.
3. The **product** of f and g is $(f \cdot g)(x) = f(x) \cdot g(x)$.
4. The **quotient** of f and g is $\left(\dfrac{f}{g}\right)(x) = \dfrac{f(x)}{g(x)}$, provided that $g(x) \neq 0$.

EXAMPLE 1   **Evaluating Operations of Functions**

Let $f(x) = x^2 + x - 3$ and $g(x) = 4x + 2$. Evaluate the following:

(a) $(f + g)(1)$ (b) $(f - g)(1)$ (c) $(f \cdot g)(1)$ (d) $\left(\dfrac{f}{g}\right)(1)$

SOLUTION

We first evaluate the two functions at $x = 1$ and get

$$f(1) = (1)^2 + 1 - 3 = -1 \quad \text{and} \quad g(1) = 4(1) + 2 = 6$$

(a) From the definition of sum of functions, we get

$$(f + g)(1) = f(1) + g(1) = -1 + 6 = 5$$

(b) From the definition of difference of functions, we get

$$(f - g)(1) = f(1) - g(1) = -1 - 6 = -7$$

(c) From the definition of the product of functions, we get

$$(f \cdot g)(1) = f(1) \cdot g(1) = (-1)(6) = -6$$

(d) From the definition of the quotient of functions, we get

$$\left(\frac{f}{g}\right)(1) = \frac{f(1)}{g(1)} = \frac{-1}{6}$$

■

✓ **CHECKPOINT 1**

Now work Exercise 11.

The critical restriction when using operations with functions occurs when using the quotient $\left(\frac{f}{g}\right)(x)$. We must make sure to exclude all values from the domain that can make the denominator zero.

Functions of Business

The operations on functions are commonly used in the functions of business, such as price–demand and cost functions. Before we extend this important family of functions, let's review a few business functions that we have already studied.

From Your Toolbox

1. The **price–demand function** p gives us the price $p(x)$ at which people buy exactly x units of product.
2. The cost $C(x)$ of producing x units of a product is given by the **cost function**

$$C(x) = \left(\begin{array}{c}\text{variable} \\ \text{costs}\end{array}\right) \cdot \left(\begin{array}{c}\text{units} \\ \text{produced}\end{array}\right) + \left(\begin{array}{c}\text{fixed} \\ \text{costs}\end{array}\right)$$

Other types of business functions are illustrated in everyday life. Let's say that a youngster has a lemonade stand in her front yard and that she charges 50 cents for each cup. If 35 cups are sold in an afternoon, the amount that the youngster makes is $0.50(35) = \$17.50$. This amount earned represents the **revenue.** We computed this revenue by multiplying the price per cup, called the **unit price,** times the quantity sold. We can generalize this notion to form the **revenue function.**

Revenue Function

The revenue generated by selling a certain quantity of a product at a certain price is

$$\left(\begin{array}{c}\text{Total} \\ \text{revenue}\end{array}\right) = \left(\begin{array}{c}\text{quantity} \\ \text{sold}\end{array}\right) \cdot \left(\begin{array}{c}\text{unit} \\ \text{price}\end{array}\right)$$

For x units sold at a price given by the price function $p(x)$, the **revenue function** R is

$$R(x) = x \cdot [p(x)]$$

EXAMPLE 2

Determining a Revenue Function

The Professional Bookstore knows from past sales records that the weekly number of units demanded, x, of a certain pack of computer tutorial software at price p is as shown in Table 1.4.1.

TABLE 1.4.1

DEMAND x	PRICE $p(x)$
10	$125
30	$75

(a) Use Table 1.4.1 to find a linear price function, p, written in CF form for the tutorial software.

(b) Use the result from part (a) to determine the revenue function R and graph R.

(c) Determine the vertex of the graph of R and interpret each coordinate.

SOLUTION

(a) To find the CF form, we must compute the slope m of the graph of the price function given by the table,

$$m = \frac{75 - 125}{30 - 10} = \frac{-50}{20} = -\frac{5}{2}$$

Using the CF form with $(x_1, y_1) = (10, 125)$, we find that the price function is

$$p(x) = -\frac{5}{2}(x - 10) + 125$$

(b) To get the revenue function R, we need to multiply the price function by the quantity x. This gives us

$$R(x) = x \cdot [p(x)] = x \cdot \left[-\frac{5}{2}(x - 10) + 125 \right]$$

$$= x \cdot \left[-\frac{5}{2}x + 25 + 125 \right] = x \cdot \left[-\frac{5}{2}x + 150 \right] = -\frac{5}{2}x^2 + 150x$$

Notice that this revenue function is a parabola that opens down (Figure 1.4.1). This is a common shape for revenue functions.

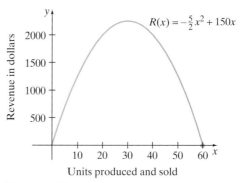

Figure 1.4.1

(c) The point where the revenue is a maximum is at the vertex. Recall that the x-coordinate of the vertex is given by $x = \frac{-b}{2a}$. For this function, we know that $a = -\frac{5}{2}$ and $b = 150$, so we get

$$x = \frac{-150}{2\left(-\frac{5}{2}\right)} = \frac{-150}{-5} = 30$$

Evaluating this number in the revenue function, we get

$$R(30) = -\frac{5}{2}(30)^2 + 150(30)$$

$$= -2250 + 4500 = 2250$$

So the weekly revenue is at its maximum of $2250 dollars when 30 packs of tutorial software are sold.   ■

Let's return to our young entrepreneur. If the cost of making the lemonade was $4.25, then the **profit** that is made during that afternoon is $17.50 − $4.25 = $13.25. This same model applies no matter how small or large the business venture may be. The profit made is just revenue minus the cost. This relationship between these two quantities allows us to form the **profit function.**

Profit Function

The profit generated after producing and selling a certain quantity of a product is given by

$$\text{Profit} = \text{revenue} - \text{cost}$$

For x units of a product, the **profit function** P is

$$P(x) = R(x) - C(x)$$

where R and C are the revenue and cost functions, respectively.

Notice that the last two definitions use operations on functions. The revenue function is a product of functions, and the profit function is the difference of functions.

EXAMPLE 3 | **Determining a Profit Function**

The FlashBroc Company makes designer ties. It determines that the weekly cost and revenue functions, in dollars, for producing and selling x designer ties are

$$C(x) = 30x + 50 \quad \text{and} \quad R(x) = 90x - x^2$$

respectively.

(a) Determine the profit function P and graph P.

(b) Find the vertex of the graph of P and interpret each coordinate.

SOLUTION

(a) Since the definition of the profit function is $P(x) = R(x) - C(x)$, we get

$$P(x) = R(x) - C(x) = (90x - x^2) - (30x + 50) = -x^2 + 60x - 50$$

So the profit function for producing and selling x units is
$P(x) = -x^2 + 60x - 50$. A graph of P is given in Figure 1.4.2.

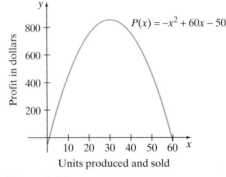

Figure 1.4.2

(b) The x-coordinate of the vertex is $x = \dfrac{-b}{2a} = \dfrac{-60}{2(-1)} = 30$. The y-coordinate is given by

$$P(30) = -30^2 + 60(30) - 50 = -900 + 1800 - 50 = 850$$

This means that a maximum profit of $850 will be realized when 30 designer ties are produced and sold each week. ✦

Intuitively, when the revenue is equal to the cost (that is, when money coming in is equal to money going out), we have reached the **break-even point.** Mathematically, this occurs at the point where the graphs of the cost and revenue functions intersect. See Figure 1.4.3.

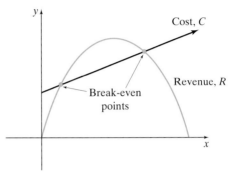

Figure 1.4.3

Algebraically, this means that the break-even point is the x-value where $R(x) = C(x)$. If the cost is more than the revenue, the result is **profit loss,** and when the revenue is greater than the cost, a **profit gain** is the result. Let's try an example of how we can use these functions in practice.

EXAMPLE 4 | **Determining the Break-even Points**

In Example 2, we showed that the revenue function for the tutorial software packs was $R(x) = -\dfrac{5}{2}x^2 + 150x$. Now assume that the fixed costs for the software are $1500, with variable costs of $8\,\dfrac{\text{dollars}}{\text{pack}}$.

(a) Write a linear cost function C for the tutorial software.

(b) Find the break-even point(s) and round to the nearest whole number. Interpret the results.

(c) Determine the profit function P.

SOLUTION

(a) Using the definition of the linear cost function with variable costs of $m = 8\,\dfrac{\text{dollars}}{\text{pack}}$ and fixed costs of $1500, we get $C(x) = 8x + 1500$.

(b) For the break-even point we need the x-values such that

$$R(x) = C(x)$$
$$-\frac{5}{2}x^2 + 150x = 8x + 1500$$
$$-\frac{5}{2}x^2 + 142x - 1500 = 0$$
$$-5x^2 + 284x - 3000 = 0$$

Since this quadratic is not easily factorable, we use the quadratic formula to get

$$x = \frac{-284 \pm \sqrt{(284)^2 - 4(-5)(-3000)}}{2(-5)}$$

$$= \frac{-284 \pm \sqrt{20,656}}{-10} \approx 14.03 \quad \text{and} \quad 42.77$$

Interactive Activity

 Graphically verify the solution to part (b) of Example 4 in two ways. First graph R and C in the same viewing window and use the INTERSECT command on your calculator to find the break-even points. Then graph the profit function P and use the ZERO or ROOT command on your calculator to find the x-intercepts.

Rounding to the nearest whole number, we get the values of 14 and 43. Since the graph of R opens down and C is increasing, this means that a profit from the tutorial software is realized when the weekly sales are between 14 and 43 packs.

(c) Using our definition of the profit function, we have

$$P(x) = R(x) - C(x)$$

$$= \left(-\frac{5}{2}x^2 + 150x\right) - (8x + 1500)$$

$$= -\frac{5}{2}x^2 + 142x - 1500$$

So the profit function for the tutorial software packs is

$$P(x) = -\frac{5}{2}x^2 + 142x - 1500.$$

We see that the break-even points on the cost and revenue functions lie directly above the zeros of the profit function P. See Figure 1.4.4.

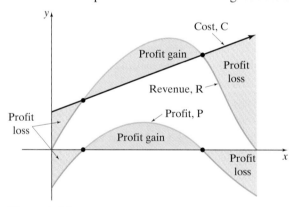

Figure 1.4.4

Thus, another way to find the break-even points is to find the zeros of the profit function. The profit gain is realized in the interval where $P(x)$ lies above the x-axis (that is, $P(x) > 0$). Let's summarize the relationship between P, R, and C before continuing.

- Break-even occurs when $R(x) = C(x)$. This is equivalent to $P(x) = 0$. Graphically, $P(x) = 0$ at the x-intercepts.
- Profit gain occurs when $R(x) > C(x)$. This is equivalent to $P(x) > 0$, when the graph of P is above the x-axis.
- Profit loss occurs when $R(x) < C(x)$. This is equivalent to $P(x) < 0$, when the graph of P is below the x-axis.

We can also apply the difference quotient to business functions to approximate the average rate of change on specific intervals.

EXAMPLE 5 | **Determining the Secant Line Slope of a Business Function**

The Midwest Manufacturing Company manufactures titanium lock washers. Its revenue function is given by

$$R(x) = -0.02x^2 + 8x$$

where x represents the number of titanium lock washers sold and $R(x)$ represents the amount of revenue generated in dollars. Find the average rate of change in revenue as the number of washers sold increases from 50 to 150 washers and interpret.

SOLUTION

To find the desired average rate of change, we need to compute the difference quotient

$$m_{\text{sec}} = \frac{R(x + \Delta x) - R(x)}{\Delta x}$$

With $x = 50$ and $\Delta x = 150 - 50 = 100$, we get

$$m_{\text{sec}} = \frac{R(50 + 100) - R(50)}{100} = \frac{R(150) - R(50)}{100}$$

$$= \frac{-0.02(150)^2 + 8(150) - (-0.02(50)^2 + 8(50))}{100} = \frac{750 - 350}{100}$$

$$= \frac{400}{100} = 4$$

This means that as the number of titanium lock washers sold increases from 50 to 150, the revenue increases at an average rate of $4 \frac{\text{dollars}}{\text{washer}}$. See Figure 1.4.5.

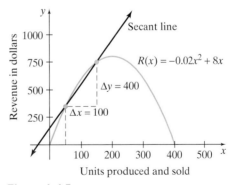

Figure 1.4.5

✓CHECKPOINT **2**

Now work Exercise 61.

Polynomial Functions

So far we have studied linear, or first-degree, functions and quadratic, or second-degree, functions. Now let's take a look at functions of higher degree. We call this general family of functions **polynomial functions.**

> **Polynomial Function**
>
> A **polynomial function** of degree n has the form
>
> $$f(x) = a_n x^n + a_{n-1} x^{n-1} + \cdots + a_1 x + a_0$$
>
> where $a_1, a_2, \ldots, a_n$ are real number constants and n is a whole number. The **degree** of the polynomial function is given by n.

Using this definition, we see that $f(x) = 2x^3 - 4x^2 - x + 5$ is a third-degree polynomial function, also called a **cubic** function. The functions $f(x) = \sqrt{x} + 3$ and $g(x) = \dfrac{4}{x} + 8$ are not polynomial functions because neither can be written with whole number exponents.

A few properties of polynomial functions make them very convenient to use mathematically.

- The domain of every polynomial function is the set of real numbers, $(-\infty, \infty)$. This means that there are no restrictions on the numbers evaluated in polynomial functions.
- There are no holes or breaks in the graphs of polynomial functions.
- We say that polynomial functions are **continuous.** This means their graphs are smooth and unbroken

EXAMPLE 6

Using Polynomial Models

The personal income per capita in the United States, in constant 1992 dollars, can be modeled by

$$f(x) = 3.75x^3 - 115.23x^2 + 1229.81x + 16{,}025.65, \qquad 1 \le x \le 15$$

where x represents the number of years since 1979 and $f(x)$ represents the personal income per capita in the United States in constant 1992 dollars.

(a) Use the difference quotient to determine the average rate of change in the personal income per capita where $x = 1$ and $\Delta x = 10$. Interpret the result.

(b) Use the difference quotient to determine the average rate of change in the personal income per capita where $x = 1$ and $\Delta x = 5$. Interpret the result. Compare it with the result from part (a).

SOLUTION

(a) Evaluating the difference quotient with $x = 1$ and $\Delta x = 10$, we get

$$m_{\text{sec}} = \frac{f(x + \Delta x) - f(x)}{\Delta x} = \frac{f(1 + 10) - f(1)}{10} = \frac{f(11) - f(1)}{10}$$

$$= \frac{[3.75(11)^3 - 115.23(11)^2 + 1229.81(11) + 16{,}025.65] - [3.75(1)^3 - 115.23(1)^2 + 1229.81(1) + 16{,}025.65]}{10}$$

$$= \frac{20{,}601.98 - 17{,}143.98}{10} = \frac{3458}{10} = 345.8$$

Here $x = 1$ corresponds to the year 1980, and $x + \Delta x = 11$ corresponds to 1990. Thus, during the period from 1980 to 1990, the personal per capita income was increasing at an average rate of $345.8 \frac{\text{dollars}}{\text{year}}$.

(b) Evaluating the difference quotient with $x = 1$ and $\Delta x = 5$, we get

$$m_{\text{sec}} = \frac{f(1 + 5) - f(1)}{5} = \frac{f(6) - f(1)}{5}$$

$$= \frac{20{,}066.23 - 17{,}143.98}{5} = \frac{2922.25}{5} = 584.45$$

This tells us that in the period between 1980 and 1985, the personal per capita income was increasing at an average rate of $584.45 \frac{\text{dollars}}{\text{year}}$. Knowing that the income data are adjusted to constant 1992 dollars, it appears that per capita income grew faster in the first half of the 1980s as compared to the decade as a whole. As can be seen in Figure 1.4.6, the personal income leveled off during the late 1980s. This is why the average rate of change is larger in part (b).

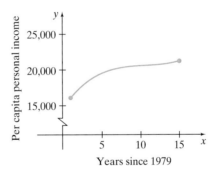

Figure 1.4.6 Graph of $f(x) = 3.75x^3 - 115.23x^2 + 1229.81x + 16{,}025.65$.

End Behavior of Functions

At times we want to know the values of a function for extremely large positive and extremely large negative values of the independent variable x. This determines what is called the **end behavior** of a function. To begin to see what happens, let's take a look at the quadratic function $f(x) = 2x^2 - 6x + 3$. Table 1.4.2 shows the function values for extremely large positive and negative values of x. From this table, it appears that as the x-values grow both negatively and positively large the function values grow positively large. Table 1.4.2 illustrates just what end behavior is all about. We commonly describe the end behavior symbolically. We write as $x \to -\infty$, $f(x) \to \infty$ to denote that, as we move from right to left on the x-axis, the $f(x)$-values grow large positively. (This is read "as x approaches negative infinity, $f(x)$ approaches infinity"). We write as $x \to \infty$, $f(x) \to \infty$ to denote that, as we move from left to right on the x-axis, the $f(x)$-values grow large positively. (This is read "as x approaches infinity, $f(x)$ approaches infinity").

To determine the end behavior of a polynomial function, all we need do is examine the leading term, $a_n x^n$, of the polynomial. We can summarize the results with the **leading coefficient test.**

TABLE 1.4.2

x	$f(x) = 2x^2 - 6x + 3$
$-100{,}000$	$20{,}000{,}600{,}003$
$-10{,}000$	$200{,}060{,}003$
-1000	$2{,}006{,}003$
-100	$20{,}603$
0	3
100	$19{,}403$
1000	$1{,}994{,}003$
$10{,}000$	$199{,}940{,}003$
$100{,}000$	$19{,}999{,}400{,}003$

Leading Coefficient Test

1. Assume that $f(x)$ is an **even-degree** polynominal function with leading term $a_n x^n$ and leading coefficient a_n.

If the leading coefficient is	as $x \to -\infty$	as $x \to \infty$	The typical graph is
Positive ($a_n > 0$)	$f(x) \to \infty$	$f(x) \to \infty$	
Negative ($a_n < 0$)	$f(x) \to -\infty$	$f(x) \to -\infty$	

2. Assume that $f(x)$ is an **odd-degree** polynominal function with leading term $a_n x^n$ and leading coefficient a_n.

If the leading coefficient is	as $x \to -\infty$	as $x \to \infty$	The typical graph is
Positive ($a_n > 0$)	$f(x) \to -\infty$	$f(x) \to \infty$	
Negative ($a_n < 0$)	$f(x) \to \infty$	$f(x) \to -\infty$	

EXAMPLE 7 | **Determining the End Behavior of a Polynomial Function**

For the cubic function $f(x) = x^3 - 3x^2 - 6x + 8$:

(a) Determine the end behavior of the function.

(b) Use the MAXIMUM and MINIMUM commands on the calculator to approximate the *peaks* and *valleys* of the graph of the function. Round the coordinates to the nearest hundredth. Use this result to determine the intervals where the function increases and where it decreases.

SOLUTION

(a) From case 2 of the Leading Coefficient Test, since f is a third, or odd, degree polynomial function and the leading coefficient, $a_n = 1$, is positive, we

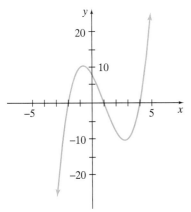

Figure 1.4.7
Graph of $f(x) = x^3 - 3x^2 - 6x + 8$.

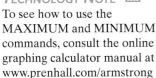

TECHNOLOGY NOTE
To see how to use the
MAXIMUM and MINIMUM
commands, consult the online
graphing calculator manual at
www.prenhall.com/armstrong

✓**CHECKPOINT 3**

know that as $x \to -\infty$, $f(x) \to -\infty$ and as $x \to \infty$, $f(x) \to \infty$. This is shown graphically in Figure 1.4.7 with the arrowheads at the beginning and end of the graph indicating that the graph continues in the same direction.

(b) Using the calculator, we obtain the MAXIMUM shown in Figure 1.4.8a and the MINIMUM shown in Figure 1.4.8b. So we estimate a peak or maximum point at $(-0.73, 10.39)$ and a valley or minimum point at $(2.73, -10.39)$. We can see from the graph that $f(x) = x^3 - 3x^2 - 6x + 8$ is increasing, or going uphill, for all the x-values to the left of $x = -0.73$ and to the right of $x = 2.73$. So, using interval notation, we write that f is increasing on the intervals $(-\infty, -0.73) \cup (2.73, \infty)$ and is decreasing on the interval $(-0.73, 2.73)$. Keep in mind that the calculator can only give us an **estimate** of the maximum and minimum values. To find the **exact** values, we need the power of calculus.

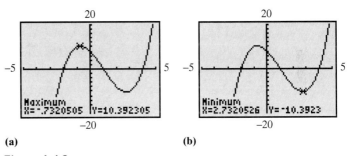

(a)   **(b)**

Figure 1.4.8

Now work Exercise 55.

Interactive Activity

Here are graphs of two cubic polynomial functions. Are they complete graphs of the functions, or is there a piece missing? How do you know this?

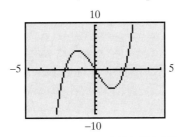 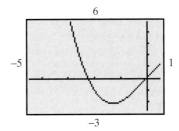

SUMMARY

This section began with operations on functions. We learned that we can add, subtract, multiply, and divide functions, much as we do with real numbers. Then we discussed the functions of business:

- The **revenue function** $R(x) = x \cdot [p(x)]$
- The **profit function** $P(x) = R(x) - C(x)$
- The **break-even** point(s) occurs when $R(x) = C(x)$

Then we turned our attention to the properties of polynomial functions. These functions all have the form

$$f(x) = a_n x^n + a_{n-1} x^{n-1} + \cdots + a_1 x + a_0$$

where $a_1, a_2, \ldots, a_n$ are real number constants and n is a whole number. The end behavior can be found by using the leading coefficient test.

LEADING COEFFICIENT TEST

	n Is Odd	n Is Even
$a_n > 0$	As $x \to -\infty$, $f(x) \to -\infty$ and as $x \to \infty$, $f(x) \to \infty$	As $x \to \pm\infty$, $f(x) \to \infty$
$a_n < 0$	As $x \to -\infty$, $f(x) \to \infty$ and as $x \to \infty$, $f(x) \to -\infty$	As $x \to \pm\infty$, $f(x) \to -\infty$

SECTION 1.4 EXERCISES

For Exercises 1–4, simplify and write the domain for the following operations.

(a) $(f + g)(x)$ (b) $(f - g)(x)$

(c) $(f \cdot g)(x)$ (d) $\left(\dfrac{f}{g}\right)(x)$

1. $f(x) = 6x + 3$, $g(x) = 4x - 1$

2. $f(x) = -2x + 9$, $g(x) = -5x - 2$

3. $f(x) = \sqrt{x + 5}$, $g(x) = x^2 + 5$

4. $f(x) = \sqrt{x - 3}$, $g(x) = x - 5$

For Exercises 5–12, let $f(x) = 2x^2 - 4x$ and $g(x) = 5x + 1$. Evaluate each of the following:

5. $(f + g)(0)$ 6. $(f + g)(-3)$

7. $(f \cdot g)(4)$ 8. $(f \cdot g)(-2)$

9. $(f - g)(2)$ 10. $(f - g)(-1)$

✓11. $\left(\dfrac{f}{g}\right)(5)$ 12. $\left(\dfrac{f}{g}\right)(-1)$

APPLICATIONS

13. The Handi-Neighbor hardware store sells PowerDriver hammers and finds that the amount made in dollars for the weekly sales is

$$f(x) = 9.75x, \qquad 0 \le x \le 50$$

The competing U-Do-It hardware store finds that the weekly sales for selling PowerDriver hammers is

$$g(x) = \frac{4}{5}x^2 + 6x, \qquad 0 \le x \le 50$$

In both functions, x represents the number of PowerDriver hammers sold.

(a) Evaluate $(f - g)(20)$ and interpret.

(b) Evaluate $(f + g)(20)$ and interpret.

14. The Nuthin'-But-Socks specialty store finds that the monthly expenses for operation during its first year of business is given by

$$f(x) = \frac{1}{5}x^2 + 300, \qquad 1 \le x \le 12$$

where $x = 1$ corresponds to January, $x = 2$ corresponds to February, and so on, and $f(x)$ is the monthly expenses in hundreds of dollars. The monthly revenue during the first year of business is given by

$$g(x) = 27.5x, \qquad 1 \le x \le 12$$

where $x = 1$ corresponds to January, $x = 2$ corresponds to February, and so on, and $g(x)$ is the monthly revenue in hundreds of dollars.

(a) Evaluate $(g - f)(6)$ and interpret.

(b) Did the store break even before the end of the first year?

15. The number of arrests for arson in the town of Waterbridge from 1960 to 1970 can be modeled by

$$f(x) = 4x + 45, \qquad 0 \le x \le 10$$

where x represents the number of years since 1960, and $f(x)$ represents the number of arrests. The number of arson convictions in Waterbridge during the same time period is given by

$$g(x) = x + 35, \qquad 0 \le x \le 10$$

(a) Determine the arson conviction rate during 1967.

(b) Determine a model for the arson *conviction rate* in Waterbridge during the decade. The conviction rate is the number of convictions divided by the number of arrests.

Exercises 16 and 17 use the following. The number of drug **convictions** in the United States can be modeled by the quartic function

$$y_1 = f(x) = 6.13x^4 - 207.23x^3 + 2127.9x^2$$
$$-6622.41x + 15{,}219.92, \qquad 1 \le x \le 11$$

where x represents the number of years since 1984, and $f(x)$ represents the number of drug convictions in the United States. The number of drug **arrests** in the United States can be modeled by

$$y_2 = g(x) = 27.82x^4 - 710.56x^3 + 5963.14x^2$$
$$-1695.09x + 27{,}423.98, \qquad 1 \le x \le 11$$

where x represents the number of years since 1984 and $g(x)$ represents the number of drug arrests in the United States.

16. (a) Evaluate $f(10)$ and $g(10)$ and interpret each.

(b) Determine $\left(\dfrac{f}{g}\right)(10)$ using the solutions to part (a). Interpret the result.

17. (a) Use your calculator to graph $y_3 = \dfrac{y_1}{y_2} = \left(\dfrac{f}{g}\right)(x)$ in the viewing window $[1, 11]$ by $[0, 1]$. This quotient of f and g represents the **conviction rate** for drug arrests in the United States from 1985 to 1995.

(b) Over which two-year period of time did the conviction rate decrease the most—$[1, 3], [5, 7],$ or $[9, 11]$?

Exercises 18 and 19 use the following. The annual Medicaid **payments** in the United States can be modeled by

$$y_1 = f(x) = 413.48x^2 + 185.72x + 24{,}031.95, \quad 0 \le x \le 15$$

where x represents the number of years since 1980 and $f(x)$ represents the amount of Medicaid payments in millions of dollars. The annual number of Medicaid **recipients** in the United States can be modeled by

$$y_2 = g(x) = 0.004x^4 + 0.02x^3 + 0.01x^2 \\ -0.24x + 21.66, \quad 0 \le x \le 15$$

where x represents the number of years since 1980 and $g(x)$ represents the number of Medicaid recipients in millions of people.

18. (a) Evaluate $f(5)$ and $g(5)$ and interpret each.

(b) Determine $\left(\dfrac{f}{g}\right)(5)$ using the solutions to part (a). Interpret the result.

19. (a) Use your calculator to graph $y_3 = \dfrac{y_1}{y_2} = \left(\dfrac{f}{g}\right)(x)$ in the viewing window $[0, 15]$ by $[400, 1400]$. This quotient of f and g is the **amount paid per recipient** of Medicaid each year from 1980 to 1995.

(b) By using the TRACE command on your calculator, find the year in which the amount paid per Medicaid recipient dropped below $900.

For Exercises 20–27, the price–demand function p is given. Determine the revenue function R.

20. $p(x) = 2.55$

21. $p(x) = 87.1$

22. $p(x) = -3.1x$

23. $p(x) = -0.19x$

24. $p(x) = -0.3x + 20$

25. $p(x) = \dfrac{-x}{2000} + 3$

26. $p(x) = -0.1x + 50$

27. $p(x) = -0.12x + 30$

For Exercises 28–31, the revenue and cost functions are given.

(a) Graph R and C in the same viewing window.

(b) Use algebra or the INTERSECT command on your calculator to determine the break-even point.

(c) Determine how much revenue must be generated to reach the break-even point.

28. $R(x) = 50x$; $C(x) = 20x + 170$

29. $R(x) = 8x$; $C(x) = 5000$

30. $R(x) = 25x - 0.25x^2$; $C(x) = 2x + 5$

31. $R(x) = 100x - x^2$; $C(x) = 20x + 4$

For Exercises 32–35, determine the profit function P, find the vertex of the graph of P, and interpret each coordinate.

32. $R(x) = 120x - 6x^2$; $C(x) = 240 + 20x$

33. $R(x) = 1200x - 37x^2$; $C(x) = 4300 + 148x$

34. $R(x) = 95.2x - 5x^2$; $C(x) = 155 + 20x$

35. $R(x) = -2.1x^2 + 500x$; $C(x) = 80x + 9500$

For Exercises 36–41, the price–demand function p, along with the fixed and variable costs, is given.

(a) Determine the profit function $P(x) = R(x) - C(x)$.
(b) Use the ZERO or ROOT command on your calculator to find the zeros of the profit function P. These are the break-even points.
(c) Find the vertex of the graph of the profit function P, and interpret each coordinate of the vertex.

36. $p(x) = 105.7 - 0.89x$; variable costs are $80 per unit and fixed costs are $61.80.

37. $p(x) = 37.8 - x$; variable costs are $29 per unit and fixed costs are $10.15.

38. $p(x) = 50 - x$; variable costs are $20 per unit and fixed costs are $200.

39. $p(x) = -10x + 1040$; variable costs are $500 per unit and fixed costs are $6650.

40. $p(x) = 40 - x$; variable costs are $20 per unit and fixed costs are $80.

41. $p(x) = -x + 30$; variable costs are $5 per unit and fixed costs are $60.

For Exercises 42–47, determine if the given function is a polynomial function. Identify each polynomial as linear, quadratic or cubic.

42. $f(x) = 7$

43. $f(x) = -30$

44. $y = 3x - \dfrac{5}{x}$

45. $g(x) = \dfrac{4}{x^2} + 3x + \dfrac{1}{2}$

46. $f(x) = x^2 + 5x + 2$

47. $f(x) = x^3 + x + 1$

In Exercises 48–57, determine the end behavior of the function.

48. $f(x) = 3x$

49. $f(x) = -17x$

50. $g(x) = -2.1x$

51. $g(x) = 3.8x$

52. $f(x) = -3x^2 + 10x$

53. $g(x) = 5.5x^2 + 4$

54. $f(x) = -x^3 + 2x - 1$ ✓ 55. $f(x) = 11x + 0.1x^3$

56. $f(x) = -x^6 - x$ 57. $g(x) = x^6 - 2x^3 - x^2$

APPLICATIONS

58. If a stone is thrown straight up from ground level with an initial velocity of 48 feet per second, then the height s of the stone above the ground after t seconds is given by

$$s(t) = -16t^2 + 48t$$

where t is the number of seconds after the stone has been thrown and $s(t)$ is the height of the stone in feet.

(a) Evaluate $s(3)$ and interpret.

(b) Find the average rate of change for $t = 1$ and $\Delta t = 2$. Interpret the result.

59. A girl throws a ball straight up from a 48-foot-tall building with an initial velocity of 30 feet per second. The height s of the ball above the ground after t seconds is given by

$$s(t) = -16t^2 + 30t + 48$$

where t is the number of seconds after the ball has been thrown and $s(t)$ is the height of the ball in feet.

(a) Evaluate $s(2)$ and interpret.

(b) Find the average rate of change for $t = 0$ and $\Delta t = 2.5$. Interpret the result.

60. The ActivLife Company finds that the revenue, in dollars, generated by selling x exercise machines is

$$R(x) = -\frac{x^2}{900} + 10x, \qquad x \geq 0$$

(a) Evaluate $R(30)$ and interpret.

(b) Find the average rate of change for $x = 10$ and $\Delta x = 90$ and interpret.

✓ 61. The BackStreet Company determines that the revenue, in dollars, generated by selling x of its Boomer headphones is given by

$$R(x) = 11x - \frac{x^2}{730}, \qquad x \geq 0$$

(a) Evaluate $R(250)$ and interpret.

(b) Find the average rate of change for $x = 100$ and $\Delta x = 200$ and interpret.

⊕ 62. The percent of people taking the SAT whose intended area of study was engineering can be modeled by the cubic function

$$f(x) = 0.002x^3 - 0.09x^2 + 1.27x + 6.76, \qquad 0 \leq x \leq 20$$

where x represents the number of years since 1975, and $f(x)$ represents the percent of people taking the SAT whose intended area of study was engineering.

(a) Evaluate $f(6)$ and interpret.

(b) Find the average rate of change for $x = 0$ and $\Delta x = 10$. Interpret the result.

(c) Find the average rate of change from 1985 to 1995. Interpret the result.

(d) Compare the results from parts (b) and (c).

⊕ 63. The percent of people taking the SAT whose intended area of study was business can be modeled by

$$f(x) = -0.09x^2 + 1.83x + 11.55, \qquad 0 \leq x \leq 20$$

where x represents the number of years since 1975, and $f(x)$ represents the percent of people taking the SAT whose intended area of study was business.

(a) Evaluate $f(7)$ and interpret.

(b) Find the average rate of change for $x = 0$ and $\Delta x = 10$. Interpret the result.

(c) Find the average rate of change from 1985 to 1995. Interpret the result.

(d) Compare the results from parts (b) and (c).

For Exercises 64–71, consider the given polynomial functions.

(a) Determine the end behavior of the function.

(b) Use the MAXIMUM and MINIMUM commands on the calculator to approximate the peaks and valleys of the graph. Estimate the points to the nearest hundredth.

(c) Use the solution from part (b) to determine the intervals where f increases and where it decreases.

64. $f(x) = 3x + 4$ 65. $f(x) = -3x + 6$

66. $y = x^2 - 6x + 7$ 67. $y = -x^2 - 10x + 21$

68. $g(x) = x^3 - 3x^2 - 9x + 15$

69. $g(x) = -2x^3 - 7x^2 + 4x - 2$

70. $f(x) = \frac{1}{2}x^4 - x^2$ 71. $f(x) = -x^4 + 8x + 12$

SECTION PROJECT

The annual total **debt** in the U.S. farming sector can be modeled by

$$y_1 = f(x) = 0.36x^3 - 4.71x^2 + 42.53x + 717.04, \quad 1 \leq x \leq 9$$

where x represents the number of years since 1986, and $f(x)$ represents the total farm debt in billions of dollars. The annual total **assets** in the U.S. farming sector can be modeled by

$$y_2 = g(x) = -0.02x^3 + 0.88x^2 - 6.14x + 149.25, \quad 1 \leq x \leq 9$$

where x represents the number of years since 1986, and $f(x)$ represents the total farm assets in billions of dollars.

(a) Evaluate $f(4)$ and $g(4)$ and interpret each.

(b) Determine $\left(\dfrac{f}{g}\right)(4)$ using the solutions to part (a). Interpret the result.

(c) Use your calculator to graph $y_3 = \dfrac{y_1}{y_2} = \left(\dfrac{f}{g}\right)(x)$ in the viewing window [1, 9] by [0, 10]. This quotient of f and g gives us what the U.S. Department of Agriculture calls the **farm debt/asset ratio.**

(d) From inspecting the graph, would you conclude that the debt/asset ratio generally increased, decreased, or remained constant from 1987 to 1996? Explain.

SECTION 1.5 — RATIONAL, RADICAL, AND POWER FUNCTIONS

Thus far, the type of functions that we have studied are polynomial functions. However, many of the functions that we study in calculus are not polynomials. In this section, we study **rational functions.** These functions are made up of polynomials, but their graphs do not resemble those that we have seen. Then we will examine **radical functions,** which are functions involving square roots, cube roots, and the like, and having restricted domains and ranges. Finally, we will investigate **power functions,** which have fractions, not only whole numbers, in their exponents. Once again we will see how these functions are used in applications, and we will calculate the slope of a secant line over an interval and interpret it as an average rate of change.

Rational Functions

Expressions for **rational functions** have fractions whose numerator and denominator are polynomials. Here are some examples of rational functions:

$$f(x) = \frac{x}{x+1}, \qquad g(x) = \frac{2x^2 - 1}{x - 2}, \qquad y = \frac{x - 3}{x^2 + x + 1}$$

A function like $f(x) = \frac{\sqrt{x+2}}{x+5}$ is not a rational function, since the numerator is not a polynomial.

Rational Function

A **rational function** has the form

$$y = \frac{f(x)}{g(x)}$$

where f and g are polynomials and $g(x) \neq 0$.

Notice that the domain of a rational function is determined by the set of all real numbers that make the denominator not equal to zero. Since there are usually x-values that make the denominator zero, rational functions frequently have breaks in their graphs.

EXAMPLE 1 | **Finding Vertical Asymptotes Using Tables**

Consider the rational function $f(x) = \frac{2x}{x - 3}$.

(a) Determine the domain.

(b) Complete Table 1.5.1 and describe what happens to the function values.

TABLE 1.5.1

x	2	2.9	2.99	2.999	2.9999	3.0001	3.001	3.01	3.1	4
$f(x)$										

SOLUTION

(a) We see that the denominator, $x - 3$, is zero when $x = 3$. So the domain of f is the set of all real numbers except 3 or, using interval notation, $(-\infty, 3) \cup (3, \infty)$.

(b)

TABLE 1.5.2

x	2	2.9	2.99	2.999	2.9999	3.0001	3.001	3.01	3.1	4
$f(x)$	−4	−58	−598	−5998	−59,998	60,002	6002	602	62	8

Table 1.5.2 shows that for x-values very close but less than 3, the function values get negatively large or, symbolically, $f(x) \to -\infty$. Also, for x-values very close but greater than 3, the function values grow positively large, or $f(x) \to \infty$. Since x cannot equal 3, the graph of $f(x) = \dfrac{2x}{x - 3}$ will never cross the vertical line $x = 3$. This line is called a **vertical asymptote.** A graph of f is given in Figure 1.5.1.

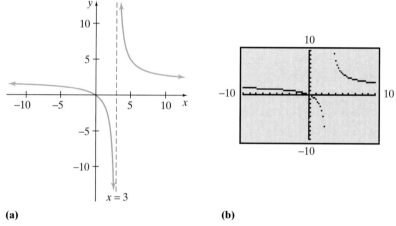

(a)   **(b)**

Figure 1.5.1   Graph of $f(x) = \dfrac{2x}{x - 3}$.

Making numerical tables is a sound, yet time-consuming, way to determine vertical asymptotes. Algebraically, we see that for the function $f(x) = \dfrac{2x}{x - 3}$, the value $x = 3$ makes the denominator equal to zero and yet makes the numerator a nonzero real number (in this case $2(3) = 6$). This is the common procedure for identifying vertical asymptotes.

Finding Vertical Asymptotes

Consider the rational function $h(x) = \dfrac{f(x)}{g(x)}$, where f and g are polynomials. If there is a value c that makes the denominator zero, yet the numerator is not zero, then the vertical line $x = c$ is a vertical asymptote.

The graphs of some rational functions have more than one vertical asymptote, as illustrated in Example 2.

EXAMPLE 2

Determining Vertical Asymptotes

Find the vertical asymptotes of the graph of the function $f(x) = \frac{x-4}{x^2+x-2}$ and graph.

SOLUTION

We start by factoring the denominator. This gives us

$$f(x) = \frac{x-4}{x^2+x-2} = \frac{x-4}{(x+2)(x-1)}$$

We see that the denominator of the rational function is zero when $x = -2$ and $x = 1$. Since neither of these values makes the numerator equal to zero, the graph of $f(x) = \frac{x-4}{x^2+x-2}$ has vertical asymptotes at the vertical lines $x = -2$ and $x = 1$. See Figure 1.5.2.

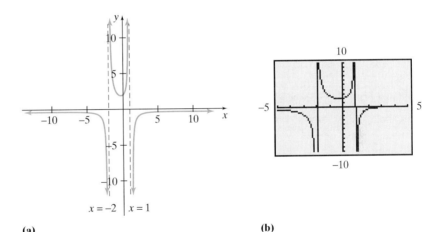

(a)   (b)

Figure 1.5.2   Graph of $f(x) = \frac{x-4}{x^2+x-2}$.

✓ CHECKPOINT 1

Now work Exercise 3.

We stated that vertical asymptotes are identified by values that make the denominator zero, yet the numerator is not zero. But what if a value makes both the numerator and denominator zero? Let's say we have the function $f(x) = \frac{x^2-4}{x-2}$. We see that the value $x = 2$ is excluded from the domain since this number makes the denominator zero. Since the factored form of the function is $f(x) = \frac{(x+2)(x-2)}{x-2}$, we can algebraically reduce the numerator and denominator by $(x-2)$ and rewrite the function as

$$f(x) = x+2, \qquad x \neq 2$$

The result is a line, $f(x) = x+2$, with a "hole" in the graph when $x = 2$. The graph of f is given in Figure 1.5.3.

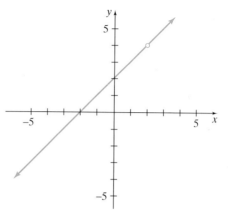

Figure 1.5.3 Graph of $f(x) = \dfrac{x^2 - 4}{x - 2}$ with a hole when $x = 2$.

End Behavior of Rational Functions

In Section 1.4, we said that the end behavior of a polynomial function describes the type of numerical values that a function takes on for extreme values of the independent variable x. To see how end behavior works for rational functions, we return to the rational function $f(x) = \dfrac{2x}{x - 3}$. The numerical output for extreme x-values is shown in Table 1.5.3. The . . . at the end of the number in the second column means that the decimal values continue on, but are not written.

From Table 1.5.3, it appears that as x gets very large positively and very large negatively, the function values are getting close to 2. Figure 1.5.4 shows a graph of $f(x) = \dfrac{2x}{x - 3}$, and we notice that $y = 2$ is a **horizontal asymptote** for the graph.

Horizontal asymptotes can be determined quickly by inspecting the degree of the numerator and the degree of the denominator. Notice that both the numerator and denominator of f have a degree of 1, and the ratio of the leading coefficients is $\dfrac{2}{1} = 2$. It is not a coincidence that this ratio is the horizontal asymptote, as we state in the following. (A more rigorous analysis of horizontal asymptotes is presented in Chapter 2.)

TABLE 1.5.3

x	$f(x)$
$-100,000$	$1.99994\ldots$
$-10,000$	$1.9994\ldots$
$-1,000$	$1.994\ldots$
-100	$1.94\ldots$
0	0
100	$2.061855\ldots$
$1,000$	$2.006018\ldots$
$10,000$	$2.000600\ldots$
$100,000$	$2.000060\ldots$

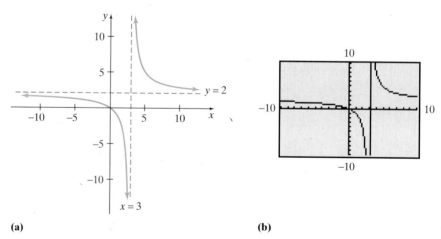

(a)   (b)

Figure 1.5.4 Graph of $f(x) = \dfrac{2x}{x - 3}$ has a vertical asymptote at $x = 3$ and a horizontal asymptote at $y = 2$.

> ### Finding Horizontal Asymptotes
>
> Consider the rational function $h(x) = \dfrac{f(x)}{g(x)}$, where f and g are polynomials.
>
> 1. If the degree of the numerator is largest, then the graph of the rational function has no horizontal asymptote.
> 2. If the degree of the denominator is largest, then the graph of the rational function has a horizontal asymptote at the line $y = 0$ (the x-axis).
> 3. If the degrees of the numerator and denominator are the same, then the horizontal asymptote is determined by the fraction made up of the coefficients of the highest-degree terms.

NOTE: The graph of a rational function can have at most one horizontal asymptote.

EXAMPLE 3 | Determining Horizontal Asymptotes

Find the horizontal asymptote of the graph of $f(x) = \dfrac{2x^2 - 5x - 3}{x^2 - 16}$.

SOLUTION

For the function $f(x) = \dfrac{2x^2 - 5x - 3}{x^2 - 16}$, the numerator and denominator are both quadratic, so the horizontal asymptote is given by the coefficients of the highest-degree terms, $y = \dfrac{2}{1} = 2$. ∎

✓ CHECKPOINT 2

Now work Exercise 13.

Another key feature of the graphs of rational functions are the intercepts. The process for determining intercepts is the same as we have shown for other functions, as illustrated in Example 4.

EXAMPLE 4 | Determining the Intercepts of the Graph of a Rational Function

Determine the x- and y-intercepts of the graph of $f(x) = \dfrac{2x^2 - 5x - 3}{x^2 - 16}$.

SOLUTION

To find the y-intercept, we substitute 0 for x. This yields

$$f(0) = \frac{2(0)^2 - 5(0) - 3}{(0)^2 - 16} = \frac{-3}{-16} = \frac{3}{16}$$

So the y-intercept is at the point $\left(0, \dfrac{3}{16}\right)$.

To find the x-intercept, we substitute 0 for y [or here $f(x)$] and solve for x. This gives

$$0 = \frac{2x^2 - 5x - 3}{x^2 - 16}$$

Factoring gives us

$$0 = \frac{(2x + 1)(x - 3)}{(x + 4)(x - 4)}$$

Now the only time a fraction equals 0 is when the numerator equals 0. Thus,

$$0 = (2x + 1)(x - 3)$$

for $x = -\frac{1}{2}$ and $x = 3$. (It is important to check that neither of these x-values makes the denominator equal 0.) We conclude that the x-intercepts occur at $\left(-\frac{1}{2}, 0\right)$ and $(3, 0)$.

Now that we know how to find vertical asymptotes, horizontal asymptotes, and intercepts, let's put it all together and sketch a graph of a rational function by hand.

EXAMPLE 5

Sketching a Rational Function

Sketch a graph of $f(x) = \frac{2x^2 - 5x - 3}{x^2 - 16}$. Label all asymptotes and intercepts.

SOLUTION

The factored form of $f(x)$ is

$$f(x) = \frac{2x^2 - 5x - 3}{x^2 - 16} = \frac{(2x + 1)(x - 3)}{(x + 4)(x - 4)}$$

Vertical asymptotes are at $x = -4$ and $x = 4$, while the horizontal asymptote is at $y = 2$. In Example 4, we determined that the y-intercept is at $\left(0, \frac{3}{16}\right)$ and the x-intercepts are at $\left(-\frac{1}{2}, 0\right)$ and $(3, 0)$. By plotting a couple of other points, $f(-5) = 8$ and $f(5) = \frac{22}{9}$, we get the graph in Figure 1.5.5.

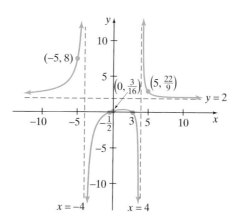

Interactive Activity

Use your calculator to graphically verify the sketch of the rational function $f(x) = \frac{2x^2 - 5x - 3}{x^2 - 16}$ in Example 5.

Figure 1.5.5 Graph of $f(x) = \frac{2x^2 - 5x - 3}{x^2 - 16}$. Vertical asymptote at $x = -4$ and $x = 4$. Horizontal asymptote at $y = 2$.

The intercepts of a rational function are often used in applications.

EXAMPLE 6

Applying Rational Functions in Economics

During the 1980s, the controversial economist Arthur Laffer promoted the idea of supply-side economics, which later gained the moniker "trickle-down theory." This theory centered on the **Laffer curve**. According to this curve, an increase in the tax rate can actually produce a reduction in government revenue. Suppose that a supply-side economist develops a model of the Laffer curve based on the rational function

$$f(x) = \frac{80x - 8000}{x - 110}, \qquad 30 \le x \le 100$$

where x represents the tax rate percentage and $f(x)$ represents the government tax revenue in tens of billions of dollars. See Figure 1.5.6.

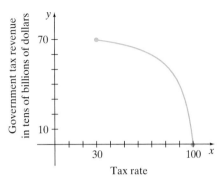

Figure 1.5.6 Graph of Laffer Curve given by the function $f(x) = \dfrac{80x - 8000}{x - 110}$.

(a) Evaluate $f(45)$ and interpret.

(b) Find the x-intercept and interpret.

(c) Find the average rate of change for $x = 45$ and $\Delta x = 30$ and interpret.

SOLUTION

(a) Evaluating the model at $x = 45$ yields

$$f(45) = \frac{80(45) - 8000}{45 - 110} = \frac{-4400}{-65} \approx 67.69$$

This means that at a tax rate of 45%, the government tax revenue generated will be 67.69 times 10 billion dollars, or $676.9 billion.

(b) Setting $f(x) = 0$ and solving gives us

$$\frac{80x - 8000}{x - 110} = 0$$

$$80x - 8000 = 0$$

$$80x = 8000$$

$$x = 100$$

This means that at a tax rate of 100%, zero dollars in tax revenue is generated. This is a common agreement among economic theorists, since no one will earn income if all of it is taken by the government.

(c) Using the values $x = 45$ and $\Delta x = 30,$ we get

$$m_{\text{sec}} = \frac{f(45 + 30) - f(45)}{30} = \frac{f(75) - f(45)}{30}$$

$$= \frac{\left[\dfrac{80(75) - 8000}{75 - 110}\right] - \left[\dfrac{80(45) - 8000}{45 - 110}\right]}{30}$$

$$\approx \frac{57.14 - 67.69}{30} = \frac{-10.55}{30} \approx -0.35$$

This means that as the tax rate increases from 45% to 75%, the tax revenue will **decrease** at an average rate of $3.5 \frac{\text{billion dollars}}{\text{percent}}$.

Radical Functions

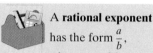
To begin to understand radical functions, let's recall a definition from algebra as shown in the From Your Toolbox to the left.

Notice that the Toolbox implies that any function written in radical form can be written as a function with a rational exponent, and vice versa. Knowing how to rewrite a function between radical and rational exponent form is essential for success in calculus. The following definition relates these two families of functions.

> ### Radical/Rational Exponent Functions
>
> A function of the form $f(x) = \sqrt[b]{[g(x)]^a}$ is called a **radical function,** where g is called the **radicand.** Rewriting $f(x) = \sqrt[b]{[g(x)]^a}$ as $f(x) = [g(x)]^{a/b}$ produces what we call the **rational exponent function.**

Many times, $g(x)$ is a polynomial expression like $5x - 1$ or $8 - x$. For example, for the function $g(x) = x^{2/3}$, the root index is 3, so we can rewrite the function in radical form as $g(x) = \sqrt[3]{x^2}$. Similarly, for the function $f(x) = (2x + 3)^{1/2}$, the function has a rational exponent of $\frac{1}{2}$, so we can rewrite f as $f(x) = \sqrt{2x + 3}$. This kind of function is called a **square root function.**

✓ CHECKPOINT 3

Now work Exercise 51.

The domains for radical/rational exponent functions are determined by the value of the root index. Let's say that we have the function $f(x) = \sqrt[3]{x - 1}$. Since we can take the cube root of any real number, the domain of this function is the set of real numbers $(-\infty, \infty)$. This can be verified visually in Figure 1.5.7.

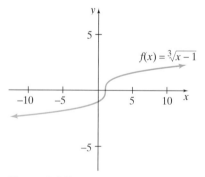

$$f(x) = \sqrt[3]{x - 1}$$

Figure 1.5.7

If the function is a square root function, we have to be more cautious. Recall from Section 1.1 that we found that the domain of $g(x) = \sqrt{2x - 5}$ is $\left[\frac{5}{2}, \infty\right)$ by determining the x-values that make $2x - 5$ greater than zero. Thus, the domain of a radical function depends on whether the root index is even or odd.

Domain of a Radical/Rational Exponent Function

1. The domain of a function of the form $f(x) = \sqrt[b]{[g(x)]^a}$ or $f(x) = [g(x)]^{a/b}$ is all real numbers $(-\infty, \infty)$ if b is an odd number.
2. The domain is determined by solving $g(x) \geq 0$ if b is even.

In practice, many of the radical functions that we study are square root functions. Let's take a look at an application of this class of radical functions.

EXAMPLE 7

Using the Square Root Function

A sapling of a certain kind of tree grows according to the mathematical model

$$g(x) = 10\sqrt{x} + 0.75$$

where x represents number of years since the sapling has been planted, and $g(x)$ represents the height of the tree in feet after x years.

(a) How tall will the sapling be 20 years after being planted? Round the solution to the nearest hundredth of a foot.

(b) How long will it take for the tree to be 30 feet tall?

SOLUTION

(a) To determine the height of the tree after 20 years, we need to evaluate $g(20)$. This gives us

$$g(20) = 10\sqrt{20} + 0.75 \approx 45.47$$

So, according to the model, the tree will be about 45.47 feet tall in 20 years.

(b) In this case, we have to find the x-values so that the $g(x)$-value is 30. Thus, we have to solve the square root equation

$$10\sqrt{x} + 0.75 = 30$$
$$10\sqrt{x} = 29.25$$
$$\sqrt{x} = 2.925$$
$$x = (2.925)^2 \approx 8.56$$

So, according to the model, the tree will be 30 feet tall in about 8.56 years.

Interactive Activity

Graphically verify the solution to part (b) of Example 7 by letting $y_1 = 10\sqrt{x} + 0.75$ and $y_2 = 30$ then determining the point of intersection.

Power Functions

Some rational exponent functions are easier to study mathematically when the exponents are written as decimals, such as $f(x) = x^{0.145}$. We will call this class of functions the **power functions.**

Power Function

A function of the form

$$f(x) = a \cdot x^b$$

is called a **power function,** where a and b are real numbers.

The power function has various properties based on the values of a and b. We will study these properties in two cases. If a is a positive number, then the

TABLE 1.5.4

x	f(x)	g(x)
0	0	0
1	1	1
2	2.82	1.62
3	5.20	2.16
4	8	2.64
5	11.18	3.09
6	14.70	3.51
7	18.52	3.90
8	22.63	4.29
9	27	4.66
10	31.62	5.01

Some values are rounded to the nearest hundredth.

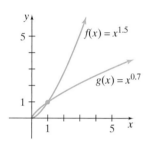

Figure 1.5.8

power function appears much like the polynomial functions that we studied in Section 1.4. Much of the shape of the graph of these functions is determined by whether b is between zero and 1 or b is greater than 1.

For example, consider the table of values and graphs of the power functions $f(x) = x^{1.5}$ and $g(x) = x^{0.7}$, where f and g are defined for $x \geq 0$ (Table 1.5.4). The graph (Figure 1.5.8) supports the numerical notion that f increases more rapidly than g. Finally, we observe that both functions are increasing.

Positive Exponent Power Functions

For the power function of the form $f(x) = a \cdot x^b$ with $a > 0$, we have

IF THE VALUE OF b IS	THE GRAPH	DIAGRAM
$0 < b < 1$	Opens down	
$b > 1$	Opens up	

EXAMPLE 8

Using Positive Exponent Power Functions in Applications

The median sale price of single-family homes in the U.S. Midwest can be modeled by

$$f(x) = 63.25 \sqrt[4]{x} \quad \text{or} \quad f(x) = 63.25 x^{0.25}, \qquad 1 \leq x \leq 14$$

where x represents the number of years since 1981, and $f(x)$ represents the sale price in thousands of dollars.

(a) Evaluate $f(11)$ and interpret.

(b) Determine the slope of the secant line of the graph of f for $x = 1$ and $\Delta x = 10$ and interpret.

SOLUTION

(a) First, note that $x = 11$ corresponds to the year $1981 + 11 = 1992$. Evaluating the model at $x = 11$, we get

$$f(11) = 63.25(11)^{0.25} \approx 63.25(1.82116) \approx 115.188$$

This means that in 1992 the median sale price of a house in the U.S. Midwest was about $115.188 thousand or $115,188.

(b) Evaluating the secant line slope, we get

$$m_{\text{sec}} = \frac{f(x + \Delta x) - f(x)}{\Delta x} = \frac{f(11) - f(1)}{10} \approx \frac{115.188 - 63.25}{10} = 5.1938$$

This means that between the years 1981 and 1991 the median sale price of a house in the Midwest increased at an average rate of about \$5193.80 per year.

In Example 8, note that when we evaluated $63.25(11)^{0.25}$ we raised 11 to the 0.25 power first and *then* multiplied by 63.25. Remember from the order of operations in algebra: exponentiation first, then multiplication.

Now let's examine power functions for which the exponent is a negative number. The domain used for these functions is $(0, \infty)$ and a is a positive number. Consider the power functions $f(x) = x^{-1.8}$ and $g(x) = x^{-0.2}$, where f and g are defined for $x > 0$. Rewriting with positive exponents, we get $f(x) = x^{-1.8} = \frac{1}{x^{1.8}}$ and $g(x) = x^{-0.2} = \frac{1}{x^{0.2}}$. Since $0.2 = \frac{1}{5}$, we could write g as $g(x) = \frac{1}{\sqrt[5]{x}}$. Notice that when $x = 0$ the denominators of f and g are zero. This is why zero is not part of the domain of these functions. The graphs of $f(x) = x^{-1.8}$ and $g(x) = x^{-0.2}$ are shown in Figure 1.5.9. Note that since $x = 0$ is not in the domain, the graphs of f and g have the y-axis as a vertical asymptote. We can see from the graph that both functions are decreasing.

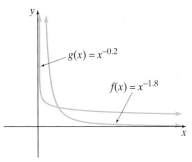

$g(x) = x^{-0.2}$

$f(x) = x^{-1.8}$

Figure 1.5.9

Interactive Activity

Contrast the graphs of $f(x) = 3 \cdot x^{2.5}$ and $g(x) = -3 \cdot x^{2.5}$. What do you think you can generally say about the relationship of the graphs of the functions $f(x) = a \cdot x^b$ and $g(x) = -a \cdot x^b$?

So when a power function has a negative exponent, the domain is the set of all positive real numbers. The functions are decreasing and their graphs have vertical asymptotes. This kind of power function is frequently used as a model for a decreasing phenomenon.

SUMMARY

In this section, we discussed the rational functions, which included finding the intercepts and asymptotes of their graphs. Then we examined radical functions and saw the relationship between radical and rational exponent functions. Finally, we examined the power functions. The investigation of all these functions is key in our understanding of calculus.

The types of functions we discussed in this section include:

- **Rational function:** $h(x) = \frac{f(x)}{g(x)}$, f and g are polynomials, $g(x) \neq 0$
- **Radical/rational exponent function:** $f(x) = \sqrt[b]{[g(x)]^a}$ or $f(x) = [g(x)]^{a/b}$
- **Square root function:** $f(x) = a\sqrt{x} + b$
- **Power function:** $f(x) = a \cdot x^b$

Vertical Asymptote Rule for the Rational Function $\dfrac{f(x)}{g(x)}$

- For $h(x) = \dfrac{f(x)}{g(x)}$, if there is a value c so that $g(c) = 0$, yet $f(c) \neq 0$, then $x = c$ is a vertical asymptote.

Horizontal Asymptote Rule for the Rational Function $\dfrac{f(x)}{g(x)}$

- If the degree of f is largest, then there is no horizontal asymptote.
- If the degree of g is largest, then the x-axis is the horizontal asymptote.
- If the degrees of f and g are the same, then the horizontal asymptote is the reduced form of the leading terms of f and g.

SECTION 1.5 EXERCISES

For Exercises 1–10:

(a) Write the domain using interval notation.
(b) Identify any holes or vertical asymptotes in the graph.

1. $f(x) = \dfrac{12x^3}{3x}$

2. $f(x) = \dfrac{15x^2}{25x^3}$

✓ 3. $f(x) = \dfrac{3x - 12}{x - 2}$

4. $f(x) = \dfrac{2x - 16}{x - 5}$

5. $f(x) = \dfrac{3x + 6}{x + 2}$

6. $g(x) = \dfrac{x^2 - 1}{x + 1}$

7. $g(x) = \dfrac{x^2 + 2x + 1}{x^2 + 4x + 3}$

8. $g(x) = \dfrac{6x^2 + x - 2}{8x^2 + 2x - 3}$

9. $f(x) = \dfrac{4x^2 + 8x + 3}{2x^2 - x - 6}$

10. $f(x) = \dfrac{x^2 + x - 30}{x^2 - x - 20}$

For Exercises 11–20, identify any horizontal asymptotes for the following:

11. The function in Exercise 1

12. The function in Exercise 2

✓ 13. The function in Exercise 3

14. The function in Exercise 4

15. The function in Exercise 5

16. The function in Exercise 6

17. The function in Exercise 7

18. The function in Exercise 8

19. The function in Exercise 9

20. The function in Exercise 10

For Exercises 21–34, determine the x- and y-intercepts.

21. $g(x) = \dfrac{x}{2x + 6}$

22. $f(x) = \dfrac{x - 1}{3x + 9}$

23. $f(x) = \dfrac{6x - 3}{2x + 4}$

24. $g(x) = \dfrac{7x + 4}{3x - 1}$

25. $f(x) = \dfrac{x^2 - x - 6}{x^2 - 7x + 12}$

26. $f(x) = \dfrac{x^2 - 5x + 4}{x^2 + 2x - 3}$

27. $g(x) = \dfrac{x - 2}{x^2 + 6x + 9}$

28. $g(x) = \dfrac{x^2 + 2x}{3x^2 + 18x + 24}$

29. $f(x) = \dfrac{x^2 - 4}{x^2 + 2x - 3}$

30. $g(x) = \dfrac{4x^2 - 9}{x^2 + 3x + 4}$

For Exercises 31–36, determine the x- and y-intercepts graphically.

31. $f(x) = \dfrac{x^2 - 2x + 5}{x + 1}$

32. $f(x) = \dfrac{x^2 - 2x + 1}{x - 2}$

33. $f(x) = \dfrac{2x^3 - x^2 + 3x - 2}{x^3 + 3}$

34. $f(x) = \dfrac{x^3 - 2x + 1}{x - 1}$

35. $f(x) = \dfrac{2x^3 - x^2 + 1}{2 - x}$

36. $f(x) = \dfrac{3x^2 - x + 5}{x^2 - 4}$

In Exercises 37–42, sketch a graph by hand for each function on the domain $1 \le x \le 10$. Label any asymptotes and intercepts.

37. $f(x) = \dfrac{x}{x+2}$

38. $f(x) = \dfrac{2x}{x-4}$

39. $f(x) = \dfrac{x+4}{x-4}$

40. $f(x) = \dfrac{x-2}{x+2}$

41. $f(x) = \dfrac{6-3x}{x-6}$

42. $f(x) = \dfrac{4-4x}{x-2}$

APPLICATIONS

43. Environmental scientists and municipal planners often are guided by **cost–benefit models.** These mathematical models estimate the cost of removing a pollutant from the atmosphere as a function of the percentage of pollutant removed. Let's suppose that a cost–benefit function is given by

$$f(x) = \frac{30x}{100-x}, \qquad 0 \le x < 100$$

where x represents the percentage of the pollutant removed and $f(x)$ represents the associated cost in millions of dollars.
(a) Evaluate $f(85)$ and interpret.
(b) Fill in the following table:

x	5	50	70	90	95
$f(x)$					

(c) Many cost–benefit functions exhibit the very high cost of removing the final percentage of a pollutant. To calculate this behavior, evaluate $f(99.9) - f(95)$. This represents the approximate cost of removing the final 5% of the pollutant.
(d) Why can we not compute $f(100) - f(95)$ to get the *actual* cost of removing the final 5% of the pollutant?

44. The cost–benefit function for removing a certain pollutant from the atmosphere is given by

$$f(x) = \frac{20x}{100-x}, \qquad 0 \le x < 100$$

where x represents the percentage of the pollutant removed and $f(x)$ represents the associated cost in millions of dollars.
(a) Evaluate $f(70)$ and interpret.
(b) Graph f in the viewing window [0, 100] by [0, 500].
(c) Evaluate $f(99.9) - f(95)$. This represents the approximate cost of removing the final 5% of the pollutant.
(d) Why can we not compute $f(100) - f(95)$ to get the *actual* cost of removing the final 5% of the pollutant?

45. The students in an anatomy class were asked to memorize a list of 20 parts of the human body. After each class, a student was chosen and asked to write as many of these anatomical parts as she could. The average number of parts remembered is given by the mathematical model

$$f(x) = \frac{20x - 18}{x}, \qquad x \ge 4$$

where x represents the number of days since the list was distributed and $f(x)$ represents the average number of items that were remembered.
(a) Evaluate $f(10)$ and interpret.
(b) Evaluate $f(100)$ and interpret.

46. The technicians at the Arp Brothers auto parts factory are given a checklist of 40 items to inspect for quality control. Over the next three months, a technician is selected and asked to write the checklist from memory. The mathematical model for the technician's performance is given by

$$f(x) = \frac{80x - 36}{2x}, \qquad 1 \le x \le 120$$

where x represents the number of days since the checklist was distributed and $f(x)$ represents the average number of inspection items that was remembered.
(a) Evaluate $f(2)$ and $f(30)$ and interpret each.
(b) Find the horizontal asymptote and interpret its meaning.
(c) The State Regulatory Agency requires technicians to remember a minimum of 25 items after three months. According to the model, will the technicians meet this requirement?

47. The Slaybaugh Satellite Company is manufacturing a new low-cost satellite dish and promotes its sales through an aggressive sales campaign. The income from sales is given by the sales function

$$S(x) = \frac{120x^2 - 600x + 3}{2x^2 - 10x + 1}, \qquad x \ge 5$$

where x represents the number of dollars spent on advertising in thousands of dollars and $S(x)$ represents the income from sales measured in hundreds of thousands of dollars.
(a) Evaluate $S(10)$ and interpret.
(b) Graph S in the viewing window [5, 10] by [50, 70].
(c) Find the horizontal asymptote and interpret its meaning.

For Exercises 48–55, rewrite the radical functions as rational exponent functions. Also, write the domain using interval notation.

48. $f(x) = \sqrt[4]{x-1}$

49. $f(x) = \sqrt{2x+3}$

50. $f(x) = \sqrt[3]{4-x}$

51. $f(x) = \sqrt[3]{5x-8}$

52. $g(x) = \sqrt{(x-4)^3}$

53. $g(x) = \sqrt[4]{(6x+1)^5}$

54. $f(x) = \dfrac{1}{\sqrt{3x+4}}$

55. $f(x) = \dfrac{3}{\sqrt{7x-2}}$

For Exercises 56–63, rewrite the rational exponent functions as radical functions. Also, write the domain using interval notation.

56. $f(x) = (4x + 3)^{1/2}$

57. $f(x) = (8x - 9)^{1/2}$

58. $f(x) = (3x + 1)^{1/3}$

59. $f(x) = (7x + 9)^{1/3}$

60. $y = (6x - 1)^{3/2}$

61. $f(x) = (4x - 5)^{3/2}$

62. $g(x) = (6x + 3)^{-1/?}$

63. $g(x) = (7x - 2)^{-1/3}$

64. A slightly banked highway corner will safely accommodate traffic at a speed given by the model

$$f(x) = \frac{29}{20}\sqrt{x}$$

where x represents the radius of the corner in feet and $f(x)$ represents the speed at which the traffic can travel safely in miles per hour.

(a) Evaluate $f(20)$ and interpret.

(b) If the highway planners expect the traffic to travel at a speed of 64 miles per hour, what radius should the corner be?

65. In a forest fire tower, the distance that an observer is able to see into the forest is related to the height of the fire tower via the function

$$g(x) = \frac{7}{5}\sqrt{x}$$

where x represents the height of the tower in feet and $g(x)$ represents the distance that the observer can see in miles.

(a) Evaluate $g(70)$ and interpret.

(b) If the observer is required to see 29 miles into the forest, how high must the tower be?

66. The distance that a person can see from an airplane to the horizon on a cloudless day is given by

$$g(x) = \frac{61}{50}\sqrt{x}$$

where x represents the altitude of the plane in feet and $g(x)$ is the distance that a person can see in miles.

(a) Evaluate $g(32,000)$ and interpret.

(b) Find the slope of the secant line for $x = 36,000$ and $\Delta x = 4000$ and interpret.

67. A biologist has shown that the number of plant species in the South American rainforest is related to the area of the land studied via the model

$$g(x) = 28.1\sqrt[3]{x}$$

where x represents the area studied in square miles and $g(x)$ represents the number of plant species.

(a) Evaluate $g(300)$ and interpret.

(b) Find the slope of the secant line for $x = 1728$ and $\Delta x = 469$ and interpret.

68. A city planner has projected that the population of a newly developed suburb will grow for the next four years according to the mathematical model

$$f(x) = 10,000 + 20\sqrt{x^3} + 30x, \qquad 0 \le x \le 48$$

where x is the number of months from the present and $f(x)$ is the suburb's population.

(a) Evaluate $f(12)$ and interpret.

(b) Graph f in the viewing window [0, 48] by [10,000, 20,000].

(c) Find the slope of the secant line for $x = 12$ and $\Delta x = 24$ and interpret.

69. The number of grocery stores in the United States can be modeled by the power function

$$f(x) = 17,311x^{-0.028}, \qquad 1 \le x \le 6$$

where x represents the number of years since 1989 and $f(x)$ represents the number of grocery stores in the United States.

(a) Evaluate $f(3)$ and interpret.

(b) Evaluate $\dfrac{f(6) - f(3)}{6 - 3}$ and interpret this result.

70. The amount of lead emitted by the United States into the atmosphere as air pollution can be modeled by

$$f(x) = \frac{13.97}{\sqrt{x}}, \qquad 1 \le x \le 10$$

where x represents the number of years since 1984 and $f(x)$ is the number of megatons of lead emitted.

(a) Evaluate $f(2)$ and interpret.

(b) Graph f in the viewing window [1, 10] by [0, 30]. From the graph, do you think that the amount of lead air pollution is generally increasing or decreasing? Explain.

(c) Compute the average rate of change of the model for $x = 1$ and $\Delta x = 5$.

71. The number of retired workers in the United States receiving Social Security benefits can be modeled by

$$f(x) = 19.14x^{0.112}, \qquad 1 \le x \le 20$$

where x represents the number of years since 1979 and $f(x)$ represents the number of retired workers, in millions, receiving Social Security benefits.

(a) Evaluate $f(11)$ and interpret.

(b) Graph f in the viewing window [1, 20] by [19, 30]. From the graph, do you think the number of retired workers receiving Social Security benefits is generally increasing or decreasing? Explain.

(c) Find the average rate of change in the number of beneficiaries for $x = 2$ and $\Delta x = 5$ and interpret.

SECTION PROJECT

The expected lifespan of African-American males in the U.S. can be modeled by

$$f(x) = 60.19 \cdot x^{0.025}, \qquad 1 \le x \le 15$$

where x represents the number of years since 1969 and $f(x)$ represents the expected lifespan of African-American males born during the xth year.

(a) Evaluate $f(10)$ and interpret.

(c) Find the average rate of change of the expected lifespan of African-American males for $x = 2$ and $\Delta x = 5$ and interpret.

The expected lifespan of African-American females in the U.S. can be modeled by

$$g(x) = 68.43 \cdot x^{0.025} \qquad 1 \le x \le 15$$

where x represents the number of years since 1969 and $g(x)$ represents the expected lifespan of African-American females born in the xth year.

(d) Evaluate $g(10)$ and interpret.

(e) During what year did the expected lifespan of African-American females exceed 70 years?

(f) Find the average rate of change of the expected lifespan of African-American females for $x = 2$ and $\Delta x = 5$ and interpret.

(g) Compare the answers of parts (c) and (f).

SECTION 1.6   EXPONENTIAL FUNCTIONS

Often business people make a proclamation like "Our sales are growing exponentially!" This may not be exactly true, but it is an effective way to communicate that the sales are growing very rapidly. In this section, we will study **general exponential functions** that typically have function values that either increase or decrease rapidly. Then we will discuss compound interest, which is a direct application of exponential functions. After introducing the irrational number **e**, we will explore additional applications in which exponential functions arise, such as population growth, radioactive decay, and learning curves. Finally, we will introduce a special type of function called the **logistic function.**

General Exponential Functions

Exponential functions are different from the functions studied so far in that the independent variable is in the exponent. The following are examples of exponential functions:

$$f(x) = 3^x, \qquad y = \left(\frac{1}{5}\right)^x, \qquad g(x) = 2 \cdot 4^x, \qquad f(x) = 1.6 \cdot (1.09)^x$$

General Exponential Function

A **general exponential function** has the form

$$f(x) = a \cdot b^x$$

where a is a real number **constant** with $a \neq 0$, and b is a real number with $b > 0, b \neq 1$. Here b is called the **base.**

NOTE: Do not confuse exponential functions with power functions. For an exponential function, the base is a constant and the exponent is the independent variable. The reverse is true for power functions. For example, $f(x) = x^3$ is a power function, whereas $f(x) = 3^x$ is an exponential function.

EXAMPLE 1   **Graphing Exponential Functions**

(a) Make a table of values for the general exponential functions $f(x) = 2^x$ and $g(x) = \left(\frac{1}{2}\right)^x$ for $x = -5, -4, -3, -2, \ldots, 4, 5$.

(b) Graph the functions f and g and write their domains using interval notation.

SOLUTION

(a) The values for the functions are shown in Table 1.6.1.

TABLE 1.6.1

x	-5	-4	-3	-2	-1	0	1	2	3	4	5
$f(x) = 2^x$	$\frac{1}{32}$	$\frac{1}{16}$	$\frac{1}{8}$	$\frac{1}{4}$	$\frac{1}{2}$	1	2	4	8	16	32
$g(x) = \left(\frac{1}{2}\right)^x$	32	16	8	4	2	1	$\frac{1}{2}$	$\frac{1}{4}$	$\frac{1}{8}$	$\frac{1}{16}$	$\frac{1}{32}$

(b) From the table and the graphs in Figures 1.6.1a and b, we see that any real number can be substituted for x, so the domain of the functions $f(x) = 2^x$ and $g(x) = \left(\frac{1}{2}\right)^x$ is $(-\infty, \infty)$.

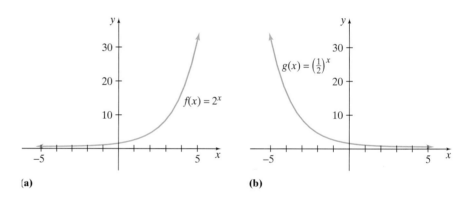

(a)   (b)

Figure 1.6.1

Interactive Activity

For Table 1.6.1, make another row of the table that calculates the **successive ratios** of the function values by dividing each value by the preceding value. For example, for the function $g(x) = \left(\frac{1}{2}\right)^x$, the ratios would begin $\frac{16}{32}, \frac{8}{16}, \ldots$ What pattern do you notice?

Notice that $f(x) = 2^x$ is increasing on its domain and as $x \to \infty, f(x) \to \infty$. Consequently, we call this kind of function an **exponential growth function.** Whereas, notice that $g(x) = \left(\frac{1}{2}\right)^x$ is decreasing on its domain and, as $x \to \infty$, $g(x) \to 0$. We call this kind of function an **exponential decay function.** (See Properties of General Exponential Functions at the top of p. 71.)

For the function $f(x) = a \cdot b^x$, the value of a can affect the end behavior. First, we see that with any value of a, $f(0) = a \cdot b^0 = a \cdot 1 = a$. This means that the y-intercept of the graph of $f(x) = a \cdot b^x$ is at $(0, a)$. To see the effect of a, compare the graphs of $f(x) = 2 \cdot 3^x$ and $g(x) = -2 \cdot 3^x$ in Figure 1.6.2.

Properties of General Exponential Functions

● The domain is the set of all real numbers $(-\infty, \infty)$.
● The y-intercept is at $(0, 1)$.
● The behavior of the function is shown in the table.

FUNCTION TYPE	DEFINITION	GRAPH
Exponential growth function	$f(x) = b^x, \quad b > 1$	
Exponential decay function	$f(x) = b^x, \quad 0 < b < 1$	

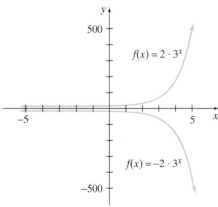

Figure 1.6.2

From Figure 1.6.2, we see that switching the sign of a from positive to negative results in a reflection about the x-axis. Now let's see how these exponential models are used in applications.

EXAMPLE 2

Using General Exponential Decay Models

The number of disabled World War I veterans receiving compensation for service-connected disabilities can be modeled by

$$f(x) = 47.55(0.78)^x, \qquad 1 \le x \le 16$$

where x represents the number of years since 1979 and $f(x)$ represents the number of World War I veterans receiving compensation, measured in thousands.

(a) Evaluate $f(6)$ and interpret.

(b) Graph f and classify the function as an exponential growth or exponential decay function.

(c) What is the behavior of f as the x-values increase? Interpret.

SOLUTION

(a) First, $x = 6$ corresponds to the year $1979 + 6 = 1985$. Evaluating, we get

$$f(6) = 47.55(0.78)^6 \approx 10.71$$

Since $f(6) \approx 10.71$, according to the model, in 1985 there were approximately 10.71 thousand, or about 10,710 disabled World War I veterans receiving compensation.

(b) The graph of f is shown in Figure 1.6.3. For the model $f(x) = 47.55(0.78)^x$, the value of $b = 0.78$ is between zero and one. This means f is an exponential decay function. This is supported by Figure 1.6.3.

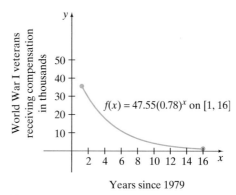

Figure 1.6.3

(c) We can see from the graph that as the x-values, representing the years, increase the function values decrease. Since x represents the number of years since 1979, this means that as we move farther away from 1979 the number of World War I veterans receiving compensation is decreasing. We would expect this trend because the number of World War I veterans has been dwindling since 1979. ∎

✓ CHECKPOINT 1

Now work Exercise 19.

Compound Interest

A common application of the general exponential function in the banking industry is *compound interest*. To understand where the compound interest formula comes from, we begin by looking at simple interest.

Simple Interest Formula

If P dollars is deposited in an account that earns interest at a rate of r (written as a decimal), then the **interest** I accumulated after time t is

$$P \cdot r \cdot t = I$$

Principal $\cdot$ rate $\cdot$ time $=$ Interest

Compound interest means that interest is paid not only on the money deposited, called the **principal,** but also on interest that has already accumulated. Usually, a bank computes the interest earned and deposits the interest in your account. The next time that the interest is computed the bank computes interest on the original principal **plus** interest! The amount in an account that earns compound interest is given by the **compound interest formula.**

> ### Compound Interest Formula
>
> If a principal of P dollars is invested in an account earning an annual interest rate of r (in decimal form) compounded n times per year, then the amount A in the account at the end of t years is given by the **compound interest formula**
>
> $$A = P\left(1 + \frac{r}{n}\right)^{nt}$$

EXAMPLE 3 | **Calculating Compound Interest**

Suppose that $20,000 is deposited into an account that yields an annual interest rate of 6.5% compounded quarterly.

(a) How much will be in the account after three years? Round the solution to the nearest cent.

(b) How much interest was made at the end of the three-year period?

SOLUTION

(a) Using the compound interest formula with $P = 20,000$, $r = 0.065$, $n = 4$, and $t = 3$ we get

$$A = P\left(1 + \frac{r}{n}\right)^{nt} = 20,000\left(1 + \frac{0.065}{4}\right)^{4(3)}$$
$$= 20,000(1.01625)^{12} \approx 24,268.15$$

So there will be about $24,268.15 in the account in three years.

(b) To determine the interest made, we take the amount A and subtract the principal P from it. This gives us $24,268.15 - 20,000 = 4268.15$. So about $4268.15 is made in interest in three years.   ◼

✓ CHECKPOINT 2

Now work Exercise 27.

If we hold the values of $P, r,$ and n constant, we can think of the compound interest formula as a function of time t. Writing this gives us the **compound interest function:** $A(t) = P\left(1 + \frac{r}{n}\right)^{nt}$. Notice here that the independent variable is time, denoted with a t. For the account in Example 3, we can write the compound interest function

$$A(t) = 20,000\left(1 + \frac{0.065}{4}\right)^{4t}$$

To find the value of the account in five years, we evaluate $A(5)$.

$$A(5) = 20,000\left(1 + \frac{0.065}{4}\right)^{4(5)} = 20,000(1.01625)^{20} \approx 27,608.40$$

This means that after five years, there will be about $27,608.40 in the account.

The Number e

At the beginning of this section, we saw that we could write an exponential function $f(x) = a \cdot b^x$ with any base b as long as $b > 0$, $b \neq 1$. However, the most frequently used base is the famous number e. This number is so widely used that it is found on nearly every scientific and graphing calculator. To see what this number is, let's numerically examine the expression $\left(1 + \dfrac{1}{n}\right)^n$, where n is an independent variable.

TABLE 1.6.2

n	$\left(1 + \dfrac{1}{n}\right)^n$
1	2
10	2.59374 ...
100	2.70481 ...
1,000	2.71692 ...
10,000	2.71815 ...
100,000	2.71827 ...
1,000,000	2.71828 ...

We see that as the n-values become very large the values of the expression $\left(1 + \dfrac{1}{n}\right)^n$ seem to get close to a single value called e. Many applications in the physical, life, and social sciences use e as the base of their exponential functions. As a matter of fact, this is referred to as *the* exponential function.

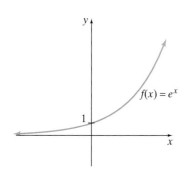

Figure 1.6.4

The Exponential Function
The **exponential function** has the form $$f(x) = ae^{bx}$$ where a and b are real number constants.

NOTE: The simplest form of the exponential function is $y = e^x$. See Figure 1.6.4.

EXAMPLE 4 | **Applying the Exponential Function**

The annual per capita consumption of whole milk in the United States can be modeled by

$$f(x) = 18.04e^{-0.046x}, \qquad 1 \leq x \leq 16$$

where x represents the number of years since 1979 and $f(x)$ represents the number of gallons consumed per capita.

(a) Evaluate $f(3)$ and interpret.

(b) Compute the difference quotient for $x = 3$ and $\Delta x = 10$ and interpret.

SOLUTION

(a) First, we notice that $x = 3$ corresponds to the year $1979 + 3 = 1982$. Evaluating the function at $x = 3$ yields

$$f(3) = 18.04e^{-0.046(3)} = 18.04e^{-0.138}$$

$$\approx 18.04(0.087109869) \approx 15.71$$

This means that in 1982 the annual per capita consumption of whole milk was about 15.71 gallons.

(b) Computing the difference quotient for $x = 3$ and $\Delta x = 10$, we get

$$m_{\text{sec}} = \frac{f(x + \Delta x) - f(x)}{\Delta x} = \frac{f(3 + 10) - f(3)}{10}$$

$$= \frac{f(13) - f(3)}{10} \approx \frac{9.92 - 15.71}{10} \approx -0.58$$

This means that from 1982 to 1992 the annual per capita whole milk consumption was decreasing at an average rate of 0.58 $\frac{\text{gallons per capita}}{\text{year}}$. See Figure 1.6.5.

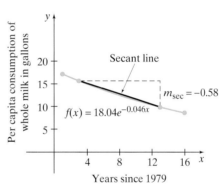

Figure 1.6.5   Slope of secant line on interval $[3, 13]$ gives the average rate of change from 1982 to 1992.

✓ CHECKPOINT 3

Now work Exercise 41.

Since the function in Example 4 is a decreasing function, we can classify it as an **exponential decay function.** The growth or decay property of an exponential function $f(x) = a \cdot e^{bx}$ is determined by the sign of the constant b.

Properties of Exponential Functions

Given the exponential function $f(x) = a \cdot e^{bx}$, with $a > 0$ and $b \neq 0$, we have

• The domain is the set of all real numbers $(-\infty, \infty)$.
• The y-intercept is at $(0, a)$.
• If $b > 0$, then f is an exponential growth function.
• If $b < 0$, then f is an exponential decay function.

The exponential function $f(x) = a \cdot e^{bx}$ is used often in the sciences, as well as the business sciences. Let's suppose that a banker with a strong entrepreneurial spirit wants to offer customers an account that accumulates interest

continuously over time. To find a formula for this kind of interest, recall that the compound interest formula is

$$A = P\left(1 + \frac{r}{n}\right)^{nt}$$

First, it can be shown that, as $n \to \infty$, $\left(1 + \frac{r}{n}\right)^{n} = e^r$. So, rewriting the compound interest formula, we get

$$A = P\left[\left(1 + \frac{r}{n}\right)^{n}\right]^{t}$$

For $n \to \infty$, we can substitute e^r for the expression in brackets and get $A = Pe^{rt}$. This is the **continuous compound interest** formula.

> ### Continuous Compound Interest
>
> If a principal of P dollars is invested into an account earning an annual interest rate r (in decimal form) compounded continuously, then the amount A in the account at the end of t years is given by the **continuous compound interest formula**
>
> $$A = Pe^{rt}$$

As a function of time, we can define the **continuous compound interest function** as $A(t) = Pe^{rt}$.

EXAMPLE 5 | **Computing Compound Interest**

Suppose that we have $5000 deposited in an account that earns 5.3% interest compounded continuously. Find the continuous compound interest function for this account, and evaluate $A(3)$. Round the solution to the nearest cent.

SOLUTION

Knowing that $r = 5.3\% = 0.053$ and $P = 5000,$ we get the function

$$A(t) = 5000e^{0.053t}$$

Evaluating $A(t)$ for $t = 3,$ we get

$$A(3) = 5000e^{0.053(3)} = 5000e^{0.159} \approx \$5861.69$$

So after 3 years, there would be $5861.69 in the account.

Logistic Curves

One type of function that is used in the study of areas such as population growth and the spread of disease is the **logistic function.** The genesis of this function will be shown in Chapter 6; for now, we simply define this new type of function.

> ### Logistic Function
>
> A **logistic function** has the form
>
> $$f(t) = \frac{L}{1 + a \cdot e^{-kLt}}$$
>
> where L is the limit of growth (that is, a horizontal asymptote) and a and k are real number constants.

Notice that the independent variable of this function is t, since the logistics function is often a function of time. The typical S-shape of the graph of a logistic function is shown in Figure 1.6.6.

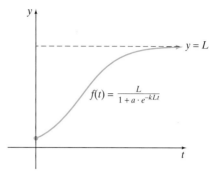

$$f(t) = \frac{L}{1 + a \cdot e^{-kLt}}$$

Figure 1.6.6

EXAMPLE 6 | **Using Logistic Functions**

The number of people affected by a flu epidemic in the isolated town of Spring Point (population 1800) can be modeled by

$$f(t) = \frac{1800}{1 + 359e^{-0.9t}}$$

where t represents the number of weeks since the flu was first detected and $f(t)$ represents the number of people who are affected (see Figure 1.6.7).

(a) Evaluate $f(1)$ and interpret.

(b) Compute the difference quotient for $t = 1$ and $\Delta t = 20$ and interpret.

SOLUTION

(a) Evaluating the function at $t = 1$, we get

$$f(1) = \frac{1800}{1 + 359e^{-0.9(1)}} = \frac{1800}{1 + 359e^{-0.9}} \approx 12.25$$

According to the model, this means that after one week about 12 people were affected.

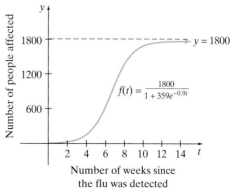

Figure 1.6.7

Interactive Activity

Use the INTERSECT command on your calculator to determine the week that the number of people who were affected by the flu epidemic in Spring Point exceeded 500.

(b) Computing the difference quotient for $t = 1$ and $\Delta t = 20$, we find that

$$m_{\text{sec}} = \frac{f(1 + 20) - f(1)}{20} = \frac{f(21) - f(1)}{20}$$

$$= \frac{\dfrac{1800}{1 + 359e^{-0.9(21)}} - \dfrac{1800}{1 + 359e^{-0.9(1)}}}{20} \approx \frac{1799.99 - 12.25}{20} \approx \frac{8}{9}.39$$

So during the interval from 1 week to 21 weeks, the number of people affected increased at an average rate of about $89.39 \ \dfrac{\text{affected people}}{\text{week}}$.   ■

SUMMARY

In this section, we examined two types of exponential functions: the general exponential function $f(x) = a \cdot b^x$ and the exponential function $f(x) = a \cdot e^{bx}$. We found that the value of b determines whether f is an exponential growth or exponential decay function. Then we reviewed the simple and compound interest formulas. The definitions that this section highlighted were the following:

- **General exponential function:** $f(x) = a \cdot b^x$
- **Simple interest formula:** $I = Prt$
- **Compound interest formula:** $A = P\left(1 + \dfrac{r}{n}\right)^{nt}$
- **The number e:** $\left(1 + \dfrac{1}{n}\right)^{n} = e$ as $n \to \infty$
- **Exponential function:** $f(x) = a \cdot e^{bx}$
- **Compound interest formula:** $A = Pe^{rt}$
- **Logistic function:** $f(t) = \dfrac{L}{1 + a \cdot e^{-kLt}}$

SECTION 1.6 EXERCISES

For Exercises 1–14, answer the following:

(a) Make a table of function values for $x = -5$ to $x = 5$, similar to Table 1.6.1.

(b) Graph f and classify the functions as exponential growth or decay.

1. $f(x) = \left(\dfrac{1}{3}\right)^{x}$ 2. $f(x) = \left(\dfrac{1}{4}\right)^{x}$ 3. $f(x) = 4^x$

4. $f(x) = 3^x$ 5. $f(x) = (2.3)^x$ 6. $f(x) = (4.5)^x$

7. $f(x) = (0.7)^x$ 8. $f(x) = (0.22)^x$ 9. $f(x) = e^{2x}$

10. $f(x) = e^{7x}$ 11. $f(x) = e^{0.3x}$ 12. $f(x) = e^{0.92x}$

13. $f(x) = e^{-1.6x}$ 14. $f(x) = e^{-3.1x}$

15. The healing of a 200-square-centimeter wound depends on the amount of time since the wound occurred and can be modeled by

$$f(x) = 200\left(\dfrac{4}{5}\right)^{x}, \qquad x \geq 0$$

where x represents the number of days since the wound occurred and $f(x)$ represents the size of the wound in square centimeters.

(a) Evaluate $f(7)$ and interpret.

(b) How many days will it take for the wound to be half its original size?

(c) Find the slope of the secant line for $x = 1$ and $\Delta x = 7$ and interpret.

16. The Desert Fox Dune Buggy Company launches a sales blitz for its new Family Buggy vehicle, offering a 30% discount during the first week. The percentage of customers who respond to the sale can be modeled by

$$f(x) = 70 - 100\left(\dfrac{3}{5}\right)^{x}, \qquad 1 \leq x \leq 7$$

where x is the day of the sales blitz and $f(x)$ represents the percentage of customers who respond to the sale.

(a) Evaluate $f(1)$ and interpret.

(b) Find the slope of the secant line for $x = 2$ and $\Delta x = 2$ and interpret.

17. The number of students at a college who hear the rumor that the college president will resign can be modeled by

$$g(x) = 8000 - 8000\left(\dfrac{43}{50}\right)^{x}, \qquad 1 \leq x \leq 14$$

where x represents the number of days since the rumor started and $g(x)$ represents the number of students who have heard the rumor.

(a) Evaluate $g(5)$ and interpret.

(b) Find the slope of the secant line for $x = 1$ and $\Delta x = 4$ and interpret.

18. The pupil–teacher ratio in U.S. elementary and secondary schools can be modeled by

$$f(x) = 25.34(0.987)^x, \qquad 1 \leq x \leq 36$$

where x represents the number of years since 1959 and $f(x)$ represents the pupil–teacher ratio.

(a) Evaluate $f(17)$ and interpret.

(b) Graph f in the viewing window $[1, 36]$ by $[0, 26]$, and classify the model as exponential growth or exponential decay.

(c) Use the INTERSECT command on your calculator to determine the year that the pupil–teacher ratio became less than $21\frac{\text{pupils}}{\text{teacher}}$.

19. The percentage of waste generated in the United States that is yard waste can be modeled by

$$f(x) = 19.12(0.944)^x, \qquad 1 \leq x \leq 6$$

where x represents the number of years since 1990 and $f(x)$ represents the percentage of waste generated that is yard waste.

(a) Evaluate $f(2)$ and interpret.

(b) Graph f in the viewing window $[1, 6]$ by $[10, 20]$, and classify the model as exponential growth or exponential decay.

(c) Compute the difference quotient for $x = 1$ and $\Delta x = 5$ and interpret.

20. The annual total assets in mutual funds in the United States can be modeled by

$$f(x) = 126.67(1.217)^x, \qquad 1 \leq x \leq 16$$

where x represents the number of years since 1979 and $f(x)$ represents the annual total assets in mutual funds in billions of dollars.

(a) Evaluate $f(9)$ and interpret.

(b) Graph f in the viewing window $[1, 16]$ by $[100, 3000]$, and classify the model as exponential growth or exponential decay.

(c) Use the INTERSECT command on your calculator to determine the year that the annual total assets in mutual funds exceeded $1000 billion (1 trillion dollars).

21. Use the simple interest formula to find the total interest earned for an account in which $500 is deposited at a 4% interest rate for 6 years.

22. Use the simple interest formula to find the total interest earned for an account in which $2000 is deposited at a 6.5% interest rate for 4 years.

23. If Kevin deposits $3000 into an account that yields 6% interest compounded annually, how much will be in the account after 3 years?

24. If Shuna deposits $10,000 into an account that yields 6.5% interest compounded semiannually, how much will be in the account after 4 years?

25. If Lucy deposits $1800 into an account that yields 7.1% interest compounded monthly, how much will be in the account after 5 years?

26. If Lana deposits $800 into an account that yields 7% interest compounded annually, how much will be in the account after 8 years?

27. If Joe deposits $2000 into an account that yields 8% annual interest, how much will be in the account after 6 years if the interest is compounded

(a) annually?

(b) semiannually (twice a year)?

(c) quarterly?

28. If Mrs. Johanson deposits $5000 into an account that yields 6.5% annual interest, how much will be in the account after 10 years if the interest is compounded

(a) annually?

(b) monthly (12 times a year)?

(c) weekly (52 times a year)?

29. If the Klien family deposits $15,000 into an account that yields 5.9% annual interest, how much will be in the account after 16 years if the interest is compounded

(a) annually?

(b) monthly (12 times a year)?

(c) weekly (52 times a year)?

30. If the LaDukes deposit $8000 into an account that yields 6.1% annual interest, how much will be in the account after 10 years if the interest is compounded

(a) annually?

(b) semiannually (twice a year)?

(c) quarterly?

For Exercises 31–34, use the compound interest function $A(t) = P\left(1 + \frac{r}{n}\right)^{nt}$ to evaluate $A(5)$ and interpret. How much interest was made at the end of the 5-year period?

31. $5500 is invested at a 5.5% interest rate compounded quarterly (four times a year).

32. $6400 is invested at a 6.5% interest rate compounded quarterly (four times a year).

33. $10,000 is invested at a 7.25% interest rate compounded monthly (12 times a year).

34. $7500 is invested at a 6.25% interest rate compounded monthly (12 times a year).

35. Suppose that Elizabeth deposits $4000 at a 5.75% annual interest rate compounded monthly.

(a) Write the compound interest function A for the given information.

(b) Graph A in the viewing window $[0, 16]$ by $[4000, 10,000]$.

(c) Use your calculator to graphically find the amount of time that it takes the account to accumulate a total balance of $8000. This is called the **doubling time** of the account.

36. Now assume that Elizabeth deposits $6000 at a 5.75% annual interest rate compounded monthly.

(a) Write and graph the compound interest function A for the given information.

(b) Find the doubling time of this new account. Compare your answer to part (c) of Exercise 35.

37. Under certain conditions, the spread of the *E. coli* strain of bacteria can be modeled by

$$f(t) = N_0 e^{0.23t}$$

where t represents time in minutes, N_0 is the size of the colony when it starts, and $f(t)$ represents the size of the colony after time t. If $N_0 = 1,200,000$, answer the following:

(a) Evaluate $f(5)$ and interpret.

(b) Find the slope of the secant line for $t = 0$ and $\Delta t = 10$ and interpret.

38. The amount of a new experimental drug present in a patient's bloodstream is modeled by

$$g(t) = 5e^{-0.3t}, \qquad t \geq 0$$

where t represents the time since the drug was administered in hours and $g(t)$ represents the amount of drug in the bloodstream, measured in milligrams.

(a) Evaluate $g(2)$ and interpret.

(b) Find the slope of the secant line for $t = 2$ and $\Delta t = 3$ and interpret.

39. The sales of the new Flashfast Cigarette Lighters can be modeled by

$$S(t) = 10,000(2 - e^{-0.2t})$$

where t represents the number of years that the lighter has been on the market and $S(t)$ represents the number of sales.

(a) Evaluate $S(5)$ and interpret.

(b) Find the slope of the secant line for $t = 0$ and $\Delta t = 5$ and interpret.

40. The annual new import car sales in the United States can be modeled by

$$f(x) = 2.72e^{-0.07x}, \qquad 1 \leq x \leq 6$$

where x represents the number of years since 1989 and $f(x)$ represents the new import car sales in millions of cars.

(a) Evaluate $f(2)$ and interpret.

(b) Graph f in the viewing window [1, 6] by [0, 3], and classify the model as exponential growth or exponential decay.

(c) Use the INTERSECT command on your calculator to determine the year that the number of new import car sales dipped below 2 million cars.

41. The number of U.S. World War II veterans receiving compensation for service-connected disabilities can be modeled by

$$f(x) = 1285.34e^{-0.037x}, \qquad 1 \leq x \leq 16$$

where x represents the number of years since 1979 and $f(x)$ represents the number of World War II veterans receiving compensation, measured in thousands.

(a) Evaluate $f(12)$ and interpret.

(b) Graph f in the viewing window [1, 16] by [600, 1300], and classify the model as exponential growth or exponential decay.

(c) Compute the difference quotient for $x = 1$ and $\Delta x = 10$ and interpret.

42. The U.S. federal government *receipts* can be modeled by

$$f(x) = 37.14e^{0.081x}, \qquad 1 \leq x \leq 46$$

where x represents the number of years since 1949 and $f(x)$ represents the federal government receipts in billions of dollars.

(a) Evaluate $f(21)$ and interpret.

(b) Graph f in the viewing window [1, 46] by [0, 1500], and classify the model as exponential growth or exponential decay.

(c) Compute the difference quotient for $x = 31$ and $\Delta x = 10$ and interpret.

43. The U.S. federal government *outlays* can be modeled by

$$g(x) = 37.52e^{0.084x}, \qquad 1 \leq x \leq 46$$

where x represents the number of years since 1949 and $g(x)$ represents the U.S. federal government outlays in billions of dollars.

(a) Classify the model as exponential growth or exponential decay.

(b) Evaluate $g(21)$ and interpret.

(c) Graph g in the viewing window [1, 46] by [0, 1500].

(d) Compute the difference quotient for $x = 31$ and $\Delta x = 10$ and interpret.

44. Using function f from Exercise 42 and the function g from Exercise 43, complete the following:

(a) Evaluate the difference $f(41) - g(41)$ and interpret.

(b) Graph $f - g$ in the viewing window [1, 46] by [−250, 50]. What does this difference of functions given by (receipts) − (outlays) represent?

For Exercises 45–50, assume that the interest is compounded continuously. Use the compound interest function to evaluate $A(7)$ and interpret.

45. $1000 is invested at a $5\frac{1}{2}$% interest rate.

46. $800 is invested at a 4.75% interest rate.

47. $20,000 is invested at a 5.9% interest rate.

48. $10,000 is invested at a 7.1% interest rate.

49. $8000 is invested at a 8.1% interest rate.

50. $75,000 is invested at a $5\frac{3}{4}$% interest rate.

⚛ 51. The proportion of U.S. households that owns a DVD player can be modeled by

$$f(x) = \frac{0.8}{1 + 6e^{-0.3x}}, \qquad x \geq 0$$

where x represents the number of years since 1997 and $f(x)$ represents the proportion of U.S. households that own a DVD player.

(a) Evaluate $f(3)$ and interpret.

(b) Compute the difference quotient for $x = 3$ and $\Delta x = 2$ and interpret.

52. The proportion of home computers sold that have the latest 3-D WOW sound technology can be modeled by

$$g(t) = \frac{0.9}{1 + 4e^{-.3t}}, \qquad t \geq 0$$

where t represents the number of months that the sound technology has been on the market and $g(t)$ represents the number of home computers sold with the sound technology.

(a) Evaluate $g(7)$ and interpret.

(b) Compute the difference quotient for $t = 5$ and $\Delta t = 10$ and interpret.

53. Suppose that the spread of a certain kind of influenza at a rural school is given by the logistic model

$$f(t) = \frac{105}{1 + 20.08e^{-t}}, \qquad 0 \leq t \leq 10$$

where t represents the number of days that healthy students are exposed to infected ones and $f(t)$ represents the number of students afflicted.

(a) Evaluate $f(0)$ and interpret.

(b) Compute the difference quotient for $t = 3$ and $\Delta t = 4$ and interpret.

54. Suppose that the spread of measles at Grover Elementary School is given by the mathematical model

$$f(t) = \frac{200}{1 + 200.34e^{-t}}, \qquad 0 \leq t \leq 10$$

where t represents the number of days since the first student was diagnosed and $f(t)$ represents the number of affected students.

(a) Evaluate $f(4)$ and interpret.

(b) Compute the difference quotient for $t = 4$ and $\Delta t = 5$ and interpret.

55. Conservationists wish to repopulate a rare breed of lizard by transporting a seed colony to a wildlife preserve. They estimate that the lizard population will grow according to the logistic model

$$f(t) = \frac{1000}{1 + 121.51e^{-0.72t}}, \qquad 0 \leq t \leq 15$$

where t represents the number of years since the seeding and $f(t)$ represents the lizard population.

(a) According to the model, what was the initial size of the seed colony?

(b) How many lizards will be in the population in 8 years?

(c) Compute the average rate of change for $t = 0$ and $\Delta t = 8$ and interpret.

56. In a biological experiment, a seed colony of paramecia were placed in a petri dish along with a nutritional medium. The number of paramecia in the dish is given by the logistic model

$$f(t) = \frac{380}{1 + 181.22e^{-2.1t}}, \qquad 0 \leq t \leq 10$$

where t represents the number of days since the start of the experiment and $f(t)$ represents the population of paramecia.

(a) According to the model, how many paramecia were initially placed in the Petri dish?

(b) How many paramecia will be in the petri dish in 2 days?

(c) Compute the average rate of change for $t = 2$ and $\Delta t = 8$ and interpret.

SECTION PROJECT

Investors compare accounts that compound interest differently using an effective rate. For example, an interest rate of 6.25% compounded *monthly* produces the same amount of interest as an interest rate of 6.43% compounded *annually.* Here 6.43% is called the **effective rate** and 6.25% is called the **nominal rate.** For the interest rate r compounded n times per year, the effective rate is

$$\text{Effective rate} = \left(1 + \frac{r}{n}\right)^n - 1$$

Compute the effective rates for the following accounts and interpret the results.

(a) A 4% nominal rate compounded weekly (52 times a year)

(b) A 7.5% nominal rate compounded monthly

(c) A 6.1% nominal rate compounded daily (365 times a year)

(d) A 5.75% nominal rate compounded semiannually

For interest compounded continuously, the effective rate is given by the formula

$$\text{Effective rate} = e^r - 1$$

Compute the effective rates for the following accounts and interpret the results.

(e) A 4% nominal rate compounded continuously

(f) A 7.5% nominal rate compounded continuously

(g) A 6.1% nominal rate compounded continuously

(h) A 5.75% nominal rate compounded continuously

SECTION 1.7 LOGARITHMIC FUNCTIONS

In the previous section, we studied exponential functions and their properties. Now we want to develop some mathematical tools that undo the process of exponentiation. This undoing is accomplished with **logarithms.** To understand logarithms, we need to introduce some other useful ideas. The first of these is the composition of functions. We will see that **composite functions** occur frequently. Next we discuss **inverse functions,** which allow us to define functions such as logarithms.

Composite Functions

In Section 1.1, we learned that when we apply a function to an independent variable value a dependent variable value is returned. Now let's suppose that the dependent value of the one function becomes the independent value of another. To illustrate, suppose that a refinery's underwater supply line ruptures, resulting in an oil spill that is fairly circular. The petroleum company, the local fisheries, and those who are concerned about the environment would like to know how much area the spill will cover as time passes.

Assume that past experiments show that the radius of the greasy, circular slick increases at a rate of about 0.7 feet per second. As a function of time t, the radius of the slick is given by

$$r(t) = 0.7t$$

Since the area of the oil spill depends on its radius, the area is given by the function

$$A(r) = \pi r^2$$

Since the radius is given by $r(t) = 0.7t$, to find the area of the slick as a function of time, we can substitute $r(t) = 0.7t$ into $A(r) = \pi r^2$. This yields

$$A(t) = \pi(0.7t)^2 = \pi \cdot 0.49 \cdot t^2 = 0.49\pi t^2$$

Figure 1.7.1 suggests that we are nesting the functions, meaning that we could write $A(t) = A(r(t))$. This is exactly what a composition of functions is—a *nesting* of one function inside another.

Composite Function

A function h is a **composition** of the functions f and g if

$$h(x) = f(g(x))$$

NOTE: The composition of functions $f(g(x))$ can also be denoted as $(f \circ g)(x)$.

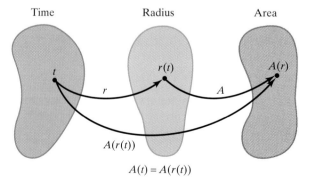

$$A(t) = A(r(t))$$

Figure 1.7.1

EXAMPLE 1 | **Evaluating Composite Functions**

Consider the functions $f(x) = \dfrac{3}{x}$ and $g(x) = 2x - 1$. Evaluate $f(g(6))$ and $g(f(6))$.

SOLUTION

Starting from the inside of the expression $f(g(6))$ and working our way out, we first evaluate

$$g(6) = 2(6) - 1 = 11$$

To evaluate $f(g(6))$, we substitute 11 for $g(6)$ to get

$$f(g(6)) = f(11) = \frac{3}{11}$$

In the case of $g(f(6))$, we first start by evaluating $f(6)$ to get

$$f(6) = \frac{3}{6} = \frac{1}{2}$$

So, to evaluate $g(f(6))$, we get

$$g(f(6)) = g\left(\frac{1}{2}\right) = 2 \cdot \frac{1}{2} - 1 = 0$$

✓CHECKPOINT 1 | Now work Exercise 3.

Notice in Example 1 that $f(g(6))$ and $g(f(6))$ were different. This shows that *generally $f(g(x)) \neq g(f(x))$*. We will soon see a special case when these compositions are the same.

When determining the domain of a composition of functions, we have to take care that the dependent values of the *inside* function are all members of the domain of the second. Knowing the range values of the inside function takes on added importance.

Building up functions through composition is a very exacting process. However, **decomposing** a function into simpler parts offers several options. This is illustrated in Example 2.

EXAMPLE 2 | **Decomposing Functions**

Determine functions f and g for each composition so that $f(g(x)) = h(x)$; then check your answer.

(a) $h(x) = \sqrt{(x + 5)^3}$ (b) $h(x) = (5x^3 + 6)^2$

SOLUTION

(a) Here we could let the *inside function* $g(x) = (x + 5)^3$ and the *outside function* $f(x) = \sqrt{x}$. Checking the solution yields

$$f(g(x)) = f\left[(x + 5)^3\right] = \sqrt{(x + 5)^3} = h(x)$$

(b) For this function we could let $g(x) = 5x^3 + 6$ and $f(x) = x^2$. Checking the solution, we get

$$f(g(x)) = f\left[(5x^3 + 6)\right] = (5x^3 + 6)^2 = h(x)$$

Inverse Functions

Addition and subtraction are examples of **inverse operations.** We know that if we start with some number like 3 and add 7, then subtract 7, we get back the original number 3. Similarly, functions such as $f(x) = \frac{x}{3}$, and $g(x) = 3x$ are inverses of one another. To show that these functions undo each other, choose a number like $x = 12$. Then in the function f we get

$$f(12) = \frac{12}{3} = 4$$

Taking this result and calculating $g(4)$, we get

$$g(4) = 3(4) = 12$$

which is the value that we started with. In terms of composition of functions, we have shown that

$$g(f(12)) = 12$$

Furthermore, we can show that for the functions $f(x) = \frac{x}{3}$ and $g(x) = 3x$ that

$$g(f(x)) = g\left(\frac{x}{3}\right) = 3\left(\frac{x}{3}\right) = x$$

These are examples of inverse functions. All inverse functions have a special property: they are all one-to-one functions.

One-to-One Function

A function f is a **one-to-one function** if, for elements a and b in the domain of f,

$$a \neq b \quad \text{implies that} \quad f(a) \neq f(b)$$

Finding these values for some functions can be rather difficult, so there is another way to determine if a function is 1-1. It is called the **horizontal line test.**

Horizontal Line Test

If every horizontal line intersects the graph of a function f in no more than one place, then f is a one-to-one function.

Now that we know what kind of functions have inverses, we present the definition of an inverse function.

Inverse Function

Two one-to-one functions f and g are **inverses** of each other if

$$f(g(x)) = x \quad \text{for every } x\text{-value in the domain of } g$$

and

$$g(f(x)) = x \quad \text{for every } x\text{-value in the domain of } f$$

A special notation is often used for the inverse function. If g is the inverse of f, then g may be written as f^{-1} (read "f inverse"). Do *not* confuse this with a negative exponent. $f^{-1}(x)$ does not represent $\dfrac{1}{f(x)}$; it represents the inverse function of f. Figure 1.7.2 shows the general relationship between a function and its inverse function.

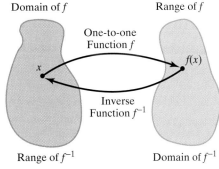

Domain of f Range of f

One-to-one Function f

x $f(x)$

Inverse Function f^{-1}

Range of f^{-1} Domain of f^{-1}

Figure 1.7.2

EXAMPLE 3

Showing That Two Functions Are Inverses

Use the definition of inverse functions to show that $f(x) = 3x - 2$ and $g(x) = \dfrac{x+2}{3}$ are inverses of each other.

SOLUTION

We need to show that $f(g(x)) = x$ and $g(f(x)) = x$. Completing the first composition, we get

$$f(g(x)) = f\left(\frac{x+2}{3}\right)$$

Replacing the expression $\dfrac{x+2}{3}$ at every occurrence of x in $f(x) = 3x - 2$, we get

$$f(g(x)) = 3\left(\frac{x+2}{3}\right) - 2 = (x+2) - 2 = x$$

In the Interactive Activity to the left, showing that $g(f(x)) = x$ will verify that $f(x) = 3x - 2$ and $g(x) = \dfrac{x+2}{3}$ are inverses of each other. ∎

Interactive Activity

Simplify the composition $g(f(x)) = x$ to verify that
$f(x) = 3x - 2$ and
$g(x) = \dfrac{x+2}{3}$ are inverses.
Then graph both functions in the viewing window $[-9.4, 9.4]$ by $[-6.2, 6.2]$. Do you see a relationship in the appearance of the graphs? Now let
$y_3 = y_1(y_2(x))$ and regraph the functions. What is this new function y_3?

The Interactive Activity to the left shows that there is not only an algebraic but also a graphical relationship between two inverse functions.

> ### Graphical Relationship between Inverse Functions *f* and *g*
>
> If *f* is a one-to-one function, then the graph of its inverse function *g* is a reflection about the line $y = x$.

Logarithmic Functions

Now we want to look at a special family of inverse functions. Their importance can be seen in the Flashback that follows.

Flashback

DISABLED WORLD WAR I VETERANS REVISITED

In Section 1.6, we saw that the number of disabled World War I veterans receiving compensation for service-connected disabilities can be modeled by

$$f(x) = 47.55(0.78)^x, \qquad 1 \leq x \leq 16$$

where x represents the number of years since 1979 and $f(x)$ represents the number of disabled World War I veterans, measured in thousands. Use tables to numerically estimate the time x when the number of disabled World War I veterans in the United States dipped below 12,000. Write the answer accurate to the half-year.

Solution

We start by making a table of values for $x = 1, 2, 3, \ldots$. The result is displayed in Table 1.7.1. Numerically, we see that at some time between the values of $x = 5$ and $x = 6$ the $f(x)$-values dipped

below 12. To get a better estimate, let's make another table using increments of 0.5 for x. This output is displayed in Table 1.7.2. This table tells us that the $f(x)$-value dipped below 12 in the interval $5.5 < x < 6.0$. Since $f(5.5) = 12.125$, our estimate accurate to 0.5 is $x = 5.5$.

TABLE 1.7.1

x	$f(x)$
1	$47.55(0.78)^1 = 37.089$
2	$47.55(0.78)^2 = 28.929$
3	$47.55(0.78)^3 = 22.565$
4	$47.55(0.78)^4 = 17.601$
5	$47.55(0.78)^5 = 13.729$
6	$47.55(0.78)^6 = 10.708$
7	$47.55(0.78)^7 = 8.352$

TABLE 1.7.2

x	$f(x)$
5.0	13.729
5.5	12.125
6.0	10.708

We could continue to get a more accurate solution to the Flashback by continuing with this numerical method or even by using the calculator's INTERSECT command. To get an accurate solution algebraically, we need a way to undo the exponential function. This is why we study the **logarithm.** The inverse association between logarithm and exponent is stressed in the definition of a logarithm.

> ### Logarithm
>
> We say that
>
> $$y = \log_b x \quad \text{if and only if} \quad b^y = x$$
>
> where $b > 0$, $b \neq 1$, $x > 0$ ($\log_b x$ is read "the log of x, base b" or "the log base b of x").

This definition tells us, for example, that

$$6^2 = 36 \quad \text{means} \quad \log_6 36 = 2$$

or $\qquad\qquad 3^{-2} = \dfrac{1}{9} \quad \text{means} \quad \log_3\left(\dfrac{1}{9}\right) = -2$

TABLE 1.7.3

x	$f(x) = 2^x$	x	$g(x) = \log_2 x$
-5	$\dfrac{1}{32}$	$\dfrac{1}{32}$	-5
-4	$\dfrac{1}{16}$	$\dfrac{1}{16}$	-4
-3	$\dfrac{1}{8}$	$\dfrac{1}{8}$	-3
-2	$\dfrac{1}{4}$	$\dfrac{1}{4}$	-2
-1	$\dfrac{1}{2}$	$\dfrac{1}{2}$	-1
0	1	1	0
1	2	2	1
2	4	4	2
3	8	8	3
4	16	16	4
5	32	32	5

To illustrate that the logarithm function is an inverse function, let's recall the table of values of $y = 2^x$ that was given in Section 1.6. These results are repeated on the left-hand side of Table 1.7.3. Now, if we use the definition of the logarithm, we obtain the values on the right-hand side. For example, for $x = \dfrac{1}{32}$, we see that since $\dfrac{1}{32} = 2^{-5}$, we can say that $\log_2 \left(\dfrac{1}{32} \right) = -5$.

The table shows that the x- and y-coordinates of $f(x) = 2^x$ and $g(x) = \log_2 x$ are interchanged. Thus, the graph of $f(x) = \log_2 x$ is a reflection of $f(x) = 2^x$ about the line $y = x$. See Figure 1.7.3. So it seems that the logarithm undoes exponentiation. That is, $f(x) = b^x$ **and** $g(x) = \log_b x$ **are inverse functions.** The table and graph also suggest that, since the range of the exponential function is the set of positive numbers, the domain of the logarithmic function is also the set of

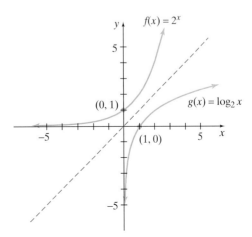

Figure 1.7.3

positive numbers. In other words, we cannot take the logarithm of a negative number. This is all summarized in the definition of a logarithm function.

Logarithm Function

$$f(x) = \log_b x$$

is the logarithm function, where $b > 0, b \neq 1,$ and the domain is $(0, \infty)$.

The definition states that the base b of a logarithm can be any positive number not equal to 1, but two bases are the most widely used.

Logarithm Notation

1. The **common logarithm** is $\log_{10} x$ and is denoted by $\log x$.
2. The **natural logarithm** is $\log_e x$ and is denoted by $\ln x$.

The properties of logarithms are much like those for exponents. Some of these properties are proved in Appendix C.

Properties of Logarithms

Let m and n be positive numbers. Then the following are true:

1. $\log_b (mn) = \log_b m + \log_b n$ 2. $\log_b\left(\dfrac{m}{n}\right) = \log_b m - \log_b n$

3. $\log_b m^n = n \cdot \log_b m$ 4. $\log_b 1 = 0$

5. $\log_b b = 1$ 6. $x = b^{\log_b x}$

7. $x = \log_b b^x$ 8. $\log_b m = \log_b n$ if and only if $m = n$

EXAMPLE 4 | **Using the Properties of Logarithms**

Use the properties of logarithms to rewrite the following:

(a) $\log(6 \cdot 17)$ (b) $\log_5\left(\dfrac{10}{3}\right)$ (c) $\log_6 \sqrt{11}$

SOLUTION

(a) Using property 1, we get $\log(6 \cdot 17) = \log 6 + \log 17$.

(b) Property 2 tells us that we can rewrite $\log_5\left(\dfrac{10}{3}\right)$ as $\log_5 10 - \log_5 3$.

(c) Before we apply property 3, we rewrite the logarithm.

$$\log_6 \sqrt{11} = \log_6 11^{1/2} = \frac{1}{2} \log_6 11$$

✓ CHECKPOINT 2

Now work Exercise 43.

In our study of calculus, we primarily use the **natural logarithm.** Since the natural logarithm is key to solving exponential equations, we note that all properties of logarithms are true for $\ln x$. These properties allow us to *take the log of* both sides of an equation so that the independent variable is no longer in the exponent. We outline this procedure to solve the general exponential equation $y = a \cdot b^x$ in Example 5.

EXAMPLE 5 | **Solving Exponential Equations**

In the Flashback, we numerically approximated the solution to the exponential equation $12 = 47.55(0.78)^x$. Use the properties of logarithms to solve the equation and round the solution to the nearest hundredth.

SOLUTION

We start solving the equation $12 = 47.55(0.78)^x$ by dividing both sides by 47.55.

$$12 = 47.55(0.78)^x$$

$$\frac{12}{47.55} = (0.78)^x \qquad \text{Divide by 47.55.}$$

$$\ln\left(\frac{12}{47.55}\right) = \ln(0.78)^x \qquad \text{Take the ln of both sides, Property 4.}$$

$$\ln\left(\frac{12}{47.55}\right) = x \cdot \ln(0.78) \qquad \text{Property 3.}$$

$$\frac{1}{\ln(0.78)} \cdot \ln\left(\frac{12}{47.55}\right) = x \qquad \text{Divide by } \ln(0.78).$$

$$5.54 \approx x$$

Note in the example that no rounding takes place until the final step. Rounding earlier in the process can throw off the final result.

Applications with Logarithmic Functions

Many phenomena that we study start with rapid growth and then the growth begins to level off. This type of behavior is well suited to the **logarithmic model** of the form **$f(x) = a + b \ln x$,** where a and b are constants. Since we cannot take the logarithm of a negative number, the domain of this function is the set of positive numbers $(0, \infty)$. In practice, we usually start with an x-value of 1.

EXAMPLE 6 | **Applying Logarithmic Models**

The percentage of U.S. households with cable TV can be modeled by

$$f(x) = 18.32 + 15.94 \ln x, \qquad 1 \le x \le 17$$

where x represents the number of years since 1979 and $f(x)$ represents the percentage of households in the United States with cable TV.

(a) Evaluate $f(6)$ and interpret.
(b) Use the result of part (a) to compute the difference quotient for $x = 6$ and $\Delta x = 5$ and interpret.

SOLUTION

(a) First, $x = 6$ corresponds to the year $1979 + 6 = 1985$. Evaluating, we get

$$f(6) = 18.32 + 15.94 \ln(6) \approx 46.88$$

Thus, according to the model, about 46.88% of the households in the United States had cable TV in 1985.

Interactive Activity

Compute the difference quotient for $x = 11$ and $\Delta x = 5$ and compare the result with part (b) of Example 6.

(b) Computing the difference quotient with $x = 6$ and $\Delta x = 5$, we get

$$m_{\text{sec}} = \frac{f(6+5) - f(6)}{5} = \frac{f(11) - f(6)}{5}$$

$$\approx \frac{56.54 - 46.88}{5} \approx 1.93$$

Thus, from 1985 to 1990 the number of U.S. households with cable TV was growing at an average rate of about $1.93 \frac{\text{percent}}{\text{year}}$. ◼

SUMMARY

This section featured several new topics, including the definition and properties of the composite and inverse functions. Then we used this information to discover that the exponential and logarithmic functions were inverses.

- A function h is a **composition** of the functions f and g if $h(x) = f(g(x))$.
- Two **one-to-one** functions f and g are **inverses** of each other if $f(g(x)) = x$ for every x-value in the domain of g and $g(f(x)) = x$ for every x-value in the domain of f.
- We say that $y = \log_b x$ if and only if $b^y = x$, where $b > 0, b \neq 1, x > 0$.
- The logarithmic model is $f(x) = a + b \ln x$.

Properties of Logarithms

- $\log_b (mn) = \log_b m + \log_b n$
- $\log_b m^n = n \cdot \log_b m$
- $\log_b b = 1$
- $x = \log_b b^x$

- $\log_b \left(\dfrac{m}{n}\right) = \log_b m - \log_b n$
- $\log_b 1 = 0$
- $x = b^{\log_b x}$
- $\log_b m = \log_b m$ if and only if $m = n$

SECTION 1.7 EXERCISES

For Exercises 1–4, answer the following:

1. Let $f(x) = 3x$ and $g(x) = 6x$. Evaluate $f(g(3))$ and $f(g(-2))$.

2. Let $f(x) = 3x - 1$ and $g(x) = x^2 + 1$. Evaluate $g(f(2))$ and $g(f(3))$.

✓ 3. Let $f(x) = 3x^2 - x$ and $g(x) = 6x + 1$. Evaluate $f(g(0))$ and $g(f(0))$.

4. Let $f(x) = \sqrt{2x}$ and $g(x) = 3x^2$. Evaluate $f(g(2))$ and $g(f(10))$.

For Exercises 5–12, find $g(f(x))$ and $f(g(x))$.

5. $f(x) = 3x - 1;\ g(x) = 8x + 2$

6. $f(x) = 6x - 9;\ g(x) = 7x + 5$

7. $f(x) = 5x - 3;\ g(x) = x^2 + 3x + 4$

8. $f(x) = 3x^2 + x + 5;\ g(x) = x - 1$

9. $f(x) = x^3;\ g(x) = \dfrac{1}{x}$

10. $f(x) = \dfrac{2}{x};\ g(x) = x + 3$

11. $f(x) = \sqrt{x + 1};\ g(x) = x^2 + 2$

12. $f(x) = \sqrt{x + 2};\ g(x) = 4x - 3$

For Exercises 13–20, determine f and g so that $f(g(x)) = h(x)$ for each composite function h; then check your answer. There is more than one correct answer.

13. $h(x) = (x + 3)^3$

14. $h(x) = (5x + 6)^4$

15. $h(x) = \left(\dfrac{1}{x + 3}\right)^2$

16. $h(x) = \left(\dfrac{5}{x - 1}\right)^4$

17. $h(x) = \sqrt[4]{x - 2}$

18. $h(x) = \sqrt[3]{2x + 3}$

19. $h(x) = 2 - 3\sqrt{x}$

20. $h(x) = 9 + 2\sqrt{x}$

21. Suppose that a spherical balloon is being inflated with the radius increasing at an average rate of 1.3 inches per second.
(a) Write the radius $r(t)$ as a function of time t. (Assume that when $t = 0$ then $r = 0$.)
(b) The volume of a sphere is a function of its radius and is given by $V(r) = \frac{4}{3}\pi r^3$. Simplify $V(t) = V(r(t))$ and interpret the meaning of this composition.
(c) Evaluate $V(t)$ when $t = 6$ and interpret.

22. The Top Drug Company has developed a new type of pill in the shape of a ball. The spherical pill is dropped into a glass of water and dissolves. Tests show that when the 2-centimeter pill is dropped into the water its radius decreases at an average rate of $0.003 \frac{\text{cm}}{\text{sec}}$.
(a) Write the radius $r(t)$ as a function of time t. In this case, $r(t) = 2$ when $t = 0$.
(b) Knowing that the surface area of a sphere as a function of its radius is $S(r) = 4\pi r^2$, simplify $S(t) = S(r(t))$ and interpret the meaning of this composition.
(c) Evaluate $S(t)$ when $t = 20$ and interpret.

For the one-to-one functions f and g in Exercises 23–30, show that $f(g(x)) = x$.

23. $f(x) = 8x;\ g(x) = \frac{1}{8}x$

24. $f(x) = \frac{3}{4}x;\ g(x) = \frac{4}{3}x$

25. $f(x) = 7x - 10;\ g(x) = \frac{x + 10}{7}$

26. $f(x) = \frac{x - 3}{4};\ g(x) = 4x + 3$

27. $f(x) = x^3 + 6;\ g(x) = \sqrt[3]{x - 6}$

28. $f(x) = x^5 - 9;\ g(x) = \sqrt[5]{x + 9}$

29. $f(x) = \sqrt{x - 1};\ g(x) = x^2 + 1,$ for $x \geq 0$

30. $f(x) = \sqrt[4]{x - 1};\ g(x) = x^4 + 1,$ for $x \geq 0$

In Exercises 31–42, rewrite the equation in the logarithmic form $y = \log_b x$.

31. $2^5 = 32$

32. $3^4 = 81$

33. $2^{-3} = \frac{1}{8}$

34. $5^{-2} = \frac{1}{25}$

35. $\left(\frac{1}{2}\right)^2 = \frac{1}{4}$

36. $\left(\frac{1}{3}\right)^{-1} = 3$

37. $81^{3/4} = 27$

38. $64^{4/3} = 256$

39. $10^5 = 100{,}000$

40. $10^3 = 1000$

41. $e^1 = e$

42. $e^0 = 1$

In Exercises 43–50, use the properties of logarithms to rewrite the following as the sum and/or difference of logarithms.

✓43. $\log_2\left(\frac{3}{5}\right)$

44. $\log_6\left(\frac{4}{7}\right)$

45. $\log(8 \cdot 20)$

46. $\log(17 \cdot 11)$

47. $\ln\sqrt{26}$

48. $\ln\sqrt[3]{11}$

49. $\log_3 \dfrac{4\sqrt{3}}{9}$

50. $\log_{16} \dfrac{8\sqrt[3]{6}}{11}$

For Exercises 51–58, solve the exponential equations algebraically and round the solution to the hundredths place.

51. $6 = 3^x$

52. $12 = 4^x$

53. $8 \cdot 2^x = 51$

54. $6 \cdot 9^x = 126$

55. $2e^x = 62$

56. $4e^x = 81$

57. $1.21(0.3)^x = 42$

58. $0.33(1.27)^x = 58$

🌐 59. The percentage of unemployed Hispanics in the U.S. labor force can be modeled by
$$f(x) = 11.39(0.94)^x, \qquad 1 \leq x \leq 7$$
where x represents the number of years since 1992 and $f(x)$ represents the percentage of unemployed Hispanics in the U.S. labor force.
(a) Evaluate $f(4)$ and interpret.
(b) Solve the exponential equation $11.39(0.94)^x = 10$ and interpret.

🌐 60. The percentage of unemployed Caucasians in the U.S. labor force can be modeled by
$$f(x) = 7.17(0.91)^x, \qquad 1 \leq x \leq 7$$
where x represents the number of years since 1989 and $f(x)$ represents the percentage of unemployed Caucasians in the U.S. labor force.
(a) Evaluate $f(4)$ and interpret.
(b) Solve the exponential equation $7.17(0.91)^x = 5.4$ and interpret.

🌐 61. The total annual personal expenditures for admission to spectator sports can be modeled by
$$f(x) = 2.05 + 1.3 \ln x, \qquad 1 \leq x \leq 10$$
where x represents the number of years since 1984 and $f(x)$ represents the total annual personal expenditures for admission to spectator sports in billions of dollars.
(a) Make a table of values for $x = 1, 3, 5, 7, 9$.
(b) Evaluate $f(6)$ and interpret.
(c) Use the table in part (a) to determine the average rate of change in personal expenditures for admission to spectator sports from 1985 to 1993.

(d) Use the table in part (a) to determine the average rate of change in personal expenditures for admission to spectator sports from 1985 to 1987.

⚙ 62. The percentage of U.S households that have video cassette recorders (VCRs) can be modeled by

$$f(x) = 21.88 + 25.34 \ln x, \qquad 1 \le x \le 17$$

where x represents the number of years since 1979 and $f(x)$ represents the percentage of households in the United States with VCRs.

(a) Make a table of values for $x = 1, 6, 11, 16$.

(b) Use the values from part (a) to compute the difference quotient for $x = 1$ and $\Delta x = 5$ and interpret.

⚙ 63. The total school expenditures in higher education by the federal government can be modeled by

$$f(x) = 16.49 + 2.85 \ln x, \qquad 1 \le x \le 15$$

where x represents the number of years since 1979 and $f(x)$ represents the total school expenditures in higher education by the federal government in billions of constant 1993–1994 dollars.

(a) Evaluate $f(10)$ and interpret.

(b) Compute the difference quotient for $x = 10$ and $\Delta x = 5$ and interpret.

⚙ 64. The total school expenditures in higher education by *state* governments can be modeled by

$$g(x) = -0.21x^2 + 2.79x + 37.97, \qquad 1 \le x \le 15$$

where x represents the number of years since 1979 and $g(x)$ represents the total school expenditures in higher education

by *state* governments in billions of constant 1993–1994 dollars.

(a) Evaluate $g(10)$ and interpret.

(b) Compute the difference quotient for $x = 10$ and $\Delta x = 5$ for the function g and compare with the result of Exercise 63 part (b).

SECTION PROJECT

The annual per capita consumption of *light* and *skim* milk in the United States can be modeled by

$$f(x) = 10.12 + 2 \ln x, \qquad 1 \le x \le 16$$

where x represents the number of years since 1979 and $f(x)$ represents the annual per capita consumption of light and skim milk in gallons.

(a) Graph f in the viewing window [1, 16] by [10, 16].

(b) Evaluate $f(12)$ and interpret.

(c) Compute the difference quotient for $x = 12$ and $\Delta x = 2$ and interpret.

The annual per capita consumption of *whole* milk in the United States can be modeled by

$$g(x) = 18.04(0.96)^x, \qquad 1 \le x \le 16$$

where x represents the number of years since 1979 and $g(x)$ represents the annual per capita consumption of whole milk in gallons.

(d) Evaluate $f(12) - g(12)$ and interpret.

(e) Compute the difference quotient for $x = 12$ and $\Delta x = 2$ for the function g and compare with part (c).

SECTION 1.8 REGRESSION AND MATHEMATICAL MODELS

Many of the applications in this book have the ⚙ icon indicating that the function is derived from data gathered on the Internet. But how is this done? The answer lies within a statistical process called **regression analysis.** In this section, we will show how computers and, particularly, graphing calculators can be used to develop a **mathematical model** to study real data. We will also discuss the shortcomings of overreliance on these regression techniques.

Mathematical Models

We often want to find a nice, compact function to explain complex events in the real world. This function, called a **mathematical model,** is a function that is derived from a real-world situation. Let's say that a local meteorologist makes an observation such as "the height of the Bend River started at 4 feet and has been steadily climbing half a foot each day for the last 10 days." We can make a simple mathematical model for the water level such as

$$w(t) = \frac{1}{2}t + 4, \qquad 0 \le t \le 10$$

where t represents the number of days and $w(t)$ represents the water level.

This simple, compact function numerically explains what the water level has been for each day of the 10-day period. Unfortunately, few phenomena in the business, life, and social sciences are so straightfoward. This is why we need to gather data and use them to create a mathematical model via **regression** techniques (see Figure 1.8.1). We can then use this model to study the phenomenon.

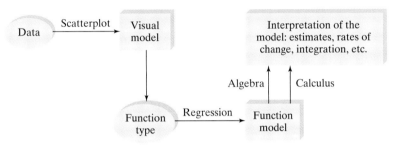

Figure 1.8.1

Regression Model Process

Finding a proper fit for data can be almost as much art and experience as it is mathematics. There are some common steps to follow. To see how such a procedure works, let's consider the Bureau of Labor Statistics data on the consumer price index (CPI) for the years 1987 to 1996 in Table 1.8.1.

As we mentioned in Section 1.1, we begin the process by **standardizing the independent variable** (Table 1.8.2). The next step is to **make a scatterplot of the** CPI data, as shown in Figure 1.8.2.

TECHNOLOGY NOTE
To see how to make scatterplots for your data, check the online calculator manual at www.prenhall.com/armstrong

TABLE **1.8.1**	
YEAR	**CPI**
1987	113.6
1988	118.3
1989	124.0
1990	130.7
1991	136.2
1992	140.3
1993	144.5
1994	148.2
1995	152.4
1996	156.9

Source: U.S. Bureau of Labor Statistics

TABLE **1.8.2**		
YEAR	**x**	**y, THE CPI**
1987	0	113.6
1988	1	118.3
1989	2	124.0
1990	3	130.7
1991	4	136.2
1992	5	140.3
1993	6	144.5
1994	7	148.2
1995	8	152.4
1996	9	156.9

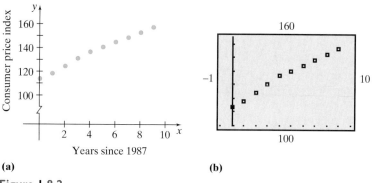

Figure 1.8.2

We can now **inspect the dependent values** to see if there is a pattern in the data. We can see from the scatterplot of the CPI data that the points appear to be fairly linear. This is a good clue that the regression model that we want to use is the linear model. But there is another way to inspect the data for a pattern. Let's compute the annual differences of the dependent values by taking a dependent value and subtracting the one before it. The results of this analysis are shown in Table 1.8.3.

TABLE 1.8.3			
YEAR	**x-VALUE**	**CPI**	**SUCCESSIVE DIFFERENCE**
1987	0	113.6	
1988	1	118.3	$118.3 - 113.6 = 4.7$
1989	2	124.0	5.7
1990	3	130.7	6.7
1991	4	136.2	5.5
1992	5	140.3	4.1
1993	6	144.5	4.2
1994	7	148.2	3.7
1995	8	152.4	4.2
1996	9	156.9	$156.9 - 152.4 = 4.5$

TECHNOLOGY NOTE
The DIAGNOSTICS ON indicator may need to be activated to get r^2 and r. Consult the on-line calculator manual at www.prenhall.com/armstrong

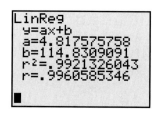

Figure 1.8.3

Even though the differences vary a little, their average value is about 4.8. We can conclude that the average rate of change is nearly constant. For each year, there is about a 4.8 increase in the CPI. This constant difference is a clue that a linear relationship exists between the independent and dependent variable values. This analysis is all part of **deciding which model is the best.**

After selecting the model, the next step is to use a computer or calculator to **compute the regression.** The common method used to compute linear regression is called the **least-squares method,** developed by mathematician G. Yule, who used a technique developed nearly a century before by Adrien Lagendre. The linear regression model takes the form $f(x) = ax + b.$ For the CPI data, the calculator output of the regression is similar to that shown in Figure 1.8.3.

Rounding to the nearest hundredth, we obtain the linear regression model $f(x) = 4.82x + 114.83$ for the data. Note that the slope of the line is $m = 4.82$, which is close to the average annual difference of 4.8. The graph of this model,

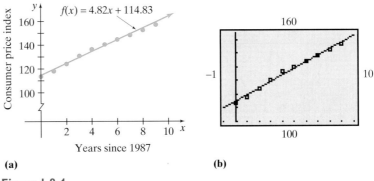

Figure 1.8.4

along with the scatterplot, is shown in Figure 1.8.4. Notice that the line does not fit exactly through the points. Some of the points are above and some are below the regression line.

The final step in the curve-fitting process is to check for **goodness of fit** of the model to the tabular data. Notice in Figure 1.8.3 that values for r and r^2 are included in the regression output. These values measure the quality of fit. The measure used for the linear regression model is the **correlation coefficient r.**

Correlation Coefficient

The **correlation coefficient** r, which ranges from -1 to 1, measures the strength and direction of the linear relationship between x and y.

The interpretation of the correlation coefficient for linear regression is summarized in Table 1.8.4. (See page 96.)

Since the CPI linear model has a correlation coefficient of $r = 0.996$, there is a strong positive relationship between x and y. Calling a relationship strong means that the data points on the scatterplot generally line up. A positive r means that as the x-values increase so do the y-values.

The r^2-value is a measure of goodness of fit for nonlinear functions, which we will discuss next. Before continuing, let's summarize the process used in curve fitting tabular data.

The Regression Model Process

1. **Standardize the independent variable** x by setting the first independent value equal to 0 or 1 and labeling all the other independent values accordingly.
2. **Make a scatterplot** of the data.
3. **Inspect the dependent values** to see if there is a pattern in the data that appears to be consistent with a familiar type of function. Use this information to **decide which curve-fitting model** to use.
4. Use a computer or graphing calculator to **compute the regression.** Record the constants along with the r and r^2 values.
5. Verify for **goodness of fit** by checking the r- and r^2-values. The closer that these numbers are to -1 and 1 the better the model fits the data.

Analyzing Nonlinear Regression Models

Many of the 🌐 mathematical models that we have seen are nonlinear models, such as quadratic, cubic, exponential, or logarithmic models. The curve-fitting process is done the same way for these models. For quadratic, cubic, and quartic

TABLE 1.8.4 INTERPRETATION OF r FOR LINEAR MODELS

IF THE SCATTERPLOT LOOKS LIKE	THEN x AND y HAVE	AND THE r VALUE IS
(scatterplot: points rising steadily, y vs x)	A strong positive association	close to 1
(scatterplot: points falling steadily, y vs x)	A strong negative association	close to –1
(scatterplot: points rising with scatter, y vs x)	A weak positive association	between 0 and 1
(scatterplot: points falling with scatter, y vs x)	A weak negative association	between –1 and 0
(scatterplot: random cloud of points, y vs x)	Little or no association	close to zero

(fourth-degree) functions, the r^2 value is used in lieu of the correlation coefficient. The closer that the r^2-value is to 1 the better the model fits the data.

EXAMPLE 1

Fitting a Polynomial Model

Table 1.8.5 shows the number of participants (in millions) in the Federal Food Stamp Program from 1990 to 1996.

TABLE 1.8.5

YEAR	PARTICIPANTS
1990	20.1
1992	25.4
1993	27
1994	27.5
1995	26.6
1996	25.5

Source: U.S. Census Bureau web site.

(a) Standardize the independent variable and enter the data in a table, make a scatterplot of the data, decide which polynomial model to use, and explain why.

(b) Compute the regression to get the polynomial model. Record the r^2-value and interpret.

SOLUTION

(a) We can standardize the independent variable values by letting x represent the number of years since 1990. Then the data table looks like Table 1.8.6. Standardizing the independent variable values this way simplifies the interpretation of the results. We see that $x = 0$ corresponds to 1990, $x = 2$ corresponds to 1992, and so on. A scatterplot of the data is shown in Figure 1.8.5. Since the scatterplot resembles a parabola, we will try a quadratic model.

TABLE 1.8.6

NUMBER OF YEARS SINCE 1990	PARTICIPANTS
0	20.1
2	25.4
3	27
4	27.5
5	26.6
6	25.5

(b) The result of the quadratic model regression is shown in Figure 1.8.6. From the output, we see that the quadratic model of the data, with all the constants rounded to the nearest hundredth, is

$$f(x) = -0.46x^2 + 3.66x + 20.08, \qquad 0 \le x \le 6$$

where x represents the number of years since 1990 and $f(x)$ represents the number of participants, in millions, in the Federal Food Stamp Program. We have $r^2 = 0.996$, which means that the model fits the data very well since 0.996 is close to 1.

Interactive Activity

Repeat Example 1, but choose a **quartic** regression model.

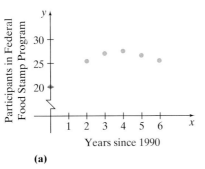

(a)

Figure 1.8.5

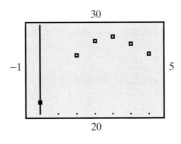

(b)

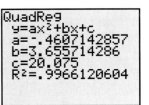

```
QuadReg
y=ax²+bx+c
a=-.4607142857
b=3.655714286
c=20.075
R²=.9966120604
```

Figure 1.8.6

Curve-fitting Polynomial Models

Most graphing calculators compute regression models for first-, second-, third-, and fourth-degree polynomials. Many times, the choice of polynomial used is determined by the number of peaks and valleys in the data. Following are some tips for each model. Remember that these are loose guidelines; when in doubt, try several models and compare the r- and r^2-values.

• **Linear Model**   The form of the linear model is $f(x) = ax + b$, where a and b are real number constants. A linear model is a good choice when the dependent values have approximately equal successive differences and the values are steadily increasing or decreasing. Figure 1.8.7a has a strong positive correlation, and Figure 1.8.7b has a strong negative association.

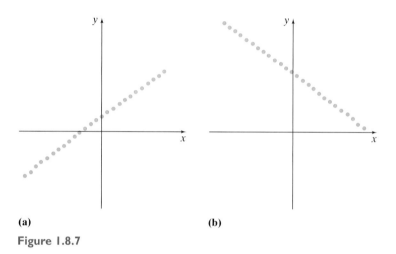

(a)   (b)

Figure 1.8.7

• **Quadratic Model**   The form of the quadratic model is $f(x) = ax^2 + bx + c$, where a, b, and c are real number constants. The quadratic model lends itself to data whose scatterplot has a distinct peak or a distinct valley. See Figure 1.8.8.

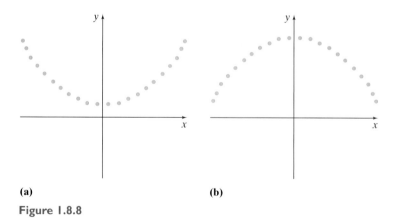

(a)   (b)

Figure 1.8.8

• **Cubic Model**   The form of the cubic model is $f(x) = ax^3 + bx^2 + cx + d$, where $a, b, c,$ and d are real number constants. The cubic model lends itself to data whose scatterplot has values with one distinct peak and one distinct valley. See Figure 1.8.9.

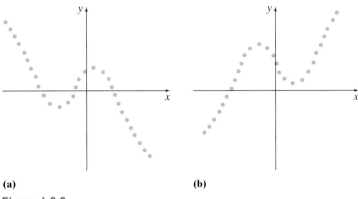

(a)   **(b)**

Figure 1.8.9

- **Quartic Model**   The form of the quartic model is

$$f(x) = ax^4 + bx^3 + cx^2 + dx + e,$$

where a, b, c, d, and e are real number constants. The cubic model lends it-self to data whose scatterplot has a distinct peak between two valleys, or vice versa. See Figure 1.8.10.

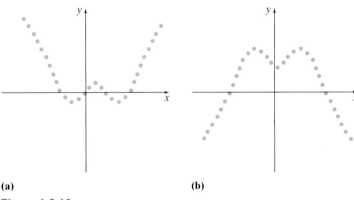

(a)   **(b)**

Figure 1.8.10

- **Power Model**   The form of the power model is $f(x) = a \cdot x^b$ where a and b are real number constants. The power model lends itself to data whose scatterplot resembles Figure 1.8.11.

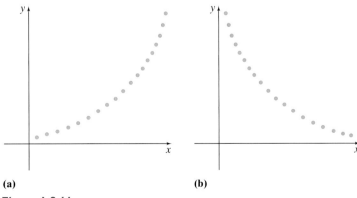

(a)   **(b)**

Figure 1.8.11

Curve-fitting Exponential and Logarithmic Models

Most graphing calculators compute regressions for general exponential, logarithmic, and logistic functions. A key step when curve fitting with this family of models is standardizing the independent variable. When computing a regression model of the form $y = a \cdot b^x$, the calculator rewrites the equation using logarithms. Since the domain of the logarithmic function is the set of positive numbers, **zero must not be a value of the independent variable.**

Following is a summary of common logarithmic and exponential models.

- **General Exponential Model**   The form of the general exponential model is $f(x) = a \cdot b^x$, where a and b are real number constants, with $b > 0$, $b \neq 1$. The exponential model lends itself to data whose scatterplot is always increasing or decreasing on the domain. See Figure 1.8.12.

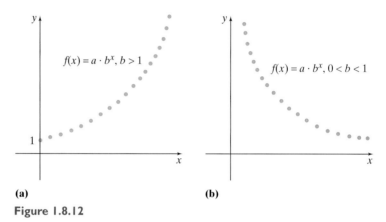

(a)   **(b)**

Figure 1.8.12

One other property of the general exponential model requires an explanation. First, we need to revisit a property of logarithms as shown in the Toolbox to the left.

Notice that we can replace x with a base like b to get $b = e^{\ln b}$. This allows us to rewrite any general exponential model $f(x) = a \cdot b^x$ in exponential form by writing $\boldsymbol{f(x) = a \cdot b^x}$ as $\boldsymbol{f(x) = a \cdot e^{\ln b \cdot x}}$.

- **Logarithmic Model**   The form of the logarithmic model is $f(x) = a + b \ln x$, where a and b are real number constants. The logarithmic model lends itself to data whose scatterplot initially increases rapidly and then begins to level off. See Figure 1.8.13.

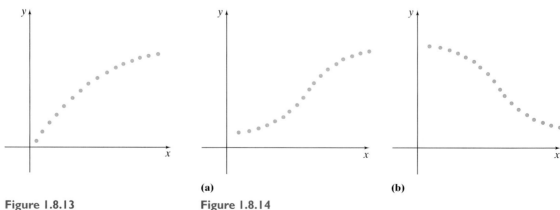

(a)   **(b)**

Figure 1.8.13   **Figure 1.8.14**

• **Logistic Model**   The form of the logistic model is $f(x) = \dfrac{c}{1 + a \cdot e^{-bx}}$ where $a, b,$ and c are real number constants. The logistic model lends itself to data whose scatterplot increases (decreases) slowly, then increases (decreases) rapidly, and then levels off. See Figure 1.8.14.

EXAMPLE 2

Fitting an Exponential/Logarithmic Model

Table 1.8.7 shows the number of employees in the environmental industry (measured in thousands of employees) in the United States.

(a) Standardize the independent variable, make a scatterplot of the data, and explain which, exponential or logarithmic, model to use.

(b) Compute the regression to determine the model. Record the r^2-value and interpret.

SOLUTION

(a) Since the data start in the year 1980, and we want to start our independent variable values at $x = 1$, let's assign x to represent the number of years since 1979. This gives us the data shown in Table 1.8.8.

   A scatterplot of the data is shown in Figure 1.8.15. Since the data in this scatterplot increase rapidly at first and then begin to level off a bit, we suspect that a logarithmic model may be a good choice.

(b) The result of the logarithmic model regression is shown in Figure 1.8.16. From the output, we see that the logarithmic model of the data, with the coefficients rounded to the nearest hundredth, is

$$f(x) = 461.87 + 299.40 \ln x, \qquad 1 \le x \le 17$$

where x represents the number of years since 1979 and $f(x)$ represents the number of employees in the environmental industry (measured in thousands of employees). Since $r^2 \approx 0.999$, this means that the model fits the data very well, since this number is close to 1. The model and the scatterplot are shown in Figure 1.8.17.

TABLE 1.8.7

YEAR	NUMBER EMPLOYED
1980	462.5
1990	1174.3
1994	1274.0
1995	1299.6
1996	1306.1

Source: U.S. Census Bureau web site.

TABLE 1.8.8

YEARS SINCE 1979	NUMBER EMPLOYED
1	462.5
11	1174.3
15	1274
16	1299.6
17	1306.1

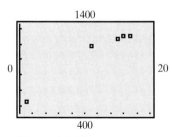

Figure 1.8.15

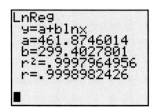

Figure 1.8.16

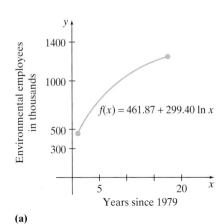

(a)

Figure 1.8.17

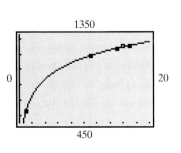

(b)

Shortcomings and Cautions

Curve fitting is a powerful tool that can be used to enhance the study of calculus. However, it does have some limitations. One of these is that regression models cannot be used to make long-term predictions. Remember that with many models, particularly polynomial models, the end behavior is given by either $-\infty$ or ∞. However, almost all phenomena in the business, life, and social sciences have some kind of physical, legal, or practical limitation. This is why almost all the regression models that we use are defined on closed intervals.

We also must be cautious of particular data points that appear to be much different from the rest of the data. Called **outliers,** they can throw off the accuracy of the resulting regression model, particularly if they occur for large independent values. If the outlier occurs as either the first or last data point, it can be excluded in order to make the model more representative of the entire set of data.

SUMMARY

This section built the framework for data modeling that is used throughout this textbook. We discussed the curve-fitting procedure that derives a function model from real-world data. Then we outlined the following types of regression models:

- **Linear model:** $f(x) = ax + b$
- **Quadratic model:** $f(x) = ax^2 + bx + c$
- **Cubic model:** $f(x) = ax^3 + bx^2 + cx + d$
- **Quartic model:** $f(x) = ax^4 + bx^3 + cx^2 + dx + e$
- **General Exponential model:** $f(x) = a \cdot b^x$
- **Exponential model:** $f(x) = a \cdot e^{\ln b \cdot x}$, $b > 0, b \neq 1$
- **Logarithmic model:** $f(x) = a + b \ln x$
- **Logistic model:** $f(x) = \dfrac{c}{1 + a \cdot e^{-bx}}$

SECTION 1.8 EXERCISES

For Exercises 1–10, determine which type of regression model would be a good choice based on the scatterplot. Answers may vary.

1.

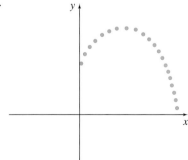

2.

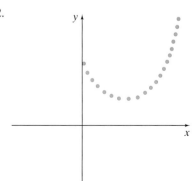

3.

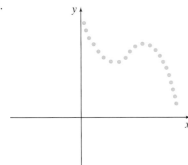

4.

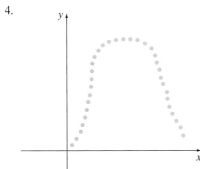

5.

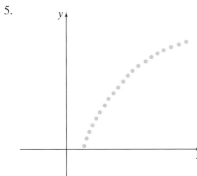

6.

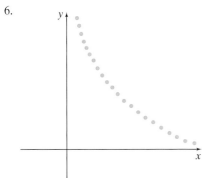

7.

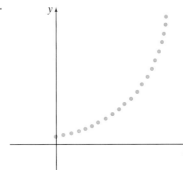

8.

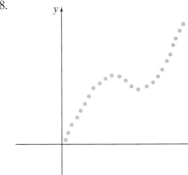

9.

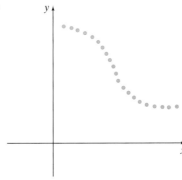

10.

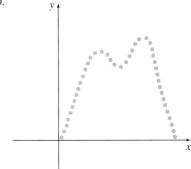

For Exercises 11–14, follow the curve-fitting process to find an appropriate model for the data.

11. Use a linear model, where x represents the number of years since 1980.

YEAR	MULTIPLE BIRTHS (TWINS, TRIPLETS, ETC.) PER 1000 LIVE BIRTHS
1980	19.3
1985	21
1986	21.6
1987	22
1988	22.4
1989	23
1990	23.3
1991	23.9
1992	24.4
1993	25.2
1994	25.7

Source: U.S. Census Bureau web site.

12. Use a cubic model, where x represents the number of years since 1975.

YEAR	NUMBER OF U.S. GOLF FACILITIES
1975	11,370
1980	12,005
1985	12,346
1990	12,846
1992	13,210
1993	13,439
1994	13,683
1995	14,074

Source: U.S. Census Bureau web site.

13. Use a logarithmic model, where x represents the number of years since 1989.

YEAR	AVERAGE EXPENDITURE FOR A NEW CAR (IN $)
1990	15,926
1991	16,650
1992	17,825
1993	18,585
1994	19,463
1995	19,757

Source: U.S. Census Bureau web site.

14. Use a power function model, where x represents the number of years since 1969.

YEAR	PERCENTAGE OF 3- TO 5-YEAR-OLDS IN PRESCHOOL
1970	37.5
1975	48.6
1980	52.5
1985	54.6
1990	59.4
1992	55.5
1993	55.1
1994	61
1995	61.8

Source: U.S. Census Bureau web site.

15. To see the shortcomings of regression modeling, consider the mathematical model of the percentage of births to unmarried mothers.

$$f(x) = 0.02x^3 - 0.23x^2 + 2.03x + 20.22, \quad 1 \le x \le 10$$

(a) Evaluate $f(10)$ and interpret.

(b) Evaluate $f(20)$ and interpret.

(c) Why does the solution to part (b) not make sense?

CHAPTER REVIEW EXERCISES

1. Use the table to construct ordered pairs and plot them in the Cartesian plane.

x	−1	2	2.8	4
$f(x)$	3.5	1	−1.7	2

2. Make a table and write the ordered pairs based on the scatterplot.

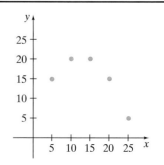

3. For the function $A = \pi r^2$, identify the independent and the dependent variable.

4. Let x represent the number of years since 1900. Standardize the values of the independent variable based on this definition, and make a scatterplot of the data.

YEAR	U.S. POPULATION (MILLIONS)
1910	92.0
1920	105.7
1930	122.8
1940	131.7
1950	150.7
1960	179.3
1970	203.3
1980	226.5
1990	248.7

5. Determine if the table represents a function.

DOMAIN	1	2	3	4	5
RANGE	5	3	1	3	5

6. Determine if the graph represents a function.

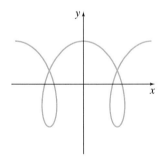

For Exercises 7 and 8, let $f(x) = x^2 - 4$ and $g(x) = 5x + 6$. Evaluate each expression and write the solution as an ordered pair.

7. $f(3)$

8. $g(-3)$

9. Use the function $y = f(x)$ to answer parts (a) through (f).

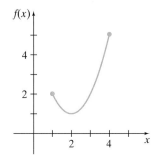

(a) What is the independent variable?

(b) What is the dependent variable?

(c) What is the value of $f(x)$ when $x = 3$?

(d) What is $f(1)$?

(e) What is the domain of f?

(f) What is the range of f?

10. Solve the linear inequality $3x - 7 \geq 20$, and write the solution using interval notation.

For Exercises 11 and 12, determine the domain of the function algebraically, and write the domain using interval notation.

11. $f(x) = \sqrt{-4 + 2x}$

12. $g(x) = \dfrac{x + 3}{x^2 + 3x - 10}$

13. Graph the function $f(x) = \dfrac{x + 3}{x^2 - 4}$ on your calculator, and determine the domain and range of the function. Write your solutions using interval notation.

14. The price–demand function for the new EasyType computer keyboard is given as

$$p(x) = 35 - 0.2x, \qquad 0 \leq x \leq 50$$

Here x represents the number of units sold per day and $p(x)$ represents the price of one unit in dollars.

(a) Complete the following table of values.

x	0	10	20	30	40	50
$p(x)$						

(b) Make a graph of p on the interval $[0, 50]$.

(c) If $[0, 50]$ is the domain, what is the range of the function?

(d) Evaluate $p(35)$ and interpret the answer.

15. Using the graph, write the domain and range of the function using interval notation.

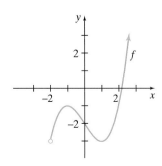

16. Find the average rate of change between the points $(3, 7)$ and $(-2, 14)$.

17. Given the graph, find the average rate of change.

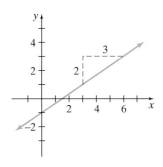

18. A line contains the points (2, 1) and (8, 19). Write the equation of the line in point–slope form and graph.

19. A line has an x-intercept at -4 and a y-intercept at 5. Write the equation of the line in point–slope form and graph.

20. A line has a slope of $\frac{3}{7}$ and passes through the point $(-3, 2)$. Write the equation of the line in CF form.

21. A linear function contains the data points in the table. Write the equation of the function in slope–intercept form.

x	y
-4	-10
12	2

22. Determine the x- and y-intercepts for the graph of the function $g(x) = \dfrac{60 - 9x}{4}$.

23. A new bicycle costs $800 and depreciates at the rate of $40 per year. Answer the following using the depreciation model.
(a) Determine the equation of the depreciation function.
(b) In how many years will the bicycle be worth $600?

24. Classify the function given by $f(x) = -3(2 - 5x)$ as increasing or decreasing, and identify the slope of the line.

25. For a certain product, the fixed costs are $2600 and it costs $3950 to produce 18 units. Write the equation of the linear cost function in the slope–intercept form $C(x) = mx + b$.

26. The manager of Pizza Pizzazz determines that the cost of preparing and delivering x pizzas is given by the cost function $C(x) = 300 + 6.5x$.
(a) What are the fixed and variable costs for producing the pizzas?
(b) Evaluate $C(36)$ and interpret the answer.
(c) How many pizzas can be prepared and delivered for $703?
(d) What is the marginal cost for the pizzas?
(e) If the current production level is 75 pizzas, what is the cost of preparing and delivering the 76th pizza?

27. Make an accurate graph of the piecewise-defined function.
$$f(x) = \begin{cases} 1 - 0.5x, & x \le 1 \\ 3, & x > 1 \end{cases}$$

28. Graph the function $g(x) = |12 - 4x|$ and then rewrite the function in piecewise form.

29. Consider the daily cost schedule for producing x Casa Verde ceiling fans.

FANS PRODUCED	COST (IN DOLLARS)
0	2,500
50	5,500
220	15,700

(a) What are the fixed costs? Explain how you determined this from the table.
(b) How much does the cost increase for each additional ceiling fan produced? How do you classify this kind of cost?
(c) Write a cost function C, where x represents the number of ceiling fans produced and $C(x)$ represents the cost.
(d) What would the cost be if 84 ceiling fans were produced?

For Exercises 30 and 31:
(a) Determine the coordinates of the vertex algebraically.
(b) Write the range of the function using interval notation.
30. $f(x) = x^2 - 4x + 3$ 31. $f(x) = -2x^2 - 6x + 12$

For Exercises 32 and 33:
(a) Determine the vertex.
(b) Use the result from part (a) to determine the intervals where the function is increasing and where it is decreasing.
32. $g(x) = 5x^2 + 20x$ 33. $g(x) = -x^2 + 3.2x - 7.8$

For Exercises 34 and 35, find the zeros of the quadratic function by factoring.
34. $f(x) = x^2 - 144$ 35. $f(x) = 2x^2 + 3x - 20$

36. Find the real number roots of the quadratic function $f(x) = 3x^2 - 6x + 2$ by using the quadratic formula. If the discriminant is a negative number, write *no real roots*.

37. Use the ZERO or ROOT capabilities of your calculator to approximate the zeros of the quadratic function $g(x) = 2.4x^2 + 1.7x - 5.3$. Round to the nearest hundredth. If the graph has no x-intercepts, write *no real roots*.

38. Consider the function $f(x) = -1.8x + 5.3$. Use the difference quotient to determine the slope of the line over the following intervals:
(a) As x changes from $x = 0$ to $x = 2$
(b) As x changes from $x = 0$ to $x = 5$
(c) As x changes from $x = 4$ to $x = 7$

39. Consider the function $g(x) = 2x^2 - 3x$. Use the difference quotient to determine the slope of the secant line over the following intervals:

(a) If $x = 0$ and $\Delta x = 1$ (b) If $x = 0$ and $\Delta x = 2$

(c) If $x = 0$ and $\Delta x = 4$

40. The average homeowner's property tax in a certain community can be modeled by

$$f(x) = -15x^2 + 300x + 1200, \qquad 0 \le x \le 9$$

where x represents the number of years since 1990 and $f(x)$ is the average property tax.

(a) Does the graph of the model open up or down? How do you know this?

(b) The domain of the model consists of x-values in the interval $[0, 9]$. What is the corresponding range?

(c) Determine the average rate of change in the property tax from 1992 to 1997.

41. The amount of time that an average American spends per year watching premium cable television programming can be modeled by

$$f(x) = 0.49x^2 - 3.41x + 88.50, \qquad 0 \le x \le 10$$

where x represents the number of years since 1990 and $f(x)$ is the number of hours spent watching premium cable television programming.

(a) Evaluate $f(4)$ and interpret.

(b) Make a graph of the model in the window $[0, 10]$ by $[75, 110]$.

(c) The domain of the model consists of x-values in the interval $[0, 10]$. What is the corresponding range?

(d) Make a table of values for $x = 0, 1, 2, 3, \ldots, 10$.

(e) According to the model, during what years did the average American spend less than 85 hours watching premium cable television programming?

For Exercises 42 and 43, simplify and write the domain for the following functions.

(a) $(f + g)(x)$ (b) $(f - g)(x)$

(c) $(f \cdot g)(x)$ (d) $\left(\dfrac{f}{g}\right)(x)$

42. $f(x) = x^2 + 4$, $g(x) = 2x + 3$

43. $f(x) = \sqrt{x + 2}$, $g(x) = x - 2$

For Exercises 44–47, let $f(x) = 3x - 5$ and $g(x) = (x + 2)^2$. Evaluate each of the following:

44. $(f + g)(2)$ 45. $(f - g)(4)$

46. $(f \cdot g)(-2)$ 47. $\left(\dfrac{f}{g}\right)(3)$

48. The population of Eastbrook from 1980 to 1990 can be modeled by

$$f(x) = 80{,}300 + 800x, \qquad 0 \le x \le 10$$

where x represents the number of years since 1980 and $f(x)$ represents the population. The population of Westbrook during the same time period is given by

$$g(x) = 62{,}600 - 200x, \qquad 0 \le x \le 10$$

(a) Evaluate $(f + g)(4)$ and interpret.

(b) Evaluate $(f - g)(8)$ and interpret.

49. For the price function $p(x) = 145 - 0.15x$, determine the revenue function R.

50. Suppose that the revenue and cost functions for a certain product are given by $R(x) = 18x - 0.08x^2$ and $C(x) = 5x + 275$.

(a) Graph R and C in the same viewing window.

(b) Use algebra or the INTERSECT command on your calculator to determine the break-even point.

(c) Determine how much revenue must be generated to reach the break-even point.

51. For the revenue and cost functions $R(x) = 32x - 0.08x^2$ and $C(x) = 16x + 400$, determine the profit function P, find the vertex, and interpret.

52. For a certain product, the price-demand function is $p(x) = 400 - 0.1x$. The variable costs are \$225 per unit, and the fixed costs are \$3000.

(a) Determine the linear cost function C.

(b) Determine the revenue function R.

(c) Determine the profit function $P(x) = R(x) - C(x)$.

(d) Use the ZERO or ROOT command on your calculator to find the zeros of the profit function P. These are break-even points.

(e) Find the vertex of the graph of the profit function P.

(f) Determine the demand level that yields the maximum profit, and find the maximum profit.

53. Determine if each function is a polynomial function. Identify each polynomial as linear, quadratic, cubic, or quartic.

(a) $f(x) = x^2 + 5x - 3$ (b) $f(x) = 17$

(c) $g(x) = \dfrac{1}{3}x - 4$ (d) $g(x) = \dfrac{3}{4}x^4 - \dfrac{2}{3}x^3 + \dfrac{1}{2}x^2 + x$

For Exercises 54 and 55, determine the end behavior of each function.

54. $f(x) = 5x - 2x^3$

55. $g(x) = 3x^2 - 5x$

56. The Disma Department Store finds that the revenue generated by selling x dresses is given by

$$R(x) = 68x - 0.3x^2, \qquad x \ge 0$$

(a) Evaluate $R(42)$ and interpret.

(b) Find the average rate of change for $x = 40$ and $\Delta x = 10$ and interpret.

57. Consider the polynomial function
$$f(x) = x^3 - 6x^2 + 6x + 4.$$

(a) Determine the end behavior of the function.

(b) Use the MAXIMUM and MINIMUM commands on the calculator to approximate the peaks and valleys of the function. Estimate the points to the nearest hundredth.

(c) Use the solution from part (b) to determine the intervals where f increases and where it decreases.

For Exercises 58–61:

(a) Write the domain using interval notation.

(b) Identify any holes or vertical asymptotes in the graph.

(c) Identify any horizontal asymptotes.

58. $f(x) = \dfrac{x+3}{x-1}$

59. $g(x) = \dfrac{x^2}{2x+3}$

60. $g(x) = \dfrac{x^2 - x - 6}{x^2 + 3x + 2}$

61. $f(x) = \dfrac{2x^2 - x - 15}{x^2 - 9}$

For Exercises 62–65, determine the x- and y-intercepts for the graph of the given function.

62. $g(x) = \dfrac{2x}{x-5}$

63. $f(x) = \dfrac{x^2 - 8x + 16}{x^2 - 9}$

64. $f(x) = \dfrac{x^2 + 5x + 6}{x^2 + x - 2}$

65. $g(x) = \dfrac{12}{x^2 + 6}$

66. For the function $f(x) = \dfrac{3x^2 - 5x + 2}{x + 1}$, determine the x- and y-intercepts graphically.

67. Sketch a graph by hand for $f(x) = \dfrac{x+3}{x-2}$ on the domain $1 \le x \le 10$. Label the asymptotes and the intercepts.

68. Consider the cost–benefit function for removing a certain pollutant from the atmosphere.

$$f(x) = \dfrac{26x}{100 - x}, \qquad 0 \le x < 100$$

where x represents the percentage of the pollutant removed and $f(x)$ represents the associated cost in millions of dollars.

(a) Evaluate $f(60)$ and interpret.

(b) Evaluate $f(95)$ and interpret.

(c) Graph f in the viewing window [0, 100] by [0, 500].

69. Rewrite the radical function $f(x) = \sqrt[4]{(x-2)^3}$ as a rational exponent function. Also, write the domain using interval notation.

70. Rewrite the rational exponent function $f(x) = (2x - 7)^{3/5}$ as a radical function. Also, write the domain using interval notation.

71. The time required for an object to fall a distance $f(x)$ is given by the model

$$f(x) = \dfrac{1}{4}\sqrt{x}$$

where x represents the distance in feet and $f(x)$ represents the time in seconds.

(a) Evaluate $f(64)$ and interpret.

(b) How long does it take an object to fall a distance of 144 feet?

72. A biologist has shown that the number of known insect species in a certain desert is related to the area of the land studied by the model

$$g(x) = 16.3\sqrt[3]{x}$$

where x represents the area studied in square miles and $g(x)$ represents the number of insect species.

(a) Evaluate $g(343)$ and interpret.

(b) Find the slope of the secant line for $x = 1000$ and $\Delta x = 331$ and interpret.

73. The total amount spent annually on research and development in the United States is given by the mathematical model

$$f(x) = 19.66x^{0.74}, \qquad 5 \le x \le 21$$

where x represents the number of years after 1975 and $f(x)$ represents the amount spent in billions of dollars.

(a) Evaluate $f(12)$ and interpret.

(b) Graph f in the viewing window [5, 21] by [0, 200].

(c) Find the average rate of change in the amount spent on R&D for $x = 8$ and $\Delta x = 4$ and interpret.

For Exercises 74–77:

(a) Complete the following table.

x	−5	−4	−3	−2	−1	0	1	2	3	4	5
$f(x)$											

(b) Graph f.

(c) Classify the function as exponential growth or decay.

74. $f(x) = \left(\dfrac{2}{5}\right)^x$

75. $f(x) = 2.3^x$

76. $f(x) = e^{0.8x}$

77. $f(x) = e^{-2.7x}$

78. The SuperValue Grocery chain announced a two-week sale in conjunction with the grand opening of its new location. The percentage of customers responding to the sale can be modeled by

$$f(x) = 65 - 80(0.71)^x$$

where x is the day of the sale and $f(x)$ represents the percentage of customers who respond to the sale.

(a) Evaluate $f(1)$ and interpret.

(b) Find the slope of the secant line for $x = 4$ and $\Delta x = 3$ and interpret.

(c) When does the percentage of customers responding to the sale first exceed 50%?

79. Suppose that Paul deposits $1700 into an account that pays interest at a rate of 5.3% compounded monthly. How much will be in the account after 7 years?

80. Suppose that Yushu deposits $20,000 into an account that pays interest at a rate of 7.5%. How much will be in the account after 12 years if the interest is compounded

(a) Annually?

(b) Monthly (12 times a year)?

(c) Weekly (52 times a year)?

81. Suppose that $8000 is invested at a 6.75% interest rate compounded quarterly (four times a year).

(a) Write the compound interest function

$$A(t) = P\left(1 + \frac{r}{n}\right)^{nt}$$

for the given information and interpret.

(b) Evaluate $A(6)$ and interpret.

(c) Calculate how much interest was made at the end of the 6-year period.

82. Suppose that Frederick deposits $5000 at a 6.2% interest rate compounded monthly.

(a) Write the compound interest function A for the given information.

(b) Graph A in the viewing window [0, 15] by [5000, 15,000].

(c) Use your calculator to graphically find doubling time of the account (that is, the amount of time that it takes the account to accumulate a total balance of $10,000).

83. The population of Granaco City can be modeled by

$$f(x) = 64{,}000e^{0.04x}, \qquad 0 \le x \le 20$$

where x represents the number of years since 1980 and $f(x)$ represents the number of people living in Granaco City.

(a) Evaluate $f(4)$ and interpret.

(b) Find the slope of the secant line for $x = 5$ and $\Delta x = 3$ and interpret.

(c) In what year did the population first exceed 110,000?

84. Suppose that $2500 is invested at a 5.7% interest rate compounded continuously.

(a) Write the compound interest function $A(t) = Pe^{rt}$ for the given information.

(b) Evaluate $A(4)$ and interpret.

(c) Use your calculator to graphically find the doubling time of the account.

85. For each account, compute the effective rate and interpret the result.

(a) An 8.2% nominal rate compounded monthly

(b) A 6.7% nominal rate compounded continuously

86. The population of leopards in a certain region is given by the logistic model

$$f(x) = \frac{160}{1 + 42.29e^{-0.1x}}, \qquad 0 \le t \le 60$$

where x represents the number of years since 1940 and $f(x)$ represents the number of leopards.

(a) Evaluate $f(35)$ and interpret.

(b) Compute the difference quotient for $x = 47$ and $\Delta x = 3$ and interpret.

For Exercises 87 and 88, let $f(x) = x^2 - 3x$ and $g(x) = 2x - 5$. Evaluate each expression.

87. $f(g(4))$

88. $g(f(-2))$

For Exercises 89 and 90:

(a) State the domain of each of the functions f and g.

(b) Algebraically simplify $f(g(x))$.

(c) Algebraically simplify $g(f(x))$.

89. $f(x) = 5x + 3$; $g(x) = x^2$

90. $f(x) = \sqrt{3x - 2}$, $g(x) = 3x^2 + 6$

For Exercises 91–94, determine the functions f and g so that $f(g(x)) = h(x)$ for each composite function h; then check your answer. There is more than one correct answer.

91. $h(x) = (x^2 + 2)^5$

92. $h(x) = x^2 + 5$

93. $h(x) = \sqrt{6 - x}$

94. $h(x) = \dfrac{3x - 4}{3x + 10}$

95. Graph the function $g(x) = \dfrac{1}{x}$, and use the horizontal line test to determine if g is a one-to-one function.

For the functions in Exercises 96 and 97:

(a) Show that $f(g(x)) = x$.

(b) Graph f, g, and the line $y = x$ in the same viewing window.

96. $f(x) = 4 - 3x$; $g(x) = \dfrac{-x + 4}{3}$

97. $f(x) = \sqrt[3]{x + 5}$; $g(x) = x^3 - 5$

For Exercises 98 and 99, rewrite the equation in logarithmic form $y = \log_b x$.

98. $3^5 = 243$

99. $10^{-4} = 0.0001$

For Exercises 100 and 101, use the properties of logarithms to rewrite the expression.

100. $\ln 18^3$

101. $\log_2 \dfrac{5^2}{7}$

For Exercises 102 and 103, solve the exponential equations algebraically and round the solution to the hundredths place.

102. $5 \cdot 4^x = 65$

103. $6.3(1.7)^x = 24$

104. The annual revenues for a certain business can be modeled by

$$f(x) = 1.38 + 2.4 \ln x, \qquad 1 \le x \le 10$$

where x represents the number of years since 1990 and $f(x)$ represents the annual revenues, in millions of dollars.

(a) Complete the following table:

x	1	3	5	10
$f(x)$				

(b) Evaluate $f(8)$ and interpret.

(c) Use the table in part (a) to determine the average rate of change in the revenues between 1991 and 1995.

(d) Use the table in part (a) to compute the difference quotient for $x = 5$ and $\Delta x = 5$ and interpret.

For Exercises 105 and 106, determine which type of regression model would be a good choice based on the scatterplot. Answers may vary.

105.

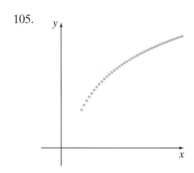

106.

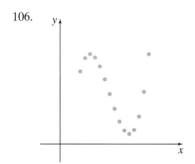

For Exercises 107–109, follow the curve-fitting process to find an appropriate model for the data.

107. Use a linear function model, where x represents the number of years since 1985.

YEAR	NUMBER OF AMERICANS COVERED BY PRIVATE OR GOVERNMENT HEALTH INSURANCE (MILLIONS)
1987	241
1989	246
1991	251
1993	260
1995	264
1997	269

Source: U.S. Census Bureau web site.

108. Use a quadratic function model, where x represents the number of years since 1960.

YEAR	NUMBER OF U.S. WORKERS WHO WORKED AT HOME (MILLIONS)
1960	4.66
1970	2.69
1980	2.18
1990	3.41

Source: U.S. Census Bureau web site.

109. Use a power function model, where x represents the number of years since 1900.

YEAR	AVERAGE COST OF A NEW HOUSE (IN $ THOUSANDS)
1963	19.3
1968	26.6
1973	35.5
1978	62.5
1983	89.8
1988	138.3
1993	147.7
1998	181.9

Source: U.S. Census Bureau web site.

Limits, Instantaneous Rate of Change, and the Derivative

The release of CFC's and other pollutants can cause smog.

The release of CFC's into the atmosphere by the U.S. from 1989–1995 can be modeled by $f(x) = -1.2x^3 + 13.37x^2 - 69.65x + 328.71$.

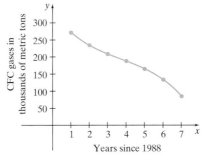

In 1994, the amount of CFC's released by the U.S. into the atmosphere was decreasing at a rate of

$$38.81 \, \frac{\text{thousand metric tons}}{\text{year}}.$$

What We Know

In Chapter 1, we reviewed algebra and also learned about an important rate of change called an *average rate of change*. We saw how an average rate of change is equivalent to the slope of a secant line over an interval.

Where Do We Go

In this chapter, we will see how the limit concept is used to introduce a new type of rate of change called an *instantaneous rate of change*. We will see how an instantaneous rate of change is equivalent to the slope of a tangent line at a specific point.

SECTION 2.1   LIMITS

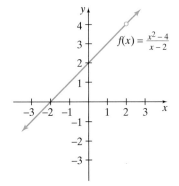

Figure 2.1.1

The two main branches of calculus, differential calculus and integral calculus, depend on the limit concept. To help grasp the ideas that calculus is based on, we first provide a practical introduction to limits. We use a combined numerical, graphical, and algebraic approach to examine the concept of a limit.

Determining Limits Numerically and Graphically

We begin our journey by considering the function $f(x) = \dfrac{x^2 - 4}{x - 2}$ and its graph shown in Figure 2.1.1. Since $x = 2$ is not in the domain of the function, in other words $f(2)$ is undefined, there appears to be a "hole" in the graph. However, what is the behavior of $f(x) = \dfrac{x^2 - 4}{x - 2}$ as x gets very, very close to the value of 2? By *behavior*, we mean what is happening to the function values (or the y-values if you prefer) as x approaches 2. One way to answer this question is by constructing a table to numerically analyze the behavior of f as x gets closer and closer to 2. Since we could approach 2 from the *left side* of 2 or from the *right side* of 2, we must include values of x less than 2 and values greater than 2. See Table 2.1.1.

TABLE 2.1.1

			x APPROACHES 2 FROM THE LEFT $\longrightarrow$					$\longleftarrow$ x APPROACHES 2 FROM THE RIGHT			
x	1	1.9	1.99	1.999	1.9999	2	2.0001	2.001	2.01	2.1	3
$f(x)$	3	3.9	3.99	3.999	3.9999		4.0001	4.001	4.01	4.1	5

NOTE: We intentionally left a blank below $x = 2$ for two reasons. First, the function is not defined at $x = 2$. Second, we wish to emphasize that in the limit process, we do not care what is happening at $x = 2$, but only in the behavior of the function as x gets close to 2.

It appears from Table 2.1.1 that if we start to the left of $x = 2$ or to the right of $x = 2$, as we allow x to approach 2, our functional values are approaching 4. We can say that,

"The limit of $\dfrac{x^2 - 4}{x - 2}$, as x approaches 2, is 4."

Using an arrow for the word *approaches* and *lim* as shorthand for the word limit, the mathematical notation for this English sentence is

$$\lim_{x \to 2} \frac{x^2 - 4}{x - 2} = 4$$

You should interpret this limit notation to mean that as x gets closer and closer to 2, from both sides of 2, $\dfrac{x^2 - 4}{x - 2}$ gets closer and closer to 4.

For a graphical perspective, let's graph $y = \dfrac{x^2 - 4}{x - 2}$. Figures 2.1.2 and 2.1.3 show the graph of $y = \dfrac{x^2 - 4}{x - 2}$, where Figure 2.1.3 is the result of utilizing the ZOOM IN command.

Notice in Figures 2.1.3b and c that we have used the TRACE command to get as close to 2 as possible from the left of 2 and from the right of 2, respectively. We now have graphical support for our numerical work. That is, numerically and graphically we believe that $\lim_{x \to 2} \dfrac{x^2 - 4}{x - 2} = 4$.

TECHNOLOGY NOTE
Due to the limitations of graphing calculators, you may not see a hole in the graph at $x = 2$. Using ZDECIMAL or selecting the x-axis window so that $x = 2$ is the midpoint of the graphing interval should show the hole. See Figure 2.1.2.

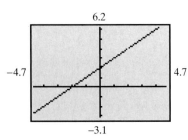

Figure 2.1.2

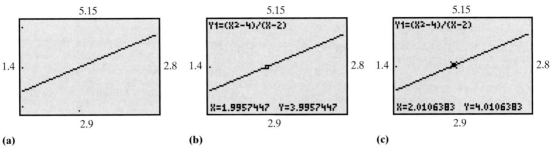

Figure 2.1.3

EXAMPLE 1 | **Analyzing a Limit**

For $f(x) = \dfrac{x^2 - 1}{x - 1}$, construct a table of values around $x = 1$ and guess the value of $f(x)$ as $x \to 1$. Graphically support your numerical work.

SOLUTION

Using a calculator, we can quickly construct Table 2.1.2. We notice that as $x \to 1$ from both sides (left and right) of 1, $f(x)$ is approaching 2. Hence, we conjecture that the limit is 2, and we write

$$\lim_{x \to 1} \frac{x^2 - 1}{x - 1} = 2$$

TECHNOLOGY NOTE
Many graphing calculators have a TABLE feature. Consult the on-line calculator manual at www.prenhall.com/armstrong

TABLE 2.1.2

	$x \to 1$ FROM LEFT					$x \to 1$ FROM RIGHT			
x	0	0.9	0.99	0.999	1	1.001	1.01	1.1	2
$f(x)$	1	1.9	1.99	1.999		2.001	2.01	2.1	3

Figure 2.1.4 shows the graph of $f(x) = \dfrac{x^2 - 1}{x - 1}$. In Figure 2.1.5 we have used the ZBOX command to capture that part of the graph where $x = 1$, and Figure 2.1.6 is the result of our ZBOX.

We now use the TRACE command to get as close to $x = 1$ as possible, from both sides of $x = 1$, and observe the corresponding y-values. See Figures 2.1.7a and b.

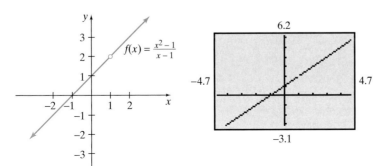

(a)

Figure 2.1.4

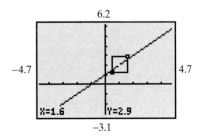

Figure 2.1.5

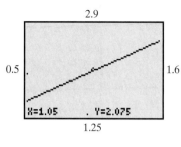

Figure 2.1.6

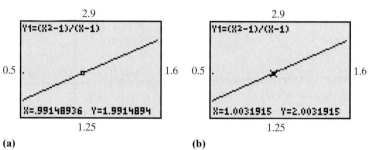

(a)   (b)

Figure 2.1.7

It appears that our graphical work indeed supports our conjecture that

$$\lim_{x \to 1} \frac{x^2 - 1}{x - 1} = 2$$

By table building and graphing (using ZOOM IN and TRACE), we can make $f(x) = \frac{x^2 - 1}{x - 1}$ as close to 2 as we like by restricting x to a sufficiently small interval *around* 1. By interval around 1, we mean to the left and right of 1. This is what the limit concept is all about! Before continuing, do Checkpoint 1 to ensure that you understand the process.

✓**CHECKPOINT 1**

Now work Exercise 13.

Left-hand and Right-hand Limits

Notice that in Example 1, our table was constructed so that we approached $x = 1$ both from the left side and from the right side. We also made sure when we zoomed in on the graph to capture x-values to the left and right of $x = 1$. Because this **left-side** and **right-side** analysis is so important in the limit process, we introduce notation for it. Specifically,

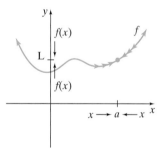

Figure 2.1.8 Illustrating the definition of a limit.

Left-hand and Right-hand Limits

$\lim\limits_{x \to a^-} f(x)$ means "the limit as x approaches a from the left side of a" and is called the **left-hand limit.**

$\lim\limits_{x \to a^+} f(x)$ means "the limit as x approaches a from the right side of a" and is called the **right-hand limit.**

Using this notation, we are ready for the following important definition.

Limit

For any function f, $\lim\limits_{x \to a} f(x) = L$ means that, as x approaches a, $f(x)$ approaches L.

Alternatively, if $\lim\limits_{x \to a^-} f(x) = L$ and $\lim\limits_{x \to a^+} f(x) = L$, then $\lim\limits_{x \to a} f(x) = L$.

NOTES: 1. If the left-hand limit does not equal the right-hand limit, then there is no limit. In this case, we say that the limit *does not exist*. See Figure 2.1.9a.

2. The existence of $\lim\limits_{x \to a} f(x)$ does *not* depend on whether $f(a)$ is defined. See Figure 2.1.9b.

3. The existence of $\lim\limits_{x \to a} f(x)$ does not depend on the value of $f(a)$ if $f(a)$ *is* defined. See Figure 2.1.9c.

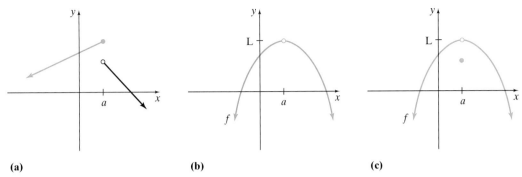

(a)   **(b)**   **(c)**

Figure 2.1.9 **(a)** $\lim\limits_{x \to a} f(x)$ does not exist since $\lim\limits_{x \to a^-} f(x) \neq \lim\limits_{x \to a^+} f(x)$. **(b)** $\lim\limits_{x \to a} f(x)$ exists even though $f(a)$ is undefined. **(c)** $f(a)$ is defined, but does *not* equal $\lim\limits_{x \to a} f(x)$.

Example 2 illustrates that the existence of $\lim\limits_{x \to a} f(x)$ does not depend on the value of $f(a)$ if $f(a)$ is defined.

EXAMPLE 2 | **Analyzing a Limit**

Estimate $\lim\limits_{x \to 2} f(x)$ for $f(x) = \begin{cases} 1 - x, & x \leq 2 \\ 4, & x > 2 \end{cases}$

SOLUTION

From Table 2.1.3 and from the graph of f in Figure 2.1.10, we conclude that

$$\lim_{x \to 2^-} f(x) = -1, \quad \text{whereas} \quad \lim_{x \to 2^+} f(x) = 4$$

Since the left-hand limit **does not equal** the right-hand limit, we conclude that $\lim\limits_{x \to 2} f(x)$ does not exist. Notice that graphically this corresponds to a jump in the graph. See Figure 2.1.10.

TABLE 2.1.3

		$x \to 2^-$					$x \to 2^+$		
x	1	1.9	1.99	1.999	2	2.001	2.01	2.1	3
$f(x)$	0	−0.9	−0.99	−0.999		4	4	4	4

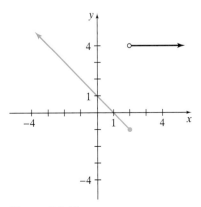

Figure 2.1.10

Again, we want to point out that in Example 2, $f(2) = -1$, yet $\lim_{x \to 2} f(x)$ does not exist. Remember that the existence of a limit does *not* depend on the value of $f(2)$.

Algebraically Determining Limits

So far, our approach to limits has been very informal. At this time, we will briefly state limit theorems to aid us in the algebraic evaluation of limits, as well as provide some mathematical validity. Later in the text, we will occasionally use some of these theorems.

Limit Theorems

If a, c, and n are real numbers, then

1. $\lim_{x \to a} c = c$

2. $\lim_{x \to a} x = a$

3. $\lim_{x \to a} [c \cdot f(x)] = c \cdot \lim_{x \to a} f(x)$

4. $\lim_{x \to a} [f(x) \pm g(x)] = \lim_{x \to a} f(x) \pm \lim_{x \to a} g(x)$

5. $\lim_{x \to a} [f(x) \cdot g(x)] = \lim_{x \to a} f(x) \cdot \lim_{x \to a} g(x)$

6. $\lim_{x \to a} \dfrac{f(x)}{g(x)} = \dfrac{\lim_{x \to a} f(x)}{\lim_{x \to a} g(x)}$, provided that $\lim_{x \to a} g(x) \neq 0$

7. $\lim_{x \to a} [f(x)]^n = [\lim_{x \to a} f(x)]^n$

NOTES:
1. The first limit theorem simply states that the limit of a constant is that constant.
2. Instead of memorizing these theorems, we suggest the following alternative, which we call the **Substitution Principle.**

Substitution Principle

When attempting to algebraically determine a limit, first try direct substitution. In other words, when attempting to find $\lim_{x \to a} f(x)$, first try to compute $f(a)$ (substitute a for x).

EXAMPLE 3

Analyzing Limits Algebraically

Determine the following limits:

(a) $\lim_{x \to 2} 7$ (b) $\lim_{x \to 1} (2x^2 - 3x + 5)$ (c) $\lim_{x \to 3} \sqrt{(2x - 1)}$

SOLUTION

(a) Since 7 is a constant, we utilize Limit Theorem 1 and have
$$\lim_{x \to 2} 7 = 7$$

(b) Here, we simply substitute and get
$$\lim_{x \to 1} (2x^2 - 3x + 5) = 2(1)^2 - 3(1) + 5 = 4$$

(c) Again, we simply substitute and have
$$\lim_{x \to 3} \sqrt{(2x - 1)} = \sqrt{(2(3) - 1)} = \sqrt{5}$$

✓ CHECKPOINT 2

Now work Exercise 31.

At this time you may be asking yourself, "When does the Substitution Principle fail?" The answer to this question can be found in Examples 1 and 2. In Example 2, the Substitution Principle fails, since we have a piecewise-defined function. In Example 1, the Substitution Principle fails, since substituting produces a fraction of the form $\frac{0}{0}$. We call the fraction $\frac{0}{0}$ an **indeterminate form.** When we try substitution and get the indeterminate form $\frac{0}{0}$, we have to use other techniques to determine the limit.

Recall that in Example 1, we determined numerically and graphically the limit

$$\lim_{x \to 1} \frac{x^2 - 1}{x - 1} = 2$$

To determine this limit algebraically, we can do the following:

$$\lim_{x \to 1} \frac{x^2 - 1}{x - 1} = \lim_{x \to 1} \frac{(x-1)(x+1)}{x-1} \qquad \text{Factor}$$

$$= \lim_{x \to 1}(x + 1) \qquad \text{Cancel, provided that } x \neq 1$$

$$= 1 + 1 = 2 \qquad \text{Substitution Principle}$$

What we have really done using algebra is to determine another function, $g(x) = x + 1$, that is equal to $f(x) = \frac{x^2-1}{x-1}$, for all values of x, except $x = 1$. In other words, we have

$$f(x) = \frac{x^2 - 1}{x - 1} = \frac{(x-1)(x+1)}{x-1} = x + 1 = g(x), \qquad \text{provided that } x \neq 1$$

The Interactive Activity to the left reinforces that the graphs of $f(x) = \frac{x^2-1}{x-1}$ and $g(x) = x + 1$ are identical for all x close to 1, but not equal to 1. We can write

$$\lim_{x \to 1} \frac{x^2 - 1}{x - 1} = \lim_{x \to 1}(x + 1) = 2$$

Interactive Activity

Complete Table 2.1.2 in Example 1 for $g(x) = x + 1$ and graph $g(x) = x + 1$ to verify that numerically and graphically $g(x) = x + 1$ and $f(x) = \frac{x^2-1}{x-1}$ are equivalent for all values of x, except $x = 1$.

EXAMPLE 4 | **Analyzing a Limit Involving $\frac{0}{0}$**

Determine $\lim_{x \to 3} g(x)$, where $g(x) = \frac{x^2-9}{x-3}$.

SOLUTION

We try substituting, which gives

$$\lim_{x \to 3} \frac{x^2 - 9}{x - 3} = \frac{(3)^2 - 9}{3 - 3} = \frac{0}{0}$$

Since we have the indeterminate form $\frac{0}{0}$, we decide to try a little algebra. We will factor the numerator and cancel as follows:

$$\lim_{x \to 3} \frac{x^2 - 9}{x - 3} = \lim_{x \to 3} \frac{(x-3)(x+3)}{x-3} \qquad \text{Factor}$$

$$= \lim_{x \to 3}(x + 3) \qquad \text{Cancel, provided that } x \neq 3$$

$$= (3 + 3) = 6 \qquad \text{Substitute}$$

Once again, notice that even though $g(3)$ is undefined, we determined that $\lim\limits_{x \to 3} g(x)$ exists. This reinforces the fact that the limit as x approaches 3 is *not* dependent on $g(3)$. See Figure 2.1.11.

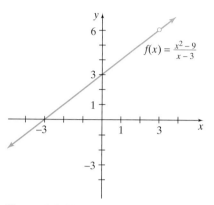

$$f(x) = \frac{x^2 - 9}{x - 3}$$

Figure 2.1.11

✓**CHECKPOINT 3**

Now work Exercise 39.

EXAMPLE 5

Analyzing a Limit Involving $\frac{0}{0}$

Determine $\lim\limits_{x \to 0} \dfrac{|x|}{x}$.

SOLUTION

If we try substituting, we get

$$\lim_{x \to 0} \frac{|x|}{x} = \frac{0}{0}$$

By letting $f(x) = \dfrac{|x|}{x}$, we can use a table and a graph to determine this limit. See Table 2.1.4 and Figure 2.1.12.

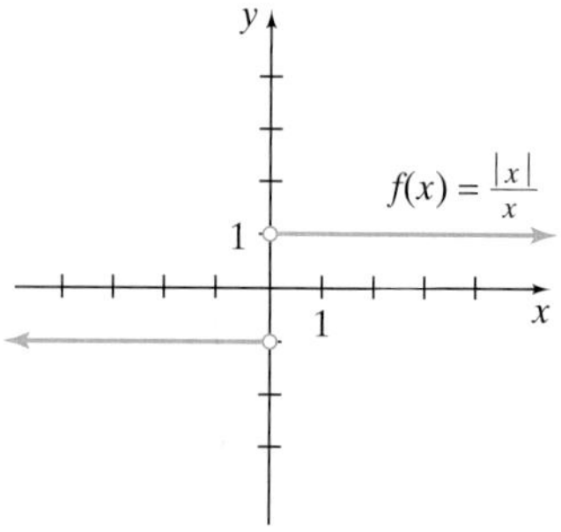

$$f(x) = \frac{|x|}{x}$$

Figure 2.1.12

Interactive Activity

Recall that $|x|$ is defined to be

$$|x| = \begin{cases} x, & \text{if } x \geq 0 \\ -x, & \text{if } x < 0 \end{cases}.$$

Use this definition to algebraically verify that $\lim\limits_{x \to 0} \dfrac{|x|}{x}$ does not exist.

TABLE 2.1.4

	$x \to 0^-$					$x \to 0^+$			
x	-1	-0.1	-0.01	-0.001	0	0.001	0.01	0.1	1
$f(x)$	-1	-1	-1	-1		1	1	1	1

As we can see numerically and graphically, since the left-hand limit does not equal the right-hand limit, we conclude that $\lim\limits_{x \to 0} \dfrac{|x|}{x}$ does not exist.

For Example 6, we need to recall Limit Theorem 1, which states that the limit of a constant is that constant.

EXAMPLE 6

Analyzing Limits Involving Two Variables

Determine the following limits:

(a) $\lim\limits_{h \to 0} (3x + 2h)$ (b) $\lim\limits_{h \to 0} \dfrac{5xh + 2h^2}{h}$

SOLUTION

(a) First, we notice that two variables, x and h, are involved. Remember that as $h \to 0$, the variable x acts as a **constant**; only the value of h changes. With this in mind, we try the Substitution Principle, which yields

$$\lim_{h \to 0} (3x + 2h) = 3x + 2(0) = 3x$$

(b) Again, two variables are involved, x and h. Proceeding as in part (a) gives

$$\lim_{h \to 0} \frac{5xh + 2h^2}{h} = \frac{5x(0) + 2(0)^2}{0} = \frac{0}{0}$$

Since this is an indeterminate form, we try some algebra.

$$\lim_{h \to 0} \frac{5xh + 2h^2}{h} = \lim_{h \to 0} \frac{h(5x + 2h)}{h} \qquad \text{Factor}$$

$$= \lim_{h \to 0} (5x + 2h) \qquad \text{Cancel, provided that } h \neq 0$$

$$= 5x + 2(0) = 5x \qquad \text{Substitute}$$

✓ CHECKPOINT 4

Now work Exercise 49.

Before we conclude this section, we would like to look at another limit that will be quite important in Section 2.3.

EXAMPLE 7 | **Analyzing a Limit of a Difference Quotient**

For $f(x) = x^2$, compute $\lim_{h \to 0} \dfrac{f(2 + h) - f(2)}{h}$.

SOLUTION

Here we have

$$\lim_{h \to 0} \frac{f(2 + h) - f(2)}{h} = \lim_{h \to 0} \frac{(2 + h)^2 - (2)^2}{h}$$

Substituting 0 for h produces the indeterminate form $\frac{0}{0}$. But, if we try some algebra, we get

$$\lim_{h \to 0} \frac{f(2 + h) - f(2)}{h} = \lim_{h \to 0} \frac{(2 + h)^2 - (2)^2}{h}$$

$$= \lim_{h \to 0} \frac{(4 + 4h + h^2) - 4}{h} \qquad \text{Recall that } (2 + h)^2 = 4 + 4h + h^2$$

$$= \lim_{h \to 0} \frac{4h + h^2}{h} \qquad \text{Still indeterminate}$$

$$= \lim_{h \to 0} \frac{h(4 + h)}{h} \qquad \text{Factor}$$

$$= \lim_{h \to 0} (4 + h) \qquad \text{Cancel, provided that } h \neq 0$$

$$= 4 + 0 = 4 \qquad \text{Substitute}$$

Applications

The main thrust of this section has been to lay the groundwork for the calculus. Although the number of applications in this section is limited, there will be more in the next section when limits at infinity and infinite limits are discussed. At this time, we offer the following application.

EXAMPLE 8 | **Applying Limits**

The graph in Figure 2.1.13 shows the cost $C(x)$, in dollars, to make x copies at Cody's Copy Center. Use the information in the graph to estimate the following:

(a) $C(10)$ (b) $\lim_{x \to 10} C(x)$ (c) $C(15)$ (d) $\lim_{x \to 15} C(x)$.

SOLUTION

(a) From the graph, we estimate that $C(10) = \$1.50$.

(b) Since $\lim_{x \to 10^-} C(x) = 1.50$ and $\lim_{x \to 10^+} C(x) = 1.50$, we estimate that $\lim_{x \to 10} C(x) = 1.50$.

(c) From the graph we estimate that $C(15) = \$2.00$.

(d) Since $\lim_{x \to 15^-} C(x) = 2.25$ and $\lim_{x \to 15^+} C(x) = 2.00$, we conclude that $\lim_{x \to 15} C(x)$ does not exist. ∎

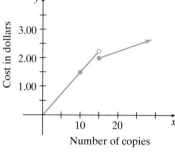

Figure 2.1.13

SUMMARY

In this introductory section on limits, we have presented the limit concept numerically, graphically, and algebraically. When analyzing a limit of a function, it is important to remember that in order for $\lim_{x \to a} f(x) = L$, we must have the **left-hand limit**, $\lim_{x \to a^-} f(x)$, and the **right-hand limit**, $\lim_{x \to a^+} f(x)$, both equal to L. We used **limit theorems** and the Substitution Principle to determine limits. In addition, we used algebra to find the limits of an indeterminate form $\frac{0}{0}$.

• The limit $\lim_{x \to a} f(x) = L$ means that as x approaches a, $f(x)$ approaches L. Alternatively, if $\lim_{x \to a^-} f(x) = L$ and $\lim_{x \to a^+} f(x) = L$, then $\lim_{x \to a} f(x) = L$.

• **Substitution Principle:** When trying to find $\lim_{x \to a} f(x)$, first try to compute $f(a)$.

SECTION 2.1 EXERCISES

For Exercises 1–10, complete the given tables to numerically estimate the following:

(a) $\lim_{x \to a^-} f(x)$ (b) $\lim_{x \to a^+} f(x)$ (c) $\lim_{x \to a} f(x)$

1. $f(x) = 2 - 3x$; $a = 1$

x	0	0.9	0.99	0.999	1	1.001	1.01	1.1	2
$f(x)$					?				

2. $f(x) = 2x - 3$; $a = -2$

x	−3	−2.1	−2.01	−2.001	−2	−1.999	−1.99	−1.9	−1
$f(x)$					?				

3. $f(x) = 2x^4 - 3x^3 + 2x - 4;$ $a = -2$

x	-3	-2.1	-2.01	-2.001	-2	-1.999	-1.99	-1.9	-1
$f(x)$					?				

4. $f(x) = -2x^3 + 3x - 2;$ $a = 1$

x	0	0.9	0.99	0.999	1	1.001	1.01	1.1	2
$f(x)$					?				

5. $f(x) = \dfrac{x^2 - 1}{x - 1};$ $a = 1$

x	0	0.9	0.99	0.999	1	1.001	1.01	1.1	2
$f(x)$					?				

6. $f(x) = \dfrac{x^2 - 3x}{x^2 - 9};$ $a = 3$

x	2	2.9	2.99	2.999	3	3.001	3.01	3.1	4
$f(x)$					?				

7. $f(x) = \dfrac{x^3 + 8}{x + 2};$ $a = -2$

x	-3	-2.1	-2.01	-2.001	-2	-1.999	-1.99	-1.9	-1
$f(x)$					?				

8. $f(x) = \dfrac{x^3 - 1}{x - 1};$ $a = 1$

x	0	0.9	0.99	0.999	1	1.001	1.01	1.1	2
$f(x)$					?				

9. $f(x) = \begin{cases} 3x + 1, & x > 0 \\ x^2 + 1, & x \le 0 \end{cases};$ $a = 0$

x	-1	-0.1	-0.01	-0.001	0	0.001	0.01	0.1	1
$f(x)$					?				

10. $f(x) = \begin{cases} x^2 + 1, & x \ge 0 \\ 2x^2 - 1, & x < 0 \end{cases};$ $a = 0$

x	-1	-0.1	-0.01	-0.001	0	0.001	0.01	0.1	1
$f(x)$					?				

For Exercises 11–18, use your calculator to graph the given function. Use the ZOOM IN and TRACE commands to graphically estimate the indicated limits. Verify your estimate numerically.

11. $f(x) = x^2 - 5x - 2;$ $\lim_{x \to 1} f(x)$

12. $f(x) = x^2 - 8x + 15;$ $\lim_{x \to 3} f(x)$

✓ 13. $g(x) = \dfrac{x^3 + 1}{x + 1};$ $\lim_{x \to -1} g(x)$

14. $g(x) = \dfrac{x + 1}{x^3 + 1};$ $\lim_{x \to -1} g(x)$

15. $f(x) = (x - 2)(x + 1);$ $\lim_{x \to 3} f(x)$

16. $f(x) = \dfrac{x - 2}{x + 1};$ $\lim_{x \to 3} f(x)$

17. $f(x) = \dfrac{\sqrt{x} - 2}{x - 4};$ $\lim_{x \to 4} f(x)$

18. $f(x) = x^3 - 2x^2 - 5x + 6;$ $\lim_{x \to 1.5} f(x)$

For Exercises 19–44, determine the indicated limit algebraically by using techniques in this section. Verify your answers numerically and graphically.

19. $\lim_{x \to -2} (3x + 1)$

20. $\lim_{x \to 2} (-2x^2 + 50x)$

21. $\lim_{x \to 2} (2x^2 + 3x - 1)$

22. $\lim_{x \to -4} \sqrt{x^2 + 9}$

23. $\lim_{x \to 3} \dfrac{x^2 - 9}{x - 3}$

24. $\lim_{x \to 1} (2 - 3x)$

25. $\lim_{x \to 0} |x|$

26. $\lim_{x \to 1} |x + 1|$

27. $\lim_{x \to -1} |x + 1|$

28. $\lim_{x \to -1} \dfrac{x + 1}{|x + 1|}$

29. $\lim_{x \to -1} \dfrac{x^2 - 1}{x + 1}$

30. $\lim_{x \to -1} \dfrac{x + 1}{x^2 - 1}$

✓ 31. $\lim_{x \to 0} \sqrt{2x + 3}$

32. $\lim_{x \to 5} \dfrac{x^2 - 25}{x - 5}$

33. $\lim_{x \to -5} \dfrac{x^2 - 25}{x - 5}$

34. $\lim_{x \to 5} \dfrac{x - 5}{x^2 - 25}$

35. $\lim_{x \to 1} \dfrac{x^2 - 1}{x + 1}$

36. $\lim_{x \to 1} \dfrac{x - 1}{x^2 - 1}$

37. $\lim_{x \to 2} (x + 1)^2 \cdot (3x - 1)^3$

38. $\lim_{x \to -1} (x + 2)^3 (3x + 2)$

✓ 39. $\lim_{x \to 3} \dfrac{x^2 - x - 6}{x - 3}$

40. $\lim_{x \to 3} \dfrac{x - 3}{x^2 - x - 6}$

41. $\lim_{x \to -2} \dfrac{x + 2}{x^2 + 5x + 6}$

42. $\lim_{x \to -2} \dfrac{x^2 + 5x + 6}{x + 2}$

43. $\lim_{x \to 4} \dfrac{\sqrt{x} - 2}{x - 4}$ (*Hint:* Algebraically, try multiplying the numerator and denominator by $\sqrt{x} + 2$.)

44. $\lim_{x \to 9} \dfrac{\sqrt{x} - 3}{x - 9}$ (*Hint:* Algebraically, try multiplying the numerator and denominator by $\sqrt{x} + 3$.)

45. Use the graph of f in Figure 2.1.14 to answer the following:
(a) $\lim_{x \to -1} f(x)$
(b) $f(-1)$
(c) $\lim_{x \to 3^-} f(x)$
(d) $\lim_{x \to 3^+} f(x)$
(e) $\lim_{x \to 3} f(x)$
(f) $f(1)$
(g) $\lim_{x \to 1} f(x)$

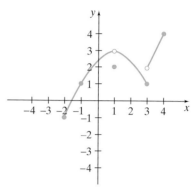

Figure 2.1.14

46. Use the graph of g in Figure 2.1.15 to answer the following:
(a) $\lim_{x \to 2^-} g(x)$
(b) $\lim_{x \to 2^+} g(x)$
(c) $\lim_{x \to 2} g(x)$
(d) $g(2)$
(e) $\lim_{x \to -2} g(x)$
(f) $g(-2)$

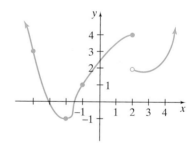

Figure 2.1.15

Determine the indicated limit in Exercises 47–53. See Example 6.

47. $\lim_{h \to 0} (2x + h)$

48. $\lim_{h \to 0} (3x^2 + 3h)$

✓ 49. $\lim_{h \to 0} \dfrac{3xh + h^2}{h}$

50. $\lim_{h \to 0} \dfrac{2xh + h^2}{h}$

51. $\lim_{h \to 0} \dfrac{4x^2 h + 2h^2}{h}$

52. $\lim_{h \to 0} \dfrac{4x^3 h^2 + 3xh^3 + 5h^4}{h}$

53. $\lim_{h \to 0} \dfrac{-3x^2 h + 6h^2}{h}$

For each function in Exercises 54–62 (see Example 7):

(a) Determine $f(1 + h)$.

(b) Compute $\lim_{h \to 0} \dfrac{f(1 + h) - f(1)}{h}$.

54. $f(x) = 2x + 3$

55. $f(x) = 3x - 1$

56. $f(x) = x^2 - 1$

57. $f(x) = x^2 + 2$

58. $f(x) = x^2 - 2x + 3$

59. $f(x) = x^2 + 3x - 1$

60. $f(x) = |x|$

61. $f(x) = \dfrac{1}{x}$

62. $f(x) = \sqrt{x}$

APPLICATIONS

63. The number of dollars spent on advertising for a product influences the number of items of the product that will be purchased by consumers. If x is the number of dollars spent, in thousands, then $N(x)$, in hundreds, is the number of items sold, as given by the function

$$N(x) = 2000 - \frac{520}{x}$$

(a) Graph N in the viewing window $[0, 20]$ by $[0, 2500]$.

(b) Determine $N(10)$ and interpret.

(c) Determine $\lim_{x \to 10} N(x)$ and interpret.

64. Repeat all parts of Exercise 63, except use

$$N(x) = 2200 - \frac{750}{x}.$$

65. The cost of producing x units of a product is given by $C(x) = 22{,}500 + 7.35x$. The **average cost,** AC, is given by

$$AC(x) = \frac{C(x)}{x} = \frac{22{,}500 + 7.35x}{x}$$

(a) Graph AC in the viewing window $[0, 25]$ by $[0, 20{,}000]$.

(b) Determine $AC(10)$ and interpret.

(c) Determine $\lim_{x \to 10} AC(x)$ and interpret.

66. Repeat all parts of Exercise 65, except use
$$C(x) = 33{,}125 + 6.38x.$$

67. Lisa's Lease-a-Car charges $25 per day plus $0.05 per mile to rent one of its medium-sized cars. Figure 2.1.16

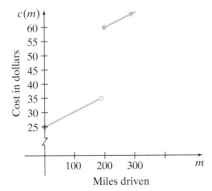

Figure 2.1.16

shows the graph of the cost in dollars, $c(m)$, realized by Sofia as a function of the miles, m, that she has driven the car.

(a) Determine $c(100)$ and interpret.

(b) Determine $\lim_{m \to 100} c(m)$.

(c) Determine $c(200)$ and interpret.

(d) Determine $\lim_{m \to 200} c(m)$.

(e) Give an explanation for the jump in the graph at $m = 200$.

68. Dylan's Truck Rental Company charges $20 per day plus $0.10 per mile to rent a moving truck. Figure 2.1.17 shows the graph of the cost in dollars, $c(m)$, realized by Austin as a function of the miles, m, that he has driven the truck.

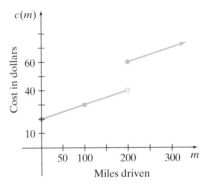

Figure 2.1.17

(a) Determine $c(50)$ and interpret.

(b) Determine $\lim_{m \to 50} c(m)$.

(c) Determine $c(100)$ and interpret.

(d) Determine $\lim_{m \to 100} c(m)$.

(e) Give an explanation for the jump in the graph at $m = 200$.

SECTION PROJECT

Finer Foods is running the following advertisement:

Hamburger

$1.50/pound for packages less than 2 pounds
$1.00/pound for packages of 2 pounds or more

(a) Let x represent the number of pounds of hamburger in a package. Determine a piecewise-defined function, $C(x)$, that models the cost of a package of hamburger.

(b) Graph C where $0 < x \leq 6$.

(c) Determine $C(1.5)$ and interpret.

(d) Determine $C(2)$ and interpret.

(e) Does $\lim_{x \to 2} C(x)$ exist? Explain.

(f) If Luther needs 2 pounds of hamburger for a cookout, how much money would Luther save by buying one 2-pound package versus two 1-pound packages?

SECTION 2.2   LIMITS AND ASYMPTOTES

In Section 2.1, we introduced the limit concept and analyzed limits numerically, graphically, and algebraically. In this section, we look at the behavior of functions via limits when the independent variable increases without bound or decreases without bound. Graphically, this is important to the understanding of *horizontal asymptotes*. We also analyze situations when the dependent variable increases and decreases without bound. This situation will be closely related to *vertical asymptotes*.

Infinite Limits and Vertical Asymptotes

Example 1 begins our discussion of the relationship between infinite limits and vertical asymptotes.

EXAMPLE 1   **Analyzing a Limit**

Consider $f(x) = \dfrac{1}{x}$. Numerically and graphically estimate $\displaystyle\lim_{x \to 0} \dfrac{1}{x}$.

SOLUTION

We begin our analysis by constructing a table of values as shown in Table 2.2.1. Numerically, it appears that the left-hand limit and the right-hand limit are not

TABLE 2.2.1

	$x \to 0^-$					$x \to 0^+$			
x	-1	-0.1	-0.01	-0.001	0	0.001	0.01	0.1	1
$f(x)$	-1	-10	-100	-1000		1000	100	10	1

equal, which means that the limit does not exist. As $x \to 0^+$, $\dfrac{1}{x}$ is positive and appears to be increasing forever or increasing without bound. We will indicate this "increasing without bound" behavior by writing

$$\lim_{x \to 0^+} \frac{1}{x} = \infty$$

Likewise, as $x \to 0^-$, $\dfrac{1}{x}$ is negative and appears to be decreasing forever or decreasing without bound. We will indicate this "decreasing without bound" behavior by writing

$$\lim_{x \to 0^-} \frac{1}{x} = -\infty$$

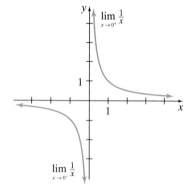

Figure 2.2.1

These behaviors are shown graphically in Figure 2.2.1. The graph seems to support our numerical work. Since both the table and the graph suggest that $\displaystyle\lim_{x \to 0^-} \dfrac{1}{x} \neq \lim_{x \to 0^+} \dfrac{1}{x}$, we conclude that $\displaystyle\lim_{x \to 0} \dfrac{1}{x}$ does not exist.   ■

NOTE: Technically speaking, in neither case does the limit, $\displaystyle\lim_{x \to 0^+} \dfrac{1}{x}$ nor $\displaystyle\lim_{x \to 0^-} \dfrac{1}{x}$, exist, since ∞ and $-\infty$ are not real numbers; they are merely concepts. To ensure that these concepts are understood, let's consider another example.

EXAMPLE 2

Analyzing an Infinite Limit

Consider $f(x) = \frac{1}{x^2}$. Use Table 2.2.2 and Figure 2.2.2 to determine the following:

(a) $\lim\limits_{x \to 0^-} \frac{1}{x^2}$ (b) $\lim\limits_{x \to 0^+} \frac{1}{x^2}$ (c) $\lim\limits_{x \to 0} \frac{1}{x^2}$

TABLE **2.2.2**

	$x \to 0^-$					$x \to 0^+$			
x	-1	-0.1	-0.01	-0.001	0	0.001	0.01	0.1	1
$f(x)$	1	100	10,000	1,000,000		1,000,000	10,000	100	1

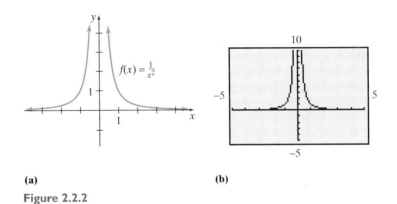

(a)   (b)

Figure 2.2.2

SOLUTION

(a) Table 2.2.2 and Figure 2.2.2 both suggest that, as $x \to 0^-$, $\frac{1}{x^2}$ increases without bound. So we write

$$\lim_{x \to 0^-} \frac{1}{x^2} = \infty$$

(b) Table 2.2.2 and Figure 2.2.2 both suggest that, as $x \to 0^+$, $\frac{1}{x^2}$ increases without bound. So we write

$$\lim_{x \to 0^+} \frac{1}{x^2} = \infty$$

(c) From our knowledge of left-hand and right-hand limits, the results from parts (a) and (b) suggest that

$$\lim_{x \to 0} \frac{1}{x^2} = \infty$$

It is worth mentioning one more time that in Example 2, none of the limits really exists, since $-\infty$ and ∞ are not real numbers. We are using the symbols ∞ and $-\infty$ to describe the behavior of the function near $x = 0$.

Notice in Examples 1 and 2 that the line $x = 0$, also known as the y-axis, serves as a **vertical asymptote** for the graphs. We now summarize the connection between infinite limits and vertical asymptotes.

Interactive Activity

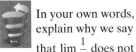

In your own words, explain why we say that $\lim\limits_{x \to 0} \frac{1}{x}$ does not exist (Example 1), yet we say that $\lim\limits_{x \to 0} \frac{1}{x^2} = \infty$ (Example 2).

> ### Infinite Limits and Vertical Asymptotes
>
> The line $x = a$ is a vertical asymptote of the graph of f if *any* of the following are true.
>
> - If as $x \to a^-$ the function values $f(x)$ increase or decrease without bound, that is, $\lim\limits_{x \to a^-} f(x) = \infty$ or $\lim\limits_{x \to a^-} f(x) = -\infty$, then $x = a$ is a vertical asymptote.
>
> - If as $x \to a^+$ the function values $f(x)$ increase or decrease without bound, that is, $\lim\limits_{x \to a^+} f(x) = \infty$ or $\lim\limits_{x \to a^+} f(x) = -\infty$, then $x = a$ is a vertical asymptote.

NOTE: If both the left- and the right-hand limits exhibit the same behavior, we say that $\lim\limits_{x \to a} f(x) = \infty$ (or $-\infty$).

EXAMPLE 3

Analyzing Limits at Excluded Domain Values

Consider $f(x) = \dfrac{x-1}{x^2-1}$. The domain for f is all real numbers except -1 and 1.

(a) Determine $\lim\limits_{x \to 1} \dfrac{x-1}{x^2-1}$ algebraically.   (b) Determine $\lim\limits_{x \to -1} \dfrac{x-1}{x^2-1}$.

SOLUTION

(a) Utilizing the Substitution Principle produces the indeterminate form $\dfrac{0}{0}$, so we need to employ some algebraic manipulation.

$$\lim_{x \to 1} \frac{x-1}{x^2-1} = \lim_{x \to 1} \frac{x-1}{(x-1)(x+1)} \qquad \text{Factor}$$

$$= \lim_{x \to 1} \frac{1}{x+1} \qquad \text{Cancel, provided that } x \neq 1$$

$$= \frac{1}{1+1} = \frac{1}{2} \qquad \text{Substitute}$$

(b) Again substitution yields $\dfrac{0}{0}$, so proceeding with the same algebra as in part (a) gives

$$\lim_{x \to -1} \frac{x-1}{x^2-1} = \lim_{x \to -1} \frac{x-1}{(x-1)(x+1)} \qquad \text{Factor}$$

$$= \lim_{x \to -1} \frac{1}{x+1} \qquad \text{Cancel, provided that } x \neq -1$$

We attempt to substitute at this time, but this gives us $\dfrac{1}{0}$, which we know is undefined. We decide to make a table and see if numerically some light can be shed on what is happening here. See Table 2.2.3.

TABLE 2.2.3

	$x \to -1^-$					$x \to -1^+$			
x	-2	-1.1	-1.01	-1.001	-1	-0.999	-0.99	-0.9	0
$f(x)$	-1	-10	-100	-1000		1000	100	10	1

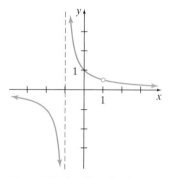

Figure 2.2.3 Graph of $f(x) = \dfrac{x-1}{x^2-1}$ has a vertical asymptote at $x = -1$.

Since the table suggests that $\lim\limits_{x \to -1^-} \dfrac{x-1}{x^2-1} \neq \lim\limits_{x \to -1^+} \dfrac{x-1}{x^2-1}$, we conclude that $\lim\limits_{x \to -1} \dfrac{x-1}{x^2-1}$ does not exist. Figure 2.2.3 supports our conjecture.

✓**Checkpoint 1**

Now work Exercise 13.

EXAMPLE 4

Analyzing a Limit in an Applied Setting

The cost, $C(x)$, in thousands of dollars of removing $x\%$ of a city's pollutants discharged into a lake is given by

$$C(x) = \frac{113x}{100 - x}$$

(a) Determine the reasonable domain for C and graph.

(b) Evaluate $C(50)$ and interpret.

(c) Determine $\lim\limits_{x \to 100^-} C(x)$ and interpret.

SOLUTION

(a) Since our independent variable, x, represents a percentage, the reasonable domain for C is $[0, 100)$. A graph of C is given in Figure 2.2.4.

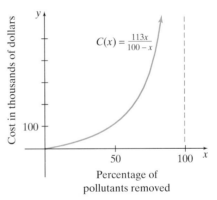

Figure 2.2.4

(b) $C(50) = \frac{113(50)}{100 - 50} = 113$. This means that it will cost the city \$113,000 to remove 50% of the pollutants being discharged into the lake.

(c) Figure 2.2.4 suggests that $\lim\limits_{x \to 100^-} C(x) = \infty$. We construct Table 2.2.4 to support our graphical belief that $\lim\limits_{x \to 100^-} C(x) = \infty$. This means that as the city tries to remove *all* the pollutants (100%) the cost to do so is "increasing without bound."

TABLE 2.2.4

			$x \to 100^-$		
x	90	99	99.9	99.99	99.999
$C(x)$	1017	11,187	112,887	1,129,887	11,299,887

✓**CHECKPOINT 2**

Now work Exercise 45.

Limits at Infinity and Horizontal Asymptotes

We now turn our attention to analyzing the behavior of a function as the independent variable values increase without bound (heads to ∞) and as the independent variable values decrease without bound (heads to $-\infty$). To introduce this concept, we return to a problem first encountered in Section 1.6.

Flashback

SPRING POINT EPIDEMIC REVISITED

In Section 1.6 we modeled a flu epidemic in Spring Point, an isolated town with a population of 1800, with the function

$$f(t) = \frac{1800}{1 + 359e^{-0.9t}}$$

where $f(t)$ represented the number of people affected by the flu after t weeks. See Figure 2.2.5. How many people were affected after 10 weeks?

Flashback Solution

To determine the number of people affected after 10 weeks, we simply substitute 10 for t and get

$$f(10) = \frac{1800}{1 + 359e^{-0.9(10)}} \approx 1723.6358$$

So after 10 weeks approximately 1724 people were affected by the flu.

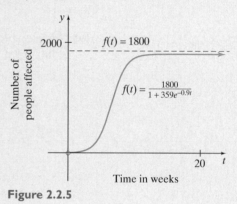

Figure 2.2.5

Interactive Activity

Construct a table of values of $f(t)$ for $t = 1$, 5, 10, 20, 50, and 100 to numerically confirm that

$$\lim_{t \to \infty} \frac{1800}{1 + 359e^{-0.9t}} = 1800$$

Notice in the Flashback that after 10 weeks 1724 out of 1800 people had been affected by the flu. Intuitively, for any value of t the largest value that $f(t)$ can have is 1800, the entire population of the town! Also, it seems reasonable that as t increases (that is, as the number of weeks since the flu outbreak increases) the number of people affected by the flu, $f(t)$, should increase. Putting this information together, namely that as t increases $f(t)$ increases, yet $f(t)$ cannot exceed Spring Point's population of 1800, leads us to claim that

$$\lim_{t \to \infty} \frac{1800}{1 + 359e^{-0.9t}} = 1800$$

Notice how this is supported by the **horizontal asymptote** in Figure 2.2.5.

EXAMPLE 5

Analyzing Limits at Infinity

Consider $f(x) = 2^x$.

(a) Use Table 2.2.5 and Figure 2.2.6 to estimate $\lim_{x \to \infty} 2^x$.

(b) Use Table 2.2.6 and Figure 2.2.6 to estimate $\lim_{x \to -\infty} 2^x$.

TABLE 2.2.5

		$x \to \infty$		
x	1	10	100	1000
$f(x)$	2	1024	1.3×10^{30}	Error

TABLE 2.2.6

		$x \to -\infty$		
x	-1	-10	-100	-1000
$f(x)$	0.5	9.8×10^{-4}	8×10^{-31}	0

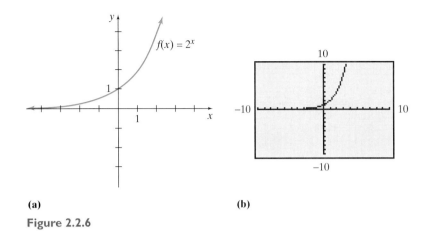

(a)   **(b)**

Figure 2.2.6

Technology Note
In Table 2.2.5, $x = 1000$ produces an error, and in Table 2.2.6, $x = -1000$ produces a value of 0 due to the limitations of our calculator. Basically, 2^{1000} is too large and 2^{-1000} is too close to 0 for our calculator.

SOLUTION

(a) Table 2.2.5 and Figure 2.2.6 suggest that $\lim\limits_{x \to \infty} 2^x = \infty$. In other words, as x increases without bound, 2^x also increases without bound.

(b) Table 2.2.6 and Figure 2.2.6 suggest that $\lim\limits_{x \to -\infty} 2^x = 0$. In other words, as x decreases without bound, 2^x gets close to 0.

Recall from Chapter 1 that Example 5 illustrates what is known as **exponential growth.** Example 6 illustrates **exponential decay.**

EXAMPLE 6   **Applying Limits at Infinity**

Pharmacological studies have determined that the amount of medication present in the body is a function of the amount given and how much time has elapsed since the medication was administered. For a certain medication, the amount present in milliliters, $A(t)$, can be approximated by the function

$$A(t) = 3e^{-0.123t}$$

where t is the number of hours since the medication was administered.

(a) Determine $A(0)$ and interpret.   (b) Determine $\lim\limits_{t \to \infty} A(t)$ and interpret.

SOLUTION

(a) $A(0) = 3e^{-0.123(0)} = 3$. This means that initially, 3 milliliters of the medication was administered.

(b) Using Table 2.2.7 and Figure 2.2.7, we conclude that $\lim\limits_{t \to \infty} A(t) = 0$. Thus, as the number of hours since administering the medication increases without bound, the amount present in the body approaches 0 milliliters. It should seem reasonable that the result is 0. Since $t = 1000$ hours is about 42 days, we would hope that, by then, our body would be rid of this medication.

TABLE 2.2.7

			$t \to \infty$		
t	0	1	10	100	1000
$A(t)$	3	2.65	0.88	1.4×10^{-5}	1×10^{-53}

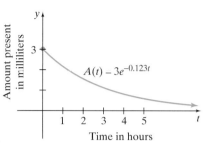

Figure 2.2.7

✓ CHECKPOINT **3**

Now work Exercise 51.

As seen in the Flashback and in Examples 5 and 6, for some functions there is a relationship between limits at infinity and horizontal asymptotes.

> ### Limits at Infinity and Horizontal Asymptotes
>
> For any function f, if $\lim\limits_{x \to \pm\infty} f(x) = L$, then the line $y = L$ is a **horizontal asymptote** for the graph of f.

Consider $f(x) = \frac{1}{x}$, $g(x) = \frac{1}{x^{5/3}}$ and $h(x) = \frac{1}{x^{1/2}}$ and their graphs in Figures 2.2.8, 2.2.9, and 2.2.10, respectively. Figure 2.2.8 suggests that $\lim\limits_{x \to \infty} \frac{1}{x} = 0$ and $\lim\limits_{x \to -\infty} \frac{1}{x} = 0$, and the line $y = 0$ is a horizontal asymptote for the graph. Figure 2.2.9 suggests that $\lim\limits_{x \to \infty} \frac{1}{x^{5/3}} = 0$ and $\lim\limits_{x \to -\infty} \frac{1}{x^{5/3}} = 0$, and the line $y = 0$ is a horizontal asymptote. Figure 2.2.10 suggests that $\lim\limits_{x \to \infty} \frac{1}{x^{1/2}} = 0$, whereas $\lim\limits_{x \to -\infty} \frac{1}{x^{1/2}}$ does not exist, since negative numbers are not in the domain. Also, the line $y = 0$ is a horizontal asymptote here as well. These three functions and their respective graphs demonstrate the following:

> ### Special Limits at Infinity
>
> 1. For n, a positive real number, $\lim\limits_{x \to \infty} \frac{1}{x^n} = 0$.
>
> 2. For n, a positive real number, $\lim\limits_{x \to -\infty} \frac{1}{x^n} = 0$, provided that x^n is a real number for negative values of x.

NOTE: All Limit Theorems from Section 2.1 are true for limits at infinity.

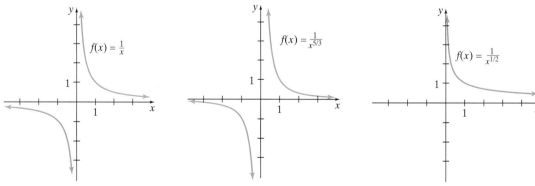

Figure 2.2.8   **Figure 2.2.9**   **Figure 2.2.10**

Example 7 illustrates the usefulness of these special limits of infinity while confirming a graphing technique first presented in Section 1.5.

EXAMPLE 7 | **Determining Horizontal Asymptotes**

For $f(x) = \dfrac{x^2 + 1}{2x^2 - 1}$, determine $\lim\limits_{x \to \infty} f(x)$ and $\lim\limits_{x \to -\infty} f(x)$ algebraically, and determine the horizontal asymptote for the graph of f.

SOLUTION

Since we have a rational function (the numerator and the denominator are polynomial functions), the algebra manipulation for problems of this type is rather straightforward:

> Divide every term in the numerator and denominator by x^b, where b is the larger degree of the numerator and the denominator.

For this problem, since both degrees are 2, we have $b = 2$. Next we divide every term in the numerator and denominator by x^2. This gives

$$f(x) = \frac{x^2 + 1}{2x^2 - 1} = \frac{(x^2 + 1) \div x^2}{(2x^2 - 1) \div x^2}$$

$$= \frac{\dfrac{x^2}{x^2} + \dfrac{1}{x^2}}{\dfrac{2x^2}{x^2} - \dfrac{1}{x^2}} = \frac{1 + \dfrac{1}{x^2}}{2 - \dfrac{1}{x^2}}$$

So, to determine the limit algebraically, we have

$$\lim_{x \to \infty} \frac{x^2 + 1}{2x^2 - 1} = \lim_{x \to \infty} \frac{1 + \dfrac{1}{x^2}}{2 - \dfrac{1}{x^2}} = \frac{\lim\limits_{x \to \infty} \left(1 + \dfrac{1}{x^2} \right)}{\lim\limits_{x \to \infty} \left(2 - \dfrac{1}{x^2} \right)}$$

$$= \frac{\lim\limits_{x \to \infty} 1 + \lim\limits_{x \to \infty} \dfrac{1}{x^2}}{\lim\limits_{x \to \infty} 2 - \lim\limits_{x \to \infty} \dfrac{1}{x^2}} = \frac{1 + 0}{2 - 0} = \frac{1}{2}$$

Interactive Activity

 Make a table and graph $y = \dfrac{x^2 + 1}{2x^2 - 1}$ to numerically and graphically support the result of Example 7.

The algebraic verification for $\lim\limits_{x \to -\infty} \dfrac{x^2 + 1}{2x^2 - 1}$ is identical. We immediately conclude that the horizontal asymptote is the line $y = \dfrac{1}{2}$.

✓ **CHECKPOINT 4**

Now work Exercise 33.

EXAMPLE 8 | **Analyzing Average Cost**

The total cost, in dollars, to produce x units of a certain product is given by $C(x) = 22{,}500 + 7.35x$. The **average cost,** AC, is given by

$$AC(x) = \frac{C(x)}{x} = \frac{22{,}500 + 7.35x}{x}$$

Determine $\lim\limits_{x \to \infty} AC(x)$ and interpret.

SOLUTION

Since $AC(x) = \dfrac{22{,}500 + 7.35x}{x}$ is already written as a fraction with a single term in the denominator, we may break it apart as follows:

$$\lim_{x \to \infty} \frac{22{,}500 + 7.35x}{x} = \lim_{x \to \infty} \left(\frac{22{,}500}{x} + \frac{7.35x}{x} \right) = \lim_{x \to \infty} \left(\frac{22{,}500}{x} + 7.35 \right)$$

$$= 0 + 7.35 = 7.35$$

This means that as the number of units produced increases without bound, the average total cost is approaching \$7.35. In other words, the cost per unit is approaching \$7.35 as the number of units produced increases without bound. ∎

Interactive Activity

Numerically and graphically verify the result in Example 8.

✓**CHECKPOINT 5**

Now work Exercise 49.

SUMMARY

In this section we continued our study of limits. We introduced the symbol ∞ (and $-\infty$) to represent an increasing without bound (decreasing without bound) behavior. We examined the relationship between **infinite limits** and **vertical asymptotes**. We also examined the relationship between **limits at infinity** and **horizontal asymptotes.**

- If as $x \to a^-$ the function values $f(x)$ increase or decrease without bound, that is, $\lim\limits_{x \to a^-} f(x) = \infty$ or $\lim\limits_{x \to a^-} f(x) = -\infty$, then the line $x = a$ is a vertical asymptote.

- If as $x \to a^+$ the function values $f(x)$ increase or decrease without bound, that is, $\lim\limits_{x \to a^+} f(x) = \infty$ or $\lim\limits_{x \to a^+} f(x) = -\infty$, then the line $x = a$ is a vertical asymptote.

- For any function f, if $\lim\limits_{x \to \pm\infty} f(x) = L$, then the line $y = L$ is a horizontal asymptote for the graph of f.

- For n, a positive real number, $\lim\limits_{x \to \infty} \dfrac{1}{x^n} = 0$.

- For n, a positive real number, $\lim\limits_{x \to -\infty} \dfrac{1}{x^n} = 0$, provided that x^n is a real number for negative values of x.

SECTION 2.2 EXERCISES

For Exercises 1–8, complete the given table to numerically estimate the following limits:

(a) $\lim\limits_{x \to a^-} f(x)$ (b) $\lim\limits_{x \to a^+} f(x)$ (c) $\lim\limits_{x \to a} f(x)$

Use the symbols ∞ and $-\infty$ where applicable.

1. $f(x) = \dfrac{1}{x^3}$; $a = 0$

x	-0.1	-0.01	-0.001	-0.00001	0	0.00001	0.001	0.01	0.1
$f(x)$									

2. $f(x) = \dfrac{2}{x^3}$; $a = 0$

x	-0.1	-0.01	-0.001	-0.00001	0	0.00001	0.001	0.01	0.1
$f(x)$									

3. $f(x) = \dfrac{1}{(x-1)^2}$; $a = 1$

x	0.9	0.99	0.999	0.99999	1	1.00001	1.001	1.01	1.1
$f(x)$									

4. $f(x) = \dfrac{3}{(x-1)^2}$; $a = 1$

x	0.9	0.99	0.999	0.99999	1	1.00001	1.001	1.01	1.1
$f(x)$									

5. $f(x) = \dfrac{x+2}{x^2-x-6}$; $a = -2$

x	-2.1	-2.001	-2.0001	-2	-1.9999	-1.999	-1.9
$f(x)$							

6. $f(x) = \dfrac{x+2}{x^2-x-6}$; $a = 3$

x	2.9	2.99	2.999	2.99999	3	3.00001	3.001	3.01	3.1
$f(x)$									

7. $f(x) = \dfrac{13{,}250 + 2.35x}{x}$; $a = 0$

x	-0.1	-0.01	-0.001	-0.00001	0	0.00001	0.001	0.01	0.1
$f(x)$									

8. $f(x) = \dfrac{4.25x - 25{,}350}{x}$; $a = 0$

x	-0.1	-0.01	-0.001	-0.00001	0	0.00001	0.001	0.01	0.1
$f(x)$									

In Exercises 9–16, determine the limit by factoring and canceling.

9. $\displaystyle\lim_{x \to 0} \frac{2}{x^3}$

10. $\displaystyle\lim_{x \to 0} \frac{13{,}250 + 2.35x}{x}$

11. $\displaystyle\lim_{x \to -2} \frac{x+2}{x^2-x-6}$

12. $\displaystyle\lim_{x \to 3} \frac{x+2}{x^2-x-6}$

✓ 13. $\displaystyle\lim_{x \to 3} \frac{x-3}{x^2-9}$

14. $\displaystyle\lim_{x \to -3} \frac{x-3}{x^2-9}$

15. $\displaystyle\lim_{x \to -1} \frac{x^2-x+1}{x^3+1}$

16. $\displaystyle\lim_{x \to 3} \frac{x^2-9}{x-3}$

For each function in Exercises 17–24:

(a) Classify f as an exponential growth or exponential decay function.

(b) Determine $\displaystyle\lim_{x \to \infty} f(x)$ and determine $\displaystyle\lim_{x \to -\infty} f(x)$.

17. $f(x) = 3^x$

18. $f(x) = 3^{-x}$

19. $f(x) = e^x$

20. $f(x) = e^{-x}$

21. $f(x) = e^{-0.215x}$

22. $f(x) = e^{0.215x}$

23. $f(x) = (0.987)^x$

24. $f(x) = (0.987)^{-x}$

In Exercises 25–32, determine the indicated limit algebraically using the method in Example 7. Verify your results numerically.

25. $\lim\limits_{x \to -\infty} \dfrac{2x + 5}{x - 1}$

26. $\lim\limits_{x \to \infty} \dfrac{2x + 5}{x - 1}$

27. $\lim\limits_{x \to \infty} \dfrac{3x^2 - x + 2}{2x^2 + x - 5}$

28. $\lim\limits_{x \to -\infty} \dfrac{3x^2 - x + 2}{2x^2 + x - 5}$

29. $\lim\limits_{x \to \infty} \dfrac{2x^2 + 2x + 1}{5x^3 + 3x - 5}$

30. $\lim\limits_{x \to \infty} \dfrac{3x^2 - 2x + 5}{2x^3 + x^2 - 2x + 3}$

31. $\lim\limits_{x \to -\infty} \dfrac{2x^2 + 2x + 1}{5x^3 + 3x - 5}$

32. $\lim\limits_{x \to -\infty} \dfrac{3x^2 - 2x + 5}{2x^3 + x^2 - 2x + 3}$

In Exercises 33–36, determine an equation for the horizontal asymptote for the graph of f. Consult your work performed in Exercises 25–32.

✓33. $f(x) = \dfrac{2x + 5}{x - 1}$

34. $f(x) = \dfrac{3x^2 - x + 2}{2x^2 + x - 5}$

35. $f(x) = \dfrac{2x^2 + 2x + 1}{5x^3 + 3x - 5}$

36. $f(x) = \dfrac{3x^2 - 2x + 5}{2x^3 + x^2 - 2x + 3}$

Use the graph of f in Figure 2.2.11 to determine the indicated limit in Exercises 37–43.

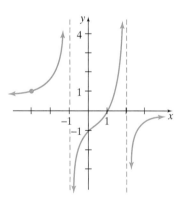

Figure 2.2.11

37. $\lim\limits_{x \to 2^+} f(x)$

38. $\lim\limits_{x \to 2^-} f(x)$

39. $\lim\limits_{x \to 2} f(x)$

40. $\lim\limits_{x \to -1} f(x)$

41. $\lim\limits_{x \to -3} f(x)$

42. $\lim\limits_{x \to \infty} f(x)$

43. $\lim\limits_{x \to -\infty} f(x)$

APPLICATIONS

44. The cost $C(x)$, in thousands of dollars, of removing $x\%$ of a city's pollutants discharged into a lake is given by

$$C(x) = \frac{223x}{100 - x}$$

(a) Determine the reasonable domain for C and graph.

(b) Evaluate $C(40)$ and interpret.

(c) Determine $\lim\limits_{x \to 100^-} C(x)$ and interpret.

✓45. The cost $C(x)$, in thousands of dollars, of removing $x\%$ of a city's pollutants discharged into a river is given by

$$C(x) = \frac{93x}{100 - x}$$

(a) Determine the reasonable domain for C and graph.

(b) Evaluate $C(40)$ and interpret.

(c) Determine $\lim\limits_{x \to 100^-} C(x)$ and interpret.

46. The number of dollars spent on advertising for a product influences the number of items of the product that will be purchased by consumers. If x is the number of dollars spent, in thousands, then $N(x)$, in hundreds, is the number of items sold, as given by the function

$$N(x) = 2000 - \frac{520}{x}$$

(a) Determine $\lim\limits_{x \to \infty} N(x)$ and interpret.

(b) Graph N in an appropriate viewing window. From the graph, estimate an interval for x that would result in $N(x)$ being between 1500 and 1800.

47. The number of dollars spent on advertising for a product influences the number of items of the product that will be purchased by consumers. If x is the number of dollars spent, in thousands, then $N(x)$, in hundreds, is the number of items sold, as given by the function

$$N(x) = 3200 - \frac{750}{x}$$

(a) Determine $\lim\limits_{x \to \infty} N(x)$ and interpret.

(b) Graph N in an appropriate viewing window. From the graph, estimate an interval for x that would result in $N(x)$ being between 2700 and 3000.

48. The total cost of producing q units of a product is given by $C(q) = 13{,}700 + 6.85q$, where $C(q)$ is in dollars. Recall that the average cost, AC, is given by

$$AC(q) = \frac{C(q)}{q} = \frac{13{,}700 + 6.85q}{q}$$

Determine $\lim\limits_{q \to \infty} AC(q)$ and interpret.

49. The total cost of producing q units of a product is given by $C(q) = 15{,}200 + 4.85q$, where $C(q)$ is in dollars. Recall that the average cost, AC, is given by

$$AC(q) = \frac{C(q)}{q} = \frac{15{,}200 + 4.85q}{q}$$

Determine $\lim\limits_{q \to \infty} AC(q)$ and interpret.

50. This exercise uses the following data for the population of Florida.

x, YEARS SINCE 1899	y, POPULATION IN THOUSANDS
1	528.542
11	752.619
21	968.470
31	1,468.211
41	1,897.414
51	2,771.305
61	4,951.560
71	6,791.418
81	9,746.961
91	12,937.926

Source: U.S. Census Bureau web site.

Bob has modeled the data for the population of Florida, where x represents the number of years since 1899 and y represents the population in thousands, with the following models:

$$y = 2.13x^2 - 65.16x + 1028.74$$
$$y = 476.33e^{0.04x}$$
$$y = \frac{32{,}785.89}{1 + 95.71e^{-0.05x}}$$

Each does a good job of modeling the given data. Aware of the dangers of using any model to predict too far into the future, Bob still wants to try to determine an upper limit for the population of Florida, but he does not know how. Help Bob by doing the following:

(a) Determine $\lim\limits_{x \to \infty} (2.13x^2 - 65.16x + 1028.74)$ and interpret.

(b) Determine $\lim\limits_{x \to \infty} (476.33e^{0.04x})$ and interpret.

(c) Determine $\lim\limits_{x \to \infty} \left(\dfrac{32{,}785.89}{1 + 95.71e^{-0.05x}} \right)$ and interpret.

(d) Which of the three models gives the most reasonable estimate for an upper limit of Florida's population and why?

51. Pharmacological studies have determined that the amount of medication present in the body is a function of the amount given and how much time has elapsed since the medication was administered. For a certain medication, the amount present in milliliters, $A(t)$, can be approximated by the function

$$A(t) = 2.5e^{-0.2t}, \quad t \geq 0$$

where t is the number of hours since the medication was administered.

(a) Determine $A(0)$ and interpret.

(b) Determine $\lim\limits_{t \to \infty} A(t)$ and interpret.

(c) Graph A in the viewing window $[0, 5]$ by $[0, 5]$.

(d) If a patient is to receive more medication when the amount present drops below 1 milliliter, graphically approximate how many hours after the initial administration of the medication more medication should be given.

52. Pharmacological studies have determined that the amount of medication present in the body is a function of the amount given and how much time has elapsed since the medication was administered. For a certain medication, the amount present in milliliters, $A(t)$, can be approximated by the function

$$A(t) = 3.5e^{-0.3t}, \quad t \geq 0$$

where t is the number of hours since the medication was administered.

(a) Determine $A(0)$ and interpret.

(b) Determine $\lim\limits_{t \to \infty} A(t)$ and interpret.

(c) Graph A in the viewing window $[0, 5]$ by $[0, 5]$.

(d) If a patient is to receive more medication when the amount present drops below 1 milliliter, graphically approximate how many hours after the initial administration of the medication more medication should be given.

53. The local game commission decided to stock a lake with bass. To do this, 100 bass were introduced into the lake. The population of the bass can be approximated by

$$P(t) = \frac{10(10 + 7t)}{1 + 0.02t}, \quad t \geq 0$$

where t is time in months since the lake was stocked and $P(t)$ is the population after t months.

(a) Graph P in the viewing window $[0, 60]$ by $[0, 3000]$.

(b) Determine the population in 1 year.

(c) Determine $\lim\limits_{t \to \infty} P(t)$ and interpret. The result is called the *limiting size* of the population.

54. The local game commission decided to stock a lake with trout. To do this, 200 trout were introduced into the lake. The population of the trout can be approximated by

$$P(t) = \frac{20(10 + 7t)}{1 + 0.02t}, \quad t \geq 0$$

where t is time in months since the lake was stocked and $P(t)$ is the population after t months.

(a) Graph P in the viewing window $[0, 60]$ by $[0, 5000]$.

(b) Determine the population in 1 year.

(c) Determine $\lim\limits_{t \to \infty} P(t)$ and interpret. The result is called the *limiting size* of the population.

55. The spread of influenza in Surfside, an isolated town with a population of 1000, can be modeled by

$$D(t) = \frac{1000}{1 + 999e^{-0.3t}}, \quad t \geq 0$$

where $D(t)$ is the number of people affected by influenza after t days.

(a) Graph D in the viewing window [0, 50] by [0, 1500]. Recall that this is a graph of a logistics function.

(b) How many people have been affected by influenza after 3 days?

(c) Determine $\lim_{t \to \infty} D(t)$ and interpret.

56. When a new product is introduced on the market, sales often follow a pattern of rapid initial growth followed by some leveling off. For a certain new candy bar, this pattern is modeled by

$$S(t) = 4000 - 4000e^{-0.25t}$$

where $S(t)$ is the number of units sold after t months.

(a) How many units were sold after 2 months?

(b) Determine $\lim_{t \to \infty} S(t)$ and interpret.

SECTION PROJECT

Use the following data for the population of California.

x, YEARS SINCE 1899	y, POPULATION IN THOUSANDS
1	1,485.053
11	2,377.549
21	3,426.861
31	5,677.251
41	6,907.387
51	10,586.223
61	15,717.204
71	19,971.069
81	23,667.764
91	29,760.021

Source: U.S. Census Bureau web site.

Let x represent the number of years since 1899 and let y represent the population, in thousands, of California. Enter the data into your calculator and determine the following regression models:

(a) Quadratic model: round coefficients to the nearest hundredth.

(b) Exponential model: round coefficients to the nearest hundredth.

(c) Logistic model: round coefficients to the nearest hundredth.

(d) Plot the data and graph each model to determine if each model gives a good fit to the data.

(e) Determine $\lim_{x \to \infty}$ for the each model and interpret.

(f) Aware of the dangers of using regression models to predict too far into the future, determine which of these three models gives the most reasonable estimate for an upper limit to the population of California.

SECTION 2.3 — INSTANTANEOUS RATE OF CHANGE AND THE DERIVATIVE

In Chapter 1 and so far with our work in Chapter 2, we have laid the foundation for our journey through calculus. We now turn our attention to an old topic first introduced in Chapter 1, **rates of change**. In Chapter 1 we discussed in detail what is known as an **average rate of change**. In this section we will use the limit concept to extend average rate of change to a new rate of change, an **instantaneous rate of change**. We will then define a new function, the **derivative**, and your journey into the **differential calculus** will have started.

Revisiting Average Rate of Change

In Chapter 1 we saw how the slope of a secant line gave us an average rate of change over an interval. Let's revisit these concepts through a Flashback.

Flashback

U.S. Imports from China Revisited

In Section 1.3, the U.S. imports from China for the years 1987 to 1996 were modeled by

$$f(x) = 0.32x^2 + 1.64x + 3.98,^* \qquad 1 \le x \le 10$$

where x represents the number of years since 1986 and $f(x)$ represents the dollar value, in billions, of goods imported. The graph of the model is shown in Figure 2.3.1.

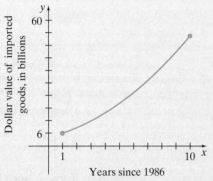

Figure 2.3.1 Graph of $f(x) = 0.32x^2 + 1.64x + 3.98$.

Determine the average rate of change for U.S. imports from China from 1990 to 1993 and interpret.

Flashback Solution

Recall that the average rate of change is simply equivalent to the *slope of the secant line* through two points. Since 1990 corresponds to an x-value of 4 and 1993 corresponds to an x-value of 7, we need to

determine the corresponding y-values for these x-values to get the two points that the secant line passes through. We determine the y-value when $x = 4$ by simply substituting $x = 4$ into our model as follows:

$$y = f(x) = 0.32x^2 + 1.64x + 3.98$$
$$= 0.32(4)^2 + 1.64(4) + 3.98 = 15.66$$

We now know one point that the secant line passes through is (4, 15.66). Similarly, when $x = 7$, we get $y = 31.14$. Hence, the second point that the secant line passes through is (7, 31.14). See Figure 2.3.2. Now the secant line slope through these two points is simply

$$m_{\text{sec}} = \frac{y_2 - y_1}{x_2 - x_1} = \frac{31.14 - 15.66}{7 - 4} = 5.16$$

So our answer is that the average rate of change for U.S. imports from China from 1990 to 1993 was $5.16 \frac{\text{billion dollars}}{\text{year}}$. Recall from Chapter 1 that this means that over the period 1990 to 1993 U.S. imports from China increased $5.16 \frac{\text{billion dollars}}{\text{year}}$ on average.

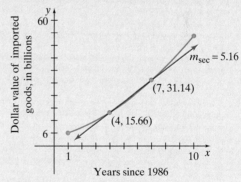

Figure 2.3.2 Slope of the secant line, m_{sec}, gives the average rate of change.

The data and model given in exercises and examples preceded by the "On the Web" icon are based on information gathered at the U.S. Census Bureau web site. These exercises and examples appear throughout this textbook.

There are a couple of important items in the Flashback that we must review:

- Since we found a *rate,* we must include *appropriate units.*
- The secant line has a positive slope indicating that from 1990 to 1993 U.S. imports from China were *increasing.*

Tangent Lines and Slopes of Curves

We now pose the following question: At what rate were U.S. imports from China changing at the beginning of 1993? We are not asking for the average rate of change over some time interval. Instead, the question asks for the rate of change at a particular *instant.* This type of rate of change is called an **instantaneous rate**

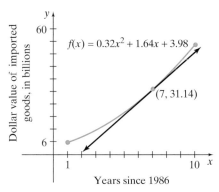

Figure 2.3.3

of change, and it is equivalent to the slope of the line **tangent** to the curve at a specific point. What do we mean by a **tangent line**?

Consider the curve in Figure 2.3.3. The steepness of the curve varies from point to point. The measure of the steepness of the curve at the point $(7, 31.14)$ is equivalent to finding the slope of the tangent line through the point. (We call the point through which the tangent line passes the **fixed point** or the **point of tangency**.) With this in mind, the following serves as a working definition of tangent line.

Tangent Line

The **tangent line** to a curve at a point A is the line through A whose slope matches the steepness of the curve at point A. We can say that the tangent line slope is equal to the slope of the curve at A.

EXAMPLE 1 | **Determining an Instantaneous Rate of Change**

Determine at what rate U.S. imports from China were changing at the beginning of 1993, that is, when $x = 7$.

SOLUTION

To answer this question requires us to find the slope of the tangent line that we considered in Figure 2.3.3. To find this, we will do the following:

1. Compute the coordinates of eight points close to the fixed point, four with an x-coordinate less than $x = 7$ and four with an x-coordinate greater than $x = 7$.
2. Compute the slope of the secant line through the fixed point and each of these eight points.

The first thing we need to observe is that $x = 6$, $x = 7$, and $x = 8$ correspond to the years 1992, 1993, and 1994, respectively. Tables 2.3.1 and 2.3.2 show x-values getting close to 7, from both sides of 7, and the corresponding y-values.

Using the coordinates of eight points relatively close to the fixed point $(7, 31.14)$, we now determine the slope of the secant line through each point and our fixed point $(7, 31.14)$. See Figure 2.3.4. The calculations for each slope are similar to the slope calculation in the Flashback. The results are given in Tables 2.3.3 and 2.3.4.

It appears that as we let $x \to 7^-$ and let $x \to 7^+$ the slope of the secant lines over these smaller and smaller intervals is getting close to 6.1. Since the slope of

TABLE 2.3.1	
$x \to 7^-$	
x	y
6	25.34
6.5	28.16
6.9	30.5312
6.99	31.078832

TABLE 2.3.2	
$x \to 7^+$	
x	y
8	37.58
7.5	34.28
7.1	31.7552
7.01	31.201232

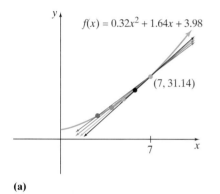

(a)

TABLE 2.3.3	
INTERVAL	**SLOPE OF SECANT LINE OVER INTERVAL**
$[6, 7]$	$m_{\sec} = \dfrac{y_2 - y_1}{x_2 - x_1} = \dfrac{31.14 - 25.34}{7 - 6} = 5.8$
$[6.5, 7]$	$m_{\sec} = \dfrac{y_2 - y_1}{x_2 - x_1} = \dfrac{31.14 - 28.16}{7 - 6.5} = 5.96$
$[6.9, 7]$	$m_{\sec} = \dfrac{y_2 - y_1}{x_2 - x_1} = \dfrac{31.14 - 30.5312}{7 - 6.9} = 6.088$
$[6.99, 7]$	$m_{\sec} = \dfrac{y_2 - y_1}{x_2 - x_1} = \dfrac{31.14 - 31.078832}{7 - 6.99} = 6.1168$

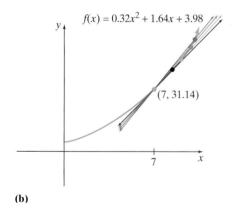

(b)

TABLE 2.3.4	
INTERVAL	**SLOPE OF SECANT LINE OVER INTERVAL**
$[7, 8]$	$m_{\sec} = \dfrac{y_2 - y_1}{x_2 - x_1} = \dfrac{37.58 - 31.14}{8 - 7} = 6.44$
$[7, 7.5]$	$m_{\sec} = \dfrac{y_2 - y_1}{x_2 - x_1} = \dfrac{34.28 - 31.14}{7.5 - 7} = 6.28$
$[7, 7.1]$	$m_{\sec} = \dfrac{y_2 - y_1}{x_2 - x_1} = \dfrac{31.7552 - 31.14}{7.1 - 7} = 6.152$
$[7, 7.01]$	$m_{\sec} = \dfrac{y_2 - y_1}{x_2 - x_1} = \dfrac{31.201232 - 31.14}{7.01 - 7} = 6.1232$

Figure 2.3.4 **(a)** Secant lines through fixed point (7, 31.14) and as $x \to 7^-$. **(b)** Secant lines through fixed point (7, 31.14) and as $x \to 7^+$.

the secant line is the average rate of change, the average rate of change seems to be getting close to $6.1 \ \dfrac{\text{billion dollars}}{\text{year}}$. So it seems reasonable to say that the slope of the tangent line is 6.1. Since the slope of the tangent line is equal to the instantaneous rate of change at $x = 7$, we believe that at $x = 7$ the instantaneous rate of change was about 6.1 billion dollars per year. This is how fast U.S. imports from China were rising at the instant 1993 began.

NOTES: 1. Graphing calculators or spreadsheets can be used for the automation of the calculations in Tables 2.3.3 and 2.3.4.

2. Notice that the rates of change, average or instantaneous, in Example 1 as well as the Flashback are measured in billions of dollars per year.

In Example 1, we said that the slope of the tangent line at the point $(7, 31.14)$ is 6.1, which means that the instantaneous rate of change is $6.1 \dfrac{\text{billion dollars}}{\text{year}}$. We were able to reach this conclusion because of the following important result.

> An instantaneous rate of change is equal to the slope of the tangent line of the curve.

Since instantaneous rate of change arises in many situations, let's summarize how to compute this rate:

- To find a slope of a tangent line, determine the point of tangency or the fixed point.
- Select points on the curve that are progressively closer to the fixed point.
- Calculate the slope of the secant line through each of these points and the fixed point. The slope of the tangent line is the limit of the slopes of the secant lines.

Now, let's generalize what we did in Example 1. Since we computed the slopes in Table 2.3.4 using small intervals such as $[7, 7.01]$ and $[7, 7.1]$, let's call the length of any such interval h so that we can write these intervals as $[7, 7 + h]$. Using this notation, we can write the slope of the secant line as

$$\frac{f(7+h) - f(7)}{(7+h) - 7} = \frac{f(7+h) - f(7)}{h}$$

This expression is nothing more than the *difference quotient* introduced in Chapter 1 with h in place of Δx. Remember that this tells us the slope of the secant line through two points, which in turn tells us an average rate of change.

Interactive Activity

 Using the notation just discussed, complete Table 2.3.5 and verify that it matches Table 2.3.4.

TABLE 2.3.5

h	INTERVAL, $[7, 7 + h]$	SLOPE OF SECANT LINE OVER INTERVAL
1	$[7, 8]$	$m_{\text{sec}} = \dfrac{f(7+h) - f(7)}{h} =$
0.5	$[7, 7.5]$	$m_{\text{sec}} = \dfrac{f(7+h) - f(7)}{h} =$
0.1	$[7, 7.1]$	$m_{\text{sec}} = \dfrac{f(7+h) - f(7)}{h} =$
0.01	$[7, 7.01]$	$m_{\text{sec}} = \dfrac{f(7+h) - f(7)}{h} =$

Difference Quotient and the Derivative

Suppose that we wish to determine the instantaneous rate of change at x for the function graphed in Figure 2.3.5. We will use the process outlined in Example 1 and now consider *any* value for x. We use the expression we just found for the difference quotient (p.140) with x in place of 7.

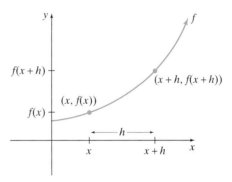

Figure 2.3.5

What happened in Example 1 should be a familiar process. By letting h get smaller and smaller, we are applying the *limit concept* introduced in Sections 2.1 and 2.2. Taking the limit of the difference quotient as $h \to 0$ yields an instantaneous rate of change and is denoted in the following definition:

> ### Instantaneous Rate of Change/Slope of a Tangent Line
>
> The **instantaneous rate of change** of a function, f, at x is equal to the **slope of the line tangent** to the graph of f at x and is given by
>
> $$\text{Instantaneous rate of change} = m_{\text{tan}} = \lim_{h \to 0} \frac{f(x+h) - f(x)}{h}$$
>
> provided that the limit exists. Its units of measurement are units of f per unit of x.

The preceding definition is so important that it has been given a special name, the *derivative*. We now present one of the most important definitions in all of mathematics, the definition of the derivative.

> ### Derivative
>
> For a function f, the **derivative of f at x,** denoted $f'(x)$ (this is read "f prime of x"), is defined to be
>
> $$f'(x) = \lim_{h \to 0} \frac{f(x+h) - f(x)}{h}$$
>
> provided that the limit exists. The units of $f'(x)$ are units of f per unit of x.

NOTE: Notice that the derivative of f at x gives an instantaneous rate of change of f at x.

EXAMPLE 2 | **Computing a Derivative**

Compute $f'(2)$ for $f(x) = x^2 + x$. Interpret the result.

SOLUTION

To compute $f'(2)$, we use the definition of the derivative

$$f'(x) = \lim_{h \to 0} \frac{f(x+h) - f(x)}{h}$$

where x has a value of 2. Proceeding, this gives

$$f'(2) = \lim_{h \to 0} \frac{f(2+h) - f(2)}{h}$$

$$= \lim_{h \to 0} \frac{[(2+h)^2 + (2+h)] - [2^2 + 2]}{h}$$

$$= \lim_{h \to 0} \frac{[4 + 4h + h^2 + 2 + h] - 6}{h}$$

$$= \lim_{h \to 0} \frac{h^2 + 5h}{h}$$

Now we rely on the algebraic techniques learned in Sections 2.1 and 2.2 to evaluate this limit. Employing the Substitution Principle yields the indeterminate form $\frac{0}{0}$, so we factor the numerator and cancel:

$$f'(2) = \lim_{h \to 0} \frac{h(h+5)}{h} = \lim_{h \to 0}(h+5) = 5$$

We have now determined that $f'(2) = 5$, which means that the instantaneous rate of change of $f(x) = x^2 + x$ at $x = 2$ is 5. Alternatively, we could say that the slope of the curve given by $f(x) = x^2 + x$ at $x = 2$ is 5, or we could even say that the slope of the line tangent to the graph of $f(x) = x^2 + x$ at $x = 2$ is 5. See Figure 2.3.6.

TECHNOLOGY NOTE

Many calculators have a DRAW TANGENT command. Figure 2.3.6b was generated using such a command. Consult the online calculator manual at www.prenhall.com/armstrong

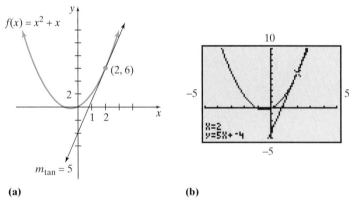

(a)   **(b)**

Figure 2.3.6   The notation $m_{\tan}$ represents "slope of tangent line."

✓**CHECKPOINT 1**

Now work Exercise 27.

EXAMPLE 3 | **Computing a Derivative in an Applied Setting**

The total cost of producing q recreational vehicles can be modeled by

$$C(q) = 100 + 60q + 3q^2$$

where q is the number of vehicles produced and $C(q)$ is in hundreds of dollars.

(a) Compute $C(5)$ and interpret.   (b) Compute $C'(5)$ and interpret.

Interactive Activity

In Example 2, we algebraically determined $f'(2) = 5$. We used the definition of the derivative at $x = 2$ to get $f'(2) = \lim_{h \to 0} \dfrac{f(2+h) - f(2)}{h}$, which we eventually simplified to $f'(2) = \lim_{h \to 0} \dfrac{h^2 + 5h}{h}$. To examine this numerically and graphically, we change the h's to x's and enter into the calculator $y = \dfrac{x^2 + 5x}{x}$. Do this, then numerically and graphically determine $\lim_{x \to 0} \dfrac{x^2 + 5x}{x}$ and see if it supports the result of Example 2. Use Table 2.3.6 for the numerical part.

TABLE **2.3.6**

	$x \to 0^-$					$x \to 0^+$			
x	-1	-0.1	-0.001	-0.0001	0	0.0001	0.001	0.01	1
$\dfrac{x^2 + 5x}{x}$									

SOLUTION

(a) To compute $C(5)$, we simply substitute 5 for q. This gives

$$C(5) = 100 + 60(5) + 3(5)^2 = 475$$

This means the total cost to produce 5 vehicles is $47,500.

(b) To compute $C'(5)$, we will use the definition of the derivative

$$C'(q) = \lim_{h \to 0} \frac{C(q+h) - C(q)}{h}$$

where q has the value of 5. Proceeding, this gives

$$C'(5) = \lim_{h \to 0} \frac{C(5+h) - C(5)}{h}$$

$$= \lim_{h \to 0} \frac{[100 + 60(5+h) + 3(5+h)^2] - [100 + 60(5) + 3(5)^2]}{h}$$

$$= \lim_{h \to 0} \frac{100 + 300 + 60h + 3(25 + 10h + h^2) - 475}{h}$$

$$= \lim_{h \to 0} \frac{100 + 300 + 60h + 75 + 30h + 3h^2 - 475}{h}$$

$$= \lim_{h \to 0} \frac{90h + 3h^2}{h}$$

As in Example 2, employing the Substitution Principle yields the indeterminate form $\dfrac{0}{0}$, so we factor the numerator and cancel.

$$C'(5) = \lim_{h \to 0} \frac{h(90 + 3h)}{h}$$

$$= \lim_{h \to 0} (90 + 3h) = 90$$

So we have determined that $C'(5) = 90$, which means that the instantaneous rate of change is $9000 per vehicle when the production level is 5 vehicles.

Alternatively, we could say that the total costs increase approximately $9000 when production increases from 5 vehicles to 6 vehicles. Figure 2.3.7 shows the graph of C along with the line tangent at the point $(5, 475)$.

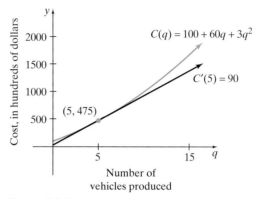

Figure 2.3.7

The derivative of a cost function such as the one in Example 3 is called **marginal cost.** The marginal concept will be discussed in full detail in Chapter 3.

Derivatives in General

Before we continue with how to compute a derivative in general, we need to present some facts about the derivative.

Facts about the Derivative

1. Recall that the derivative of a function f can be found for any value x where the derivative exists. Thus, the derivative of a function, f', is itself a function. This means that it has a domain for which it is defined.
2. For any c in the domain of f', $f'(c)$ is a number that gives the slope of a line tangent to the graph of f at $x = c$. Also, $f'(c)$ gives the instantaneous rate of change of f at $x = c$.
3. The process of determining $f'(x)$ is called **differentiation.**

By taking full advantage of the definition of the derivative coupled with the fact that the derivative is a function, we can compute a formula for the derivative. In other words, recall the total cost function in Example 3, which was given by $C(q) = 100 + 60q + 3q^2$. Suppose that we were asked to compute $C'(3)$, $C'(7)$, $C'(13)$, and $C'(15)$. At first glance, it appears that we would have the long and tedious task of duplicating our work in Example 3 for these four different values of q. However, the definition of the derivative tells us that the derivative is a *function,* which means that we should be able to determine a rule for this function. In Example 4, we illustrate how to use the definition of the derivative to do this.

EXAMPLE 4 | **Computing a Derivative Formula**

Compute $C'(q)$ for $C(q) = 100 + 60q + 3q^2$.

SOLUTION

The process is the same as in Example 3, except here we have our independent variable q in place of 5.

$$
\begin{aligned}
C'(q) &= \lim_{h \to 0} \frac{C(q+h) - C(q)}{h} \\
&= \lim_{h \to 0} \frac{[100 + 60(q+h) + 3(q+h)^2] - [100 + 60(q) + 3(q)^2]}{h} \\
&= \lim_{h \to 0} \frac{[100 + 60q + 60h + 3(q^2 + 2qh + h^2)] - (100 + 60q + 3q^2)}{h} \\
&= \lim_{h \to 0} \frac{100 + 60q + 60h + 3q^2 + 6qh + 3h^2 - 100 - 60q - 3q^2}{h} \\
&= \lim_{h \to 0} \frac{60h + 6qh + 3h^2}{h} = \lim_{h \to 0} \frac{h(60 + 6q + 3h)}{h} \\
&= \lim_{h \to 0} (60 + 6q + 3h) = 60 + 6q.
\end{aligned}
$$

So $C'(q) = 60 + 6q$. A huge advantage in having this formula for the derivative is that we can now evaluate it for different values of our independent variable very quickly. For example,

$$
\begin{aligned}
C'(3) &= 60 + 6(3) = 78 \\
C'(5) &= 60 + 6(5) = 90 \\
C'(7) &= 60 + 6(7) = 102
\end{aligned}
$$

✔ **CHECKPOINT 2**

Now work Exercise 47.

EXAMPLE 5

Computing and Using a Derivative Formula

Determine $f'(x)$ for $f(x) = x^2 + 3x$. Also, determine the slope of the line tangent to the graph of f at $x = -2$.

SOLUTION

Utilizing the definition of the derivative gives

$$
\begin{aligned}
f'(x) &= \lim_{h \to 0} \frac{f(x+h) - f(x)}{h} \\
&= \lim_{h \to 0} \frac{[(x+h)^2 + 3(x+h)] - (x^2 + 3x)}{h} \\
&= \lim_{h \to 0} \frac{x^2 + 2xh + h^2 + 3x + 3h - x^2 - 3x}{h} \\
&= \lim_{h \to 0} \frac{2xh + h^2 + 3h}{h} = \lim_{h \to 0} \frac{h(2x + h + 3)}{h} \\
&= \lim_{h \to 0} (2x + h + 3) = 2x + 3
\end{aligned}
$$

Interactive Activity

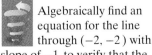 Algebraically find an equation for the line through $(-2, -2)$ with slope of -1 to verify that the equation of the tangent line in Figure 2.3.8 is $y = -x - 4$.

So $f'(x) = 2x + 3$. Knowing that the derivative gives the slope of the tangent line at any point, we can quickly compute that at $x = -2$

$$
m_{\tan} = f'(-2) = 2(-2) + 3 = -1
$$

Figure 2.3.8 has a graph of f and the line tangent at $x = -2$.

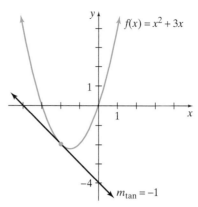

Figure 2.3.8

✓ CHECKPOINT **3**

Now work Exercise 59.

Example 5 communicates that the derivative of a function, in its most general form, is very useful when determining tangent line slopes or instantaneous rates of change at *any* point on the graph of the given function.

We conclude this section with the same example that started it. However, we now compute the derivative in general.

EXAMPLE 6 | **Computing a Derivative Formula**

Compute $f'(x)$ for $f(x) = 0.32x^2 + 1.64x + 3.98$. Recall that f models U.S. imports from China over the years 1987 to 1996, where x is the number of years since 1986 and $f(x)$ is the dollar value, in billions, of the imported goods.

SOLUTION

By definition, $f'(x) = \lim\limits_{h \to 0} \dfrac{f(x+h) - f(x)}{h}$. So we have

$$f'(x) = \lim_{h \to 0} \frac{[0.32(x+h)^2 + 1.64(x+h) + 3.98] - [0.32x^2 + 1.64x + 3.98]}{h}$$

$$= \lim_{h \to 0} \frac{[0.32(x^2 + 2xh + h^2) + 1.64x + 1.64h + 3.98] - 0.32x^2 - 1.64x - 3.98}{h}$$

$$= \lim_{h \to 0} \frac{[0.32x^2 + 0.64xh + 0.32h^2 + 1.64x + 1.64h + 3.98] - 0.32x^2 - 1.64x - 3.98}{h}$$

$$= \lim_{h \to 0} \frac{0.64xh + 0.32h^2 + 1.64h}{h}$$

$$= \lim_{h \to 0} \frac{h(0.64x + 0.32h + 1.64)}{h}$$

$$= \lim_{h \to 0} (0.64x + 0.32h + 1.64) = 0.64x + 1.64$$

So $f'(x) = 0.64x + 1.64$. Notice that $f'(7) = 6.12$, which is very close to the same result that we numerically determined in Example 1. However, we could now quickly compute the instantaneous rate of change for *any* value of x using $f'(x) = 0.64x + 1.64$.

Summary

We have presented some very powerful mathematical concepts in this section. The main ideas have been rates of change, specifically introducing the **instantaneous rate of change.** In Table 2.3.7, we compare and contrast average rate of change with instantaneous rate of change.

TABLE 2.3.7

AVERAGE RATE OF CHANGE	INSTANTANEOUS RATE OF CHANGE
Change over an interval	Change at a point
Slope of a secant line	Slope of a tangent line; slope of a curve
$$\dfrac{f(x+h) - f(x)}{h}$$	$$\lim_{h \to 0} \dfrac{f(x+h) - f(x)}{h}$$
Measured in units of f per unit of x	Measured in units of f per unit of x

We calculated **derivatives** for specific values of x and for any arbitrary x. The derivative is a function, which means that it has a domain and can be evaluated at any value c in the domain. Remember when computing a derivative, whether at a specific value of c or for an arbitrary x, that we are finding a limit. Therefore, the limit techniques learned in Sections 2.1 and 2.2 can be employed.

Section 2.3 Exercises

In Exercises 1–10, compute the average rate of change of the function over the indicated interval.

1. $f(x) = 2x^2 + x - 3$, $[-2, 1]$

2. $f(x) = 3x^2 - 2x + 2$, $[-3, 0]$

3. $g(x) = 3\sqrt{x}$, $[1, 9]$ 4. $g(x) = 2\sqrt{x}$, $[1, 16]$

5. $f(x) = 2.1e^{0.2x}$, $[0, 20]$ 6. $f(x) = 1.3e^{0.3x}$, $[0, 25]$

7. $f(x) = 2.1e^{-0.2x}$, $[0, 15]$ 8. $f(x) = 3.2e^{-0.4x}$, $[0, 5]$

9. $f(x) = 1 + \ln x$, $[1, 10]$ 10. $g(x) = 2.1 + \ln x$, $[1, 8]$

11. Between which pairs of consecutive points on the given curve is the average rate of change positive? Negative? Zero?

(a)
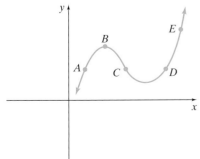

(b)
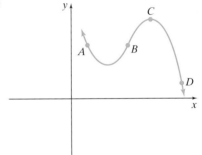

12. At which points on the given curve is the tangent line slope positive? Negative? Zero?

(a)
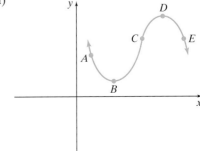

(b)

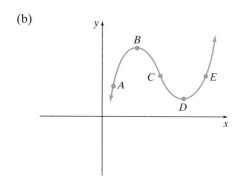

In Exercises 13–22, estimate the instantaneous rate of change of the function at the indicated value by using $\lim\limits_{h\to0}\dfrac{f(x+h)-f(x)}{h}$ and by completing the following tables:

		$h \to 0^-$			
h	-1	-0.1	-0.01	-0.001	-0.0001
$\dfrac{f(x+h)-f(x)}{h}$					

		$h \to 0^+$			
h	1	0.1	0.01	0.001	0.0001
$\dfrac{f(x+h)-f(x)}{h}$					

13. $f(x) = 3x; \quad x = 2$

14. $f(x) = 4x; \quad x = 2$

15. $f(x) = x^2 + x; \quad x = 3$

16. $f(x) = x^2 - x; \quad x = -3$

17. $f(x) = 2x^3 + 3x + 1; \quad x = 4$

18. $f(x) = x^3 - 2x - 3; \quad x = -1$

19. $f(x) = 3\sqrt{x}; \quad x = 4$

20. $f(x) = 2\sqrt{x}; \quad x = 9$

21. $f(x) = 2.1e^{0.2x}; \quad x = 2$

22. $f(x) = 2.1e^{-0.2x}; \quad x = 3$

23. Based on the result of Exercise 15, estimate the slope of the line tangent to the graph of $f(x) = x^2 + x$ at $x = 3$.

24. Based on the result of Exercise 20, estimate the slope of the line tangent to the graph of $f(x) = 2\sqrt{x}$ at $x = 9$.

In Exercises 25–34, compute $f'(c)$ by using the definition of the derivative:

$$f'(c) = \lim_{h\to0}\frac{f(c+h)-f(c)}{h}$$

See Example 2.

25. $f(x) = 2x + 1; \quad c = 2$

26. $f(x) = 3x - 4; \quad c = 3$

✓ 27. $f(x) = x^2 - 2x; \quad c = -2$

28. $f(x) = x^2 - 3x; \quad c = 2$

29. $f(x) = -x^2 + 2x; \quad c = 4$

30. $f(x) = -2x^2 + x; \quad c = 4$

31. $f(x) = 2\sqrt{x}; \quad c = 4$

32. $f(x) = 2.3x^2 - 0.2x + 3.2; \quad c = 8$

33. $f(x) = -1.2x^2 + 3.2x - 2.1; \quad c = 7$

34. $f(x) = \dfrac{1}{x}; \quad c = 3$

Exercises 35–40 use the results of Exercises 25–34. For each function and given value for x:

(a) Determine an equation for the line tangent to the graph of the function at the indicated value for x.

(b) Graph the function and the tangent line in the same viewing window.

35. $f(x) = x^2 - 2x$ at $x = -2$. See Exercise 27.

36. $f(x) = x^2 - 3x$ at $x = 2$. See Exercise 28.

37. $f(x) = -x^2 + 2x$ at $x = 4$. See Exercise 29.

38. $f(x) = 2\sqrt{x}$ at $x = 4$. See Exercise 31.

39. $f(x) = -1.2x^2 + 3.2x - 2.1$ at $x = 7$. See Exercise 33.

40. $f(x) = \dfrac{1}{x}$ at $x = 3$. See Exercise 34.

In Exercises 41–56, compute a formula for the derivative by using the definition of the derivative

$$f'(x) = \lim_{h\to0}\frac{f(x+h)-f(x)}{h}$$

as was illustrated in Examples 4, 5, and 6.

41. $f(x) = 2x - 5$ 42. $f(x) = 3x + 1$

43. $g(x) = -2x + 3$ 44. $g(x) = -3x - 1$

45. $f(x) = x^2$ 46. $f(x) = x^2 - 3x$

✓ 47. $h(x) = x^2 - 2x + 3$ 48. $h(x) = x^2 - 3x + 1$

49. $g(x) = 2.1x^2 + 3.2x$ 50. $g(x) = 1.3x^2 - 2.1x$

51. $f(x) = -2x^2 + 3x$ 52. $f(x) = -3x^2 + 2x$

53. $f(x) = -2.3x^2 + 3.1x$ 54. $f(x) = x^3$

55. $g(x) = x^3 + x^2$ 56. $g(x) = \sqrt{x}$

Exercises 57–66 use the results of Exercises 41–56. Determine the slope of the line tangent to the graph of the given function at the indicated values of x. See Example 5.

57. $f(x) = 2x - 5$ at $x = -1$ and $x = 2$. See Exercise 41.

58. $g(x) = -3x - 1$ at $x = -1$ and $x = 2$. See Exercise 44.

✓ 59. $f(x) = x^2 - 2x + 3$ at $x = -3$ and $x = 2$. See Exercise 47.

60. $h(x) = x^2 - 3x + 1$ at $x = -2$ and $x = 3$. See Exercise 48.

61. $g(x) = 2.1x^2 + 3.2x$ at $x = 0$ and $x = 2$. See Exercise 49.

62. $g(x) = 1.3x^2 - 2.1x$ at $x = 1$ and $x = 4$. See Exercise 50.

63. $f(x) = -2x^2 + 3x$ at $x = 3$ and $x = 6$. See Exercise 51.

64. $f(x) = -3x^2 + 2x$ at $x = 2$ and $x = 7$. See Exercise 52.

65. $f(x) = x^3 + x^2$ at $x = -1$ and $x = 1$. See Exercise 55.

66. $g(x) = \sqrt{x}$ at $x = 4$ and $x = 9$. See Exercise 56.

APPLICATIONS

67. The Network Standards Company has determined from data that it has collected that the profit function is given by

$$P(q) = 1000q - 2q^2, \qquad 0 \le q \le 300$$

where q is the number of modems produced and sold and $P(q)$ is the profit in dollars.

(a) Use the definition of the derivative to determine $P'(q)$. Here you will use

$$P'(q) = \lim_{h \to 0} \frac{P(q + h) - P(q)}{h}$$

(b) Use the result from part (a) to compute $P'(200)$ and $P'(300)$, and interpret each using appropriate units.

68. The Fore Link Company has determined from data that it has collected that the profit function is given by

$$P(q) = 1000q - q^2, \qquad 0 \le q \le 800$$

where q is the number of golf clubs produced and sold and $P(q)$ is the profit in dollars.

(a) Use the definition of the derivative to determine $P'(q)$.

(b) Use the result from part (a) to compute $P'(400)$ and $P'(600)$, and interpret each using appropriate units.

69. The Fore Link Company has determined that the revenue, in dollars, from the sale of q sets of Junior Golf Clubs can be estimated by

$$R(q) = 200q - q^2, \qquad 0 \le q \le 200$$

where $R(q)$ is in dollars.

(a) Compute $R(50)$ and $R(150)$.

(b) Compute $R'(q)$ by using the definition of the derivative. Here you will use

$$R'(q) = \lim_{h \to 0} \frac{R(q + h) - R(q)}{h}$$

(c) Use the result from part (b) to determine $R'(50)$ and $R'(150)$, and interpret each using appropriate units.

70. The Network Standards Company has determined that the revenue, in dollars, from the sale of q modems can be estimated by

$$R(q) = 300q - q^2, \qquad 0 \le q \le 300$$

where $R(q)$ is in dollars.

(a) Compute $R(100)$ and $R(200)$.

(b) Compute $R'(q)$ by using the definition of the derivative.

(c) Use the result from part (b) to determine $R'(100)$ and $R'(200)$, and interpret each using appropriate units.

71. The FrezMore Company has determined that the cost of producing x refrigerators is modeled by

$$C(x) = x^2 + 15x + 1500, \qquad 0 \le x \le 200$$

where x is the number of refrigerators produced and $C(x)$ is the weekly cost in dollars.

(a) Use the definition of the derivative to determine $C'(x)$. Here you will use

$$C'(x) = \lim_{h \to 0} \frac{C(x + h) - C(x)}{h}$$

(b) Use the result from part (a) to compute $C'(40)$ and $C'(100)$, and interpret each using appropriate units.

72. The FrezMore Company has determined that the cost of producing x dorm room-sized refrigerators is modeled by

$$C(x) = \frac{1}{2}x^2 + 3x + 15, \qquad 0 \le x \le 300$$

where x is the number of refrigerators produced and $C(x)$ is the weekly cost in dollars.

(a) Use the definition of the derivative to determine $C'(x)$.

(b) Use the result from part (a) to compute $C'(40)$ and $C'(100)$, and interpret each using appropriate units.

73. The U.S. imports of all merchandise from all countries can be modeled by

$$f(x) = 2.39x^2 + 47.05x + 432.2, \qquad 1 \le x \le 6$$

where x is the number of years since 1990 and $f(x)$ is the dollar value in billions of all U.S. imports from all countries.

(a) Determine $f'(x)$.

(b) Use the result from part (a) to compute $f'(3)$, $f'(4)$, and $f'(5)$, and interpret each.

(c) Using your knowledge that the derivative gives the slope of a tangent line, does the result from part (b) indicate that the function is increasing or decreasing? Explain.

74. The U.S. exports of all merchandise to all countries can be modeled by

$$f(x) = 5.18x^2 + 5.86x + 410.6, \qquad 1 \le x \le 6$$

where x is the number of years since 1990 and $f(x)$ is the dollar value in billions of all U.S. exports to all countries.

(a) Determine $f'(x)$.

(b) Use the result from part (a) to compute $f'(3)$, $f'(4)$, and $f'(5)$, and interpret each.

(c) Using your knowledge that the derivative gives the slope of a tangent line, does the result from part (b) indicate that the function is increasing or decreasing? Explain.

75. The U.S. water consumption in gallons per day per capita can be modeled by

$$f(x) = -0.41x^2 + 13.53x + 330.69, \qquad 1 \le x \le 31$$

where x is the number of years since 1959 and $f(x)$ is the number of gallons of water consumption per day per capita.

(a) Compute $f'(x)$.

(b) Use the result from part (a) to compute $f'(6), f'(16),$ and $f'(26)$, and interpret each.

(c) Using your knowledge that the derivative gives the slope of a tangent line and the result from part (b), at what values of x was water consumption per day per capita increasing? Decreasing? What years do these x-values correspond to?

76. If $1000 is deposited in a bank account earning 5% interest compounded annually, then the amount in the account after t years is given by

$$A(t) = 1000(1.05)^t$$

a) Determine $A(10)$ and interpret.

b) To determine $A'(10)$ to the nearest cent, use the definition of the derivative,

$$A'(10) = \lim_{h \to 0} \frac{A(10 + h) - A(10)}{h}$$

and employ tables as outlined in the Interactive Activity on p. 143. Interpret $A'(10)$.

77. If $2000 is deposited in a bank account earning 6.5% interest compounded annually, then the amount in the account after t years is given by

$$A(t) = 2000(1.065)^t$$

(a) Determine $A(7)$ and interpret.

(b) To determine $A'(7)$ to the nearest cent, use the definition of the derivative,

$$A'(7) = \lim_{h \to 0} \frac{A(7 + h) - A(7)}{h}$$

and employ tables as outlined in the Interactive Activity on p. 143. Interpret $A'(7)$.

78. In marketing, the number of items sold can depend on the amount of money spent on advertising. Suppose that

$$N(x) = 2000 - \frac{520}{x}, \qquad [0.26, 100]$$

is a function where x is the amount spent in thousands of dollars to sell $N(x)$ items, in hundreds.

(a) Given that $N'(x) = \frac{520}{x^2}$, compute $N(20)$ and $N'(20)$ and interpret each.

(b) Compute $N(36)$ and $N'(36)$ and interpret each.

(c) As the amount of dollars, x, spent on advertising increases, what happens to $N'(x)$, the rate of change of items

sold per thousands of dollars spent? Is it increasing or decreasing?

(d) Graph N and N', and determine if your result in part (c) seems reasonable.

79. In marketing, the number of items sold can depend on the amount of money spent on advertising. Suppose that

$$N(x) = 1500 - \frac{630}{x}, \qquad [0.42, 100]$$

is a function where x is the amount spent in thousands of dollars to sell $N(x)$ items, in hundreds.

(a) Given that $N'(x) = \frac{630}{x^2}$, compute $N(15)$ and $N'(15)$ and interpret each.

(b) Compute $N(25)$ and $N'(25)$ and interpret each.

(c) As the amount of dollars, x, spent on advertising increases, what happens to $N'(x)$, the rate of change of items sold per thousands of dollars spent? Is it increasing or decreasing?

(d) Graph N and N', and determine if your result in part (c) seems reasonable.

80. The total inmate population in the Federal Bureau of Prisons can be modeled by

$$f(x) = \frac{1}{4}x^2 + \frac{9}{25}x + 24.5,^* \qquad 1 \le x \le 17$$

where x is the number of years since 1979 and $f(x)$ represents the total inmate population in thousands.

(a) Determine $f(11)$ and $f(16)$ and interpret each.

(b) Determine $f'(x)$.

(c) Compute $f'(11)$ and $f'(16)$ and interpret each.

81. Assume that the mathematical model for the growth of a locust tree in its first century of life is given by

$$h(t) = \sqrt{3t}, \qquad [0, 100]$$

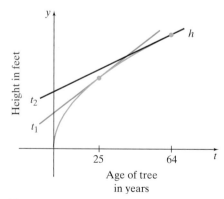

Figure 2.3.9

*Source: *Based on data gathered at the Federal Bureau of Prisons web site.*

where t is the age of the tree in years and $h(t)$ is the height of the tree in feet. Figure 2.3.9 is a graph of h along with two tangent lines labeled t_1 and t_2. t_1 is tangent to the graph of h at $t = 25$, and t_2 is tangent to the graph of h at $t = 64$. Which tangent line is associated with the faster growth rate? Explain in a complete sentence.

SECTION PROJECT

Use Figure 2.3.10 to answer the following:

(a) For what value of q is $P(q)$ a maximum?

(b) For what value of q does $P'(q) = 0$?

(c) What is the maximum profit?

(d) Using interval notation, determine for what values of q is $P(q)$ increasing.

(e) Using interval notation, determine for what values of q is $P'(q) > 0$.

(f) Using interval notation, determine for what values of q is $P(q)$ decreasing.

(g) Using interval notation, determine for what values of q is $P'(q) < 0$.

(h) In your own words, describe the connection between the results of parts (a) and (b).

(i) In your own words, describe the connection between the results of parts (d) and (e).

(j) In your own words, describe the connection between the results of parts (f) and (g).

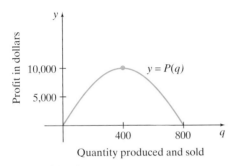

Figure 2.3.10

SECTION 2.4 DERIVATIVES OF CONSTANTS, POWERS, AND SUMS

In Section 2.3 we defined the *derivative* of a function and used it to find slopes of tangent lines as well as instantaneous rates of change. However, computing the derivative of a function from the definition can be quite involved. Calculus would not be very useful if all derivatives had to be calculated from the definition.

In this section and the next, we present several **rules of differentiation,** or shortcuts if you like, that will greatly simplify differentiation. So why did we painstakingly compute derivatives via the definition if these shortcut rules exist? Quite simply, these rules are *derived* from the definition! Also, we want to make sure that the *concept* of instantaneous rate of change/tangent line slope was understood for what it is . . . a limit.

Before we delve into differentiation rules, we need to present alternative ways that can be used to represent the derivative.

Alternative Notation for the Derivative

For $y = f(x),$ the following may be used to represent the derivative:

$$f'(x), \quad y', \quad \frac{dy}{dx}, \quad \frac{d}{dx}[f(x)]$$

Each notation has its own advantage, and we will use each of these where appropriate.

Derivative of a Constant Function

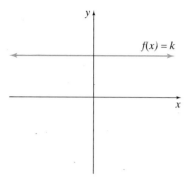

Figure 2.4.1

The first rule that we present shows how to differentiate a constant function. Geometrically, this rule is fairly obvious. The graph of a constant function, $f(x) = k$, is a horizontal line like the one shown in Figure 2.4.1. We know that the slope of a horizontal line is 0. This leads us to believe that the instantaneous rate of change at any point on the graph of a horizontal line is 0 or, in other words, for $f(x) = k$ in Figure 2.4.1 we believe that $f'(x) = 0$.

Rule 1: Constant Function Rule

For any constant k, if $f(x) = k$, then $f'(x) = 0$. Equivalently, we have $\dfrac{d}{dx}[k] = 0$, and if $y = k$, then $\dfrac{dy}{dx} = 0$.

In English, this rule states:

The derivative of a constant is 0.

To show that the Constant Function Rule is true, we simply apply the definition of the derivative as follows. If $f(x) = k$, $f(x + h) = k$ and $f(x) = k$. From the definition of the derivative, we have

$$f'(x) = \lim_{h \to 0} \frac{f(x+h) - f(x)}{h} = \lim_{h \to 0} \frac{k - k}{h} = \lim_{h \to 0} \frac{0}{h} = \lim_{h \to 0} 0 = 0$$

EXAMPLE 1 | **Differentiating Constant Functions**

Differentiate the following:

(a) $f(x) = 7$ (b) $g(x) = \sqrt[3]{2}$

SOLUTION

(a) Since 7 is a constant, we have $f'(x) = 0$.

(b) Since $\sqrt[3]{2}$ is a constant, we have $g'(x) = 0$.

Power Rule

We now present one of the most useful differentiation rules in calculus. It is used to differentiate functions of the form $f(x) = x^n$, where n is any real number. Functions of this form are collectively called **power functions.** The definition of the derivative can be used to determine derivatives for power functions, although the algebra may become cumbersome quickly. For example, the definition of the derivative can be used to derive the following:

$$\text{For } f(x) = x^3, \text{ we have } f'(x) = 3x^2.$$
$$\text{For } f(x) = x^6, \text{ we have } f'(x) = 6x^5.$$
$$\text{For } f(x) = x^7, \text{ we have } f'(x) = 7x^6.$$

There appears to be a pattern with the derivatives of power functions. It appears that the power of $f(x)$ becomes the coefficient of $f'(x)$, and the power of $f'(x)$

is 1 less than the power of $f(x)$. This observation leads us to the Power Rule. The proof of the Power Rule is in Appendix C.

> ### Rule 2: Power Rule
>
> If $f(x) = x^n$, where n is any real number, then $f'(x) = nx^{n-1}$. Equivalently,
>
> $\frac{d}{dx}[x^n] = nx^{n-1}$, and if $y = x^n$, then $\frac{dy}{dx} = nx^{n-1}$.

In English, this rule states:

> To differentiate x^n, simply bring the exponent out front as a coefficient and then decrease the exponent by 1.

To use the Power Rule, we must have the function in the form $f(x) = x^n$. To do this, sometimes we need some rules from algebra, as Example 2 illustrates.

EXAMPLE 2 | **Differentiating Power Functions**

Differentiate the following:

(a) $f(x) = x^4$ (b) $g(x) = x^{1.32}$ (c) $y = \sqrt{x}$ (d) $g(x) = \frac{1}{x^3}$

SOLUTION

(a) Thanks to the Power Rule, we quickly compute the derivative to be

$$f'(x) = 4x^{4-1} = 4x^3$$

(b) Again, using the Power Rule, we have

$$g'(x) = 1.32x^{1.32-1} = 1.32x^{0.32}$$

(c) Here we need to use a little algebra before differentiating. We need to recall that $\sqrt[n]{x} = x^{1/n}$. This allows us to rewrite $y = \sqrt{x}$ as $y = x^{1/2}$. Now we can apply the Power Rule to get

$$y' = \frac{1}{2}x^{1/2-1} = \frac{1}{2}x^{-1/2}$$

Utilizing the algebraic fact that $x^{-n} = \frac{1}{x^n}$ allows us to write the simplified version of the derivative as

$$y' = \frac{1}{2x^{1/2}} = \frac{1}{2\sqrt{x}}$$

(d) We begin by using a little algebra to rewrite the function. That is, by utilizing the algebraic fact that $x^{-n} = \frac{1}{x^n}$, the function can be rewritten as

$$g(x) = \frac{1}{x^3} = x^{-3}$$

Now we can apply the Power Rule and get

$$g'(x) = -3x^{-3-1} = -3x^{-4} = -\frac{3}{x^4}$$

✓ CHECKPOINT 1

Now work Exercise 5.

Constant Multiple Rule

The Constant Multiple Rule extends the Power Rule to differentiating functions that are of the form $k \cdot g(x)$, a constant times a function. The definition of the derivative can be used to derive the following:

For $f(x) = 2x^3$, we have $f'(x) = 6x^2$.

For $f(x) = 3x^5$, we have $f'(x) = 15x^4$.

For $f(x) = -2x^4$, we have $f'(x) = -8x^3$.

Again, there appears to be a pattern in these derivatives. It appears that the power of $f(x)$ gets multiplied by the coefficient of $f(x)$, and this product becomes the coefficient of $f'(x)$, while the power of $f'(x)$ is 1 less than the power of $f(x)$. This observation leads to the Constant Multiple Rule.

> ### Rule 3: Constant Multiple Rule
>
> If $f(x) = k \cdot g(x)$, where k is any real number, then $f'(x) = k \cdot g'(x)$, assuming that g is differentiable. Equivalently, $\dfrac{d}{dx}[k \cdot g(x)] = k \cdot \dfrac{d}{dx}[g(x)] = k \cdot g'(x)$.

In English, this rule states:

> The derivative of a constant times a function is simply the constant times the derivative of the function.

The proof of this rule is given in Appendix C.

EXAMPLE 3 | **Differentiating a Constant Times a Function**

Differentiate the following:

(a) $g(x) = 1.2x^5$ 　　　　(b) $y = \dfrac{1}{7x^3}$ 　　　　(c) $f(x) = \dfrac{2}{3}\sqrt[5]{x}$

SOLUTION

(a) Applying the Constant Multiple Rule yields

$$g'(x) = 1.2 \cdot (5x^{5-1}) = 6x^4$$

(b) Rewriting the function via algebra gives

$$y = \frac{1}{7}x^{-3}$$

Now we apply the Constant Multiple Rule to get

$$y' = \frac{1}{7} \cdot (-3x^{-3-1}) = \frac{1}{7}(-3x^{-4})$$

$$= -\frac{3}{7}x^{-4} = \frac{-3}{7x^4}$$

(c) We begin by rewriting the function, using some algebra, as

$$f(x) = \frac{2}{3}x^{1/5}$$

Now we apply the Constant Multiple Rule, which gives

$$f'(x) = \frac{2}{3}\left(\frac{1}{5}x^{1/5-1}\right) = \frac{2}{15}x^{-4/5}$$

Simplifying this derivatives yields

$$f'(x) = \frac{2}{15x^{4/5}} \quad \text{or} \quad \frac{2}{15\sqrt[5]{x^4}}$$

✓ **CHECKPOINT 2**

Now work Exercise 13.

Sum and Difference Rule

Consider the two functions $f(x) = 2x^2$ and $g(x) = 3x + 1$. We define two new functions, S and D, as follows:

$$S(x) = f(x) + g(x) = 2x^2 + 3x + 1$$
$$D(x) = f(x) - g(x) = 2x^2 - (3x + 1) = 2x^2 - 3x - 1$$

From differentiation rules developed so far in this section, we know that

$$f'(x) = 4x \quad \text{and} \quad g'(x) = 3$$

Evaluating $f'(x)$ and $g'(x)$ at $x = 1$ gives

$$f'(1) = 4 \quad \text{and} \quad g'(1) = 3$$

Recall that $f'(1)$ and $g'(1)$ give the slope of the graphs of f and g, respectively, at $x = 1$.

In Figure 2.4.2, we have graphed S and used the calculator's $\frac{dy}{dx}$ feature to numerically compute the derivative of $S(x)$ at $x = 1$. In Figure 2.4.3, we have graphed and numerically computed the derivative of $D(x)$ at $x = 1$. Notice that the slope of the graph of f at $x = 1$ is 4, and the slope of the graph of g at $x = 1$ is 3, while the slope of the graph of S at $x = 1$ is 7 and the slope of the graph of D at $x = 1$ is 1. This seems to indicate the following relationship:

$$S'(1) = f'(1) + g'(1)$$
$$D'(1) = f'(1) - g'(1)$$

This is not a coincidence! There is a rule that handles functions that are sums or differences of other functions, the Sum and Difference Rule.

TECHNOLOGY NOTE

Most calculators have a $\frac{dy}{dx}$ command. Check the online calculator manual at www.prenhall.com/armstrong

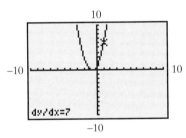

Figure 2.4.2
$\frac{dy}{dx} = 7$ at $x = 1$

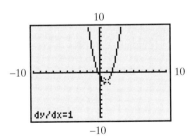

Figure 2.4.3
$\frac{dy}{dx} = 1$ at $x = 1$

Rule 4: Sum and Difference Rule

If $h(x) = f(x) \pm g(x)$, where f and g are both differentiable functions, then $h'(x) = f'(x) \pm g'(x)$. Equivalently,

$$\frac{d}{dx}[f(x) \pm g(x)] = \frac{d}{dx}[f(x)] \pm \frac{d}{dx}[g(x)] = f'(x) \pm g'(x).$$

In English this rule states:

To differentiate a sum/difference of two (or more) functions, just differentiate the functions separately and add/subtract the results.

A proof of the Sum and Difference Rule is given in Appendix C.

EXAMPLE 4 | **Differentiating Sums and Differences**

Differentiate the following functions:

(a) $f(x) = 3x^2 + 2x - 1$ (b) $g(x) = \frac{1}{2}x^3 - \frac{3}{2}x^{-1}$ (c) $y = 3x^4 + 2\sqrt{x} - \frac{2}{x^2}$

SOLUTION

(a) According to the Sum and Difference Rule,

$$f'(x) = \frac{d}{dx}[3x^2 + 2x - 1] = \frac{d}{dx}[3x^2] + \frac{d}{dx}[2x] - \frac{d}{dx}[1]$$

$$= 6x + 2 - 0 = 6x + 2$$

(b) Applying the Sum and Difference Rule yields

$$g'(x) = \frac{d}{dx}\left[\frac{1}{2}x^3 - \frac{3}{2}x^{-1}\right] = \frac{d}{dx}\left[\frac{1}{2}x^3\right] - \frac{d}{dx}\left[\frac{3}{2}x^{-1}\right]$$

$$= \frac{3}{2}x^2 - \left(-\frac{3}{2}x^{-2}\right) = \frac{3}{2}x^2 + \frac{3}{2}x^{-2}$$

(c) First, we rewrite the function as

$$y = 3x^4 + 2x^{1/2} - 2x^{-2}$$

Now we differentiate to get

$$\frac{dy}{dx} = \frac{d}{dx}[3x^4 + 2x^{1/2} - 2x^{-2}]$$

$$= \frac{d}{dx}[3x^4] + \frac{d}{dx}[2x^{1/2}] - \frac{d}{dx}[2x^{-2}]$$

$$= 12x^3 + x^{-1/2} + 4x^{-3}$$

Notice in Example 4 that we not only used the Sum and Difference Rule, but we also used the Constant Function Rule, the Power Rule, and the Constant Multiple Rule. In short, we used all the rules presented in this section.

✓ CHECKPOINT 3

Now work Exercise 23.

Interactive Activity

Algebraically simplify the derivatives in Example 4b and 4c so that there are no negative exponents. Then determine the domain of each derivative. What is the behavior of the original function at these excluded domain values?

EXAMPLE 5 | **Analyzing Greenhouse Gas Emissions**

Table 2.4.1 displays data for the amount of chlorofluorocarbons (CFCs), in thousands of metric tons, released into the atmosphere by the United States from 1989 to 1995.

TABLE **2.4.1**

YEAR	1989	1990	1991	1992	1993	1994	1995
CFC GASES	272	231	210	187	166	133	87

Source: U.S. Census Bureau web site.

If we let x represent the number of years since 1988 and $f(x)$ represent the CFCs in thousands of metric tons released into the atmosphere, using the regression capabilities of your calculator, the data can be modeled by

$$f(x) = -1.2x^3 + 13.37x^2 - 69.65x + 328.71, \qquad 1 \le x \le 7$$

Figure 2.4.4 gives a graph of the data along with the cubic model. Determine $f'(x)$. Compute $f'(2)$ and $f'(6)$ and interpret each. Use appropriate units in the answer.

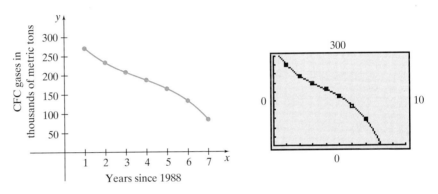

Figure 2.4.4 Graph of $f(x) = -1.2x^3 + 13.37x^2 - 69.65x + 328.71$ along with data points.

SOLUTION

Using the techniques learned in this section, we get

$$f'(x) = -1.2(3x^2) + 13.37(2x) - 69.65(1) + 0$$
$$f'(x) = -3.6x^2 + 26.74x - 69.65$$

To compute $f'(2)$, we simply substitute $x = 2$ into the derivative for x:

$$f'(2) = -3.6(2)^2 + 26.74(2) - 69.65 = -30.57 \frac{\text{thousand metric tons}}{\text{year}}$$

This means that in 1990 U.S. emission of CFCs into the atmosphere was decreasing at a rate of 30.57 $\frac{\text{thousand metric tons}}{\text{year}}$. Similarly, to determine $f'(6)$, we substitute $x = 6$ into the derivative.

$$f'(6) = -3.6(6)^2 + 26.74(6) - 69.65 = -38.81 \frac{\text{thousand metric tons}}{\text{year}}$$

Thus, in 1994 U.S. emission of CFCs into the atmosphere was decreaasing at a rate of 38.81 $\frac{\text{thousand metric tons}}{\text{year}}$. See Figure 2.4.5 and Figure 2.4.6.

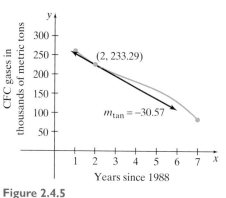

Figure 2.4.5

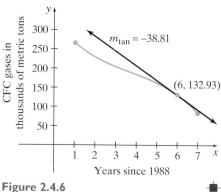

Figure 2.4.6

✓ **CHECKPOINT 4**

Now work Exercise 69.

The next example deals with what is known as a *position function* and a *velocity function*. These functions are commonly used in physics.

EXAMPLE 6 | **Analyzing Velocity**

If a coconut falls from a tree that is 75 feet tall, its height above the ground after t seconds is given by

$$s(t) = 75 - 16t^2$$

where $s(t)$ is measured in feet. This function is called a **position function** since it gives the position of the coconut above the ground as a function of time.

(a) The derivative of a position function, $s'(t)$, is called the **velocity function** and is denoted by $v(t)$. Determine $v(t)$.

(b) Compute $s(2)$ and $v(2)$ and interpret each.

(c) When does the coconut hit the ground?

SOLUTION

(a) Since we have been told that $s'(t) = v(t)$, we determine $v(t)$ as follows:

$$v(t) = s'(t) = -32t$$

(b) We have

$$s(2) = 75 - 16(2)^2 = 11 \text{ feet}$$

This means that after 2 seconds the coconut is 11 feet above the ground. We also have

$$v(2) = -32(2) = -64 \frac{\text{feet}}{\text{sec}}$$

This means that after 2 seconds the velocity of the falling coconut is $-64 \frac{\text{feet}}{\text{sec}}$. The negative sign indicates that the coconut is falling toward the ground.

(c) The coconut hits the ground when it is 0 feet above the ground. Hence, we need to solve $s(t) = 0$. Setting $s(t) = 0$ and solving yields

$$75 - 16t^2 = 0$$
$$75 = 16t^2$$
$$\frac{75}{16} = t^2$$
$$\pm\sqrt{\frac{75}{16}} = t$$
$$t \approx \pm 2.17 \text{ seconds}$$

Since -2.17 seconds does not make sense (we cannot yet travel back in time), we conclude that the coconut hits the ground in about 2.17 seconds. Figure 2.4.7 shows a graphical solution for this problem.

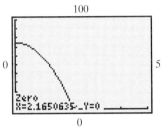

Figure 2.4.7   Result of ZERO command.

| EXAMPLE 7 | **Analyzing a Profit Function** |

The Cool Air company has determined that the total cost function for the production of its refrigerators can be described by

$$C(x) = 2x^2 + 15x + 1500$$

where x is the weekly production of refrigerators and $C(x)$ is the total cost in dollars. The revenue function for these refrigerators is given by

$$R(x) = -0.3x^2 + 460x$$

where x is the number of refrigerators sold and $R(x)$ is in dollars.

(a) Determine P, the profit function, and determine $P'(x)$.

(b) Evaluate $P(20)$ and $P'(20)$ and interpret each.

SOLUTION

(a) In Chapter 1 we learned that $P(x) = R(x) - C(x)$. Thus,

$$P(x) = (-0.3x^2 + 460x) - (2x^2 + 15x + 1500)$$
$$= -2.3x^2 + 445x - 1500$$

Since $P(x) = -2.3x^2 + 445x - 1500$, $P'(x)$ is

$$P'(x) = -2.3(2x) + 445(1) - 0 = -4.6x + 445$$

(b) We have $P(20) = -2.3(20)^2 + 445(20) - 1500 = 6480$, or $6480. This means that when the Cool Air company makes and sells 20 refrigerators, a profit of $6480 is realized. On the other hand, $P'(20) = -4.6(20) + 445 = 353$, or $353. This means that the instantaneous rate of change is $353 per refrigerator when the production level is 20 refrigerators (Figure 2.4.8). In other words, at a weekly production of 20 refrigerators, the profit is increasing at a rate of $353 per refrigerator. Alternatively, we could say that it means the profit increases approximately $353 to make and sell the 21st refrigerator.

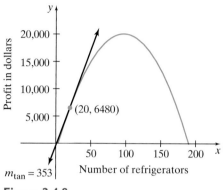

Figure 2.4.8

The derivative of a profit function is called **marginal profit,** which will be discussed more thoroughly in Chapter 3.

SUMMARY

In this section we introduced four differentiation rules to allow us to compute derivatives more quickly:

NAME	RULE	EXAMPLE
Constant Function Rule	If $f(x) = k$, then $f'(x) = 0$.	If $f(x) = 7$, then $f'(x) = 0$.
Power Rule	If $f(x) = x^n$, then $f'(x) = nx^{n-1}$.	If $f(x) = x^8$, then $f'(x) = 8x^7$.
Constant Multiple Rule	If $f(x) = k \cdot g(x)$, then $f'(x) = k \cdot g'(x)$.	If $f(x) = 3x^4$, then $f'(x) = 12x^3$.
Sum and Difference Rule	If $h(x) = f(x) \pm g(x)$, then $h'(x) = f'(x) \pm g'(x)$.	If $h(x) = x^3 + x^2$, then $h'(x) = 3x^2 + 2x$.

Even though we now have some tools to allow us to calculate a derivative more quickly than using the definition of the derivative, do not lose sight of what the derivative tells us: It is a limit that gives an instantaneous rate of change as well as the slope of a tangent line.

SECTION 2.4 EXERCISES

In Exercises 1–16, determine the derivative for the given function. Where appropriate, simplify the derivative so that there are no negative or fractional exponents. Some algebra rules to keep in mind are $x^{-n} = \dfrac{1}{x^n}$ and $\sqrt[n]{x^m} = x^{m/n}$.

1. $f(x) = 3$

2. $g(x) = 5$

3. $f(t) = -2$

4. $g(t) = -7$

✓ 5. $f(x) = x^6$

6. $f(x) = x^{10}$

7. $y = -3x^4$

8. $y = -5x^7$

9. $g(x) = 2x^{2/3}$

10. $g(x) = 3x^{4/5}$

11. $f(x) = -3x^{-1/3}$

12. $f(x) = -2x^{-2/5}$

✓ 13. $h(x) = \frac{2}{3}x^4$

14. $h(x) = \frac{3}{4}x^8$

15. $y = -\frac{2}{5}x^{5/3}$

16. $y = -\frac{5}{6}x^{6/5}$

In Exercises 17–48, determine the derivative for the given function. Where appropriate, simplify the derivative so that there are no negative or fractional exponents.

17. $f(x) = 2x^3 + 4x^2 - 7x + 1$

18. $f(x) = 4x^3 - 3x^2 + 5x - 3$

19. $g(x) = 3x^2 - 2x + 6$

20. $g(x) = 3x^2 + 5x - 2$

21. $y = -5x^2 - 6x + 2$

22. $y = -3x^2 + 9x - 1$

✓ 23. $y = -5x^3 + 7x - 5$

24. $y = -8x^3 - 8x + 3$

25. $f(x) = \frac{1}{2}x^3 + \frac{3}{5}x^2 - \frac{2}{3}x + \frac{2}{5}$

26. $f(x) = \frac{2}{5}x^3 - \frac{2}{3}x^2 + \frac{1}{2}x - 5$

27. $g(x) = 1.31x^2 + 2.05x - 3.9$

28. $g(x) = 3.15x^2 - 1.13x + 5.2$

29. $d(x) = -0.2x^2 + 3.5x^3 - 0.4x^4$

30. $d(x) = 0.3x^2 - 0.67x^3 + 0.8x^4$

31. $y = 1.15x^3 - 2.3x^2 + 2.53x - 7.1$

32. $y = 2.35x^3 + 3.56x^2 - 63.25x + 365.3$

33. $f(x) = 3\sqrt{x} + \frac{1}{2}x - 5x^2$

34. $f(x) = 5\sqrt{x} - \frac{1}{2}x + 7x^2$

35. $y = \sqrt[3]{x} + x^2 - 3x^3$

36. $y = \sqrt[3]{x} - 2x^3 + 5x^4$

37. $g(x) = \sqrt[3]{x^2} - \frac{4}{\sqrt{x}}$

38. $g(x) = \sqrt[3]{x^2} + \frac{3}{\sqrt{x}}$

39. $f(x) = 2.35x^{1.35}$

40. $f(x) = 3.2x^{1.14}$

41. $AC(x) = 2000 + \frac{5}{x^2}$

42. $AC(x) = 3300 + \frac{13}{x^2}$

43. $y = 2x^2 + \frac{1}{x}$

44. $y = -3x^2 - \frac{1}{x}$

45. $f(x) = 3x^{-3/2} - 4x^{1/2} + 5$

46. $f(x) = 4x^{-3/2} - 2x^{1/2} + x^{-1/2}$

47. $y = 2.35x^{-1/2} - 2.3x^{-2/3}$

48. $y = 3.52x^{-2/5} + 3.2x^{-1/2}$

For Exercises 49–56:

(a) Determine the domain of the given function. Write in interval notation.

(b) Use the addition rule for fractions, $\frac{a+b}{c} = \frac{a}{c} + \frac{b}{c}$, to rewrite the function and simplify. For example,

$$f(x) = \frac{x^3 + 3x^2 - x + 2}{x} = \frac{x^3}{x} + \frac{3x^2}{x} - \frac{x}{x} + \frac{2}{x}$$

$$= x^2 + 3x - 1 + 2x^{-1}$$

(c) Compute the derivative using differentiation rules.

(d) Determine the domain of the derivative. Write in interval notation.

49. $f(x) = \frac{3x^3 - 9x^2 + 4}{12}$

50. $f(x) = \frac{x^2 + 4x + 3}{2}$

51. $f(x) = \frac{2x^3 + 3x^2 - x + 3}{x}$

52. $f(x) = \frac{-3x^3 - 4x^2 + 2x - 7}{x}$

53. $y = \frac{7x^4 - 50x^2 + x}{x^2}$

54. $y = \frac{-5x^4 + 25x^2 - x}{x^2}$

55. $h(x) = \frac{4x^3 - 14x^2 + 3}{2\sqrt{x}}$

56. $g(x) = \frac{x^3 + 2x^2}{\sqrt[3]{x^2}}$

For Exercises 57–62:

(a) Determine the derivative for the given function using differentiation rules.

(b) Determine the slope of the line tangent to the graph of the function at the indicated x-value.

(c) Determine an equation for the line tangent to the graph of the function at the indicated x-value.

(d) Graph the function and the tangent line in the same viewing window.

57. $f(x) = x^3$ at $x = -1$

58. $f(x) = \sqrt{x}$ at $x = 4$

59. $y = \frac{1}{x}$ at $x = 3$

60. $y = \frac{1}{x^2}$ at $x = 3$

61. $f(x) = x^{2/3}$ at $x = 8$

62. $g(x) = x^{-1/3}$ at $x = 1$

Applications

63. The number of dollars spent on advertising for a new CD influences the number of CDs consumers purchase. If x is the number of dollars spent, in thousands, then $N(x)$, measured in hundreds, is the number of CDs sold, as given by

$$N(x) = 2500 - \frac{520}{x}, \qquad x \geq 0.208$$

(a) Compute $N'(x)$.

(b) Compute $N(10)$ and $N'(10)$ and interpret each.

64. The number of dollars spent on advertising for a new surround-sound stereo receiver influences the number of these stereo receivers that consumers purchase. If x is the number of dollars spent, in thousands, then $N(x)$, measured in hundreds, is the number of stereo receivers sold, as given by

$$N(x) = 3600 - \frac{700}{x}, \qquad x \geq 0.195$$

(a) Compute $N'(x)$.

(b) Compute $N(5)$ and $N'(5)$ and interpret each.

Exercises 65 and 66 use the following: Recall that if the total cost of making x units of a product is given by $C(x)$ then the average total cost, $AC(x)$, is simply given by

$$AC(x) = \frac{C(x)}{x}$$

Suppose that from past experience a manufacturer of coats knows that the total cost in dollars of making x coats per week is given by

$$C(x) = 3000 + 11x - 7\sqrt{x} + 0.03x^{3/2}$$

65. Determine $C(300)$ and $C'(300)$ and interpret each.

66. (a) Determine $AC(x)$.

(b) Determine $AC(300)$ and $AC'(300)$ and interpret each.

67. An electric generating plant burns coal, and data collected from air samples downwind from the plant indicate that the amount of sulfur dioxide, $SD(x)$, x miles downwind is given by

$$SD(x) = \frac{93.21}{x^2}, \qquad x > 0$$

where $SD(x)$ is in parts per million (ppm). Compute $SD(1)$ and $SD'(1)$ and interpret each.

68. An electric generating plant burns coal, and data collected from air samples downwind from the plant indicate that the amount of sulfur dioxide, $SD(x)$, x miles downwind is given by

$$SD(x) = \frac{78.35}{x^2}, \qquad x > 0$$

where $SD(x)$ is in parts per million (ppm). Compute $SD(2)$ and $SD'(2)$ and interpret each.

69. The amount of sulfur dioxide emitted into the atmosphere by the United States from 1985 to 1995 can be modeled by

$$SD(t) = -0.02t^3 + 0.27t^2 - 1.22t + 24.12, \qquad 1 \leq t \leq 11$$

where t is the number of years since 1984 and $SD(t)$ is the amount of sulfur dioxide, in millions of tons, released into the atmosphere by the United States.

a) Compute $SD(3)$ and $SD(7)$ and interpret each.

b) Compute $SD'(3)$ and $SD'(7)$ and interpret each. Use appropriate units.

70. The size of a certain bacteria culture at time t, in minutes, is approximated by $N(t) = 4t^{7/2}$, where $N(t)$ is in milligrams. Compute $N'(9)$ and interpret using appropriate units.

71. The size of a certain bacteria culture at time t, in minutes, is approximated by $N(t) = 6t^{5/2}$, where $N(t)$ is in milligrams.

(a) Compute the average rate of change from $t = 1$ to $t = 4$. Interpret using appropriate units.

(b) Compute $N'(4)$ and interpret using appropriate units.

72. If a watermelon is dropped from a building 200 feet tall, its height above the ground after t seconds is given by $s(t) = 200 - 16t^2$, where $s(t)$ is measured in feet.

(a) Determine an expression for the velocity function, $v(t)$. (Recall that $v(t) = s'(t)$.)

(b) Determine $s(1)$ and $s(3)$ and interpret each using appropriate units.

(c) Determine the average velocity from $t = 1$ to $t = 3$.

(d) Determine $v(1)$ and $v(3)$ and interpret each.

(e) When does the watermelon hit the ground? Round your answer to the nearest tenth of a second.

73. If an egg is dropped from a building 150 feet tall, its height above the ground after t seconds is given by $s(t) = 150 - 16t^2$, where $s(t)$ is measured in feet.

(a) Determine an expression for the velocity function, $v(t)$. (Recall that $v(t) = s'(t)$.)

(b) Determine $s(1)$ and $s(3)$ and interpret each using appropriate units.

(c) Determine the average velocity from $t = 1$ to $t = 3$.

(d) Determine $v(1)$ and $v(3)$ and interpret each.

(e) When does the egg hit the ground? Round your answer to the nearest tenth of a second.

TABLE 2.4.2

YEAR	1980	1985	1989	1990	1991	1992	1993	1994	1995
TUBERCULOSIS	27.7	22.2	23.5	25.7	26.3	26.7	25.3	24.4	22.9

Source: U.S. Census Bureau web site.

Exercises 74 and 75 use the data in Table 2.4.2, which gives the number of cases of tuberculosis (in thousands) reported from 1980 to 1995. If we let x represent the number of years since 1979, then the number of cases of tuberculosis reported can be modeled by

$$f(x) = -0.02x^3 + 0.57x^2 - 4.25x + 31.47, \qquad 1 \le x \le 16$$

where $f(x)$ is the number of cases in thousands.

74. (a) Using the model, compute the average rate of change of cases of tuberculosis reported from 1985 to 1992. Interpret the result.

(b) Using the model, compute the average rate of change of cases of tuberculosis reported from 1992 to 1995. Interpret the result.

75. (a) Determine $f'(x)$.

(b) Compute $f'(12)$ and $f'(15)$ and interpret each. Make sure to use appropriate units.

76. Consider any linear function of the form $f(x) = mx + b$, where m and b are real numbers. Compute $f'(x)$ and interpret.

77. The number of calories consumed per day per capita for the U.S. civilian population can be modeled by

$$C(t) = 1.65t^2 - 9.14t + 3306.88, \qquad 1 \le t \le 20$$

where t is the number of years since 1974 and $C(t)$ is the number of calories consumed per day per capita.

(a) Compute $C(11)$ and $C(16)$ and interpret each.

(b) Compute $C'(11)$ and $C'(16)$ and interpret each using appropriate units.

Exercises 78 and 79 utilize the following: Double D Cola ran a 10-week ad campaign for its new caffeine-free cola. The total weekly sales, $S(t)$, where $S(t)$ is in thousands of dollars, t weeks after the ad campaign began can be modeled by

$$S(t) = -0.51t^2 + 20.20t + 49.49, \qquad 1 \le t \le 23$$

78. (a) Compute $S(10)$ and interpret.

(b) Compute $S'(10)$ and interpret using appropriate units.

(c) Using the results from parts (a) and (b), complete the following:

Ten weeks after the ad campaign began, weekly sales of the new cola were_____. Weekly sales were _____ (increasing/decreasing) at a rate of _____.

79. (a) Compute $S(23)$ and interpret.

(b) Compute $S'(23)$ and interpret using appropriate units.

(c) Using the results from parts (a) and (b), complete the following:

Twenty-three weeks after the ad campaign began, weekly sales of the new cola were_____. Weekly sales were_____ (increasing/decreasing) at a rate of _____.

80. Psychologists have found that, when learning a new task, people learn quickly in the beginning and then the rate of learning slows. Eventually, an individual reaches a level that cannot be exceeded. For example, to test learning, a psychologist asks people to memorize a long sequence of digits and checks with each person every few minutes to see how many digits have been memorized. Assume that the learning can be described by the equation $y = 22(1 - e^{-0.2x})$, where y is the number of digits memorized and x is the time in minutes. A graph of this equation is called a **learning curve** and will be studied in Chapter 4.

(a) Determine $\lim\limits_{x \to \infty} [22(1 - e^{-0.2x})]$ and interpret.

(b) Graph $y = 22(1 - e^{-0.2x})$, use your calculator's $\dfrac{dy}{dx}$ or DRAW TANGENT command to determine $\dfrac{dy}{dx}$ at $x = 3$ and $x = 10$, and interpret each.

(c) Does the result from part (b) indicate that the rate of learning slows down? Explain.

81. Membership in the Girl Scouts of America can be modeled by

$$f(t) = 9.25t^3 - 123.81t^2 + 484.8t + 2882.57, \qquad 1 \le t \le 7$$

where t is the number of years since 1989 and $f(t)$ is the membership in thousands.

(a) Determine $f(4)$ and interpret.

(b) Determine $f'(t)$.

(c) Determine $f'(4)$ and interpret.

82. Membership in the Boy Scouts of America can be modeled by

$$g(t) = 21.82t^2 - 143.75t + 5555.57, \qquad 1 \le t \le 7$$

TABLE 2.4.3

YEAR	1980	1985	1987	1988	1989	1990	1991	1992	1993	1994
EXPENDITURES	42.4	68.4	76.3	84.1	96.3	115.1	136.2	159.7	181.1	190.6

Source: U.S. Census Bureau web site.

where t is the number of years since 1989 and $g(t)$ is the membership in thousands.

(a) Determine $g(4)$ and interpret.

(b) Determine $g'(t)$.

(c) Determine $g'(4)$. and interpret.

83. Comparing the results from Exercises 81c and 82c, whose membership was increasing at a faster rate at the beginning of 1993? Explain.

SECTION PROJECT

In Table 2.4.3, expenditures refers to federal government expenditures, in billions of dollars, for health services and supplies.

(a) Let x represent the number of years since 1979, and let y represent the amount, in billions, spent on health services and supplies by the federal government. Enter the data into your calculator, and perform a cubic regression on the data to model the data on the interval [1, 15]. Plot the data and graph the model in the same viewing window.

(b) Use the model to compute the average rate of change of federal government expenditures for health services and supplies from 1983 to 1990.

(c) Compute the derivative of the model.

(d) Evaluate the derivative at $x = 4$, and interpret using appropriate units.

(e) Evaluate the derivative at $x = 11$, and interpret using appropriate units.

(f) To the nearest tenth, graphically determine a value for x in the interval [4, 11], where the derivative at this value is equivalent to the average rate of change found in part (b). In other words, find a value for x in [4, 11], where the slope of the tangent line is equal to the slope of the secant line that passes through the points determined by $x = 4$ and $x = 11$. (*Hint:* Two lines have the same slope if they are parallel.)

SECTION 2.5   DERIVATIVES OF PRODUCTS AND QUOTIENTS

In Section 2.4 we learned four useful rules for computing derivatives. In this section, we illustrate how to differentiate the **product** and **quotient** of two functions. These two rules are not as simple as the ones in Section 2.4.

Product Rule

To differentiate the product of two functions, we use the Product Rule.

> ### Rule 5: Product Rule
>
> If $h(x) = f(x) \cdot g(x)$, where f and g are differentiable functions, then the derivative is $h'(x) = f'(x) \cdot g(x) + f(x) \cdot g'(x)$.

In English, the Product Rule states:

Take the derivative of the first function times the second function plus the first function times the derivative of the second function.

A proof of the Product Rule is given in Appendix C.

EXAMPLE 1 | **Differentiating Products**

Differentiate $h(x) = 3x^3(x^4 + 2)$ by using the Product Rule.

Interactive Activity

 The derivative of h in Example 1 could have been computed if we had first multiplied out the function and then applied the rules learned in Section 2.4. Do this to verify our result in Example 1.

SOLUTION

Let $f(x) = 3x^3$ and $g(x) = x^4 + 2$. By the Product Rule, we compute the derivative to be

$$h'(x) = f'(x) \cdot g(x) + f(x) \cdot g'(x)$$
$$= 9x^2(x^4 + 2) + 3x^3(4x^3)$$
$$= 9x^6 + 18x^2 + 12x^6$$
$$= 21x^6 + 18x^2$$
$$= 3x^2(7x^4 + 6)$$

EXAMPLE 2 | **Differentiating Products**

Differentiate the following by using the Product Rule.

(a) $f(x) = (2x^2 + 4x + 5)(5x - 4)$ (b) $y = \sqrt{x}(3x^3 - 4x^2 + 8x)$

SOLUTION

(a) Applying the Product Rule yields

$$f'(x) = (4x + 4)(5x - 4) + (2x^2 + 4x + 5)(5)$$
$$= (20x^2 - 16x + 20x - 16) + (10x^2 + 20x + 25)$$
$$= 30x^2 + 24x + 9$$

(b) Before differentiating, we rewrite the function as

$$y = x^{1/2}(3x^3 - 4x^2 + 8x)$$

Now, using the Product Rule, we get

$$y' = \left(\frac{1}{2}x^{-1/2}\right)(3x^3 - 4x^2 + 8x) + x^{1/2}(9x^2 - 8x + 8)$$

Writing the derivative without negative or fractional exponents yields

$$y' = \frac{3x^3 - 4x^2 + 8x}{2\sqrt{x}} + \sqrt{x} \cdot (9x^2 - 8x + 8)$$

✓ CHECKPOINT **1**

Now work Exercise 13.

EXAMPLE 3

Differentiating Products

Differentiate $f(x) = (6x^{4/3} + 2x)(3x^{5/3} + 4x - 1)$.

SOLUTION

By the Product Rule, we immediately compute the derivative to be

$$f'(x) = (8x^{1/3} + 2)(3x^{5/3} + 4x - 1) + (6x^{4/3} + 2x)(5x^{2/3} + 4)$$

EXAMPLE 4

Determining an Equation for a Tangent Line

Determine an equation for the line tangent to $y = (x^4 - x^2)(x^3 - x + 2)$ at $x = 1$.

SOLUTION

Since the derivative gives the slope of a tangent line at any point, our first task is to determine the derivative of the function by using the Product Rule. This gives

$$\frac{dy}{dx} = (4x^3 - 2x)(x^3 - x + 2) + (x^4 - x^2)(3x^2 - 1)$$

We then evaluate $\frac{dy}{dx}$ at $x = 1$ to get the slope of the tangent line. This yields, at $x = 1$,

$$\frac{dy}{dx} = (4 \cdot 1^3 - 2 \cdot 1)(1^3 - 1 + 2) + (1^4 - 1^2)(3 \cdot 1^2 - 1)$$

$$= 4$$

So, at $x = 1$, $m_{\text{tan}} = 4$. Also, when $x = 1$, we determine the point of tangency by substituting 1 for x in the original function. This gives

$$y = (1^4 - 1^2)(1^3 - 1 + 2) = 0$$

So our point of tangency is $(1, 0)$. Since we know a point on the tangent line and we know that the slope of the tangent line is 4, we use the point–slope form of a line to determine an equation for the tangent line. This results in

$$y - y_1 = m(x - x_1)$$
$$y - 0 = 4(x - 1)$$
$$y = 4x - 4$$

TECHNOLOGY NOTE

In Figure 2.5.1, the DRAW TANGENT command does not give the same equation for the tangent line. This is a limitation of the calculator. Even though it is very close, we needed calculus for the **exact** answer.

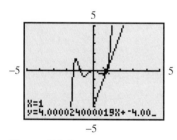

Figure 2.5.1

Figure 2.5.1 shows a graph of the function along with the tangent line.

✓ CHECKPOINT **2**

Now work Exercise 39.

EXAMPLE 5

Applying the Product Rule

Extensive market research has determined that for the next five years the price of a certain mountain bike is predicted to vary according to $p(t) = 300 - 30t + 7.5t^2$, where t is time in years and $p(t)$ is the price in dollars. The number of mountain bikes sold annually by Skinner's Bikes is expected to follow $q(t) = 3000 + 90t - 15t^2$, where $q(t)$ is the number sold and t is time in years.

(a) Determine $R(t)$ and $R'(t)$.

(b) Compute $R'(1)$ and interpret.

(c) Compute $R'(4)$ and interpret.

SOLUTION

(a) Since revenue equals price times quantity, we have

$$R(t) = p(t) \cdot q(t) = (300 - 30t + 7.5t^2) \cdot (3000 + 90t - 15t^2)$$

By the Product Rule, we compute the derivative to be

$$R'(t) = \underbrace{(-30 + 15t)}_{p'(t)}\underbrace{(3000 + 90t - 15t^2)}_{q(t)} + \underbrace{(300 - 30t + 7.5t^2)}_{p(t)}\underbrace{(90 - 30t)}_{q'(t)}$$

We will not simplify $R'(t)$ so that we may see separately the effects on $R(t)$ caused by changing prices and sales. Notice in the derivative that the first term is $p'(t) \cdot q(t)$. This term describes the rate at which $R(t)$ is changing as the price changes. The second term in the derivative is $p(t) \cdot q'(t)$. This term describes the rate at which $R(t)$ is changing as the number of bikes sold changes.

(b) We compute $R'(1)$ to be

$$R'(1) = (-30 + 15)(3000 + 90 - 15) + (300 - 30 + 7.5)(90 - 30)$$
$$= (-15)(3075) + (277.5)(60)$$
$$= -46,125 + 16,650 = -29,475$$

The negative sign indicates that after one year the revenue is decreasing at a rate of $29,475 per year. Notice that the effect of falling prices represented by the first term, $p'(1) \cdot q(1) = -46,125,$ is greater than the effect of rising sales given by the second term, $p(1) \cdot q'(1) = 16,650.$

(c) We have

$$R'(4) = (-30 + 60)(3000 + 360 - 240) + (300 - 120 + 120)(90 - 120)$$
$$= (30)(3120) + (300)(-30)$$
$$= 93,600 - 9000 = 84,600$$

Since this is positive, we claim that after four years the revenue is increasing at a rate of $84,600 per year. Notice that the effect of rising prices represented by the first term, $p'(4) \cdot q(4) = 93,600,$ is greater than the effect of falling sales given by the second term, $p(4) \cdot q'(4) = -9000.$ Figure 2.5.2 shows a graph of R.

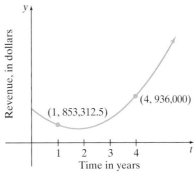

Figure 2.5.2 Graph of
$R(t) = (300 - 30t + 7.5t^2)(3000 + 90t - 15t^2).$

Quotient Rule

The final differentiation rule of this section, the Quotient Rule, shows how to differentiate the quotient of two functions.

Rule 6: Quotient Rule

If $h(x) = \dfrac{f(x)}{g(x)}$, where f and g are differentiable functions, then

$$h'(x) = \frac{f'(x) \cdot g(x) - f(x) \cdot g'(x)}{[g(x)]^2}, \text{ where } g(x) \neq 0.$$

In English, the Quotient Rule states:

Take the derivative of the numerator times the denominator minus the numerator times the derivative of the denominator, all of this over the denominator squared.

A proof of the Quotient Rule is given in Appendix C.

EXAMPLE 6 | **Differentiating Quotients**

Differentiate the following using the Quotient Rule.

(a) $h(x) = \dfrac{x + 3}{x - 2}$ (b) $y = \dfrac{x^4 - 3x}{x^2 + 1}$

SOLUTION

(a) Using the Quotient Rule, we have

$$h'(x) = \frac{f'(x)g(x) - f(x)g'(x)}{[g(x)]^2}$$

$$= \frac{1 \cdot (x - 2) - (x + 3) \cdot 1}{(x - 2)^2}$$

$$= \frac{x - 2 - x - 3}{(x - 2)^2} = \frac{-5}{(x - 2)^2}$$

(b) Again, by the Quotient Rule, the derivative is

$$y' = \frac{f'(x)g(x) - f(x)g'(x)}{[g(x)]^2}$$

$$= \frac{(4x^3 - 3)(x^2 + 1) - (x^4 - 3x)(2x)}{(x^2 + 1)^2}$$

$$= \frac{(4x^5 + 4x^3 - 3x^2 - 3) - 2x^5 + 6x^2}{(x^2 + 1)^2} = \frac{2x^5 + 4x^3 + 3x^2 - 3}{(x^2 + 1)^2}$$

✓ CHECKPOINT 3

Now work Exercise 27.

The next example requires us to use both the Product Rule and the Quotient Rule.

EXAMPLE 7 | **Differentiating Using the Product Rule and Quotient Rule**

Differentiate $y = \frac{(2x+1)(3x-2)}{x+1}$.

SOLUTION

Notice that when applying the Quotient Rule we must use the Product Rule when differentiating the numerator. To aid us in this computation, let $f(x) = 2x + 1$ and $g(x) = 3x - 2$. We then have

$$f'(x) = 2 \quad \text{and} \quad g'(x) = 3$$

Now, when applying the Quotient Rule, the derivative of the numerator is

$$2 \cdot (3x - 2) + (2x + 1) \cdot 3 = 6x - 4 + 6x + 3 = 12x - 1$$

Putting this all together using the Quotient Rule gives

$$
\begin{aligned}
\frac{dy}{dx} &= \frac{(12x - 1)(x + 1) - (2x + 1)(3x - 2) \cdot 1}{(x + 1)^2} \\
&= \frac{12x^2 + 11x - 1 - (6x^2 - x - 2)}{(x + 1)^2} \\
&= \frac{12x^2 + 11x - 1 - 6x^2 + x + 2}{(x + 1)^2} \\
&= \frac{6x^2 + 12x + 1}{(x + 1)^2}
\end{aligned}
$$

✓ CHECKPOINT 4

Now work Exercise 31.

EXAMPLE 8 | **Applying the Quotient Rule**

Researchers have determined through experimentation that the percent concentration of a certain medication can be approximated by

$$p(t) = \frac{200t}{2t^2 + 5} - 4, \qquad [0.25, 20]$$

where t is the time in hours after administering the medication and $p(t)$ is the percent concentration.

(a) Graph p and approximate its range over the given interval.

(b) Evaluate $p'(1)$ and $p'(6)$ and interpret each.

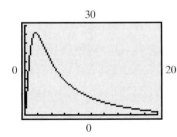

Figure 2.5.3

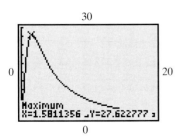

Figure 2.5.4

SOLUTION

(a) Since $p(t)$ represents percent concentration of a medication, it should be obvious that $p(t)$ cannot be negative. In fact, from the graph in Figure 2.5.3 we can see that $p(t) > 0$ on the interval [0.25, 20]. We can determine the maximum value of $p(t)$ on the interval [0.25, 20] in several ways. (In Chapter 5 we will see how calculus allows us to determine *exactly* where it is.) For now, we rely on the calculator and the MAXIMUM command.

As we can see in Figure 2.5.4, the maximum value for $p(t)$ occurs when $t \approx 1.6$ and is approximately 27.6. We conclude that the range of p is fairly close to (0, 27.6).

(b) Utilizing the Quotient Rule, along with the Sum and Difference Rule, we compute the derivative to be

$$p'(t) = \frac{200(2t^2 + 5) - (200t)(4t)}{(2t^2 + 5)^2} = \frac{-200(2t^2 - 5)}{(2t^2 + 5)^2}$$

Evaluating $p'(1)$ gives

$$p'(1) = \frac{-200(2(1)^2 - 5)}{(2(1)^2 + 5)^2} \approx 12.24$$

This means that at the end of one hour the concentration of the medication is increasing at a rate of about 12.24% per hour. Likewise,

$$p'(6) = \frac{-200(2(6)^2 - 5)}{(2(6)^2 + 5)^2} \approx -2.26$$

Thus, at the end of six hours the concentration of the medication is decreasing at the rate of about 2.26% per hour. Do these results seem reasonable based on Figure 2.5.4? ◼

SUMMARY

In this section we presented the **Product Rule** and the **Quotient Rule** for computing derivatives. The differentiation rules learned so far are listed in Table 2.5.1. The letters k and n represent constants, while f and g represent differentiable functions of x.

TABLE 2.5.1

NAME	RULE
Constant Function	$\dfrac{d}{dx}[k] = 0$
Power	$\dfrac{d}{dx}[x^n] = nx^{n-1}$
Constant Multiple	$\dfrac{d}{dx}[k \cdot f(x)] = k \cdot f'(x)$
Sum and Difference	$\dfrac{d}{dx}[f(x) \pm g(x)] = f'(x) \pm g'(x)$
Product	$\dfrac{d}{dx}[f(x) \cdot g(x)] = f'(x) \cdot g(x) + f(x) \cdot g'(x)$
Quotient	$\dfrac{d}{dx}\left[\dfrac{f(x)}{g(x)}\right] = \dfrac{f'(x) \cdot g(x) - f(x) \cdot g'(x)}{[g(x)]^2}, \quad g(x) \neq 0$

SECTION 2.5 EXERCISES

In Exercises 1–22, use the Product Rule to determine the derivative for the given function.

1. $f(x) = x^2(2x + 1)$

2. $f(x) = x^2(3x - 5)$

3. $f(x) = x^3(3x^2 + 2x - 5)$ 4. $f(x) = x^3(5x^2 - 6x + 3)$

5. $y = 3x^4(2x^2 - 9x + 1)$ 6. $y = 5x^3(3x^2 - 6x + 2)$

7. $y = -5x^2(3x^3 + 5x - 7)$ 8. $y = -7x^3(2x^3 - 3x^2 + 8)$

9. $f(x) = (3x + 4)(2x - 1)$ 10. $f(x) = (4x - 1)(x + 6)$

11. $y = (5x + 3)(3x^3 + 2x^2 + 1)$

12. $y = (2x - 1)(x^2 - 2x + 3)$

✓ 13. $g(x) = (3x^2 - 2x + 1)(2x^2 + 5x - 7)$

14. $g(x) = (2x^2 + 5x - 1)(3x^2 - 7x + 3)$

15. $y = (2\sqrt{x} + 4x - 3)(3x - 4)$

16. $y = (3\sqrt{x} - 2x + 1)(5x + 2)$

17. $f(x) = (3x^{6/5} - 5x)(4x^{5/3} + 2x - 5)$

18. $f(x) = (2x^{4/3} + 3x)(-2x^{7/3} + 2x - 5)$

19. $f(x) = (3\sqrt{x} - 5)\left(2\sqrt{x} - \dfrac{1}{x^3}\right)$

20. $f(x) = (4\sqrt{x} + 2x - 6)(3\sqrt{x} + 6x)$

21. $h(x) = (x^{2/3} + x + 1)(x^{-1} + x^{-2})$

22. $h(x) = (6x^{4/3} - 2x + 3)(3x^{-1} - 4x^{-2})$

In Exercises 23–36, determine the derivative for the given function.

23. $f(x) = \dfrac{x + 2}{x + 1}$ 24. $f(x) = \dfrac{3x - 4}{x - 1}$

25. $y = \dfrac{4x - 3}{2x + 1}$ 26. $y = \dfrac{5x - 11}{3x - 4}$

✓ 27. $f(x) = \dfrac{3x^2 - 5x + 1}{5x^2 + 3x + 2}$ 28. $f(x) = \dfrac{-2x^2 + 6x - 5}{3x^2 + 5x + 2}$

29. $f(x) = \dfrac{3\sqrt{x} - 5}{6x - 1}$ 30. $f(x) = \dfrac{4\sqrt{x} + 3}{2x + 7}$

✓ 31. $y = \dfrac{(x^2 + 2)(x - 3)}{x - 1}$ 32. $y = \dfrac{(x + 2)(x^3 - 3x^2 + 1)}{x - 2}$

33. $f(x) = \dfrac{(5x^4 + 2)(x^2 + 3)}{x - 4}$

34. $f(x) = \dfrac{(6x^3 - 2x^2 + 1)(3x - 5)}{2x + 1}$

35. $g(x) = \dfrac{1}{2}(4x^3 + 2x^2 - 3x - 5)$

36. $g(x) = \dfrac{2x^3 + 3x^2 - 7x + 1}{5}$

For Exercises 37–44:

(a) Determine the derivative.

(b) Determine an equation of the line tangent to the graph of the function at the indicated x-value.

(c) Graph the function and the tangent line in the same viewing window.

(d) Use the $\dfrac{dy}{dx}$ or DRAW TANGENT command of your calculator to check your work.

37. $f(x) = x^2(x^2 - 5)$ at $x = 1$

38. $y = -3x^2(2x + 3)$ at $x = 3$

✓ 39. $f(x) = (x^2 + 1)(x^3 + 1)$ at $x = 1$

40. $f(x) = (2x - 3)(x^2 + 3)$ at $x = 5$

41. $y = \dfrac{x + 2}{x - 1}$ at $x = 2$

42. $y = \dfrac{x^2 + 1}{x}$ at $x = -1$

43. $g(x) = \dfrac{3x^2 - 2x}{-2x + 3}$ at $x = -1$

44. $g(x) = \dfrac{-2x^2 + 3x}{3x - 5}$ at $x = 3$

The Product Rule presented in this section can be extended to functions made up of any finite number of differentiable functions. For example, if $k(x) = f(x) \cdot g(x) \cdot h(x)$, then $k'(x) = f'(x) \cdot g(x) \cdot h(x) + f(x) \cdot g'(x) \cdot h(x) + f(x) \cdot g(x) \cdot h'(x)$. Use this extension of the Product Rule to compute derivatives for each function in Exercises 45–50.

45. $f(x) = (x + 1)(x - 2)(x + 5)$

46. $f(x) = x^2(x^3 - 3x^2 + 1)(x - 4)$

47. $f(x) = (x + 1)(2x^2 - 3)(3x + 4)$

48. $f(x) = (x - 4)(3x^2 - 5)(2x - 9)$

49. $y = \sqrt{x} \cdot (2x - 1)(3x^2 + 2)$

50. $y = \sqrt[3]{x} \cdot (3x + 1)(2x^3 - 3)$

APPLICATIONS

51. The monthly sales of a new computer are given by $q(t) = 30t - 0.5t^2$ hundred units per month t months after it hits the market, where $0 \le t \le 7$. Compute $q(3)$ and $q'(3)$ and interpret.

52. Suppose that the computer described in Exercise 51 has a retail price, in dollars, given by $p(t) = 2200 - 34t^2$ in t months after it hits the market, where $0 \le t \le 7$. Compute $p(3)$ and $p'(3)$ and interpret.

53. For the computer described in Exercises 51 and 52:
(a) Determine the revenue function $R(t)$. Do not simplify.
(b) Determine the rate of change of revenue $R'(t)$.
(c) Compute $R(3)$ and $R'(3)$ and interpret each.

54. The monthly sales of a new computer are given by $q(t) = 30t - 0.5t^2$ hundred units per month t months after it hits the market, where $0 \le t \le 7$. Compute $q(6)$ and $q'(6)$ and interpret.

55. Suppose that the computer described in Exercise 54 has a retail price, in dollars, given by $p(t) = 2200 - 34t^2$ in t months after it hits the market, where $0 \le t \le 7$. Compute $p(6)$ and $p'(6)$ and interpret.

56. For the computer described in Exercises 54 and 55:
(a) Determine the revenue function $R(t)$. Do not simplify.
(b) Determine the rate of change of revenue $R'(t)$.
(c) Compute $R(6)$ and $R'(6)$ and interpret each.

57. (a) Verify that upon multiplying and simplifying $R'(t)$ in Exercise 53 we get

$$R'(t) = 68t^3 - 3060t^2 - 2200t + 66{,}000$$

(b) Graph $R'(t)$ and use the ZERO or ROOT or SOLVE command to find where $R'(t) = 0$ in the interval $[0, 7]$.

(c) Graph R and use the $\dfrac{dy}{dx}$ or DRAW TANGENT command to calculate the slope of the line tangent to the graph of R at the value found in part (b). (*Hint:* It should be close to zero.)

(d) Is *revenue* maximized or minimized at the t-value determined in part (b)? Explain.

58. The monthly sales of a new CD-ROM drive are given by $q(t) = 30t - 0.5t^2$ hundred units per month t months after being introduced on the market, where $0 \le t \le 5$. Compute $q(3)$ and $q'(3)$ and interpret.

59. The retail price, in dollars, of the CD-ROM drive in Exercise 58 is given by $p(t) = 220 - t^2$, where t is the number of months after being introduced on the market, where $0 \le t \le 5$. Compute $p(3)$ and $p'(3)$ and interpret.

60. For the CD-ROM drive described in Exercises 58 and 59:
(a) Determine the revenue function $R(t)$. Do not simplify.
(b) Determine the rate of change of revenue $R'(t)$.
(c) Compute $R(3)$ and $R'(3)$ and interpret each.

61. The average hourly earnings, in dollars, of employees in finance, insurance, and real estate in the United States are given by

$$g(t) = 0.48t + 9.44, \qquad 1 \le t \le 7$$

where t is the number of years since 1989. The average weekly earnings, in dollars, of employees in finance, insurance, and real estate in the United States are given by

$$f(t) = 17.5t + 336.86, \qquad 1 \le t \le 7$$

where t is the number of years since 1989.

(a) Compute $\dfrac{f(3)}{g(3)}$ and interpret. (*Hint:* Watch the units.)

(b) Let $h(t) = \dfrac{f(t)}{g(t)}$. What does this function represent?

(c) Compute $h(3)$ and $h'(3)$ and interpret each.

62. The cost $C(x)$ in thousands of dollars of removing x % of a city's pollutants discharged into a lake is given by

$$C(x) = \frac{113x}{100 - x}, \qquad 0 \le x < 100$$

Compute $C(50)$ and $C'(50)$ and interpret each.

63. The cost $C(x)$ in thousands of dollars of removing x % of a city's pollutants discharged into a lake is given by

$$C(x) = \frac{50x}{100 - x}, \qquad 0 \le x < 100$$

Compute $C(50)$ and $C'(50)$ and interpret each.

64. The local game commission decides to stock a lake with bass. To do this, 100 bass are introduced into the lake. The population of the bass is approximated by

$$P(t) = \frac{10(10 + 7t)}{1 + 0.02t}, \qquad t \ge 0$$

where t is time in months. Compute $P(5)$ and $P'(5)$ and interpret each.

65. The local game commission decides to stock a lake with bass. To do this, 200 bass are introduced into the lake. The population of the bass is approximated by

$$P(t) = \frac{20(10 + 7t)}{1 + 0.02t}, \qquad t \ge 0$$

where t is time in months. Compute $P(5)$ and $P'(5)$ and interpret each.

66. A cost analyst for Jetway Airlines is trying to estimate the rate at which the total fuel cost will change during the next three years. The number of gallons of jet fuel that Jetway Airlines expects to use annually is given by $N(t) = 300 + 60t$, where $N(t)$ is in millions of gallons per year for t, the time in years, in the interval $[0, 3]$. The price per gallon of the jet fuel is expected to decrease according to $p(t) = \dfrac{100}{100 + 2t^2}$, where $p(t)$ is dollars per gallon for t, the time in years, in the interval $[0, 3]$. Determine the rate at which the airline's annual fuel costs are changing when
(a) $t = 0$ (b) $t = 1$ (c) $t = 2$

67. The amount of sulfur dioxide emitted into the atmosphere by the United States from 1985 to 1995 can be modeled by

$$SD(t) = -0.02t^3 + 0.27t^2 - 1.22t + 24.12, \qquad 1 \le t \le 11$$

where t is the number of years since 1984 and $SD(t)$ is the amount of sulfur dioxide, in millions of tons, released into the atmosphere by the United States. The U.S. civilian population from 1985 to 1995 can be modeled by

$$CP(t) = 2.57t + 232.91, \qquad 1 \le t \le 11$$

where t is the number of years since 1984 and $CP(t)$ is the U.S. civilian population in millions.

(a) Compute $SD(5)$ and $CP(5)$ and interpret each.

(b) Compute $\dfrac{SD(5)}{CP(5)}$ and interpret.

(c) Define $H(t) = \dfrac{SD(t)}{CP(t)}$. This gives the per capita emission of sulfur dioxide. Compute $H'(t)$.

(d) Compute $H'(5)$ and interpret.

(e) Graph H in the window $[0, 12]$ by $[0.05, 0.10]$. Does H appear to be increasing or decreasing?

SECTION PROJECT

The table gives the U.S. national debt in billions of dollars and the U.S. civilian population in millions from 1980 to 1996.

YEAR	NATIONAL DEBT	U.S. CIVILIAN POPULATION
1980	909.1	225.621
1982	1137.3	229.995
1984	1564.7	234.110
1986	2120.6	238.412
1988	2601.3	242.817
1990	3206.6	247.758
1992	4002.1	253.426
1994	4643.7	258.960
1996	5207.3	263.998

Source: U.S. Census Bureau web site.

(a) Let x represent the number of years since 1980 and y represent the national debt in billions. Enter the data into your calculator, and determine a quadratic regression model for the data.

(b) Round the coefficients to the nearest hundredth and store the regression model in y_1. Plot the data points with the regression model to see how well the curve fits the data.

(c) Compute the derivative of the quadratic equation. Evaluate the derivative at $x = 6$, $x = 12$, and $x = 16$, and interpret each.

(d) Let x represent the number of years since 1980 and y the U.S. civilian population in millions. Enter the data into your calculator, and determine a linear regression model for the data.

(e) Round the coefficients to the nearest hundredth and store the linear model in y_2. Plot the data points with the linear model to see how well the line fits the data.

(f) Compute the derivative for the linear equation. Evaluate the derivative at $x = 6$, $x = 12$, and $x = 16$ and interpret each.

(g) If we take the national debt and divide it by the U.S. civilian population, we get the per capita national debt. Define $y_3 = \dfrac{y_1}{y_2}$ and graph y_3. Notice its increasing behavior.

(h) Determine the derivative for y_3. Evaluate the derivative at $x = 6$, $x = 12$, and $x = 16$ and interpret each.

SECTION 2.6   CONTINUITY AND NONDIFFERENTIABILITY

In this final section of Chapter 2, we address what it means for a function to be *continuous* at a point (and over an interval) and when a function does not have a derivative at a point. These two topics nicely tie together the main concepts in this chapter—the limit concept and the derivative.

Continuity

A function is said to be **continuous** if its graph has no breaks in it such as holes, gaps, or jumps. This means that the graph of the function can be drawn without lifting a pencil off the paper. If the graph of a function has a hole, gap, or jump at $x = a$, we say that the function is **discontinuous** at $x = a$. There are many ways for a function to be discontinuous at $x = a$. Rather than list every possibility, we now supply a definition to determine whether a function is continuous at $x = a$.

Continuity

A function, f, is said to be continuous at the point $x = a$ if all the following are true:

(1) $f(a)$ is defined   (2) $\lim\limits_{x \to a} f(x)$ exists   (3) $\lim\limits_{x \to a} f(x) = f(a)$

NOTE: A function is continuous on open interval (b, c) if it is continuous for all x in the interval.

Figure 2.6.1 shows a graph of three different functions, each discontinuous at $x = a$. Notice that the function in Figure 2.6.1a violates condition 1 of our continuity definition, the function in Figure 2.6.1b violates condition 2, and the function in Figure 2.6.1c violates condition 3.

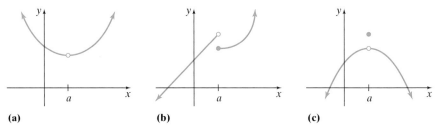

(a)   (b)   (c)

Figure 2.6.1   **(a)** $f(a)$ is not defined. **(b)** $\lim\limits_{x \to a} f(x)$ does not exist. **(c)** $\lim\limits_{x \to a} f(x) \neq f(a)$.

EXAMPLE 1

Determining Continuity

Determine the continuity of $g(x) = \dfrac{x^2 - 4}{x - 2}$ at $x = 1$ and at $x = 2$.

SOLUTION

According to the definition, three items must be true for continuity to exist. For $x = 1$,

1. Is $g(1)$ defined? Yes; $g(1) = \dfrac{1^2 - 4}{1 - 2} = 3$.
2. Does $\lim\limits_{x \to 1} \dfrac{x^2 - 4}{x - 2}$ exist? Yes; the Substitution Principle shows that

$$\lim\limits_{x \to 1} \frac{x^2 - 4}{x - 2} = \frac{(1)^2 - 4}{1 - 2} = 3$$

3. Does $\lim\limits_{x \to 1} \dfrac{x^2 - 4}{x - 2} = g(1)$? Yes; $3 = 3$.

So we conclude that $g(x) = \dfrac{x^2 - 4}{x - 2}$ is continuous at $x = 1$.
For $x = 2$,

1. Is $g(2)$ defined? No!

Since $g(2)$ is not defined, hence violating part 1 of the definition of continuity, we conclude that $g(x) = \dfrac{x^2 - 4}{x - 2}$ is not continuous at $x = 2$.   ◼

✓ CHECKPOINT 1

Now work Exercise 13.

Notice in Example 1 that $\lim\limits_{x \to 2} \dfrac{x^2 - 4}{x - 2}$ exists, as seen in Figure 2.6.2. In fact, in Section 2.1 we determined that $\lim\limits_{x \to 2} \dfrac{x^2 - 4}{x - 2} = 4$. But since $g(2)$ is not defined,

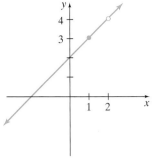

Figure 2.6.2 Graph of $g(x) = \dfrac{x^2 - 4}{x - 2}$.

there is a hole in the graph at the point $(2, 4)$. This type of discontinuity is said to be *removable*.

Continuity for Specific Functions

From Chapter 1 and our work thus far in Chapter 2, you may have a hunch that continuity can be determined for certain families of functions just by recognition. In fact, we can determine continuity for certain functions as indicated in the following theorem.

> **Theorem 2.1**
>
> 1. A polynomial function is continuous for all real x.
> 2. A rational function is continuous for all real x except those values of x for which the denominator is 0.
> 3. An exponential function is continuous for all real x.
> 4. The natural logarithmic function is continuous for all real x in its domain.
> 5. A radical function, or rational exponent function, is continuous for all real x in its domain.

EXAMPLE 2 | **Determining Intervals of Continuity**

Use Theorem 2.1 to determine intervals where the following functions are continuous.

(a) $f(x) = \dfrac{2x + 1}{(x - 1)(2x + 1)}$ (b) $g(x) = (1.21)^x$ (c) $h(x) = \sqrt{x - 7}$

SOLUTION

(a) Since this is a rational function, we conclude that it is continuous everywhere except at $x = 1$ and at $x = -\dfrac{1}{2}$, the values of x that make the denominator 0. See Figure 2.6.3. The function is continuous on $\left(-\infty, -\dfrac{1}{2}\right) \cup \left(-\dfrac{1}{2}, 1\right) \cup (1, \infty)$.

(b) Since this is an exponential function, it is continuous for all reals. See Figure 2.6.4.

(c) The domain of this radical function is $[7, \infty)$, so we conclude that the function is continuous on $[7, \infty)$. See Figure 2.6.5.

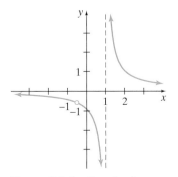

Figure 2.6.3 Graph of $f(x) = \dfrac{2x + 1}{(x - 1)(2x + 1)}$.

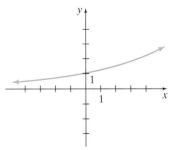

Figure 2.6.4 Graph of $g(x) = (1.21)^x$.

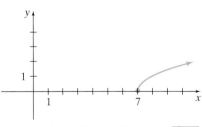

Figure 2.6.5 Graph of $h(x) = \sqrt{x - 7}$.

✓ CHECKPOINT 2

Now work Exercise 21.

Nondifferentiable Functions

So far we have presented six basic rules for differentiation, and we will see a few more in Chapter 4. In spite of all the rules for differentiation, some functions are **nondifferentiable.** That is, some functions cannot be differentiated at certain values. (You may have noticed in some of the rules presented so far the phrase, "where f is differentiable.") We conclude this section by determining where these functions are not differentiable. We begin by analyzing the **absolute value function,** $f(x) = |x|$, whose graph is shown in Figure 2.6.6.

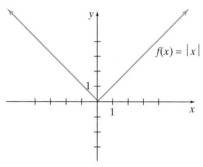

Figure 2.6.6

Recall that the definition for $f(x) = |x|$ is

$$f(x) = |x| = \begin{cases} x, & x \geq 0 \\ -x, & x < 0 \end{cases}$$

EXAMPLE 3 | **Analyzing Nondifferentiability at a Point**

Show that $f(x) = |x|$ is not differentiable at $x = 0$.

SOLUTION

Since we have not given a rule for differentiating the absolute value function, we appeal to the definition of the derivative that was presented in Section 2.3. (See the Toolbox to the left.) Let's try to compute $f'(0)$ from the definition. We have

$$f'(0) = \lim_{h \to 0} \frac{f(0 + h) - f(0)}{h} = \lim_{h \to 0} \frac{f(h) - f(0)}{h}$$

$$= \lim_{h \to 0} \frac{|h| - 0}{h} = \lim_{h \to 0} \frac{|h|}{h}$$

In Section 2.1, we learned that if we try substitution here it produces the indeterminate form $\frac{0}{0}$. In Table 2.6.1, we numerically compute the limit.

Table 2.6.1 suggests that the limit does not exist. We confirm this belief algebraically by appealing to the definition of absolute value. Specifically,

$$\lim_{h \to 0^-} \frac{|h|}{h} = \lim_{h \to 0^-} \frac{-h}{h} \qquad |h| = -h, \text{ for } h < 0$$

$$= \lim_{h \to 0^-} (-1) = -1 \qquad \text{Limit of a constant}$$

From Your Toolbox

For a function f, the derivative of f at x, denoted $f'(x)$, is defined to be

$$f'(x) = \lim_{h \to 0} \frac{f(x + h) - f(x)}{h}$$

provided that the limit exists.

TABLE 2.6.1

h	$h \to 0^-$				0	$h \to 0^+$			
h	-1	-0.1	-0.01	-0.001	0	0.001	0.01	0.1	1
$\dfrac{|h|}{h}$	-1	-1	-1	-1		1	1	1	1

Interactive Activity

Graphically support that $\lim\limits_{h \to 0} \dfrac{|h|}{h}$ does not exist by graphing $y = \dfrac{|x|}{x}$ and analyzing the graph as $x \to 0$.

Similarly,

$$\lim_{h \to 0^+} \frac{|h|}{h} = \lim_{h \to 0^+} \frac{h}{h} \qquad \text{\small $|h| = h$, for $h \geq 0$}$$

$$= \lim_{h \to 0^+} (1) = 1 \qquad \text{\small Limit of a constant}$$

Since the *left-hand limit* does not equal the *right-hand limit*, $\lim\limits_{h \to 0^-} \dfrac{|h|}{h} \neq \lim\limits_{h \to 0^+} \dfrac{|h|}{h}$, we conclude that $\lim\limits_{h \to 0} \dfrac{|h|}{h}$ *does not exist*. Since $\lim\limits_{h \to 0} \dfrac{|h|}{h}$ does not exist and since $f'(0) = \lim\limits_{h \to 0} \dfrac{|h|}{h}$, we conclude that the derivative does not exist. This is exactly what we wanted to show—that the absolute value function is not differentiable at $x = 0$. Notice that $f(x) = |x|$ is continuous at $x = 0$. ∎

Geometrically, the reason $f(x) = |x|$ is not differentiable at $x = 0$ is due to its graph consisting of two lines with slopes of -1 and $+1$ that meet at the origin. See Figure 2.6.7. The two conflicting slopes make it impossible to define a single slope at the origin. To gain a better understanding, consider the sequence of graphs in Figure 2.6.8a through e.

Notice that after four applications of the ZOOM IN command the graph of $f(x) = |x|$ still has a corner at the origin. It has not straightened out. If a function is differentiable at a point, several applications of the ZOOM IN command will show that the graph "straightens" out and practically becomes its own tangent line.

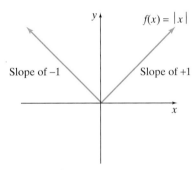

Figure 2.6.7

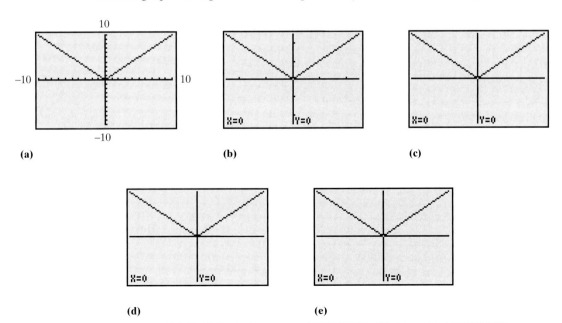

Figure 2.6.8 **(b)** After one ZOOM IN. **(c)** After second ZOOM IN. **(d)** After third ZOOM IN. **(e)** After fourth ZOOM IN.

Example 3 and the preceding graphical discussion show that if a graph has a sharp turn or a corner at a point, then the function is not differentiable at that point. In Example 4, we analyze another scenario in which a function is non-differentiable.

EXAMPLE 4

Analyzing Nondifferentiability at a Point

Show that $f(x) = \sqrt[3]{x}$ is not differentiable at $x = 0$.

SOLUTION

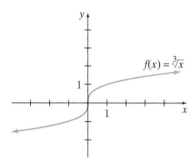

Figure 2.6.9

By looking at the graph of $f(x) = \sqrt[3]{x}$ in Figure 2.6.9, it appears that the tangent line at $x = 0$ would be a vertical line. We know the slope of a vertical line is undefined, which leads us to believe that here, $f'(0)$ is undefined. To analytically verify this, we compute the derivative by using the Power Rule.

$$f(x) = \sqrt[3]{x} = x^{1/3}$$

$$f'(x) = \frac{1}{3}x^{-2/3} = \frac{1}{3x^{2/3}} = \frac{1}{3\sqrt[3]{x^2}}$$

Remembering that f' is a function, we see that its domain is $(-\infty, 0) \cup (0, \infty)$. In other words, the domain of the derivative *excludes* zero. Since $f'(0)$ is not defined, the derivative is not defined at zero, and so our function is not differentiable at zero. Notice that $f(x) = \sqrt[3]{x}$ is continuous at $x = 0$. ◼

Example 4 reminds us that the derivative is also a function. Using that fact (and being able to take a function's derivative) can aid us in determining where functions are nondifferentiable.

So far we have seen that a function is not differentiable at $x = c$ if

- The graph of the function has a sharp turn or corner at $x = c$.
- The graph of the function has a vertical tangent at $x = c$.

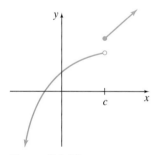

Figure 2.6.10

The third situation for which a function is not differentiable is wherever the function is *discontinuous*. The function graphed in Figure 2.6.10 is not differentiable at $x = c$ since it is discontinuous at $x = c$.

Now both functions analyzed in Examples 3 and 4 were continuous at $x = 0$, but in both cases $f'(0)$ was undefined. At this time we present the following astounding theorem.

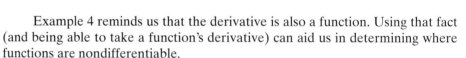

Theorem: Differentiability Implies Continuity

If f is differentiable at $x = c$, then f is continuous at $x = c$.

A proof of this theorem is given in Appendix C. We could condense the contents of this theorem to the following:

Differentiability implies continuity, but continuity does *not* imply differentiability.

Note that Examples 3 and 4 illustrate continuous functions that are not differentiable.

So we have determined that a function is not differentiable at $x = c$ if

- The graph of the function has a sharp turn or corner at $x = c$.
- The graph of the function has a vertical tangent at $x = c$.
- The graph of the function is not continuous at $x = c$.

EXAMPLE 5 | **Determining Points of Nondifferentiability**

For what values of x is the function graphed in Figure 2.6.11 not differentiable?

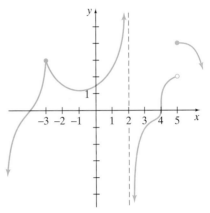

Figure 2.6.11

SOLUTION

$f(x)$ is not differentiable at

$x = -3$	Sharp turn or corner
$x = 2$	Not continuous
$x = 4$	Vertical tangent
$x = 5$	Not continuous

✓**CHECKPOINT 3**   Now work Exercise 35.

SUMMARY

In this section we formalized the concept of **continuity.** We then presented Theorem 2.1, which tells us the continuity for certain functions quickly. We concluded this section by looking at **nondifferentiable functions** or, more specifically, by locating values where a function is not differentiable.

- A function, f, is said to be continuous at the point $x = a$ if all the following are true: (1) $f(a)$ is defined; (2) $\lim_{x \to a} f(x)$ exists; (3) $\lim_{x \to a} f(x) = f(a)$.

We then presented Theorem 2.1, which tells us the continuity for certain functions quickly.

- Theorem 2.1

  1. A polynomial function is continuous for all real x.
  2. A rational function is continuous for all real x except those values of x where the denominator is 0.

3. An exponential function is continuous for all real x.
4. The natural logarithmic function is continuous for all real x in its domain.
5. A radical function, or rational exponent function, is continuous for all real x in its domain.

• A function is not differentiable at $x = c$ if

1. The function has a sharp turn or corner at $x = c$.
2. The function has a vertical tangent at $x = c$.
3. The function is not continuous at $x = c$.

SECTION 2.6 EXERCISES

For Exercises 1–4, use the graph of f in Figure 2.6.12 to determine if f is continuous at the indicated point. If f is not continuous at the indicated point, state which of the three conditions given in the definition of continuity is violated.

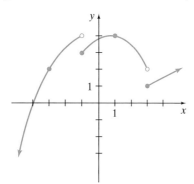

Figure 2.6.12

1. Is f continuous at $x = 1$?   2. Is f continuous at $x = -1$?

3. Is f continuous at $x = 3$?   4. Is f continuous at $x = -3$?

For Exercises 5–8, use the graph of f in Figure 2.6.13.

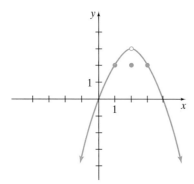

Figure 2.6.13

5. Is f continuous at $x = 1$? Explain.

6. Is f continuous at $x = 2$? Explain.

7. Is f continuous at $x = 3$? Explain.

8. Determine $f(2)$.

In Exercises 9–18, determine the continuity of the given function at the indicated point.

9. $g(x) = 3x - 2$, $x = -2$

10. $f(x) = 2x + 6$, $x = -3$

11. $h(x) = x^2 - x - 6$, $x = 3$

12. $h(x) = 2x^3 + 3x^2 - 5$, $x = 2$

✓ 13. $g(x) = \dfrac{x+1}{x-3}$, $x = 3$

14. $f(x) = \dfrac{x-5}{x+2}$, $x = -2$

15. $f(x) = \dfrac{x^2 - 25}{x - 5}$, $x = 5$

16. $g(x) = \dfrac{x^2 - 25}{x - 5}$, $x = 0$

17. $f(x) = \begin{cases} x + 2, & x \le 1 \\ x^2 + 3, & x > 1 \end{cases}$, $x = 1$

18. $f(x) = \begin{cases} x - 3, & x \le 0 \\ x^2 + x - 3, & x > 0 \end{cases}$, $x = 0$

In Exercises 19–32, determine intervals where the following functions are continuous. Refer to Theorem 2.1 and Example 2.

19. $f(x) = 7x^2 - 3.2x + 10.5$

20. $f(x) = 4x^3 - 2x^2 + 1.3x - 5$

✓ 21. $g(x) = \dfrac{2x - 3}{x + 5}$ 22. $g(x) = \dfrac{3x - 7}{2x + 1}$

23. $f(x) = \dfrac{x + 1}{(x + 1)(x - 3)}$ 24. $f(x) = \dfrac{2x - 3}{(2x - 3)(x + 4)}$

25. $f(x) = (1.221)^x$ 26. $g(x) = (0.958)^x$

27. $f(x) = -2.1 - \ln(x + 3)$ 28. $f(x) = 3.2 + \ln(x + 1)$

29. $h(x) = e^{-0.254x}$ 30. $g(x) = e^{0.21x}$

31. $f(x) = \sqrt{2x + 3}$ 32. $g(x) = \sqrt{3x - 2}$

For each function graphed in Exercises 33–40, state the *x*-values for which the derivative does not exist and explain why.

33.

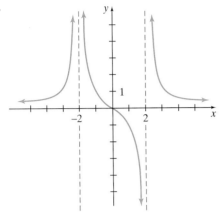

34.

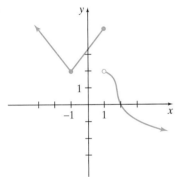

✓ 35.

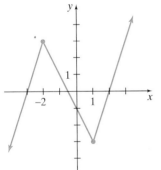

36.

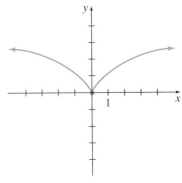

37.

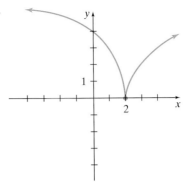

38.

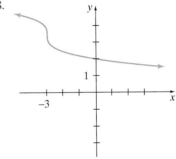

39.

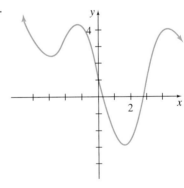

40.

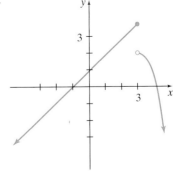

For Exercises 41–48:

(a) Graph the function.

(b) Compute the derivative.

(c) Determine the x-values for which the derivative is undefined.

(d) For the x-values where the derivative is undefined, does the graph of the function have a sharp turn or a vertical tangent or is it discontinuous?

41. $f(x) = \sqrt{x}$

42. $f(x) = \sqrt[4]{x}$

43. $f(x) = x^{2/3}$

44. $f(x) = x^{3/5}$

45. $f(x) = \dfrac{x^2 - 1}{x + 1}$

46. $f(x) = \dfrac{x^2 - 1}{x - 1}$

47. $f(x) = \dfrac{x^2 - 4}{x - 2}$

48. $f(x) = \dfrac{x^2 - 4}{x + 2}$

APPLICATIONS

49. Lisa's Lease-a-Car charges $25 per day plus $0.05 per mile to rent one of its medium-sized cars. Figure 2.6.14 shows the graph of the cost in dollars, $c(m)$, realized by Sofia as a function of the miles, m, that she has driven the car. Give a reason for the discontinuity at $m = 200$.

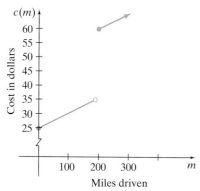

Figure 2.6.14

50. Dylan's Truck Rental Company charges $20 per day plus $0.10 per mile to rent a moving truck. Figure 2.6.15 shows the graph of the cost in dollars, $c(m)$, realized by Austin as a function of the miles, m, that he has driven the truck.

(a) Determine $c(100)$ and interpret.

(b) Is c continuous at $m = 100$?

(c) Determine $c(200)$ and interpret.

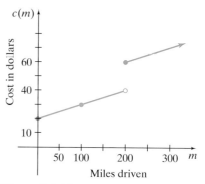

Figure 2.6.15

(d) Is c continuous at $m = 200$?

(e) Give a reason for the discontinuity at $m = 200$.

51. Finer Foods is running the following advertisement:

<u>Hamburger</u>

$1.50/pound for packages less than 2 pounds

$1.00/pound for packages of 2 pounds or more

(a) Let x represent the number of pounds of hamburger in a package. Determine a piecewise-defined function, C, that models the cost of a package of hamburger.

(b) Graph C, where $0 < x \le 6$.

(c) Determine $C(1.5)$ and interpret.

(d) Determine $C(2)$ and interpret.

(e) Does $\lim\limits_{x \to 2} C(x)$ exist? Explain.

(f) Is C continuous at $x = 2$?

SECTION PROJECT

Consider $f(x) = \sqrt{x}$.

(a) Use the definition of the derivative to show that
$$f'(0) = \lim_{h \to 0} \frac{\sqrt{h}}{h}.$$

(b) Numerically, attempt to find the limit in part (a) by making a table for $h = 0.1, 0.001, 0.00001, 0.0000001$, and 0.00000001.

(c) Recall from Section 2.3 that what you did in part (b) was compute secant line slopes. The limit of the secant line slope is the slope of the tangent line. From part (b), do you believe that the limit exists?

(d) Graphically support your conjecture in part (b) by graphing $y = \dfrac{\sqrt{x}}{x}$ and use the ZOOM IN and

TRACE commands. Does this graphical analysis support your numerical conjecture?

(e) Using the results of parts (b), (c), and (d), does the derivative of $f(x) = \sqrt{x}$ at $x = 0$ exist? Explain.

(f) Graph $f(x) = \sqrt{x}$. Explain why the slope at $x = 0$ does not exist.

CHAPTER REVIEW EXERCISES

1. For $f(x) = \dfrac{1 - \sqrt{1 - 2x - x^2}}{x}$, complete the table to numerically estimate the following.

(a) $\lim\limits_{x \to 0^-} f(x)$ (b) $\lim\limits_{x \to 0^+} f(x)$ (c) $\lim\limits_{x \to 0} f(x)$

x	−0.1	−0.01	−0.001	0	0.001	0.01	0.1
$f(x)$				?			

2. For $f(x) = (\sqrt{4 - x})^5$, complete the table to numerically estimate the following.

(a) $\lim\limits_{x \to 4^-} f(x)$ (b) $\lim\limits_{x \to 4^+} f(x)$ (c) $\lim\limits_{x \to 4} f(x)$

x	3	3.9	3.99	4	4.01	4.1	5
$f(x)$				?			

For Exercises 3 and 4, use your calculator to graph the given function. Use the ZOOM IN and TRACE commands to graphically estimate the indicated limits. Verify your estimate numerically.

3. $f(x) = x^3 - 2x$; $\lim\limits_{x \to 3.1} f(x)$

4. $f(x) = \sqrt{x} + \dfrac{1}{|x|}$; $\lim\limits_{x \to 2.5} f(x)$

For Exercises 5–10, determine the indicated limit algebraically.

5. $\lim\limits_{x \to 2} (7x^3 - 10x)$

6. $\lim\limits_{x \to -1} \dfrac{x^2 - 1}{x + 1}$

7. $\lim\limits_{x \to -3} |x - 5|$

8. $\lim\limits_{x \to 0} \sqrt{36 - 8x}$

9. $\lim\limits_{x \to 10} \dfrac{x + 10}{x^2 - 100}$

10. $\lim\limits_{x \to 2.2} (x + 9)^3$

11. Use the graph of f to find the following:

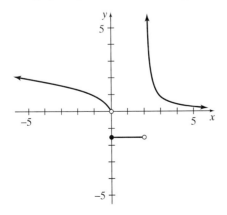

(a) $\lim\limits_{x \to -4} f(x)$ (b) $\lim\limits_{x \to 0^-} f(x)$ (c) $f(0)$

(d) $\lim\limits_{x \to 2} f(x)$ (e) $\lim\limits_{x \to 3} f(x)$ (f) $f(3)$

Determine the indicated limit in Exercises 12–14.

12. $\lim\limits_{h \to 0} (x + 2h)^2$ 13. $\lim\limits_{h \to 0} \dfrac{2x^2 h - 9h}{h}$ 14. $\lim\limits_{h \to 0} \dfrac{6x^3 h^2 + h}{h}$

15. For $f(x) = 3x^2$, find

(a) $f(2 + h)$ (b) $\lim\limits_{h \to 0} \dfrac{f(2 + h) - f(2)}{h}$

16. For $f(x) = \dfrac{9}{x}$, find

(a) $f(4 + h)$ (b) $\lim\limits_{h \to 0} \dfrac{f(4 + h) - f(4)}{h}$

17. Acme Stuffed Animals, Inc, is introducing a new line of teddy bears. The total cost of producing Scare Bear (with glowing eyes) is projected to be $C(x) = 36{,}000 + \sqrt{10{,}000x}$, where x is the number of units made and $C(x)$ is the cost in dollars.

(a) Find and interpret $C(400)$.

(b) Find and interpret $\lim\limits_{x \to 100} C(x)$.

(c) Find and interpret $AC(x) = \dfrac{C(x)}{x}$ for $x = 25$.

18. A toy speedboat moves away from a dock, and its distance from the dock is given by

$$d(t) = \begin{cases} 2t^2, & 0 \le t \le 3 \\ 12t - 18, & 3 < t \end{cases}$$

where $d(t)$ is in feet and t is in seconds.

(a) Graph d for $0 \le t \le 6$.

(b) Find $\lim\limits_{t \to 3^-} d(t)$.

(c) Find $\lim\limits_{t \to 3^+} d(t)$.

(d) Find $\lim\limits_{t \to 3} d(t)$.

(e) Describe the behavior of the toy speedboat.

19. A pastry chef in a commercial test kitchen is fine-tuning a pie-filling recipe. Colleagues acting as tasters have rated various recipes on a scale of 1 to 10. The average rating $R(s)$ appears to be a function of the sugar content s.

$$R(s) = 8 - \dfrac{s^2 - 48s + 512}{50}$$

where s is the number of tablespoons of sugar.

(a) Graph R in the viewing window $[0, 40]$ by $[0, 10]$.

(b) Find $R(20)$, $R(25)$, and $R(30)$.

(c) Find $\lim\limits_{h \to 0} \dfrac{R(24 + h) - R(24)}{h}$.

(d) How much sugar seems to be optimal?

20. A furnace switches on in such a way that the temperature of the heating element is a function of time, as shown by the graph.

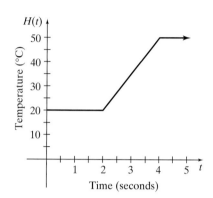

(a) Write an algebraic definition of $H(t)$.

(b) Find $\lim\limits_{h \to 0} \dfrac{H(5+h) - H(5)}{h}$, if possible.

(c) Find $\lim\limits_{h \to 0} \dfrac{H(2+h) - H(2)}{h}$, if possible.

(d) Explain why the middle segment of the curve is not vertical.

21. For $f(x) = \dfrac{2}{(x+3)^2}$, complete the table to numerically estimate

(a) $\lim\limits_{x \to -3^-} f(x)$ (b) $\lim\limits_{x \to -3^+} f(x)$ (c) $\lim\limits_{x \to -3} f(x)$

Use the symbols ∞ and $-\infty$ as appropriate.

x	-3.1	-3.01	-3.001	-3	-2.999	-2.99	-2.9
$f(x)$							

22. For $f(x) = \dfrac{x+1}{x-2}$, complete the table to numerically estimate

(a) $\lim\limits_{x \to 2^-} f(x)$ (b) $\lim\limits_{x \to 2^+} f(x)$ (c) $\lim\limits_{x \to 2} f(x)$

Use the symbols ∞ and $-\infty$ as appropriate.

x	1.9	1.99	1.999	2	2.001	2.01	2.1
$f(x)$							

For the functions given in Exercises 23 and 24:

(a) Determine the domain.

(b) Determine the indicated limit.

23. For $f(x) = \dfrac{1250 + 3.2x}{x}$; $\lim\limits_{x \to 0} f(x)$

24. For $f(x) = \dfrac{x+5}{x^2 - 25}$; $\lim\limits_{x \to -5} f(x)$

For the functions in Exercises 25–27:

(a) Classify f as modeling exponential growth or decay.

(b) Determine $\lim\limits_{x \to \infty} f(x)$.

(c) Determine $\lim\limits_{x \to -\infty} f(x)$.

25. $f(x) = 4^x$ 26. $f(x) = (0.72)^x$ 27. $f(x) = (2.72)^x$

In Exercises 28–31, determine the indicated limit algebraically. Verify your results numerically.

28. $\lim\limits_{x \to -\infty} \dfrac{x+2}{2x-3}$ 29. $\lim\limits_{x \to \infty} \dfrac{-2x^2 + 5x - 1}{x^2 - 13}$

30. $\lim\limits_{x \to \infty} \dfrac{-4x^2 - 3x + 11}{8x^3 - 5}$ 31. $\lim\limits_{x \to -\infty} \dfrac{3x^4 - 27}{x+3}$

32. Find an equation for the horizontal asymptote of the graph of $f(x) = \dfrac{2x^2 - 9x + 9}{6x^2 + x + 11}$.

33. Find an equation for the horizontal asymptote of the graph of $f(x) = \dfrac{7x+9}{3x^2+2}$.

For Exercises 34–36, use the following graph of f:

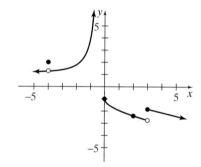

34. Is f continuous at $x = -4$? If not, why not?

35. Is f continuous at $x = 0$? If not, why not?

36. Is f continuous at $x = 2$? If not, why not?

37. Is $g(x) = \dfrac{x^2 - 9}{x + 3}$ continuous at $x = 9$? At $x = -3$? Explain.

38. Is $h(x) = \begin{cases} x - 8, & x \le 2 \\ \dfrac{6}{x-3}, & x > 2 \end{cases}$ continuous at $x = 2$? At $x = 3$? Explain.

39. Find the intervals where the function $f(x) = \dfrac{x-2}{(x-5)(x+3)}$ is continuous.

40. Find the intervals where the function $g(x) = 6.4 - \ln(x-9) + \sqrt{x}$ is continuous.

41. The annual cost $C(x)$, in thousands of dollars, of getting the reliability rate of an automated production quality

control system up to $x\%$ is given by

$$C(x) = \begin{cases} 0, & 0 \le x \le 80 \\ \dfrac{400 - 5x}{x - 100}, & 80 < x < 100 \end{cases}$$

(a) Graph and interpret the function.

(b) Evaluate $\lim\limits_{x \to 80} C(x)$.

(c) Evaluate $\lim\limits_{x \to 100^-} C(x)$.

42. A woman creates lithographs depicting the healing auras surrounding crystals. Producing q copies of a lithograph costs $C(q) = 238 + 5q$ dollars. The average cost, $AC(q)$, is given by $AC(q) = \dfrac{C(q)}{q}$. Find and interpret $\lim\limits_{q \to 40} AC(q)$ and $\lim\limits_{q \to \infty} AC(q)$.

43. In a swampy area, researchers model the springtime mosquito population using the function $N(t) = 2.4e^{0.008t}$, where t is the number of days after April 1 and $N(t)$ is population in millions.

(a) Find $N(0)$ and interpret.

(b) Find $\lim\limits_{t \to \infty} N(t)$ and interpret. Comment on this result.

(c) Graph N in the viewing window $[0, 100]$ by $[0, 10]$.

(d) How long does it take for the population to reach 5 million?

44. A bowl of soup taken out of a microwave has a temperature given by the function $C(t) = 70e^{-0.065t} + 20$, where t is time in minutes and $C(t)$ is in degrees Celsius.

(a) What is the initial temperature?

(b) What is the temperature at $t = 15$ minutes?

(c) What is the temperature of the surrounding air? (*Hint:* What temperature does the soup tend toward in the long run?)

45. A peach grower takes steps to improve the quality of his product over several years. The annual profits follow the function

$$P(t) = \frac{25(t + 4)}{t + 5}$$

where t is time in years and $P(t)$ is profit in thousands of dollars.

(a) Graph P in the viewing window $[0, 10]$ by $[0, 30]$.

(b) Find the annual profit 3 years after the start of the improvement program.

(c) Find and interpret $\lim\limits_{t \to \infty} P(t)$.

In Exercises 46–49, compute the average rate of change of the given function over the indicated interval.

46. $f(x) = x^3 - 6x + 1$; $[3, 7]$

47. $g(x) = 2\sqrt{4x}$; $[4, 9]$

48. $h(x) = 1.7e^{0.2x}$; $[-2, 7]$

49. $f(x) = \dfrac{3}{x - 2}$; $[6.2, 7.9]$

50. Consider the curve shown.

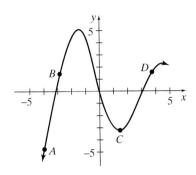

(a) Between which pairs of points is the average rate of change positive? Negative? Zero?

(b) At which points shown is the tangent line slope positive? Negative? Zero?

In Exercises 51–53, estimate the instantaneous rate of change of the function at the indicated value by using $\lim\limits_{h \to 0} \dfrac{f(x + h) - f(x)}{h}$ and completing the following tables:

			$h \to 0^-$		
h	-1	-0.1	-0.01	-0.001	-0.0001
$\dfrac{f(x + h) - f(x)}{h}$					

			$h \to 0^+$		
h	1	0.1	0.01	0.001	0.0001
$\dfrac{f(x + h) - f(x)}{h}$					

51. $f(x) = 2.5x$; $x = 4$

52. $f(x) = 3x^3 + 3x - 25$; $x = -2$

53. $f(x) = 1.4e^{0.6x}$; $x = -1$

54. Find the slope of the tangent line to the graph of $f(x) = 3\sqrt{x + 2}$ at $x = 4$ by estimating $\lim\limits_{h \to 0} f(x)$ using successively smaller values for h.

In Exercises 55–57, compute $f'(c)$ by using the definition of the derivative

$$f'(c) = \lim_{h \to 0} \frac{f(c + h) - f(c)}{h}$$

55. $f(x) = x^2 - 2x$; $c = 7$

56. $f(x) = x^3 + 1$; $c = 4$

57. $f(x) = \dfrac{1}{x^2}$; $c = 3$

58. For $f(x) = -3x^2 - 2x$:

(a) Find an equation for the line tangent to the graph of the function at $x = -1$.

(b) Graph the function and tangent line in the same window.

In Exercises 59–62, find a formula for the general derivative by using the definition of the derivative

$$f'(x) = \lim_{h \to 0} \frac{f(x+h) - f(x)}{h}$$

59. $f(x) = 3x - 7$ 60. $g(x) = x^4$

61. $h(x) = 3.4x^2 + 1.9x$ 62. $f(x) = \frac{1}{x^2}$

63. Find the slope of the tangent line to the graph of $f(x) = 6x$ at $x = 4$ by first finding a general formula for the derivative.

64. Find the slope of the tangent line to the graph of $g(x) = 3x^2$ at $x = 9$ by first finding a general formula for the derivative.

65. Find the slope of the tangent line to the graph of $h(x) = \frac{1}{x}$ at $x = 7$ by first finding a general formula for the derivative.

66. The graph of a profit function is shown.

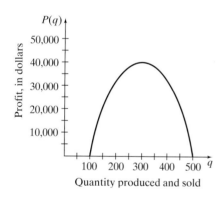

P(q)

Profit, in dollars

50,000

40,000

30,000

20,000

10,000

100   200   300   400   500   q

Quantity produced and sold

(a) For what value of q is $P(q)$ a maximum?

(b) For what value of q does $P'(q) = 0$?

(c) What is the maximum profit?

67. If \$3000 is deposited in a bank account earning 8.5% interest compounded annually, then the amount in the account after t years is given by $A(t) = 3000(1.085)^t$.

(a) Find and interpret $A(6)$.

(b) Use the definition of the derivative to find and interpret $A'(6)$.

68. The number of people attending a high school play is a function of the effort put into publicity, according to the equation $N(w) = 800 - \frac{8000}{w + 10}$, where w is the number of person-hours (1 person working 1 hour = 1 person-hour).

(a) Given $N'(w) = \frac{8000}{(w + 10)^2}$, compute $N(20)$ and $N'(20)$ and interpret each.

(b) How many people will attend if no effort at all is put into publicity?

(c) What is the maximum number of people that can be expected to attend with heavy publicity?

69. Assume that a mathematical model for the growth over time of a fish population is given by $P(t) = 20 + \sqrt{400t}$, where t is in years.

(a) Use an approximation method to find $P'(t)$ at $t = 4$.

(b) Write an equation for the line tangent to the graph of the population model at $t = 4$.

(c) Interpret the slope of the tangent line.

In Exercises 70–81, determine the derivative for the given function. Where appropriate, simplify the derivative so that there are no negative or fractional exponents.

70. $f(x) = x^8$ 71. $g(x) = 3x^{1/5}$

72. $h(x) = -4x^{-3/7}$ 73. $f(x) = -\frac{4}{5}x^{7/6}$

74. $g(x) = 6x^2 + 5x - 11$ 75. $h(x) = -9x^3 + x + 12$

76. $f(x) = \frac{1}{2}x^3 - 3x^2 + \frac{2}{3}x + 5$

77. $g(x) = 6.23x^2 + 1.98x - 3.34$

78. $h(x) = -7.10x^3 - 5.02x^2 + 11.19x - 16.07$

79. $f(x) = 2\sqrt{x} + 7x^2 - \frac{1}{2}x^3$

80. $g(x) = \sqrt[5]{x} - \sqrt{x} + \frac{1}{x}$ 81. $h(x) = 2.08x^{3.79}$

For Exercises 82 and 83:

(a) Write the domain of f in interval notation.

(b) Use the addition rule for fractions to rewrite and simplify f.

(c) Find $f'(x)$ using the differentiation rules.

(d) Write the domain of f' in interval notation.

82. $f(x) = \frac{3x^3 - 2x^2 + 6x + 2}{x}$

83. $f(x) = \frac{6x^4 + 25x^3 - 9x + 11}{x^2}$

For Exercises 84 and 85:

(a) Determine the derivative for the given function using the differentiation rules.

(b) Determine the slope of the line tangent to the graph of the function at the indicated x-value.

(c) Determine an equation for the line tangent to the graph of the function at the indicated x-value.

(d) Graph the function and the tangent line in the same viewing window.

84. $f(x) = \frac{1}{x^2}$ at $x = 4$ 85. $f(x) = x^{3/4}$ at $x = 16$

86. The total amount of energy that an appliance has used since time $t = 0$ is given by

$$E(t) = 6t + 9\sqrt{t} + 0.02t^{3/2}$$

where t is in hours and $E(t)$ is in watt-hours. Find and interpret $E'\left(\frac{1}{3600}\right)$.

87. A rocket taking off has an altitude from the ground of $h(t) = 200t^2$, where $h(t)$ is in feet and t is in seconds. Find and interpret $h'(4)$.

88. Over the course of one year, the average price of homes for sale in a certain area follows the function $P(t) = -30t^2 + 2100t + 163{,}250$, where $P(t)$ is in dollars and t is in weeks after January 1.

(a) Find the average rate of change of the price from $t = 10$ weeks to $t = 30$ weeks.

(b) Graph the function on a calculator and find the maximum price and the time at which that price holds.

89. It has been said that from 1980 through the late 1990s the clock speed of commercially available home computer microprocessors doubled every 18 months. (This is called Moore's law.)

(a) Assuming that the clock speed was 1.172 MHz in 1985 ($t = 0$), write an equation for the clock speed $S(t)$ at t years after 1985.

(b) Find and interpret $S'(13)$ by using successively smaller values of h.

In Exercises 90–98, find the derivative of the given function.

90. $f(x) = x^3(6x^2 - 3x + 8)$

91. $g(x) = -3x^2(6x^3 - 2x^2 + 9x + 10)$

92. $h(x) = (2x^2 - 5x + 1)(3x^2 + 4x - 1)$

93. $f(x) = (3x^{1/2} + 5x)(-2x^{2/5} + x - 9)$

94. $g(x) = (5\sqrt{x} - 3x - 1)(4\sqrt{x} - 7x)$

95. $h(x) = \dfrac{3x + 2}{x - 1}$

96. $f(x) = \dfrac{2x^2 + 2x - 5}{3x^2 - x + 9}$

97. $g(x) = \dfrac{3\sqrt{x} - 2}{3x + 1}$

98. $h(x) = \dfrac{4}{3}(6x^3 - x^2 + 2x + 11)$

For Exercises 99 and 100:

(a) Determine the derivative.

(b) Determine an equation of the line tangent to the graph of the function at the indicated x-value.

(c) Graph the function and the tangent line in the same viewing window.

(d) Use the $\dfrac{dy}{dx}$ or DRAW TANGENT command on your calculator to check your work.

99. $f(x) = -3x^2(3x + 5)$ at $x = 2$

100. $g(x) = \dfrac{2x^2 - 3}{x}$ at $x = 2$

For each function graphed in Exercises 101 and 102, state the x-values for which the derivative does not exist.

101.

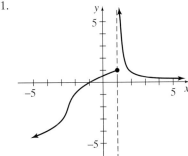

102.

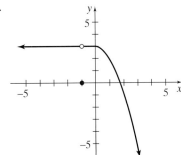

For Exercises 103 and 104:

(a) Graph the function.

(b) Compute the derivative.

(c) Determine the x-values for which the derivative is undefined.

(d) Compare the result from part (c) with the behavior of the graph of the function in part (a). That is, for the x-values for which the derivative is undefined, does the graph of the function have a sharp turn or a vertical tangent or is it discontinuous?

103. $f(x) = x^{5/8}$ 104. $g(x) = \dfrac{x^2 - 9}{3 - x}$

105. For $f(x) = \sqrt[4]{x^3} \cdot (3x - 1)(2x + 9)$, find $f'(x)$.

106. A musical instrument manufacturer's monthly sales of clarinets are given by $q(t) = 200e^{-0.3t}$, where t is in months after January 1.

(a) Compute $q(4)$ and interpret.

(b) Compute $q'(4)$ and interpret.

Applications of the Derivative

Graph of profit function
$P(x) = -2.3x^2 + 445x - 1500.$

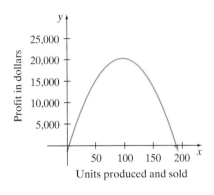

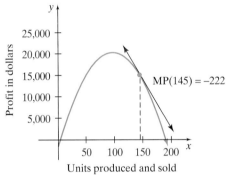

$MP(145) = -222$

Profit lost by producing and selling
the 146th refrigerator.

What We Know

In Chapter 2, we found that the instantaneous rate of change gave us the slope of the line tangent to the graph of a function. The way we found this new rate of change was through the derivative. We also saw applications of the derivative.

Where Do We Go

In this chapter, we will use the derivative to study a new quantity called the *differential*. We will use the differential to approximate solutions, to study the functions of business, and to measure errors.

SECTION 3.1   THE DIFFERENTIAL AND LINEAR APPROXIMATIONS

How often have you heard on TV or read in the newspapers a statement such as "If crime keeps growing at this rate. . . ." The problem with this statement is that rates do indeed change. This is the reason that we study calculus—to see the impact of these fluctuating rates of change. In this section, we will examine the mathematics of the statement made above, and how to use the **differential** to make short-term conclusions about applications. Then we will study the idea of **linear approximations,** which can be used to approximate the change in the dependent variable as changes in the independent variable values are made.

The Differential

The differential is a tool used to study the relationship between changes in the independent and dependent variable values. So let's begin by revisiting some tools that we introduced with the difference quotient.

From Your Toolbox

 Recall that for a function f on a closed interval $[x_1, x_2]$:

- $\Delta x = x_2 - x_1$ is called the *increment in x.* (Remember, in Chapter 2, we called this quantity h.)
- $\Delta y = f(x_2) - f(x_1)$ is called the *increment in y.*
- $m_{\text{sec}} = \dfrac{\Delta y}{\Delta x} = \dfrac{f(x + \Delta x) - f(x)}{\Delta x}$ is called the *difference quotient.*

Now we can establish a context for our discussion through a Flashback.

Flashback   **PRISON INMATE POPULATION REVISITED**

In Section 2.3, we modeled the total U.S. Federal Bureau of Prisons inmate population with the function

$$f(x) = \frac{1}{4}x^2 + \frac{9}{25}x + 24.5,^* \qquad 1 \le x \le 17$$

where x represents the number of years since 1979 and $f(x)$ represents the total inmate population in thousands.

(a) Evaluate $f(7) - f(6)$ and interpret this result.

(b) Determine the difference quotient with $x = 6$ and $\Delta x = 1$ and interpret. Compare this result with part (a).

---
*Source: *Based on data gathered at the Bureau of Prisons web site.*

Solution

(a) First, note that the values $x = 6$ and $x = 7$ correspond to the years 1985 and 1986, respectively.

$$\Delta y = f(7) - f(6)$$

$$= \left(\frac{1}{4}(7)^2 + \frac{9}{25}(7) + 24.5 \right)$$

$$- \left(\frac{1}{4}(6)^2 + \frac{9}{25}(6) + 24.5 \right)$$

$$= 39.27 - 35.66 = 3.61$$

So from 1985 to 1986 the U.S. federal prison inmate population increased by 3.61 thousand (or 3610 inmates).

(*Continued*)

(b) Computing the difference quotient with $x = 6$ and $\Delta x = 1$,

$$m_{sec} = \frac{f(x + \Delta x) - f(x)}{\Delta x} = \frac{f(6+1) - f(6)}{1}$$

$$= \frac{f(7) - f(6)}{1} = \frac{39.27 - 35.66}{1} = 3.61$$

This means that between the years 1985 and 1986, the U.S. federal prison inmate population increased by an average rate of $3610 \frac{\text{inmates}}{\text{year}}$. Notice that the result is the same as in part (a). In other words, **when $\Delta x = 1$, then $\Delta y = 3.61 = m_{sec}$**, where m_{sec} is the slope of the secant line.

Now let's mathematically try the scenario that we suggested at the beginning of this section. What if the prison population continued to grow at a rate constant with that of 1985? Will this assumption give an accurate prediction of the 1986 U.S. federal prison population? Using the derivative and the tangent line, we can find out. First, we find the growth rate at the beginning of 1985 ($x = 6$) using the derivative f', since we know that $f'(6)$ is the instantaneous rate of change at $x = 6$.

$$f'(x) = \frac{d}{dx}\left(\frac{1}{4}x^2 + \frac{9}{25}x + 24.5\right) = \frac{1}{2}x + \frac{9}{25}$$

$$f'(6) = \frac{1}{2}(6) + \frac{9}{25} = \frac{84}{25} = 3.36$$

Thus, the U.S. federal prison population was growing at a rate of $3.36 \frac{\text{thousand inmates}}{\text{year}}$ at the beginning of 1985. Assuming that the rate of change is constant during 1985, we use the equation of the line tangent to the curve at $x = 6$ to predict the 1986 federal inmate population. Example 1 illustrates the process.

EXAMPLE 1 **Using the Tangent Line Equation**

Determine an equation of the line tangent to the graph of the model $f(x) = \frac{1}{4}x^2 + \frac{9}{25}x + 24.5$ when $x = 6$. On the tangent line, determine y when $x = 7$ and interpret what this means.

SOLUTION

Knowing that $f(6) = \frac{1}{4}(6)^2 + \frac{9}{25}(6) + 24.5 = \frac{1783}{50} = 35.66$ and $f'(6) = 3.36$, we use the CF form of a line to find an equation of the tangent line at $x = 6$.

$$y = m(x - x_1) + y_1$$
$$y = 3.36(x - 6) + 35.66$$

Evaluating the tangent line equation when $x = 7$ (that is, the year 1986), we get

$$y = 3.36(7 - 6) + 35.66 = 3.36(1) + 35.66 = 39.02$$

This means that if the U.S. federal prison inmate population continued growing at the 1985 rate, the 1986 inmate population would be about 39.02 thousand. The number of inmates in 1986 using the model was $f(7) = 39.27$, meaning that the difference is only about 0.25 or 250 inmates. See Figure 3.1.1.

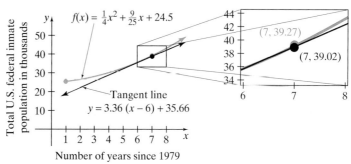

Figure 3.1.1 Graph of the line tangent to the graph of
$$f(x) = \frac{1}{4}x^2 + \frac{9}{25}x + 24.5 \text{ at } x = 6.$$

Interactive Activity

What would the prison population have been in 1994 if it kept growing at the 1985 rate? To find out, graph the model $f(x) = \frac{1}{4}x^2 + \frac{9}{25}x + 24.5$ and the tangent line equation $y = 3.36(x - 6) + 35.66$ in the viewing window [0, 17] by [0, 105]. Evaluate the model and the tangent line value y when $x = 15$. Compare the results. Explain the shortcoming of using the tangent line to make long-term predictions.

✓ **CHECKPOINT 1**

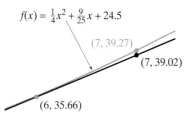

Figure 3.1.2

Now work Exercise 47.

We can make two important observations from the results of the Flashback and from Example 1.

1. From part (a) of the Flashback, we found that the actual change as x changed from 6 to 7 ($\Delta x = 1$) was $\Delta y = 3.61$ thousand prisoners. We also found that $f'(6)$, the instantaneous rate of change at $x = 6$, was $3.36 \frac{\text{thousand inmates}}{\text{year}}$. Thus, for $\Delta x = 1$, Δy is approximately equal to $f'(6)$.
2. Example 1 showed that $f(7) = 39.27$ and the y-value on the tangent line at $x = 7$ ($y = 39.02$) are very close. We display these observations graphically in Figure 3.1.2.

The concept of the *differential* is imbedded in observation 1, while the concept of a *linear approximation* is imbedded in part 2. We will discuss the linear approximation later in this section, but before that, let's examine the differential and its applications.

The Differential

If $y = f(x)$ where f is a differentiable function then we define the following:

- The **differential in x,** denoted by dx, of the independent variable is given by

$$dx = \Delta x$$

- The **differential in y,** denoted by dy, of the dependent variable is given by

$$dy = f'(x)\, dx$$

NOTE: Δy and dy are not the same. Δy is the *actual* change in the dependent variable values, whereas dy is an *approximation* of this change. If dx is small, then $dy \approx \Delta y$.

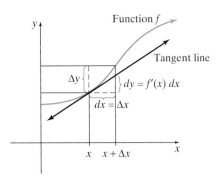

Figure 3.1.3

Many disciplines think of dy as the change in y on the tangent line. We show the difference between Δy and dy in Figure 3.1.3.

EXAMPLE 2 | **Computing the Differential**

For the function $f(x) = x^2 + 3x - 8$, evaluate Δy and dy for $x = 2$ and $\Delta x = dx = 0.1$.

SOLUTION

We can compute the change in y, Δy, by evaluating $f(x + \Delta x) - f(x)$.

$$\Delta y = f(2 + 0.1) - f(2) = f(2.1) - f(2)$$
$$= ((2.1)^2 + 3(2.1) - 8) - ((2)^2 + 3(2) - 8)$$
$$= 2.71 - 2 = 0.71$$

Since $f'(x) = 2x + 3$, the differential in y for the function is

$$dy = f'(x)\,dx$$
$$dy = (2x + 3)\,dx$$

So at $x = 2$ and $\Delta x = dx = 0.1$ we get

$$dy = (2(2) + 3)(0.1) = 7(0.1) = 0.7 \qquad \blacksquare$$

✓CHECKPOINT 2

Now work Exercise 17.

Our next example illustrates how differentials can be used in business applications.

EXAMPLE 3 | **Using Differentials in Applications**

The Garland Toddler Company determined that the price–demand function for their new pacifier/thermometer is given by

$$p(x) = 15 - 0.2\sqrt{x}$$

where x represents the quantity demanded and $p(x)$ represents the unit price in dollars.

(a) Compute Δp, the actual change in price, for $x = 100$ and $\Delta x = dx = 1$.

(b) Determine the differential dp for the price–demand function. Use the differential dp to approximate the change in price that would cause the quantity demanded to increase from 100 to 101 units.

SOLUTION

(a) Using $x = 100$ and $\Delta x = dx = 1$, we get the increment in p as

$$\Delta p = p(x + \Delta x) - p(x)$$
$$= p(100 + 1) - p(100) = p(101) - p(100)$$
$$\approx 12.99002 - 13 = -0.00998$$

(b) For the function $p(x) = 15 - 0.2\sqrt{x}$, the differential in the dependent variable p is

$$dp = p'(x)\,dx = \frac{d}{dx}(15 - 0.2\sqrt{x})\,dx$$
$$= -0.2\left(\frac{1}{2}x^{-1/2}\right)dx$$
$$= (-0.1x^{-1/2})\,dx$$

Evaluating dp when $x = 100$ and $\Delta x = dx = 101 - 100 = 1$, we get

$$dp = (-0.1(100)^{-1/2})(1)$$
$$= (-0.1)(0.1)(1) = -0.01$$

This means that as the quantity demanded changes from 100 to 101 units, the unit price of the pacifier/thermometer would decrease by about 1 cent.

Notice how close the numerical solutions to parts (a) and (b) are in Example 3. This is evidence that **if dx is small, then $dy \approx \Delta y$.**

Linear Approximations

To get a better picture of why linear approximations are used, let's return to the model for the U.S. Federal Bureau of Prisons total inmate population

$$f(x) = \frac{1}{4}x^2 + \frac{9}{25}x + 24.5, \qquad 1 \le x \le 17$$

Let's investigate the graph of f around $x = 16$. We graph f in smaller and smaller viewing windows as shown in Figure 3.1.4. Notice that as we zoom in on f for values close to $x = 16$ the graph appears to straighten out and look like a line. Let's see if we can take advantage of this characteristic to approximate function values.

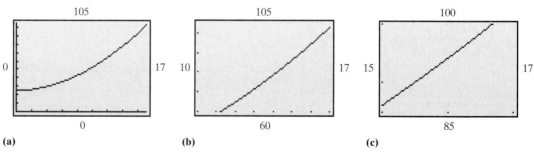

Figure 3.1.4

EXAMPLE 4

Computing a Linear Approximation

For the model of the total U.S. Federal Bureau of Prisons inmate population:

(a) Determine dy and evaluate when $x = 15$ and $\Delta x = dx = 1$.

(b) Evaluate $f(16)$ and interpret.

(c) Add $f(15)$ to the result of part (a) and compare to part (b).

SOLUTION

(a) The differential in y for the model is

$$dy = f'(x)dx = \frac{d}{dx}\left(\frac{1}{4}x^2 + \frac{9}{25}x + 24.5\right)dx = \left(\frac{1}{2}x + \frac{9}{25}\right)dx$$

So for $x = 15$ and $dx = 1$, the differential in y is

$$dy = f'(15)(1)$$

$$= \left(\frac{1}{2}(15) + \frac{9}{25}\right)(1) = \frac{393}{50} = 7.86$$

(b) First, note that $x = 16$ corresponds to the year $1979 + 16 = 1995$. Evaluating, we get

$$f(16) = \frac{1}{4}(16)^2 + \frac{9}{25}(16) + 24.5 = 94.26$$

Thus, according to the model, in 1995 there were about 94.26 thousand inmates in U.S. federal prisons.

(c) Evaluating the model at $x = 15$ gives

$$f(15) = \frac{1}{4}(15)^2 + \frac{9}{25}(15) + 24.5 = 86.15$$

Adding this result to part (a) gives us

$$f(15) + dy = f(15) + f'(15)(1) = 86.15 + 7.86 = 94.01$$

Notice how close this result is to part (b). In other words, $f(16) \approx f(15) + f'(15)(1)$. Numerically, the difference between this result and the result of part (b) is

$$f(16) - [f(15) + f'(15)(1)] = 94.26 - 94.01 = 0.25$$

Notice in Example 4 that we can rewrite $f(16)$ as $f(15 + 1)$. So it appears, for a small value of dx, that $f(15 + 1) \approx f(15) + f'(15)(1)$. This is exactly what a linear approximation is. We generalize this new tool in the following definition.

Linear Approximation

For a differentiable function f, where $y = f(x)$, the **linear approximation** of f is given by

$$f(x + dx) \approx f(x) + dy = f(x) + f'(x)\,dx$$

when the value of dx is small.

Both the linear approximation and the differential will be the cornerstone of the remainder of this chapter. Let's now try a classic application of linear approximations.

EXAMPLE 5 | **Using Linear Approximations to Make Estimates**

Use a linear approximation to estimate $\sqrt[3]{63}$, and compare with the calculator value rounded to four decimal places.

SOLUTION

At first, it appears that we have nothing to work with to get the differential. But, on closer inspection, we see that a number very close to $\sqrt[3]{63}$ has an integer cube root, that is, $\sqrt[3]{64} = 4$. Thus, we can use the linear approximation to make an estimate of $\sqrt[3]{63}$ by letting $f(x) = \sqrt[3]{x}$, $x = 64$ (a perfect cube), and $dx = 63 - 64 = -1$. So we want to find

$$f(x + dx) \approx f(x) + f'(x)\,dx$$
$$= f(64 + (-1)) \approx f(64) + f'(64)(-1)$$

Knowing that $f(x) = \sqrt[3]{x} = x^{1/3}$, we get $f'(x) = \frac{1}{3}x^{-(2/3)} = \frac{1}{3\sqrt[3]{x^2}}$. So the linear approximation is

$$\sqrt[3]{63} = f(64 + (-1)) \approx f(64) + f'(64)(-1)$$

$$= \sqrt[3]{64} + \frac{1}{3\sqrt[3]{(64)^2}}(-1)$$

$$= 4 + \frac{1}{3(16)}(-1)$$

$$= 4 - \frac{1}{48} = 3\frac{47}{48}$$

As a decimal number, $3\frac{47}{48} \approx 3.9792$, while $\sqrt[3]{63} \approx 3.9791$. Thus, we find that the difference using the linear approximation is only about one ten-thousandth. ✦

✓CHECKPOINT 3 | Now work Exercise 31.

SUMMARY |

This section began with a discussion of the **differential in x** denoted by dx and the **differential in y,** where

- $dy = f'(x)\,dx$

We said that if dx is small in value then $\Delta y \approx dy$. Then we introduced the **linear approximation** and stated that, when dx is small,

- $f(x + dx) \approx f(x) + dy = f(x) + f'(x)\,dx$

SECTION 3.1 EXERCISES

In Exercises 1–16, find dy for the given function by determining $f'(x)\,dx$.

1. $y = 6x$

2. $y = -3x$

3. $f(x) = -3x^2 + 2x$

4. $f(x) = 7x^3 + 3x^2 - 13$

5. $y = \dfrac{5}{x-1}$

6. $y = \dfrac{-2}{x+3}$

7. $f(x) = \dfrac{x}{x+1}$

8. $f(x) = \dfrac{2x}{x-3}$

9. $y = \sqrt{x} + \dfrac{2}{x}$

10. $y = \sqrt[3]{x} - \dfrac{3}{x^2}$

11. $y = \dfrac{1}{\sqrt{x}} + \sqrt[3]{x^2}$

12. $y = \dfrac{1}{2\sqrt{x}} - \sqrt{x}$

13. $f(x) = \dfrac{x^2+1}{x^2-1}$

14. $f(x) = \dfrac{x^2+3}{x^2-3}$

15. $y = 3x^{1.7} + 7x^{0.8} + 3$

16. $y = 4x^{2.2} - 6x^{0.7} + 7$

In Exercises 17–28 evaluate Δy and dy for each function at the indicated values.

✓ 17. $y = x^2 - 2x - 1$, $x = 2$, $\Delta x = dx = 0.1$

18. $y = x^2 + 5x$, $x = 1$, $\Delta x = dx = 0.2$

19. $y = 750 + 5x - 2x^2$, $x = 50$, $\Delta x = dx = 2$

20. $y = 1000 + 2x - 3x^2$, $x = 100$, $\Delta x = dx = 1$

21. $y = 100 - \dfrac{270}{x}$, $x = 9$, $\Delta x = dx = 0.5$

22. $y = 75 - \dfrac{150}{x}$, $x = 5$, $\Delta x = dx = 0.5$

23. $y = \sqrt{x}$, $x = 2$, $\Delta x = dx = 0.1$

24. $y = 3\sqrt{x}$, $x = 1.5$, $\Delta x = dx = 0.1$

25. $f(x) = \dfrac{x^2+1}{x^2-1}$, $x = 2$ and $\Delta x = dx = 0.1$

26. $f(x) = \dfrac{x^2-5}{2x^2+1}$, $x = 3$ and $\Delta x = dx = 0.1$

27. $y = 2x^2(3x^2 - 2x)$, $x = 1$ and $\Delta x = dx = 0.1$

28. $y = x^3(x^2 - 1)$, $x = 2$ and $\Delta x = dx = 0.1$

For Exercises 29–36, use the linear approximation to estimate the values of the given numbers. Compare to the calculator value when rounded to four decimal places.

29. $\sqrt{26}$

30. $\sqrt{8.9}$

✓ 31. $\sqrt[3]{26}$

32. $\sqrt[3]{124}$

33. $\sqrt[4]{15.8}$

34. $\sqrt[4]{15}$

35. $\dfrac{2}{\sqrt{50}}$

36. $\dfrac{3}{\sqrt[3]{7}}$

APPLICATIONS

37. The radius of a circle increases from an initial value of $r = 5$ inches by an amount $\Delta r = dr = 0.2$ inch. Estimate the corresponding increase in the circle's area by evaluating dA. The area of a circle is given by the function $A(r) = \pi r^2$.

38. The radius of a circle increases from an initial value of $r = 8$ inches by an amount $\Delta r = dr = 0.1$ inch. Estimate the corresponding increase in the circle's area by evaluating dA. The area of a circle is given by the function $A(r) = \pi r^2$.

39. The Dakorn Company determines that the annual cost of covering its employees' eye and dental insurance can be modeled by the function

$$y = 1000 + 110x^{1/2}$$

where x represents the number of employees covered and y represents the annual insurance cost in dollars. Determine dy and use the differential to approximate the increase in cost if the number of employees increases from $x = 250$ to 254.

40. In a study conducted by the Northern Aid Organization, the number of people identified as having incomes below the poverty level in a certain region of the country can be modeled by the function

$$y = 10 + 707x^{1/2}$$

where x is the total population of the region in thousands and y represents the number of people identified as having incomes below the poverty level in thousands. Determine dy and use the differential to approximate the increase in the number of people below the poverty level if the population in the region increases from $x = 20$ to 22.

41. The Paulson Motor Company estimates that the weekly cost of producing its most popular custom sport utility vehicle can be modeled by

$$y = 0.22x^3 - 2.35x^2 + 14.32x + 10.22$$

where x represents the number of custom sport utility vehicles produced each week and y represents the total cost in thousands of dollars. Determine dy and use the differential to approximate the increase in cost if the weekly production is increased from $x = 30$ to 33 vehicles.

42. The A & D Publishing Company finds that its cost of printing textbooks can be modeled by

$$y = 0.02x^3 - 0.6x^2 + 9.15x + 98.43$$

where x is the number of textbooks printed in thousands and y is the cost of printing the textbooks in thousands. Determine dy and use the differential to approximate the increase in printing cost if the number of textbooks printed each day is increased from $x = 19$ to 20.

43. Using past records, the PowerSet Company has estimated that the association between its monthly sales of volleyballs and its advertising can be modeled by

$$y = 120x - 2.4x^2$$

where x represents the amount spent on advertising in thousands of dollars and y is the number of volleyballs sold in hundreds. Determine dy and use the differential to approximate the increase in sales caused by increasing advertising from $x = 10$ to 11.

44. Repeat Exercise 43, except use the model

$$y = 189.24x - 3.5x^2.$$

45. Compute the actual change Δy in sales in Exercise 43 and compare Δy to the approximation dy.

For each of the models in Exercises 46–48:

(a) Write an equation of the tangent line at $x = 2$.

(b) Find the y-value on the tangent line when $x = 3$. Interpret what this estimate means. See Example 1.

46. The percentage of people living alone in Arizona can be modeled by

$$f(x) = -0.62x^2 + 3.14x + 64.66, \qquad 0 \le x \le 6$$

where x represents the number of years after 1990 and $f(x)$ represents the percentage of people living alone in Arizona.

✓ 47. The percentage of people living alone in Florida can be modeled by

$$f(x) = 0.08x^2 - 0.11x + 65.1, \qquad 0 \le x \le 6$$

where x represents the number of years after 1990 and $f(x)$ represents the percentage of people living alone in Florida.

48. The percentage of people living alone in Ohio can be modeled by

$$f(x) = 0.11x^3 - 0.94x^2 + 1.73x + 68.71, \qquad 0 \le x \le 6$$

where x represents the number of years after 1990 and $f(x)$ represents the percentage of people living alone in Ohio.

SECTION PROJECT

The average salary of major league baseball players can be modeled by

$$f(x) = -0.5x^4 + 9.03x^3 - 39.04x^2 + 70.03x + 335.12,$$
$$1 \le x \le 11$$

where x represents the number of years since 1984 and $f(x)$ represents average salary of major league baseball players in thousands of dollars.

(a) Compute $f(7) - f(6)$ and interpret.

(b) Determine the differential dy for the salary function.

(c) Use the differential to approximate the change in salary as x changes from 6 to 7 and interpret.

(d) Determine the equation of the tangent line at $x = 6$.

(e) Find the y-value of the tangent line when $x = 7$ and interpret.

SECTION 3.2   MARGINAL ANALYSIS

Many decisions made by managers in business involve analyzing the effect on the dependent variable when a small change is made to a specific independent value. For example, a company may wish to consider changing the price of an item and examining how this change affects the revenue or profit of the product. **Marginal analysis** can be defined as the study of the amount of change in the dependent variable that results from a single unit change in an independent variable. A **unit change** means a change of one single unit. This change in the dependent variable is a direct application of our now familiar tool—the derivative.

Marginal Analysis Concept

Let's start this discussion of business functions by reviewing their definitions.

From Your Toolbox

1. The *price–demand function p* gives us the price $p(x)$ at which people buy exactly x units of product.

2. The cost of producing x units of product with variable costs m and fixed costs b is given by the *cost function:*

$$C(x) = mx + b = \left(\begin{array}{c}\text{variable}\\\text{costs}\end{array}\right)x + \left(\begin{array}{c}\text{fixed}\\\text{cost}\end{array}\right)$$

Note that since variable costs are often expressed as a function, $C(x)$ may be a higher-order polynomial function.

3. The total revenue R generated by producing and selling x units of product at price $p(x)$ is given by the *revenue function:*

$$R(x) = x \cdot [p(x)] = \left(\begin{array}{c}\text{quantity}\\\text{sold}\end{array}\right) \cdot \left(\begin{array}{c}\text{unit}\\\text{price}\end{array}\right)$$

4. The profit P generated after producing and selling x units of a product is given by the *profit function:*

$$P(x) = R(x) - C(x) = \text{revenue} - \text{cost}$$

We referred to marginal analysis as the study of the dependent variable if the independent variable had a single unit change. Let's say that we want to study the marginal cost at a production level x, given a cost function C. We can start by determining the *actual change* in cost, denoted by ΔC, when the number of units produced is increased by 1.

$$\begin{pmatrix} \text{actual change} \\ \text{in cost} \end{pmatrix} = \begin{pmatrix} \text{cost to produce} \\ x + 1 \text{ units} \end{pmatrix} - \begin{pmatrix} \text{cost to produce} \\ x \text{ units} \end{pmatrix}$$

In terms of the cost function C, this is

$$\Delta C = C(x + 1) - C(x)$$

This relationship is illustrated in Figure 3.2.1.

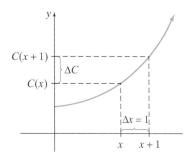

Figure 3.2.1 ΔC is the actual change in cost.

Let's try an example of computing this actual change. In Section 2.3 we found that the cost of producing q units of a recreation vehicle can be modeled by

$$C(q) = 100 + 60q + 3q^2$$

where q represents the number of vehicles produced and $C(q)$ is the cost in hundreds of dollars. To find ΔC for $q = 5$ and $\Delta q = 1$, we seek the difference in cost where

$$\begin{aligned} \Delta C &= C(q + \Delta q) - C(q) \\ &= C(5 + 1) - C(5) \\ &= C(6) - C(5) \\ &= 568 - 475 = 93 \text{ hundred dollars} \end{aligned}$$

Since $C(6)$ is the cost of producing 6 vehicles and $C(5)$ is the cost of producing the first 5 vehicles, then $\Delta C = C(6) - C(5)$ must represent the cost of producing the sixth vehicle. Thus, the cost of producing the sixth vehicle is $9300.

An exact value can always be computed by evaluating $\Delta C = C(x + 1) - C(x)$, but in Section 3.1 we found that, when dx is small, $\Delta C \approx dy$, where dy is the differential in y given by $dy = C'(x)\, dx$. Thus, for dx being small,

$$\begin{pmatrix} \text{actual change} \\ \text{in cost, } \Delta C \end{pmatrix} \approx \begin{pmatrix} \text{differential} \\ \text{in } C, \text{ called } dC \end{pmatrix}$$

But for marginal analysis, $dx = 1$. This gives us

$$\begin{pmatrix} \text{marginal cost at} \\ \text{production level } x \end{pmatrix} \approx \begin{pmatrix} \text{differential} \\ \text{in } C, \text{ where } dx = 1 \end{pmatrix} = C'(x) \cdot (1) = C'(x)$$

Consequently, the marginal cost function, denoted by MC is simply the derivative of the cost function. Also note that MC is the differential in the cost

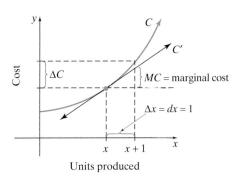

Figure 3.2.2

function where $dx = 1$. This is sometimes called a *unit differential*. The relationship of $C(x)$, ΔC, and $MC(x)$ is shown in Figure 3.2.2.

Marginal Cost Function

The **marginal cost function**, MC, given by $MC(x) = C'(x)$, is the approximate cost of producing one additional unit at a production level x.

Normally, when we write the units for a rate of change, we write them as $\frac{\text{dependent units}}{\text{independent units}}$. But, since $dx = 1$ in these cases, we just use the dependent units, which are usually measured in dollars.

EXAMPLE 1

Computing a Marginal Cost Function

The cost of producing q units of a certain recreational vehicle can be modeled by

$$C(q) = 100 + 60q + 3q^2$$

where q represents the number of vehicles produced and $C(q)$ is the cost in hundreds of dollars. Compute the marginal cost $MC(q) = C'(q)$. Evaluate $MC(5)$ and interpret.

SOLUTION

Computing the derivative of $C(q)$ with respect to q, we get

$$MC(q) = C'(q) = \frac{d}{dq}\left(100 + 60q + 3q^2\right) = 60 + 6q$$

Evaluating the marginal cost function at $q = 5$ yields

$$MC(5) = 60 + 6(5) = 90$$

So the approximate cost of producing the next, or sixth vehicle is 90 hundred dollars, or $9000. Earlier in this section, we showed that the *exact cost* of producing the sixth vehicle is $9300, so the error of the marginal cost approximation is $300. ■

✓ CHECKPOINT 1

Now work Exercise 3.

Other Marginal Business Functions

At the beginning of this section, we stated that marginal analysis focused on the change in cost, revenue, and profit with a single unit change of the independent

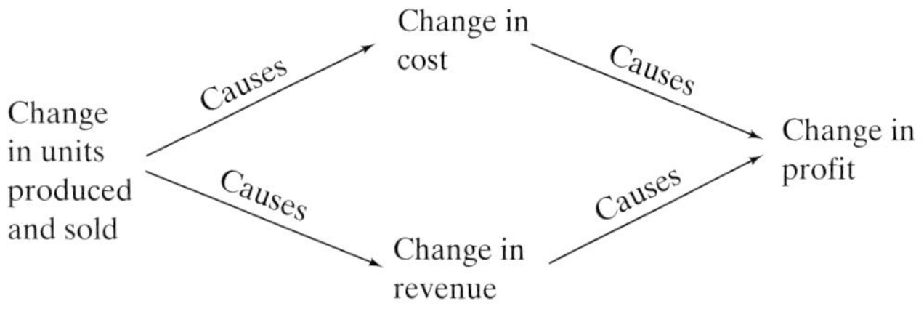

Figure 3.2.3

variable. This small change can have a significant impact on profit. This is because a change in price can cause changes in the quantity produced and sold, the cost and revenue, and the profit. See Figure 3.2.3.

So it seems logical that we need to examine the marginal functions that are associated with the remainder of the business functions.

Marginal Business Functions

- The **marginal revenue function** MR, given by $MR(x) = R'(x)$, is the approximate loss or gain in revenue by producing one additional unit at a production level x.
- The **marginal profit function** MP, given by $MP(x) = P'(x)$, is the approximate loss or gain in profit by producing one additional unit at a production level x.

These definitions show that differentiating the business functions gives us the marginal business functions. Let's see how these functions can be used to affect profit.

EXAMPLE 2 | **Using the Marginal Profit Function**

The FrezMore Company has determined that its cost of producing x refrigerators can be modeled by

$$C(x) = 2x^2 + 15x + 1500, \qquad 0 \le x \le 200$$

where x is the number of refrigerators produced each week and $C(x)$ represents the weekly cost in dollars. The company also determines that the price–demand function for the refrigerators is

$$p(x) = -0.3x + 460$$

(a) Determine the profit function P for the refrigerators.

(b) Determine the marginal profit function.

(c) Compute $MP(60)$ and $MP(145)$ and interpret the results.

SOLUTION

(a) Before we determine the profit function, we must first find the revenue function. Since the price–demand function for producing and selling x re-frigerators is $p(x) = -0.3x + 460$, the revenue from selling x refrigerators is

$$R(x) = x[p(x)] = x(-0.3x + 460) = -0.3x^2 + 460x$$

The profit function can now be written as

$$
\begin{aligned}
P(x) &= R(x) - C(x) \\
&= (-0.3x^2 + 460x) - (2x^2 + 15x + 1500) \\
&= -0.3x^2 + 460x - 2x^2 - 15x - 1500 \\
&= -2.3x^2 + 445x - 1500
\end{aligned}
$$

(b) Differentiating to determine the marginal profit function yields

$$MP(x) = \frac{d}{dx}(-2.3x^2 + 445x - 1500) = -4.6x + 445$$

(c) Evaluating the marginal profit at $x = 60$ gives

$$MP(60) = -4.6(60) + 445 = 169$$

This means that at a production level of $x = 60$ refrigerators each week there is about \$169 profit for manufacturing and selling the 61st refrigerator. Finally, evaluating $MP(x)$ at $x = 145$, we get

$$MR(145) = -4.6(145) + 445 = -222$$

Thus, at a production level of $x = 145$ refrigerators each week there is about \$222 loss in profit for manufacturing and selling the 146th refrigerator. Figure 3.2.4 shows that the profit at $x = 145$ is decreasing at a rate of $222 \frac{\text{dollars}}{\text{refrigerator}}$ because the slope of the tangent line at $x = 145$ is -222. ◼

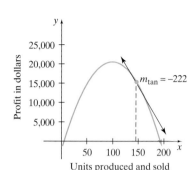

Figure 3.2.4

An extension of the marginal concept is the use of linear approximation for business functions. Recall from Section 3.1 that for a differentiable function f where $y = f(x)$, the *linear approximation* of f is given by

$$f(x + dx) \approx f(x) + dy = f(x) + f'(x)\,dx$$

when the value of dx is small.

In the case of marginal analysis, $dx = 1$. This means that we can write a linear approximation for a cost function with $f(x) = C(x)$ as

$$C(x + 1) \approx C(x) + C'(x)$$

or, in terms of the marginal cost function, as

$$C(x + 1) \approx C(x) + MC(x)$$

Consequently, we can apply the linear approximation concept to marginal analysis to easily make approximations for cost, revenue, and profit functions with information that is already known. This is particularly useful if the data given are tabular or some partial results are already available. Let's apply this new piece of information on differentials to find a linear approximation for a revenue function.

Interactive Activity

For Example 2, use your calculator to:

- Approximate the maximum number of refrigerators that a company can make in a week and still realize a profit.
- Determine the production level x at which the profit is at a maximum. Now find the x-intercept of the graph of MP. Make a conjecture about the association between the production level that yields maximum profit and the x-intercept of the graph of MP.

EXAMPLE 3 | **Computing Linear Approximations for a Revenue Function**

The revenue function for the production of x refrigerators per week in Example 2 was given as $R(x) = -0.3x^2 + 460x$.

(a) Compute $R(110)$ and $MR(110)$ and interpret these results.

(b) Use the solution from part (a) to get a linear approximation for $R(111)$. Compare your answer to the exact value of $R(111)$.

SOLUTION

(a) For the revenue function $R(x) = -0.3x^2 + 460x$, we evaluate to get

$$R(110) = -0.3(110)^2 + 460(110) = -0.3(12,100) + 50,600$$
$$= -3630 + 50,600 = 46,970$$

So, when producing and selling 110 refrigerators per week, the revenue realized is $46,970. Knowing that $R(x) = -0.3x^2 + 460x$, we differentiate to get the marginal revenue function as

$$MR(x) = R'(x) = \frac{d}{dx}(-0.3x^2 + 460x)$$

$$= -0.6x + 460$$

So the marginal revenue at a production level of $x = 110$ is

$$MR(110) = -0.6(110) + 460$$

$$= -66 + 460 = 394$$

This means that the revenue gained from producing and selling one additional refrigerator at a production level of $x = 110$ is about $394.

(b) To get an approximation for $R(111)$, use the linear approximation for the revenue function

$$R(x + 1) \approx R(x) + MR(x)$$

Using this, along with the result from part (a), we can get an approximation for $R(111)$ as

$$R(111) = R(110 + 1) \approx R(110) + MR(110)$$

$$= 46,970 + 394 = 47,364$$

The exact revenue using the given revenue function is $R(111) = \$47,363.70$, so the error from using the approximation is only $0.30! The key to the closeness of this linear approximation is that the change in the independent variable is relatively small. ◆

Average Business Functions

Many times in business situations, financial reports are simplified so that the numerical results can be easily understood. Often managers are interested in the *per unit* cost of a product, which is usually easier to work with than the total cost for the production of x units of a product. For example, it is much easier for a manager to think of the production costs at a music production company to be $5.25 per CD or for a cycling company to think of it costing $215 to produce each bike.

To find this per unit or *average* for the cost function, we take the total cost and divide it by the number of items produced. In words,

$$\text{per unit cost} = \frac{\text{total cost to produce } x \text{ items}}{\text{number of items produced, } x}$$

Let's see how this works through a Flashback.

Flashback

FASHION MYSTIQUE REVISITED

In Section 1.2, we found that the cost function of copying promotional flyers for The Fashion Mystique was given by

$$C(x) = 0.10x + 5$$

where x represents the number of copies produced and $C(x)$ represents the cost of producing x flyers.

(a) Compute $C(100)$ and interpret.

(b) Evaluate $\frac{C(100)}{100}$ and interpret.

(Continued)

Solution

(a) Evaluating the cost function at $x = 100$, we get

$$C(100) = 0.10(100) + 5 = 10 + 5 = 15 \text{ dollars}$$

This means that the cost of producing 100 promotional flyers is $15.

(b) We use the solution from part (a) to evaluate the expression $\frac{C(100)}{100}$.

$$\frac{C(100)}{100} = \frac{15}{100} = 0.15$$

Thus, at a production level of $x = 100$ flyers, the average cost is 15 cents per flyer.

The Flashback leads us to the following definition:

Average Business Functions

The **average cost function,** AC, which gives the per unit cost of producing x items, is given by

$$AC(x) = \frac{C(x)}{x}$$

The **average profit function,** AP, which gives the per unit profit of producing and selling x, is given by

$$AP(x) = \frac{P(x)}{x}$$

NOTE: Statistics students may remember that the average of a set of data is $\bar{x} = \dfrac{\text{sum of data}}{\text{number of data items}}$. This is why some textbooks denote the average cost function as $\overline{C}$.

Notice that the average revenue function is omitted from the average business functions definiton. This is because the average revenue function is $AR(x) = \frac{R(x)}{x} = \frac{x \cdot p(x)}{x} = p(x)$. So we see that the average revenue funtion is just another name for the price function. The average profit function is discussed in more detail in Exercises 17 through 20. Now let's take a look at an application of an average cost function.

EXAMPLE 4 | **Analyzing an Average Cost Function**

The Ventoux Athletic Shoe Manufacturing Company knows that, for its Stampeder model basketball shoes, the daily cost function can be modeled by

$$C(x) = 700\sqrt{x} + 5000, \qquad 0 \le x \le 500$$

where x is the number of pairs of shoes produced daily and $C(x)$ is the daily cost in dollars.

(a) Determine AC, the average cost function.

(b) Evaluate and interpret $C(400)$ and $AC(400)$.

SOLUTION

(a) Using the definition of average cost function, we get

$$AC(x) = \frac{C(x)}{x} = \frac{700\sqrt{x} + 5000}{x}$$

(b) Evaluating the cost function when $x = 400$ yields

$$C(400) = 700\sqrt{400} + 5000 = 700(20) + 5000 = 19,000$$

This means that the total cost of producing 400 pairs of shoes a day is $19,000. Evaluating $AC(x)$ when $x = 400$, we get

$$AC(400) = \frac{700\sqrt{400} + 5000}{400} = 47.5$$

Thus, the average cost of producing 400 pairs of shoes in a day is $47.50 for each pair.

✓**CHECKPOINT 2**

Now work Exercise 15a.

A manager may also wish to apply marginal analysis techniques to average business functions. The result is the **marginal average business functions.** These are found by computing the derivative of each of the average business functions, respectively. The result is a function that approximates the per unit cost (or profit) of producing one more item. Now let's define these marginal average business functions.

> ### Marginal Average Business Functions
>
> The **marginal average cost function** approximates the per unit cost for producing an additional item of a product and is given by
>
> $$MAC(x) = \frac{d}{dx}\left(\frac{C(x)}{x}\right)$$
>
> The **marginal average profit function** approximates the per unit profit for producing and selling an additional item of a product and is given by
>
> $$MAP(x) = \frac{d}{dx}\left(\frac{P(x)}{x}\right)$$

NOTE: To determine these marginal average profit functions, we first determine the average business function and then compute the derivative *in that order.*

The marginal average profit function is discussed in more detail in Exercises 17 through 20. Now let's take a look at an application of marginal average business functions.

EXAMPLE 5 | **Computing a Marginal Average Cost Function**

In Example 4, we found that the cost function for producing x pairs of Stampeder model basketball shoes was

$$C(x) = 700\sqrt{x} + 5000, \quad 0 \le x \le 500$$

(a) Compute the marginal average cost function MAC.

(b) Evaluate $MAC(400)$. Round your answer to the nearest hundredth and interpret.

SOLUTION

(a) In Example 4 we computed the average cost function as $AC(x) = \frac{700\sqrt{x} + 5000}{x}$. Before differentiating, we simplify to

$AC(x) = \dfrac{700\sqrt{x}}{x} + \dfrac{5000}{x} = \dfrac{700}{\sqrt{x}} + \dfrac{5000}{x} = 700x^{-1/2} + 5000x^{-1}$. Now, differentiating this average cost function with respect to x, we get

$$MAC(x) = \frac{d}{dx}(700x^{-1/2} + 5000x^{-1})$$
$$= -350x^{-3/2} - 5000x^{-2}$$

Simplifying the rational and negative exponents, we get the marginal average cost function

$$MAC(x) = -\frac{350}{\sqrt{x^3}} - \frac{5000}{x^2}$$

(b) Evaluating the result in part (a) at $x = 400$, we get

$$MAC(400) = -\frac{350}{\sqrt{(400)^3}} - \frac{5000}{(400)^2} = -0.075 \approx -0.08$$

This means that when 400 pairs of basketball shoes have been produced, the average cost per pair decreases by about \$0.08 for an additional pair produced. Note that the negative answer tells us that the per unit cost is decreasing.

SUMMARY

In this section, we revisited the business functions and from them derived the marginal business functions. These functions are found by differentiation, and they approximate the cost, revenue, or profit for producing one more item of a product. Then we discussed the average business functions, which were found by taking the business function and dividing by the independent variable. Finally, we discussed the marginal average business functions, which were the derivatives of the average business functions.

Important Functions

- **Marginal cost function:** $MC(x) = C'(x)$
- **Marginal revenue function:** $MR(x) = R'(x)$
- **Marginal profit function:** $MP(x) = P'(x)$
- **Average cost function:** $AC(x) = \dfrac{C(x)}{x}$
- **Average profit function:** $AP(x) = \dfrac{P(x)}{x}$
- **Marginal average cost function:** $MAC(x) = \dfrac{d}{dx}\left(\dfrac{C(x)}{x}\right)$
- **Marginal average profit function:** $MAP(x) = \dfrac{d}{dx}\left(\dfrac{P(x)}{x}\right)$

SECTION 3.2 EXERCISES

For Exercises 1–6, assume $C(x)$ is in dollars and complete the following:

(a) Determine the marginal cost function MC.

(b) For the given production level x, evaluate $MC(x)$ and interpret.

(c) Evaluate the actual change in cost by evaluating $C(x + 1) - C(x)$ and compare with the answer to part (b).

1. $C(x) = 23x + 5200;$ $x = 10$

2. $C(x) = 14x + 870;$ $x = 12$

✓ 3. $C(x) = \frac{1}{2}x^2 + 12.7x + 2100$; $x = 11$

4. $C(x) = \frac{1}{2}x^2 + 27x + 1200$; $x = 20$

5. $C(x) = 0.2x^3 - 3x^2 + 50x + 20$; $x = 30$

6. $C(x) = 0.08x^3 - 2x^2 + 10x + 70$; $x = 90$

In Exercises 7–12, the cost function C and the price–demand function p are given. Assume $C(x)$ and $p(x)$ are in dollars.

(a) Determine the revenue function R and the profit function P.

(b) Determine the marginal cost function MC and the marginal profit function MP.

7. $C(x) = 5x + 500$; $p(x) = 6$

8. $C(x) = 12x + 4500$; $p(x) = 15$

9. $C(x) = \frac{x^2}{100} + 7x + 1000$; $p(x) = \frac{-x}{20} + 15$

10. $C(x) = \frac{1}{100}x^2 + \frac{1}{2}x + 8$; $p(x) = \frac{-x}{200} + 1$

11. $C(x) = -0.001x^3 + 4x + 100, 0 \le x \le 70$; $p(x) = -0.005x + 7$

12. $C(x) = -0.002x^3 + 0.01x^2 + 2x + 50, 0 \le x \le 40$; $p(x) = -x^{-1/2} + 5$

APPLICATIONS

13. The Country Day Company determines that the daily cost of producing lawn tractor tires is given by

$$C(x) = 100 + 40x - 0.001x^2, \quad 0 \le x \le 300$$

where x represents the number of tires produced each day and $C(x)$ is the total cost, in dollars, of producing the tires. Determine MC, the marginal cost function. Evaluate and interpret $MC(200)$.

14. The Kelomata Company determines that the daily cost of producing patio swings can be modeled by

$$C(x) = 15{,}000 + 100x - 0.001x^2, \quad 0 \le x \le 200$$

where x represents the number of patio swings produced each day and $C(x)$ represents the daily production cost in dollars. Determine MC, the marginal cost function. Evaluate and interpret $MC(100)$.

✓ 15. For the cost function in Exercise 13:

✓ (a) Determine AC, the average cost function. Evaluate and interpret $AC(200)$.

(b) Determine MAC, the marginal average cost function. Evaluate and interpret $MAC(200)$.

16. For the cost function in Exercise 14:

(a) Determine AC, the average cost function. Evaluate and interpret $AC(100)$.

(b) Determine MAC, the marginal average cost function. Evaluate and interpret $MAC(100)$.

17. A telemarketer determines that the monthly profit from selling magazine subscriptions can be modeled by

$$P(x) = 5x + x^{1/2}, \quad 0 \le x \le 100$$

where x is the number of magazine subscriptions sold per month and $P(x)$ is the profit in dollars. Determine MP, the marginal profit function. Evaluate and interpret $MP(55)$.

18. A newspaper courier determines that the monthly profit from a typical newspaper route can be modeled by

$$P(x) = 2x - x^{1/2}, \quad 0 \le x \le 200$$

where x represents the number of subscribers and $P(x)$ represents the monthly profit in dollars. Determine MP, the marginal profit function. Evaluate and interpret $MP(110)$.

19. For the profit function in Exercise 17:

(a) Determine AP, the average profit function. Evaluate and interpret $AP(55)$.

(b) Determine MAP, the marginal average profit function. Evaluate and interpret $MAP(55)$.

20. For the profit function in Exercise 18:

(a) Determine AP, the average profit function. Evaluate and interpret $AP(110)$.

(b) Determine MAP, the marginal average profit function. Evaluate and interpret $MAP(110)$.

21. The GlobalTalk Company manufactures pocket pagers and finds that the fixed and variable costs to produce q pagers are $1200 and $12, per pager respectively.

(a) Find the cost function C in the form $C(q) = mq + b$, and use calculus to compute the marginal cost function $MC(q) = \frac{d}{dq}C(q)$.

(b) Evaluate $MC(100)$ and $MC(150)$ and interpret these results.

(c) Why are the answers in part (b) equal?

22. The NewJoy Company toy manufacturer has just produced a new doll action set that it sells to wholesalers for $20 each. The cost $C(x)$ in dollars to produce x action sets is given by the function $C(x) = 0.001x^2 + 4x + 5000$.

(a) Algebraically derive the profit function P and simplify it.

(b) Evaluate $P(1000)$ and interpret.

(c) Evaluate $MP(1000)$ and interpret.

23. The NewJoy Company hires a consulting firm to audit their books and consequently revise their price and cost functions to $p(x) = 23$ and $C(x) = \frac{x^2}{95} + \frac{7}{2}x + 5500$.

(a) Algebraically derive the profit function P and simplify it.

(b) Evaluate $P(500)$ and interpret.

(c) Evaluate $MP(500)$ and interpret.

24. The CraftEz Company determines that the price–demand function (in dollars) for their new make-at-home

picture frame is

$$p(x) = \frac{-x}{30} + 200$$

with a cost function of

$$C(x) = 60x + 72{,}000$$

where x represents the number of frames produced and $C(x)$ is the cost in dollars.

(a) Determine the revenue function R.

(b) Determine the profit function P. Find the smallest and largest production levels x so that the company realizes a profit. (That is, find the smallest and largest independent variable values so that the revenue is greater than the cost.)

(c) Evaluate $P'(3000)$ and interpret.

25. The Vroncom Company determines that the price–demand function for their handheld computer device is

$$p(x) = \frac{-x}{30} + 300$$

They know that their fixed costs are \$150,000 and variable cost is $30 \dfrac{\text{dollars}}{\text{device}}$.

(a) Determine the revenue function R and the cost function C in the form $C(x) = mx + b$.

(b) Determine the profit function P. Find the smallest and largest production levels x so that the company realizes a profit. (That is, find the smallest and largest independent values so that the revenue is greater than the cost.)

(c) Evaluate $P'(1000)$ and interpret.

For Exercises 26 and 27, consider a stockholder's report that lists the following information:

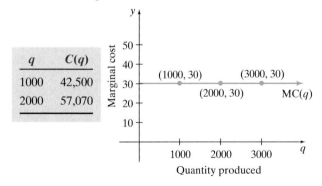

q	$C(q)$
1000	42,500
2000	57,070

Use the table and the graph of $MC(q)$ to get a linear approximation for the given cost function. Use the approximation $C(q + 1) \approx C(q) + MC(q)$.

26. $C(1001)$ 27. $C(2001)$

SECTION PROJECT

Consider the following data table for the cost and revenues at various production levels for a new brand of fuel injector.

NUMBER OF INJECTORS PRODUCED, x	COST, $C(x)$	REVENUE, $R(x)$
100	\$5,500	\$3,050
200	\$9,600	\$24,100
300	\$13,500	\$81,200
400	\$17,600	\$192,200

(a) Use your calculator to determine a linear regression model for the cost of producing the fuel injector in the form

$$C(x) = ax + b, \qquad 100 \le x \le 400$$

where x represents the number of fuel injectors produced and $C(x)$ represents the cost in dollars.

(b) Use your calculator to determine a cubic regression model for the revenue of producing and selling the fuel injectors in the form

$$R(x) = ax^3 + bx^2 + cx + d, \qquad 100 \le x \le 400$$

where x represents the number of fuel injectors produced and sold and $R(x)$ represents the revenue in dollars.

(c) Compute $MAC(x)$ and simplify the result.

(d) Evaluate $MAC(250)$ and interpret.

(e) Compute $MAP(x)$ and simplify the result.

(f) Evaluate $MAP(250)$ and interpret.

SECTION 3.3   MEASURING RATES AND ERRORS

Many times when computing values with differentials, particularly when those values are computed with regression models, there is a difference between the actual measurement and the computed one. In this section, we will quantify these differences using the tools of differential calculus by examining the association between the instantaneous rate of change and the error measurement by using the **relative error** and **percentage error**. We will also measure the amount of error resulting from using a linear approximation. Then we will introduce the **relative rate of change**, a measure that can be used to compare rates of different mathematical models.

Relative Error

How often do you see public opinion polls in the newspapers, on television, or on the Internet and wonder just what is meant by the *margin of error* measurement displayed below the polling results? To see how the margin of error works, let's say that the Concord Computer Company announces that its sales forecast for the next quarter is about 600,000 computers. In reality, the company does not really expect to generate *exactly* this number of sales. To account for uncertainty and volatility in the market, they assume that the actual number of sales will be in the range from 555,000 to 645,000 computers. Computing the difference between the range values and the given forecast, we get

$$555,000 - 600,000 = -45,000$$

and

$$645,000 - 600,000 = 45,000$$

This means that the error in the forecast can be written as $\pm 45,000$. So, if the sales forecast is 600,000 units, the error in measurement relative to the sales forecast can be computed as

$$\frac{\left(\begin{array}{c}\text{maximal error in} \\ \text{the forecast}\end{array}\right)}{\left(\begin{array}{c}\text{given value of} \\ \text{the forecast}\end{array}\right)} = \pm \frac{45,000}{600,000} = \pm 0.075$$

We call this number the **relative error**. To make this error measurement more informative to managers, analysts and stockholders, we can write this as a percentage by multiplying by 100%.

$$\pm 0.075 \cdot 100\% = \pm 7.5\%$$

So the error in the forecast is read as "plus or minus 7.5%." Thus, the predicted sales could be 7.5% more than or 7.5% less than the 600,000 units forecast. This relationship is illustrated in Figure 3.3.1.

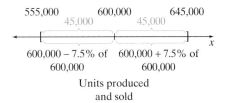

Figure 3.3.1   Sales Forecast of the Concord Computer Company

Notice in this example that we were measuring the error in the quantity produced and sold (that is, the independent variable x). By generalizing the idea of the computer company sales forecast, we can define the *relative error* and *percentage error* of the independent variable.

Relative and Percentage Error

• The **relative error** in the independent variable x, denoted by ε_x, is defined as

$$\varepsilon_x = \pm \frac{\left(\begin{array}{c}\text{maximal error in}\\\text{the forecast}\end{array}\right)}{\left(\begin{array}{c}\text{given value of}\\\text{the forecast}\end{array}\right)} = \pm \frac{dx}{x}$$

Note that, just as in Section 3.1, dx represents a change in the independent variable x.

• The **percentage error** or margin of error in the independent variable is given by

$$\varepsilon_x \cdot 100\%$$

Note: The ε is the Greek letter epsilon.

Let's take a look at an example of everyday use of this relative error in x.

| EXAMPLE 1 | **Determining a Range of Values from the Relative Error** |

TABLE 3.3.1

| Should | 31% |
| Should not | 62% |

Source: Data gathered at CNN web site.

Interactive Activity

Using Example 1, determine the highest and lowest values of the measure of those who think that the government *should* pass legislation that officially apologizes to American blacks for the fact that slavery was practiced before the Civil War in this country.

A Gallup poll conducted in June 1997 asked the question, "Do you think the U.S. government should pass legislation that officially apologizes to American blacks for the fact that slavery was practiced before the Civil War in this country?" The results of the poll are shown in Table 3.3.1.

The methodology of the poll states that the margin of error for the poll was $\pm 3.5\%$. Determine the highest and lowest values of the measure of those who think the government should not pass this legislation. Interpret the result.

SOLUTION

Since the figures are already written as percentages, we see that in this case the relative and percentage error are the same. We are given $x = 62\%$ and a maximal error of $dx = 3.5\%$. This gives us the lowest value, x_a, determined by the margin of error as

$$x_a = x - dx = 62\% - 3.5\% = 58.5\%$$

and the highest value, x_b, is

$$x_b = x + dx = 62\% + 3.5\% = 65.5\%$$

This means that, based on the results of the poll, somewhere between 58.5% and 65.5%, or [58.5%, 65.5%], of the American public think that the government should not pass the legislation.

Since we have defined the relative error for the independent variable, it makes sense to examine how the relative error affects the corresponding dependent variable. Let's say that the relative change in the dependent variable is the differential in y divided by the y-value. This quantity is

$$\varepsilon_y = \frac{dy}{y} = \frac{f'(x) \cdot dx}{f(x)}$$

Many times, in practice, the value of the relative error in x is already given, so to write this expression in terms of ε_x, we can multiply the numerator and denominator of the fraction by x to get

$$\varepsilon_y = \frac{f'(x) \cdot dx \cdot x}{f(x) \cdot x} = \frac{f'(x) \cdot x}{f(x)} \cdot \frac{dx}{x} = \frac{f'(x) \cdot x}{f(x)} \cdot \varepsilon_x$$

Now we can define the relative error in the dependent variable y in terms of ε_x.

Relative Error of the Dependent Variable

With a given value x and the relative error in the independent variable ε_x, for a differentiable function f, the **relative error in the dependent variable** $y = f(x)$, denoted by ε_y, is given by

$$\varepsilon_y = \frac{f'(x) \cdot x}{f(x)} \cdot \varepsilon_x$$

or, alternatively, without the given relative error ε_x, we can write

$$\varepsilon_y = \frac{f'(x) \cdot dx}{f(x)}$$

The **percentage error** in the dependent variable is given by

$$\varepsilon_y \cdot 100\%$$

Let's revisit the Concord Computer Company and see if we can compute a relative error for a dependent variable.

EXAMPLE 2 | **Computing the Relative Error of the Dependent Variable**

Assume that the revenue function for the sale of x computers is

$$R(x) = 0.0001x^2 + 2x$$

where $R(x)$ is in dollars.

(a) Compute the predicted revenue from the predicted sales forecast of $x = 600,000$ and interpret.

(b) Knowing that $\varepsilon_x = \pm 0.075$, compute ε_y, the percentage error, and interpret.

SOLUTION

(a) The predicted revenue based on the forecast of $x = 600,000$ units is
$$R(600,000) = 0.0001(600,000)^2 + 2(600,000) = \$37,200,000$$

This means that if the sales forecast is correct, the company will generate $37.2 million in revenue.

(b) Knowing that the derivative of the function is $R'(x) = 0.0002x + 2$, we get the relative error based on the predicted sales.

$$\varepsilon_y = \frac{R'(x) \cdot x}{R(x)} \cdot \varepsilon_x = \frac{R'(600,000) \cdot 600,000}{R(600,000)} (\pm 0.075)$$

$$= \pm \frac{(0.0002(600,000) + 2)(600,000)}{0.0001(600,000)^2 + 2(600,000)} (0.075)$$

$$= \pm \frac{(122)(600,000)}{37,200,000} (0.075) = \pm \frac{73,200,000}{37,200,000} (0.075) \approx \pm 0.148$$

As a percentage error, this value is $\pm 0.148 \cdot 100\% = \pm 14.8\%$. This indicates that the revenues will range from 14.8% below to 14.8% above the predicted $37.2 million, that is, in the interval $[31{,}694{,}400, 42{,}705{,}600]$, if the sales are in the predicted range from 7.5% above to 7.5% below the 600,000 forecast, that is, in the interval $[555{,}000, 645{,}000]$.

NOTE: We use the "plus or minus" symbol, $\pm$, to denote the numerical *range* of the error. This sign is not used in exactly the same way that it is used in the quadratic formula. In that case, one term is added and then subtracted from another.

Many relative errors in the dependent variable result from basing estimates of cost, revenue, and profit on models, many of which are computed through means of regression. Such models lend themselves to the study of the relative error of these dependent variable values.

Flashback

REFRIGERATOR COST FUNCTION REVISITED

In Section 3.2, we saw that the FrezMore Company finds that its cost of producing x refrigerators can be modeled by

$$C(x) = 2x^2 + 15x + 1500, \qquad 0 \le x \le 200$$

where x is the number of refrigerators produced and $C(x)$ represents the weekly cost in dollars. Assume that this cost function was modeled based on regression methods and the company's analysts determine that the percentage error in the quantity produced is 4% (that is, $\varepsilon_x = \pm 0.04$). Find the relative and percentage errors in the cost when $x = 150$ and interpret.

Solution

The relative error for the cost function is $\varepsilon_y = \dfrac{C'(x) \cdot x}{C(x)} \cdot \varepsilon_x$. Knowing that $\varepsilon_x = \pm 0.04$ and

$C'(x) = 4x + 15$ with $x = 150$, we find that

$$\varepsilon_y = \frac{C'(x) \cdot x}{C(x)} \cdot \varepsilon_x = \frac{C'(150) \cdot (150)}{C(150)}(\pm 0.04)$$

To avoid confusion, we can move the $\pm$ sign to the front of the fraction to get

$$= \pm \frac{C'(150) \cdot (150)}{C(150)}(0.04) = \pm \frac{(615)(150)}{48{,}750}(0.04)$$

$$= \pm \frac{92{,}250}{48{,}750}(0.04) \approx \pm 0.076$$

As a percentage error, this value is $\pm 7.6\%$, which means that if the production is 4% more than predicted the production costs will rise 7.6%. On the other hand, if the production is 4% less than predicted, the production costs will decrease 7.6%.

Relative Errors and Linear Approximations

In the previous section on marginal analysis, we found that we could use a linear approximation to estimate the cost, revenue, or profit of producing the $(x + 1)^{\text{st}}$ unit of a product. We can use the relative error here to see just how "approximate" this linear approximation really is. Remember, in the beginning of this section we stated that

$$\text{Relative error} = \frac{\left(\begin{array}{c}\text{maximal error in}\\\text{measurement}\end{array}\right)}{\left(\begin{array}{c}\text{given value of}\\\text{measurement}\end{array}\right)}$$

We apply this idea in Example 3.

EXAMPLE 3 | **Computing the Relative Error of a Linear Approximation**

The StopFirst Brake Company determines that their annual profit function for producing and selling brakes is given by

$$P(x) = 4\sqrt[5]{x}, \qquad 0 \le x \le 100$$

where x represents the number of brakes sold in thousands and $P(x)$ represents the profit in hundred thousand dollars.

(a) Evaluate $P(55)$ and $MP(55)$ and interpret. Round to the nearest ten-thousandth.

(b) Use the solutions from part (a) to get a linear approximation for $P(56)$.

(c) Evaluate $P(56)$ and compute the relative error in measure of the approximation found in part (b).

SOLUTION

(a) Evaluating $P(x) = 4\sqrt[5]{x}$ when $x = 55$, we get

$$P(55) = 4\sqrt[5]{55} \approx 8.9152$$

This means that when 55 thousand brakes are produced and sold the profit realized is about 8.9152 hundred thousand dollars, or \$891,520. The marginal profit function for the brake company is

$$MP(x) = \frac{d}{dx}(4\sqrt[5]{x}) = \frac{d}{dx}(4x^{1/5}) = \frac{4}{5}x^{-4/5} = \frac{4}{5x^{4/5}} = \frac{4}{5\sqrt[5]{x^4}}$$

Evaluating this function at $x = 55$ yields

$$MP(55) = \frac{4}{5\sqrt[5]{(55)^4}} \approx 0.0324$$

This means that at a production level of 55 thousand units the profit realized from producing and selling one additional thousand units is about 0.0324 hundred thousand dollars, or \$3240.

(b) A linear approximation for the profit function (with $dx = 1$) has the form

$$P(x + 1) \approx P(x) + MP(x)$$

So for $x = 55$ we get the approximation of $P(56)$ as

$$P(56) = P(55 + 1) \approx P(55) + MP(55)$$
$$\approx 8.9152 + 0.0324 = 8.9476$$

(c) Since the actual value at $x = 56$ is $P(56) = 4\sqrt[5]{56} \approx 8.9474$, the error in measurement is

$$\Delta P = 8.9476 - 8.9474 = 0.0002$$

So, knowing that relative error $= \dfrac{\left(\begin{smallmatrix}\text{maximal error in}\\\text{measurement}\end{smallmatrix}\right)}{\left(\begin{smallmatrix}\text{given value of}\\\text{measurement}\end{smallmatrix}\right)}$, we find the relative

error is $\frac{0.0002}{8.9474} \approx 0.00002$. This tells us that the relative error made by the linear approximation is 0.00002 or 0.002%. ∎

✓ **CHECKPOINT 1**

Now work Exercise 23.

Relative Rates of Change

Ideas about rates of change make more sense at times if they are given as a percentage. If someone says that the amount of ozone in the atmosphere decreased at an annual rate of $7\frac{1}{2}\%$ in 1992, this seems to make more sense than saying, "the amount of ozone in the atmosphere decreased at an annual rate of 0.008 parts per million." (*Source:* www.census.gov). This annual rate of $7\frac{1}{2}\%$ in 1992 can be found by dividing the instantaneous rate of change by the current function value. We call this result the **relative rate of change**.

Relative Rate of Change

If f is a differentiable function with a range $(0, \infty)$, then the **relative rate of change,** denoted by $Rel(x)$, is given by

$$Rel(x) = \frac{f'(x)}{f(x)}, \qquad \text{provided that } f(x) \neq 0$$

We will see in Chapter 4 why the range of f must be restricted to the positive numbers. Relative rates of change can be compared between different functions and models, as shown in Example 4.

EXAMPLE 4

Comparing Relative Rates of Change

The membership in the Boy Scouts of America between 1980 and 1996 can be modeled by

$$f(x) = 21.82x^2 - 100.11x + 5433.64, \qquad 0 \leq x \leq 16$$

where x represents the number of years since 1980 and $f(x)$ represents Boy Scout membership in thousands. During the same time period, the membership in the Girl Scouts of America can be modeled by

$$g(x) = 9.25x^3 - 96.06x^2 + 264.93x + 3252.81, \qquad 0 \leq x \leq 16$$

where x represents the number of years since 1980 and $g(x)$ represents Girl Scout membership in thousands.

(a) Compute $Rel(10)$ for the model f and interpret.

(b) Compute $Rel(10)$ for the model g and compare to the result of part (a).

SOLUTION

(a) We need to compute $Rel(x) = \frac{f'(x)}{f(x)}$ for $x = 10$. Knowing that the derivative of the model f is

$$f'(x) = \frac{d}{dx}(21.82x^2 - 100.11x + 5433.64) = 43.64x - 100.11$$

we get for $x = 10$

$$Rel(10) = \frac{f'(10)}{f(10)} = \frac{336.29}{6614.54} \approx 0.05$$

Interactive Activity

 Graph the models f and g in the viewing window $[0, 16]$ by $[0, 9500]$. At $x = 10$, how can you tell which function will have the greater relative rate? According to the model, when did Girl Scout membership exceed that of the Boy Scouts?

Since $x = 10$ corresponds to 1990, we see that during that year the membership in the Boy Scouts of America was growing at a rate of about 5%.

(b) Now we compute $Rel(x) = \frac{g'(x)}{g(x)}$ for $x = 10$. The derivative of this model is

$$g'(x) = \frac{d}{dx}(9.25x^3 - 96.06x^2 + 264.93x + 3252.81)$$

$$= 27.75x^2 - 192.12x + 264.93$$

So for $x = 10$ we have

$$Rel(10) = \frac{g'(10)}{g(10)} = \frac{1118.73}{5546.11} \approx 0.20$$

Thus, in 1990 the membership in the Girl Scouts of America was growing at a rate of about 20%. So Girl Scout membership was growing about four times faster than that of the Boy Scouts. ▪

SUMMARY

In this section, we studied errors in estimates and the impact that calculus has on them. We defined the relative and percentage errors and how to determine the range of values based on the maximal error. We applied the relative error to linear approximations of business functions. Then we discussed the relative rates of change and how they can be used to compare rates of growth.

Important Formulas

- **Relative error (independent variable):** $\varepsilon_x = \pm \dfrac{\left(\begin{smallmatrix}\text{maximal error in}\\\text{the forecast}\end{smallmatrix}\right)}{\left(\begin{smallmatrix}\text{given value of}\\\text{the forecast}\end{smallmatrix}\right)} = \pm \dfrac{dx}{x}$

- **Percentage error (independent variable):** $\varepsilon_x \cdot 100\%$

- **Relative error (dependent variable):** $\varepsilon_y = \dfrac{f'(x) \cdot x}{f(x)} \cdot \varepsilon_x$

- **Percentage error (dependent variable):** $\varepsilon_y \cdot 100\%$

- **Relative rate of change:** $Rel(x) = \dfrac{f'(x)}{f(x)}$

SECTION 3.3 EXERCISES

In Exercises 1–6, the range of values $[x_a, x_b]$ is given.

(a) Determine the maximal error of measurement.
(b) Compute the relative error in the independent variable ε_x.
(c) Determine the percentage error.

1. $[10, 14]$ 2. $[200, 210]$ 3. $[4000, 4300]$

4. $[6000, 6100]$ 5. $[42,000, 45,000]$ 6. $[91,000, 92,500]$

For Exercises 7–16, find the relative error in the dependent variable ε_y, given $f(x)$, x, and the relative error in the dependent variable ε_x.

7. $f(x) = 6x^3$; $x = 1$; $\varepsilon_x = \pm 0.10$

8. $f(x) = 4x^2$; $x = 2$; $\varepsilon_x = \pm 0.10$

9. $f(x) = \dfrac{5}{x - 1}$; $x = -3$; $\varepsilon_x = \pm 0.15$

10. $f(x) = \dfrac{3}{x - 2}$; $x = 4$; $\varepsilon_x = \pm 0.15$

11. $f(x) = x^2 - 5x - 2$; $x = 2$; $\varepsilon_x = \pm 0.05$

12. $f(x) = x^2 - 6x - 1$; $x = 3$; $\varepsilon_x = \pm 0.08$

13. $f(x) = 3\sqrt{x}$; $x = 2$; $\varepsilon_x = \pm 0.07$

14. $f(x) = 2\sqrt{x^2}$; $x = -3$; $\varepsilon_x = \pm 0.15$

15. $f(x) = \dfrac{3 - 2x}{x}$; $x = 1$; $\varepsilon_x = \pm 0.19$

16. $f(x) = \dfrac{x^2 + 1}{2}$; $x = 5$; $\varepsilon_x = \pm 0.15$

APPLICATIONS

17. A television newscast highlights a pre-election poll showing that two candidates have garnered 52.5% and 47.5% of the vote, respectively. The poll has a margin of error of $\pm 3\%$. The anchorperson states that the race is a statistical deadheat. Explain her statement.

18. A newspaper poll shows that a school levy is passing 56% to 44%, with a margin of error measured at $\pm 3\%$. What is the closest that the vote could actually be in the poll and stay within the margin of error?

19. Sales projections for the Griffort Company's new model of ballet shoe have targeted a projected sales of 24,000 pairs of shoes sold in the first year. They know that the sales could actually range from 18,000 to 30,000. Compute the percentage error in the independent variable for the number of pairs of ballet shoes sold in the next year.

20. Sales projections for an author's new romance novel is 70,000. The marketing department knows that the sales could actually range from 65,000 to 75,000. Compute the percentage error in the independent variable for the number of books sold.

For Exercises 21–23, consider the following scenario: The New Alliance Tool and Die factory sells bolts through large government contracts. In the next quarter, the target sales for the bolts are 60,000, but the accounting office knows that the sales could actually range from 55,000 to 65,000.

21. Compute the percentage error in the independent variable for the number of bolts sold in the next quarter.

22. If the cost function for the bolts is $C(x) = 11x^2 + 20x$, where x represents the number of bolts sold in thousands and $C(x)$ is in dollars:

(a) Find the percentage error in the dependent variable ε_y for the cost in next quarter's forecast.

(b) Determine $C(5)$ and $MC(5)$ and interpret.

(c) Use the solutions from part (b) to get a linear approximation for $C(6)$ by evaluating $C(5) + MC(5)$.

(d) Evaluate $C(6)$ and compute the relative error in measure of the approximation found in part (c).

✓ 23. If the revenue function for the bolts is $R(x) = 200\sqrt{x^3}$, where x represents the number of bolts sold in thousands and $R(x)$ is in dollars:

(a) Find the percentage error in the dependent variable ε_y for the revenue in next quarter's forecast.

(b) Determine $R(5)$ and $MR(5)$ and interpret.

(c) Use the solutions from part (b) to get a linear approximation for $R(6)$.

(d) Evaluate $R(6)$ and compute the relative error in measure of the approximation found in part (c).

🌐 24. The results of an August 23, 1998, Gallup/CNN poll follow:

"The United States recently launched military attacks against suspected terrorist sites in Afghanistan and Sudan. Do you approve or disapprove of these attacks?"

Approve:	75 percent
Disapprove:	18 percent

Source: Data gathered at CNN web site.

Knowing that the margin of error of the poll was $\pm 3.5\%$, determine the lowest and highest number of those who approved of the military attacks.

For Exercises 25–32, find the relative rate of change and then evaluate $Rel(x)$ at the given value of x.

25. $f(x) = 2x^2$; $x = 5$

26. $f(x) = 5x^3$; $x = 2$

27. $f(x) = \dfrac{x - 100}{10}$; $x = 125$

28. $f(x) = \dfrac{x - 1}{x^2}$; $x = 2$

29. $f(x) = 20\sqrt{x}$; $x = 8$

30. $f(x) = 5\sqrt[3]{x}$; $x = 1$

31. $f(x) = 3x^2 + 5x$; $x = 2$

32. $f(x) = 4 + \dfrac{3}{x}$; $x = -3$

🌐 33. The poverty threshold in the Unites States has followed a quadratic model since 1960 and is modeled by

$$f(x) = 0.004x^2 + 0.057x + 1.280, \qquad 0 \le x \le 35$$

where x represents the number of years since 1960 and $f(x)$ represents the U.S. poverty threshold in thousands of dollars. Use this model to compute $Rel(25)$ and interpret the answer.

🌐 34. The population of Florida can be modeled by the quadratic function

$$f(x) = 0.003x^2 + 0.168x + 4.908, \qquad 0 \le x \le 35$$

where x represents the number of years since 1960 and $f(x)$ represents the population in millions of people.

(a) Evaluate $f(30)$ and interpret.

(b) Compute $f'(x)$. Evaluate $f'(30)$ and interpret.

(c) Evaluate $Rel(30)$ and interpret.

SECTION PROJECT

The number of employees (in thousands) at business establishments that have less than 20 employees and between 20 and 99 employees, from the years 1980 and 1994, are displayed in Table 3.3.1.

TABLE 3.3.1

YEAR	x	LESS THAN 20 EMPLOYEES	20–99 EMPLOYEES
1980	1	19,423	21,168
1985	6	21,810	23,539
1988	9	23,583	25,930
1989	10	23,992	26,829
1990	11	24,373	27,414
1991	12	24,482	26,906
1992	13	25,000	27,030
1993	14	25,233	27,443
1994	15	25,373	28,138

Source: U.S. Census Bureau web site.

(a) Use the regression capabilities of your calculator to find a power model for the number of business establishments that have less than 20 employees in the form

$$f(x) = a \cdot x^b, \qquad 1 \le x \le 15$$

where x represents the number of years since 1979 and $f(x)$ represents the number of business establishments that have less than 20 employees, in thousands.

(b) Use the regression capabilities of your calculator to find a quartic model for the number of business establishments that have between 20 and 99 employees in the form

$$g(x) = ax^4 + bx^3 + cx^2 + dx + e, \qquad 1 \le x \le 15$$

where x represents the number of years since 1979 and $g(x)$ represents the number of business establishments that have between 20 and 99 employees, in thousands.

(c) Compute $Rel(14)$ for $f(x)$.

(d) Compute $Rel(14)$ for $g(x)$.

(e) Compare the rates found in parts (c) and (d).

CHAPTER REVIEW EXERCISES

In Exercises 1–12, find dy for the given function by determining $f'(x)$.

1. $y = 4x + 2$

2. $f(x) = 5x^2 - 3x + 2$

3. $y = \dfrac{x+3}{x-5}$

4. $y = \dfrac{7}{x+5}$

5. $y = \sqrt{x} - \dfrac{3}{x^4}$

6. $f(x) = \dfrac{8}{x^2} + \sqrt[3]{x}$

7. $f(x) = x^4 - 2x + \sqrt[3]{x^2}$

8. $y = x^3 - 5x^2 + 2x + 3$

9. $y = 4x^5 - 21x + 4$

10. $y = \dfrac{x^2 - 5}{x^2 + 5}$

11. $y = 2x^{1.7} - 5x^{0.8} + 4$

12. $f(x) = 3x^{4.1} + 7x^{0.6} - 12$

In Exercises 13–18, evaluate Δy and dy for each function for the indicated values.

13. $y = 4x^2 - x + 6$; $x = 3, \Delta x = dx = 0.1$

14. $y = \dfrac{18}{x} + 5$; $x = 2, \Delta x = dx = 0.5$

15. $f(x) = \sqrt[3]{x}$; $x = 8, \Delta x = dx = 1.261$

16. $f(x) = 2x^3 - 7x^2 + 2x$; $x = 4, \Delta x = dx = 0.2$

17. $y = \dfrac{x^2 + 2}{x^2 - 2}$; $x = 2, \Delta x = dx = 0.1$

18. $y = 4x(2x + 5)$; $x = 1, \Delta x = dx = 0.1$

For Exercises 19–22, use the linear approximation to estimate the values of the given numbers. Compare to the calculator value when rounded to four decimal places.

19. $\sqrt{65}$ 20. $\sqrt{24.6}$ 21. $\sqrt[4]{16.3}$ 22. $\sqrt[3]{62}$

23. The Wild & Wacky T-shirt company has estimated that the association between its monthly T-shirt sales and its advertising can be modeled by

$$y = 90x - 2.7x^2, \qquad 0 \le x \le 8$$

where x represents the amount spent on advertising in hundreds of dollars and y is the number of T-shirts sold in hundreds.

(a) Determine dy.

(b) Approximate the increase in sales if the advertising is increased from $400 to $500.

24. Repeat Exercise 23 using the model $y = 82.76x - 1.87x^2$.

25. For Exercises 23 and 24, compute the actual change Δy in sales and compare to the approximation.

26. The number of 21-year-olds living in New York can be modeled by

$$f(x) = 0.83x^3 - 8.11x^2 + 7.66x + 283.96, \qquad 0 \le x \le 7$$

where x represents the number of years since 1990 and $f(x)$ represents the number of 21-year-olds, in thousands.

(a) Evaluate $f(3)$ and $f(5)$ and interpret.

(b) Determine $f'(x)$

(c) Write the equation of the tangent line at $x = 2$.

(d) Find the y-value on the tangent line when $x = 3$. Interpret what this estimate means and compare to the value of $f(3)$ in part (a).

For Exercises 27–32:

(a) Determine the marginal cost function $MC(x)$.

(b) For the given production level x, evaluate $MC(x)$ and interpret.

(c) Determine the actual change in cost by evaluating $C(x + 1) - C(x)$ and compare with the answer to part (b).

27. $C(x) = 18x + 642$; $x = 12$

28. $C(x) = 9x + 1460$; $x = 27$

29. $C(x) = 26.7x + 87.4$; $x = 8$

30. $C(x) = \frac{1}{2}x^2 + 3x + 16$; $x = 15$

31. $C(x) = \frac{1}{4}x^2 + 12x + 47$; $x = 31$

32. $C(x) = \frac{1}{3}x^2 + 318x + 1783$; $x = 23$

In Exercises 33–38, the cost function C and the price–demand function p are given.

(a) Determine the revenue function R.

(b) Determine the profit function P.

(c) Differentiate P in part (b) to get the marginal profit function MP.

(d) Determine the marginal cost function MC and the marginal revenue function MR.

(e) Subtract the solutions found in part (d) to get $MR - MC$ and simplify. Compare with the result of part (c).

33. $C(x) = 7x + 250$; $p(x) = 11$

34. $C(x) = 14x + 1380$; $p(x) = 21$

35. $C(x) = \frac{1}{10}x^2 + 3x + 850$; $p(x) = -\frac{x}{15} + 50$

36. $C(x) = \frac{1}{50}x^2 + \frac{1}{4}x + 70$; $p(x) = -\frac{x}{50} + 5$

37. $C(x) = -0.001x^3 + 8x + 100$; $p(x) = -0.005x + 10$

38. $C(x) = -0.01x^3 + 0.1x^2 + 4x + 18$
    $p(x) = -0.6x + 15$

For Exercises 39–43, the Wheelex Company manufactures bicycles and finds the price function for the bicycles to be

$$p(q) = -0.02q + 150$$

where q represents the number of bicycles produced and sold and $p(q)$ is the price of the bicycle. Furthermore, the fixed and variable costs to produce q bicycles are $5600 and $85 per bicycle, respectively.

39. The cost function follows the linear form $C(q) = mq + b$. Answer the following:

(a) Write the cost C in the linear form $C(q) = mq + b$.

(b) Use calculus to compute the marginal cost function $MC(q) = \frac{d}{dq}C(q)$.

40. Using the information found in Exercise 39, complete part (a) through part (d).

(a) Evaluate $MC(250)$ and $MC(500)$ and interpret these answers.

(b) Why are the answers in part (a) equal?

(c) Algebraically find $AC(q)$ and simplify.

(d) Evaluate $AC(250)$ and interpret.

41. Use the solution from part (c) of Exercise 40 to answer parts (a) and (b).

(a) Use calculus to compute the marginal average cost function $MAC(q)$.

(b) Evaluate $MAC(250)$ and interpret.

42. Use $p(q)$ and the cost function information given in Exercise 39 to complete parts (a) through (c).

(a) Derive the revenue function $R(q)$.

(b) Use calculus to compute $MR(q) = \frac{d}{dq}R(q)$.

(c) Evaluate $MR(500)$ and interpret.

43. Use the solution from part (a) of Exercise 39 with the solution to part (a) of Exercise 42 to complete parts (a) and (b).

(a) Derive the profit function $P(q)$.

(b) If the bicycles must be manufactured in lots of 250, how many bicycles should be manufactured so that the profit is as large as possible? Verify your answer by filling out the table.

PRODUCED, q	TOTAL PROFIT, $P(q)$
0	
250	
500	
750	
1000	
1250	
1500	
1750	
2000	

For Exercises 44–48, repeat Exercises 39–43 all parts if the price–demand function is $p(q) = -0.022q + 180$ and the variable costs are $125 per bicycle and the fixed costs are $7300.

49. Consider the following scenario. The Colorama Company, a paint manufacturer, has just produced a watercolor set that it sells to wholesalers for $6 each. The cost $C(x)$ to produce x watercolor sets is given by the function $C(x) = 0.0002x^2 + 2x + 1250$.

(a) Algebraically derive the profit function $P(x)$ and simplify it.

(b) Evaluate $P(3000)$ and interpret the answer.

(c) Use calculus to compute the marginal profit function.

(d) Evaluate $MP(3000)$ and interpret.

(e) Use the solutions from parts (b) and (d) to get a linear approximation for the value of $P(3001)$.

(f) Compute the error of the approximation for $P(3001)$ in part (e).

50. The Colorama Company hires a consulting firm to assess its work and consequently revises its price and cost functions to $p(x) = 6.5$ and $C(x) = \dfrac{x^2}{5500} + \dfrac{7}{3}x + 1500$. Redo parts (a) to (f) in Exercise 49 using these revisions.

51. Knowing that $AC(x) = \dfrac{C(x)}{x}$, use the quotient rule to show that the marginal average cost function can be written as

$$MAC(x) = \frac{MC(x) - AC(x)}{x}$$

52. The Between the Lines Publishing Company determines that the price–demand function for a new book is

$$p(x) = \frac{-x}{500} + 20$$

with fixed costs of $12,000 and variable costs of $4.5 \dfrac{\text{dollars}}{\text{book}}$.

(a) Find the cost function C and the revenue function R.

(b) Graph C and R in the same viewing window. Graphically find the production level x that represents the break-even point.

(c) Find the profit function P.

(d) Find the smallest and largest production levels x so that the company realizes a profit. (That is, find the smallest and largest independent values so that the revenue is greater than the cost.)

(e) Compute the marginal profit function $P'(x)$.

(f) Evaluate $P'(2500)$ and interpret the result.

53. The Deluxe Furniture Company determines that the price–demand function for its new bookshelf is

$$p(x) = \frac{-x}{75} + 250$$

The fixed costs are $10,000 and variable costs are $150 \dfrac{\text{dollars}}{\text{unit}}$. Redo parts (a) through (f) in Exercise 52.

For Exercises 54–59, consider a stockholder's report that lists the following information:

q	$C(q)$	$R(q)$
100	22,830	38,000
200	30,830	72,000
300	38,830	102,000

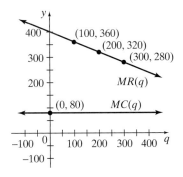

Use the table and the graphs of $MC(q)$ and $MR(q)$ to get a linear approximation for the given cost and revenue functions.

54. $C(101)$ 55. $C(201)$ 56. $C(301)$

57. $R(101)$ 58. $R(201)$ 59. $R(301)$

60. Consider the following data table for the cost and revenues at various production levels for a new brand of computer printer.

Number of Printers Produced, x	**Cost, $C(x)$**	**Revenue, $R(x)$**
1000	243,600	296,950
2000	363,000	575,800
3000	482,800	818,550
4000	603,300	1,007,200

(a) Use your calculator to determine a linear regression model for the cost of producing the printer in the form

$$C(x) = ax + b, \quad 1 \le x \le 4$$

where x represents the number of printers produced in thousands and $C(x)$ represents the cost of production.

(b) Use your calculator to determine a cubic regression model for the revenue of producing and selling the printers in the form

$$R(x) = ax^3 + bx^2 + cx + d, \quad 1 \le x \le 4$$

where x represents the number of printers produced and sold in thousands and $R(x)$ represents the resulting revenue.

(c) Compute $MAC(x)$ and simplify the result.

(d) Evaluate $MAC(1.5)$ and interpret.

61. Use the models found in Exercise 60 parts (a) and (b) to answer parts (a) and (b).

(a) Compute $MAP(x)$ and simplify the result.

(b) Evaluate $MAP(1.5)$ and interpret the answer.

For Exercises 62–67, find ε_y, the relative error in the dependent variable, given $f(x)$, x, and ε_x, the relative error in the independent variable.

62. $f(x) = 4x^2$; $x = 2$; $\varepsilon_x = \pm 0.15$

63. $f(x) = x^3 - 2x + 3$; $x = 1$; $\varepsilon_x = \pm 0.10$

64. $f(x) = \sqrt{5}x$; $x = 5$; $\varepsilon_x = \pm 0.05$

65. $f(x) = \dfrac{2x + 3}{3x - 5}$; $x = -4$; $\varepsilon_x = \pm 0.17$

66. $f(x) = \dfrac{2x^2 - 7}{15}$; $x = 4$; $\varepsilon_x = \pm 0.02$

67. $f(x) = 8\sqrt[4]{x^3}$; $x = 2$; $\varepsilon_x = \pm 0.15$

In Exercises 68–73, the range of values $[x_a, x_b]$ is given.

(a) Determine the forecast value in the independent variable x.

(b) Compute ε_x, the relative error in the independent variable.

(c) Determine the percentage error.

68. [13, 17]

69. [180, 250]

70. [99, 100]

71. [246, 254]

72. [56,000, 72,000]

73. [94,150, 94,350]

74. A local newspaper prints pre-election poll results showing that a ballot initiative is losing 48% to 52%. If the margin of error is ±2.5%, could the race be a tie? Explain.

75. A magazine publishes a poll regarding an upcoming mayoral election. The poll shows that 58% of the voters support Candidate Evans and 42% support Candidate Hawthorne, with a margin of error of 4%. If the actual vote falls within the margin of error, what is the closest that the vote could be?

For Exercises 76–78, consider the following scenario:

A high-tech company sells modems to computer manufacturers. In the next quarter, the sales target for the modems is 88,000, but the accounting office knows that the sales could actually range from 85,000 to 91,000.

76. Compute the percentage error in the independent variable for the number of modems sold in the next quarter.

77. The cost function for the modems is $C(x) = 350x^2 + 30,000x + 24,000$, where x represents the number of modems sold in thousands.

(a) Find ε_y, the percentage error in the dependent variable, for the cost in next quarter's forecast.

(b) Determine $C(88)$ and $MC(88)$ and interpret.

(c) Use the solutions from part (b) to get a linear approximation for $C(89)$.

(d) Evaluate $C(89)$ and compute the relative error in measure of the approximation found in part (c).

78. The revenue function for the modems is $R(x) = 340,000\sqrt[3]{x^2}$, where x represents the number of modems sold in thousands. Find the percentage error in the dependent variable ε_y for the revenue in next quarter's forecast.

(a) Find ε_y, the percentage error in the dependent variable, for the revenue in next quarter's forecast.

(b) Determine $R(88)$ and $MR(88)$ and interpret.

(c) Use the solutions from part (b) to get a linear approximation for $R(89)$.

(d) Evaluate $R(89)$ and compute the relative error in measure of the approximation found in part (c).

79. According to a Gallup poll conducted on October 23, 1998, 78% of Americans were generally satisfied with their standard of living, while 22% were generally dissatisfied. The margin of error for this poll was 2%. Determine the low-

est and highest values of the measure of those who were generally *satisfied* with their standard of living. (*Source:* Based on data gathered at the Gallup poll web site.)

For Exercises 80–87, find the relative rate of change and then evaluate the $Rel(x)$ at the given value of x.

80. $f(x) = 4x^2$; $x = 7$

81. $f(x) = 7x^3 - 2x$; $x = 5$

82. $f(x) = \dfrac{x + 18}{5}$; $x = 12$

83. $f(x) = \dfrac{x - 4}{x^2}$; $x = 8$

84. $f(x) = 5\sqrt[3]{x}$; $x = 27$

85. $f(x) = 3\sqrt{x}$; $x = 2$

86. $f(x) = x^3 + 7x^2 - 3x$; $x = 1$

87. $f(x) = \dfrac{7}{x} - 1$; $x = 2$

88. The median income of U.S. one-person households consisting of a female under 65 years old can be modeled by

$$f(x) = -1.33x^3 + 42.80x^2 - 32.31x + 16,318.98,$$
$$0 \le x \le 26$$

where x represents the number of years since 1960 and $f(x)$ represents the median income in dollars. Use this model to compute $Rel(22)$ and interpret the answer.

89. The number of U.S. households with a male householder (no spouse) and related children under 18 years of age can be modeled by the quadratic function

$$f(x) = 1.52x^2 + 19.90x + 437.52, \quad 0 \le x \le 26$$

where x represents the number of years since 1970 and $f(x)$ represents the number of households in thousands.

(a) Evaluate $f(18)$ and interpret the answer.

(b) Compute $f'(x)$, evaluate $f'(18)$, and interpret the answer.

(c) Evaluate $Rel(18)$ and interpret the answer.

90. The following table shows the estimated total annual retail sales (in million of dollars) for apparel and accessory stores and for automotive dealers, for the years 1986 through 1997.

Year	x	Apparel and Accessory Stores	Automotive Dealers
1986	1	75,626	326,138
1987	2	79,322	342,896
1988	3	85,307	372,570
1989	4	92,341	386,011
1990	5	95,819	387,605
1991	6	97,441	372,647
1992	7	104,212	406,935
1993	8	107,199	457,797
1994	9	109,976	521,768
1995	10	110,936	556,708
1996	11	114,635	599,667
1997	12	117,826	625,682

Source: U.S. Census Bureau web site.

(a) Use the regression capabilities of your calculator to find a power model for the annual retail sales of apparel and accessory stores in the form

$$f(x) = a \cdot x^b, \qquad 1 \le x \le 12$$

where x represents the number of years since 1985 and $f(x)$ represents the annual retail sales of apparel and accessory stores in millions of dollars.

(b) Use the regression capabilities of your calculator to find a quartic model for the annual retail sales of automotive

dealers in the form

$$g(x) = ax^4 + bx^3 + cx^2 + dx + e, \qquad 1 \le x \le 12$$

where x represents the number of years since 1985 and $g(x)$ represents the annual retail sales of automotive dealers in millions of dollars.

(c) Compute $Rel(5)$ for $f(x)$.

(d) Compute $Rel(5)$ for $g(x)$.

(e) Compare the rates found in parts (c) and (d).

Additional Differentiation Techniques

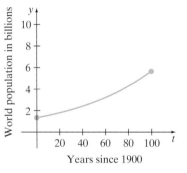

Graph of world population model $p(t) = 1.419e^{0.014t}$.

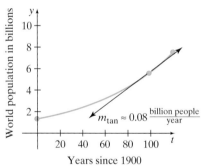

In 1999, the world's population was growing at a rate of about 80 million people per year.

What We Know

In the previous two chapters, we learned the basic rules of differentiation and their applications. We also learned that the derivative can be used to study marginal analysis of business functions. We found that the role of the differential was central to marginal analysis.

Where Do We Go

In this chapter, we will learn how to differentiate other families of functions for which the rules we have learned so far may not apply. These include composite functions, exponential and logarithmic functions, and implicit functions.

SECTION 4.1   THE CHAIN RULE

So far, the type of functions that we have differentiated are polynomial functions, rational functions, and power functions. But one family of functions that we have not differentiated are the **composite** functions. If the composite function $f(x) = 1000\sqrt{180 - 2x}$ models the number of college graduates surviving to x years of age, we must find a way to compute the derivative of f in order to determine the rate of change of this function. In this section, we introduce the **Chain Rule,** a powerful technique used to differentiate composite functions.

Chain Rule

We will discuss the differentiation of composite functions that have the form $h(x) = f(g(x))$. A reasonable question to ask at this point is "Can we even determine the derivative of a composite function in the first place?" To answer that question, let's take a Flashback to Chapter 1.

Flashback   **OIL SPILL FUNCTION REVISITED**

In Section 1.7 we saw that a refinery's underwater supply line ruptures, resulting in a fairly circular oil spill. The radius was modeled by

$$r(t) = 0.7t$$

where t represents the number of seconds since the spill occurred and $r(t)$ represents the radius of the oil slick in feet. The area of the spill was given by

$$A(r) = \pi r^2$$

where r is the radius of the slick in feet and $A(r)$ is the area of the slick in square feet. Simplify the composition $A(r(t))$, which yields $A(t)$, and find the derivative $\frac{d}{dt}(A(t))$. Interpret the resulting function.

Solution

Simplifying, we get

$$A(r(t)) = A(0.7t)$$

Substituting $0.7t$ for r in the function $A(r) = \pi r^2$, we get the area of the oil slick as a function of time.

$$A(r(t)) = \pi(0.7t)^2 = \pi(0.49 \cdot t^2) = 0.49\pi t^2$$

Differentiating this result yields

$$\frac{d}{dt}(A(t)) = \frac{d}{dt}(0.49\pi t^2) = 0.98\pi t$$

The derivative, $\frac{d}{dt}(A(t)) = 0.98\pi t$, represents the instantaneous rate of change of the area of the oil slick with respect to time.

The Flashback shows that we can differentiate a composite function, yet it does not explicitly show how this is done. Let's examine the functions in the Flashback again and see whether there is another way to find the derivative of $A(t) = A(r(t))$. If we compute the derivatives of the original functions $r(t) = 0.7t$ and $A(r) = \pi r^2$, with respect to t and r respectively, we have

$$r'(t) = 0.7 \quad \text{and} \quad A'(r) = 2\pi r$$

Since we know that $A'(r(t)) = 2\pi(0.7t) = 1.4\pi t$ and $r'(t) = 0.7$, it appears that the derivative $A'(t)$ is equivalent to

$$A'(t) = A'(r(t)) \cdot r'(t) = (1.4\pi t) \cdot (0.7) = 0.98\pi t$$

We have just differentiated a composite function using the Chain Rule! We now state this rule for functions made up of a composition of functions f and g.

Chain Rule

If $y = f(u)$ and $u = g(x)$ are used to define $h(x)$, where $h(x) = f(g(x))$, then

$$h'(x) = f'(g(x)) \cdot g'(x)$$

provided that $f'(g(x))$ and $g'(x)$ exist. Equivalently, using Leibniz notation, we have

$$h'(x) = \frac{dy}{du} \cdot \frac{du}{dx}$$

provided that $\dfrac{dy}{du}$ and $\dfrac{du}{dx}$ exist.

To differentiate $h(x) = f(g(x))$, it appears that we can differentiate the "outside" function f, and then chain it to the derivative of the "inside" function g. Many functions that we will study have the form $h(x) = (\text{function})^{\text{power}}$, where the power is some real number. For these types of functions, we can rely on an extension, or corollary, of this rule, called the **Generalized Power Rule.** Notice that this rule looks much like our Power Rule from Chapter 2.

Generalized Power Rule

If u is a differentiable function of x and n is any real number with $f(x) = [u(x)]^n$, then

$$f'(x) = n[u(x)]^{n-1} \cdot u'(x)$$

EXAMPLE 1 | **Applying the Generalized Power Rule**

Use the Generalized Power Rule to determine the derivatives.

(a) $f(x) = (5x^3 + 3x)^4$

(b) $g(x) = (x^2 + 1)^{15}$

SOLUTION

(a) For the function $f(x) = (5x^3 + 3x)^4$, consider $u(x) = 5x^3 + 3x$ and $n = 4$. Applying the Generalized Power Rule gives

$$f'(x) = \frac{d}{dx}[(5x^3 + 3x)^4] = 4(5x^3 + 3x)^{4-1} \cdot \frac{d}{dx}(5x^3 + 3x)$$

$$= 4(5x^3 + 3x)^3(15x^2 + 3) = (5x^3 + 3x)^3(60x^2 + 12)$$

(b) For this 15th-degree polynomial, we can consider $u(x) = x^2 + 1$ and $n = 15$ and apply the Generalized Power Rule to get

$$g'(x) = \frac{d}{dx}[(x^2 + 1)^{15}] = 15(x^2 + 1)^{14} \cdot \frac{d}{dx}(x^2 + 1)$$

$$= 15(x^2 + 1)^{14}(2x) = 30x(x^2 + 1)^{14}$$

✓CHECKPOINT 1

Now work Exercise 11.

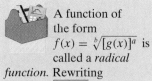
Notice the power of the Generalized Power Rule in part (b) of Example 1. It would have been possible, yet not at all practical, to algebraically expand the binomial $(x^2 + 1)^{15}$ in order to apply the differentiation techniques from Chapter 2.

We can use this new technique of differentiation to readily determine derivatives of new families of functions, including the radical and rational exponent functions. To use the Generalized Power Rule with these functions, let's review how these functions can be rewritten as shown in the Toolbox to the left.

The key in differentiating radical functions is to rewrite them in the rational exponent form so that we can apply the Generalized Power Rule. We illustrate this in Example 2.

EXAMPLE 2 | **Determining Derivatives of Radical Functions**

(a) Use the Generalized Power Rule to determine $f'(x)$ for $f(x) = \sqrt[3]{2x - 4}$.

(b) Find an equation of the line tangent to the graph of f at the point $(6, 2)$.

SOLUTION

(a) Before we can differentiate, we rewrite $f(x) = \sqrt[3]{2x - 4}$ using rational exponents as

$$f(x) = \sqrt[3]{2x - 4} = (2x - 4)^{1/3}$$

Using the Generalized Power Rule with $u(x) = (2x - 4)$ and $n = \frac{1}{3}$ gives us

$$f'(x) = \frac{d}{dx}[(2x - 4)^{1/3}]$$

$$= \frac{1}{3}(2x - 4)^{1/3 - 1} \cdot \frac{d}{dx}(2x - 4) = \frac{1}{3}(2x - 4)^{-2/3}(2) = \frac{2}{3}(2x - 4)^{-2/3}$$

Writing the derivative without negative or rational exponents, this simplifies to

$$f'(x) = \frac{2}{3} \cdot (2x - 4)^{-2/3} = \frac{2}{3(2x - 4)^{2/3}} = \frac{2}{3\sqrt[3]{(2x - 4)^2}}$$

(b) Since $f'(6)$ gives the slope of the tangent line at $x = 6$, we have

$$f'(6) = \frac{2}{3\sqrt[3]{(2(6) - 4)^2}} = \frac{2}{3\sqrt[3]{(8)^2}}$$

$$= \frac{2}{3\sqrt[3]{64}} = \frac{2}{3 \cdot 4} = \frac{1}{6}$$

With a slope of $\frac{1}{6}$ and point $(6, 2)$, the tangent line equation is

$$y - 2 = \frac{1}{6}(x - 6)$$

or, in CF form,

$$y = \frac{1}{6}(x - 6) + 2$$

Interactive Activity

For what x-value is the function $f(x) = \sqrt[3]{2x-4}$ not differentiable? Which of the three nondifferentiability characteristics (sharp turn or corner, vertical tangent, or discontinuity) does the graph of f exhibit? To check visually, graph f in the viewing window $[1.95, 2.05]$ by $[-0.3, 0.3]$.

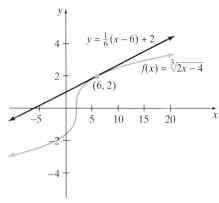

Figure 4.1.1

The graphs of $f(x) = \sqrt[3]{2x-4}$ and $y = \frac{1}{6}(x-6) + 2$ are shown in Figure 4.1.1.

✓ **CHECKPOINT 2**

Now work Exercise 37.

Another family of functions that can be differentiated using the Generalized Power Rule are rational functions. Even though the Generalized Power Rule can be used on any rational function, it is particularly useful on rational functions that have constants in their numerators.

EXAMPLE 3 | **Determining Derivatives of Rational Functions Two Ways**

For the function $h(x) = \dfrac{5}{(2x-3)^2}$, determine $h'(x)$ using the Quotient Rule and then determine $h'(x)$ using the Generalized Power Rule.

SOLUTION

Using the Quotient Rule, we get

$$h'(x) = \frac{\dfrac{d}{dx}[5] \cdot (2x-3)^2 - (5) \cdot \dfrac{d}{dx}[(2x-3)^2]}{[(2x-3)^2]^2}$$

$$= \frac{0 \cdot (2x-3)^2 - 5 \cdot [2(2x-3)(2)]}{(2x-3)^4}$$

$$= \frac{-20(2x-3)}{(2x-3)^4} = \frac{-20}{(2x-3)^3}$$

When using the Generalized Power Rule, we rewrite the rational function as $h(x) = 5(2x-3)^{-2}$. With help from the Constant Multiple Rule, we determine

$$h'(x) = 5 \cdot \frac{d}{dx}[(2x-3)^{-2}]$$

$$= 5 \cdot (-2)(2x-3)^{-3} \cdot (2)$$

$$= -20(2x-3)^{-3} = \frac{-20}{(2x-3)^3}$$

✓CHECKPOINT **3**

Now work Exercise 33.

Applications

In our first application, we apply the Generalized Power Rule to a rational exponent function.

EXAMPLE 4

Applying the Generalized Power Rule to Rational Exponent Functions

The death rate caused by heart disease in the United States can be modeled by

$$f(x) = 336.18(x + 1)^{-0.06}, \qquad 0 \le x \le 15$$

where x represents the number of years since 1980 and $f(x)$ represents the death rate (measured in deaths per 100,000 people) caused by heart disease. Determine $f'(x)$. Evaluate $f'(2)$ and interpret.

SOLUTION

Using the Generalized Power Rule, we get

$$\begin{aligned} f'(x) &= \frac{d}{dx}[336.18(x + 1)^{-0.06}] \\ &= (336.18) \cdot (-0.06)(x + 1)^{-0.06-1} \cdot (1) \\ &= -20.1708(x + 1)^{-1.06} \end{aligned}$$

Notice that $x = 2$ corresponds to the year 1982. Continuing, we have

$$f'(2) = -20.1708(3)^{-1.06} \approx -6.29$$

So in 1982 the death rate caused by heart disease was decreasing at a rate of about $6.29 \frac{\text{deaths per 100,000 people}}{\text{year}}$.

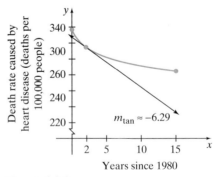

Figure 4.1.2

✓CHECKPOINT **4**

Now work Exercise 83.

The final example demonstrates how the Generalized Power Rule can be used with business functions.

EXAMPLE 5 | **Applying the Generalized Power Rule to Business Functions**

The RadRadio Company produces and sells personal stereo devices. Market research has found that the price–demand function is

$$p(x) = 100 - \sqrt{x^2 + 20}, \qquad 0 \le x \le 35$$

where x is the number of devices demanded in thousands and $p(x)$ represents the unit price in dollars.

(a) Evaluate $p'(30)$ and interpret.

(b) Determine the marginal revenue function. Evaluate $MR(30)$ and interpret.

SOLUTION

(a) The rational exponent form of the price–demand function is

$$p(x) = 100 - (x^2 + 20)^{1/2}$$

Using the Generalized Power Rule, we get the derivative

$$p'(x) = -\frac{1}{2}(x^2 + 20)^{-1/2} \cdot \frac{d}{dx}(x^2 + 20)$$

$$= -\frac{1}{2}(x^2 + 20)^{-1/2} \cdot (2x)$$

$$= -x(x^2 + 20)^{-1/2}$$

$$= \frac{-x}{(x^2 + 20)^{1/2}} = \frac{-x}{\sqrt{x^2 + 20}}$$

Evaluating $p'(30)$ yields

$$p'(30) = \frac{-(30)}{\sqrt{(30)^2 + 20}} = \frac{-30}{\sqrt{920}} \approx -0.99 \frac{\text{dollars}}{\text{thousand devices}}$$

This means that at a demand level of 30 thousand personal stereo devices the price is decreasing at a rate of about 99 cents per 1,000 devices sold.

(b) Knowing that the revenue function is $R(x) = x \cdot [p(x)]$, we use the price–demand function to get

$$R(x) = x \cdot [p(x)] = x \cdot \left(100 - \sqrt{x^2 + 20}\right)$$

$$= 100x - x\sqrt{x^2 + 20}$$

To determine the marginal revenue function, we determine the derivative $R'(x)$. We must apply the Product Rule to the second term, which gives us

$$MR(x) = R'(x) = \frac{d}{dx}\left(100x - x\sqrt{x^2 + 20}\right)$$

$$= 100 - \frac{d}{dx}(x) \cdot \sqrt{x^2 + 20} - x \cdot \frac{d}{dx}[(x^2 + 20)^{1/2}]$$

$$= 100 - (1)\sqrt{x^2 + 20} - x\left(\frac{1}{2}\right)(x^2 + 20)^{-1/2}(2x)$$

$$= 100 - \sqrt{x^2 + 20} - x^2(x^2 + 20)^{-1/2}$$

$$= 100 - \sqrt{x^2 + 20} - \frac{x^2}{\sqrt{x^2 + 20}}$$

Evaluating $MR(x)$ at $x = 30$ produces

$$MR(30) = 100 - \sqrt{(30)^2 + 20} - \frac{(30)^2}{\sqrt{(30)^2 + 20}}$$

$$= 100 - \sqrt{920} - \frac{900}{\sqrt{920}} \approx 40.0 \frac{\text{dollars}}{\text{thousand devices}}$$

Thus, when the production level is 30 thousand, an increase of 1000 personal stereo devices produced and sold increases revenue by about \$40. ∎

SUMMARY

In this section, we reviewed composite functions and then introduced the Chain Rule, which is a technique used to differentiate composite functions. We then focused on the Generalized Power Rule, which is an extension of the Chain Rule. It can be used when a composite function has the form $f(x) = (\text{function})^{\text{power}}$, where the power is a real number.

- **Chain Rule:** If $y = f(u)$ and $u = g(x)$ are used to define $h(x)$, where $h(x) = f(g(x))$, then $h'(x) = f'(g(x)) \cdot g'(x)$, provided that $f'(g(x))$ and $g'(x)$ exist.
- **Generalized Power Rule:** If u is a differentiable function of x and n is any real number, with $f(x) = [u(x)]^n$, then $f'(x) = n[u(x)]^{n-1} \cdot u'(x)$.

SECTION 4.1 EXERCISES

In Exercises 1–6, differentiate using the Generalized Power Rule.

1. $f(x) = (x + 1)^2$
2. $f(x) = (x + 3)^2$
3. $f(x) = (x - 5)^3$
4. $f(x) = (x - 2)^3$
5. $f(x) = (2 - x)^2$
6. $f(x) = (5 - x)^2$

In Exercises 7–30, differentiate using the Generalized Power Rule.

7. $g(x) = (2x + 4)^3$
8. $g(x) = (3x + 3)^3$
9. $f(x) = (5 - 2x)^5$
10. $f(x) = (10 - 5x)^4$
✓ 11. $f(x) = (3x^2 + 7)^5$
12. $f(x) = (4x^2 - 3)^3$
13. $f(x) = (x^3 - 2x^2 + x)^2$
14. $f(x) = (2x^3 + 4x + 3)^3$
15. $g(x) = 3(x^3 - 4)^3$
16. $g(x) = 5(4x^2 - 10)^6$
17. $f(x) = 5(5x^2 - 3x - 1)^{10}$
18. $f(x) = 10(10x^3 + x - 9)^8$
19. $g(x) = (4x^2 - x - 4)^{55}$
20. $g(x) = (8x^2 - 2x + 5)^{94}$
21. $g(x) = (2x - 4)^{1/2}$
22. $f(x) = (7x + 6)^{1/2}$
23. $g(x) = (x^2 + 2x)^{1/3}$
24. $g(x) = (x^3 + 5x)^{1/3}$
25. $f(x) = (5x - 2)^{-2}$
26. $f(x) = (4x + 3)^{-3}$
27. $g(x) = (x^2 + 2x + 4)^{-1/2}$
28. $g(x) = (3x^2 + 5x + 6)^{-1/2}$
29. $g(x) = (3x^3 - x)^{-1/4}$
30. $g(x) = (4x^5 + 5x^3)^{-1/3}$

For the rational functions in Exercises 31–36:

(a) Determine the derivative using the Quotient Rule.
(b) Determine the derivative using the Generalized Power Rule.

31. $f(x) = \dfrac{1}{3x + 4}$
32. $f(x) = \dfrac{1}{7x - 5}$
✓ 33. $f(x) = \dfrac{5}{(x - 2)^2}$
34. $f(x) = \dfrac{10}{(2x - 1)^3}$
35. $f(x) = \dfrac{2}{x^2 + 2x + 3}$
36. $f(x) = \dfrac{9}{x^3 + 2x + 10}$

In Exercises 37–44, determine an equation of the line tangent to the graph of f at the indicated point.

✓ 37. $f(x) = (2x - 1)^3$; $(1, 1)$
38. $f(x) = (3x - 4)^3$; $(1, -1)$
39. $f(x) = (2 - x)^4$; $(1, 1)$
40. $f(x) = (x^2 - 1)^4$; $(1, 0)$
41. $f(x) = (x^3 - 4x + 2)^4$; $(2, 16)$
42. $f(x) = (4x - 3)^{1/2}$; $(3, 3)$
43. $f(x) = (2x - 4)^{1/2}$; $(2, 0)$
44. $f(x) = (2x + 8)^{1/2}$; $(4, 4)$

Use the Generalized Power Rule to differentiate the functions in Exercises 45–52.

45. $f(x) = \sqrt{x^2 + 5}$

46. $f(x) = \sqrt{3x + 6}$

47. $g(x) = \sqrt[3]{2x - 1}$

48. $g(x) = \sqrt[3]{4x - 3}$

49. $f(x) = \dfrac{5}{\sqrt{2x - 8}}$

50. $f(x) = \dfrac{10}{\sqrt{5x + 8}}$

51. $f(x) = \dfrac{64}{\sqrt[3]{5x^2 - 6x + 3}}$

52. $f(x) = \dfrac{27}{\sqrt[3]{3x^3 + x}}$

In Exercises 53–62, use the Generalized Power Rule, along with the Product and Quotient Rules to find the derivatives of the given functions.

53. $g(x) = x(x - 4)^3$

54. $g(x) = x(10 - x)^3$

55. $g(x) = x\sqrt{x^2 + 3x}$

56. $g(x) = x^2\sqrt{2x^2 - 11}$

57. $f(x) = \dfrac{x^3}{(3x - 8)^2}$

58. $f(x) = \dfrac{x^2}{(4x^2 - x + 5)^3}$

59. $f(x) = (x + 3)^3(2x - 1)^2$

60. $f(x) = (3x - 3)^4(2x - 2)^3$

61. $g(x) = \sqrt{\dfrac{x + 3}{x - 3}}$

62. $g(x) = \sqrt{\dfrac{2x + 1}{2x - 1}}$

In Exercises 63–68, find the derivatives using the Generalized Power Rule.

63. $y = (4x^2 + 5x + 6)^{0.23}$

64. $y = 3(0.7x^3 - 0.02x^2)^{0.09}$

65. $g(x) = \left(\dfrac{1}{x + 3}\right)^{-1.03}$

66. $g(x) = \left(\dfrac{1}{0.2x + 1.7}\right)^{-1.1}$

67. $f(x) = 1.44(x + 1)^{1.22}$

68. $f(x) = 67.41(x + 1)^{0.97}$

APPLICATIONS

69. An actuary has determined that for a certain demographic group the number of people surviving over the duration of a century can be modeled by

$$f(x) = 400\sqrt{100 - x}, \qquad 0 \le x \le 100$$

where x represents the age of the person in years in the group and $f(x)$ represents the number of people surviving. Evaluate and interpret $f'(70)$.

70. During its first season, the number of viewers who watched the new television series "It Ain't Me!" can be modeled by

$$g(x) = \sqrt[3]{(50 + 2x)^2}, \qquad 1 \le x \le 26$$

where x represents the number of weeks that the series has been airing and $g(x)$ is the number of viewers in millions. Evaluate and interpret $g'(13)$.

71. A study by the bursar at Clarksman College determines that the number of students enrolled in the Arts and Science

programs during the past decade can be modeled by

$$f(t) = -\frac{10{,}000}{\sqrt{1 + 0.18t}} + 11{,}000, \qquad 1 \le t \le 11$$

where t represents the number of years since the beginning of the study and $f(t)$ represents the number of students enrolled in the Arts and Science programs. Evaluate and interpret $f'(10)$.

72. Medical researchers studying arteriosclerosis have found that, if the radius of a person's artery is currently 1 centimeter, the amount of fatty tissue called plaque that will build up in the artery can be modeled by

$$g(t) = 0.5t^2(t^2 + 10)^{-1}, \qquad 0 \le t \le 10$$

where t represents the number of years since the present time and $g(t)$ represents the thickness of the plaque in the wall of the artery in centimeters. Evaluate and interpret $g'(7)$.

73. The StopCop Company determines that the cost to produce auto antitheft devices is modeled by

$$C(x) = (3x + 6)^{1.5} + 30, \qquad 0 \le x \le 50$$

where x represents the number of auto antitheft devices produced in hundreds and $C(x)$ represents the production costs in thousands of dollars.

(a) Determine the marginal cost function.

(b) Evaluate and interpret $MC(5)$.

74. (Continuation of Exercise 73)

(a) Determine the average cost function AC.

(b) Determine the marginal average cost function $MAC(x) = \dfrac{d}{dx}(AC(x))$.

(c) Evaluate and interpret $MAC(5)$.

75. The SnapPic Company determines that its cost for producing disposable cameras is modeled by

$$C(x) = 60 + \sqrt{3x + 5}, \qquad 0 \le x \le 40$$

where x represents the number of disposable cameras produced during each shift and $C(x)$ represents the cost of production in hundreds of dollars.

(a) Determine the marginal cost function.

(b) Evaluate and interpret $MC(15)$.

76. (Continuation of Exercise 75)

(a) Determine the average cost function AC.

(b) Evaluate and interpret $AC(15)$.

(c) Determine the marginal average cost function $MAC(x) = \dfrac{d}{dx}(AC(x))$.

77. The price–demand function for a collectable doll is found to be

$$p(x) = \sqrt{22{,}500 - 50x}, \qquad 0 \le x \le 30$$

where x represents the number of collectable dolls produced in hundreds and $p(x)$ is the price of the dolls in dollars.

(a) Determine $p'(x)$ using the Generalized Power Rule.

(b) Evaluate $p'(25)$ and interpret.

78. (Continuation of Exercise 77)

(a) Determine the revenue function R.

(b) Determine the marginal revenue function using the Generalized Power Rule.

(c) Evaluate and interpret $MR(15)$.

79. The HotSpark Company has assumed that the price–demand function for their spark plug is

$$p(x) = \frac{125}{\sqrt{2x+5}}, \qquad 0 \le x \le 20$$

where x represents the number of spark plugs manufactured in hundreds and $p(x)$ is the price of the spark plug.

(a) Determine $p'(x)$ using the Generalized Power Rule.

(b) Evaluate $p'(20)$ and interpret.

(c) Determine the revenue function R.

(d) Determine the marginal revenue function using the Generalized Power Rule.

(e) Evaluate and interpret $MR(20)$.

For Exercises 80 and 81, consider the following: In the early 1930s, psychologist L.L. Thurstone determined that the time needed to learn a list of a certain length is given by the model

$$f(x) = ax\sqrt{x-b}, \qquad x \ge b$$

where x represents the number of items on the list and $f(x)$ represents the time needed to learn the list, measured in minutes. The constants a and b are different for each subject and are determined by pretesting.

80. Suppose that it has been determined that a freshman subject in a psychology class learns the items on a list according to the model

$$f(x) = \frac{5x}{2}\sqrt{x-6}, \qquad x \ge 6$$

(a) Evaluate $f(20)$ and interpret.

(b) Determine $f'(x)$ using the Generalized Power Rule.

(c) Evaluate $f'(20)$ and interpret.

81. Suppose that it has been determined that a sophomore subject in a psychology class learns the items on a list according to the model

$$f(x) = 2x\sqrt{x-3}, \qquad x \ge 3$$

(a) Evaluate $f(12)$ and interpret.

(b) Determine $f'(x)$ using the Generalized Power Rule.

(c) Evaluate $f'(12)$ and interpret.

82. The amount of toxic material entering Lake Formica is related to the number of years that the Bristine Chemical

Company has been operating by the model

$$g(t) = \left(\frac{4}{5}t^{1/5} + 2\right)^4, \qquad 0 \le t \le 30$$

where t represents the number of years that the company has been operating and $g(t)$ represents the amount of toxic material entering the lake in thousands of gallons.

(a) Evaluate $g(15)$ and interpret.

(b) Determine $g'(t)$ using the Generalized Power Rule.

(c) Evaluate $g'(15)$ and interpret.

✓ 83. Suppose that an actuary has determined that in Tribble Township the number of people surviving a certain number of years is given by the model

$$P(t) = 400\sqrt{101-t}, \qquad 0 \le t \le 101$$

where t represents the age of the person and $P(t)$ represents the number of people in the township who are still surviving.

(a) Determine $P(75)$ and interpret.

(b) Determine $P'(t)$ using the Generalized Power Rule.

(c) Determine $P'(75)$ and interpret.

84. Suppose that the rural town of Rufusville decides to relax its zoning laws so that more land can be made eligible for commercial use. They find that the town's annual tax base after the zoning changes can be modeled by

$$f(x) = 5x\sqrt{2x+2}, \qquad x \ge 0$$

where x represents the number of years since the zoning laws have changed and $f(x)$ represents the annual tax base in thousands of dollars.

(a) Determine $f(15)$ and interpret.

(b) Determine $f'(x)$ using the Generalized Power Rule.

(c) Determine $f'(15)$ and interpret.

85. Suppose that a lab technician finds that the number of bacteria present in an unidentified culture can be modeled by

$$g(t) = 70(10 + 0.5t)^{2.1}, \qquad t \ge 0$$

where t represents the time since the first observation in hours and $g(t)$ represents the number of bacteria present.

(a) Determine $g(8)$ and interpret.

(b) Determine $g'(t)$ using the Generalized Power Rule.

(c) Determine $g'(8)$ and interpret.

SECTION PROJECT

The data in Table 4.1.1 give the percentage of 3- to 5-year-olds enrolled in preschool from 1970 to 1995.

(a) Use your calculator to determine a power regression model of the form

$$f(x) = a \cdot x^b, \qquad 1 \le x \le 26$$

TABLE **4.1.1**	
YEAR	PERCENT ENROLLED
1970	37.5
1975	48.6
1980	52.5
1985	54.6
1990	59.4
1992	55.5
1993	55.1
1994	61.0
1995	61.8

Source: U.S. Census Bureau web site.

where x represents the number of years since 1969 and $f(x)$ represents the percentage of 3- to 5-year-olds enrolled in preschool. Round the constants a and b to the nearest hundredth.

(b) Use the constants a and b found in part (a) to rewrite the model in the form

$$g(x) = a \cdot (x + 1)^b, \qquad x_1 \le x \le x_2$$

where x represents the number of years since 1970. What are the values of x_1 and x_2 for which the model is valid?

(c) Use your calculator to make a table of values for f and g for $x = 0, 1, \ldots, 25$. What pattern do you notice?

(d) Determine $g'(x)$ using the Generalized Power Rule.

(e) Evaluate and interpret $g'(7)$.

(f) Evaluate and interpret $f'(8)$ and compare to part (e).

SECTION 4.2   DERIVATIVES OF LOGARITHMIC FUNCTIONS

Now we wish to focus our attention on a function first introduced in Section 1.7, the *logarithmic function.* This function has applications in business, social and life sciences. To discuss its application in calculus, in particular for determining rates of change, we first must learn its derivative. Since the logarithmic function is not a polynomial function, we must develop a new rule for its derivative.

From Your Toolbox

• The natural logarithm is $\log_e x$ and is denoted by $\ln x$ for $x > 0$.
• Let m and n be positive numbers. Then the following are true:

1. $\ln(mn) = \ln m + \ln n$
2. $\ln\left(\dfrac{m}{n}\right) = \ln m - \ln n$
3. $\ln m^n = n \cdot \ln m$
4. $\ln 1 = 0$
5. $\ln e = 1$
6. $e^{\ln x} = x$
7. $\ln e^x = x$

Derivative of the Natural Logarithm Function

To set the stage for studying the natural logarithm function, let's review some of its properties that we first encountered in Section 1.7. (See the Toolbox to the left.)

To determine the derivative of $f(x) = \ln x$, we cannot rely on the differentiation rules that we have learned so far. To determine the derivative, we return to the definition of the derivative, which gives

$$f'(x) = \lim_{h \to 0} \frac{\ln(x + h) - \ln x}{h}$$

The algebra behind computing this derivative is challenging. The proof of this derivative using the difference quotient is given in Appendix C. But to gain some insight into what the derivative is, we can use our calculator.

EXAMPLE 1

Analyzing $\dfrac{d}{dx}(\ln x)$ Numerically

For the function $f(x) = \ln x$, use your calculator to make a table of values for $y_1 = \ln(x)$ for $x = 1, 2, \ldots, 7$ and compare it to the values of the derivative of $\ln x$ by using the NDERIV (numerical derivative) command on your calculator.

SOLUTION

The output is shown in Table 4.2.1. In the calculator screen, Table 4.2.1b, $y_1 = \ln x$ and $y_2 = \text{NDERIV}(\ln x)$.

TABLE **4.2.1**

x	$y_1 = \ln x$	$y_2 = $ NDERIV $(\ln x)$
1	0	1
2	0.69	$\dfrac{1}{2}$
3	1.10	$\dfrac{1}{3}$
4	1.39	$\dfrac{1}{4}$
5	1.61	$\dfrac{1}{5}$
6	1.79	$\dfrac{1}{6}$
7	1.95	$\dfrac{1}{7}$

X	Y₁	Y₂
1	0	1
2	.69315	.5
3	1.0986	.33333
4	1.3863	.25
5	1.6094	.2
6	1.7918	.16667
7	1.9459	.14286

X=7

(a)

TECHNOLOGY NOTE
For more on NDERIV and other ways to find derivative values with your graphing calculator, see the online calculator manual at www.prenhall.com/armstrong

Do you see a pattern? It appears that we can get the values of the derivative by taking the reciprocal of x, the independent variable value. We now state this surprising result.

Derivative of the Natural Logarithm Function

For $f(x) = \ln x$, with $x > 0$, the **derivative of the natural logarithm function** is given by

$$f'(x) = \frac{d}{dx}(\ln x) = \frac{1}{x}$$

EXAMPLE 2 | **Computing Derivatives of Natural Logarithmic Functions**

Compute the derivatives of the given functions.

(a) $f(x) = 2 \ln x$ (b) $g(x) = \ln x^3$ (c) $y = 7 - 4 \ln x$

SOLUTION

(a) We can apply the Constant Multiple Rule here to get

$$f'(x) = \frac{d}{dx}(2 \ln x) = 2 \cdot \frac{d}{dx}(\ln x)$$

$$= 2 \cdot \frac{1}{x} = \frac{2}{x}$$

(b) Before we differentiate, we first take advantage of the logarithm property $\ln m^n = n \cdot \ln m$ and rewrite $g(x) = \ln x^3$ as $g(x) = 3 \ln x$. Differentiation gives

$$g'(x) = \frac{d}{dx}(\ln x^3) = \frac{d}{dx}(3 \ln x)$$

$$= 3 \cdot \frac{d}{dx}(\ln x)$$

$$= 3 \cdot \frac{1}{x} = \frac{3}{x}$$

(c) For this derivative, we use the Sum and Difference Rule. This gives

$$y' = \frac{d}{dx}(7 - 4\ln x) = \frac{d}{dx}(7) - \frac{d}{dx}(4\ln x)$$

$$= 0 - 4 \cdot \frac{d}{dx}(\ln x)$$

$$= -4 \cdot \frac{1}{x} = \frac{-4}{x}$$

✓**CHECKPOINT 1**

Now work Exercise 3.

Example 2c illustrates a special form that we will frequently use for modeling data. The function $y = 7 - 4\ln x$ is an example of the **natural logarithm model** because it has the form $f(x) = a + b \cdot \ln x$, where a and b represent constants. Example 3 illustrates how to calculate the rate of change of this type of function.

EXAMPLE 3

Differentiating and Interpreting a Natural Logarithm Model

The life expectancy of women born in the United States from 1951 to 1991 can be modeled by

$$f(x) = 66.11 + 3.29 \ln x, \qquad 1 \leq x \leq 41$$

where x represents the number of years since 1950 and $f(x)$ represents the life expectancy, in years, of a woman born in the year corresponding to x. Find $f'(x)$ and evaluate $f'(5)$ and $f'(40)$. Interpret the results.

SOLUTION

The derivative of the model is

$$f'(x) = \frac{d}{dx}(66.11 + 3.29\ln x) = \frac{d}{dx}(66.11) + \frac{d}{dx}(3.29\ln x)$$

$$= 0 + 3.29 \cdot \frac{1}{x} = \frac{3.29}{x}$$

First, note that $x = 5$ corresponds to the year 1955. Evaluating $f'(5)$, we get

$$f'(5) = \frac{3.29}{5} \approx 0.66$$

This means that the life expectancy of a woman born in the United States in 1955 was increasing at a rate of about 0.66 year of life per year.

The value of $x = 40$ corresponds to the year 1990. Evaluating $f'(40)$, we get

$$f'(40) = \frac{3.29}{40} \approx 0.08$$

Thus, the life expectancy of a woman born in the United States in 1990 was increasing at a rate of approximately 0.08 year of life per year, about one-eighth of the 1955 rate of change (see Figure 4.2.1). Perhaps the increase in life expectancy of women in 1955 is greater than the increase in 1990 because many medical advancements were made during the 1950s and 1960s, resulting in a fairly rapid increase in life expectancy.

Interactive Activity

For the model in Example 3, determine an equation of the line tangent to the graph of f at $x = 5$. Graph the tangent line and the model f in the same viewing window. Determine the value of the tangent line at $x = 40$, and interpret in terms of the life expectancy of women.

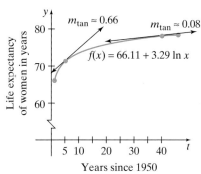

Figure 4.2.1 Comparison of $f'(5)$ and $f'(40)$.

✓ CHECKPOINT 2

Now work Exercise 43.

NOTE: One word of caution when computing logarithmic regressions: Since the domain of the logarithmic function $y = \ln x$ is $(0, \infty)$, we must have all nonzero entries for our independent variable values. In particular, when standardizing the independent variable, be careful not to begin with $x = 0$.

Sometimes we need to differentiate functions that involve more than just $\ln x$. Let's say that the argument of the logarithmic function was another function, which we will call $u = g(x)$. If we let $f(x) = \ln x$, then we have a composite function $h(x) = f(g(x)) = \ln (g(x)) = \ln u$. We must use the Chain Rule to determine the derivative of $h(x) = f(g(x))$. Since $f'(x) = \dfrac{1}{x}$ and $g'(x) = \dfrac{du}{dx}$, we have

$$h'(x) = f'(g(x)) \cdot g'(x)$$

$$= \frac{1}{g(x)} \cdot g'(x) = \frac{g'(x)}{g(x)}$$

Chain Rule for the Natural Logarithm Function

If g is a differentiable function of x and the range of g is $(0, \infty)$, the derivative of the composition of functions $h(x) = \ln [g(x)]$ is given by

$$h'(x) = \frac{1}{g(x)} \cdot g'(x) = \frac{g'(x)}{g(x)}$$

NOTE: We saw this derivative once before in Section 3.3. There we defined the *relative rate of change* of a function f as $Rel(f(x)) = \dfrac{\left(\substack{\text{derivative} \\ \text{of } f(x)}\right)}{\left(\substack{\text{given} \\ \text{function } f(x)}\right)} = \dfrac{f'(x)}{f(x)}$. This is exactly what the preceding rule states, with $g(x)$ in place of $f(x)$. So we can redefine the relative rate of change of a function as $Rel(f(x)) = \dfrac{d}{dx}[\ln (f(x))]$. (This is why we had to restrict the domain of f in our definition of $Rel(f(x))$ to $(0, \infty)$.)

EXAMPLE 4 | **Differentiating the Natural Logarithm Function with the Chain Rule**

Determine the derivative of each function.

(a) $y = \ln (x^4 - 2x)$

(b) $h(x) = \ln \sqrt{6x - 1}$

(c) $f(x) = (\ln x)^3$

SOLUTION

(a) Here the inside, or g, function is $g(x) = x^4 - 2x$. So applying the Chain Rule gives us

$$y' = \frac{1}{x^4 - 2x} \cdot \frac{d}{dx}(x^4 - 2x)$$

$$= \frac{1}{x^4 - 2x} \cdot (4x^3 - 2) = \frac{4x^3 - 2}{x^4 - 2x}$$

(b) Here, we begin by using some algebra to rewrite the function,

$$h(x) = \ln \sqrt{6x - 1} = \ln (6x - 1)^{1/2} = \frac{1}{2} \ln (6x - 1)$$

Now, we differentiate using the Chain Rule.

$$h'(x) = \frac{d}{dx}\left[\frac{1}{2} \ln (6x - 1)\right]$$

$$= \frac{1}{2} \cdot \frac{d}{dx} \ln (6x - 1)$$

$$= \frac{1}{2} \cdot \frac{1}{6x - 1} \cdot \frac{d}{dx}(6x - 1)$$

$$= \frac{1}{2} \cdot \frac{1}{6x - 1} \cdot 6 = \frac{3}{6x - 1}$$

(c) This is a little different from the previous two parts in that we must apply the Generalized Power Rule to get the derivative.

$$f(x) = (\ln x)^3$$

$$f'(x) = \frac{d}{dx}[(\ln x)^3]$$

$$= 3(\ln x)^2 \cdot \frac{d}{dx}(\ln x)$$

$$= 3(\ln x)^2 \cdot \frac{1}{x} = \frac{3(\ln x)^2}{x}$$

The difference between the functions in Example 2b and Example 4c is an important one. In Example 2b, $g(x) = \ln x^3$ means that x is taken to the third power and *then* the logarithm is taken. In Example 4c, $f(x) = (\ln x)^3$ means that the logarithm of x is taken first and *then* the result is cubed. The parentheses are used to make the difference clear. (Some textbooks denote a function like $f(x) = (\ln x)^3$ by writing $f(x) = \ln^3 x$.)

Derivatives of General Logarithmic Functions

The most frequently used logarithm function for applications is the natural logarithm function $f(x) = \ln x$. But what about the derivatives of the logarithm function with other bases? In other words, what is $\frac{d}{dx}(\log_b x)$? The following definition shows how to differentiate a general logarithmic function. The derivation of this result is left to you. (See Interactive Activity to the left.)

Interactive Activity

The general logarithmic function $f(x) = \log_b x$ with $b > 0$, $b \neq 1$ may be rewritten as

$$f(x) = \log_b x = \frac{\ln x}{\ln b}$$

via what is known as the change of base formula. Differentiate $f(x) = \frac{\ln x}{\ln b}$ to show that for

$f(x) = \log_b x$, $f'(x) = \frac{1}{\ln b} \cdot \frac{1}{x}$.

> ### Derivative of the General Logarithmic Function
>
> - For the general logarithmic function $f(x) = \log_b x$ with $b > 0, b \neq 1$,
>
> $$f'(x) = \frac{1}{\ln b} \cdot \frac{1}{x}$$
>
> - If g is a differentiable function, where the range of g is $(0, \infty)$, then the **Chain Rule** for the general logarithmic function $f(x) = \log_b [g(x)]$ is
>
> $$f'(x) = \frac{1}{\ln b} \cdot \frac{1}{g(x)} \cdot g'(x)$$

EXAMPLE 5

Determining Derivatives of General Logarithmic Functions

Differentiate the following:

(a) $y = \log_3 x$ (b) $f(x) = \log (x^3 + 9)$

SOLUTION

(a) Here we have a general logarithm function with base 3. So the derivative is

$$y' = \frac{d}{dx}(\log_3 x) = \frac{1}{\ln 3} \cdot \frac{1}{x} = \frac{1}{x \ln 3}$$

(b) Since the base is not written, we know that this is a common logarithm with base 10. Applying the Chain Rule, with $g(x) = x^3 + 9$, gives

$$f'(x) = \frac{d}{dx}[\log (x^3 + 9)]$$

$$= \frac{1}{\ln 10} \cdot \frac{1}{(x^3 + 9)} \cdot \frac{d}{dx}(x^3 + 9)$$

$$= \frac{1}{\ln 10} \cdot \frac{1}{(x^3 + 9)} \cdot (3x^2) = \frac{1}{\ln 10} \cdot \frac{3x^2}{(x^3 + 9)}$$

✓ CHECKPOINT 3

Now work Exercise 31.

SUMMARY

In this section we learned how to differentiate the natural logarithmic function, as well as the logarithmic function with any base.

- **Derivative of the natural logarithm function:** For $f(x) = \ln x$,
  $f'(x) = \frac{d}{dx}(\ln x) = \frac{1}{x}$.

- **Chain rule for the natural logarithm function:** If $h(x) = \ln (g(x))$ then
  $h'(x) = \frac{1}{g(x)} \cdot g'(x) = \frac{g'(x)}{g(x)}$.

- **Derivative of the general logarithmic function:** For the general logarithmic function $f(x) = \log_b x$, $f'(x) = \frac{1}{\ln b} \cdot \frac{1}{x}$. The **Chain Rule** for the general logarithmic function $f(x) = \log_b [g(x)]$ is $f'(x) = \frac{1}{\ln b} \cdot \frac{1}{g(x)} \cdot g'(x)$.

SECTION 4.2 EXERCISES

In Exercises 1–10, determine the derivative for the following functions.

1. $f(x) = 5\ln x$

2. $f(x) = -8\ln x$

✓ 3. $f(x) = \ln x^6$

4. $f(x) = \ln x^4$

5. $f(x) = 4x^3 \cdot \ln x$

6. $f(x) = 12x^3 \cdot \ln x$

7. $f(x) = \dfrac{3x^5}{\ln x}$

8. $f(x) = \dfrac{12}{\ln x}$

9. $f(x) = 10 - 12\ln x$

10. $f(x) = -2 + 8\ln x$

In Exercises 11–24, determine the derivative for the following functions.

11. $g(x) = \ln(x + 7)$

12. $g(x) = \ln(2 - x)$

13. $g(x) = \ln(2x - 5)$

14. $g(x) = \ln(3x + 4)$

15. $g(x) = \ln(x^2 + 3)$

16. $g(x) = \ln(3x^3 - 11)$

17. $g(x) = \ln\left(\sqrt{2x + 5}\right)$

18. $g(x) = \ln\left(\sqrt[3]{4x + 2}\right)$

19. $g(x) = (\ln x)^6$

20. $g(x) = (\ln x)^4$

21. $g(x) = \sqrt{x} \cdot \ln\left(\sqrt{x}\right)$

22. $g(x) = 4x^5 \cdot \ln(3x^3)$

23. $g(x) = \dfrac{x^2 + 2x + 3}{\ln(x + 5)}$

24. $g(x) = \dfrac{4x^3 - x + 2}{\ln(x + 7)}$

For Exercises 25–34, determine the derivative for the following functions.

25. $f(x) = \log_{10} x$

26. $f(x) = \log_5 x$

27. $f(x) = 6\log_3 x$

28. $f(x) = 11\log_4 x$

29. $f(x) = x^2 \log_9 x$

30. $f(x) = 2x^5 \log_8 x$

✓ 31. $f(x) = \log_2(5x + 3)$

32. $f(x) = \log_5(3x + 9)$

33. $f(x) = \log_{10}\left(\dfrac{x + 3}{x^2 + 1}\right)$

34. $f(x) = \log_2\left(\dfrac{x^3}{x^2 - 1}\right)$

For Exercises 35–42, determine an equation for the line tangent to the graph of the function at the given point.

35. $f(x) = \ln x$; $(2, \ln 2)$

36. $f(x) = \ln x$; $(1, 0)$

37. $f(x) = \ln\sqrt{2x - 1}$; $(1, 0)$

38. $f(x) = \ln(3x)$; $(2, \ln 6)$

39. $f(x) = 4x^3 \cdot \ln x$; $(1, 0)$

40. $f(x) = 12x^3 \cdot \ln x$; $(2, 96\ln 2)$

41. $y = (\ln x)^6$; $(e, 1)$

42. $y = \ln x^6$; $(e, 6)$

APPLICATIONS

✓ 43. A research assistant in biology finds in an experiment that at low temperatures the growth of a certain bacteria culture can be modeled by

$$f(t) = 750 + 12\ln t, \qquad t \geq 1$$

where t represents the number of hours since the start of the experiment and $f(t)$ represents the number of bacteria present.

(a) Determine $f'(t)$.

(b) Evaluate and interpret $f(12)$ and $f'(12)$.

44. The city of Plantersville has enacted new zoning laws in order to curb the growth of the city's population. They find that the population can be modeled by

$$P(x) = 10{,}000 + 100\ln x, \qquad x \geq 1$$

where x represents the number of years since the laws were adopted and $P(x)$ represents the city's population.

(a) Determine $P'(x)$.

(b) Evaluate and interpret $P(20)$ and $P'(20)$.

45. Prescription drug companies have found that the popularity of the new drug Vectrum has dwindled and can be modeled by

$$f(x) = 150 + 5\log_2 x, \qquad x \geq 1$$

where x represents the number of years that the drug has been on the market and $f(x)$ represents the number of prescriptions written for the drug annually in thousands.

(a) Determine $f'(x)$.

(b) Evaluate $f'(2)$ and $f'(10)$ and interpret each.

46. The urban school district of Molisburg has started a new educational campaign in an attempt to reduce the increase in lice found in the elementary school student population. The number of children who contracted lice can be modeled by

$$g(t) = 200 + 8\log_3 t, \qquad t \geq 1$$

where t represents the number of years since the new educational campaign has been enacted and $g(t)$ represents the number of students who are diagnosed with lice annually.

(a) Determine $g'(t)$.

(b) Evaluate $g'(2)$ and $g'(7)$ and interpret each.

47. The life expectancy for African-American females in the United States can be modeled by

$$f(x) = 68.41 + 1.75\ln x, \qquad 1 \leq x \leq 26$$

where x represents the birth year since 1969 and $f(x)$ represents the life expectancy in years.

(a) Determine $f'(x)$.

(b) Evaluate and interpret $f(3)$ and $f'(3)$.

(c) Write the equation of the tangent line at $x = 3$, and determine y on the tangent line when $x = 15$. Interpret and compare to $f(15)$.

48. The life expectancy for white females in the United States can be modeled by

$$f(x) = 75.32 + 1.29 \ln x, \qquad 1 \le x \le 26$$

where x represents the birth year since 1969 and $f(x)$ represents the life expectancy in years.

(a) Determine $f'(x)$.

(b) Evaluate and interpret $f(3)$ and $f'(3)$. Compare to part (b) in Exercise 47.

(c) Write the equation of the tangent line at $x = 3$, and determine y on the tangent line when $x = 15$. Interpret and compare to $f(15)$. Compare to part (c) in Exercise 47.

49. The annual per capita consumption of light and skim milk in the United States can be modeled by

$$f(x) = 10.12 + 2 \ln x, \qquad 1 \le x \le 16$$

where x represents the number of years since 1979 and $f(x)$ represents the annual per capita consumption of light and skim milk in gallons.

(a) Graph f in the viewing window $[1, 16]$ by $[10, 17]$.

(b) Determine $f'(x)$.

(c) Evaluate and interpret $f'(5)$ and compare to $f'(10)$.

50. The average expenditure for a new domestic car in the United States can be modeled by

$$f(x) = 15{,}302.93 + 1685.66 \ln x, \qquad 1 \le x \le 7$$

where x represents the number of years since 1980 and $f(x)$ represents the average expenditure for a new domestic car in dollars.

(a) Graph f in the viewing window $[1, 7]$ by $[15{,}000, 19{,}000]$.

(b) Determine $f'(x)$.

(c) Evaluate and interpret $f'(2)$ and compare to $f'(6)$.

SECTION PROJECT

The revenue, in billions of dollars, generated in the solid waste management industry in the United States from 1980 to 1996 are displayed in Table 4.2.2.

TABLE 4.2.2

YEAR	x	REVENUE (IN $BILLIONS)
1980	1	52.0
1990	11	146.4
1994	15	172.5
1995	16	180.0
1996	17	184.3

Source: U.S. Census Bureau web site.

(a) Use your calculator to determine a logarithmic regression model of the form

$$f(x) = a + b \ln x, \qquad 1 \le x \le 17$$

where x represents the number of years since 1979 and $f(x)$ represents the revenue generated in the solid waste management industry, in billions of dollars.

(b) Determine $f'(x)$.

(c) Evaluate and interpret $f'(5)$.

(d) Write an equation of the line tangent to the graph of f at $x = 3$. Determine y on the tangent line when $x = 10$ and interpret.

(e) Compare the value found in part (d) to $f(10)$ and interpret.

SECTION 4.3 DERIVATIVES OF EXPONENTIAL FUNCTIONS

In this section, we discuss how to differentiate the **exponential function**, $f(x) = e^x$, and the **general exponential function**, $f(x) = b^x$. The exponential function has applications in business as well as in the social and life sciences. One application first introduced in Section 1.6 is exponential growth and decay. Here we analyze the rate of change of the growth or the decay. To discuss these applications, in particular when determining rates of change, we need to learn the derivative. As in Section 4.2, we must determine the derivative utilizing the definition of the derivative.

Derivatives of Exponential Functions with Base e

An important function that has not yet been differentiated is the exponential function with base e, $f(x) = e^x$. To determine this derivative, we must return to the definition of the derivative. By definition, if $f(x) = e^x$, then

$$f'(x) = \lim_{h \to 0} \frac{e^{x+h} - e^x}{h}$$

Using the properties of exponents gives us

$$f'(x) = \lim_{h \to 0} \frac{e^x \cdot e^h - e^x}{h}$$

$$= \lim_{h \to 0} \frac{e^x(e^h - 1)}{h}$$

Since we are computing the limit with respect to h, e^x can be treated as constant.

$$= \lim_{h \to 0} e^x \frac{e^h - 1}{h}$$

$$= e^x \cdot \lim_{h \to 0} \frac{e^h - 1}{h}$$

But what is $\lim_{h \to 0} \frac{e^h - 1}{h}$? If we simply substitute zero for h, we get the indeterminate form $\frac{0}{0}$. To find this limit, we rely on our numerical and graphical methods.

From the graph of $\frac{e^h - 1}{h}$ in Figure 4.3.1, we believe that $\lim_{h \to 0} \frac{e^h - 1}{h} = 1$. Checking numerically, Table 4.3.1 shows the values of $\frac{e^h - 1}{h}$ for h-values close to zero. From the graph and the table, we believe that

$$\lim_{h \to 0} \frac{e^h - 1}{h} = 1$$

We can now take the limit of our difference quotient.

$$f'(x) = e^x \cdot \lim_{h \to 0} \frac{e^h - 1}{h}$$

$$= e^x \cdot 1 = e^x$$

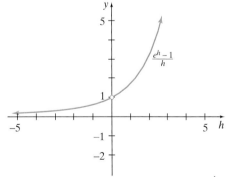

Figure 4.3.1 Graphical determination that $\lim_{h \to 0} \dfrac{e^h - 1}{h} = 1$.

TABLE 4.3.1							
h	-0.1	-0.001	-0.0001	0	0.0001	0.001	0.1
$\dfrac{e^h - 1}{h}$	0.9516	0.9995	0.9999		1.0001	1.0005	1.0517

Interactive Activity

 Use the rule to the right along with the constant multiple rule to determine the derivatives for $h(x) = 5e^x$, $g(x) = \frac{1}{2}e^x$, and $y = \sqrt{3} \cdot e^x$. What rule could we conclude for the derivative of $f(x) = ce^x$, where c is any real number?

This surprising result means that the exponential function $f(x) = e^x$ is its own derivative! (The only other function that had a property like this up to now is the trivial function $f(x) = 0$.)

Derivative of the Exponential Function

The **derivative** of the **exponential function** $f(x) = e^x$ is

$$f'(x) = e^x$$

EXAMPLE 1 | **Differentiating Exponential Functions**

Determine derivatives for the following functions:

(a) $y = \dfrac{x^2}{e^x}$ (b) $g(x) = x^4 \cdot e^x$ (c) $f(x) = \log e^x$

SOLUTION

(a) Applying the Quotient Rule, we get

$$y' = \frac{d}{dx}\left(\frac{x^2}{e^x}\right) = \frac{\frac{d}{dx}[x^2]\cdot(e^x) - (x^2)\cdot\frac{d}{dx}[e^x]}{(e^x)^2}$$

$$= \frac{2xe^x - x^2e^x}{e^{2x}} = \frac{e^x(2x - x^2)}{e^{2x}} = \frac{2x - x^2}{e^x}$$

(b) Applying the Product Rule for this function gives

$$g'(x) = \frac{d}{dx}(x^4 \cdot e^x) = \frac{d}{dx}[x^4]\cdot(e^x) + (x^4)\cdot\frac{d}{dx}(e^x)$$

$$= 4x^3e^x + x^4e^x$$

(c) For this common logarithm function, we need to recall the differentiation rule for the general logarithm function from Section 4.2.

$$f'(x) = \frac{d}{dx}(\log e^x) = \frac{1}{\ln 10}\cdot\frac{1}{e^x}\cdot\frac{d}{dx}(e^x)$$

$$= \frac{1}{\ln 10}\cdot\frac{1}{e^x}\cdot(e^x) = \frac{1}{\ln 10}$$

Notice that this derivative is actually a constant.

The derivative of e^x by itself has limited value. But the Chain Rule applied to the exponential function is very useful, as it lends itself to many applications. Let's add this extension to our list of differentiation rules.

Chain Rule for the Exponential Function

If f is a differentiable function of x, the derivative of the composition of functions $h(x) = e^{f(x)}$ is given by

$$h'(x) = e^{f(x)}\cdot f'(x)$$

EXAMPLE 2 | **Differentiating Composite Exponential Functions**

Determine derivatives for the following functions:

(a) $h(x) = e^{3x-3}$ 　　　　(b) $g(x) = e^{6x-(1/2)x^6}$

(c) $y = e^{1-\ln x}$ 　　　　(d) $f(x) = \ln(e^{3x} - 8)$

SOLUTION

(a) In the composite form $h(x) = e^{f(x)}$, $f(x)$ is $3x - 3$.

$$h'(x) = \frac{d}{dx}(e^{3x-3}) = e^{3x-3} \cdot \frac{d}{dx}(3x-3)$$

$$= e^{3x-3} \cdot (3) = 3e^{3x-3}$$

(b) Applying the Chain Rule for the Exponential Function gives

$$g'(x) = \frac{d}{dx}(e^{6x-(1/2)x^6}) = (e^{6x-(1/2)x^6}) \cdot \frac{d}{dx}\left(6x - \frac{1}{2}x^6\right)$$

$$= (e^{6x-(1/2)x^6}) \cdot (6 - 3x^5)$$

(c) This function has a natural logarithm term in the exponent.

$$y' = \frac{d}{dx}(e^{1-\ln x}) = (e^{1-\ln x}) \cdot \frac{d}{dx}(1 - \ln x)$$

$$= (e^{1-\ln x}) \cdot \left(-\frac{1}{x}\right) = -\frac{e^{1-\ln x}}{x} = -\frac{e}{x^2}$$

(d) This function requires us to start with the Chain Rule for the natural logarithm function.

$$f'(x) = \frac{d}{dx}[\ln(e^{3x} - 8)] = \frac{1}{e^{3x} - 8} \cdot \frac{d}{dx}(e^{3x} - 8)$$

$$= \frac{1}{e^{3x} - 8} \cdot (e^{3x} \cdot 3) = \frac{3e^{3x}}{e^{3x} - 8}$$

✓CHECKPOINT **1**

Now work Exercise 19.

The properties that we studied in earlier chapters can be applied to the exponential functions as well. This is illustrated in Example 3.

EXAMPLE 3 | **Applying an Exponential Growth Model**

The world's population during the 20th century closely follows the mathematical model

$$p(t) = 1.419e^{0.014t}, \quad 0 \le t \le 100$$

where t represents the number of years since 1900 and $p(t)$ represents the world's population in billions of people. Evaluate and interpret $p'(10)$ and compare to $p'(99)$.

SOLUTION

Differentiating the population model, we get

$$p'(t) = \frac{d}{dt}\left(1.419e^{0.014t}\right) = 1.419\frac{d}{dt}(e^{0.014t})$$

$$= 1.419e^{0.014t} \cdot \frac{d}{dt}(0.014t)$$

$$= 1.419e^{0.014t} \cdot (0.014) \approx 0.02e^{0.014t}$$

The year 1910 corresponds to $t = 10$. Evaluating $p'(10)$ gives

$$p'(10) = 0.02e^{(0.014)(10)} \approx 0.023\frac{\text{billion people}}{\text{year}}$$

This means that in 1910 the world's population was increasing at a rate of about 0.023 billion (that is, 23 million) people per year.

The year 1999 corresponds to $t = 99$. Evaluating $p'(99)$ yields

$$p'(99) = 0.02e^{(0.014)(99)} \approx 0.08\frac{\text{billion people}}{\text{year}}$$

Thus, in 1999 the world's population was increasing at a rate of about 0.08 billion (that is, 80 million), people per year. The growth rate in 1999 was nearly 3.5 times the growth rate in 1910. See Figure 4.3.2. ■

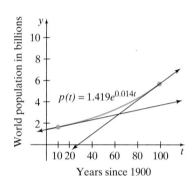

Figure 4.3.2 Growth rate in 1999 was about 3.5 times the growth rate in 1910 as shown by tangent lines.

From Your Toolbox

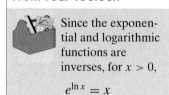

Since the exponential and logarithmic functions are inverses, for $x > 0$,

$$e^{\ln x} = x$$

Derivatives of General Exponential Functions

One type of mathematical model we have not yet differentiated is the **general exponential function**, $f(x) = b^x$. Remember that for these functions b is any positive number other that 1. To find this derivative, let's return to a property that we covered in Chapter 1 as shown in the Toolbox to the left.

This fact in the Toolbox means that we can substitute the value of b for x and write $b = e^{\ln b}$. Thus,

$$b^x = (e^{\ln b})^x = e^{\ln b \cdot x}$$

Using this property, we can now differentiate the function $f(x) = b^x$ using the Chain Rule for the exponential function.

$$f'(x) = \frac{d}{dx}(b^x)$$

$$= \frac{d}{dx}(e^{\ln b \cdot x})$$

$$= e^{\ln b \cdot x} \cdot \frac{d}{dx}(\ln b \cdot x)$$

Since b is a constant, so is $\ln b$, which gives us

$$f'(x) = e^{\ln b \cdot x} \cdot \ln b$$

$$f'(x) = b^x \cdot \ln b \qquad \text{Rewrite } b \text{ for } e^{\ln b}$$

> ### Derivative of the General Exponential Function
>
> If $f(x) = b^x$, with $b > 0$ and $b \neq 1$, then the **derivative** of the **general exponential function** is
> $$f'(x) = b^x \cdot \ln b$$

EXAMPLE 4 | **Determining Derivatives of General Exponential Functions**

Determine the derivatives for the given functions.

(a) $f(x) = 10^x$ (b) $g(x) = \dfrac{3^x}{4^x}$

SOLUTION

(a) Here we use the general exponential function differentiation rule with $b = 10$. So we get

$$f'(x) = \frac{d}{dx}(10^x) = 10^x \cdot \ln 10$$

(b) Instead of using the quotient rule, we can use the properties of exponents to rewrite $g(x) = \dfrac{3^x}{4^x}$ as $g(x) = \left(\dfrac{3}{4}\right)^x$. Now we have a general exponential function with $b = \dfrac{3}{4}$. So the derivative is

$$g'(x) = \frac{d}{dx}\left[\left(\frac{3}{4}\right)^x\right]$$

$$= \left(\frac{3}{4}\right)^x \cdot \ln\left(\frac{3}{4}\right) = \left(\frac{3}{4}\right)^x (\ln 3 - \ln 4)$$

It seems natural at this point to extend our differentiation capabilities of the general exponential function by adding in the Chain Rule so that we can differentiate composite general exponential functions.

> ### Chain Rule for the General Exponential Function
>
> If f is a differentiable function of x, and $b > 0, b \neq 1$, the derivative of the composition of functions $h(x) = b^{f(x)}$ is given by
> $$h'(x) = b^{f(x)} \cdot \ln b \cdot f'(x)$$

EXAMPLE 5 | **Differentiating Composite General Exponential Functions**

Determine derivatives for the given functions.

(a) $y = 5^{9x-5}$ (b) $g(x) = 3^{\ln x + 5}$

SOLUTION

(a) In the form $h(x) = b^{f(x)}$, we have $b = 5$ and $f(x) = 9x - 5$. So the derivative is

$$y' = \frac{d}{dx}(5^{9x-5})$$

$$= 5^{9x-5} \cdot \ln 5 \cdot \frac{d}{dx}(9x - 5)$$

$$= 5^{9x-5} \cdot \ln 5 \cdot 9$$

Interactive Activity

For $y = \ln(3^{2x})$, use the Chain Rule for natural logarithmic functions along with a property of logarithms to show that $y' = \ln 9$.

(b) Applying the Chain Rule for the general exponential function yields

$$y' = \frac{d}{dx}(3^{\ln x + 5})$$

$$= 3^{\ln x + 5} \cdot \ln 3 \cdot \frac{d}{dx}(\ln x + 5)$$

$$= 3^{\ln x + 5} \cdot \ln 3 \cdot \left(\frac{1}{x}\right) = \frac{3^{\ln x + 5} \cdot \ln 3}{x}$$

✓ **CHECKPOINT 2**

Now work Exercise 35.

Applications

From Your Toolbox

• The exponential function $f(x) = b^x$, with $b > 1$ (or $f(x) = b^{-x}$, where $0 < b < 1$) models **exponential growth.**

• The exponential function $f(x) = b^x$, where $0 < b < 1$ (or $f(x) = b^{-x}$, where $b > 1$) models what is known as **exponential decay.**

Recall that models such as $p(t) = 1.419e^{0.014t}$ from Example 3 are called **exponential growth models**. Let's take another look at the definition of exponential growth and exponential decay by consulting the Toolbox to the left.

Before we look at one final application, we would like to point out that, since $f(x) = b^x$ can be written as $f(x) = e^{\ln b \cdot x}$, any general exponential model may be written as follows:

> ### Rewriting the General Exponential Model
>
> The general exponential model $f(x) = a \cdot b^x$ can be written in the exponential form
>
> $$g(x) = a \cdot e^{kx}$$
>
> where $k = \ln b$.

For example, the exponential model $f(x) = 3.1(1.92)^x$ can be written as $g(x) = 3.1e^{(\ln 1.92) \cdot x} \approx 3.1e^{0.65x}$. This rewriting technique is useful for modeling, since many calculators will only model data in the general exponential form $f(x) = a \cdot b^x$.

Flashback

YARD WASTE MODEL REVISITED

In Section 1.6, we modeled the percentage of waste generated that is yard waste with the general exponential function

$$f(x) = 19.12(0.944)^x, \qquad 1 \le x \le 6$$

where x represents the number of years since 1990 and $f(x)$ represents the percentage of waste generated that is yard waste. Determine $f'(2)$ and interpret.

Solution

The derivative is computed to be

$$f'(x) = \frac{d}{dx}[19.12(0.944)^x]$$

$$= 19.12 \cdot (0.944)^x \cdot \ln(0.944)$$

Evaluating the derivative when $x = 2$ gives us

$$f'(2) = 19.12 \cdot (0.944)^2 \cdot \ln(0.944) \approx -0.98$$

Since $x = 2$ corresponds to 1992, we conclude that in 1992 the percentage of waste generated that is yard waste was decreasing at a rate of about $0.98 \frac{\text{percent}}{\text{year}}$. See Figure 4.3.3.

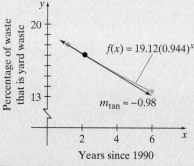

Figure 4.3.3

| ✓ CHECKPOINT 3 | Now work Exercise 45. |

SUMMARY

In this section, we examined the **derivatives of exponential functions.** We also saw how the property $b = e^{\ln b}$ can be used to rewrite $f(x) = b^x$ as $f(x) = e^{\ln b \cdot x}$. This rewriting technique is useful for modeling since many calculators only model data in the general exponential form $f(x) = a \cdot b^x$. We determined the following differentiation rules:

- $\dfrac{d}{dx}[e^x] = e^x$

- $\dfrac{d}{dx}[e^{f(x)}] = e^{f(x)} \cdot f'(x),$ where f is a differentiable function of x

- $\dfrac{d}{dx}[b^x] = b^x \cdot \ln b$

- $\dfrac{d}{dx}[b^{f(x)}] = b^{f(x)} \cdot \ln b \cdot f'(x),$ where f is a differentiable function of x

SECTION 4.3 EXERCISES

In Exercises 1–16, determine the derivative for the following functions:

1. $g(x) = 7e^x$
2. $g(x) = 10e^x$
3. $g(x) = 2x(4 + e^x)$
4. $g(x) = 10x(e^x + 40)$
5. $g(x) = \dfrac{10}{5 - e^x}$
6. $g(x) = \dfrac{15}{2 + e^x}$
7. $g(x) = 4x^2 e^x$
8. $g(x) = 10x^2 e^x$
9. $g(x) = \sqrt{12 - e^x}$
10. $g(x) = \sqrt[3]{e^x + 5}$
11. $g(x) = \dfrac{e^x - 10}{x^3 - 1}$
12. $g(x) = \dfrac{x^2 + 5}{2 - e^x}$
13. $g(x) = \dfrac{e^x + 1}{e^x - 1}$
14. $g(x) = \dfrac{4 - e^x}{4 + e^x}$
15. $g(x) = 2xe^x - x$
16. $g(x) = 4x^2 e^x - e^x$

In Exercises 17–28, determine the derivative for the following functions:

17. $f(x) = e^{2x-1}$
18. $f(x) = e^{6x+9}$
✓ 19. $f(x) = e^{\sqrt{x}}$
20. $f(x) = e^{\sqrt[3]{x}}$
21. $f(x) = e^{\ln x}$
22. $f(x) = e^{2\ln x}$
23. $f(x) = 5x \cdot e^{2x}$
24. $f(x) = 2x^3 \cdot e^{6x^2}$
25. $f(x) = \dfrac{e^{x-1}}{e^{x+1}}$
26. $f(x) = \dfrac{e^{2+x}}{e^{2-x}}$
27. $f(x) = \ln(x^2 + e^{-x})$
28. $f(x) = \ln(x^3 - e^{5x})$

In Exercises 29–34, determine the derivative for the following functions:

29. $g(x) = 10^x$
30. $g(x) = 4^x$
31. $g(x) = \dfrac{5^x}{15^x}$
32. $g(x) = \dfrac{8^x}{2^x}$
33. $g(x) = x^3 \cdot 0.3^x$
34. $g(x) = x^2 \cdot 0.2^x$

In Exercises 35–42, determine the derivative for the following functions:

✓ 35. $f(x) = 10^{x+3}$
36. $f(x) = 2^{9-x}$
37. $f(x) = 9^{1/x}$
38. $f(x) = 4^{\sqrt{x}}$
39. $f(x) = x \cdot e^x - 5^{2x}$
40. $f(x) = 2x \cdot \ln x + 4^{2x-1}$
41. $f(x) = \ln 5x \cdot 5^{x^2}$
42. $f(x) = \log_3 4x \cdot 4^{x-e}$

APPLICATIONS

43. A veterinarian finds that when a lab animal specimen is exposed to a new pesticide, the growth of a tumor in the specimen can be modeled by

$$f(t) = 2.1e^{0.2t}, \qquad t > 0$$

where t represents the number of days since exposure to the pesticide and $f(t)$ represents the diameter of the tumor in millimeters.

(a) Determine $f'(t)$.
(b) Evaluate and interpret $f'(3)$.

44. Since adding a new herbicide to wheat seed, a farmer finds that the total yield of the crop can be modeled by

$$f(x) = 1500\, e^{0.15x}, \qquad x \geq 0$$

where x represents the number of growing seasons since the farmer started using the herbicide and $f(x)$ represents the crop yield in bushels.

(a) Determine $f'(x)$.

(b) Evaluate and interpret $f'(5)$.

✓ 45. A zoologist has studied the population growth of a certain species of fish near a recently built lake shore factory. The population can be modeled by

$$p(t) = 12 \cdot (0.8)^t, \qquad 0 \leq t \leq 10$$

where t represents the number of years since the factory opened and $p(t)$ represents the population of the fish in hundreds.

(a) Determine $p'(t)$.

(b) Evaluate and interpret $p'(1)$ and compare to $p'(8)$. See Example 3.

46. The director of health services at a large university enacts an immunization drive in order to decrease the number of students contracting the flu. The number of students getting the flu can be modeled by

$$g(x) = 7000 \cdot (0.92)^x, \qquad 0 \leq x \leq 12$$

where x represents the number of years since the immunization drive began and $g(x)$ represents the number of students who contract the flu each year.

(a) Determine $g'(x)$.

(b) Evaluate and interpret $g'(2)$ and compare to $g'(11)$. See Example 3.

47. The average salary of players in the National Basketball Association from 1983 to 1993 can be modeled by

$$f(x) = 264.1e^{0.19x}, \qquad 1 \leq x \leq 11$$

where x represents the number of years since 1982 and $f(x)$ represents the average salary in thousands of dollars.

(a) Classify the function as an exponential growth or decay model. Explain.

(b) Determine $f'(x)$.

(c) Evaluate and interpret $f'(3)$ and compare to $f'(9)$.

48. The tax liability of a single U.S. resident with no dependents who earned at most $35,000 can be modeled by

$$f(x) = 7042.12e^{-0.03x}, \qquad 1 \leq x \leq 11$$

where x represents the number of years since 1984 and $f(x)$ represents the tax liability in dollars.

(a) Classify the function as an exponential growth or decay model. Explain.

(b) Determine $f'(x)$.

(c) Evaluate and interpret $f'(1)$ and compare to $f'(10)$.

49. Suppose that $2000 is invested in an account that earns 6.5% interest compounded continuously. The amount accumulated after t years is given by

$$A(t) = 2000e^{0.065t}, \qquad t \geq 0$$

(a) Graph A in the viewing window [0, 10] by [2000, 4000].

(b) Determine the derivative $A'(t)$.

(c) Evaluate and interpret $A(5)$ and $A'(5)$.

50. Suppose that $5000 is invested in an account that earns 5% interest compounded continuously. The amount accumulated after t years is given by

$$A(t) = 5000e^{0.05t}, \qquad t \geq 0$$

(a) Graph A in the viewing window [0, 10] by [5000, 8500].

(b) Determine the derivative $A'(t)$.

(c) Evaluate and interpret $A(9)$ and $A'(9)$.

51. The total amount of individual retirement accounts (IRAs) that are held by mutual funds can be modeled by

$$f(x) = 28.69 \cdot (1.28)^x, \qquad 1 \leq x \leq 12$$

where x represents the number of years since 1984 and $f(x)$ represents the total amount of individual retirement accounts invested in mutual funds, in billions of dollars.

(a) Classify the function as an exponential growth or decay model. Explain.

(b) Determine $f'(x)$ for the model. Evaluate and interpret $f'(1)$ and compare to $f'(11)$.

(c) Rewrite the model in the form $g(x) = a \cdot e^{kx}$, where a and k are constants rounded to the nearest hundredth.

52. The number of civil cases going to trial in U.S. district courts can be modeled by

$$f(x) = 12.4 \cdot (0.95)^x, \qquad 1 \leq x \leq 12$$

where x represents the number of years since 1984 and $f(x)$ represents the number of civil cases going to trial in U.S. district courts, in thousands.

(a) Classify the function as an exponential growth or decay model. Explain.

(b) Determine $f'(x)$ for the model. Evaluate and interpret $f'(3)$.

(c) Rewrite the model in the form $g(x) = a \cdot e^{kx}$, where a and k are constants rounded to the nearest hundredth.

SECTION PROJECT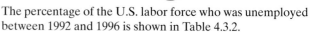

The percentage of the U.S. labor force who was unemployed between 1992 and 1996 is shown in Table 4.3.2.

(a) Use your calculator to determine a general exponential regression model of the form

$$f(x) = a \cdot b^x, \qquad 1 \leq x \leq 5$$

where x represents the number of years since 1991 and $f(x)$ represents the percentage of the U.S. labor force who was unemployed.

(b) Classify the function as an exponential growth or decay model.

(c) Graph f in the viewing window [1, 5] by [4, 8].

(d) Determine $f'(x)$ for the model.

(e) Evaluate and interpret $f'(1)$ and compare to $f'(5)$.

(f) Rewrite the model in the form $g(x) = a \cdot e^{kx}$, where a and k are constants rounded to the nearest thousandth.

(g) Evaluate and interpret $g'(1)$ and $g'(5)$ and compare to part (e).

TABLE 4.3.2		
YEAR	x	PERCENT UNEMPLOYED
1992	1	7.5
1993	2	6.9
1994	3	6.1
1995	4	5.6
1996	5	5.4

Source: U.S. Census Bureau web site.

SECTION 4.4   IMPLICIT DIFFERENTIATION AND RELATED RATES

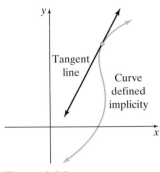

Figure 4.4.1

Until now, all the graphs of functions that we have differentiated are ones that have passed the vertical line test. But what about other graphs like the one in Figure 4.4.1. We see that these graphs have tangent lines, too.

In this section we will learn a new method of differentiation that can be used to find derivatives and, subsequently, rates of change of graphs that may not pass the vertical line test. This method is called **implicit differentiation.** Then we examine applications called **related rates** that use implicit differentiation.

Implicit Differentiation

Thus far, all the functions that we have studied have the form $y = f(x)$. Algebraically, this means that we can solve for y *explicitly* in terms of x, such as $y = x^3 + 3x^2 - 1$. However, not all equations are expressed in this form. For example, for an equation such as

$$xy^4 + y^2 - xy + 3x^2 - 5 = 0$$

solving for y for this implicit function would be both difficult and impractical. The natural question is, "How do we find the derivative $\frac{dy}{dx}$ in this case?" The key is to use implicit differentiation. Since this differentiation method is really an extension of the Chain Rule, let's recall what this rule states.

From Your Toolbox   **THE CHAIN RULE**

 If $y = f(u)$ and $u = g(x)$ are used to define $h(x)$, where $h(x) = f(g(x))$, then

$$h'(x) = f'(g(x)) \cdot g'(x)$$

provided that $f'(g(x))$ and $g'(x)$ exist.

For a function like $y = (x^3 + 3)^4$, we can use the Chain Rule to determine the derivative because our inside function, $g(x) = x^3 + 3$, is given. But what if we were asked to determine

$$\frac{d}{dx}[f(x)^4]?$$

Notice here that our inside function is not explicitly given; it is just represented as $f(x)$. We could still perform the differentiation using the Chain Rule, as follows:

$$\frac{d}{dx}[f(x)^4] = 4 \cdot [f(x)^3] \cdot f'(x)$$

Now, if we replace $f(x)$ with y and $f'(x)$ with $\frac{dy}{dx}$, we have

$$\frac{d}{dx}[f(x)^4] = \frac{d}{dx}[y^4] = 4y^3 \frac{dy}{dx}$$

This suggests the following rule:

> ### Implicit Differentiation Rule for y^n
>
> If n is any real number and y is differentiable, then
>
> $$\frac{d}{dx}[y^n] = n(y^{n-1}) \cdot \frac{dy}{dx}$$

This rule allows us to determine derivatives of expressions that involve powers of x, as well as powers of y. The process we use is the following:

- Use the usual differentiation rules for terms in x.
- Use implicit differentiation for terms in y.
- Solve for $\frac{dy}{dx}$.

EXAMPLE 1 | **Determining Derivatives of Expressions Containing y^n**

Use implicit differentiation to determine $\frac{dy}{dx}$ for the following:

(a) $y^3 - 4x^2 = 7$ (b) $x^{1/3} + y^{1/3} - 1 = 0$

SOLUTION

(a) We start by differentiating each term with respect to x.

$$\frac{d}{dx}(y^3) - \frac{d}{dx}(4x^2) = \frac{d}{dx}(7)$$

To differentiate $\frac{d}{dx}(y^3)$, we use the Implicit Differentiation Rule for y^n to get

$$3y^2 \cdot \frac{dy}{dx} - 8x = 0$$

Solving for $\frac{dy}{dx}$ yields

$$3y^2 \cdot \frac{dy}{dx} = 8x$$

$$\frac{dy}{dx} = \frac{8x}{3y^2}$$

(b) Again, we begin by differentiating each term with respect to x.

$$\frac{d}{dx}(x^{1/3}) + \frac{d}{dx}(y^{1/3}) - \frac{d}{dx}(1) = 0$$

$$\frac{1}{3}x^{-2/3} + \frac{1}{3}y^{-2/3} \cdot \frac{dy}{dx} - 0 = 0$$

When writing the equation with positive exponents and in radical form, we get

$$\frac{1}{3\sqrt[3]{x^2}} + \frac{1}{3\sqrt[3]{y^2}} \cdot \frac{dy}{dx} = 0$$

Solving for $\frac{dy}{dx}$ gives

$$\frac{1}{3\sqrt[3]{y^2}} \cdot \frac{dy}{dx} = -\frac{1}{3\sqrt[3]{x^2}}$$

$$\frac{dy}{dx} = -\frac{3\sqrt[3]{y^2}}{3\sqrt[3]{x^2}}$$

$$\frac{dy}{dx} = -\sqrt[3]{\frac{y^2}{x^2}}$$

✓**CHECKPOINT 1**

Now work Exercise 5.

Notice that the derivatives of these implicit functions are written in terms of both x and y.

EXAMPLE 2 | **Determining $\dfrac{dy}{dx}$ Implicitly**

Determine $\frac{dy}{dx}$ for the following:

(a) $4e^y = x^2$ \qquad\qquad (b) $2x + 1 = \sqrt{2 - y^2}$

SOLUTION

(a) We begin by differentiating each term with respect to x.

$$\frac{d}{dx}[4e^y] = \frac{d}{dx}[x^2]$$

Using techniques from Section 4.3 on the left side of the equation gives us

$$4e^y \frac{dy}{dx} = 2x$$

Solving for $\frac{dy}{dx}$ yields

$$\frac{dy}{dx} = \frac{2x}{4e^y} = \frac{x}{2e^y}$$

(b) Again, we start by differentiating each term with respect to x.

$$\frac{d}{dx}[2x] + \frac{d}{dx}[1] = \frac{d}{dx}[\sqrt{2 - y^2}]$$

$$\frac{d}{dx}[2x] + \frac{d}{dx}[1] = \frac{d}{dx}[(2 - y^2)^{1/2}]$$

$$2 + 0 = \frac{1}{2}(2 - y^2)^{-1/2} \cdot (-2y)\frac{dy}{dx}$$

$$2 = -y(2 - y^2)^{-1/2}\frac{dy}{dx}$$

$$2 = \frac{-y}{\sqrt{2 - y^2}}\frac{dy}{dx}$$

Solving for $\frac{dy}{dx}$ gives

$$\frac{dy}{dx} = \frac{-2\sqrt{2-y^2}}{y}$$

✓ **CHECKPOINT 2**

Now work Exercise 25.

Some functions also contain mixed terms, such as $x^2 y^3$. To apply our implicit differentiation technique to these mixed terms, we must apply the Product Rule and then the Implicit Differentiation Rule for y^n when we differentiate the factor containing y^n. For example, suppose that we want to determine $\frac{d}{dx}(x^2 y^3)$. Using the Product Rule, we get

$$\frac{d}{dx}(x^2 y^3) = \frac{d}{dx}(x^2) \cdot y^3 + x^2 \cdot \frac{d}{dx}(y^3)$$

$$= 2x \cdot y^3 + x^2 \cdot 3y^2 \frac{dy}{dx}$$

EXAMPLE 3 | **Determining $\frac{dy}{dx}$ of an Expression Containing x and y Terms**

Use implicit differentiation to determine $\frac{dy}{dx}$ for $x^2 y^2 - 7x^3 - 5 = 0$.

SOLUTION

Differentiating each term with respect to x gives

$$\frac{d}{dx}(x^2 y^2) - \frac{d}{dx}(7y^3) - \frac{d}{dx}(5) = \frac{d}{dx}(0)$$

Using the Product Rule for $\frac{d}{dx}(x^2 y^2)$ yields

$$\frac{d}{dx}(x^2) \cdot y^2 + x^2 \cdot \frac{d}{dx}(y^2) - \frac{d}{dx}(7y^3) - \frac{d}{dx}(5) = \frac{d}{dx}(0)$$

$$2x \cdot y^2 + x^2 \cdot 2y\frac{dy}{dx} - 21y^2\frac{dy}{dx} - 0 = 0$$

$$2xy^2 + 2x^2 y\frac{dy}{dx} - 21y^2\frac{dy}{dx} = 0$$

Now we need to solve for $\frac{dy}{dx}$ algebraically.

$$2x^2 y\frac{dy}{dx} - 21y^2\frac{dy}{dx} = -2xy^2$$

$$\frac{dy}{dx}(2x^2 y - 21y^2) = -2xy^2$$

$$\frac{dy}{dx} = \frac{-2xy^2}{2x^2 y - 21y^2}$$

✓ **CHECKPOINT 3**

Now work Exercise 13.

When finding the equation for the line tangent to the graph of an implicit function, we use a procedure that is much the same as what we have always done.

- Compute the derivative.
- Evaluate the derivative at a specified point.
- Write an equation of the tangent line.

The only difference is that for implicit functions we need both the x- and y-coordinates for the tangent point, since many derivatives are expressed in terms of both x and y.

NOTE: When we are evaluating a derivative of an implicit expression by replacing both x- and y-values, we write $\left.\dfrac{dy}{dx}\right|_{(x,y)}$

Determining an Equation of a Line Tangent to the Graph of an Implicit Function

EXAMPLE 4

Consider the equation $x^2 + y^2 = 25$. This equation represents the graph of a circle centered at the origin and has a radius of 5, as shown in Figure 4.4.2.

(a) Use implicit differentiation to determine $\dfrac{dy}{dx}$ for $x^2 + y^2 = 25$. Evaluate $\left.\dfrac{dy}{dx}\right|_{(4,-3)}$ and interpret.

(b) Write an equation of the line tangent to the graph of $x^2 + y^2 = 25$ at the point $(4, -3)$.

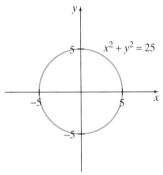

Figure 4.4.2

SOLUTION

(a) Using the Implicit Differentiation Rule for y^n, we get

$$\frac{d}{dx}(x^2) + \frac{d}{dx}(y^2) = \frac{d}{dx}(25)$$

$$2x + 2y\frac{dy}{dx} = 0$$

$$2y\frac{dy}{dx} = -2x$$

$$\frac{dy}{dx} = \frac{-2x}{2y} = \frac{-x}{y}$$

To evaluate $\left.\dfrac{dy}{dx}\right|_{(4,-3)}$, we replace 4 for x and -3 for y in the derivative $\dfrac{dy}{dx} = \dfrac{-x}{y}$ to get

$$\left.\frac{dy}{dx}\right|_{(4,-3)} = \frac{-(4)}{-3} = \frac{4}{3}$$

This means that the slope of the line tangent to the graph of $x^2 + y^2 = 25$ at the point $(4, -3)$ is $m_{\tan} = \dfrac{4}{3}$.

(b) Using the slope of $\dfrac{4}{3}$ and the point–slope form of a line, we find that the tangent line equation at the point $(4, -3)$ is

$$y - (-3) = \frac{4}{3}(x - 4)$$

$$y + 3 = \frac{4}{3}(x - 4) \quad \text{or} \quad y = \frac{4}{3}(x - 4) - 3$$

The graph of $x^2 + y^2 = 25$ with tangent line $y = \dfrac{4}{3}(x - 4) - 3$ is shown in Figure 4.4.3.

Interactive Activity

Verify the result to part (b) of Example 4 on your calculator by defining

$y_1 = \sqrt{(25 - x^2)}$

$y_2 = -y_1$

$y_3 = (4/3)(x - 4) - 3$

Do you think all implicit functions can be defined in this way? Explain.

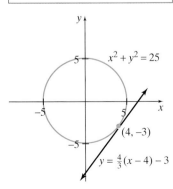

Figure 4.4.3

We can also use implicit differentiation in applications. In the next example we are given a price–demand equation in terms of both demand x and price $p(x)$. To avoid confusion, we will denote the price in such applications by just writing p.

EXAMPLE 5 | **Applying Implicit Differentiation**

Suppose that the VeeCam Company determines that the price–demand equation for their economy tripod is given by

$$p + 2xp + x^2 = 125, \qquad 0 < x \le 100$$

where x represents the demand for the tripod in thousands and p represents the price in dollars. Use implicit differentiation to determine $\frac{dp}{dx}$. Evaluate and interpret $\left.\frac{dp}{dx}\right|_{(2.5,19.5)}$.

SOLUTION

To differentiate $p + 2xp + x^2 = 125$ with respect to x implicitly, we need to use the Product Rule for the term $2xp$.

$$\frac{d}{dx}(p) + \frac{d}{dx}(2xp) + \frac{d}{dx}(x^2) = \frac{d}{dx}(125)$$

$$\frac{d}{dx}(p) + \frac{d}{dx}(2x) \cdot p + 2x \cdot \frac{d}{dx}(p) + \frac{d}{dx}(x^2) = \frac{d}{dx}(125)$$

$$\frac{dp}{dx} + 2p + 2x\frac{dp}{dx} + 2x = 0$$

Solving for $\frac{dp}{dx}$ gives

$$\frac{dp}{dx} + 2x\frac{dp}{dx} = -2x - 2p$$

$$\frac{dp}{dx}(1 + 2x) = -2x - 2p$$

$$\frac{dp}{dx} = \frac{-2x - 2p}{1 + 2x} = \frac{-2(x + p)}{1 + 2x}$$

Evaluating the derivative, with $x = 2.5$ and $p = 19.5$, yields

$$\left.\frac{dp}{dx}\right|_{(2.5,19.5)} = \frac{-2(2.5 + 19.5)}{1 + 2(2.5)} = \frac{-2(22)}{6} \approx -7.33$$

Thus, at a demand level of 2.5 thousand tripods and a price of $19.50, the price is decreasing at a rate of 7.33 $\frac{\text{dollars}}{\text{thousand tripods}}$. ◼

Related Rates

To begin to understand the concept behind related rates, let's suppose that the TastyMix Company determines that the daily revenue function, R for their cupcake mix is given by

$$R(x) = 40x - x^2$$

where x represents the number of boxes of mix produced and sold in hundreds. But instead of having a price–demand equation giving price as a function of x, they find that the amount produced and sold is a function of *time t* or, mathematically, $x = x(t)$. Can we write an expression for $\frac{dR}{dt}$ even if we do not know how the quantity, x, or the revenue, R, are related to t? The answer lies again with the Chain Rule. Since the quantity x is a function of time t and the revenue $R(x)$ is a function of the quantity x, we can write

$$\left(\begin{array}{c} \text{change in revenue} \\ \text{with respect to time} \end{array} \right) = \left(\begin{array}{c} \text{change in revenue} \\ \text{with respect to amount} \\ \text{produced and sold} \end{array} \right) \cdot \left(\begin{array}{c} \text{change in amount} \\ \text{produced and sold} \\ \text{with respect to time} \end{array} \right)$$

$$\frac{dR}{dt} = \frac{dR}{dx} \cdot \frac{dx}{dt}$$

Since $\frac{dR}{dx} = \frac{d}{dx}(40x - x^2) = 40 - 2x$, we get

$$\frac{dR}{dt} = (40 - 2x) \cdot \frac{dx}{dt}$$

$$\frac{dR}{dt} = 40\frac{dx}{dt} - 2x\frac{dx}{dt}$$

This equation is called a **rate of change equation.** It expresses the rate of change of revenue with respect to time, $\frac{dR}{dt}$, in terms of $\frac{dx}{dt}$, the rate of change in the amount produced and sold with respect to time. Hence, we say that $\frac{dR}{dt}$ and $\frac{dx}{dt}$ are **related rates.** Thus, if we know the rate $\frac{dx}{dt}$, then we can find the rate $\frac{dR}{dt}$ using the rate of change equation.

EXAMPLE 6 | **Determining a Rate of Change Equation**

Determine the rate of change equation by implicitly differentiating $p^2 = x^3 - 300x$ with respect to time, t.

SOLUTION

Using implicit differentiation, we $\frac{d}{dx}$ each term to get

$$2p \cdot \frac{dp}{dt} = 3x^2 \cdot \frac{dx}{dt} - 300 \cdot \frac{dx}{dt}$$

✓ CHECKPOINT 4

Now work Exercise 47.

In the related rates applications that follow, it is necessary to determine a rate of change equation, using implicit differentiation, as we did in Example 6. We then solve for one of the rates, given some information, as illustrated in Example 7.

EXAMPLE 7 | **Solving General Related Rates Applications**

Assuming that x and p are both functions of t, determine $\frac{dp}{dt}$ for $p^2 = x^3 - 300x$, given $\frac{dx}{dt} = 5$, $x = 2$, and $p = 3$.

SOLUTION

In Example 6 we determined that the rates $\frac{dp}{dt}$ and $\frac{dx}{dt}$ were related by the equation

$$2p \cdot \frac{dp}{dt} = 3x^2 \cdot \frac{dx}{dt} - 300 \cdot \frac{dx}{dt}$$

Substituting the given values and solving for $\frac{dp}{dt}$ gives

$$2(3)\frac{dp}{dt} = 3(2)^2(5) - 300(5)$$

$$6\frac{dp}{dt} = 60 - 1500$$

$$6\frac{dp}{dt} = -1440$$

$$\frac{dp}{dt} = -\frac{1440}{6} = -240$$

Applications

The applications of related rates are varied. To aid us in solving related rates applications, we offer the following five-step process.

Solving Related Rates Applications

1. Sketch a diagram, if it can be helpful.
2. Write down all the variables, along with the rates that are given. Be sure to include proper units.
3. Write an equation that relates the variables given.
4. Assuming that the variables are functions of time, t, differentiate the equation implicitly with respect to time, t.
5. Solve for the unknown rate.

EXAMPLE 8

Applying Related Rates in Geometry Applications

The FreshDay Company delivers its bread in electric-powered trucks. The delivery area is restricted by the range of the trucks to a radius r miles from the company bakery. Through improvements in battery efficiency, the company finds that the radius for the delivery area is increasing at a rate of $3\frac{\text{miles}}{\text{year}}$. At the time that the radius is 50 miles, how fast is the delivery area increasing?

SOLUTION

For clarity, we will follow the five steps listed in the preceding box.

1. The first step is to sketch a diagram for the application, as shown in Figure 4.4.4.
2. Next, we can label t as the time in years, r as the radius in miles, and A as the delivery area in square miles. When the radius $r = 50$, we know that the radius is increasing at a rate of $\frac{dr}{dt} = 3\frac{\text{miles}}{\text{year}}$.
3. The radius and the area are related by the equation for a circle, $A = \pi r^2$, where A is the area in square miles and r is the radius in miles.
4. Since r is changing with respect to time t, which means that A is a function of time t, we differentiate the equation $A = \pi r^2$ implicitly with respect to t.

$$\frac{d}{dt}[A] = \frac{d}{dt}[\pi r^2]$$

$$\frac{dA}{dt} = 2\pi r \frac{dr}{dt} \qquad (\pi \text{ is a constant})$$

Figure 4.4.4

$r = 50$

$\frac{dr}{dt} = 3\frac{\text{miles}}{\text{year}}$

bakery

5. Since we want to know how fast the delivery area is increasing, we seek $\frac{dA}{dt}$ when $r = 50$ and $\frac{dr}{dt} = 3$. Solving for the unknown rate $\frac{dA}{dt}$ yields

$$\frac{dA}{dt} = 2\pi r \frac{dr}{dt}$$

$$= 2\pi(50)(3) = 300\pi \approx 942.48$$

This means that when the radius is 50 miles, the delivery area is increasing at a rate of about $942.48 \frac{\text{square miles}}{\text{year}}$.

Many times, the equation for a related rates application is not given directly. In the previous example, we saw that the geometry formula for the area of a circle was necessary to describe the delivery area. Now let's see how related rates are used in business applications.

EXAMPLE 9 | **Applying Related Rates in Business Applications**

Past records of the TechTop Company determine that the revenue for the number of software suites produced and sold is given by

$$R = 90x - x^2$$

where x is the number of units produced and sold daily and R is the generated revenue in dollars. The company also finds that the software is selling at a rate of 5 suites per day. How fast is the revenue changing when 40 suites are being produced and sold?

SOLUTION

Again, we will follow the steps for **solving related rates applications.**

1. In this application, a diagram is not needed.
2. We have the number of units sold, x, the revenue generated, R, and the time, t. We know that the suites are selling at a rate of $5\frac{\text{suites}}{\text{day}}$, so $\frac{dx}{dt} = 5$ when $x = 40$.
3. In this case, we are given the revenue equation $R = 90x - x^2$.
4. Since we are being asked how revenue, R, is changing with respect to time, t, we differentiate the equation for R with respect to time t.

$$\frac{d}{dt}(R) = \frac{d}{dt}(90x - x^2)$$

$$\frac{dR}{dt} = 90\frac{dx}{dt} - 2x\frac{dx}{dt}$$

5. Since we want to know how fast the revenue is changing, we seek $\frac{dR}{dt}$ when $\frac{dx}{dt} = 5$ and $x = 40$. This gives us

$$\frac{dR}{dt} = 90\frac{dx}{dt} - 2x\frac{dx}{dt}$$

$$= 90(5) - 2(40)(5) = 50$$

Thus, when 40 suites are produced and sold daily, the revenue is increasing at a rate of $50 \frac{\text{dollars}}{\text{day}}$.

SUMMARY

In this section, we focused on two primary topics. We saw that **implicit differentiation** was a method that could be used to determine derivatives of expressions that were written in terms of x, as well as in terms of both x and y. We introduced a new rule for implicit differentiation. Next we turned our attention to **related rates**. By using implicit differentiation, we found that we could differentiate two quantities with respect to time, t, a third variable. We saw applications of related rates that used geometric formulas and also applications in the business sciences that used our five-step process for solving related rates applications.

- **Implicit differentiation rule for y^n:** $\dfrac{d}{dx}[y^n] = n(y^{n-1}) \cdot \dfrac{dy}{dx}$.

- **Solving related rates applications**

  1. Sketch a diagram, if it can be helpful.
  2. Write down all the variables, along with the rates that are given. Be sure to include proper units.
  3. Write an equation that relates the variables given.
  4. Assuming that the variables are functions of time, t, differentiate the equation implicitly with respect to time t.
  5. Solve for the unknown rate.

SECTION 4.4 EXERCISES

IMPLICIT DIFFERENTIATION

In Exercises 1–10, use implicit differentiation to determine $\dfrac{dy}{dx}$.

1. $2x + y = 5$
2. $x + 3y = 0$
3. $x + 3y^2 = 4$
4. $7x^2 - 5y^2 - 100 = 0$
✓ 5. $2x^2 + 2y^2 = 32$
6. $3x^2 + y^2 = 81$
7. $5x^3 + y^3 - x^4 = 0$
8. $4x^2 + 2y^3 = x^3$
9. $x^{1/4} - y^{1/4} = 1$
10. $x^{2/3} - y^{2/3} = 4$

In Exercises 11–20, use implicit differentiation to determine $\dfrac{dy}{dx}$.

11. $x^2 + xy = 6$
12. $xy + y^2 = 4$
✓ 13. $x^3 - y^3 + 12xy = 0$
14. $2x^3 + 3xy + y^3 = 0$
15. $(x + y)^2 - 1 = 7x^2$
16. $(y + 3)^2 = x^2 - 5$
17. $y \cdot \ln x = 10 - y$
18. $x + \ln y + 11 = 0$
19. $xe^y + x^2 = y^2$
20. $e^x y + y = 4x$

In Exercises 21–26, determine $\dfrac{dy}{dx}$ for the following:

21. $5^y = x^3$
22. $6^y = 2x^4$
23. $10^{y-2} = x - 3$
24. $7^{3y-1} = x + 1$

✓ 25. $\sqrt[3]{y} = x^2 + 6$
26. $\sqrt{y} = 3x^4 - 9$

In Exercises 27–32, complete the following:

(a) Use implicit differentiation to determine $\dfrac{dy}{dx}$.

(b) Solve the equation explicitly for y and differentiate to determine $\dfrac{dy}{dx}$.

27. $x + 2y = 6$
28. $3x - 4y = 12$
29. $xy + 4 = x^4$
30. $xy = 2$
31. $\dfrac{x}{y} - x^2 = 1$
32. $\dfrac{1}{x} + \dfrac{1}{y} = 3$

For Exercises 33–40, determine an equation of the line tangent to the graph of the given equation at the indicated point.

33. $x^2 + y^2 = 13$; $(3, 2)$
34. $x^3 + y^3 = 9$; $(1, 2)$
35. $4x^2 + 9y^2 = 36$; $(0, 2)$
36. $x^2 + y^2 = 25$; $(3, 4)$
37. $x \ln y = 2x^3 - 2y$; $(1, 1)$
38. $x + \ln y - 2y^2 = 0$; $(2, 1)$
39. $x^2 + y^2 = e^y$; $(1, 0)$
40. $4e^y + y = x^2$; $(2, 0)$

APPLICATIONS

41. The Crabtree Company determines that the price–demand equation for their mini picture frames is given by

$$p + x^2 = 150$$

where x represents the demand for the frames in hundreds and p represents the price.

(a) Use implicit differentiation to determine $\frac{dp}{dx}$.

(b) Evaluate and interpret $\frac{dp}{dx}\Big|_{(11,29)}$

42. Suppose that Pottery Plus determines that the price–demand equation for their glazed presidential inauguration mugs is given by

$$\frac{p}{2} + \sqrt{x} = 30$$

where x represents the demand for the mugs and p represents the price.

(a) Use implicit differentiation to determine $\frac{dp}{dx}$.

(b) Evaluate and interpret $\frac{dp}{dx}\Big|_{(400,20)}$

43. The WoodedLife Company has determined that the price–demand equation for their new single-occupant tent is given by

$$px + 2x = 1000, \qquad 10 \le x \le 30$$

where x represents the demand for the tents in hundreds and p represents the price.

(a) Use implicit differentiation to determine $\frac{dp}{dx}$.

(b) Evaluate and interpret $\frac{dp}{dx}\Big|_{(20,48)}$

44. The CorpStyle Company determines that the price–demand equation for their all-silk power ties is given by

$$x^2 = 1000 - p^2, \qquad 10 \le x \le 30$$

where x represents the demand for the ties in thousands and p represents the price.

(a) Use implicit differentiation to determine $\frac{dp}{dx}$.

(b) Evaluate and interpret $\frac{dp}{dx}\Big|_{(18,26)}$

RELATED RATES

In Exercises 45–52, determine the rate of change equation by implicitly differentiating each of the following with respect to time t.

45. $2x + 3y = 20$

46. $x^2 + 2y = 11$

✓ **47.** $x^2 - 3y = 1$

48. $y = x^2 - 3x + 5$

49. $x^2 + y^2 = 5x$

50. $x^3 - y^3 = 10y$

51. $5xy + y^4 = x$

52. $xy = 4x + 5y + 9$

For Exercises 53–58, assume that x and y are both functions of t and determine the indicated rate.

53. Determine $\frac{dy}{dt}$ for $xy = 7$ given $x = 7$, $y = 1$, and $\frac{dx}{dt} = 2$.

54. Determine $\frac{dx}{dt}$ for $x^2 + 9y^2 = 18$ given $x = 3$, $y = 1$, and $\frac{dy}{dt} = 10$.

55. Determine $\frac{dx}{dt}$ for $y^2 + x = 3$ given $x = 2$, $y = 1$, and $\frac{dy}{dt} = 2$.

56. Determine $\frac{dy}{dt}$ for $y = x^2 - 2x + 3$ given $x = 5$, $y = 18$, and $\frac{dx}{dt} = 3$.

57. Determine $\frac{dy}{dt}$ for $x^2 + y^2 = 25$ given $x = 3$, $y = -4$, and $\frac{dx}{dt} = 2$.

58. Determine $\frac{dx}{dt}$ for $y^2 + xy - 3x = -1$ given $x = 1$, $y = -2$, and $\frac{dy}{dt} = -2$.

APPLICATIONS

59. The area of a circle with radius r is increasing at a rate of 12 square inches per minute. Find the rate at which the radius is increasing when the radius is 2 inches. The area of a circle is $A = \pi r^2$.

60. The area of a circle with radius r is increasing at a rate of 5 square inches per minute. Find the rate at which the radius is increasing when the radius is 0.8 inch. The area of a circle is $A = \pi r^2$.

61. The area of a square with sides x inches is increasing at a rate of 10 square inches per minute. Find the rate at which a side is increasing when the sides are 3 inches. The area of a square is $A = x^2$.

62. The area of a square with sides x inches is increasing at a rate of 25 square inches per minute. Find the rate at which a side is increasing when the sides are 4 inches. The area of a square is $A = x^2$.

63. A circular pan is being heated and the radius of the heated area increases at a rate of 0.02 inch per minute. Find the rate at which the area is increasing when the radius is 8 inches.

64. A 5-foot-tall man is walking away from a 20-foot street lamp at a speed of 6 feet per second. How fast is the tip of his shadow moving along the ground?

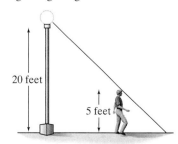

20 feet

5 feet

65. A 6-foot-tall woman is walking away from a 24-foot street lamp at a speed of 8 feet per second. How fast is the tip of her shadow moving along the ground?

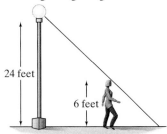

24 feet

6 feet

66. The SoftSkirt Company determines that the monthly revenue for a new style of skirt is given by

$$R = 60 - \frac{1}{2}x^2$$

where x is the number of skirts produced and sold in hundreds and R is the revenue, in thousands of dollars, generated from the skirts.

(a) Differentiate the revenue function with respect to time t.

(b) Determine the rate of change in the revenue with respect to time at a production level of $x = 3$ and $\frac{dx}{dt} = 20 \frac{\text{hundred skirts}}{\text{month}}$.

67. The SharpSuit Company determines that monthly revenue for a new type of casual suit is given by

$$R = 250x - \frac{2}{5}x^2$$

where x is the number of suits produced and sold and R is the revenue in dollars.

(a) Differentiate the revenue function with respect to time t.

(b) Determine the rate of change in the revenue with respect to time at a production level of $x = 100$ and $\frac{dx}{dt} = 200 \frac{\text{suits}}{\text{month}}$.

68. A 24-foot ladder is leaning against the wall as shown in the figure below. Assume that the values of x and y are related by the Pythagorean Theorem equation

$$x^2 + y^2 = 24^2$$

(a) Differentiate each side of the equation with respect to time t.

(b) If the base of the ladder is sliding away from the wall at a rate of 2 $\frac{\text{feet}}{\text{second}}$, find the rate at which the top of the ladder

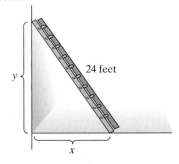

24 feet

y

x

is sliding down the wall when the top of the ladder is 12 feet from the ground.

69. A 20-foot ladder is leaning against the wall as shown in the figure below. Assume that the values of x and y are related by the Pythagorean Theorem equation

$$x^2 + y^2 = 20^2$$

(a) Differentiate each side of the equation with respect to time t.

(b) If the base of the ladder is sliding away from the wall at a rate of 3 $\frac{\text{feet}}{\text{second}}$, find the rate at which the top of the ladder is sliding down the wall when the top of the ladder is 8 feet from the ground.

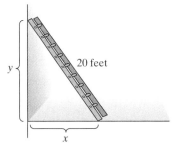

y

20 feet

x

70. A surgeon has a patient with a tumor in the shape of a sphere. Assume that the volume of the tumor is given by the equation

$$V = \frac{4}{3}\pi r^3$$

where r is the radius of the tumor measured in millimeters and V is its volume, measured in cubic millimeters.

(a) Differentiate each side of the equation with respect to time t.

(b) If tests show that the radius of the tumor is growing at a rate of 0.3 $\frac{\text{millimeter}}{\text{week}}$ determine the rate at which the volume is changing when the radius is 15 millimeters.

71. Assume that the number of bass in a pond is related to the level of polychlorinated biphenyls (or PCBs) in the pond. (PCBs are a group of industrial chemicals used in plasticizers, fire retardants, and other materials.) The bass population is modeled by

$$y = \frac{2500}{1 + x}$$

where x represents the PCB level in parts per million (ppm) and y represents the number of bass in the pond.

(a) Differentiate each side of the equation with respect to time t.

(b) If the level of PCBs is increasing at rate of 40 $\frac{\text{ppm}}{\text{year}}$, find the rate at which y is changing when there are 100 bass in the pond.

72. The price–demand equation for oranges in the town of Southburg is given by

$$p = \frac{40,000}{x^{1.5}}$$

where x is the demand for oranges in bushels per month and p is the wholesale price per bushel. The local farmer's market finds that the demand for oranges is currently 900 bushels each month and is increasing at a rate of $100 \frac{bushels}{month}$. How fast is the price changing?

73. The price–demand equation of the number of DVD players sold by a local electronics store is given by

$$p = 0.2x^2 + 2x$$

where x is the demand for the DVD players per week and p is the price per player. The store finds that the demand for the players is currently 150 players each week and is increasing at a rate of $10 \frac{players}{week}$. How fast is the price changing?

SECTION PROJECT

The strategic planning department of the Higbrize Company is given the data in Table 4.4.1 on the cost of producing a new lapel pager. The department is asked to answer the following scenario: The current production level is 7500 pagers and is

TABLE **4.4.1**	
UNITS PRODUCED, x	**COST, C**
1500	14,725
2500	18,125
3500	21,725
4000	23,600
5500	29,525
7000	35,900

increasing at a rate of 50 units per day. How is the average cost changing?

(a) Use the regression features of your calculator to determine a linear cost model for the lapel pagers.

(b) Determine the average cost function AC.

(c) Differentiate AC with respect to time t.

(d) What do the numbers 7500 and 50 represent in terms of production x and time t?

(e) Use related rates to find the rate of change for the average cost with respect to time and interpret.

SECTION 4.5   ELASTICITY OF DEMAND

In Chapter 3, we discussed the concept of marginal analysis, which focused on the effect that a small change in the quantity produced and sold had on cost, revenue, and profit. Now we will take a different approach. We wish to know what effect a change in price has on the quantity demanded. Let's take a look at an illustration. In the 1980s and early 1990s, Japanese microprocessor manufacturers flooded the technology market with low-price microchips in an attempt to force other overseas manufacturers out of business. This huge increase in microchips caused lower prices. The low prices, in turn, stimulated consumer demand, but reduced manufacturers' revenues dramatically. **Elasticity of demand** is a mathematical tool that can be used to measure the impact that a change in price has on the demand for a product. The term **elasticity** generally refers to how sensitive the demand is to a change in price.

Arc Elasticity

Let's begin by looking at some examples of elasticity. We call a product **elastic** if a small change in price produces a significant change in demand. The furniture market is an example of demand that is elastic. This market is very sensitive to changes in price, as furniture shoppers are always seeking value and quality. On the other hand, we call a product **inelastic** if a change in price generally does not affect demand. The heating oil market is an example of a commodity that is inelastic. Since homeowners who use heating oil must have the product in the winter months, changes in price have little effect on the demand.

> ### Elastic and Inelastic Products
> - If small changes in the unit price of a product result in significant changes in demand, the product is **elastic.**
> - If small changes in the unit price of a product do not result in significant changes in demand, the product is **inelastic.**

Generally, there are two types of elasticity. The first type that economists commonly use is **arc elasticity.** This type of elasticity is computed when the actual changes in price and quantity are known and a price–demand function may not be given. Arc elasticity measures the ratio of relative change in quantity to the relative change in price.

$$\begin{pmatrix} \text{Arc} \\ \text{elasticity} \end{pmatrix} = -\frac{\begin{pmatrix} \text{relative change} \\ \text{in quantity demanded} \end{pmatrix}}{\begin{pmatrix} \text{relative change} \\ \text{in price} \end{pmatrix}} = -\frac{\dfrac{\Delta q}{q}}{\dfrac{\Delta p}{p}}$$

The relative changes are simply the change in the quantity or price divided by their original value. The reason for the negative sign is purely a convention used by economists.

> ### Arc Elasticity
> For finite changes in quantity and price, the **arc elasticity,** denoted by E_a, is
>
> $$E_a = -\frac{\dfrac{\Delta q}{q}}{\dfrac{\Delta p}{p}} = -\frac{\dfrac{q_2 - q_1}{q_1}}{\dfrac{p_2 - p_1}{p_1}}$$
>
> where q_1 and q_2 are the original and new quantities, respectively, and p_1 and p_2 are the original and new prices, respectively.

EXAMPLE 1 | **Determining Arc Elasticity**

The GreenLawn Company can sell 300 mulching lawn mowers each week when the unit price of the mowers is $200. They find through a promotion that when the price is reduced by 10% to $180 the number sold rises by 20% to 360 mowers each week. Determine the arc elasticity.

SOLUTION

Here we have the original quantity of $q_1 = 300$ and original price of $p_1 = 200$. So the arc elasticity is

$$E_a = -\frac{\dfrac{q_2 - q_1}{q_1}}{\dfrac{p_2 - p_1}{p_1}} = -\frac{\dfrac{360 - 300}{300}}{\dfrac{180 - 200}{200}} = -\frac{\dfrac{60}{300}}{\dfrac{-20}{200}} = -\frac{0.2}{-0.1} = 2$$

Thus, for the lawn mowers a $0.1 = 10\%$ change in price causes a $0.2 = 20\%$ change in sales. Since the percentage change in sales exceeds the percentage change in price, we say that the demand is *elastic*. That is, if $E_a > 1$ then the demand is elastic.

✓ CHECKPOINT 1

Now work Exercise 3.

Arc elasticity is applicable when we have finite changes in quantity and price. However, we cannot determine the elasticity at a single quantity demanded. For this, we would need the *instantaneous* change in price and quantity. To compute this type of elasticity, we need the derivative. This results in a new type of elasticity, called **point elasticity.**

Point Elasticity

From Your Toolbox

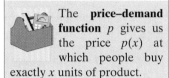

The **price–demand function** p gives us the price $p(x)$ at which people buy exactly x units of product.

To determine this new type of elasticity, we need to rewrite the price–demand function. Let's recall what this function tells us as shown in the Toolbox to the left.

Since the price–demand function is always decreasing on its domain, we know that the function is one-to-one. Thus, we can solve the price–demand function for x to get what economists call the **demand function**. This function plays a key role in determining point elasticity.

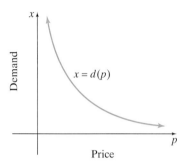

Figure 4.5.1
Graph of demand function $x = d(p)$.

> ### Demand Function
> The **demand function,** denoted by $d(p)$, is given by
> $$x = d(p)$$
> The demand function gives the quantity x of a product demanded by consumers at price p.

A graph of a typical demand function is displayed in Figure 4.5.1. Notice that the price p is now on the horizontal axis and the demand x is on the vertical axis. To determine the demand function, we simply need to solve the price–demand function for x. Since p will become the independent variable, we will express $p(x)$ simply as p. Example 2 illustrates this.

EXAMPLE 2

Determining the Demand Function

Determine the demand function $x = d(p)$ for the price–demand function

$$p = \frac{1000}{x^2}, \qquad 10 \le x \le 50$$

SOLUTION

We can get the demand function by solving $p = \frac{1000}{x^2}$ for x.

$$p = \frac{1000}{x^2}$$

$$x^2 \cdot p = 1000$$

$$x^2 = \frac{1000}{p}$$

$$x = \sqrt{\frac{1000}{p}}$$

So the demand function is $x = d(p) = \sqrt{\frac{1000}{p}}$. Notice that we only use the positive square root since both price and demand are represented by positive values. ◾

✓ **CHECKPOINT 2**

Now work Exercise 13.

To determine point elasticity, suppose that for a demand function the unit price of a product is increased by h dollars to $p + h$ dollars, as shown in Figure 4.5.2. This causes the quantity demanded to decrease from $d(p)$ to $d(p + h)$ units. So we see that the relative change in demand is

$$\left(\begin{array}{c} \text{relative change} \\ \text{in quantity demanded} \end{array}\right) = \frac{d(p+h) - d(p)}{d(p)}$$

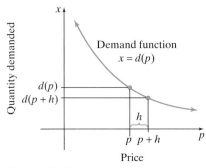

Figure 4.5.2

The corresponding change in unit price is

$$\left(\begin{array}{c} \text{relative change} \\ \text{in price} \end{array}\right) = \frac{(p+h) - p}{p} = \frac{h}{p}$$

Knowing from the beginning of this section that elasticity is the ratio of relative change in quantity to the relative change in price, we get

$$-\frac{\left(\begin{array}{c} \text{relative change} \\ \text{in quantity demanded} \end{array}\right)}{\left(\begin{array}{c} \text{relative change} \\ \text{in price} \end{array}\right)} = -\frac{\dfrac{d(p+h) - d(p)}{d(p)}}{\dfrac{h}{p}}$$

As the change in price becomes small, that is, $h \to 0$, we get the point elasticity formula.

$$\left(\begin{array}{c} \text{Point} \\ \text{elasticity} \end{array}\right) = \lim_{h \to 0} -\frac{\dfrac{d(p+h) - d(p)}{d(p)}}{\dfrac{h}{p}}$$

$$= \lim_{h \to 0} -\frac{\dfrac{d(p+h) - d(p)}{h}}{\dfrac{d(p)}{p}} = -\frac{\displaystyle\lim_{h \to 0} \dfrac{d(p+h) - d(p)}{h}}{\dfrac{d(p)}{p}}$$

$$= -\frac{d'(p)}{\dfrac{d(p)}{p}} = -\frac{p \cdot d'(p)}{d(p)}$$

We will simply call this type of elasticity the **elasticity of demand**.

Elasticity of Demand

Given the demand function $x = d(p)$, the **elasticity of demand,** denoted by $E(p)$, of a product at a price p is given by

$$E(p) = \frac{-p \cdot d'(p)}{d(p)}$$

EXAMPLE 3 | **Determining the Elasticity of Demand**

Determine the elasticity of demand $E(p)$ for each price–demand function.

(a) $p = 12 - 0.4x$ 　　　　　　　(b) $p = \dfrac{50}{2 - x}$

SOLUTION

(a) First, we must solve this price–demand function for x to get the demand function, $x = d(p)$.

$$p = 12 - 0.4x$$
$$p - 12 = -0.4x$$
$$\frac{p - 12}{-0.4} = x$$
$$-2.5p + 30 = x = d(p)$$

Now, to find $d'(p)$, we differentiate $d(p)$ with respect to p to get

$$d'(p) = \frac{d}{dp}(-2.5p + 30) = -2.5$$

Substituting into the elasticity of demand formula gives

$$E(p) = \frac{-p \cdot d'(p)}{d(p)} = \frac{-p(-2.5)}{-2.5p + 30} = \frac{2.5p}{-2.5p + 30}$$

(b) Solving $p = \dfrac{50}{2-x}$ for x, we get

$$p = \frac{50}{2 - x}$$
$$p(2 - x) = 50$$
$$2p - xp = 50$$
$$-xp = 50 - 2p$$
$$x = \frac{50 - 2p}{-p} = -\frac{50}{p} + 2 = 2 - \frac{50}{p} = 2 - 50p^{-1} = d(p)$$

Differentiating with respect to p yields

$$d'(p) = \frac{d}{dp}(2 - 50p^{-1}) = (-1)(-50p^{-2})$$
$$= 50p^{-2} = \frac{50}{p^2}$$

This gives the elasticity of demand $E(p)$ as

$$E(p) = \frac{-p \cdot d'(p)}{d(p)} = \frac{-p\left(\dfrac{50}{p^2}\right)}{2 - \dfrac{50}{p}}$$

$$= \frac{\dfrac{-50}{p}}{\dfrac{2p - 50}{p}} = \frac{-50}{2p - 50}$$

✓ **CHECKPOINT 3**

Now work Exercise 25.

Notice that the value of $E(p)$ has no units, since the units of demand are canceled out when computing the ratio of the relative changes. Also notice that the point elasticity function only requires one price to compute the elasticity coefficient. The interpretation of the values of $E(p)$ is summarized in Table 4.5.1.

TABLE 4.5.1 INTERPRETING THE $E(p)$ VALUE

IF THE $E(p)$ VALUE IS:	THEN THE DEMAND IS:	AND THIS MEANS A CHANGE IN PRICE WILL CAUSE:
Between 0 and 1	Inelastic	Relatively small changes in demand
Greater than 1	Elastic	Relatively large changes in demand
Equal to 1	Unitary	A relatively equal change in demand

EXAMPLE 4

Interpreting the Elasticity of Demand

A school's junior business club is holding its annual raffle. Data collected from raffles in the past indicate that the demand function for the tickets follows the model

$$x = d(p) = 36 - p^2$$

where p represents the price of a ticket and $x = d(p)$ represents the number of tickets each member sells each day.

(a) Find the elasticity of demand $E(p)$.

(b) Evaluate $E(3)$ and $E(4)$ and interpret the results.

SOLUTION

(a) To determine $d'(p)$, we differentiate the demand function with respect to p to get

$$d'(p) = \frac{d}{dp}(36 - p^2) = -2p$$

Using the definition of elasticity of demand yields

$$E(p) = \frac{-p \cdot d'(p)}{d(p)} = \frac{-p(-2p)}{36 - p^2} = \frac{2p^2}{36 - p^2}$$

(b) First, evaluating the elasticity of demand at $3, we get

$$E(3) = \frac{2(3)^2}{36 - (3)^2} = \frac{18}{27} = \frac{2}{3}$$

Since $E(3) = \frac{2}{3} < 1$, we see that the demand for the raffle ticket is **inelastic** at a price of $3 a ticket. A small increase in price will cause a negligible change in demand. Evaluating the function at $4, we get

$$E(4) = \frac{2(4)^2}{36 - (4)^2} = \frac{32}{20} = \frac{8}{5}$$

Here $E(4) = \frac{8}{5} > 1$, so the demand for the raffle tickets is **elastic** at a price of $4 a ticket. So a small increase in price will cause a significant change in demand.

✓ **CHECKPOINT 4**

Now work Exercise 39.

In Example 4, we found that at a price of $3 the demand for a ticket was inelastic, so demand is only negligibly sensitive to price changes. Increasing prices slightly will increase the revenue $R(x)$ because the higher price will turn away relatively few buyers. Conversely, when the price was $4, the demand was elastic, so the demand is sensitive to price changes. Thus, to increase revenue in this case, the club should slightly lower its price. The lower price will increase demand for tickets and will generate more revenue, which will offset the lower price. Furthermore, if the point elasticity was $E(p) = 1$, then prices should not be changed, because an increase in price will result in exactly the same percentage decrease in demand. Thus, when $E(p) = 1$, revenue is at a maximum.

Since $R'(p)$ is the rate of change in the revenue as the price p changes, the equation $R'(p) = d(p)(1 - E(p))$ expresses this relationship between $R'(p)$ and $d(p)$. This is summarized in Table 4.5.2.

TABLE 4.5.2 ELASTICITY–REVENUE RELATIONSHIP			
IF THE DEMAND IS:	**THEN $E(p)$ IS:**	**AND $R'(p)$ IS:**	**SO THE REVENUE IS:**
Elastic	Greater than 1	Negative	Decreasing
Inelastic	Less than 1	Positive	Increasing
Unitary	1	Zero	At a maximum

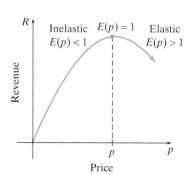

Figure 4.5.3

From Table 4.5.2 and Figure 4.5.3, we can see the following:

- If demand is elastic, then prices should be lowered to increase revenue.
- If demand is inelastic, then prices should be raised to increase revenue.
- If demand is unitary, then the revenue is at its maximum and the price should not be changed.

In Chapter 5, we will focus more on the sign of the derivative and finding maximum values.

EXAMPLE 5

Determining Price Adjustment to Increase Revenue

The PowerTi Company produces neckties and sells them at a price of $15. The demand function for the ties is given by

$$x = d(p) = 170{,}000\sqrt{20 - p}$$

where p is the price of a necktie and $d(p)$ is the demand for the ties.

(a) Determine $E(15)$ and state whether the manufacturer should lower or raise the price to increase revenue.

(b) Set $E(p) = 1$ and solve for p to determine at what price the revenue is greatest.

SOLUTION

(a) First, we need the elasticity of demand $E(p)$. Computing the derivative of the demand function $d(p) = 170{,}000(20 - p)^{1/2}$, we get

$$d'(p) = \frac{d}{dp}[170{,}000(20 - p)^{1/2}]$$

$$= 170{,}000\left(\frac{1}{2}\right)(20 - p)^{-1/2}(-1) = \frac{-85{,}000}{\sqrt{20 - p}}$$

Now, using this result in the point elasticity formula,

$$E(p) = \frac{-p \cdot d'(p)}{d(p)} = \frac{-p\left(\dfrac{-85{,}000}{\sqrt{20 - p}}\right)}{170{,}000\sqrt{20 - p}}$$

$$= \frac{85{,}000p}{\sqrt{20 - p}} \cdot \frac{1}{170{,}000\sqrt{20 - p}} = \frac{0.5p}{20 - p}$$

So at a price of $p = 15$, we get

$$E(15) = \frac{0.5(15)}{20 - 15} = \frac{7.5}{5} = 1.5$$

Since $E(15) = 1.5 > 1$, the demand is relatively elastic and the prices should be *lowered* to increase revenue.

(b) To find the price that maximizes revenue, we need to find the value of p that makes $E(p) = 1$. So, solving this equation, we get

$$E(p) = 1$$

$$\frac{0.5p}{20 - p} = 1$$

$$0.5p = 20 - p$$

$$1.5p = 20$$

$$p = \frac{20}{1.5} \approx 13.33$$

Thus, to maximize revenue, the price of the neckties should be $13.33. ◼

✓ **CHECKPOINT 5**

Now work Exercise 43.

SUMMARY

In this section we discussed a topic that is specific to economics, **elasticity of demand.** We first computed elasticity for finite changes in price and demand using the **arc elasticity** formula. We saw that products could be classified as **elastic, inelastic** or **unitary.** Then we examined elasticity for instantaneous changes in price and quantity, called **point elasticity.** Finally, we examined the relationship between the **elasticity of demand** and maximizing the revenue.

- **Arc elasticity:** $E_a = -\dfrac{\dfrac{q_2 - q_1}{q_1}}{\dfrac{p_2 - p_1}{p_1}}$

- **Point elasticity:** $E(p) = \dfrac{-p \cdot d'(p)}{d(p)}$

SECTION 4.5 EXERCISES

In Exercises 1–6, compute the arc elasticity E_a.

1. $q_1 = 50, q_2 = 55, p_1 = 110, p_2 = 100$

2. $q_1 = 10, q_2 = 13, p_1 = 8, p_2 = 5$

✓ 3. $q_1 = 300, q_2 = 260, p_1 = 40, p_2 = 45$

4. $q_1 = 5000, q_2 = 5800, p_1 = 60, p_2 = 54$

5. $q_1 = 5700, q_2 = 5600, p_1 = 390, p_2 = 400$

6. $q_1 = 420, q_2 = 419, p_1 = 389, p_2 = 395$

7. The 4Com Company determines that the demand function for their mouse pad is

$$x = d(p) = 20 - 2p$$

where p is the unit price of the mouse pad and $d(p)$ is the quantity demanded in hundreds.

(a) Determine $d(5)$ and $d(6)$ and interpret.

(b) Using $q_1 = d(5)$, $q_2 = d(6)$, $p_1 = 5$, and $p_2 = 6$, determine the arc elasticity E_a and interpret.

8. The WetLands Company determines that the demand function for their rain poncho is

$$x = d(p) = 50 - 5p$$

where p is the unit price of the rain poncho and $d(p)$ is the quantity demanded in thousands.

(a) Determine $d(5)$ and $d(7)$ and interpret.

(b) Using $q_1 = d(5)$, $q_2 = d(7)$, $p_1 = 5$, and $p_2 = 7$, determine the arc elasticity E_a and interpret.

9. The 725th Street Movie Theater finds that it can sell 320 tickets per day when the price of each ticket is $8.50. During a promotion week, the theater finds that by reducing the ticket price to $7.25, the number of tickets sold increases to 410 tickets per day. Determine the arc elasticity and interpret.

10. The HappyFace Portrait Studio sells 325 family packs of photos monthly at a unit price of $49. When they offer a sale price of $39, the number of family packs increases to 330 sold per month. Determine the arc elasticity and interpret.

11. The JemJam Jewelry Store finds that they sell 52 pairs of emerald earrings per week at a price of $149 each. Under new management, the price is increased to $159 and the number sold decreases to 45 pairs per week. Determine the arc elasticity and interpret.

12. The Fleetwood Company sells 600 stained glass window ornaments monthly at a unit price of $19. When they increased the price to $24 in order to absorb higher fixed costs, they found that the monthly sales decreased to 450 ornaments. Determine the arc elasticity and interpret.

In Exercises 13–22, the price–demand function p is given. Determine the demand function $x = d(p)$.

✓ 13. $p = 600 - 100x$

14. $p = 100 - x$

15. $p = \dfrac{300}{x^2} + 10$

16. $p = \dfrac{700}{2x + 1}$

17. $p = 12 - 0.04x$

18. $p = \sqrt{250 - x}$

19. $p = \sqrt{300 - x^2}$

20. $p = 1200 - x^2$

21. $p = 100e^{-0.1x}$

22. $p = 250e^{-0.3x}$

In Exercises 23–38, the demand function $x = d(p)$ and the price p are given. Determine the point elasticity of demand $E(p)$, and classify the demand as elastic, inelastic, or unitary.

23. $d(p) = 220 - 5p; \quad p = 10$

24. $d(p) = 50 - 4p; \quad p = 5$

✓ 25. $d(p) = 200 - p^2; \quad p = 8$

26. $d(p) = 250 - p^2;$ $p = 10$

27. $d(p) = \dfrac{200}{p};$ $p = 15$

28. $d(p) = \dfrac{1000}{7p};$ $p = 20$

29. $d(p) = \sqrt{150 - 3p};$ $p = 30$

30. $d(p) = \sqrt{100 - 2p};$ $p = 25$

31. $d(p) = \dfrac{100}{p^2};$ $p = 30$

32. $d(p) = \dfrac{500}{p^3};$ $p = 20$

33. $d(p) = 4500e^{-0.02p};$ $p = 200$

34. $d(p) = 6500e^{-0.04p};$ $p = 100$

35. $d(p) = 100 \ln (1000 - 10p);$ $p = 19$

36. $d(p) = 150 \ln (100 - 2p);$ $p = 5$

37. $d(p) = 100e^{-0.05p};$ $p = 40$

38. $d(p) = 3000e^{-0.03p};$ $p = 100$

APPLICATIONS

✓ 39. Currently, about 1800 people ride the MetroTram local public transportation system per day and pay \$4 for each ticket. The number of people x willing to take the public transportation at price p is given by the demand function

$$x = 600(5 - p^{1/2})$$

(a) Is the demand elastic or inelastic at the current \$4 ticket price?

(b) Should the ticket price be raised or lowered to increase revenue?

40. The EconoCommute bus system charges 65 cents for each ride and serves approximately 57,800 riders each day. The demand function for x riders at p cents for each ride is given by

$$x = 2000\sqrt{900 - p}$$

(a) Is the demand elastic or inelastic at the current 65-cent fare?

(b) Should the rider's price be raised or lowered to increase revenue?

41. The SilkTop clothing manufacturer sells blouses at \$15 each and calculates the demand function for the blouses to be

$$d(p) = 60 - 3p$$

where p is the price of a blouse and $d(p)$ represents the demand in hundreds.

(a) Is the demand elastic or inelastic at the current \$15 price?

(b) Should the blouse price be raised or lowered to increase revenue?

42. Suppose that the demand function for a certain brand of cigarettes is

$$d(p) = 4.5p^{-0.73}$$

where p is the price of a pack of cigarettes and $d(p)$ is the demand in thousands.

(a) Determine the point elasticity of demand $E(p)$.

(b) Show that the demand is inelastic for any price $p > 0$. Interpret what this means in terms of raising cigarette prices through manufacturer and tax price increases.

✓ 43. The IceDream Company determines that the demand function for their frozen yogurt is

$$d(p) = 50 - 2p$$

where p is the price, in dollars, of a quart of frozen yogurt and $d(p)$ represents the demand in hundreds.

(a) Determine $E(4)$ and state whether the manufacturer should lower or raise the price to increase revenue.

(b) Set $E(p) = 1$ and solve for p to determine at what price the revenue is greatest.

44. The PackIt Company determines that the demand function for their lightweight daypack is given by

$$d(p) = -p^2 + 400$$

where p is the unit price, in dollars, of a daypack and $d(p)$ is the demand.

(a) Determine $E(10)$ and state whether the manufacturer should lower or raise the price to increase revenue.

(b) Set $E(p) = 1$ and solve for p to determine at what price the revenue is greatest.

45. The EverGlo Company determines that the demand for their new night light is given by

$$d(p) = 2 - \dfrac{p^2}{5}$$

where p is the unit price of a night light, in dollars, and $d(p)$ is the demand in thousands.

(a) Determine $E(1.50)$ and state whether the manufacturer should lower or raise the price to increase revenue.

(b) Set $E(p) = 1$ and solve for p to determine at what price the revenue is greatest.

46. The SuperChip Company determines that the demand function for their gourmet potato chips is given by

$$d(p) = 3 - 0.1p - 0.1p^2$$

where p is the unit price, in dollars, of a bag of potato chips and $d(p)$ is the demand in thousands.

(a) Determine $E(4)$ and state whether the manufacturer should lower or raise the price to increase revenue.

(b) Set $E(p) = 1$ and solve for p to determine at what price the revenue is greatest.

47. The MicroTV Company determines that the demand for their 3-inch color TV is

$$d(p) = 100 \ln (150 - p)$$

where p is the unit price, in dollars, of a 3-inch color TV and $d(p)$ is the demand in hundreds.

(a) Determine $E(100)$ and state whether the manufacturer should lower or raise the price to increase revenue.

(b) Set $E(p) = 1$ and solve for p to determine at what price the revenue is greatest.

SECTION PROJECT

Some who study the theory of economics want to generalize the trends that they see relative to demand functions. To begin to understand how this is done, answer the following:

(a) If the price–demand for a product is $d(p) = 300p^{-0.3}$ show that the elasticity of demand $E(p)$ is a constant.

(b) Consider the price–demand function $d(p) = c \cdot p^{-n}$, where c and n are fixed constants. Show that the elasticity $E(p)$ is also a fixed constant.

(c) If the price–demand for a product is $d(p) = \dfrac{100}{e^{0.2x}}$ show that the elasticity of demand $E(p)$ is a linear function.

(d) Consider the price–demand function $d(p) = \dfrac{a}{e^{cp}}$, where a and c are fixed constants. Show that the elasticity $E(p)$ is a linear function of the form $E(p) = cp$.

CHAPTER REVIEW EXERCISES

In Exercises 1–10, differentiate using the Generalized Power Rule.

1. $f(x) = (x + 2)^3$

2. $f(x) = (x - 5)^2$

3. $f(x) = (8 - x)^3$

4. $f(x) = (4x - 3)^3$

5. $f(x) = (2x + 5)^4$

6. $g(x) = (4x^2 + 7)^5$

7. $g(x) = 3(x^2 - 5x + 3)^2$

8. $f(x) = (3x^2 + 7x - 2)^{63}$

9. $f(x) = (2x^2 - 5x + 7)^{1/3}$

10. $g(x) = (3x^2 - 9x - 4)^{-6}$

For the rational functions in Exercises 11–14:

(a) Determine the derivative using the Quotient Rule.

(b) Determine the derivative using the Generalized Power Rule.

11. $f(x) = \dfrac{3}{2x + 9}$

12. $f(x) = \dfrac{7}{(x - 7)^2}$

13. $f(x) = \dfrac{2}{(3x + 5)^3}$

14. $f(x) = \dfrac{9}{x^2 - 6x + 18}$

In Exercises 15–18, determine an equation of the line tangent to the graph of f at the indicated point.

15. $f(x) = (5x + 3)^5$; $(-1, -32)$

16. $f(x) = (7x - 6)^3$; $(1, 1)$

17. $f(x) = (x^2 - 5x + 8)^4$; $(3, 16)$

18. $f(x) = (6x - 11)^{1/2}$; $(6, 5)$

Use the Generalized Power Rule to differentiate the functions in Exercises 19–22.

19. $g(x) = \sqrt{7x - 12}$

20. $f(x) = \sqrt[3]{8x + 1}$

21. $f(x) = \dfrac{3}{\sqrt{4x + 5}}$

22. $g(x) = \dfrac{8}{\sqrt[3]{2x^3 - 5x + 4}}$

In Exercises 23–26, use the Generalized Power Rule, along with the Product and Quotient Rules, to find the derivatives of the given functions.

23. $f(x) = x(x^2 + 5)^3$

24. $g(x) = 4x\sqrt{x^2 - 2x}$

25. $f(x) = \dfrac{3x - 7}{\sqrt{5x - 6}}$

26. $g(x) = (6x - 5)^7(3x + 2)^4$

In Exercises 27–30, find the derivatives using the Generalized Power Rule.

27. $y = (3x^2 - x + 1)^{0.67}$

28. $y = \left(\dfrac{1}{8.3x - 5.7}\right)^{-2.4}$

29. $f(x) = (x^3 - x^2 + 5x + 1)^{-0.7}$

30. $g(x) = 4.96(x + 1)^{2.78}$

31. The ClearView Window Company determines that the cost to produce a certain kind of window is modeled by

$$C(x) = (10x - 8)^{1.5} + 480, \qquad 1 \le x \le 100$$

where x represents the number of windows produced in hundreds and $C(x)$ represents the production costs in hundreds of dollars.

(a) Determine the marginal cost function, MC.

(b) Evaluate and interpret $MC(7)$.

(c) Determine the average cost function, AC.

(d) Determine the marginal average cost function, MAC.

(e) Evaluate and interpret $MAC(7)$.

In Exercises 32–37, determine the derivatives of the natural logarithm functions.

32. $f(x) = -6\ln x$

33. $f(x) = \ln x^5$

34. $f(x) = 4x^5 \cdot \ln x$

35. $f(x) = \ln(3x^2 - 5)$

36. $f(x) = (\ln x)^{-2}$

37. $f(x) = \dfrac{x^2 + 5x - 2}{\ln(x + 4)}$

In Exercises 38–41, determine the derivatives of the general logarithm functions.

38. $g(x) = 3 \log_5 x$

39. $g(x) = 4x^3 \log_2 x$

40. $g(x) = \log_{10}(6x - 5)$

41. $g(x) = \log_4 \left(\dfrac{x^2 + 5}{2x - 3} \right)$

In Exercises 42–47, determine the derivatives of the exponential functions.

42. $f(x) = 3e^x + 5$

43. $f(x) = \sqrt{e^x - 5}$

44. $f(x) = 7x^3 e^x - 3e^x$

45. $f(x) = e^{2x - 13}$

46. $f(x) = e^{4 \ln x + 5}$

47. $f(x) = \dfrac{e^{2x}}{\ln(x - 4)}$

For Exercises 48–51:

(a) Determine the tangent line equation for the function at the given point.

(b) Check by graphing the function and the tangent line in the same viewing window.

48. $f(x) = 3e^x$; $(2, 3e^2)$

49. $f(x) = \ln x^4$; $(e^4, 16)$

50. $f(x) = \ln \sqrt{2x + 3}$; $(3, \ln 3)$

51. $f(x) = e^x - e^{-x}$; $(0, 0)$

In Exercises 52–55, determine the derivatives of the general exponential functions.

52. $g(x) = 4^x$

53. $g(x) = \dfrac{3^x}{15^x}$

54. $g(x) = x^4 \cdot 0.7^x$

55. $g(x) = \sqrt[3]{10^x}$

In Exercises 56–59, determine the derivatives of the given functions.

56. $f(x) = 5^{x^2 - 1}$

57. $f(x) = 0.4^{\sqrt{x}}$

58. $f(x) = 3xe^{5x} - \ln(x + 4)$

59. $f(x) = \log_7 3x - \log_3 7x$

60. The amount of annual air pollution emissions in the United States can be modeled by

$$f(x) = 13.44 - 3.20 \ln x, \qquad 1 \le x \le 26$$

where x represents the number of years since 1969 and $f(x)$ represents the number of millions of tons of particulate matter of less than 10 micrometers.

(a) Graph f in the viewing window $[1, 26]$ by $[0, 15]$.

(b) Determine f' for the model.

(c) Evaluate and interpret $f'(10)$.

(d) Write the equation of the tangent line at $x = 10$ and determine y on the tangent line when $x = 15$. Interpret and compare to $f(15)$.

61. The number of commercial FM radio stations in the United States can be modeled by

$$f(x) = 2149.6 \cdot (1.036)^x, \qquad 1 \le x \le 26$$

where x represents the number of years since 1969 and $f(x)$ represents the number of commercial FM radio stations.

(a) Classify the function as an exponential growth or decay model.

(b) Graph f in the viewing window $[1, 26]$ by $[2000, 6000]$.

(c) Determine f' for the model.

(d) Evaluate and interpret $f'(7)$ and compare to $f'(12)$.

62. The number of subscribers (in millions) to cellular phone service in the United States between 1989 and 1996 is shown in the following table.

YEAR	x	SUBSCRIBERS (MILLIONS)
1989	1	3.509
1990	2	5.283
1991	3	7.557
1992	4	11.033
1993	5	16.009
1994	6	24.134
1995	7	33.786
1996	8	44.043

(a) Use your calculator to determine a general exponential regression model of the form

$$f(x) = a \cdot b^x, \qquad 1 \le x \le 8$$

where x represents the number of years since 1988 and $f(x)$ represents the number of subscribers (in millions) to cellular phone service in the United States.

(b) Classify the function as an exponential growth or decay model.

(c) Graph f in the viewing window $[1, 8]$ by $[0, 50]$.

(d) Determine f' for the model.

(e) Evaluate and interpret $f'(3)$ and compare to $f'(7)$.

(f) Rewrite the model in the form $g(x) = a \cdot e^{kx}$, where a and k are constants rounded to the nearest hundredth.

In Exercises 63–70, use implicit differentiation to find $\dfrac{dy}{dx}$.

63. $3x - 5y = 7$

64. $9x^2 + 4y^2 = 36$

65. $5x^6 + y^3 = x^2$

66. $x^{1/3} - y^{1/3} = 8$

67. $y^3 - x^2 y = 5$

68. $x^2 + 6xy + 9y^2 = 4$

69. $x \ln y = y + 4$

70. $\ln(x + y) = e^x + 3y$

For Exercises 71–74:

(a) Use implicit differentiation to determine $\dfrac{dy}{dx}$.

(b) Solve the equation explicitly for y and determine $\dfrac{dy}{dx}$.

71. $2x + 3y = 6$ 72. $7x - xy = 5x^2$

73. $x^3 y - y = 2x + 2$ 74. $\dfrac{3}{x} - \dfrac{4}{y} = 5$

For Exercises 75–78, determine the equation of the tangent line for the given equation at the indicated point.

75. $x^2 - y^2 = 9$; $(5, 4)$

76. $x^2 y + y^2 x = 12$; $(3, -4)$

77. $2x \ln y = 4x^2 - 4y$; $(1, 1)$

78. $3x + xe^y = 4 + y$; $(1, 0)$

79. The BriteWite Light Company has determined that the price–demand equation for a new lamp is

$$px + 10x = 600, \qquad 10 \le x \le 25$$

where x represents the demand for lamps in thousands and p represents the price of one lamp in dollars.

(a) Use implicit differentiation to determine $\dfrac{dp}{dx}$.

(b) Evaluate and interpret $\dfrac{dp}{dx}\Big|_{(15,30)}$

In Exercises 80–83, determine the rate of change equation by implicitly differentiating each side of the equation with respect to time t.

80. $x^2 - 3y = 4$ 81. $x^3 + y^3 = 6x$

82. $2xy + 5x = 18y$ 83. $xy - y^3 = 2x$

For Exercises 84–87, assume that x and y are both functions of t and determine the indicated rate.

84. Determine $\dfrac{dy}{dt}$ for $2x + y^2 = 7$, given $y = 3$ and $\dfrac{dx}{dt} = 12$.

85. Determine $\dfrac{dx}{dt}$ for $xy + 4y - 3x = 7$, given $x = 1$ and $\dfrac{dy}{dt} = 4$.

86. Determine $\dfrac{dx}{dt}$ for $y = 2x + 3^{x-4}$, given $x = 4$ and $\dfrac{dy}{dt} = -6$.

87. Determine $\dfrac{dy}{dt}$ for $x^2 + 2xy + y^2 = 25$, given $x = 2$, $y = 3$, and $\dfrac{dx}{dt} = 11$.

88. The Zoot Soot Chimney Service has determined that its monthly revenue is given by

$$R = 150x - \frac{x^2}{300}$$

where x is the number of chimneys cleaned and R is the revenue in dollars.

(a) Differentiate the revenue function with respect to time t.

(b) Determine the rate of change in the revenue if $x = 45$ and $\dfrac{dx}{dt} = 3 \dfrac{\text{chimneys}}{\text{month}}$.

89. The surface area of a spherical balloon is given by the equation

$$S = 4\pi r^2$$

where r is the radius of the balloon and S is the voloume.

(a) Differentiate the equation with respect to time t.

(b) If the balloon is being blown up so that the radius is increasing at a rate of $0.7 \dfrac{\text{centimeter}}{\text{second}}$, determine the rate at which the surface area is changing when the radius is 8 centimeters.

In Exercises 90–93, compute the arc elasticity E_a.

90. $p_1 = 18$, $p_2 = 20$, $q_1 = 200$, $q_2 = 195$

91. $p_1 = 50$, $p_2 = 55$, $q_1 = 80$, $q_2 = 70$

92. $p_1 = 630$, $p_2 = 650$, $q_1 = 1300$, $q_2 = 1250$

93. $p_1 = 15{,}000$, $p_2 = 15{,}250$, $q_1 = 75$, $q_2 = 73$

94. The Macro Software Company determines that the price–demand function for a new FileFinder program is

$$x = d(p) = 350 - 2p$$

where p is the unit price in dollars of the program and x is the quantity demanded in thousands.

(a) Determine $d(8)$ and $d(9)$ and interpret.

(b) Using $q_1 = d(8)$, $q_2 = d(9)$, $p_1 = 8$, and $p_2 = 9$, determine the arc elasticity E_a and interpret.

95. The Thai Way Buffet Restaurant typically has 45 customers for lunch each day when the price for lunch is \$6.95. During a special promotion week, the price is reduced to \$5.50 and the number of lunch customers grows to 58 per day. Determine the arc elasticity and interpret.

In Exercises 96–99, the price–demand function $p(x)$ is given. Determine the demand function $x = d(p)$.

96. $p = 18 - 0.3x$ 97. $p = \dfrac{12}{x} + 5$

98. $p = \sqrt{500 - x^2}$ 99. $p = 800e^{-0.02x}$

In Exercises 100–103, the demand function $x = d(p)$ and the price p are given.

(a) Determine the point elasticity function $E(p)$.

(b) Classify the demand as elastic, inelastic, or unitary.

100. $d(p) = 180 - 5p$; $p = 20$

101. $d(p) = 400 - p^2$; $p = 12$

102. $d(p) = \sqrt{200 - p}$; $p = 100$

103. $d(p) = 18 \ln(120 - 3p)$; $p = 5$

104. An amusement park charges $15 per admission and serves approximately 6400 patrons per day. The demand function for the x patrons at p dollars per admission is given by

$$x = 130(35 - p)^{1.3}$$

(a) Determine the point elasticity function $E(p)$.

(b) Is the demand elastic or inelastic at the current $15 admission price?

(c) Should the ticket price be raised or lowered in order to increase revenue?

105. The FlyingThyme Watch Company determines that the demand for a new watch is given by

$$d(p) = 35 \ln(150 - 5p)$$

where p is the unit price, in dollars, of a watch and $d(p)$ is the demand in thousands.

(a) Determine $E(25)$ and state whether the manufacturer should raise or lower the price in order to increase revenue.

(b) Set $E(p) = 1$ and solve for p to determine the watch price that maximizes revenue.

CHAPTER

Further Applications of the Derivative

Cola waiting to be distributed to stores.

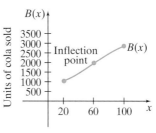

$B(x)$ represents the number of units of cola sold after spending x thousand dollars on advertising.

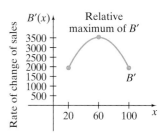

The rate of change of sales, $B'(x)$, is maximized at $x = 60$. At $x = 60$ on the graph of $B(x)$, we find an inflection point. In business, this is also known as the *point of diminishing returns*.

What We Know

In Chapter 4, we continued our study of differentiation and how the derivative gives an instantaneous rate of change. We also examined more applications utilizing the power of the derivative.

Where Do We Go

In this chapter, we will see how the rate of change concept and the derivative tell us the behavior of a function. Specifically, we will see how the derivative tells us where a function is increasing (or decreasing) and where its graph reaches a high point (or low point). We then examine how the increasing (or decreasing) behavior of the derivative gives us the *concavity* of the graph of the function.

275

SECTION 5.1 FIRST DERIVATIVES AND GRAPHS

In this section we examine more deeply the relationship that exists between a function and its derivative. Our study will reveal how the sign of a derivative, whether it is positive or negative, on an interval indicates whether the function is increasing or decreasing. Also, we will see how the derivative can help locate the **relative extrema** for a function.

Intervals of Increase and Decrease for Functions

So far, we have looked at the graphs of many different functions. The graphs of functions generally have portions that are *increasing* and portions that are *decreasing,* such as the graph in Figure 5.1.1.

TECHNOLOGY NOTE
You can verify the results in the Flashback by graphing the model on your calculator and using the MAXIMUM command. Consult the online calculator manual at www.prenhall.com/armstrong

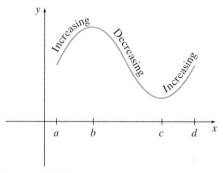

Figure 5.1.1

SAT DATA REVISITED

Flashback

In Section 1.3, we found that the percentage of people taking the SAT who intended to study business and commerce can be modeled by

$$f(x) = -0.09x^2 + 2.01x + 9.63, \qquad 1 \le x \le 22$$

where $f(x)$ is the percentage of people taking the SAT and x is the number of years since 1974. See Figure 5.1.2.

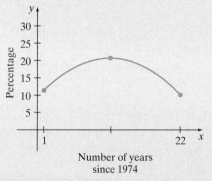

Figure 5.1.2 Graph of $f(x) = -0.09x^2 + 2.01x + 9.63$.

Determine intervals where f is increasing and decreasing. Round to the nearest hundredth and interpret the results.

Flashback Solution

Since the graph of the model is a parabola, to determine intervals where the model is increasing or decreasing, we must find the coordinates of the vertex, the maximum point of the graph. In Chapter 1 it was shown that the x-coordinate of the vertex can be determined algebraically by using $x = -\dfrac{b}{2a}$. Using this formula, the x-coordinate is

$$x = -\frac{b}{2a} = -\frac{2.01}{2(-0.09)} = -\frac{2.01}{-0.18} \approx 11.17$$

The y-coordinate is found by substituting $x = 11.17$ back into $f(x)$. This gives

$$f(11.17) = -0.09(11.17)^2 + 2.01(11.17)$$
$$+ 9.63 \approx 20.85, \quad \text{or } 20.85\%$$

So we conclude that f is increasing on the interval $(1, 11.17)$ and decreasing on the interval $(11.17, 22)$. Rounding 11.17 to the nearest year gives a value of 11. Thus, the percentage of people taking the SAT who intended to study business and commerce in- creased from 1975 to 1985 and then it decreased from 1985 to 1996. There was a peak of about 20.85% who intended to study business and commerce, and it occurred in 1985.

In the Flashback, we found the intervals where f is increasing and where it is decreasing. Another way to find these intervals is to analyze the sign of f'. (Since $f'(x)$ gives the rate of change of f at x, f' is called a **rate function.**) In Example 1, we observe the relationship between the graph of f and the sign of f'.

EXAMPLE 1 | **Analyzing the Behavior of a Function and Its Derivative**

The graphs of $f(x) = x^3 - 12x + 4$ and its derivative $f'(x) = 3x^2 - 12$ are shown in Figure 5.1.3 and 5.1.4. Use the graphs to determine where:

(a) f is increasing   (b) $f' > 0$ (c) f is decreasing

(d) $f' < 0$ (e) f is constant   (f) $f' = 0$

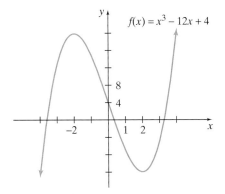

Figure 5.1.3

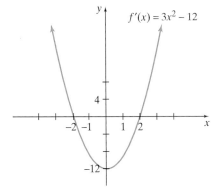

Figure 5.1.4

SOLUTION

(a) f is increasing on $(-\infty, -2) \cup (2, \infty)$.

(b) Recall that a function is greater than zero whenever its graph is *above* the x-axis. Thus, $f' > 0$ on $(-\infty, -2) \cup (2, \infty)$ since this is where the graph of f' is above the x-axis.

(c) f is decreasing on $(-2, 2)$.

(d) Recall that a function is less than zero whenever its graph is *below* the x-axis. Thus, $f' < 0$ on $(-2, 2)$ since this is where the graph of f' is below the x-axis.

(e) f is constant at $x = -2$ and $x = 2$.

(f) Recall that a function is equal to zero whenever its graph *crosses* the x-axis. Thus, $f'(x) = 0$ at $x = -2$ and $x = 2$ since this is where the graph of f' crosses the x-axis.

Notice in Example 1 that there appears to be a connection between our answers to parts (a) and (b) as well as parts (c) and (d) and parts (e) and (f). Pairing the results in this manner leads us to make the following observation, which is true for any function and its derivative.

Increasing and Decreasing Functions

On an open interval (a, b) on which f is differentiable and continuous

1. If $f'(x) > 0$ for all x in (a, b), then f is increasing on (a, b).
2. If $f'(x) < 0$ for all x in (a, b), then f is decreasing on (a, b).
3. If $f'(x) = 0$ for all x in (a, b), then f is constant on (a, b).

In English this means that *wherever the derivative of a function is positive, the function is increasing; wherever the derivative of a function is negative, the function is decreasing; and wherever the derivative of a function is zero, the function is constant.* See Figure 5.1.5.

Knowing that the derivative tells us the slope of a tangent line at a point, the preceding properties should come as no surprise. Basically, wherever tangent lines have a positive slope, $f'(x) > 0$, we see that a function is increasing (Figure 5.1.6), and wherever tangent lines have a negative slope, $f'(x) < 0$, we see that a function is decreasing (Figure 5.1.7). Also, where we have *horizontal tangent lines*, $f'(x) = 0$, we say that the function is constant (Figure 5.1.8).

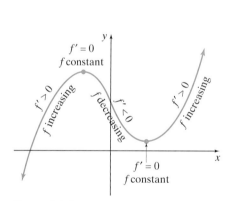

Figure 5.1.5

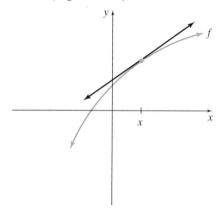

Figure 5.1.6 Tangent line has positive slope; that is, $f'(x) > 0$.

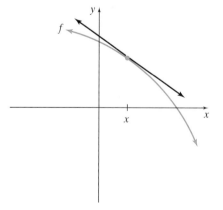

Figure 5.1.7 Tangent line has negative slope; that is, $f'(x) < 0$.

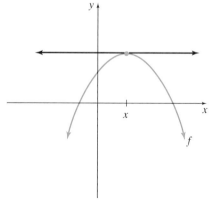

Figure 5.1.8 Tangent line has zero slope; that is, $f'(x) = 0$.

Relative Extrema and Critical Values

In Figure 5.1.5, notice that the graph has a *valley* and a *hill*. In mathematics, the hill is called a **relative maximum**, while the valley is called a **relative minimum**. By using the word *relative* we mean *relative to points nearby*. (In Section 5.4 we will discuss *absolute maximum* and *absolute minimum*). These terms are expressed more formally in the following definition.

Relative Extrema

If f is a continuous function on an open interval containing c, then the point $(c, f(c))$:

- Is a **relative maximum** if $f(c) \geq f(x)$ for all x in the interval.
- Is a **relative minimum** if $f(c) \leq f(x)$ for all x in the interval.

NOTE: The plural of relative maximum is relative maxima and the plural of relative minimum is relative minima. Collectively, the maxima and minima are called **extrema.**

EXAMPLE 2	**Locating Relative Extrema**

Determine where the graphs in Figure 5.1.9 and 5.1.10 have a relative maximum and a relative minimum.

(a)   (b)

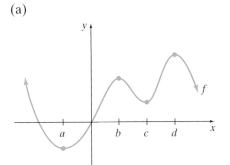

Figure 5.1.9   **Figure 5.1.10**

SOLUTION

(a) f has a relative maximum at $x = b$ and $x = d$, and f has a relative minimum at $x = a$ and $x = c$.

(b) g has a relative maximum at $x = f$ and a relative minimum at $x = e$.

    In Example 2a, the relative extrema occurred where the graph has horizontal tangents, that is, where $f'(x) = 0$. In Example 2b the relative extrema occurred where the graph had a corner or made a sharp turn, that is, where $f'(x)$ is undefined. We call points on a curve where $f'(x) = 0$ or where $f'(x)$ is undefined the **critical points** of the function. The x-coordinate of the critical points are called **critical values** or **critical numbers**.

Critical Values

A **critical value** for f is an x-value in the domain of f for which (1) $f'(x) = 0$ *or* (2) $f'(x)$ is undefined.

EXAMPLE 3 | **Determining Critical Values**

Determine the critical values for $f(x) = 5x^3 + 4x^2 - 12x - 25$.

SOLUTION

The key to locating critical values is to first determine the derivative. The derivative is

$$f'(x) = 15x^2 + 8x - 12$$

Since $f'(x)$ is defined for all x, the only critical values occur where $f'(x) = 0$. So we need to solve

$$15x^2 + 8x - 12 = 0$$
$$(5x + 6)(3x - 2) = 0$$
$$x = -\frac{6}{5} \quad \text{or} \quad x = \frac{2}{3}$$

Thus, the critical values occur at $x = -\frac{6}{5}$ and $x = \frac{2}{3}$. See Figure 5.1.11.

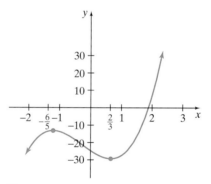

Interactive Activity

Solve $f'(x) = 0$ graphically to support the result in Example 3.

Figure 5.1.11

EXAMPLE 4 | **Determining Critical Values**

Determine the critical values for the following functions.

(a) $f(x) = \sqrt[3]{x}$ (b) $y = x^3$

SOLUTION

(a) Again, our first step is to compute the derivative. This yields

$$f(x) = \sqrt[3]{x} = x^{1/3}$$
$$f'(x) = \frac{1}{3}x^{-2/3} = \frac{1}{3x^{2/3}} \quad \text{or} \quad \frac{1}{3\sqrt[3]{x^2}}$$

Here we notice that $f'(x)$ is undefined at $x = 0$ and that $f'(x)$ cannot equal zero; that is, $f'(x) = 0$ has no solution. So we conclude that the only critical value is $x = 0$. See Figure 5.1.12.

(b) The derivative is $\frac{dy}{dx} = 3x^2$ and it is defined for all x. We see that $\frac{dy}{dx} = 0$ at $x = 0$, which means that the only critical value is $x = 0$. See Figure 5.1.13.

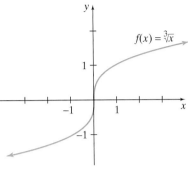

Figure 5.1.12

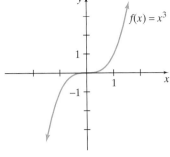

Figure 5.1.13

✓ CHECKPOINT 1

Now work Exercise 17.

Notice that in Example 3 our critical values produced a relative maximum and a relative minimum, whereas in Example 4 our critical values produced neither a relative maximum nor a relative minimum. What this means is that *every relative extreme occurs at a critical value, but not every critical value produces a relative extreme.* So how do we know if a critical value is going to produce a relative maximum or a relative minimum? The answer to this question is found in the First Derivative Test.

First Derivative Test

The First Derivative Test pulls together all the ideas in this section.

> ### First Derivative Test
>
> Let $x = c$ be a critical value of a function f that is continuous on an open interval containing $x = c$. The point $(c, f(c))$ can be called:
>
> 1. A **relative minimum** if $f' < 0$ to the left of c and $f' > 0$ to the right of c.
> 2. A **relative maximum** if $f' > 0$ to the left of c and $f' < 0$ to the right of c.

Recalling that the sign of the derivative tells us the increasing-decreasing behavior of the function, we can construct a **sign diagram** that communicates the First Derivative Test in a more visual manner. In the following sign diagrams, the bottom row gives the sign of f' and the top row gives the behavior of f.

> ### First Derivative Test Using a Sign Diagram
>
> Let $x = c$ be a critical value of a function f that is continuous on an open interval containing $x = c$. Place c on a number line.
>
> 1. **Relative minimum**
>
>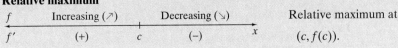
>
> 2. **Relative maximum**

NOTE: If f' has the same sign on both sides of $x = c$, then f has neither a relative maximum nor a relative minimum at $x = c$.

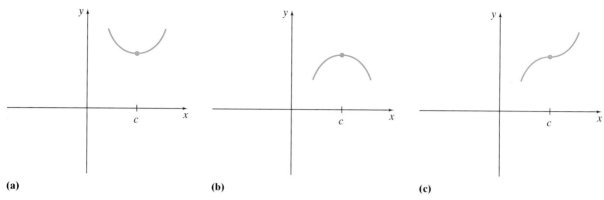

(a)   **(b)**   **(c)**

Figure 5.1.14   **(a)** Relative minimum at $x = c$. **(b)** Relative maximum at $x = c$. **(c)** Neither a relative maximum nor a relative minimum at $x = c$.

Figure 5.1.14(a–c) illustrates the First Derivative Test when $x = c$ is a critical value, where $f'(c) = 0$, that is, a horizontal tangent.

Figure 5.1.15(a–c) illustrates the First Derivative Test when $x = c$ is a critical value, where $f'(c)$ is undefined.

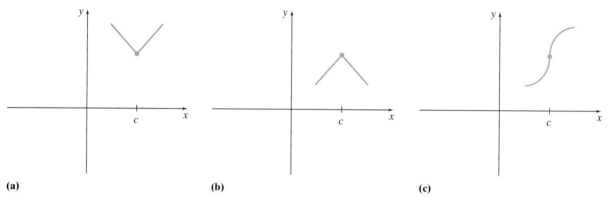

(a)   **(b)**   **(c)**

Figure 5.1.15   **(a)** Relative minimum at $x = c$. **(b)** Relative maximum at $x = c$. **(c)** Neither relative maximum nor relative minimum at $x = c$.

To understand why the sign diagram and the method that we will use in constructing sign diagrams works, we need to quickly review **continuous functions.** Suppose that a function f is continuous on the open interval $(2, 7)$, and $f(x) \neq 0$ for any x in $(2, 7)$. If $f(x)$ is positive for some x, say, $f(3) = 4$, then $f(x)$ is positive for every x in $(2, 7)$. Why? Well, suppose we assume that $f(6)$ is negative, say $f(6) = -3$. Since f is *continuous,* the graph of f could not connect the points $(3, 4)$ and $(6, -3)$ without crossing the x-axis. See Figure 5.1.16. But recall that $f(x) = 0$

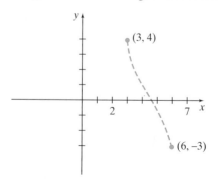

Figure 5.1.16

at the point where the graph of f crosses the x-axis. Since we agreed that $f(x) \neq 0$ for any x in $(2, 7)$, this assumption has been contradicted, and thus f cannot be negative in $(2, 7)$. This property of continuous functions is stated in the following theorem, which we will use frequently in this chapter.

Theorem 5.1

If f is continuous and $f(x) \neq 0$ for any x in the interval (a, b), then f cannot change sign on (a, b).

EXAMPLE 5

Applying the First Derivative Test

For $f(x) = 5x^3 + 4x^2 - 12x - 25$, determine intervals where f is increasing and where f is decreasing by making a sign diagram. Locate all relative extrema.

SOLUTION

To find intervals where f is increasing and where it is decreasing, we need to determine the critical values for this function. In Example 3, we determined that the derivative is $f'(x) = 15x^2 + 8x - 12$ and the critical values are

$$x = -\frac{6}{5} \quad \text{and} \quad x = \frac{2}{3}$$

To determine intervals where the function is increasing and decreasing, we place the critical values on a number line representing the x-axis in Figure 5.1.17.

Figure 5.1.17

Since f' is a polynomial function, it is continuous on the open intervals determined by the critical values. Thanks to Theorem 5.1, f' can only change sign at the critical values. We can determine the sign of the derivative within each interval by evaluating $f'(x)$ for some x within the interval. We call such x-values **test numbers.** In the interval $\left(-\infty, -\frac{6}{5}\right)$, we choose $x = -2$ as our test number.

$$f'(-2) = 15(-2)^2 + 12(-2) - 12 = 24$$

Since $f'(-2) = 24$, which is positive, we conclude that on the interval $\left(-\infty, -\frac{6}{5}\right)$, $f(x) = 5x^3 + 4x^2 - 12x - 25$ is *increasing*. Choosing $x = 0$ as our test number in the interval $\left(-\frac{6}{5}, \frac{2}{3}\right)$ gives

$$f'(0) = 15(0)^2 + 12(0) - 12 = -12$$

So $f'(0) = -12$, a negative result. In a similar fashion, we select $x = 1$ as our test number in the interval $\left(\frac{2}{3}, \infty\right)$ and determine that

$$f'(1) = 15(1)^2 + 12(1) - 12 = 15$$

So $f'(1) = 15$, a positive number. We now list the signs of the derivative on the sign diagram in Figure 5.1.18.

Figure 5.1.18

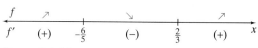

Figure 5.1.19

Now using our knowledge of how the sign of the derivative tells us the behavior of the function, we complete the sign diagram in Figure 5.1.19. This sign diagram indicates that f is increasing on $\left(-\infty, -\frac{6}{5}\right) \cup \left(\frac{2}{3}, \infty\right)$ and decreasing on $\left(-\frac{6}{5}, \frac{2}{3}\right)$. To find the relative extrema, we use the First Derivative Test with a sign diagram. Since f is increasing on $\left(-\infty, -\frac{6}{5}\right)$ and then decreasing on $\left(-\frac{6}{5}, \frac{2}{3}\right)$, we conclude that there is a relative maximum at $x = -\frac{6}{5}$. Since f is decreasing on $\left(-\frac{6}{5}, \frac{2}{3}\right)$ and then increasing on $\left(\frac{2}{3}, \infty\right)$, we conclude that there is a relative minimum at $x = \frac{2}{3}$. Specifically, we have a relative maximum at the point $\left(-\frac{6}{5}, -\frac{337}{25}\right)$ and a relative minimum at the point $\left(\frac{2}{3}, -\frac{803}{27}\right)$.

Example 5 demonstrates the following process, which we can use to locate relative extrema.

> *To locate relative extrema, we find the critical values of the function and apply the First Derivative Test.*

Interactive Activity

Graph $f(x) = 5x^3 + 4x^2 - 12x - 25$ and use the MAXIMUM and MINIMUM commands to verify the result of Example 5.

EXAMPLE 6 | **Applying the First Derivative Test**

For $f(x) = x + \dfrac{1}{x}$, determine intervals where f is increasing and decreasing. Locate all relative extrema.

SOLUTION

To determine critical values, we need to first determine the derivative.

$$f(x) = x + \frac{1}{x} = x + x^{-1}$$

$$f'(x) = 1 - x^{-2} = 1 - \frac{1}{x^2}$$

Next we set the derivative equal to zero and solve.

$$1 - \frac{1}{x^2} = 0$$

$$1 = \frac{1}{x^2}$$

$$x^2 = 1$$

$$x = \pm 1$$

Notice that, although the derivative is undefined at $x = 0$, we do not list $x = 0$ as a critical value, because $x = 0$ is not in the domain of f. So the only critical values are $x = 1$ and $x = -1$. We proceed as in Example 5 and construct the sign diagram in Figure 5.1.20. Notice that we placed $x = 0$, where the derivative was

Figure 5.1.20

undefined, in our sign diagram. We did this to take advantage of Theorem 5.1. In the interval $(-\infty, -1)$, we select the test number -2; in $(-1, 0)$, we select $-\frac{1}{2}$; in $(0, 1)$, we pick $\frac{1}{2}$; and in $(1, \infty)$, we choose 2. (Notice that in each of these open intervals our function is continuous, thus allowing us to use Theorem 5.1.) Substituting these test numbers into the derivative yields

$$x = -2 \qquad\qquad x = -\frac{1}{2} \qquad\qquad x = \frac{1}{2} \qquad\qquad x = 2$$

$$f'(-2) = 1 - \frac{1}{(-2)^2} \qquad f'\left(-\frac{1}{2}\right) = 1 - \frac{1}{\left(-\frac{1}{2}\right)^2} \qquad f'\left(\frac{1}{2}\right) = 1 - \frac{1}{\left(\frac{1}{2}\right)^2} \qquad f'(2) = 1 - \frac{1}{(2)^2}$$

$$= 1 - \frac{1}{4} \qquad\qquad = 1 - \frac{1}{\frac{1}{4}} \qquad\qquad = 1 - \frac{1}{\frac{1}{4}} \qquad\qquad = 1 - \frac{1}{4}$$

$$= \frac{3}{4} \qquad\qquad = -3 \qquad\qquad = -3 \qquad\qquad = \frac{3}{4}$$

Finally, using these results and our knowledge of how the sign of the derivative determines the behavior of the function, we complete the sign diagram as shown in Figure 5.1.21.

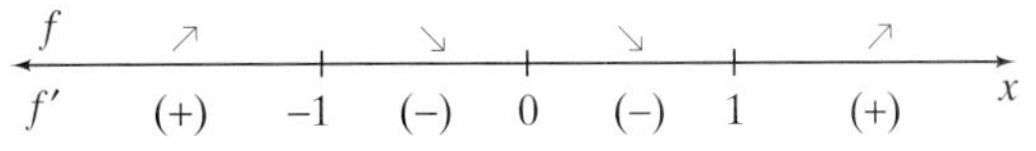

$$
\begin{array}{c}
f \quad\quad \nearrow \quad\quad \searrow \quad\quad \searrow \quad\quad \nearrow \\
\overline{f' \quad (+) \quad -1 \quad (-) \quad 0 \quad (-) \quad 1 \quad (+)} \quad x
\end{array}
$$

Figure 5.1.21

So $f(x) = x + \frac{1}{x}$ is increasing on the intervals $(-\infty, -1) \cup (1, \infty)$. It is decreasing on the intervals $(-1, 0) \cup (0, 1)$. Using the sign diagram and the First Derivative Test, we have a relative maximum at $x = -1$ and a relative minimum at $x = 1$. Specifically, we have a relative maximum at $(-1, -2)$ and a relative minimum at $(1, 2)$. See Figure 5.1.22.

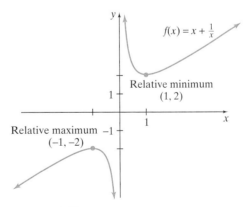

Figure 5.1.22

✓ **Checkpoint 2**   Now work Exercise 31.

Now let's see how the First Derivative Test can be used in an application.

EXAMPLE 7   **Generating Revenue from Marginal Revenue**

The owner of Everything Fit to Print has received a shipment of new computer tutorial software. The manufacturer of the software claims that their market

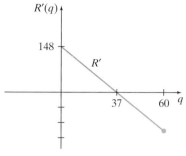

Figure 5.1.23

research has determined that a retail price of \$74 maximizes revenue. The manufacturer supplied the graph of the marginal revenue function shown in Figure 5.1.23.

(a) Determine what value of q, quantity, will maximize the revenue. Also, determine the maximum revenue.

(b) Sketch a possible graph of R.

SOLUTION

(a) To find the relative maximum, we determine the critical values of the function and apply the First Derivative Test. Since the critical values of R occur where $R'(q) = 0$, we know from the graph of R' that $q = 37$ is a critical value. Now we determine the sign of R' to the left and to the right of this critical value.

- The graph R' is above the q-axis on the interval $(0, 37)$. This means that R' is positive and thus R is increasing on this interval.

- The graph R' is below the q-axis on the interval $(37, 60)$. This means that R' is negative and thus R is decreasing on this interval.

We use this information to construct the sign diagram in Figure 5.1.24.

Figure 5.1.24

Since revenue is increasing on the interval $(0, 37)$ and decreasing on the interval $(37, 60)$, we conclude, by the First Derivative Test, that revenue is maximized at $q = 37$. The maximum revenue is then simply

$$p \cdot q = 74 \cdot 37 = \$2738$$

(b) We know two points on the graph, $(0, 0)$ (if nothing is sold, then revenue is zero), and the relative maximum $(37, 2738)$. Using this information, a possible graph is given in Figure 5.1.25.

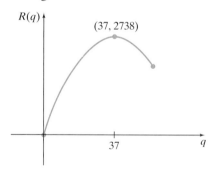

Figure 5.1.25

SUMMARY

In this section we saw how the sign of the derivative tells us the behavior of a function. Specifically, on an open interval:

- If $f'(x) > 0$ for all x in (a, b), then f is increasing on (a, b).
- If $f'(x) < 0$ for all x in (a, b), then f is decreasing on (a, b).
- If $f'(x) = 0$ for all x in (a, b), then f is constant on (a, b).

We defined **critical values** as *x*-values in the domain of the function such that $f'(x) = 0$ or $f'(x)$ is undefined. To find relative extrema, determine critical values and apply the **First Derivative Test.** Use the following process:

1. Determine the derivative and use it to find critical values.
2. Make a sign diagram to determine intervals of increase or decrease.
3. Use the sign diagram and the First Derivative Test to locate relative extrema.
4. Use the calculator as an effective tool to verify your work.

SECTION 5.1 EXERCISES

For each function graphed in Exercises 1–6, determine where the function is increasing, where it is decreasing, and where it is constant.

1.

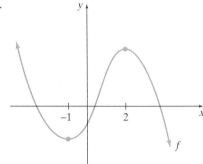

2.

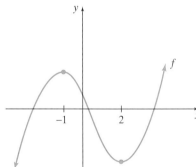

3.

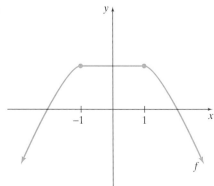

4.

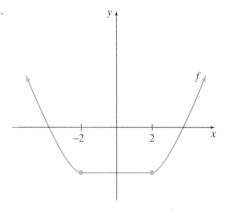

5.

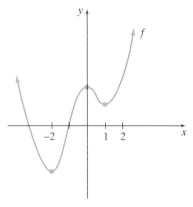

6.

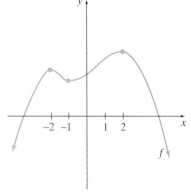

For each **function** graphed in Exercises 7–12:

(a) Determine intervals where the derivative is positive.

(b) Determine intervals where the derivative is negative.

(c) List where the derivative is undefined and where the derivative equals zero.

7.

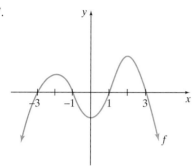

8.

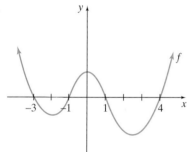

9.

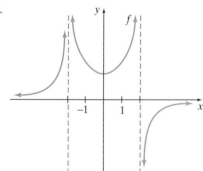

10.

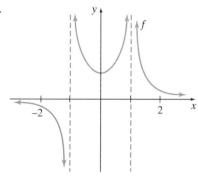

11.

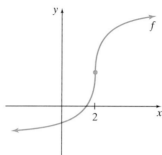

12.

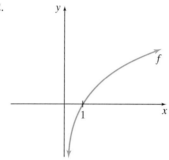

Determine the critical values of each function in Exercises 13–28.

13. $f(x) = 2x - 3$

14. $f(x) = -3x + 5$

15. $g(x) = x^2 - 5x + 6$

16. $f(x) = 2x^2 + 3x - 5$

✓ 17. $f(x) = \dfrac{1}{3}x^3 + x^2 - 15x + 3$

18. $g(x) = -x^3 - 3x^2 + 45x - 5$

19. $y = \sqrt{2x + 1}$

20. $y = \sqrt{3x - 7}$

21. $f(x) = e^x$

22. $g(x) = e^{3x}$

23. $y = x + \ln x$

24. $y = x - \ln x$

25. $f(x) = \sqrt[3]{x + 1}$

26. $g(x) = \sqrt[3]{x - 1}$

27. $f(x) = \dfrac{x - 2}{2x - 3}$

28. $f(x) = \dfrac{2x + 1}{5 - 3x}$

Exercises 29–38 use the results of Exercises 13–28. For each function:

(a) Determine intervals where the function is increasing and decreasing.

(b) Locate the relative extrema.

(c) Use a calculator to graph the function, and verify your results for parts (a) and (b).

29. The function in Exercise 15.

30. The function in Exercise 16.

✓ 31. The function in Exercise 17.

32. The function in Exercise 18.

33. The function in Exercise 19.

34. The function in Exercise 20.

35. The function in Exercise 23.

36. The function in Exercise 24.

37. The function in Exercise 25.

38. The function in Exercise 26.

Use the derivatives given in Exercises 39–44 to determine the x-values where f has a relative maximum and/or a relative minimum.

39. $f'(x) = -2x + 5$

40. $f'(x) = -2x - 3$

41. $f'(x) = 3x(x + 1)$

42. $f'(x) = -2x(x - 4)$

43. $f'(x) = 2(x + 1)^2(x - 1)(x + 3)^3$

44. $f'(x) = 4(x - 1)(x + 1)^2(x + 2)^3$

For Exercises 45–48, sketch a graph of a continuous function that satisfies the given data.

45.

x	$f(x)$	$f'(x)$
0	3	−1
1	2	−1
2	1	Undefined
3	2	1
4	3	1
5	4	1

46.

x	$f(x)$	$f'(x)$
0	2	−2
1	1	−1
2	0.25	0
3	1.25	1
4	2.5	1.5
5	2	−0.5

47.

x	$f(x)$	$f'(x)$
0	0	Undefined
1	1	$\frac{1}{2}$
2	1.4	0.35
3	1.7	0.29
4	2	0.25
5	2.2	0.22

48

x	$f(x)$	$f'(x)$
0	−7	12
1	0	3
2	1	0
3	2	3
4	9	12
5	28	27

APPLICATIONS

49. In Table 5.1.1, x represents the number of years since 1984 and $f(x)$ represents the average number of patient visits per week to a General/Family Practice doctor's office.
(a) Sketch a graph of a continuous function, on the interval [1, 10], that satisfies the data.
(b) What are the units of $f'(x)$?

50. In Table 5.1.2, x represents the number of years since 1980 and $f(x)$ represents the number of cases of tuberculosis, in thousands, reported in the United States.
(a) Sketch a graph of a continuous function, on the interval [0, 16], that satisfies the data.
(b) What are the units of $f'(x)$?

51. The Fore Link Company has determined from data that it has collected that the profit function P is given by

$$P(q) = 1000q - q^2$$

where q is the number of golf clubs produced and sold and $P(q)$ is the profit in dollars.

TABLE 5.1.1

x	1	2	3	4	5	6	7	8	9	10
$f(x)$	138	141	143	144	145	144	143	140	137	133
$f'(x)$	3.5	2.6	1.7	0.8	−0.1	−1	−2	−2.8	−3.7	−4.6

Source: U.S. Census Bureau web site.

TABLE 5.1.2

x	0	2	4	6	8	10	12	14	16
$f(x)$	31	25	22.2	21.8	23	24.7	25.9	25.6	22.9
$f'(x)$	−4.3	−2.2	−0.7	0.3	0.8	0.81	0.3	−0.7	−2.2

Source: U.S. Census Bureau web site.

(a) Determine intervals where P is increasing and decreasing.

(b) Determine the relative maximum and interpret each coordinate.

52. The Network Standards Company has determined that the revenue, in dollars, from the sale of 56K modems can be estimated by

$$R(q) = 300q - q^2$$

where $R(q)$ is in dollars and q is the number of modems.

(a) Determine intervals where R is increasing and decreasing.

(b) Determine the relative maximum and interpret each coordinate.

⬢ 53. The U.S. water consumption in gallons per day per capita can be modeled by

$$f(x) = -0.41x^2 + 13.53x + 330.69, \qquad 1 \le x \le 31$$

where x is the number of years since 1959 and $f(x)$ is the number of gallons of water consumed per day per capita.

(a) Determine intervals where f is increasing and decreasing.

(b) Determine the relative maximum and interpret each coordinate.

⬢ 54. The U.S. per capita consumption of beef can be modeled by

$$B(t) = 0.03t^3 - 0.76t^2 + 4.54t + 68.34, \qquad 1 \le t \le 16$$

where t is the number of years since 1979 and $B(t)$ is the per capita consumption, in pounds, of beef.

(a) Determine intervals where B is increasing and decreasing.

(b) Determine the relative extrema and interpret each coordinate.

⬢ 55. The U.S. total petroleum production can be modeled by

$$G(t) = 0.69t^2 - 9.42t + 62.82, \qquad 1 \le t \le 6$$

where t is the number of years since 1989 and $G(t)$ is the U.S. total petroleum production in billions of dollars. Show that from 1990 to 1995 U.S. total petroleum production has been decreasing.

56. Linguini's Pizza Palace is starting an all-you-can-eat pizza buffet from 5:00 to 9:00 P.M. on Friday evenings. A survey of local residents produced the price–demand function

$$p(x) = -0.02x + 8.3$$

where x represents the quantity demanded and $p(x)$ represents the price in dollars.

(a) Use the model to determine the price if the demand is 250. Round to the nearest cent.

(b) Determine $R(x)$, revenue as a function of the quantity x demanded.

(c) Use the techniques of this section to determine intervals where R is increasing and where R is decreasing.

(d) Determine the relative maximum and interpret each coordinate.

57. Linguini's Pizza Palace is introducing a line of specialty pizzas, such as a chicken and garlic pizza. Taste tests of this type of pizza produced the price–demand function

$$p(x) = 15.22e^{-0.015x}$$

where x represents the quantity demanded and $p(x)$ represents the price in dollars.

(a) Use the model to determine the price if the demand is 55. Round to the nearest cent.

(b) Determine $R(x)$, revenue as a function of the quantity x demanded.

(c) Use the techniques of this section to determine intervals where R is increasing and where R is decreasing.

(d) Determine the relative maximum. Round the x-coordinate to the nearest whole number and the y-coordinate to the nearest hundredth, and interpret each coordinate.

58. Researchers have determined through experimentation that the percent concentration of a certain medication during the first 20 hours after it has been administered is approximated by

$$p(t) = \frac{200t}{2t^2 + 5} - 4, \qquad [0.25, 20]$$

where t is the time in hours after administration of the medication and $p(t)$ is the percent concentration.

(a) Compute $p'(t)$ and determine the critical value(s).

(b) Determine intervals where p is increasing and decreasing.

(c) Determine the relative maximum and interpret each coordinate.

59. Researchers have determined through experimentation that the percent concentration of a certain medication during the first 20 hours after it has been administered is approximated by

$$p(t) = \frac{230t}{t^2 + 6t + 9}, \qquad [0, 20]$$

where t is the time in hours after administration of the medication and $p(t)$ is the percent concentration.

(a) Compute $p'(t)$ and determine the critical values.

(b) Determine intervals where p is increasing and decreasing.

(c) Determine the relative maximum and interpret each coordinate.

60. A manufacturer of Digital Pet, a virtual pet, has the following costs when producing x pets in one day, where $0 \le x \le 200$: Fixed costs are \$150, unit production costs

are $3 per pet, and equipment maintenance is $\dfrac{2x^2}{30}$. This yields a cost function for manufacturing x pets in one day of

$$C(x) = 150 + 3x + \frac{2x^2}{30}, \qquad 0 \le x \le 200$$

(a) Determine the average cost function, AC, and find the critical value(s) for AC.

(b) Determine intervals where AC is increasing and decreasing.

(c) Determine the relative minimum and interpret each coordinate.

61. A manufacturer of coffee mugs has the following costs when producing x mugs in one day, where $0 \le x \le 100$: Fixed costs are $50, unit production costs are $1 per mug, and equipment maintenance is $\dfrac{x^2}{40}$. This yields a cost function for manufacturing x coffee mugs in one day of

$$C(x) = 50 + x + \frac{x^2}{40}, \qquad 0 \le x \le 100$$

(a) Determine the average cost function, AC, and find the critical value(s) for AC.

(b) Determine intervals where AC is increasing and decreasing.

(c) Determine the relative minimum and interpret each coordinate.

62. Consider a quadratic function of the form $f(x) = ax^2 + bx + c$, where a, b, and c are real numbers with $a \ne 0$. Use $f'(x)$ to determine the x-coordinate where a horizontal tangent line occurs. (The result should look familiar.)

63. Recall that profit equals revenue minus cost; that is, $P(x) = R(x) - C(x)$.

(a) At a production level x, show that $P(x)$ is maximized when $R'(x) = C'(x)$, that is, when marginal revenue equals marginal cost.

(b) For all x on production interval (a, b), show that profit is increasing when $R'(x) > C'(x)$, that is, when marginal revenue is greater than marginal cost.

(c) For all x on production interval (a, b), show that profit is decreasing when $R'(x) < C'(x)$, that is, when marginal revenue is less than marginal cost.

SECTION PROJECT

Figure 5.1.26 shows a graph of the rate of change for catfish sold in the United States that were raised by aquaculture. Given that CF is the function that represents the number of catfish sold, in millions, in the United States that were raised by aquaculture and t represents the number of years since 1989, $1 \le t \le 7$, answer the following:

(a) What are the units for $CF'(t)$?

(b) Determine intervals where CF is increasing and decreasing.

(c) Determine the values of t where the rate of change of $CF(t) = 0$.

(d) Determine the t-values that produce relative extrema. Classify each as producing a relative maximum or a relative minimum.

(e) Sketch a possible graph for CF.

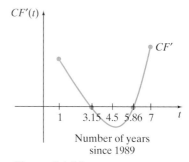

Figure 5.1.26

SECTION 5.2   SECOND DERIVATIVES AND GRAPHS

In Section 5.1, we saw how the derivative of a function, f', allowed us to determine where the function f is increasing and where it is decreasing. Recall that $f'(x)$ tells us an *instantaneous rate of change* at x and that f' is itself a *function*. Since the derivative is a function, we can compute the derivative of $f'(x)$ and use it to determine where f' is increasing and decreasing. The derivative of a derivative is called the **second derivative.**

We just stated that the second derivative tells us where the derivative is increasing and decreasing. This analysis will also give us more information about the behavior of the original function. Specifically, it will give us the intervals of **concavity** and **inflection points,** and other characteristics of the graphs of functions. We will interpret these concepts in the context of rates of change. Before we go much further, let's consider **higher-order derivatives.**

Higher-order Derivatives

If f is some function, $f'(x)$ is the derivative at x which gives the instantaneous rate of change for f at x or, equivalently, the slope of a tangent line at x. We know that f' is used to tell where f is increasing and where it is decreasing. Since f' is itself a function, computing the derivative of f' tells where the derivative is increasing and where it is decreasing. The derivative of f', denoted f'' and read "f double prime", is called the **second derivative.** Table 5.2.1 gives some different notations for the second derivative.

TABLE 5.2.1

FUNCTION	FIRST DERIVATIVE	SECOND DERIVATIVE
$f(x)$	$f'(x)$	$f''(x)$
y	y'	y''
y	$\dfrac{dy}{dx}$	$\dfrac{d^2 y}{dx^2}$

For most applications, we have no need for a derivative beyond the second derivative. However, any derivative beyond the first derivative is called a **higher-order derivative.**

EXAMPLE 1 | **Computing Higher-order Derivatives**

Determine the first three derivatives for the following functions.

(a) $f(x) = 2x^5 + 6x^3 - 7x + 1$ (b) $y = e^{2x}$

SOLUTION

(a) Applying the rules for differentiation yields

$$f'(x) = 10x^4 + 18x^2 - 7$$

$$f''(x) = \frac{d}{dx}(10x^4 + 18x^2 - 7) = 40x^3 + 36x$$

$$f'''(x) = \frac{d}{dx}(40x^3 + 36x) = 120x^2 + 36$$

(b) Again, applying the rules for differentiation yields

$$y' = 2e^{2x}$$

$$y'' = \frac{d}{dx}(2e^{2x}) = 4e^{2x}$$

$$y''' = \frac{d}{dx}(4e^{2x}) = 8e^{2x}$$

As we can see in Example 1, computing higher-order derivatives is no different than just computing a derivative.

✓ CHECKPOINT 1 | Now work Exercise 3.

FIRST DERIVATIVE TEST REVISITED

Flashback

Determine intervals where $f(x) = 5x^3 + 4x^2 - 12x - 25$ is increasing and where it is decreasing. Also, locate any relative extrema.

Flashack Solution

To find intervals where f is increasing and where it is decreasing, we need to determine the critical values for this function. As we saw in Section 5.1, Example 5, the derivative is given by $f'(x) = 15x^2 + 8x - 12$, and the critical values are $x = \frac{-6}{5}$ and $x = \frac{2}{3}$. Placing the critical values on a number line and choosing test numbers produces the sign diagram in Figure 5.2.1. The sign diagram indicates that $f(x) = 5x^3 + 4x^2 - 12x - 25$ is increasing on $\left(-\infty, \frac{-6}{5}\right) \cup \left(\frac{2}{3}, \infty\right)$ and

decreasing on $\left(\frac{-6}{5}, \frac{2}{3}\right)$. To locate the relative extrema, we use the First Derivative Test using a Sign Diagram. This gives a relative maximum at $\left(\frac{-6}{5}, \frac{-337}{25}\right)$ and a relative minimum at $\left(\frac{2}{3}, \frac{-803}{27}\right)$, as shown in Figure 5.2.2.

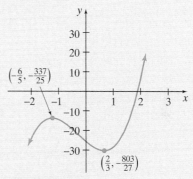

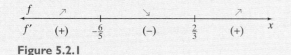

Figure 5.2.1

Figure 5.2.2

Recall that, since f' is also a function, we should be able to compute the *second derivative*, f'', and use it to determine the behavior of the derivative, f'. Later we will see how this information is used to determine the shape of the graph of f. First, let's see how we can use f'' to find information about f'.

EXAMPLE 2   **Using f'' to Graph f'**

For the function in the Flashback, compute $f''(x)$ and use it to determine where f' is increasing and decreasing. Graph f'.

SOLUTION

In the Flashback we computed the derivative of $f(x) = 5x^3 + 4x^2 - 12x - 25$ to be $f'(x) = 15x^2 + 8x - 12$. So the second derivative is

$$f''(x) = \frac{d}{dx}(15x^2 + 8x - 12) = 30x + 8$$

Since $f''(x)$ is defined for all x, the only critical value for f' is when $f''(x) = 0$. So we solve

$$30x + 8 = 0$$

$$x = \frac{-8}{30} = \frac{-4}{15}$$

So the only critical value for f' is $x = \frac{-4}{15}$.

We employ the same process from Section 5.1 and the Flashback; we place $x = \frac{-4}{15}$ on a number line, select some test numbers, and make a sign diagram. See Figure 5.2.3.

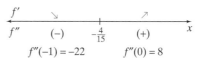

Figure 5.2.3

Notice that f'' and f' are used in the sign diagram. This is exactly what we did in Section 5.1, since we simply have a function and its derivative listed here! From the sign diagram we determine that f' is decreasing on $\left(-\infty, \frac{-4}{15}\right)$ and increasing on $\left(\frac{-4}{15}, \infty\right)$.

Figure 5.2.4 shows a graph of $f'(x) = 15x^2 + 8x - 12$.

Interactive Activity

From the work in Example 2, notice that a relative minimum for f' occurs at $x = \frac{-4}{15}$. Verify that the y-coordinate is $\frac{-196}{15}$. We know that $f'(x)$ tells us the slope of a line tangent to the graph of f for any x, so at $x = \frac{-4}{15}$ on the graph of f we know that the slope of the tangent line is $\frac{-196}{15}$. What does the fact that this is also the relative minimum for f' mean about the graph of f?

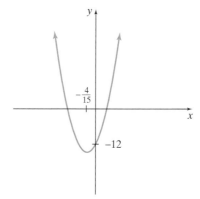

Figure 5.2.4   Graph of $f'(x) = 15x^2 + 8x - 12$.

Concavity

In Chapter 2 we saw how the derivative gives the instantaneous rate of change of a function at any point. Example 2 showed how the second derivative tells us where the first derivative is increasing and where it is decreasing. Since the first derivative gives the instantaneous rate of change, the second derivative tells us the increasing or decreasing behavior of the instantaneous rate of change. Knowing intervals where the instantaneous rate of change, f', is increasing and intervals where it is decreasing then tells us the type of **concavity** that the graph of a function, f, has on the interval. Since concavity is important in our continuing analysis of functions and their graphs, we offer the following definition.

Concavity

On an open interval (a, b) where f is differentiable:

1. If f' is increasing, then the graph of f is **concave up**.
2. If f' is decreasing, then the graph of f is **concave down**.

Notice in Figure 5.2.5, on an interval where the graph of a function is concave up, tangent lines to the curve for any point in the interval are below the

Concave up: Tangent lines lie below the curve.

Concave down: Tangent lines lie above the curve.

Figure 5.2.5

curve. Also, on an interval where the graph of a function is concave down, tangent lines to the curve for any point in the interval are above the curve.

Now let's return to our work in the Flashback and Example 2. Recall that

$$f(x) = 5x^3 + 4x^2 - 12x - 25 \qquad \text{(see Figure 5.2.6)}$$
$$f'(x) = 15x^2 + 8x - 12 \qquad \text{(see Figure 5.2.7)}$$
$$f''(x) = 30x + 8 \qquad \text{(see Figure 5.2.8)}$$

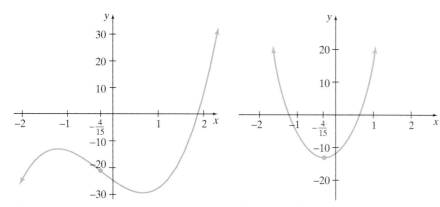

Figure 5.2.6

Figure 5.2.7

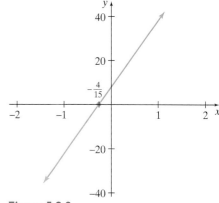

Figure 5.2.8

The Flashback, Example 2, and Figures 5.2.6 through 5.2.8 demonstrate the following:

- On the interval $\left(-\infty, \frac{-4}{15}\right)$, $f'' < 0$ (Figure 5.2.8) and f' is decreasing (Figure 5.2.7 and Example 2).
- On the same interval $\left(-\infty, \frac{-4}{15}\right)$, the graph of f is concave down. See Figure 5.2.6.
- On the interval $\left(\frac{-4}{15}, \infty\right)$, $f'' > 0$ (Figure 5.2.8) and f' is increasing (Figure 5.2.7 and Example 2).
- On the same interval $\left(\frac{-4}{15}, \infty\right)$, the graph of f is concave up. See Figure 5.2.6.

These observations can be condensed into the following Tests for Concavity.

Tests for Concavity

For a function f whose second derivative exists on open interval (a, b):

1. If $f''(x) > 0$ for all x on (a, b), then the graph of f is concave up on (a, b).
2. If $f''(x) < 0$ for all x on (a, b), then the graph of f is concave down on (a, b).

NOTE: In Section 5.1 we saw how the sign of f' determines where f is increasing or decreasing. Here we need to determine the sign of f'' in order to determine the concavity of the graph of f. Hence, the process outlined in Section 5.1 will be used here, as shown in Example 3.

EXAMPLE 3 | **Determining Intervals of Concavity**

Determine intervals where the graph of $f(x) = x^3 + 3x^2 - 4$ is concave up and where the graph is concave down.

SOLUTION

To determine the concavity, we need to determine the sign of the second derivative. We compute the second derivative to be

$$f'(x) = 3x^2 + 6x$$
$$f''(x) = 6x + 6$$

We note that $f''(x)$ is defined for all values of x, so, using the same process as in Section 5.1, we need to determine where $f''(x) = 0$. Solving this equation yields

$$6x + 6 = 0$$
$$x = -1$$

We place $x = -1$ on a number line and make a sign diagram in Figure 5.2.9 using the same steps as in Section 5.1.

Figure 5.2.9

Selecting $x = -2$ as a test number from the interval $(-\infty, -1)$ and substituting this into $f''(x)$ gives $f''(-2) = 6(-2) + 6 = -6$. Selecting $x = 0$ from the interval $(-1, \infty)$ and substituting into $f''(x)$ gives $f''(0) = 6(0) + 6 = 6$. Putting this information on the sign diagram yields Figure 5.2.10.

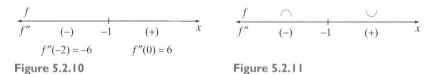

Figure 5.2.10   **Figure 5.2.11**

Knowing that the sign of the second derivative tells the concavity of the graph, we complete the sign diagram as shown in Figure 5.2.11. We use ⌢ to mean concave down and ⌣ to mean concave up.

The graph of $f(x) = x^3 + 3x^2 - 4$ is concave down on $(-\infty, -1)$ and concave up on $(-1, \infty)$. ∎

✓ CHECKPOINT **2**

Now work Exercise 11.

The graph of $f(x) = x^3 + 3x^2 - 4$ is shown in Figure 5.2.12. At the point $(-1, -2)$, the graph switches concavity from concave down to concave up. A

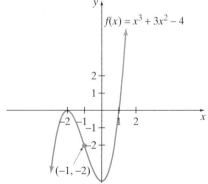

Figure 5.2.12

point on the graph where concavity changes from concave up to concave down (or from concave down to concave up) is called an **inflection point.** It is also worth noting that the x-coordinate of an inflection point is a critical value for f'.

EXAMPLE 4 | **Determining Intervals of Concavity**

Determine intervals where the graph of $f(x) = x^4$ is concave up and where it is concave down. Also, locate any inflection points.

SOLUTION

We can solve problems that ask for the concavity of the graph of a function by using the following procedure.

> *To determine intervals of concavity of the graph of a function f, we determine the second derivative f″ and apply the Tests for Concavity.*

We begin by computing the second derivative to be

$$f'(x) = 4x^3$$
$$f''(x) = 12x^2$$

Since $f''(x)$ is defined for all x, just like in Example 3, we need to solve $f''(x) = 0$. This results in

$$12x^2 = 0$$
$$x = 0$$

We place $x = 0$ on a number line and make a sign diagram as in Figure 5.2.13. Selecting $x = -1$ as a test number from the interval $(-\infty, 0)$ and substituting this into $f''(x)$ gives $f''(-1) = 12(-1)^2 = 12$. Selecting $x = 1$ from the interval $(0, \infty)$ and substituting into $f''(x)$ gives $f''(1) = 12(1)^2 = 12$. Knowing that the sign of the second derivative tells the concavity of the graph, we complete the sign diagram as shown in Figure 5.2.14.

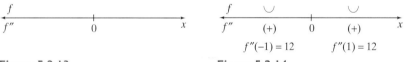

Figure 5.2.13   Figure 5.2.14

We conclude that the graph of $f(x) = x^4$ is concave up on $(-\infty, 0)$ and concave up on $(0, \infty)$. Since there is no change in concavity, the graph of the function $f(x) = x^4$ has no inflection points.

Example 3 showed that $f''(x)$ may equal zero at an inflection point. There is only one other possibility for an inflection point. If a continuous function has an inflection point at $x = d$, then either $f''(d) = 0$ or $f''(d)$ is undefined. We use the following process to locate inflection points.

Locating Inflection Points

1. Determine the values $x = d$ where $f''(d) = 0$ or where $f''(d)$ is undefined.
2. Place these values on a number line and make a sign diagram.
3. The point $(d, f(d))$ is an inflection point if f'' changes sign at $x = d$ **and** if $x = d$ is in the domain of f.

NOTE: Example 4 indicates that not every value of x that satisfies $f''(x) = 0$ produces an inflection point. It is worth noting one more time that the process outlined in Examples 3 and 4 is similar to the process that we used in Section 5.1 to locate relative extrema.

EXAMPLE 5 | **Locating Inflection Points**

Determine any inflection points for the graph of $f(x) = 5x^3 + 4x^2 - 12x - 25$.

SOLUTION

This is the same function from the Flashback and Example 2. In Example 2 we determined that $f''(x) = 30x + 8$ and $f''(x) = 0$ at $x = \frac{-4}{15}$. Placing $x = \frac{-4}{15}$ on a number line and constructing a sign diagram gives Figure 5.2.15. The sign diagram indicates that the graph of f changes concavity at $x = \frac{-4}{15}$. So the point $\left(\frac{-4}{15}, f\left(\frac{-4}{15}\right)\right)$, which is the same as $\left(\frac{-4}{15}, \frac{-14,587}{675}\right)$, is an inflection point. See Figure 5.2.16.

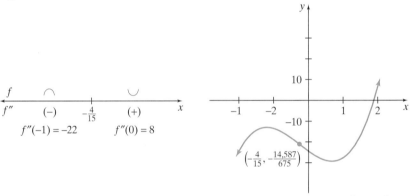

Figure 5.2.15

Figure 5.2.16   Graph of $f(x) = 5x^3 + 4x^2 - 12x - 25$ has an inflection point at $\left(\frac{-4}{15}, \frac{-14,587}{675}\right)$.

TABLE 5.2.2

FIRST AND SECOND DERIVATIVE	SHAPE OF GRAPH OF f	BEHAVIOR OF f
$f' > 0$ implies f increasing $f'' > 0$ implies f concave up	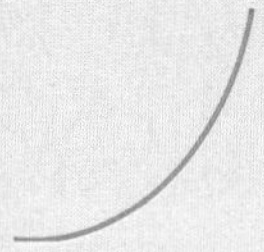	f increasing at a faster rate
$f' > 0$ implies f increasing $f'' < 0$ implies f concave down	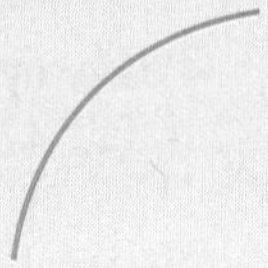	f increasing at a slower rate
$f' < 0$ implies f decreasing $f'' < 0$ implies f concave down	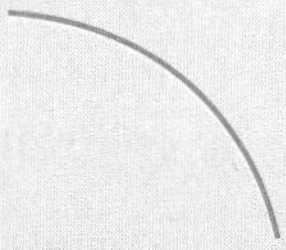	f decreasing at a faster rate
$f' < 0$ implies f decreasing $f'' > 0$ implies f concave up		f decreasing at a slower rate

✓ CHECKPOINT **3**

Now work Exercise 19.

Inflection Points, Rates of Change, and Applications

In this section, we have learned that the second derivative tells us the intervals where the rate of change is increasing and where it is decreasing. In addition, we know that:

- f' gives the intervals where f is increasing and where it is decreasing.
- f'' gives the intervals where the graph of f is concave up and where it is concave down.

We summarize the results of all this information in Table 5.2.2.

EXAMPLE 6 | **Maximizing a Rate of Change**

The marketing research department for the Spritz Cola company analyzed data on the number of units of a certain cola that sold after spending x dollars on advertising. They estimate that the company will sell $B(x)$ units of a diet cola after spending x thousand dollars on advertising, according to

$$B(x) = -\frac{1}{3}x^3 + 60x^2 - 110x + 5200, \qquad 20 \leq x \leq 100.$$

(a) Determine where the rate of change of sales is increasing and where it is decreasing.

(b) What level of spending maximizes the rate of change of sales?

SOLUTION

(a) We must first determine the rate of change of sales, $B'(x)$. Differentiating $B(x)$ gives

$$B'(x) = -x^2 + 120x - 110$$

To determine where B' is increasing and decreasing, we compute $B''(x)$, the derivative of $B'(x)$.

$$B''(x) = -2x + 120$$

Since $B''(x)$ is defined for all x, we solve $B''(x) = 0$ and get

$$-2x + 120 = 0$$
$$x = 60$$

Making a sign diagram and analyzing B' and B'' gives Figure 5.2.17. From this diagram we conclude that the rate of change of sales, B', is increasing on the interval $(20, 60)$ and decreasing on the interval $(60, 100)$.

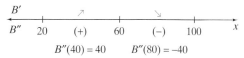

Figure 5.2.17

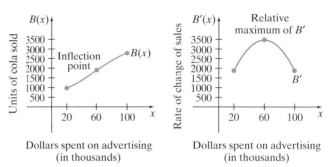

Figure 5.2.18

(b) The sign diagram indicates that a spending level of 60, which is $60,000, for advertising maximizes the rate of change of sales. Figure 5.2.18 shows graphs of B and B' that demonstrate this result. ▆

Example 6 and Figure 5.2.18 nicely summarize the key concepts of this section. That is, where B' is increasing, the graph of B is concave up, and where B' is decreasing, the graph of B is concave down. Also, where the rate of change of sales is maximized, at $x = 60$, the graph of B has an inflection point. In Example 6, the inflection point is also known as the **point of diminishing returns**. At the inflection point in Example 6, concavity changes from concave up to concave down. As we analyzed in Example 6, this is exactly where the rate of change of sales started to decrease.

EXAMPLE 7

Interpreting Concavity and Inflection Point

From 1985 to 1994, the number of births to women in the United States who were 20 to 24 years old can be modeled by

$$b(x) = -1.1x^3 + 17.55x^2 - 86.76x + 1211, \qquad 1 \le x \le 10$$

where x is the number of years since 1984 and $b(x)$ is the number of births in thousands.

(a) On the interval $(1, 10)$, determine where the graph of b is concave up and where it is concave down.

(b) Locate and interpret the inflection point.

SOLUTION

(a) To determine intervals of concavity, we find the second derivative and apply the tests for concavity.

$$b'(x) = -3.3x^2 + 35.1x - 86.76$$
$$b''(x) = -6.6x + 35.1$$

Since $b''(x)$ is defined for all x, we must solve $b''(x) = 0$. This gives

$$-6.6x + 35.1 = 0$$
$$x = 5.32, \quad \text{rounded to the nearest hundredth}$$

We place $x = 5.32$ on a number line and construct a sign diagram as in Figure 5.2.19. So the graph of b is concave up on $(1, 5.32)$ and concave down on $(5.32, 10)$.

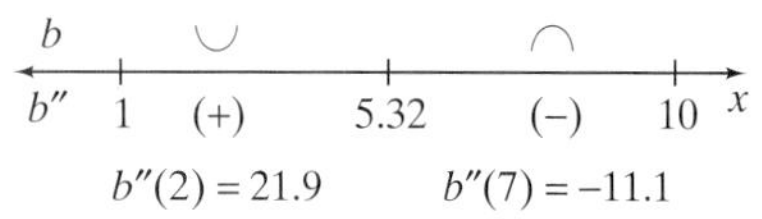

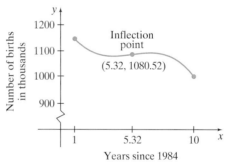

Figure 5.2.19

(b) Since the graph of b changes from concave up to concave down at $x \approx 5.32$, we conclude that there is an inflection point at $(5.32, b(5.32))$ or, specifically, at $(5.32, 1080.52)$. Note that both coordinates have been rounded to the nearest hundredth.

Since we know that the graph of b is concave up on $(1, 5.32)$ from part (a), the rate of change of births is *increasing* on this interval. Also from part (a), we know that the graph of b is concave down on $(5.32, 10)$. This means that the rate of change of births is *decreasing* on this interval. Hence, we interpret the inflection point as indicating that the rate of change of births was maximized at $(5.32, 1080.52)$. See Figure 5.2.20.

Figure 5.2.20

✓ **CHECKPOINT 4**

Now work Exercise 63.

SUMMARY

In this section we saw how the second derivative can be used to determine the increasing and decreasing behavior of the rate of change, or the derivative, of a function. We examined the intervals where the graph of a function is concave up and where it is concave down, and we saw that a point where a graph switches concavity is called an **inflection point.**

To determine the **concavity** of the graph of a function on an interval, apply the Tests for Concavity.

• If $f'' > 0$ on an interval, then the graph of f is concave up on the interval.
• If $f'' < 0$ on an interval, then the graph of f is concave down on the interval.

To locate inflection points:

1. Compute $f''(x)$.
2. Determine where $f''(x)$ is undefined and where $f''(x) = 0$.
3. Place these values on a number line and construct a sign diagram.
4. If concavity changes at any point, then there is an inflection point.

SECTION 5.2 EXERCISES

In Exercises 1–10, compute the first three derivatives for the given function.

1. $f(x) = -4x^5 - 6x^3 + 7x$

2. $f(x) = 5x^4 + 3x^2 - 7x + 1$

✓ 3. $y = 7x^3 - 3x^2 + 4x + 5$

4. $y = -8x^3 - 7x^2 + 5x + 6$

5. $f(x) = e^x$

6. $f(x) = e^{x^2}$

7. $f(x) = \sqrt{x}$

8. $f(x) = \sqrt[3]{x}$

9. $y = \ln x$

10. $y = \ln 2x$

For each function graphed in Exercises 11–16:

(a) Determine intervals where the graph of function is concave up and where it is concave down.

(b) Determine intervals where the derivative of the function is increasing and where the derivative is decreasing.

✓ 11.

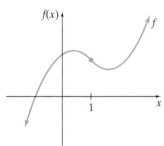

12.

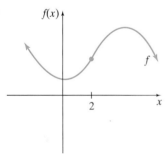

13.

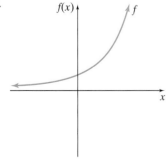

14.

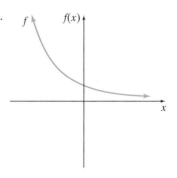

15.

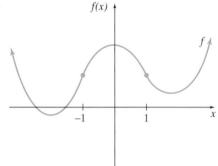

16.

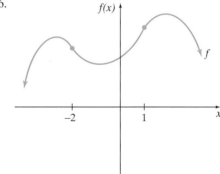

For each function in Exercises 17–34, determine intervals where the graph of the function is concave up and where it is concave down, and locate any inflection points.

17. $y = x^3 + 6x^2 + 18x - 5$

18. $f(x) = x^3 - 6x^2 + 6x - 3$

✓ 19. $g(x) = -12x^3 + 6x^2 - 24x - 11$

20. $y = -3x^3 + 5x^2 - 2x + 4$

21. $f(x) = 2x^2 + 3x + 1$

22. $f(x) = 3x^2 - 2x - 5$

23. $f(x) = -3x^2 + 3x - 2$

24. $f(x) = -2x^2 - 5x + 3$

25. $f(x) = x + \dfrac{2}{x}$

26. $f(x) = x + \dfrac{5}{x}$

27. $f(x) = x^{2/3}$

28. $f(x) = (x-1)^{2/3}$

29. $y = e^{2x}$

30. $y = e^{1.2x}$

31. $y = e^{-2x}$

32. $y = e^{-1.5x}$

33. $g(x) = \ln(x+1)$

34. $g(x) = \ln(x-3)$

For each function in Exercises 35–48, determine the following:

(a) Intervals where the function is increasing and where it is decreasing.

(b) The relative extrema.

(c) Intervals where the graph of the function is concave up and where it is concave down.

(d) Any inflection points.

35. $f(x) = 2x^2 + 3x + 1$

36. $f(x) = 3x^2 - 2x - 5$

37. $y = x^3 + 6x^2 - 15x - 5$

38. $f(x) = x^3 - 6x^2 - 15x + 3$

39. $y = -3x^3 + 5x^2 + 2x - 9$

40. $y = -2x^3 + 3x^2 + 6x - 5$

41. $f(x) = x + \dfrac{5}{x}$

42. $f(x) = x + \dfrac{2}{x}$

43. $f(x) = (x-1)^{2/3}$

44. $f(x) = (x+2)^{2/3}$

45. $y = e^{3x}$

46. $y = e^{-3x}$

47. $g(x) = \ln(x+1)$

48. $g(x) = \ln(x-3)$

49. Sketch the graph of a function that satisfies the following conditions:

Domain: All real numbers
Range: All real numbers less than or equal to 5
Continuous for all real numbers
$f' > 0$ on $(-\infty, 2)$
$f' < 0$ on $(2, \infty)$
$f'' < 0$ on $(-\infty, \infty)$

50. Sketch the graph of a function that satisfies the following conditions:

Domain: All real numbers
Range: All real numbers greater than or equal to 5
Continuous for all real numbers
$f' < 0$ on $(-\infty, -3)$
$f' > 0$ on $(-3, \infty)$
$f'' > 0$ on $(-\infty, \infty)$

51. Sketch the graph of a function that satisfies the following conditions:

Domain: All real numbers
Range: All real numbers
Continuous for all real numbers
$f' > 0$ on $(-\infty, -1) \cup (3, \infty)$
$f' < 0$ on $(-1, 3)$
$f'' < 0$ on $(-\infty, 1)$
$f'' > 0$ on $(1, \infty)$

52. Sketch the graph of a function that satisfies the following conditions:

Domain: All real numbers
Range: All real numbers
Continuous for all real numbers
$f' > 0$ on $(-\infty, 0) \cup (0, \infty)$
$f'(0)$ is undefined
$f'' > 0$ on $(-\infty, 0)$
$f'' < 0$ on $(0, \infty)$

53. In Chapter 1 we saw that the graph of the quadratic function $f(x) = ax^2 + bx + c$, where a, b, and c are real numbers and $a \neq 0$, is concave up if $a > 0$ and concave down if $a < 0$. Show that this is true.

54. Show that the x-coordinate of the inflection point on the graph of any cubic polynomial of the form $f(x) = ax^3 + bx^2 + cx + d$, where $a, b, c,$ and d are real numbers and $a \neq 0$, is given by $x = \dfrac{-b}{3a}$.

Applications

55. The Beagle Works Company has determined that its cost, in hundreds of dollars, for producing x items of its best selling product is given by

$$C(x) = x^3 - 6x^2 + 13x + 10$$

(a) Determine $C(5)$ and $C'(5)$ and interpret each.

(b) Determine intervals where *marginal cost* is increasing and where it is decreasing. Determine the relative minimum for the marginal cost function.

(c) Determine the inflection point for the graph of C.

56. The Chug-a-Mug Company has determined that its cost, in hundreds of dollars, for producing x items of its best selling product is given by

$$C(x) = x^3 - 6x^2 + 15x$$

(a) Determine $C(5)$ and $C'(5)$ and interpret each.

(b) Determine intervals where *marginal cost* is increasing and where it is decreasing. Determine the relative minimum for the marginal cost function.

(c) Determine the inflection point for the graph of C.

57. The Cool Air refrigerator company has determined that its monthly profit, in dollars, for producing and selling x items of its best-selling refrigerator, measured in hundreds, is given by

$$P(x) = -2.3x^3 + 445x^2 - 1500x - 200$$

(a) Determine $P(35)$ and $P'(35)$ and interpret each.

(b) Determine intervals where the *marginal profit* is increasing and where it is decreasing. Determine the relative extremum for marginal profit.

58. (Continuation of Exercise 57.)

(a) Determine the inflection point for the graph of P.

(b) Explain why the relative extremum for the marginal profit and the inflection point for the graph of P have the same x-value.

59. For the profit function, P, given in Exercise 57, explain why the intervals where the *marginal profit* is increasing and where it is decreasing and the intervals where the graph of P is concave up and where it is concave down are the same.

60. The Sucre Cola Company estimates that total sales of its new cola, $TS(x)$, when spending x million dollars on advertising, can be modeled by

$$TS(x) = -\frac{5}{2}x^3 + 112.5x^2 + 150x + 10{,}000, \quad 10 \le x \le 30$$

Locate the point of diminishing returns for $TS(x)$ and interpret its meaning. (*Hint:* Recall the discussion following Example 6, that the *point of diminishing returns* is an inflection point.)

61. The Big Cola Company estimates that total sales of its new diet cola, $TS(x)$, when spending x million dollars on advertising, can be modeled by

$$TS(x) = -2x^3 + 90x^2 - 1200x + 10{,}000, \quad 10 \le x \le 25$$

Locate the point of diminishing returns for $TS(x)$ and interpret its meaning. (*Hint:* Recall the discussion following Example 6, that the *point of diminishing returns* is an inflection point.)

62. Fill in the blanks.

For $B(x)$ on the interval $[20, 100]$ in Example 6, the rate of change of sales, $B'(x)$, is increasing. This rate of change increases at a _____ (faster/slower) rate on the interval $(20, 60)$, whereas this rate of change increases at a _____ (faster/slower) rate on the interval $(60, 100)$.

✓ ⊕ 63. From 1985 to 1994, the number of births, in thousands, to women 35 to 39 years old in the United States can be modeled by

$$b(x) = 200x^{1/4}, \quad 1 \le x \le 10$$

where $b(x)$ is the number of births in thousands and x represents the number of years since 1984.

(a) Compute $b(5)$ and $b'(5)$ and interpret each.

(b) On the interval $[1, 10]$, show that b is increasing.

(c) Determine $b''(x)$ and show that the graph of b is concave down on the interval $[1, 10]$.

(d) Complete the following sentence:

From 1985 to 1994, the number of births to women 35 to 39 years old was _____ (increasing/decreasing), and it was _____ (increasing/decreasing) at a _____ (slower/faster) rate.

64. As arterial pressure increases, blood flow through the artery also increases. This can be modeled by

$$f(x) = 0.267e^{0.0256x}, \quad 20 \le x \le 120$$

where x represents the arterial pressure, measured in millimeters of mercury (mm Hg), and $f(x)$ is blood flow measured in milliliters per minute (mL/min).

(a) Evaluate $f(50)$ and $f'(50)$ and interpret each.

(b) Show that the graph of f is concave up on the interval $[20, 120]$. Explain what this means.

65. As arterial pressure increases, blood flow through the artery also increases. This can also be modeled by

$$f(x) = 0.278(1.026)^x, \quad 20 \le x \le 120$$

where x represents the arterial pressure, measured in mm Hg, and $f(x)$ is blood flow measured in mL/min.

(a) Evaluate $f(50)$ and $f'(50)$ and interpret each.

(b) Show that the graph of f is concave up on the interval $[20, 120]$. Explain what this means.

66. For males, the blood volume in milliliters (mL) can be approximated by

$$f(x) = -10{,}822 + 3800 \ln x, \quad 40 \le x \le 90$$

where x is weight in kilograms (kg).

(a) Compute $f(60)$ and $f'(60)$ and interpret each.

(b) Show that on the interval $[40, 90]$ the graph of f is concave down.

67. The *cardiac index* is the cardiac output per square meter of body surface area, and its units of measure are $\dfrac{\text{liters per minute}}{\text{square meters}}$. The cardiac index can be modeled by

$$CI(x) = \frac{7.644}{\sqrt[4]{x}}, \quad 10 \le x \le 80$$

where x is an individual's age in years.

(a) Compute $CI(20)$ and $CI'(20)$ and interpret each.

(b) Determine the intervals where CI is increasing and where it is decreasing.

(c) Show that the graph of CI is concave up on $10 \le x \le 80$ and interpret what this means.

68. A certain product has a demand curve modeled by

$$q(t) = \frac{5000}{1 + 2e^{-0.2t}}$$

where $q(t)$ represents the number sold t weeks after the product was introduced to the market.

(a) Compute $q(26)$ and $q'(26)$ and interpret each.

(b) Determine $q''(t)$.

(c) On the interval $[0, 52]$, determine where the graph of q is concave up and where it is concave down. (*Hint:* Graph $q''(t)$ and use the ZERO/ROOT command.)

(d) Assuming that advertising increases sales, based on the result from part (c), when would be a good time to begin an advertising campaign for this product? Explain.

69. The U.S. federal debt, in billions of dollars, from 1980 to 1992 can be modeled by

$$f(x) = 10.28x^2 + 113.98x + 749.6, \qquad 1 \leq x \leq 13$$

(a) Use $f'(x)$ to show that on [1, 13] the federal debt is increasing.

(b) Use $f''(x)$ to show that the rate of change of the U.S. federal debt was increasing from 1980 to 1992.

SECTION PROJECT

The following data give the number of births, in thousands, to women in the United States under 20 years old from 1985 to 1994.

By letting x represent the number of years since 1984 and $b(x)$ represent the number of births in thousands to women under 20 years of age, we can model the data with the following:

$$b(x) = 0.24x^4 - 5.66x^3 + 43.38x^2 - 113.32x + 557.17,$$
$$1 \leq x \leq 10$$

(a) Compute $b'(4)$ and interpret.

(b) Use $b'(x)$ to determine intervals where b is increasing and where it is decreasing.

(c) Determine intervals where the rate of change of b is increasing and where it is decreasing.

(d) Determine intervals where the graph of b is concave up and where it is concave down.

(e) Use the results from part (b) to determine any relative extrema for b, and interpret each coordinate.

(f) Use the result from part (d) to determine any inflection points for the graph of b, and interpret each coordinate.

YEAR	1985	1986	1987	1988	1989	1990	1991	1992	1993	1994
BIRTHS	478	472	473	489	518	533	532	518	501	518

Source: U.S. Census Bureau web site.

SECTION 5.3   GRAPHICAL ANALYSIS AND CURVE SKETCHING

In Section 5.1 we learned how the rate function f' can be used to determine the behavior of a continuous function f. To find relative maxima and minima, determine the critical values of f and use the First Derivative Test.

- The point $(c, f(c))$ is a relative maximum if $f' > 0$ to the left of c and $f' < 0$ to the right of c.
- The point $(c, f(c))$ is a relative minimum if $f' < 0$ to the left of c and $f' > 0$ to the right of c.

In Section 5.2 we learned how the second derivative f'' gives important information about the behavior of the function f. To find intervals of concavity for a function f, determine the second derivative f'' and apply the Tests for Concavity.

- If $f'' > 0$ on an open interval, then the graph of f is concave up.
- If $f'' < 0$ on an open interval, then the graph of f is concave down.

In this section, we pull together all these ideas to analyze the behavior of functions in greater depth. We will also analyze how the **Second Derivative Test** may be used to determine relative extrema.

The Second Derivative Test

In Section 5.1 we saw how the First Derivative Test was used to locate relative extrema. At this time, we explore another method for locating relative extrema by what is called the **Second Derivative Test**.

EXAMPLE 1 | **Determining Relative Extrema**

Determine the critical values of the function shown in Figure 5.3.1. Classify each as giving a relative maximum or a relative minimum, and determine the concavity of the graph of f at each relative extrema.

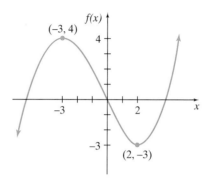

Figure 5.3.1

SOLUTION

Since the graph of f has no breaks or sharp turns, $f'(x)$ is defined for all x. It appears that $f'(x) = 0$ at $x = -3$ and $x = 2$. So the critical values are $x = -3$ and $x = 2$. We have a relative maximum at $x = -3$, specifically at $(-3, 4)$, and a relative minimum at $x = 2$, specifically at $(2, -3)$. From the graph, it appears that at the relative maximum $(-3, 4)$ the graph is *concave down,* whereas at the relative minimum $(2, -3)$ the graph is *concave up.* ■

In Example 1, since the graph is concave down at the relative maximum $(-3, 4)$, we know that $f''(-3) < 0$. Since the graph is concave up at the relative minimum $(2, -3)$, we know that $f''(2) > 0$. This observation is formally given as the **Second Derivative Test.**

The Second Derivative Test

For a function whose second derivative exists on an open interval containing c and has a critical value $x = c$, where $f'(c) = 0$, the point $(c, f(c))$ is a

1. **Relative minimum** if $f''(c) > 0$ 2. **Relative maximum** if $f''(c) < 0$

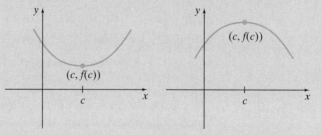

The Second Derivative Test fails if $f''(c) = 0$ or if $f''(c)$ is undefined. In this case, the First Derivative Test may be used.

EXAMPLE 2 | **Using the Second Derivative Test to Locate Extrema**

Use the Second Derivative Test to locate the relative extrema for
$$f(x) = x^3 + 3x^2 - 9x + 5$$

SOLUTION

To apply the Second Derivative Test, we must first determine the critical values for f, which means that we need to determine $f'(x)$.

$$f'(x) = 3x^2 + 6x - 9$$

Since $f'(x)$ is defined for all x, the only critical values are where $f'(x) = 0$. Solving $f'(x) = 0$ yields

$$3x^2 + 6x - 9 = 0$$
$$3(x^2 + 2x - 3) = 0$$
$$3(x + 3)(x - 1) = 0$$
$$x = -3 \quad \text{or} \quad x = 1$$

So the critical values are $x = -3$ and $x = 1$. We now compute $f''(x)$ to be

$$f''(x) = 6x + 6$$

We substitute each critical value into $f''(x)$ to obtain

$$f''(-3) = 6(-3) + 6 = -12$$
$$f''(1) = 6(1) + 6 = 12$$

Since $f''(-3) < 0$, there is a relative maximum at $x = -3$. Since $f''(1) > 0$, there is a relative minimum at $x = 1$. Figure 5.3.2 shows a graph of f.

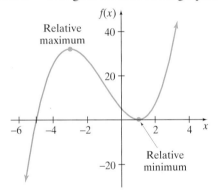

Figure 5.3.2

Interactive Activity

Use the First Derivative Test to verify the results of Example 2.

✓**CHECKPOINT 1**

Now work Exercise 19.

Synthesis

At this time, we are ready to pull together all the concepts studied so far in Chapter 5.

EXAMPLE 3 | **Analyzing a Function and Its Graph**

Consider $f(x) = 5x^4 - x^5$.

(a) Locate all relative extrema.

(b) Determine intervals where the graph of f is concave up and where it is concave down.

(c) Locate any inflection points.

(d) Using the information gathered in parts (a)–(c), sketch a graph of f and label the points found in part (a) and part (c).

SOLUTION

(a) To locate relative extrema, we determine the critical values of f and apply the First Derivative Test. The derivative is

$$f'(x) = 20x^3 - 5x^4$$

Since $f'(x)$ is defined for all x, we need to solve $f'(x) = 0$ to determine the critical values.

$$20x^3 - 5x^4 = 0$$
$$5x^3(4 - x) = 0$$
$$x = 0 \quad \text{or} \quad x = 4$$

We place the critical values on a number line and make a sign diagram using the test values $x = -1$, $x = 1$, and $x = 5$. Recall that we evaluate the derivative at these test values and place the sign of the result in our sign diagram as in Figure 5.3.3. Since $f(x) = 5x^4 - x^5$ is decreasing on $(-\infty, 0)$ and increasing on $(0, 4)$, there is a relative minimum at $x = 0$. Since $f(x) = 5x^4 - x^5$ is increasing on $(0, 4)$ and decreasing on $(4, \infty)$, there is a relative maximum at $x = 4$. Specifically, there is a relative minimum at $(0, 0)$ and a relative maximum at $(4, 256)$.

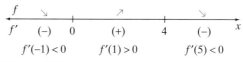

Figure 5.3.3

(b) To find intervals of concavity, we determine the second derivative and apply the Tests for Concavity. The second derivative is

$$f''(x) = 60x^2 - 20x^3$$

Since $f''(x)$ is defined for all x, we need to solve $f''(x) = 0$.

$$60x^2 - 20x^3 = 0$$
$$20x^2(3 - x) = 0$$
$$x = 0 \quad \text{or} \quad x = 3$$

We place $x = 0$ and $x = 3$ on a number line and make a sign diagram (Figure 5.3.4) using the test values $x = -1$, $x = 1$, and $x = 5$. Recall that we evaluate the second derivative at these test values and place the sign of the result in our sign diagram.

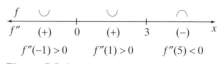

Figure 5.3.4

So the graph of $f(x) = 5x^4 - x^5$ is concave up on $(-\infty, 0) \cup (0, 3)$ and concave down on $(3, \infty)$.

(c) Recall that at an inflection point the sign of f'' changes. Since the sign of f'' changes at $x = 3$, the only inflection point occurs when $x = 3$. Specifically, it is at the point $(3, 162)$.

(d) Figure 5.3.5 is a sketch of the function incorporating all the information gathered in parts (a)–(c).

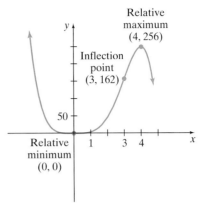

Figure 5.3.5

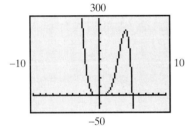

Figure 5.3.6

The final step in any problem dealing with the graphical analysis of a function, such as we did in Example 3, should be to use the calculator to confirm the sketch. In doing so, notice how the calculus is helpful in setting an appropriate viewing window. For example, in Example 3, part (a), we determined that the function has a relative maximum at $(4, 256)$ and a relative minimum at $(0, 0)$. This tells us that a viewing window of $[-10, 10]$ by $[-50, 300]$ should show the important features of the graph. See Figure 5.3.6.

✓ **CHECKPOINT 2**

Now work Exercise 33.

Before proceeding to Example 4, let's summarize the steps employed in Example 3.

Graphical Analysis of Function f

1. Use $f'(x)$ to:
   - Determine critical values of f.
   - Construct a sign diagram to determine intervals where f is increasing and where it is decreasing.
   - Locate relative extrema by the First Derivative Test.
2. Use $f''(x)$ to:
   - Construct a sign diagram to determine intervals where the graph of f is concave up and where it is concave down.
   - Locate inflection points.

EXAMPLE 4 | **Analyzing a Function and Its Graph**

Consider $f(x) = \dfrac{x - 1}{2x - 3}$.

(a) Determine any relative extrema.

(b) Determine intervals where the graph of f is concave up and where it is concave down. Also, locate any inflection points.

(c) Identify any vertical and horizontal asymptotes.

(d) Sketch a graph of f.

SOLUTION

(a) To locate relative extrema, we determine the critical values of f and apply the First Derivative Test. Differentiating this function utilizing the **Quotient Rule** gives

$$f'(x) = \frac{1 \cdot (2x-3) - (x-1) \cdot 2}{(2x-3)^2} = \frac{2x-3-2x+2}{(2x-3)^2} = \frac{-1}{(2x-3)^2}$$

Notice that $f'(x)$ is undefined at $x = \frac{3}{2}$ since this value makes the denominator 0. Even though $x = \frac{3}{2}$ makes the derivative undefined, it is not a critical value, because $x = \frac{3}{2}$ is not in the domain of f. Also notice that there are no values of x such that $f'(x) = 0$, that is,

$$\frac{-1}{(2x-3)^2} = 0$$

has no real solution. Thus, f has no critical values. However, we include $x = \frac{3}{2}$ on a number line when constructing our sign diagram so that we can exploit the power of Theorem 5.1. See Figure 5.3.7.

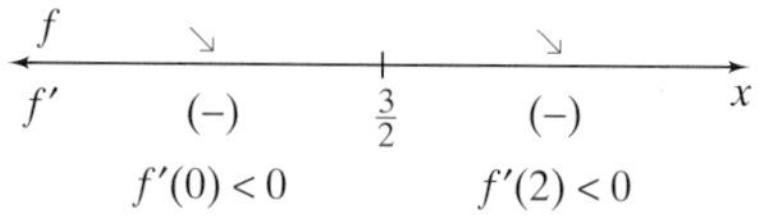

Figure 5.3.7

So $f(x) = \frac{x-1}{2x-3}$ is decreasing on $\left(-\infty, \frac{3}{2}\right) \cup \left(\frac{3}{2}, \infty\right)$. Utilizing the First Derivative Test using a sign diagram indicates that f has no relative extrema.

(b) To find intervals of concavity, we determine the second derivative and apply the Tests for Concavity. We have from part (a) that the first derivative is $f'(x) = \frac{-1}{(2x-3)^2}$. To compute $f''(x)$, we will rewrite $f'(x)$ as

$$f'(x) = -1(2x-3)^{-2}$$

and employ the Power Rule and Chain Rule. This gives

$$f''(x) = -1[-2(2x-3)^{-3} \cdot 2]$$

$$= 4(2x-3)^{-3} = \frac{4}{(2x-3)^3}$$

As in part (a), $f''(x)$ is undefined at $x = \frac{3}{2}$, and $f''(x) = 0$ has no real solution. Thus, the only value that we place on the number line is $x = \frac{3}{2}$. See Figure 5.3.8.

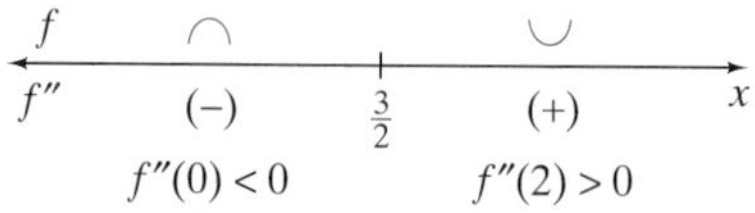

Figure 5.3.8

So the graph of $f(x) = \frac{x-1}{2x-3}$ is concave down on $\left(-\infty, \frac{3}{2}\right)$ and concave up on $\left(\frac{3}{2}, \infty\right)$. Even though the graph changes concavity at $x = \frac{3}{2}$, as stated earlier, this number is not in the domain of f. Hence, there are no inflection points for the graph of f.

(c) Utilizing information from Section 2.2, since $\lim\limits_{x \to (3/2)^-} \dfrac{x-1}{2x-3} = -\infty$ and $\lim\limits_{x \to (3/2)^+} \dfrac{x-1}{2x-3} = \infty$, we have a vertical asymptote at $x = \dfrac{3}{2}$. Also from Section 2.2, we determine the horizontal asymptote by computing

$$\lim_{x \to \infty} \frac{x-1}{2x-3} = \lim_{x \to \infty} \frac{\dfrac{x}{x} - \dfrac{1}{x}}{\dfrac{2x}{x} - \dfrac{3}{x}} = \lim_{x \to \infty} \frac{1 - \dfrac{1}{x}}{2 - \dfrac{3}{x}} = \frac{1-0}{2-0} = \frac{1}{2}$$

Similarly,

$$\lim_{x \to -\infty} \frac{x-1}{2x-3} = \frac{1}{2}$$

So the horizontal asymptote is $y = \dfrac{1}{2}$.

(d) Figure 5.3.9 shows a graph of f using the information gathered in parts (a)–(c). We recommend that you graph the function on your calculator to check this result.

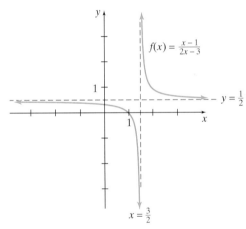

Figure 5.3.9

✓ **CHECKPOINT 3**   Now work Exercise 39.

Applications

Example 5 illustrates the power of using the Second Derivative Test in locating relative extrema in some applications.

EXAMPLE 5 | **Analyzing Executive Branch Employment**

The number of civilians employed by the executive branch of the U.S. government can be modeled by

$$f(x) = 0.09x^4 - 3.78x^3 + 50.06x^2 - 212.32x + 3101.16, \qquad 1 \le x \le 17$$

where x represents the number of years since 1979 and $f(x)$ represents the number of civilians employed in thousands. Use the Second Derivative Test to locate any relative extrema and interpret each coordinate.

SOLUTION

The derivative is

$$f'(x) = 0.36x^3 - 11.34x^2 + 100.12x - 212.32$$

Since the derivative is defined for all x, the only critical value is when $f'(x) = 0$. Hence, we must solve

$$0.36x^3 - 11.34x^2 + 100.12x - 212.32 = 0$$

To solve this equation, we graph $y = 0.36x^3 - 11.34x^2 + 100.12x - 212.32$ and utilize the ZERO command. Figure 5.3.10 shows the result of utilizing the ZERO command.

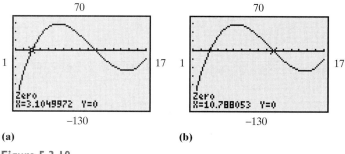

(a)   **(b)**

Figure 5.3.10

Figure 5.3.10a shows that f' has a zero at $x = 3.10$, rounded to the nearest hundredth. Figure 5.3.10b shows that f' has another zero at $x = 10.79$, rounded to the nearest hundredth. We now need to evaluate the second derivative at these values. The second derivative is computed to be

$$f''(x) = 1.08x^2 - 22.68x + 100.12$$

Evaluating $f''(x)$ at $x = 3.10$ and $x = 10.79$ gives

$$f''(3.10) = 1.08(3.10)^2 - 22.68(3.10) + 100.12 \approx 40.19$$
$$f''(10.79) = 1.08(10.79)^2 - 22.68(10.79) + 100.12 \approx -18.86$$

Since $f''(3.10) \approx 40.19$, we have a relative minimum at $x = 3.10$, specifically at the point (3.10, 2819.75). Since $f''(10.79) \approx -18.86$, we have a relative maximum at $x = 10.79$, specifically at the point (10.79, 3109.83). (The y-coordinates have been rounded to the nearest hundredth.) We interpret these relative extrema to mean that early in 1982 the number of civilians employed by the executive branch of the U.S. government hit a relative minimum of 2,819,750 people. On the other hand, about three-fourths of the way through 1989 the number of civilians employed by the executive branch of the U.S. government reached a relative maximum of 3,109,830 people. ◀

Interactive Activity

 Give a possible explanation for why the relative extrema in Example 5 occurred in the years that they did.

EXAMPLE 6 | **Analyzing Executive Branch Employment**

In Example 5 we saw that the number of civilians employed by the executive branch of the U.S. government can be modeled by

$$f(x) = 0.09x^4 - 3.78x^3 + 50.06x^2 - 212.32x + 3101.16, \qquad 1 \le x \le 17$$

where x represents the number of years since 1979 and $f(x)$ represents the number of civilians employed in thousands. Locate any inflection points and interpret each coordinate.

SOLUTION

From Example 5 we determined that

$$f''(x) = 1.08x^2 - 22.68x + 100.12$$

The second derivative is defined for all x, so we must solve $f''(x) = 0$. This gives

$$1.08x^2 - 22.68x + 100.12 = 0$$

We opt to solve this equation by the quadratic formula. This produces

$$x = \frac{-b \pm \sqrt{b^2 - 4ac}}{2a} = \frac{22.68 \pm \sqrt{(-22.68)^2 - 4(1.08)(100.12)}}{2(1.08)}$$

$$= \frac{22.68 \pm \sqrt{81.864}}{2.16}$$

So we have, rounded to the nearest hundredth, $x = 14.69$ or $x = 6.31$. Placing these values on a number line, we construct the sign diagram in Figure 5.3.11. We use $x = 3$, $x = 7$, and $x = 15$ as our test values. Evaluating the second derivative at these test values gives the diagram shown.

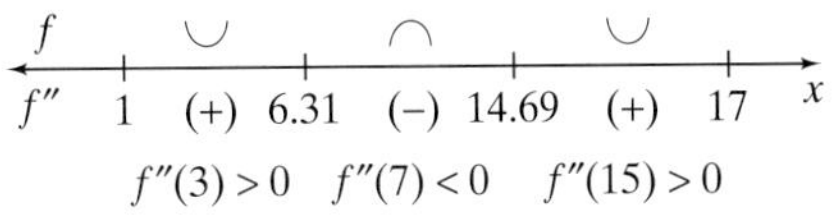

$$
\begin{array}{c}
f \qquad \cup \qquad \cap \qquad \cup \\
\hline
f'' \quad 1 \quad (+) \quad 6.31 \quad (-) \quad 14.69 \quad (+) \quad 17 \qquad x \\
f''(3) > 0 \quad f''(7) < 0 \quad f''(15) > 0
\end{array}
$$

Figure 5.3.11

From Figure 5.3.11 we see that the graph of f changes concavity at $x = 6.31$ and at $x = 14.69$. Thus, we have inflection points at $(6.31, 2947.61)$ and $(14.69, 2993.28)$. (The y-coordinates have been rounded to the nearest hundredth.) The interpretation of these points, which includes the interpretation from Example 5, is as follows:

> The rate at which the number of civilians employed by the executive branch was increasing the greatest occurred about a third of the way into 1985. The rate at which the number of civilians employed by the executive branch was decreasing the greatest occurred about two-thirds of the way into 1993.

✓ CHECKPOINT 4

Now work Exercise 67.

SUMMARY

Section 5.3 has brought together the many concepts studied in Sections 5.1 and 5.2. For a function f:

$f'(x)$	$f''(x)$

$f'(x)$
- Gives the slope of a tangent line at any point
- Gives an instantaneous rate of change
- Determines critical values
- Determines intervals of increase or decrease for f
- Determines relative extrema

$f''(x)$
- Gives the increasing or decreasing behavior of the rate of change
- Determines concavity for the graph of f
- Determines inflection points of the graph of f

In this section, we also examined the **Second Derivative Test** as a way to classify critical values as a **relative maximum** or a **relative minimum.**

SECTION 5.3 EXERCISES

Use the graph of f shown in Figure 5.3.12 to answer Exercises 1–6.

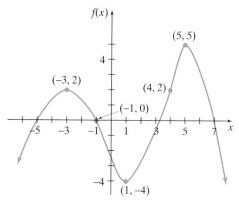

Figure 5.3.12

1. Determine intervals where f is positive and where f is negative.

2. Determine intervals where f is increasing and where f is decreasing.

3. Determine intervals where f' is positive and where f' is negative.

4. Determine intervals where f' is increasing and where f' is decreasing.

5. Determine intervals where the graph of f is concave up and where the graph of f is concave down.

6. Locate any relative extrema, and locate any inflection points.

Use the graph of f shown in Figure 5.3.13 to answer Exercises 7–12.

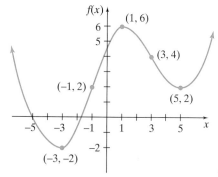

Figure 5.3.13

7. Determine intervals where f is positive and where f is negative.

8. Determine intervals where f is increasing and where f is decreasing.

9. Determine intervals where f' is positive and where f' is negative.

10. Determine intervals where f' is increasing and where f' is decreasing.

11. Determine intervals where the graph of f is concave up and where the graph of f is concave down.

12. Locate any relative extrema, and locate any inflection points.

In Exercises 13–28, use the Second Derivative Test to locate any relative extrema, if they exist.

13. $f(x) = 3x^2 - 2x - 3$

14. $f(x) = -2x^2 + 3x + 2$

15. $y = x^3 - 2x^2 - 13x - 10$

16. $y = x^3 + 3x^2 - x - 3$

17. $y = \frac{1}{3}x^3 + \frac{5}{2}x^2 + 6x - 2$

18. $y = \frac{1}{3}x^3 - \frac{1}{2}x^2 - 6x + 2$

✓ 19. $f(x) = x^3 + \frac{3}{2}x^2 - 6x - 3$

20. $f(x) = -x^3 + 3x^2 - 3x + 5$

21. $y = x^4 + x^3 - 7x^2 - x + 6$

22. $y = x^4 - 3x^3 - 8x^2 + 12x + 16$

23. $g(x) = 3x^4 - 24x^2 + 16$

24. $g(x) = 2x^4 - 36x^2 - 23$

25. $f(x) = \dfrac{1}{x^2 + 1}$ 26. $f(x) = \dfrac{1}{x^2 - 1}$

27. $y = 3x^6 + 9x^4 - 5$ 28. $y = -\dfrac{1}{3}x^6 - 2x^4 + 3$

In Exercises 29–48, sketch the graph of the given function using the techniques in this section. Label all relative extrema, inflection points, and asymptotes.

29. $y = x^3 + 3x^2 - x - 3$ 30. $y = x^3 - 2x^2 - 13x - 10$

31. $y = \frac{1}{3}x^3 - \frac{1}{2}x^2 - 6x + 2$

32. $f(x) = -x^3 + 3x^2 - 3x + 5$

✓ 33. $f(x) = x^3 - 6x^2 + 5$

34. $f(x) = \frac{1}{3}x^3 - 3x^2 + 5x - 2$

35. $y = x^4 - 3x^3 - 8x^2 + 12x + 16$

36. $y = x^4 + x^3 - 7x^2 - x + 6$

37. $f(x) = x^4 - 2x^2 + 1$

38. $f(x) = x^4 - 9x^2 + 2$

✓ 39. $y = \dfrac{x-1}{x+2}$ 40. $y = \dfrac{3x+1}{5x-2}$

41. $f(x) = \dfrac{1}{x^2 + 1}$ 42. $f(x) = \dfrac{x}{x^2 - 1}$

43. $y = 0.2x + 40 + \dfrac{20}{x}$ 44. $y = 0.3x + 20 + \dfrac{10}{x}$

45. $f(x) = x - \ln x$ 46. $y = e^x - x$

47. $f(x) = \sqrt[3]{x^2}$ 48. $f(x) = \sqrt[3]{(x-1)^2}$

In Exercises 49–52, the graph of f'' is given. Use the graph to sketch a graph of (a) f' and (b) f. There are many correct answers.

49.

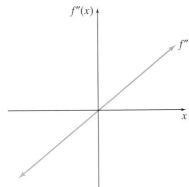

50.

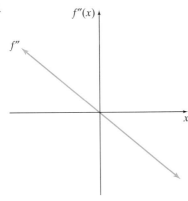

51.

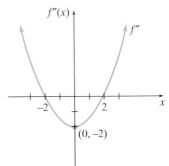

52.

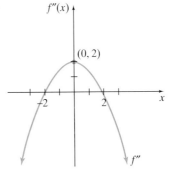

APPLICATIONS

53. The marketing research department for Music Time, a manufacturer of integrated amplifiers, used a large metropolitan area to test market their new product. They determined that price, $p(x)$, in dollars per unit, and quantity demanded per week, x, was approximated by

$$p(x) = 216 - 0.08x^2, \qquad 0 \le x \le 50$$

So the weekly revenue can be approximated by

$$R(x) = p(x) \cdot x = 216x - 0.08x^3, \qquad 0 \le x \le 50$$

(a) Compute $R'(x)$ and use it to determine intervals where R is increasing and where it is decreasing.

(b) Determine intervals where the graph of R is concave up and where it is concave down.

54. Suppose that Music Time, the manufacturer in Exercise 53, has a weekly cost function, in dollars given by $C(x) = 20x + 1000$.

(a) Determine P, the weekly profit function.

(b) Using the techniques of this section, graph P. Label all extrema and inflection points, if they exist.

55. The Cool Air refrigerator company has determined that its monthly profit, in dollars, for producing and selling x items of its best-selling refrigerator is given by

$$P(x) = -0.023x^3 + 4.45x^2 - 15x - 2, \qquad 0 \le x \le 130$$

(a) The *average profit function*, denoted AP, is defined to be $AP(x) = \dfrac{P(x)}{x}$. Determine $AP(x)$.

(b) Using the techniques of this section, sketch a graph of AP on the interval $(0, 130]$. Label all extrema and inflection points.

56. (Continuation of Exercise 55.)

(a) Determine P', the *marginal profit function*.

(b) In the viewing window $[0, 130]$ by $[0, 300]$, graph both P' and AP.

(c) Using either algebraic techniques or the INTERSECT command on your calculator, locate the point where the two curves cross.

(d) Compare the x-coordinate of the result from part (c) with the x-coordinate of the relative maximum for AP. Are they the same?

57. (Continuation of Exercises 55 and 56.) Fill in the blanks illustrating an important fact from economics:

The maximum average profit occurs when _____ _____ is equal to _____.

58. The Digital Pet Company has determined that its daily cost, in dollars, for producing x virtual pets is given by

$$C(x) = 150 + 3x + \frac{2x^2}{30}, \qquad 0 \le x \le 200$$

(a) Determine the *average cost function*, $AC(x) = \dfrac{C(x)}{x}$.

(b) Using the techniques of this section, sketch a graph of AC on the interval $(0, 200]$. Label all extrema and inflection points.

59. (Continuation of Exercise 58.)

(a) Determine C', the *marginal cost function*.

(b) In the viewing window $[0, 200]$ by $[0, 50]$, graph both C' and AC.

(c) Using either algebraic techniques or the INTERSECT command on your calculator, locate the point where the two curves cross.

(d) Compare the x-coordinate of the result from part (c) with the x-coordinate of the relative minimum for AC. Are they the same?

60. (Continuation of Exercises 58 and 59.) Fill in the blanks illustrating an important fact from economics.

The minimum average cost occurs when _____ _____ is equal to _____.

61. Miranda's New and Used Cars has a special program for individuals who purchase a new car. At the time of purchase, buyers can also purchase a maintenance and service contract with the dealership. This contract covers all recommended servicing of the vehicle. Tony just purchased a new car for

$11,000. The cost for the maintenance and service contract for his new car is $500 the first year and increases $200 per year thereafter, as long as Tony owns the car. Using the techniques of regression, he found that the total cost of the car (excluding items not covered by the service contract such as gasoline) after t years is given by

$$C(t) = 100t^2 + 400t + 11{,}000$$

(a) Determine a function for the average cost per year, $AC(t) = \dfrac{C(t)}{t}$.

(b) When is the average cost per year a minimum? Round to the nearest tenth.

(c) What is the minimum average cost per year rounded to the nearest dollar?

62. Elisa just purchased a new car at Miranda's New and Used Cars (see Exercise 61). The cost was $15,000 and the cost for the maintenance and service contract for her new car is $700 the first year and increases $200 per year thereafter, as long as Elisa owns the car. Using the techniques of regression, she found that the total cost of the car (excluding items not covered by the service contract such as gasoline) after t years is given by

$$C(t) = 100t^2 + 600t + 15{,}000$$

(a) Determine a function for the average cost per year, $AC(t) = \dfrac{C(t)}{t}$.

(b) When is the average cost per year a minimum? Round to the nearest tenth.

(c) What is the minimum average cost per year rounded to the nearest dollar?

63. For males, blood volume, in milliliters (mL), can be approximated by

$$f(x) = -10{,}822 + 3800 \ln x, \qquad 40 \le x \le 90$$

where x is weight in kilograms (kg).

(a) Show that f is increasing on the interval $[40, 90]$.

(b) Show that the graph of f is concave down on the interval $[40, 90]$.

64. For females, blood volume (in milliliters), can be approximated by

$$f(x) = -11{,}456 + 3888 \ln x, \qquad 40 \le x \le 60$$

where x is weight in kilograms.

(a) Show that f is increasing on the interval $[40, 60]$.

(b) Show that the graph of f is concave down on the interval $[40, 60]$.

65. The movement of air is known as convection, and the removal of heat from the human body by convection air currents is called *heat loss by convection*. If we let x represent wind velocity, in miles per hour, then $f(x)$ represents the percent of total heat loss by convection and is approximated by

$$f(x) = 12 + 16.694 \ln (x + 1), \qquad 0 \le x \le 60$$

(a) Compute $f(20)$ and $f'(20)$ and interpret each.

(b) Show that f is increasing on the interval $[0, 60]$.

(c) Show that the graph of f is concave down on the interval $[0, 60]$.

(d) Sketch a graph of f on the interval $[0, 60]$.

66. In the 1990s, many municipalities stopped hauling yard waste to land fills due to lack of space. The percent of the waste generated in the United States that is yard waste is approximated by

$$f(t) = 19.12e^{-0.06t}, \qquad 1 \le t \le 5$$

where t is the number of years since 1990 and $f(t)$ is the percent of the waste generated in the United States that is yard waste.

(a) Compute $f(3)$ and $f'(3)$ and interpret each.

(b) Compute $f'(t)$ and use it to show that on the interval $[1, 5]$ the percent of waste generated that is yard waste is decreasing.

67. The total carbon content, in millions of metric tons, of carbon dioxide emissions released into the atmosphere by the United States is given by

$$f(t) = -1.32t^3 + 19.66t^2 - 72.76t + 1441.47, \qquad 1 \le t \le 7$$

where t is the number of years since 1988 and $f(t)$ is the total carbon content in millions of metric tons.

(a) Compute $f(4)$ and $f'(4)$ and interpret each.

(b) Determine where f is increasing and where it is decreasing. Also, determine the relative minimum.

(c) Determine where the graph of f is concave up and where it is concave down. Also, determine the inflection point.

(d) Sketch a graph of f on the interval $[1, 7]$. Label the relative minimum and the inflection point.

68. Evaluate $f'(t)$ in Exercise 67 at the t-coordinate of the inflection point and interpret.

SECTION PROJECT

Referring to Table 5.3.1, participation in the labor force can be interpreted to mean "had a job."

(a) Let x represent the number of years since 1959, and let y represent the married female participation rate as a percent-

TABLE **5.3.1**	
YEAR	**MARRIED FEMALE PARTICIPATION RATE IN THE LABOR FORCE (%)**
1960	31.9
1970	40.5
1975	44.3
1980	49.8
1985	53.8
1990	58.4
1991	58.5
1992	59.3
1993	59.4
1994	60.7
1995	61.0

Source: U.S. Census Bureau web site.

age in the U.S. labor force. Plot the data and determine an exponential regression model for the data.

(b) Compute the derivative of the model. Evaluate it at $x = 25$ and interpret.

(c) Use the derivative of the model to show that the married female participation rate was increasing from 1960 to 1995.

(d) Use the second derivative to show that the graph of the model is concave up on the indicated interval. What does this result mean?

(e) The participation rate in the U.S. labor force, as a percentage, for married males from 1960 to 1995 can be modeled by

$$f(t) = 89.18e^{-0.004t}, \qquad 1 \le t \le 36$$

where t represents the number of years since 1959 and $f(t)$ represents the married male participation rate. Compute $f'(25)$ and compare with part (b).

(f) Use $f'(t)$ to show that from 1960 to 1995 the married male participation rate was decreasing.

(g) Use $f''(t)$ to show that the graph of f is concave up on $[1, 36]$. What does this result mean?

SECTION 5.4   OPTIMIZING FUNCTIONS ON A CLOSED INTERVAL

In Section 5.1 we used the first derivative and the First Derivative Test to determine the location of the *relative maximum* and *relative minimum* (plural is *relative extrema*) for a function. In this section we go one step further and determine the location of the **absolute maximum** and **absolute minimum** for a function. Intuitively, these points are on the "highest hill" and in the "lowest valley." Once we know where these **absolute extrema** are located, we will start to apply our new knowledge to a wide array of applications. Thus, we begin our voyage into what is known as **optimization**. We begin by taking a look at the tools used to find the absolute maximum and absolute minimum values on closed intervals.

Absolute Extrema on a Closed Interval

In Section 5.1 we stated that the word *relative* in *relative maximum* and *relative minimum* meant *relative to points near by*. An **absolute maximum** is the largest value that the function attains on its domain or, stated another way, the highest point on the graph. Similarly, an **absolute minimum** is the smallest value that the function attains on its domain or, stated another way, the lowest point on the graph. For any given function, the **absolute extrema** may or may not exist. The graphs in Figure 5.4.1 show the possibilities for some functions that are defined for all values of x. These graphs suggest that absolute extrema may occur at critical points, that is, where $f'(x) = 0$ or where $f'(x)$ is undefined, which is quite similar to what we found in Section 5.1. Notice that Figure 5.4.1 (d) indicates that a critical point does not necessarily give an absolute extrema.

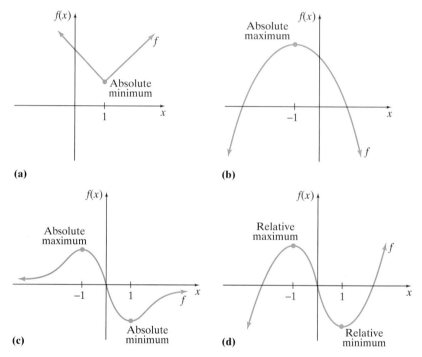

Figure 5.4.1 **(a)** Absolute minimum at $x = 1$. No absolute maximum. **(b)** Absolute maximum at $x = -1$. No absolute minimum. **(c)** Absolute maximum at $x = -1$. Absolute minimum at $x = 1$. **(d)** Relative maximum at $x = -1$. Relative minimum at $x = 1$. No absolute extrema.

Now, if we consider any continuous function on a *closed interval* $[a, b]$, the function will have an absolute maximum and an absolute minimum on the interval. In fact, these absolute extrema will be located at either (1) critical points that are *within* the interval (a, b), or (2) at the endpoints of the interval $[a, b]$. Figures 5.4.2 through 5.4.4 illustrate some of the possibilities.

The prior discussion and figures lead to the following:

Extreme Value Theorem

If a function is continuous on closed interval $[a, b]$, then the function must have both an absolute maximum and an absolute minimum on $[a, b]$.

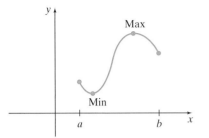

Figure 5.4.2 Absolute extrema occur at critical points.

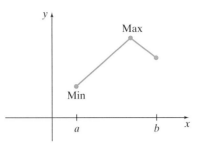

Figure 5.4.3 Absolute maximum at critical point. Absolute minimum at endpoint.

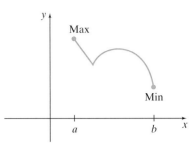

Figure 5.4.4 Absolute extrema occur at endpoints.

The Extreme Value Theorem tells us when an absolute maximum and absolute minimum are guaranteed to exist. Since we know that absolute extrema occur at critical points within the closed interval or at the endpoints, we supply the following procedure to locate these absolute extrema. When we determine the absolute extrema for a function, we are said to be **optimizing** the function.

> **Four-step Process for Locating the Absolute Extrema of f on $[a, b]$**
>
> 1. Verify that f is continuous on $[a, b]$.
> 2. Determine the critical values of f in (a, b).
> 3. Evaluate $f(x)$ at the critical values and at the endpoints of the interval.
> 4. The largest value that you obtain from step 3 is the absolute maximum of f on $[a, b]$, and the smallest value that you obtain from step 3 is the absolute minimum of f on $[a, b]$.

EXAMPLE 1 | **Determining Absolute Extrema**

Determine the absolute extrema for $f(x) = 2x^3 + 3x^2 - 12x + 1$ on $[-3, 2]$.

SOLUTION

Step 1: Since f is a polynomial function, we know that it is continuous on the closed interval $[-3, 2]$. By the Extreme Value Theorem, it must have an absolute maximum and an absolute minimum.

Step 2: To determine the critical values, we first need the derivative

$$f'(x) = 6x^2 + 6x - 12$$

Since $f'(x)$ is defined for all x, the only critical values are where $f'(x) = 0$. So we need to solve

$$6x^2 + 6x - 12 = 0$$
$$6(x^2 + x - 2) = 0$$
$$6(x + 2)(x - 1) = 0 \qquad \text{Factor}$$
$$x = -2 \quad \text{or} \quad x = 1$$

So the critical values are $x = -2$ and $x = 1$.

Step 3: Here, we evaluate $f(x)$ at the critical values and the endpoints.

Critical values: $\quad x = -2 \quad f(-2) = 2(-2)^3 + 3(-2)^2 - 12(-2) + 1 = 21$
$\qquad\qquad\qquad\quad x = 1 \qquad f(1) = 2(1)^3 + 3(1)^2 - 12(1) + 1 = -6$

Endpoints: $x = -3$ $f(-3) = 2(-3)^3 + 3(-3)^2 - 12(-3) + 1 = 10$

$x = 2$ $f(2) = 2(2)^3 + 3(2)^2 - 12(2) + 1 = 5$

Step 4: Since 21 is the largest value from step 3, we conclude that the absolute maximum is 21 and it occurs at $x = -2$. The smallest value from step 3 is -6, so -6 is the absolute minimum and occurs at $x = 1$. See Figure 5.4.5.

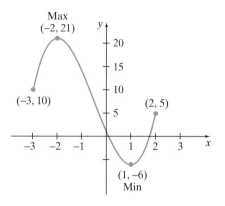

Figure 5.4.5

EXAMPLE 2 | **Determining Absolute Extrema**

Determine the absolute extrema for $f(x) = 3x^4 + 4x^3 - 36x^2 + 1$ on $[-1, 1]$.

SOLUTION

Step 1: Since f is a polynomial function, we know that it is continuous on the closed interval $[-1, 1]$. By the Extreme Value Theorem, it must have an absolute maximum and an absolute minimum.

Step 2: To determine the critical values, we first need the derivative

$$f'(x) = 12x^3 + 12x^2 - 72x$$

Since $f'(x)$ is defined for all x, the only critical values are where $f'(x) = 0$. So we need to solve

$$12x^3 + 12x^2 - 72x = 0$$
$$12x(x^2 + x - 6) = 0$$
$$12x(x + 3)(x - 2) = 0 \qquad \text{Factor}$$
$$x = 0, \quad x = -3, \quad \text{or} \quad x = 2$$

So the critical values are $x = 0$, $x = -3$, and $x = 2$.

Step 3: Since $x = 0$ is the only critical value in the interval $[-1, 1]$, we evaluate $f(x)$ at this critical value and at the endpoints.

Critical value: $x = 0$ $f(0) = 3(0)^4 + 4(0)^3 - 36(0)^2 + 1 = 1$

Endpoints: $x = -1$ $f(-1) = 3(-1)^4 + 4(-1)^3 - 36(-1)^2 + 1$
$$= -36$$

$x = 1$ $f(1) = 3(1)^4 + 4(1)^3 - 36(1)^2 + 1 = -28$

Step 4: The absolute maximum is 1 and it occurs at $x = 0$. The absolute minimum is -36 and occurs at $x = -1$. See Figure 5.4.6.

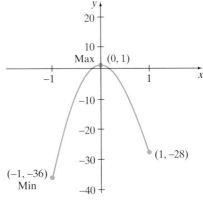

Figure 5.4.6

<hr>

✓**CHECKPOINT 1**

Now work Exercise 9.

Applications

We now turn our attention to problems where we can maximize revenue, profit, area, or volume, as well as minimize costs, average costs, and pollution. This is a small sample of the types of problems that we may encounter. The examples that follow supply a general strategy for solving any optimization problem on a closed interval.

EXAMPLE 3

Optimizing NASA's Budget

The budget for NASA from 1962 to 1980 can be modeled by

$$y = -0.76x^4 + 34.15x^3 - 502.77x^2 + 2607.19x + 926.66, \qquad 1 \le x \le 19$$

where x represents the number of years since 1961 and y is the budget in millions of dollars. Determine the absolute extrema on [1, 19], that is, from 1962 to 1980. Give your answer to the nearest year.

SOLUTION

Even though our function models a real-world situation, we still employ the four-step process.

Step 1: Since we have a polynomial function, it is continuous on the interval [1, 19]. Thus, the Extreme Value Theorem applies, and we can expect to find an absolute maximum and an absolute minimum on the interval.

Step 2: To determine the critical values, we need the derivative, which is

$$y' = -3.04x^3 + 102.45x^2 - 1005.54x + 2607.19$$

The derivative is defined for all x, so the only critical values are where $y' = 0$. We need to solve $-3.04x^3 + 102.45x^2 - 1005.54x + 2607.19 = 0$ and we opt to solve this graphically. Figure 5.4.7 shows the result of using the ZERO command. The critical values are, rounded to the nearest hundredth, $x = 4.09$, $x = 11.72$, and $x = 17.89$.

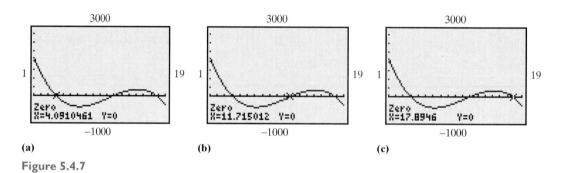

Figure 5.4.7

> **Step 3:** Next we evaluate the function at the critical values, rounded to the nearest hundredth, and at the endpoints.

Critical values: $x = 4.09$

$y = -0.76(4.09)^4 + 34.15(4.09)^3 - 502.77(4.09)^2 + 2607.19(4.09) + 926.66 \approx 5303.48$

$x = 11.72$

$y = -0.76(11.72)^4 + 34.15(11.72)^3 - 502.77(11.72)^2 + 2607.19(11.72) + 926.66 \approx 3060.12$

$x = 17.89$

$y = -0.76(17.89)^4 + 34.15(17.89)^3 - 502.77(17.89)^2 + 2607.19(17.89) + 926.66 \approx 4341.10$

Endpoints: $x = 1$

$y = -0.76(1)^4 + 34.15(1)^3 - 502.77(1)^2 + 2607.19(1) + 926.66 \approx 3064.47$

$x = 19$

$y = -0.76(19)^4 + 34.15(19)^3 - 502.77(19)^2 + 2607.19(19) + 926.66 \approx 4154.19$

> **Step 4:** Thus, 5303.48 is the absolute maximum and it occurs at $x = 4.09$. Rounding x to the nearest year gives $x = 4$. We conclude that from 1962 to 1980 the largest budget for NASA was \$5,303,480,000 and it occurred in 1965. Also, 3060.12 is the absolute minimum and it occurs at $x = 11.72$. Rounding x to the nearest year gives $x = 12$. We conclude that from 1962 to 1980 the smallest budget for NASA was \$3,060,120,000 and it occurred in 1973. ◼

✓**CHECKPOINT 2**

Now work Exercise 19.

In many optimization problems, we must first determine the function that we are trying to optimize. Example 4 illustrates how we handle these types of problems.

EXAMPLE 4 | **Optimizing Area**

One of the authors needs to build a rectangular enclosure for his beloved beagle. He has 200 feet of fence. Determine the dimensions of the rectangle that will make the area of the enclosure as large as possible.

SOLUTION

Two rectangular regions are shown in Figure 5.4.8 that each use the 200 feet of fencing. Our goal here is to find the one rectangle that has the maximum area.

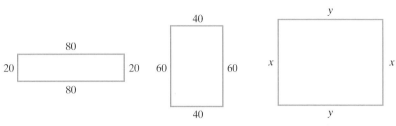

Figure 5.4.8   **Figure 5.4.9**

Instead of continuing in a "guess-and-check" fashion, let's consider a generic rectangle with a width of x and a length of y, as shown in Figure 5.4.9. The area, A, of this generic rectangle is given by

$$A = x \cdot y$$

We want to maximize the area, A, but it is written as a function of two variables. To use the tools we currently have, we need to rewrite $A = x \cdot y$ as a function of *one* variable. Since we have 200 feet of fence, we conclude that the perimeter, P, of the rectangular enclosure is 200 feet. That is,

$$P = x + y + x + y = 200$$
$$2x + 2y = 200$$
$$2y = 200 - 2x$$
$$y = 100 - x$$

We can substitute this into our area equation and get $A = x \cdot (100 - x)$ or, more precisely,

$$A(x) = 100x - x^2$$

Since this enclosure is for an animal that must be able to move around in it and turn around in it, a reasonable domain is $10 \leq x \leq 90$. So we now have the situation of determining the absolute maximum of

$$A(x) = 100x - x^2, \qquad \text{on } [10, 90]$$

We now employ our four-step process just as we did in Examples 1, 2, and 3.

Step 1: Since A is a polynomial function, it is continuous on the stated interval, which means that the Extreme Value Theorem applies.

Step 2: Next, determine the critical values. The derivative of $A(x)$ is

$$A'(x) = 100 - 2x$$

Since $A'(x)$ is defined for all x, the only critical value is where $A'(x) = 0$.

Solving this equation yields

$$100 - 2x = 0$$
$$x = 50$$

Step 3: Evaluating $A(x)$ at the critical value and endpoints results in

Critical value: $x = 50$ $A(50) = 100(50) - (50)^2 = 2500$

Endpoints: $x = 10$ $A(10) = 100(10) - (10)^2 = 900$

$x = 90$ $A(90) = 100(90) - (90)^2 = 900$

Step 4: The largest number from step 3 is 2500 and occurs at $x = 50$. So the dimensions of the rectangular enclosure that maximizes the area for the

beagle are

$$x = 50$$
$$y = 100 - x = 100 - 50 = 50$$

Thus, a rectangle with length of 50 feet and width of 50 feet maximizes the area. ◼

Notice that in Example 4 we needed to do some preliminary work *before* we could use our four-step process. Example 4 suggests the following procedure when solving these types of applied optimization problems.

<table>
<tr><td>

Interactive Activity

 Reread Example 4 and describe how the strategy to the right was used in the solution.

</td></tr>
</table>

Strategy for Solving Applied Optimization Problems on a Closed Interval

1. Read the question carefully and, if possible, sketch a picture that represents the problem.
2. Select variables to represent the quantity to be maximized or minimized and all other unknowns.
3. Write an equation for the quantity to be maximized or minimized. If necessary, eliminate extra variables so that the quantity to be optimized is a function of one variable.
4. Apply the four-step process to determine the absolute extrema and answer the question posed in the problem.

In Example 5 we return to a problem first encountered in the Section 5.1 Exercises.

EXAMPLE 5 | **Determining a Maximum Medication Concentration**

Researchers have determined through experimentation that the percent concentration of a certain medication t hours after it has been administered can be approximated by

$$p(t) = \frac{230t}{t^2 + 6t + 9}, \qquad 1 \le t \le 20$$

where $p(t)$ is the percent concentration. How many hours after the administration of this medication is the concentration at a maximum? What is the maximum concentration? See Figure 5.4.10.

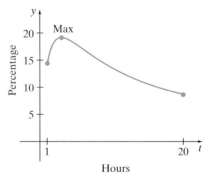

Figure 5.4.10

SOLUTION

Step 1: The function p is continuous for all values of t except $t = -3$. But, since $t = -3$ is not in the stated interval, p is continuous on the stated interval. Thus, the Extreme Value Theorem guarantees an absolute maximum.

Step 2: By the Quotient Rule, we compute the derivative to be

$$p'(t) = \frac{230(t^2 + 6t + 9) - 230t(2t + 6)}{(t^2 + 6t + 9)^2} = \frac{-230t^2 + 2070}{(t^2 + 6t + 9)^2} = \frac{-230(t^2 - 9)}{(t^2 + 6t + 9)^2}$$

Notice that $p'(t)$ is undefined at $t = -3$, but, since this is not in the stated interval, we do not consider it. We now solve $p'(x) = 0$, which gives

$$\frac{-230(t^2 - 9)}{(t^2 + 6t + 9)^2} = 0$$

Since the only time that a fraction can equal zero is when the numerator equals zero, we have

$$-230(t^2 - 9) = 0$$
$$t^2 - 9 = 0$$
$$t^2 = 9$$
$$t = \pm 3$$

Disregarding $t = -3$, we have a critical value at $t = 3$.

Step 3: Evaluating $p(t)$ at the critical value $t = 3$ and at the endpoints and rounding to the nearest hundredth gives

$$\text{Critical value:} \quad p(3) = \frac{230(3)}{(3)^2 + 6(3) + 9} \approx 19.17$$

$$\text{Endpoints:} \quad p(1) = \frac{230(1)}{(1)^2 + 6(1) + 9} \approx 14.38$$

$$p(20) = \frac{230(20)}{(20)^2 + 6(20) + 9} \approx 8.70$$

Step 4: The largest result from step 3 is 19.17 and it occurs at $t = 3$. This means that 3 hours after administration of this medication the percent concentration is maximized at about 19.17%.

✓ **CHECKPOINT 3**

Now work Exercise 37.

In our final example, we return to the functions of business, as well as determining an equation for a line.

EXAMPLE 6 | **Maximizing Profit**

Only Beef, a local sandwich store, estimates that it can sell 500 roast beef sandwiches per week if it sets the price at $2.50, but its weekly sales will rise by 50 sandwiches for each $0.05 decrease in price. The company has fixed costs each week of $525.00, and variable costs for making a roast beef sandwich are $0.55. Let x represent the number of roast beef sandwiches made and sold each week. Determine the value of x that maximizes weekly profit, assuming that $500 \leq x \leq 1500$.

SOLUTION

We need to employ the strategy supplied after Example 4 to attack this problem. We want to maximize profit, $P(x)$. Recall that $P(x) = R(x) - C(x)$, $R(x) = $ price $\cdot$ quantity, and $C(x) = $ fixed costs $+$ variable costs. Since fixed costs are \$525 and x is the number of sandwiches made and sold each week, the variable costs are $0.55 \cdot x$. Hence,

$$C(x) = \text{fixed costs} + \text{variable costs}$$
$$C(x) = 525 + 0.55x$$

To determine $R(x)$, we need price, $p(x)$. We know that when $x = 500$ price is \$2.50, and for each \$0.05 *decrease* in price, weekly sales *increase* by 50. To aid us in analyzing this situation, we make Table 5.4.1 with this information.

TABLE 5.4.1

x, NUMBER OF SANDWICHES	$p(x)$, PRICE OF SANDWICH
500	2.50
550	2.45
600	2.40
650	2.35

Notice that, for each change in x of 50, $p(x)$ changes by 0.05. As x increases at a constant rate, $p(x)$ seems to be decreasing at a constant rate, which means that we should be able to model $p(x)$ with a *linear function*. We can write

$$\frac{\text{Change in } p(x)}{\text{Change in } x} = \frac{-0.05}{50} = -0.001$$

which tells us that the slope of the line is $m = -0.001$. Using the point–slope form of a line with the point $(500, 2.50)$ gives

$$y - y_1 = m(x - x_1)$$
$$y - 2.50 = -0.001(x - 500)$$
$$y = -0.001(x - 500) + 2.50 \quad \text{or} \quad y = -0.001x + 3$$

So the price function is given by

$$p(x) = -0.001x + 3$$

The revenue function is then given by

$$\text{Revenue} = \text{price} \cdot \text{quantity}$$
$$R(x) = p(x) \cdot x$$
$$R(x) = (-0.001x + 3) \cdot x = -0.001x^2 + 3x$$

Now that we have $R(x)$ and $C(x)$, profit, $P(x)$, is given by

$$P(x) = R(x) - C(x)$$
$$P(x) = -0.001x^2 + 3x - (525 + 0.55x) = -0.001x^2 + 2.45x - 525$$

We have determined the function that is to be optimized. In other words, our task is to determine the absolute maximum for

$$P(x) = -0.001x^2 + 2.45x - 525, \quad \text{on } [500, 1500]$$

Step 1: Since P is a polynomial function, it is continuous on the stated interval and the Extreme Value Theorem guarantees an absolute maximum.

Step 2: The derivative is

$$P'(x) = -0.002x + 2.45$$

Since $P'(x)$ is defined for all x, the only critical value is where $P'(x) = 0$. Solving this equation gives

$$-0.002x + 2.45 = 0$$
$$x = 1225$$

Step 3: Evaluating $P(x)$ at the critical value $x = 1225$ and the endpoints results in

Critical value: $P(1225) = -0.001(1225)^2 + 2.45(1225) - 525 = 975.625$

Endpoints: $P(500) = -0.001(500)^2 + 2.45(500) - 525 = 450$

$P(1500) = -0.001(1500)^2 + 2.45(1500) - 525 = 900$

Step 4: The largest value from step 3 is 975.625 and it occurs at $x = 1225$. Thus, to maximize weekly profit, Only Beef should sell 1225 roast beef sandwiches each week.

In Example 6, the price function, p, was determined from the information given in the problem. We recommend that you check its validity. One way of doing this is to substitute values into the price function that correspond with the table. For example, we could substitute $x = 500$ or $x = 600$ into the price function to verify that we get a price of \$2.50 and \$2.40, respectively. This type of checking is an invaluable tool in helping to secure the correct result.

✓ **CHECKPOINT 4**

Now work Exercise 47.

SUMMARY

The main concept in this section was **optimizing,** that is, determining absolute extrema for a function on a closed interval. The **Extreme Value Theorem** stated that if f is a continuous function on a closed interval then f is guaranteed to have an absolute maximum and an absolute minimum. These absolute extrema occur at either the endpoints of an interval or at a critical point that is in the interior of the interval. We then presented a **four-step process** to locate these absolute extrema and the **Strategy for Solving Applied Optimization Problems on a Closed Interval.**

SECTION 5.4 EXERCISES

In Exercises 1–18, determine the absolute extrema of each function on the indicated interval.

1. $f(x) = x^2 - 2x - 7$ on $[-2, 1]$

2. $f(x) = 2x^2 + 3x - 1$ on $[-1, 2]$

3. $f(x) = x^3 - 2x^2 - 5x + 6$ on $[-2, 2]$

4. $f(x) = x^3 + 4x^2 + x - 6$ on $[-2, 0]$

5. $f(x) = x^3 - 3x^2$ on $[-1, 3]$

6. $f(x) = x^3 - 12x$ on $[0, 4]$

7. $f(x) = x^4 - 15x^2 - 10x + 24$ on $[-3, 3]$

8. $y = x^4 + 2x^3 - 13x^2 - 14x + 24$ on $[0, 4]$

✓ 9. $h(x) = x^4 - x^3 + 5$ on $[-2, 2]$

10. $f(x) = 2x^3 - 6x^2 + 4$ on $[-1, 4]$

11. $y = (2x^2 - 1)^4$ on $[0, 2]$

12. $y = (3x + 1)^3$ on $[-2, 1]$

13. $y = \sqrt[3]{x}$ on $[-1, 1]$

14. $f(x) = \sqrt[3]{x^2}$ on $[-1, 8]$

15. $f(x) = \dfrac{1}{x - 2}$ on $[0, 1]$

16. $f(x) = \dfrac{x}{x - 2}$ on $[3, 5]$

17. $y = \dfrac{1}{x^2 + 1}$ on $[1, 4]$

18. $y = \dfrac{1}{x^2 + 1}$ on $[-1, 1]$

APPLICATIONS

✓ 19. From past records, the owner of the Sleep Cheap Motel has determined that when $\$x$ per day is charged to rent a room the daily profit, $P(x)$, is given by

$$P(x) = -x^2 + 92x - 180, \qquad 40 \le x \le 60$$

What should the owner charge to maximize profit?

20. The Wax and Wick Company sells jumbo-sized holiday scented candles with 10 in a box. The price–demand function is given by

$$p(x) = 102 - 3x$$

where x is the number of boxes and $p(x)$ is in dollars per box.

(a) Determine the revenue function, R.

(b) Determine the number of boxes sold that maximizes revenue.

21. The 40-Acre Pond is a local swimming pond for residents of Point King. Periodically, the pond is treated to control the growth of harmful bacteria. The concentration, $HB(x)$, of harmful bacteria per cubic centimeter is given by

$$HB(x) = 15x^2 - 210x + 750, \qquad 0 \le x \le 14$$

where x is the number of days after a treatment.

(a) How many days after a treatment is the concentration minimized?

(b) What is the minimum concentration?

22. The Camp-n-Swim campgrounds has a small lake for recreational swimming. Periodically, the lake is treated to control the growth of harmful bacteria. The concentration, $HB(x)$, of harmful bacteria per cubic centimeter is given by

$$HB(x) = 3x^2 - 42x + 150, \qquad 0 \le x \le 14$$

where x is the number of days after a treatment.

(a) How many days after a treatment is the concentration minimized?

(b) What is the minimum concentration?

23. From 1986 to 1995, the percent of the U.S. labor force unemployed can be modeled by

$$y = -0.037x^3 + 0.605x^2 - 2.776x + 9.333, \qquad 1 \le x \le 10$$

where x is the number of years since 1985 and y is the percent of the labor force that is unemployed. To the nearest year, determine when the United States had the highest unemployment percentage and when it had the lowest unemployment percentage. Determine the percentage for each.

24. The total number of public elementary and secondary school districts in the United States from 1980 to 1996 can be modeled by

$$f(x) = -0.04x^3 - 2.12x^2 - 23.08x + 16,063, \qquad 1 \le x \le 17$$

where x is the number of years since 1979 and $f(x)$ represents the number of districts. Determine the absolute maximum and the absolute minimum and interpret each.

25. From 1974 to 1995, the number of bowling establishments in the United States can be modeled by

$$f(x) = -4.21x^2 + 10.6x + 8613.57, \qquad 1 \le x \le 22$$

where x is the number of years since 1973 and $f(x)$ represents the number of bowling establishments in the United States. Determine the absolute maximum and the absolute minimum and interpret each. Round each to the nearest whole number.

26. From 1985 to 1994, the average salary, in thousands of dollars per year, for an NBA player can be modeled by

$$f(x) = 264e^{0.18x}, \qquad 1 \le x \le 10$$

where x is the number of years since 1984 and $f(x)$ is the average salary. Determine the absolute maximum and the absolute minimum and interpret each. Round the salary to the nearest thousand.

27. The total number of Catholic elementary schools in the United States from 1960 to 1995 can be modeled by

$$f(x) = 10,560e^{-0.012x}, \qquad 1 \le x \le 36$$

where x represents the number of years since 1959 and $f(x)$ represents the number of Catholic elementary schools. Determine the absolute maximum and the absolute minimum and interpret each.

28. The total number of Catholic secondary schools in the United States from 1960 to 1995 can be modeled by

$$f(x) = 2410e^{-0.019x}, \qquad 1 \le x \le 36$$

where x is the number of years since 1959 and $f(x)$ is the number of Catholic secondary schools. Determine the absolute maximum and the absolute minimum and interpret each.

29. Let y_1 equal the model in Exercise 27 and y_2 equal the model in Exercise 28. Define y_3 as follows:

$$y_3 = y_1 + y_2$$

(a) What does y_3 represent?

(b) Determine the absolute extrema for y_3 and interpret.

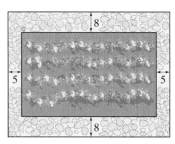

Figure 5.4.11

the garden and y is the width of the garden. Also assume that $10 \leq x \leq 100$.

30. From 1980 to 1993, the total number of postneonatal (infants from 28 days to 11 months old) deaths in the United States can be modeled by

$$y = -1.69x^4 + 39.88x^3 - 277.12x^2 + 432.56x + 14{,}715.83,$$
$$1 \leq x \leq 14$$

where x is the number of years since 1979 and y represents the total number of postneonatal deaths. Determine the absolute maximum and the absolute minimum and interpret each.

31. From 1990 to 1996, the average price, in dollars per 1000 cubic feet, of natural gas in the United States can be modeled by

$$f(x) = 0.02x^4 - 0.31x^3 + 1.6x^2 - 3.09x + 3.51, \quad 1 \leq x \leq 7$$

where x represents the number of years since 1989 and $f(x)$ is the average price. Determine the absolute extrema and interpret each.

32. In Example 4 we learned that one of the authors needs to build a rectangular enclosure for his beloved beagle. His wife suggested that, if the rectangular enclosure was built so that one side was along the outside wall of the house, the 200 feet of fencing would make an even bigger enclosure. Follow the wife's advice and determine the dimensions of the rectangular enclosure that will make the enclosure as large as possible. Assume that the author has 200 feet of fencing and the side along the wall needs no fencing.

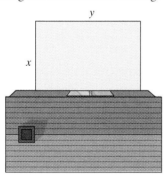

33. The recreation department in Crown Point has been authorized to construct a rectangular playground whose area is 10,000 square feet. Due to a city ordinance, the playground must be enclosed on all four sides by a fence. Let x be the length and y be the width. Given that $50 \leq x \leq 200$, determine the dimensions of the rectangle that minimize the number of feet of fencing required to enclose the playground.

34. One of the authors of this text recently purchased a new home. He has an extra acre of land to grow a garden. He wants the garden to be rectangular, and he wants the garden to have an area of 1440 square feet. The design of his garden incorporates an 8-foot-wide walkway on the north and south sides of the garden and a 5-foot-wide walkway on the east and west sides. See Figure 5.4.11. Determine the dimensions of the garden that will minimize the *total* area for the garden *and* the walkways. Assume that x is the length of

35. The marketing research department of *Shank,* a quarterly magazine for beginning golfers, has determined that the price–demand equation for the magazine is approximated by

$$p(x) = 2.75 - 0.01x, \qquad 0 \leq x \leq 275$$

where x represents the number of magazines printed and sold each quarter, in hundreds, and $p(x)$ is the price, in dollars, of the magazine. The cost of printing, distributing, and advertising is given by

$$C(x) = 5 + 0.5x + 0.003x^2$$

where $C(x)$ is in hundreds of dollars.

(a) Determine the level of sales that maximizes profit.

(b) Determine the price that the magazine should sell at in order to maximize profit.

36. The concentration of a certain medication in a patient's bloodstream can be given by

$$C(t) = \frac{2.5t}{2t^2 + 7t + 4}, \qquad 0 \leq t \leq 8$$

where $C(t)$ is in milligrams per cubic centimeter and t is the number of hours after the medication has been administered.

(a) How many hours after the medication has been administered is the concentration at a maximum?

(b) What is the maximum concentration?

37. The concentration of a certain medication in a patient's bloodstream can be given by

$$C(t) = \frac{5.3t}{t^2 + 4t + 5}, \qquad 0 \leq t \leq 8$$

where $C(t)$ is in milligrams per cubic centimeter and t is the number of hours after the medication has been administered.

(a) How many hours after the medication has been administered is the concentration at a maximum?

(b) What is the maximum concentration?

38. The Double D Corporation analyzed the production costs for one of its products and determined that the daily

cost function can be given by

$$C(x) = 0.03x^3 - 2.4x^2 + 73.16x + 102.27, \qquad 5 \le x \le 55$$

where x represents the number of units produced each day and $C(x)$ is the daily cost in dollars.

(a) Determine the average cost function, AC.

(b) Determine the production level that minimizes the average cost.

(c) Determine the average cost and the total cost at the production level found in part (b).

39. For the product in Exercise 38, the Double D Corporation gathered data for the price of a unit, in dollars, and the number of units demanded per day. The price–demand function that models these data is given by

$$p(x) = 0.14x^2 - 15.52x + 561.21, \qquad 5 \le x \le 55$$

where x represents the number of units demanded each day and $p(x)$ is the price, in dollars, that consumers pay to buy exactly x units per day.

(a) Determine R, revenue as a function of x.

(b) Determine the maximum revenue.

(c) Determine the demand that maximizes revenue.

(d) Determine the price that maximizes revenue.

40. Refer to Exercises 38 and 39. Using the model for C from Exercise 38 and the function for R from Exercise 39, determine the following:

(a) P, profit as a function of x.

(b) Maximum profit.

(c) The demand that maximizes profit.

(d) The price that maximizes the profit.

41. From 1970 to 1994, the total energy consumption in the United States, measured in quadrillions of Btu is given by

$$y = -0.0007x^4 + 0.045x^3 - 0.868x^2 + 6.347x + 60.574,$$
$$1 \le x \le 25$$

where x represents the number of years since 1969 and y is the total consumption in quadrillions of Btu.

(a) Determine the relative maximum and relative minimum and interpret each.

(b) Determine the absolute maximum and absolute minimum and interpret each.

42. The United States per capita consumption of beef, in pounds, can be modeled by

$$B(t) = 0.03t^3 - 0.76t^2 + 4.54t + 68.34, \qquad 1 \le t \le 16$$

where t is the number of years since 1979 and $B(t)$ is the per capita consumption of beef, in pounds. Determine the absolute maximum and the absolute minimum and interpret each.

43. The U.S. water consumption, in gallons per day per capita, can be modeled by

$$f(x) = -0.41x^2 + 13.53x + 330.69, \qquad 1 \le x \le 31$$

where x is the number of years since 1959 and $f(x)$ is the number of gallons of water consumed per day per capita. Determine the absolute maximum and the absolute minimum and interpret each.

44. As arterial pressure increases, blood flow through the artery also increases. This can be modeled by

$$f(x) = 0.267e^{0.0256x}, \qquad 20 \le x \le 120$$

where x represents arterial pressure, measured in millimeters of mercury (mm Hg), and $f(x)$ is blood flow measured in milliliters per minute (mL/min). Determine the absolute extrema and interpret each.

45. The hematocrit of blood is the percent of blood that is cells. For example, a hematocrit of 40 means that 40% of the blood volume is cells and the rest is plasma. Viscosity is the internal friction of a fluid caused by molecular attraction, which makes it resist a tendency to flow. As hematocrit increases, the viscosity of blood increases. If we arbitrarily consider water to have a viscosity of 1, then the viscosity of whole blood at normal hematocrit is about 3 or 4. This means that three to four times as much pressure is needed to force whole blood through a given tube than to force water through the same tube. The relationship between hematocrit and viscosity is given by

$$v(h) = 0.0015h^2 - 0.019h + 1.563, \qquad 10 \le h \le 80$$

where h is the hematocrit and $v(h)$ represents the viscosity. (The hematocrit of a normal human is about 40.) Determine the absolute maximum and the absolute minimum and interpret each.

46. The Boogie-All-Night Company manufactures 1970s-style black lights. The monthly fixed costs are $100 and variable costs are $12 for each light. The company estimates that 20 lights can be sold each month if the price for a light is $20 and that 2 more lights can be sold for each decrease of $1 in the price. Let x represent the number of lights produced and sold each month.

(a) Determine the monthly cost function, C.

(b) Determine the monthly revenue function, R.

(c) Determine the value of x that maximizes the monthly profit, assuming that $10 \le x \le 40$.

47. The Boogie-All-Night Company also manufactures authentic 1970s-style lava lamps. The monthly fixed costs are $1000 and variable costs are $60 for each lamp. The company estimates that 100 lamps can be sold each month if the price for a lamp is $100 and that 7 more lamps can be sold for each decrease of $5 in the price. Let x represent the number of lamps produced and sold each month.

(a) Determine the monthly cost function, C.

(b) Determine the monthly revenue function, R.

(c) Determine the value of x that maximizes the monthly profit, assuming that $50 \leq x \leq 200$.

48. The Tool Shack manufactures ratchet sets. The monthly fixed costs are \$15,000 and variable costs are \$180 for each ratchet set produced. The company estimates that 300 sets can be sold each month if the unit price is \$280 and that 20 more units can be sold for each decrease of \$14 in the price. Let x represent the number of ratchet sets produced and sold each month.

(a) Determine the monthly cost function, C.

(b) Determine the monthly revenue function, R.

(c) Determine the value of x that maximizes the monthly profit, assuming that $200 \leq x \leq 580$.

49. The VirtualPet Company makes handheld electronic virtual pets. The marketing department is confident that it can sell 3000 virtual pets per week at a price of \$10 each and also believes that reducing the price by \$0.17 each will increase the weekly sales by 180. Let x represent the number of virtual pets manufactured and sold each week.

(a) Determine the weekly revenue, R.

(b) Determine the value of x that maximizes weekly revenue, assuming that $3000 \leq x \leq 10,000$.

SECTION PROJECT

Lisa's Bakery collected the data shown in Table 5.4.2 on the price of a dozen brownies, in dollars, and the number of dozens demanded per day.

(a) A price–demand function that models these data is

$$p(x) = -0.07x + 5.04, \qquad 14 \leq x \leq 40$$

where x is the number of dozens of brownies demanded per day and $p(x)$ represents the price at which people buy exactly x dozen brownies per day. Determine R, revenue as a function of x.

TABLE 5.4.2

NUMBER OF DOZENS DEMANDED	PRICE, IN DOLLARS
14	4.00
19	3.75
21	3.50
23	3.25
26	3.00
31	2.75
35	2.50

(b) Determine the maximum revenue and determine what price and what demand will result in the maximum revenue.

(c) If each dozen brownies costs Lisa's Bakery \$1.50 to make, determine P, profit as a function of x.

(d) Determine the maximum profit and determine what price and what demand will result in the maximum profit.

(e) A price–demand function for the data can also be modeled by

$$p(x) = 5.64e^{-0.023x}, \qquad 14 \leq x \leq 40$$

where x is the number of dozens of brownies demanded per day and $p(x)$ represents the price at which people buy exactly x dozen brownies per day. Determine R, revenue as a function of x.

(f) Using the result from part (e), determine the maximum revenue and determine what price and what demand will result in the maximum revenue.

(g) If each dozen brownies costs Lisa's Bakery \$1.50 to make, using the result from part (f), determine P, profit as a function of x.

(h) Determine the maximum profit and determine what price and what demand will result in the maximum profit.

(i) If you owned Lisa's Bakery, which model would you use to determine the price to maximize profit, the model from part (a) or the model from part (e)? Explain.

SECTION 5.5   THE SECOND DERIVATIVE TEST AND OPTIMIZATION

In Section 5.4 we learned that if a function is continuous on a closed interval then it has an absolute maximum and an absolute minimum. We also saw how to locate the absolute extrema by a four-step process. The key to the last section was that we considered functions on *closed intervals*. In this section we consider functions on **open** and **half-open** intervals. This means that we cannot use the four-step process and must rely on something else. What we tend to rely on when optimizing a function on an open (or half-open) interval is a graph of the function, a table of values, and **the Second Derivative Test.** What we use depends on the number of critical values on the open interval.

Absolute Extrema on Open Intervals

Recall, in Section 5.4, that we used the Extreme Value Theorem to find the absolute maximum and absolute minimum on a closed interval. The key to this theorem was that it told us on a *closed interval,* a function is *guaranteed* to have an absolute maximum and an absolute minimum. If the interval is not closed, there is no guarantee that absolute extrema exist.

Figures 5.5.1 through 5.5.3 demonstrate that when the interval is open we may or may not have absolute extrema. Probably one of the best methods to determine if a function has absolute extrema on an open interval is to look at the graph of the function. However, when the function in question is continuous and has only one critical value in the indicated interval, we can employ the **Second Derivative Test** to determine if an absolute maximum or an absolute minimum exists.

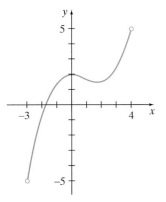

Figure 5.5.1 No absolute extrema on $(-3, 4)$.

Figure 5.5.2 Absolute minimum, but no absolute maximum on $(-2, 2)$.

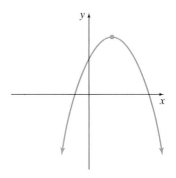

Figure 5.5.3 Absolute maximum, but no absolute minimum on $(-\infty, \infty)$.

The Second Derivative Test may be used on *any* interval (closed, open, or half-open) that contains only one critical value. So when we have a function, f, that is continuous on an interval and $x = c$ is the only critical value in the interval, we can modify the Second Derivative Test from Section 5.3 to locate **absolute extrema** as follows:

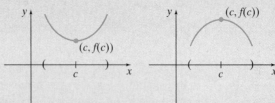

Second Derivative Test for Absolute Extrema

For a continuous function, f, on *any* interval (open, closed, or half-open), and if $x = c$ is the *only* critical value in the interval where $f'(c) = 0$, if $f''(c)$ exists, then the point $(c, f(c))$ is

1. an **absolute minimum** if $f''(c) > 0$ 2. an **absolute maximum** if $f''(c) < 0$

If $f''(c) = 0$, then this test fails.

EXAMPLE 1 | **Determining an Absolute Extremum on an Interval**

Determine the absolute minimum for $f(x) = 3x + \dfrac{27}{x}$ on $(0, \infty)$.

SOLUTION

Since $(0, \infty)$ is an open interval, the techniques of Section 5.4 do not apply here. Applying the new techniques of this section requires us to first determine the derivative. Rewriting the function algebraically and then differentiating yields

$$f(x) = 3x + \frac{27}{x} = 3x + 27x^{-1}$$

$$f'(x) = 3 - 27x^{-2} = 3 - \frac{27}{x^2}$$

Now, according to the Second Derivative Test for Absolute Extrema, we need to find the critical value $x = c$, where $f'(c) = 0$. Setting the derivative equal to zero and solving gives

$$3 - \frac{27}{x^2} = 0$$

$$\frac{3x^2 - 27}{x^2} = 0$$

$$\frac{3(x^2 - 9)}{x^2} = 0$$

$$\frac{3(x + 3)(x - 3)}{x^2} = 0$$

$$x = 3 \quad \text{or} \quad x = -3$$

So the only critical value in $(0, \infty)$, where $f'(x) = 0$, is $x = 3$. Since there is only one critical value that makes $f'(x) = 0$, we can apply the Second Derivative Test for Absolute Extrema. To apply this requires the second derivative.

$$f'(x) = 3 - 27x^{-2}$$

$$f''(x) = 54x^{-3} = \frac{54}{x^3}$$

We now evaluate the second derivative at the critical value and get

$$f''(3) = \frac{54}{(3)^3} = \frac{54}{27} = 2$$

Since $f''(3) = 2$, $f''(3) > 0$, we conclude by the Second Derivative Test for Absolute Extrema that the absolute minimum for f on $(0, \infty)$ occurs at $x = 3$, and the absolute minimum is $f(3) = 3(3) + \frac{27}{3} = 18$. See Figure 5.5.4.

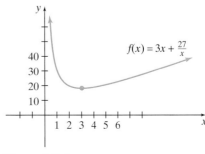

Figure 5.5.4

✓ **CHECKPOINT 1**

Now work Exercise 3.

Applications

In contrast to the problems in Section 5.4, the problems in this section have intervals that are not closed. Hence, we need the techniques presented in this section.

EXAMPLE 2

Maximizing Volume

Lydia would like to make an open-top box out of a square piece of corrugated cardboard that measures 12 inches on each side. To do this, she needs to cut small squares from each corner and then turn up the edges. Determine the dimensions of the resulting box if its volume is to be a maximum.

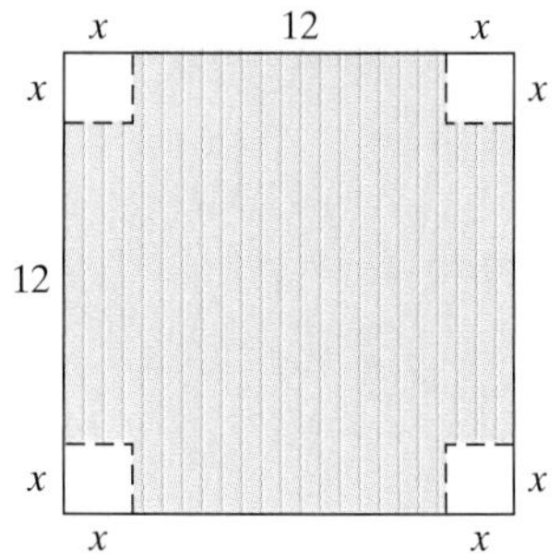

SOLUTION

We begin the solution by sketching the square piece of corrugated cardboard and labeling the length and width of the cutout pieces with an x. See Figure 5.5.5.

The formula for volume of a box is

$$V = l \cdot w \cdot h$$

Figure 5.5.5

Volume is what we want to maximize, but as it is currently written, we do not have the tools in our calculus arsenal to solve it. We need to represent the volume as a function of a single variable. To do this, we now draw a figure of what the box would look like. See Figure 5.5.6.

The height, h, of the box is the same as x. The length, l, of our box is given by $(12 - 2x)$. This is due to the fact that the 12-inch square had x inches cut out from each corner. Hence, $12 - x - x$ results in the length being $12 - 2x$. By a similar argument, the width, w, of the box also measures $12 - 2x$. So we now have for the volume of the box

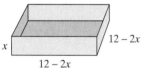

Figure 5.5.6

$$V = l \cdot w \cdot h$$
$$V(x) = (12 - 2x) \cdot (12 - 2x) \cdot x = (12 - 2x)^2 \cdot x$$

Now that we have the volume as a function of a single variable, we can apply techniques that are in our calculus arsenal to solve the problem. But first we need to consider the interval, or reasonable domain, for V. Since we want to make a box, $x \neq 0$, because $x = 0$ corresponds to making no cuts at all. Also, $x < 6$ since a square cutout of 6 inches on each side would produce four square pieces of corrugated cardboard and we could not make a box. (There would be nothing left to turn up!) So we claim that the length and width of the cutout, x, is in the interval $0 < x < 6$. Thus, we need to maximize

$$V(x) = (12 - 2x)^2 \cdot x, \quad \text{on } (0, 6)$$

The interval is an open interval, so we need the techniques of this section to solve it. We first determine the critical value as follows:

$$V'(x) = 2(12 - 2x)(-2) \cdot x + (12 - 2x)^2 \cdot 1 \quad \text{Product and Chain Rules}$$
$$= -4x(12 - 2x) + (12 - 2x)^2$$
$$= (12 - 2x)[-4x + (12 - 2x)]$$
$$= (12 - 2x)(-6x + 12)$$

Setting the derivative equal to zero and solving gives

$$(12 - 2x)(-6x + 12) = 0$$
$$x = 6 \quad \text{or} \quad x = 2$$

The only critical value that is *in* the interval $(0, 6)$ is $x = 2$. We now use the Product Rule to determine $V''(x)$.

$$V''(x) = \frac{d}{dx}[(12 - 2x)(-6x + 12)]$$
$$= -2(-6x + 12) + (12 - 2x)(-6)$$
$$= 12x - 24 - 72 + 12x = 24x - 96$$

Evaluating $V''(x)$ at the critical value $x = 2$ yields

$$V''(2) = 24(2) - 96 = 48 - 96 = -48$$

Since $V''(2) = -48$, $V''(2) < 0$, we have an absolute maximum at $x = 2$. Thus, the dimensions of the box that has maximum volume is

Length: $12 - 2x = 12 - 2(2) = 12 - 4 = 8$

Width: $12 - 2x = 12 - 2(2) = 12 - 4 = 8$

Height: $x = 2$

So a length of 8 inches, width of 8 inches, and height of 2 inches maximize the volume at 128 cubic inches. ▆

Example 2 suggests the following procedure when solving applied optimization problems on an open interval.

Strategy for Solving Applied Optimization Problems on an Open Interval

1. Read the question carefully and, if possible, sketch a picture that represents the problem.
2. Select variables to represent the quantity to be maximized or minimized and other unknowns.
3. Write an equation for the quantity to be maximized or minimized and determine the interval (reasonable domain) over which the function is to be optimized. If necessary, eliminate extra variables so that the quantity to be optimized is a function of one variable.
4. If the interval is open (or half-open), you must use the techniques of this section. If the interval is closed, you can use the techniques of Section 5.4.
5. Answer the question posed in the problem.

Interactive Activity

In Example 2, expand the equation for $V(x) = (12 - 2x)^2 \cdot x$ by multiplication. Compute $V'(x)$ and solve the problem to verify that the same result is attained. Notice that by doing this, the differentiation may be easier.

✓ **CHECKPOINT 2**

Now work Exercise 21.

In many settings it is very easy to become too greedy. For example, if a farmer plants too much corn on an acre of ground, the yield can actually be reduced. If a restaurant has too many tables and they are too close together, patrons may actually stay away. If an accounting firm places too many accountants in an office, their productivity may decrease. Calculus and optimization can allow us to handle the problem of overcrowding, as illustrated in Example 3.

EXAMPLE 3   **Maximizing a Harvest**

Laurie's Lemons has determined that the annual yield per lemon tree is fairly constant at 320 pounds when the number of trees per acre is 50 or fewer. The owner of Laurie's Lemons would like to maximize the annual yield per acre. To do this, she wants to plant more lemon trees. Research has shown that for each additional tree over 50 the annual yield per tree decreases by 4 pounds due to overcrowding. How many trees should be planted on each acre to maximize the annual yield from an acre?

SOLUTION

The total yield for an acre is represented by

$$\text{Total yield} = (\text{number of trees per acre}) \cdot (\text{yield per tree})$$

Let x represent the number of trees per acre. To aid us in determining the yield per tree, we make the chart shown in Table 5.5.1.

TABLE 5.5.1

NUMBER OF TREES	YIELD PER TREE	TOTAL YIELD
10	320	3200
20	320	6400
40	320	12,800
50	320	16,000
51	$320 - 4(51 - 50) = 316$	16,116
52	$320 - 4(52 - 50) = 312$	16,224
53	$320 - 4(53 - 50) = 308$	16,324
x	$320 - 4(x - 50)$	$x \cdot [320 - 4(x - 50)]$

The "yield per tree" column was unchanged until we hit 51 trees per acre. *Each additional tree beyond 50 results in a decrease in yield of 4 pounds per tree.* After listing a few values beyond 50, we generalized by placing an x in for the number of trees and followed the pattern that was found for 51, 52, and 53 trees. If we now let $TY(x)$ be the total yield in pounds, we can represent this scenario by

$$TY(x) = x \cdot [320 - 4(x - 50)]$$

The reasonable domain for TY is $[50, 130)$. We do not include $x = 130$ since a quick check shows that at $x = 130$ the yield would be zero! So our problem is to maximize

$$TY(x) = x \cdot [320 - 4(x - 50)], \qquad \text{on } [50, 130)$$

The derivative is computed after algebraically simplifying $TY(x)$ to be

$$TY(x) = x \cdot [520 - 4x] = 520x - 4x^2$$
$$TY'(x) = 520 - 8x$$

Determining the critical value x such that $TY'(x) = 0$, gives

$$520 - 8x = 0$$
$$x = 65$$

We compute the second derivative to be

$$TY''(x) = -8$$

Evaluating the second derivative at $x = 65$ gives

$$TY''(65) = -8$$

So $TY''(65) < 0$, which means, according to the techniques of this section, that at $x = 65$ we have an absolute maximum. So Laurie's Lemons should plant 65 lemon trees per acre to maximize the total yield. ✦

✓ **CHECKPOINT 3**

Now work Exercise 31.

The method outlined in Example 3 is used in any situation where increasing one item decreases total output by some amount. For example, adding additional work stations in a machine shop may decrease the output per station. This scenario would follow the Example 3 process.

Sustainable Harvest

Our next example shows how to **maximize a sustainable harvest**. If one is in the business of harvesting fish, possibly by aquaculture, animals, or even trees, the ability to resist overharvesting is fundamental to the long-term success of the business. Overharvesting can inhibit the ability of the population to reproduce and sustain itself. This is illustrated in Example 4.

EXAMPLE 4 | **Maximizing a Sustainable Harvest**

Sussex County allows hunters to shoot deer during a limited open hunting season. The length of the season is carefully determined by the State Department of Natural Resources to ensure a harvest that is sustainable year after year. A state biologist has determined that, in Sussex County, the deer population after one year is given by

$$f(x) = 2.1x - 0.001x^2$$

where x is the original population size of the deer measured in hundreds. Determine the optimal population size and the yearly kill that it will sustain.

SOLUTION

We start by determining $h(x)$, the yearly harvest. The harvest is simply the difference between the deer population, $f(x)$, after one year and the original population, x. That is,

$$\begin{aligned} h(x) &= f(x) - x \\ &= 2.1x - 0.001x^2 - x \\ &= 1.1x - 0.001x^2 \end{aligned}$$

Since we have no idea of the population size, the interval on which we consider $h(x)$ is $(0, \infty)$. (We know that the population cannot be negative!) We wish to maximize $h(x)$ using the techniques in this section. Computing the derivative gives

$$h'(x) = 1.1 - 0.002x$$

The critical value x, such that $h'(x) = 0$, is found to be

$$\begin{aligned} 1.1 - 0.002x &= 0 \\ x &= 550 \end{aligned}$$

The second derivative is $h''(x) = -0.002$. Evaluating the second derivative at the critical value gives

$$h''(550) = -0.002$$

Thus, we have a maximum at $x = 550$. So the population of deer in Sussex County should grow to 55,000. It will then sustain a yearly harvest of

$$h(550) = 1.1(550) - 0.001(550)^2 = 302.5$$

that is, a yearly sustainable harvest of 30,250 deer per year.

Inventory Costs

A successful retail store must pay attention to the size of its inventory. Overstocking can lead to extra interest costs, excessive warehouse rental charges, and possibly the danger of the product becoming obsolete or damaged. On the other hand, too small an inventory involves additional paperwork in reordering and extra delivery charges. There is also the danger of running out of stock.

Let's take a look at a real inventory problem. Records from previous years show that Linger's Luxury Office Furniture sells 300 executive desks a year. To plan their inventory accurately, their analysts assume that the desks sell steadily throughout the year. They could order these desks in lots of size 300, 150, or 100 or in general lots of size x. Regardless of the lot size, on average this store will have $\frac{x}{2}$ executive desks in stock on which it must pay *inventory costs*. See Figure 5.5.7. Ordering all 300 at once could result in high *storage costs*, whereas reordering several times throughout the year could result in high *reorder costs*. In Example 5 we determine the best lot size that minimizes the total of storage costs and reorder costs.

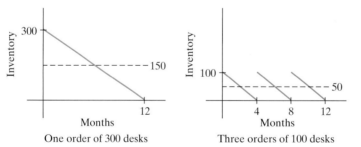

One order of 300 desks   Three orders of 100 desks   Orders of x desks

Figure 5.5.7

EXAMPLE 5 | **Minimizing Inventory Costs**

Linger's Luxury Office Furniture expects to sell 300 executive desks a year. Each desk costs the store $400, and there is a fixed charge of $800 per order. If it costs $200 to store an executive desk for a year, how large should each order be and how often should orders be placed to minimize the inventory costs?

SOLUTION

We begin by letting x represent the lot size, the number in each order. Our total inventory costs are represented by

$$\text{Total costs} = \text{storage costs} + \text{reorder costs}$$

We begin by determining the storage costs. We assume that the desks sell steadily throughout the year and that Linger's reorders x more when the stock is depleted. The inventory throughout the year would look like the graph of inventory as a function of the months in a year, as shown in Figure 5.5.8.

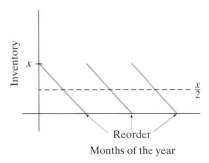

Figure 5.5.8

Since the number in stock varies from x to 0, we consider an *average inventory* of $\frac{x}{2}$. Since it costs $200 to store a desk for a year, we determine the storage costs as follows:

$$\text{Storage costs} = (\text{storage per item}) \cdot (\text{average number of items})$$

$$= 200 \cdot \left(\frac{x}{2}\right) = 100x$$

The next piece to determine is the reorder costs. In general, reorder costs are determined by

$$\text{Reorder costs} = (\text{cost per order}) \cdot (\text{number of orders}).$$

Since each desk costs $400, we know that ordering x desks costs $400x$. There is a fixed charge of $800 per order, which gives the cost per order as

$$\text{Cost per order} = 400x + 800$$

A yearly supply is 300 desks, and with x desks in each order, we find the number of orders in one year to be

$$\text{Number of orders} = \frac{300}{x}$$

This means that our reorder costs are given by

$$\text{Reorder costs} = (400x + 800) \cdot \left(\frac{300}{x}\right)$$

Thus, our total inventory costs, $IC(x)$, are given by

$$IC(x) = 100x + \frac{300}{x}(400x + 800), \qquad 0 < x \le 300$$

To minimize $IC(x)$, we employ the methods of this section. The derivative is

$$IC(x) = 100x + 120{,}000 + \frac{240{,}000}{x} = 100x + 120{,}000 + 240{,}000x^{-1}$$

$$IC'(x) = 100 - 240{,}000x^{-2} = 100 - \frac{240{,}000}{x^2}$$

We now find the critical value x, where $IC'(x) = 0$.

$$100 - \frac{240{,}000}{x^2} = 0$$

$$100 = \frac{240{,}000}{x^2}$$

$$x^2 = 2400$$

$$x \approx \pm 48.99$$

or, since x represents the lot size, we round to the nearest integer and get

$$x = \pm 49$$

We can immediately reject $x = -49$ and begin to verify that $x = 49$ minimizes inventory costs. Computing the second derivative gives

$$IC''(x) = 480{,}000 x^{-3} = \frac{480{,}000}{x^3}$$

Evaluating the second derivative at the critical value $x = 49$ gives

$$IC''(49) = \frac{480{,}000}{(49)^3} \approx 4.0799$$

Since $IC''(49) > 0$, we conclude that inventory costs, $IC(x)$, are minimized at $x = 49$. If there are 49 desks per order, the yearly total of 300 would require $\frac{300}{49} \approx 6.1224$ orders per year. Since this number of orders per year is impossible, it seems reasonable to round x to 50 (the number of orders would then be 6) and answer the question as follows:

> To minimize inventory costs, each lot size should have 50 desks with orders placed 6 times per year.

✓ **CHECKPOINT 4**

Now work Exercise 39.

SUMMARY

The main theme in this section was determining absolute extrema on intervals that are not closed. We illustrated how the second derivative handles this case, provided that there is only one critical value, $x = c$, on the interval where $f'(c) = 0$. Specifically, we presented the **Second Derivative Test for Absolute Extrema** and the **Strategy for Solving Applied Optimization Problems on an Open Interval.**

- **Second Derivative Test for Absolute Extrema:** For a continuous function, f, on *any* interval (open, closed, or half-open), and if $x = c$ is the *only* critical value in the interval where $f'(c) = 0$, if $f''(c)$ exists, then the point $(c, f(c))$ is an (1) **absolute minimum** if $f''(c) > 0$ or (2) **absolute maximum** if $f''(c) < 0$. If $f''(c) = 0$, then this test fails.

SECTION 5.5 EXERCISES

In Exercises 1–7, determine the absolute minimum of each function on the indicated interval.

1. $f(x) = 2x + \dfrac{6}{x}$ on $(0, 10)$

2. $f(x) = 4x + \dfrac{2}{x}$ on $(0, 10)$

✓ 3. $f(x) = 4x - 3 + \dfrac{2}{x}$ on $(0, 20)$

4. $f(x) = 2x - 5 + \dfrac{6}{x}$ on $(0, 20)$

5. $y = 3x^2 - 2 + \dfrac{6}{x^2}$ on $(0, 5)$

6. $y = 2x^2 - 3 + \dfrac{2}{x^2}$ on $(0, 5)$

7. $g(x) = 3x^2 - 1 + \dfrac{2}{x^2}$ on $(0, 10)$

In Exercises 8–15, determine the absolute maximum of each function on the given interval.

8. $f(x) = -2x - \dfrac{9}{x}$ on $(0, 20)$

9. $f(x) = -3x - \dfrac{2}{x}$ on $(0, 20)$

10. $y = -3x - \dfrac{5}{x}$ on $(0, 10)$

11. $y = -2x - \dfrac{3}{x}$ on $(0, 10)$

12. $g(x) = 3 - 2x^2 - \dfrac{3}{x^2}$ on $(0, 10)$

13. $g(x) = 2 - 3x^2 - \dfrac{2}{x^2}$ on $(0, 7)$

14. $f(x) = 2 - 3x - \dfrac{2}{x^2}$ on $(0, 10)$

15. $y = 3 - 2x - \dfrac{5}{x}$ on $(0, 10)$

APPLICATIONS

16. Melissa has 400 feet of fencing with which to enclose two adjacent lots as shown in Figure 5.5.9. Determine the dimensions x and y that maximize the total area. What is the maximum area?

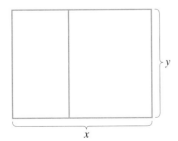

Figure 5.5.9

17. Melissa decided to put the two adjacent lots along a canal in such a way that the side labeled x in Figure 5.5.9

requires no fencing. Assuming that she has 400 feet of fencing, determine the dimensions x and y that maximize the total area. What is the maximum area?

18. A 1320-foot track encloses a rectangular region and the adjoining semicircular ends, as shown in Figure 5.5.10. Determine the dimensions x and y of the rectangle of maximum area. What is the maximum area?

Figure 5.5.10

19. Juan is an avid gardener. He wishes to enclose two identical rectangular plots each with an area of 1500 square feet, as shown in Figure 5.5.11. To keep raccoons out of his garden, the outer boundary requires a heavy fence costing $8.50 per foot, but the fence for the partition costs only $3.20 per foot. Determine the dimensions x and y that minimize the total cost of the fencing.

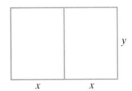

Figure 5.5.11

20. Richelle is an avid gardener. She wishes to enclose two identical rectangular plots each with an area of 1200 square feet. (Consult Figure 5.5.11 here, too.) To keep skunks out of her garden, the outer boundary requires a heavy fence costing $6 per foot, but the fence for the partition costs only $4 per foot. Determine the dimensions x and y that minimize the total cost of the fencing.

✓ 21. A serving tray is made from a 6-inch by 16-inch sheet of tin by cutting identical squares from the corners and then folding up the flaps. Determine the dimensions of the tray of maximum volume.

22. An open-top box used to carry small toys is to be made from a 10-inch by 12-inch piece of corrugated cardboard by cutting identical squares from the corners and then folding up the flaps. Determine the dimensions of the box of maximum volume.

Figure for Exercises 23 and 24.

23. An open-top box is made with a square base and should have a volume of 6000 cubic inches. If the material for the sides costs $0.20 per square inch and the material for the base costs $0.30 per square inch, determine the dimensions of the box that minimize the cost of the materials. (*Hint:* The

surface area is equal to the area of the base plus the area of the four sides.)

24. A closed box is made with a square base and must have a volume of 343 cubic inches. The material for the sides and the top cost $0.02 per square inch, and the material for base costs $0.04 per square inch. Determine the dimensions of the box that minimize the cost of the materials.

25. The marketing research department of *Shank,* a quarterly magazine for beginning golfers, has determined that the cost of printing, distributing, and advertising is given by

$$C(x) = 5 + 0.5x + 0.003x^2, \qquad x > 0$$

where x represents the number of magazines printed and sold each quarter, in hundreds, and $C(x)$ is in hundreds of dollars.

(a) Determine AC, the average cost function.

(b) Determine the value of x that minimizes $AC(x)$.

26. It has been determined that the cost $C(t)$, in dollars, of owning a new car (including a service and maintenance contract) that sells for $11,000 after t years can be approximated by

$$C(t) = 100t^2 + 400t + 11,000$$

Determine $AC(t)$ and the value for t that minimizes $AC(t)$.

27. It has been determined that the cost $C(t)$ in dollars, of owning a new car (including a service and maintenance contract) that sells for $15,000 after t years can be approximated by

$$C(t) = 100t^2 + 600t + 15,000$$

Determine $AC(t)$ and the value for t that minimizes $AC(t)$.

28. A 200-room hotel is filled when the room rate is $50 per day. For each $1.50 increase in the room rate, two fewer rooms are rented. Determine the room rate that maximizes daily revenue.

29. For the hotel described in Exercise 28, it costs $4.50 per day to clean and maintain each room occupied. Determine the room rate that maximizes daily profit.

30. Dudley's Delicious Apples determines that the annual yield per apple tree is fairly constant at 352 pounds when the number of trees per acre is 55 or fewer. For each additional tree over 55, the annual yield decreases by 5 pounds due to overcrowding. How many trees should be planted on each acre to maximize the annual yield from an acre?

✓ 31. Pear Paradise has determined that the annual yield per pear tree is fairly constant at 160 pounds when the number of trees per acre is 40 or fewer. For each additional tree over 40, the annual yield per tree decreases by 2 pounds due to overcrowding. How many trees should be planted on each acre to maximize the annual yield from an acre?

32. The Orange Works has determined that the annual yield per orange tree is fairly constant at 270 pounds per tree when the number of trees per acre is 30 or fewer. For each additional tree over 30, the annual yield per tree decreases by 3 pounds due to overcrowding. How many trees should be planted on each acre to maximize the annual yield from an acre?

33. Walt's Walnut Grove has determined that the annual yield per walnut tree is fairly constant at 50 pounds per tree when the number of trees per acre is 30 or fewer. For each additional tree over 30, the annual yield per tree decreases by 1.25 pounds due to overcrowding. How many trees should be planted on each acre to maximize the annual yield from an acre?

34. In a certain state, hunters are allowed to shoot deer during a limited hunting season. The length of the season is carefully determined by the State Department of Natural Resources to ensure a harvest that is sustainable year after year. A state biologist has determined that the yearly growth curve for the deer population is given by

$$f(x) = 1.5x - 0.002x^2$$

where x is measured in thousands. Determine the optimal population size and the yearly kill that it will sustain.

35. In a western U.S. state, it has been estimated that the yearly growth curve for elk can be approximated by

$$f(x) = 1.75x - 0.003x^2$$

where x is measured in hundreds. Determine the optimal population size and the yearly kill that it will sustain.

36. A certain state has determined that the yearly growth curve for rabbits can be approximated by

$$f(x) = 1.3x - 0.0003x^2$$

where x is measured in hundreds. Determine the optimal population size and the yearly kill that it will sustain.

37. Essex County is renowned for its pheasant hunting. It has been determined that the growth curve for pheasant can be approximated by

$$f(x) = 2.2x - 0.01x^2$$

where x is in thousands. Determine the optimal population size and the yearly kill that it will sustain.

In Exercises 38–46, find the lot size and how often orders should be placed to minimize the inventory costs. See Example 5. Round to the nearest lot size that produces an integer value for the number of orders placed.

38. Linger's Luxury Office Furniture expects to sell 400 junior executive desks a year. Each desk costs the store $300,

and there is a fixed charge of $600 per order. If it costs $200 to store a junior executive desk for a year, how large should each order be and how often should orders be placed to minimize the inventory costs?

✓ 39. Linger's Luxury Office Furniture expects to sell 200 presidential desks in a year. Each desk costs the store $600, and there is a fixed charge of $600 per order. If it costs $300 to store a presidential desk for a year, how large should each order be and how often should orders be placed to minimize inventory costs?

40. The VideoPhile Camera Store expects to sell 480 video camera carrying cases in a year. Each carrying case costs the store $26, and there is a fixed charge of $30 per order. If it costs $1 to store a carrying case for a year, how large should each order be and how often should orders be placed to minimize inventory costs?

41. The VideoPhile Camera Store described in Exercise 40 has recently found a new place to store the carrying cases. Here it costs $0.50 to store a carrying case for a year. Given this new amount, rework Exercise 40 and determine how large each order should be and how often orders should be placed to minimize inventory costs.

42. Sandkuhl's Appliances expects to sell 2500 gas stoves in a year. Each gas stove costs the store $500, and there is a fixed charge of $20 per order. If it costs $10 to store a gas stove for a year, how large should each order be and how often should orders be placed to minimize inventory costs?

43. Sandkuhl's Appliances expects to sell 2400 microwave ovens in a year. Each microwave oven costs the store $120, and there is a fixed charge of $54 per order. If it costs $9.00 to store a microwave oven for a year, how large should each order be and how often should orders be placed to minimize inventory costs?

44. Beagle's department store expects to sell 1200 pairs of blue jeans in a year. Each pair of blue jeans costs the store $12, and there is a fixed charge of $100 per order. If it costs $4 to store a pair of blue jeans for a year, how large should each order be and how often should orders be placed to minimize the inventory costs?

45. Gene's New and Used Cars expects to sell 300 cars in a year. On average, the cars cost $8500 each, and there is a fixed charge of $750 per order. If it costs $1500 to store a car for a year, how large should each order be and how often should orders be placed to minimize the inventory costs?

46. Simpson's Shoes expects to sell 1000 pairs of a certain running shoe in a year. Each pair of shoes costs the store $15, and there is a fixed charge of $150 per order. If it costs $6 to store a pair of shoes for a year, how large should each order be and how often should orders be placed to minimize inventory costs?

Exercises 47–49 examine that aspect of manufacturing known as *production runs*. These exercises can be handled in a manner similar to Exercises 38–46.

47. A publisher estimates that the annual demand for a book will be 5000 copies. Each book costs $17 to print, and setup costs are $1800 for each printing. If storage costs are $3 per book per year, determine how many books should be printed per run and how many printings will be needed in order to minimize costs.

48. A manufacturer estimates the demand for a new CD-ROM game to be 8000 per year. Each CD-ROM game costs $14 to produce, plus it costs $800 in setup costs. If a CD-ROM game can be stored for a year at a cost of $3, how many should be produced at a time and how many production runs will be needed in order to minimize costs?

49. A toy manufacturer estimates the demand for a toy car to be 50,000 per year. Each toy car costs $3 to make, plus setup costs of $500 for each production run. If it costs $2 to store a toy car for a year, how many toy cars should be manufactured at a time and how many production runs will be needed in order to minimize costs?

SECTION PROJECT

The source of the data for this project is the Federal Express web site (www.fedex.com).

(a) *Federal Express* states that any package that exceeds 165 inches in length plus girth are to be considered U.S. Domestic Freight. See Figure 5.5.12. Assuming that the front of the package is a square, determine the largest-volume package that *Federal Express* will accept and not declare it to be U.S. Domestic Freight.

(b) *Federal Express* states that the maximum length plus girth for its FedEx Overnight Freight and its FedEx 2Day Freight shipments is 300 inches. (See Figure 5.5.12). Assuming that the front of the package is a square and the maximum length plus girth is 300 inches, determine the dimensions that will maximize the volume and determine the maximum volume.

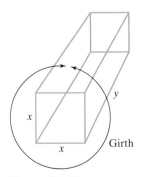

Figure 5.5.12

CHAPTER REVIEW EXERCISES

For each function graphed in Exercises 1–4:

(a) Determine where the function is increasing.

(b) Determine where the function is decreasing.

(c) Determine where the function is constant.

For the *derivatives* graphed in Exercises 5–8:

(a) List intervals where the function is increasing.

(b) List intervals where the function is decreasing.

(c) List where the function is constant.

1.

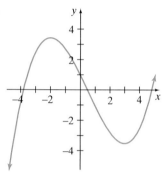

2.

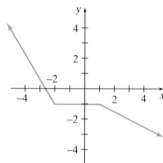

3.

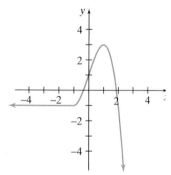

4.

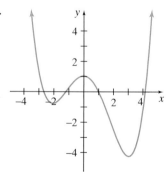

5.

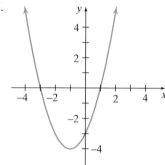

6.

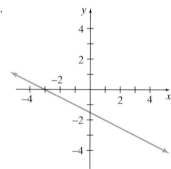

7.

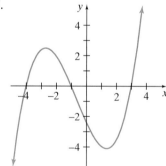

8.

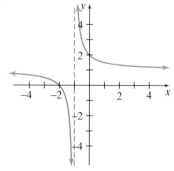

For each function in Exercises 9–14:

(a) Determine the critical values of the function.

(b) Determine intervals where the function is increasing and decreasing.

(c) Locate the relative extrema.

(d) Use a calculator to graph the function and verify your results to parts (a) and (b).

9. $f(x) = \dfrac{1}{2}x - 4$

10. $g(x) = -x^2 + 3x + 8$

11. $y = x^3 + 3x^2 - 24x + 20$

12. $f(x) = \sqrt{9 - x}$

13. $g(x) = (3x - 2)^4$

14. $y = \dfrac{x + 4}{3 - x}$

Use the derivatives given in Exercises 15–18 to determine the x-values where f has a relative maximum and/or a relative minimum.

15. $f'(x) = -3x + 8$

16. $f'(x) = 5x(x - 1)$

17. $f'(x) = x^2 - 9$

18. $f'(x) = (x - 4)^2(x - 1)(x + 2)^3(x + 5)$

For Exercises 19 and 20, sketch a graph of a continuous function that satisfies the given data.

19.

x	$f(x)$	$f'(x)$
0	1	−4
1	−2	−2
2	−3	0
3	−2	2
4	1	4
5	6	6

20.

x	$f(x)$	$f'(x)$
0	3	17
1	12	2
2	9	−7
3	0	−10
4	−9	−7
5	−12	2

21. The LuvYerPet Company is introducing a new brand of dog shampoo. The company has determined that the demand for the product can be modeled by

$$p(x) = 4.5e^{-0.03x}$$

where x represents the quantity demanded in thousands of containers of dog shampoo and $p(x)$ represents the price in dollars.

(a) Use the model to determine the price if there is demand for 30,000 containers of dog shampoo (that is, $x = 30$).

(b) Determine R, the revenue in thousands of dollars as a function of the quantity demanded in thousands of containers.

(c) Use the techniques of Section 5.1 to determine intervals where R is increasing and decreasing.

(d) Determine the relative maximum. Round the x- and y-coordinates to the nearest hundredth and interpret.

(e) Graph R in the viewing window [0, 100] by [0, 100] and verify your work.

22. Researchers have determined through experimentation that the percent concentration of a certain medicine during the first 20 hours after it has been administered can be approximated by

$$p(t) = \frac{150t}{t^2 + 4t + 4}, \qquad 0 \le t \le 20$$

where t is the time in hours after administration of the medication and $p(t)$ is the percent concentration.

(a) Compute $p'(t)$.

(b) Determine the critical values.

(c) Determine intervals where p is increasing and decreasing.

(d) Determine the relative maximum and interpret.

23. A manufacturer of telephones has the following costs when producing x telephones in one day, where $0 < x < 500$:

Fixed costs: $800

Unit production cost: $15 per telephone

Equipment maintenance: $\dfrac{x^2}{18}$

Therefore, the total cost of manufacturing x telephones in one day is given by

$$C(x) = 800 + 15x + \frac{x^2}{18}, \qquad 0 < x < 500$$

(a) Determine the average total cost, AC.

(b) Determine the critical value(s) for AC.

(c) Determine intervals where AC is increasing and decreasing.

(d) Determine the relative minimum and interpret.

24. The average value of farmland in the United States (excluding Alaska and Hawaii) can be modeled by

$$F(t) = 0.813t^3 - 2.63t^2 + 0.815t + 741.676, \qquad 1 \le t \le 17$$

where t is the number of years since 1979 and $F(t)$ is the average value of farmland in dollars per acre.

(a) Determine the critical value(s).

(b) Determine intervals where F is increasing and decreasing.

(c) Determine the relative extrema and interpret.

In Exercises 25–28, compute the first three derivatives of the given function.

25. $f(x) = x^3 - 7x^2 + 5$

26. $y = 1.3x^5 + 2.7x^4 - 6.1x^2 + 5.8x$

27. $y = \sqrt[3]{x - 5}$

28. $f(x) = x^2 + 5x - \ln 3x$

For each function in Exercises 29–35:

(a) Determine intervals where the function is concave up and where it is concave down.

(b) Determine any inflection points.

29. $f(x) = x^3 - x^2 + 5x - 7$ 30. $y = x^2 - 18x + 15$

31. $y = 12 - \sqrt{x}$ 32. $f(x) = \sqrt{25 - x^2}$

33. $y = 3x - \dfrac{5}{x}$ 34. $g(x) = e^{-0.7x}$

35. $y = e^{5x}$

For each function in Exercises 36–39:

(a) Determine intervals where the function is increasing and where it is decreasing.

(b) Determine the relative extrema.

(c) Determine intervals where the graph of the function is concave up and where it is concave down.

(d) Determine any inflection points.

36. $y = 3x^2 + 2x - 8$

37. $f(x) = 2x^3 + 3x^2 - 36x$

38. $g(x) = 2x + \dfrac{18}{x}$

39. $y = \ln (2 - x)$

40. Sketch the graph of a function that satisfies the following conditions.

Domain: All reals
Range: All reals greater than or equal to 2
Continuous for all reals

$f' > 0$ on $(-3, 0) \cup (3, \infty)$

$f' < 0$ on $(-\infty, -3) \cup (0, 3)$

$f'(0)$ is undefined

$f'' > 0$ on $(-\infty, 0) \cup (0, \infty)$

41. The Sweet Truth Cookie Company has determined that its daily profit is given by

$$P(x) = -0.5x^3 + 3x^2 + 68x - 133$$

where x is the number of hundreds of cookies baked and sold and $P(x)$ is the daily profit in dollars.

(a) Determine $P(4)$ and $P'(4)$ and interpret each.

(b) Determine intervals where P is increasing and where it is decreasing. Determine the relative maximum and interpret each coordinate.

(c) Determine intervals where the *marginal profit* is increasing and where it is decreasing. What can you say about the graph of P on these intervals?

(d) Determine the inflection point for the graph of P.

42. Suppose that blood flow through the artery of a certain animal species can be modeled by

$$f(x) = 0.317e^{0.0103x}, \qquad 20 \le x \le 120$$

where x represents arterial pressure, measured in millimeters of mercury, and $f(x)$ is blood flow, measured in milliliters per minute.

(a) Evaluate $f(40)$ and interpret.

(b) Evaluate $f'(40)$ and interpret.

(c) Show that the graph of f is concave up on the interval [20, 120]. Explain what this means.

43. Suppose that the number of mountain goats in a certain region can be modeled by

$$p(t) = \frac{1400}{1 + 43e^{-0.09t}}$$

where $p(t)$ represents the number of goats t years after the goats were first introduced to the region.

(a) Graph p in the viewing window [0, 100] by [0, 1500].

(b) Compute $p(84)$ and $p'(84)$ and interpret each.

(c) Determine $p''(t)$.

(d) On the interval [0, 100], determine intervals where the graph of p is concave up and where it is concave down. (*Hint:* Graph p'' and use the ZERO/ROOT command.)

(e) When is the population of goats growing fastest?

44. The following table gives the annual number of complaints against U.S. airlines from 1988–1996.

YEAR	COMPLAINTS
1988	21,493
1989	10,553
1990	7,703
1991	6,106
1992	5,639
1993	4,438
1994	5,179
1995	4,629
1996	5,778

Let x represent the number of years since 1987 and let y represent the annual number of complaints against U.S. airlines.

(a) Determine a cubic regression equation for the data in the form

$$y_1 = ax^3 + bx^2 + cx + d, \qquad 1 \le x \le 9$$

(b) Determine intervals where y_1 is increasing and where it is decreasing.

(c) Determine intervals where the graph of y_1 is concave up and where it is concave down.

(d) According to the model, when was the number of complaints lowest?

Use the graph of f shown to answer Exercises 45–51.

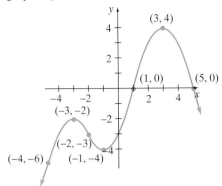

45. Determine intervals where f is positive and where f is negative.

46. Determine intervals where f is increasing and where f is decreasing.

47. Determine intervals where f' is positive and where f' is negative.

48. Determine intervals where f' is increasing and where f' is decreasing.

49. Determine intervals where the graph of f is concave up and where the graph of f is concave down.

50. Locate any relative extrema.

51. Locate any inflection points.

In Exercises 52–55, use the Second Derivative Test to locate any relative extrema, if they exist.

52. $f(x) = x^2 + 12x + 5$ 53. $y = x^3 - 5x^2 - 8x$

54. $y = 3x^4 - 2x^3 - 3x^2$ 55. $g(x) = \dfrac{1}{x^2 + 9}$

In Exercises 56–61, sketch the graph of the given function. Label all relative extrema, inflection points, and any asymptotes.

56. $y = x^3 - 9x^2 + 4$ 57. $f(x) = 2x^3 - \dfrac{3}{2}x^2 - 9x + 5$

58. $g(x) = x^4 - 14x^2 - 24x$ 59. $y = \dfrac{1}{x^2 - 4}$

60. $f(x) = 0.4x + 16 + \dfrac{8}{x}$ 61. $g(x) = e^x - 4x$

In Exercises 62 and 63, the graph of f'' is given. Use the graph to sketch graphs of f' and f. There are many correct answers.

62.

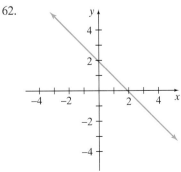

63.

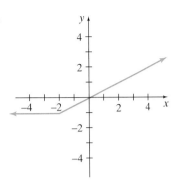

64. Frank is an artist who sells rock sculptures at local art festivals. He has determined that the relationship between price $p(x)$, in dollars per sculpture, and the quantity x demanded per day can be approximated by $p(x) = 85 - 0.12x^2$. He also knows that the cost function for the sculptures is given by $C(x) = 15x + 200$.

(a) Determine the daily revenue, R.

(b) Determine the daily profit, P.

(c) Compute $P'(x)$ and use it to determine intervals where P is increasing and where it is decreasing.

(d) Determine intervals where the graph of P is concave up and where it is concave down.

(e) Sketch the graph of P. Label all relative extrema and inflection points, if they exist.

65. (Continuation of Exercise 64.)

(a) Determine the average cost function, $AC(x)$.

(b) For what value of x is $AC(x)$ a minimum?

(c) Determine the average profit function, AP.

(d) For what value of x is $AP(x)$ a maximum?

66. The Honest Otis Auto Center offers a special maintenance and service contract for individuals who purchase a used car. This contract covers all recommended servicing of the vehicle, such as oil changes and lube jobs. Penelope just purchased a used car for $6000. The maintenance and service contract for her car costs $300 at the end of the first year and increases $150 per year thereafter, as long as Penelope owns the car. Using techniques of regression, she found that the total cost of the car (excluding items not covered by the service contract, such as gasoline) after t years is given by

$$C(t) = 75t^2 + 225t + 6000$$

(a) Define a function for the average cost per year, AC.

(b) When is the average cost per year a minimum? Round to the nearest tenth.

(c) What is the minimum average cost per year? Round to the nearest dollar.

67. The number of African American members of the U.S. House of Representatives can be modeled by

$$f(t) = 15.7e^{0.058t}, \qquad 1 \le t \le 15$$

where t is the number of years since 1980 and $f(t)$ is the number of African American members of the U.S. House of Representatives.

(a) Compute $f(9)$ and $f'(9)$ and interpret each.

(b) Compute $f'(t)$ and use it to show that, according to the model, the number of African American members of the U.S. House of Representatives was increasing from 1981 to 1995.

(c) Compute $f''(t)$ and use it to show that the graph of f is concave up on the indicated interval. Interpret what this means.

(d) Sketch a graph of f.

In Exercises 68–74, determine the absolute extrema of each function on the indicated interval.

68. $f(x) = x^2 - 5x + 3$ on $[0, 10]$

69. $y = x^3 + 6x^2 + 9x$ on $[-5, 5]$

70. $f(x) = x^3 - 6x$ on $[0, 5]$

71. $h(x) = \dfrac{1}{5}x^5 - \dfrac{5}{3}x^3 + 4x$ on $[-5, 5]$

72. $g(x) = \sqrt{x^3}$ on $[0, 5]$

73. $f(x) = x + \dfrac{4}{x}$ on $[1, 10]$

74. $y = x + \dfrac{4}{x}$ on $[-10, -5]$

75. From 1973 to 1997, the amount of money spent by the U.S. government on national defense can be modeled by

$$y = -0.0668x^3 + 2.04x^2 - 2.41x + 71.29, \qquad 1 \leq x \leq 25$$

where x is the number of years since 1972 and y is the amount spent on national defense, in billions of dollars. Determine the absolute maximum and the absolute minimum and interpret each.

76. Mike has 60 feet of fencing that he intends to use to build a rectangular enclosure for his Scottish terrier. He plans to build the enclosure against one side of his house, so the fencing is needed on just three sides of the enclosure. Determine the dimensions that will make the enclosure have the largest possible area.

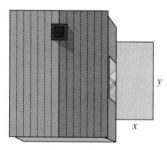

77. Your friend's family is planning to build a swimming pool. The water surface is to have an area of 1200 square feet. The pool will also have a 6-foot-wide walkway on the

east and west sides and an 8-foot walkway on the north and south sides. Let x be the length of the pool (east to west), and let y be the width (north to south). Given that $10 \leq x \leq 100$, find the dimensions that will minimize the *total* area of the swimming pool and walkways.

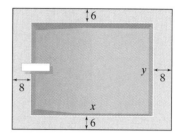

78. The GruntWorks Company analyzed the production costs for its new weight-training equipment and determined that its weekly cost function can be given by

$$C(x) = 0.001x^3 - 0.6x^2 + 217x + 7200, \qquad 50 \leq x \leq 550$$

where x represents the number of units produced each week and $C(x)$ is the weekly cost in dollars.

(a) Determine the average cost function, AC.

(b) Determine the production level that minimizes the average cost.

(c) Determine the average cost and the total cost at the production level found in part (b).

79. For the weight-training equipment in Exercise 78, the GruntWorks company gathered data for the price of a unit and the number of units demanded per week. The price–demand function that models these data is given by

$$p(x) = -0.002x^2 + 1.5x - 1.7, \qquad 50 \leq x \leq 550$$

where x represents the number of units demanded each week and $p(x)$ is the price, in dollars, that consumers pay for each unit to buy exactly x units per week.

(a) Determine R, the weekly revenue as a function of x.

(b) Determine the maximum weekly revenue.

(c) Determine the demand that maximizes weekly revenue.

(d) Determine the price that maximizes weekly revenue.

80. (Continuation of Exercise 79.) Using the functions C and R from Exercises 78 and 79, respectively:

(a) Determine P, the profit as a function of x.

(b) Determine the maximum weekly profit.

(c) Determine the demand that maximizes weekly profit.

(d) Determine the price that maximizes weekly profit.

In Exercises 81–84, determine the absolute minimum of each function on the indicated interval.

81. $y = 3x + \dfrac{6}{x}$ on $(0, 10)$

82. $f(x) = 5x - 9 + \dfrac{15}{x}$ on $(0, 6)$

83. $g(x) = 4x^2 - 2 + \dfrac{9}{x^2}$ on $(0, 5)$

84. $y = \dfrac{x^2 + 4}{x}$ on $(0, 4)$

In Exercises 85–88, determine the absolute maximum of each function on the given interval.

85. $f(x) = -4x - \dfrac{3}{x}$ on $(0, 10)$

86. $y = 3x - 5 + \dfrac{6}{x}$ on $(-5, 0)$

87. $f(x) = 5 - x^2 - \dfrac{9}{x^2}$ on $(0, 5)$

88. $g(x) = (9 - x)\sqrt{x}$ on $(2, 7)$

89. A window has the shape of a rectangle with an adjoining semicircle, as shown. The perimeter of the window is 12 feet. Determine the dimensions x and y if the area of the window is as large as possible. What is the maximum area?

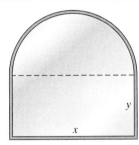

90. A newspaper publisher has determined that the daily cost of printing, distributing, and advertising is

$$C(x) = 0.0015x^2 + 0.6x + 4.5, \qquad x > 0$$

where x represents the daily circulation, in thousands, and $C(x)$ is the cost in hundreds of dollars.
(a) Determine AC, the average cost function.
(b) Determine the value of x that minimizes $AC(x)$.

91. An open-top box is made by cutting identical squares from the corners of a 15-inch by 24-inch piece of cardboard and then folding up the flaps. Determine the dimensions of the box of maximum volume.

92. The IzataFax store expects to sell 800 fax machines in a year. Each fax machine costs the store $80, and there is a fixed cost of $24 per order. If it costs $62 to store a fax machine for a year, how large should each order be and how often should orders be placed to minimize inventory costs? Round to the nearest lot size that produces an integer value for the number of orders placed each year.

93. Freddy's Fruit Orchard has determined that the annual yield per fruit tree is 150 pounds per tree when the number of trees per acre is 35 or fewer. For each additional tree over 35, the annual yield per tree decreases by 4 pounds due to overcrowding. How many trees should be planted on each acre to maximize the annual yield from an acre?

94. The United Parcel Service states that a lightweight package (under 30 pounds) is *oversize* if the sum of its length and girth is 84 inches. (Data from the UPS web site, www.ups.com.) Assuming that the front of the package is a square and the maximum length plus girth is 84 inches, determine the dimensions that will maximize the volume without requiring the customer to pay the oversize charge. What is the maximum volume?

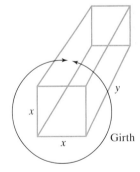

Integral Calculus

CHAPTER 6

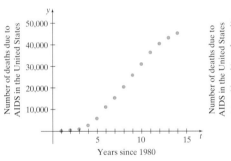

U.S. deaths due to AIDS, 1981 to 1994.

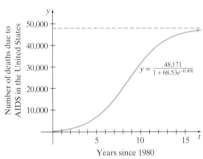

$y = \dfrac{48{,}171}{1 + 68.53e^{-0.49t}}$

Model of U.S. deaths due to AIDS.

What We Know

Our focus thus far has been on the study of the derivative and instantaneous rates of change. Among other things, we have seen how we can use the derivative to determine maximum and minimum values on open and closed intervals.

Where Do We Go

We now wish to discover techniques for undoing the process of differentiation. This technique, called *antidifferentiation*, can be used to recover functions if the rate of change function is given. We will then study the *Fundamental Theorem of Calculus*, which is the glue that holds differential and integral calculus together.

SECTION 6.1   THE INDEFINITE INTEGRAL

Until now, we have focused almost exclusively on how to determine the derivative and interpret the resulting rate of change function. For example, we found the derivative of a profit function and learned that this derivative is a rate of change function that measures the marginal profit. But what if we are given a *rate of change function* (that is, a rate function or derivative) and wish to find the original function? In this section, we will learn the basics of *integration,* a process that finds the original function, given its derivative, by reversing the differentiation. Then we will consider applications of integration.

The Indefinite Integral

To see how differentiation can be reversed, let's start with a common rate function from business. The WorkSharp Company determines that the marginal cost function for their designer suspenders is

$$MC(x) = C'(x) = 2x, \qquad 0 \le x \le 10$$

where x is the number of designer suspenders produced in thousands and $MC(x)$ is the marginal cost in thousands of dollars.

To find the company's cost function, we need to determine a cost function with the property that $\frac{d}{dx}(C(x)) = MC(x)$. In words, we seek a cost function so that when we take its derivative we get $MC(x) = 2x$. A function such as this is referred to as an **antiderivative** of the given function MC. To gain insight into how this antidifferentiation process is done, let's review our basic differentiation rules.

From Your Toolbox

 For differentiable functions f and g with real numbers n and k, we have:
- If $f(x) = x^n$, then $f'(x) = n \cdot x^{n-1}$.   (Power Rule)
- If $f(x) = kx$, then $f'(x) = k$.   (Constant Rule)
- If $h(x) = f(x) \pm g(x)$, then $h'(x) = f'(x) \pm g'(x)$.   (Sum and Difference Rule)

Since the WorkSharp Company's marginal cost function is $2x$, we need an expression whose derivative is $2x$. Since the term $2x$ is linear (first degree), the Power Rule tells us that we need a quadratic function (second degree) for the antiderivative. Notice that if we choose the quadratic x^2 then $\frac{d}{dx}(x^2) = 2x$. This is just what we wanted! So is $C(x) = x^2$ the cost function we want? Not quite, because there are many functions whose derivatives are $2x$, including $x^2 + 3$, $x^2 + \frac{11}{2}$, and $x^2 - 10$. In fact, if C is any real number constant, then $x^2 + C$ is the antiderivative that we seek. We call C an **arbitrary constant.** To determine the WorkSharp Company's actual cost function, it would appear that we need a little more information.

Before continuing, let's generalize what we have done.

Antiderivative

We call F an antiderivative of f on an interval if $F'(x) = f(x)$ for all x in the interval. In other words,

$$\frac{d}{dx}[F(x)] = f(x)$$

We call $F(x) + C$, where C is any real number constant, the **general antiderivative** of f on an interval if

$$\frac{d}{dx}[F(x) + C] = f(x)$$

for all x in the interval.

NOTE: The process of determining an antiderivative is called **antidifferentiation.** Antidifferentiation is the **inverse operation** of differentiation.

Our definition provides an easy way to determine if a function is an antiderivative of another function. Example 1 illustrates how we can do this.

EXAMPLE 1 | **Verifying General Antiderivatives**

Determine if the function F is the general antiderivative of the function f.

(a) $F(x) = \frac{2}{3}x^{3/2} + 4x + C$; $f(x) = \sqrt{x} + 4$

(b) $F(x) = 2x^4 - x + C$; $f(x) = \frac{2}{3}x^3 - 1$

SOLUTION

For each, we need to determine if $\frac{d}{dx}F(x) = f(x)$.

(a) Differentiating, we get

$$\frac{d}{dx}F(x) = \frac{d}{dx}\left(\frac{2}{3}x^{3/2} + 4x + C\right) = \frac{2}{3} \cdot \frac{3}{2}x^{1/2} + 4 + 0$$

$$= x^{1/2} + 4 = \sqrt{x} + 4 = f(x)$$

So $F(x) = \frac{2}{3}x^{3/2} + 4x + C$ *is* the general antiderivative of $f(x) = \sqrt{x} + 4$.

(b) Again, computing the derivative of $F(x)$ yields

$$\frac{d}{dx}F(x) = \frac{d}{dx}(2x^4 - x + C) = 8x^3 - 1 \neq f(x)$$

Since $\frac{d}{dx}F(x) \neq f(x)$, we find that $F(x) = 2x^4 - x + C$ is *not* the general antiderivative of f. ∎

Another way to represent the general antiderivative of a function f is by

$$\int f(x)\, dx$$

which is called the **indefinite integral** of f.

Indefinite Integral

If $F'(x) = f(x)$, then

$$\int f(x)\, dx = F(x) + C$$

In this notation, $\int$ is the **integral sign,** $f(x)$ is the **integrand,** and C is any real number constant. dx tells us the variable of integration.

NOTE: If we can determine the indefinite integral of a function, we say that the function is **integrable**. The process of determining a general antiderivative is called **integration**.

With this new notation, we can write the result of part (a) of Example 1 as

$$\int \left(\sqrt{x} + 4 \right) dx = \frac{2}{3} x^{3/2} + 4x + C$$

Now let's look for a method to determine the indefinite integral for an expression involving x^n. First, we will consider the expression x^3. We know from the Power Rule for differentiation that the antiderivative of x^3 must be fourth degree. Since we know that $\frac{d}{dx} x^4 = 4x^3$, then $\frac{1}{4} \frac{d}{dx}(x^4) = \frac{1}{4}(4x^3) = x^3$. Thus,

$$\int x^3\, dx = \frac{1}{4} x^4 + C$$

Notice in the antiderivative $\frac{1}{4} x^4 + C$ that the power of x is one more than in x^3. Also, the antiderivative includes a constant multiple that is the reciprocal of the exponent 4.

On p. 353 we said that antidifferentiation is the inverse operation of differentiation. Here we are using the inverse operations of the Power Rule for differentiation to find an antiderivative. Instead of *subtracting* 1 from the exponent, we *add* 1, and instead of *multiplying* by the constant, we *divide*. We call this rule the **Power Rule for Integration.**

Power Rule for Integration

For any real number n, where $n \neq -1$, the indefinite integral of x^n is

$$\int x^n\, dx = \frac{1}{n+1} x^{n+1} + C$$

NOTE: The restriction that n cannot be -1 is because a value of -1 for n would make the denominator of the coefficient zero. In Section 6.5 we will learn the rule for the indefinite integral $\int x^{-1}\, dx = \int \frac{1}{x}\, dx$.

EXAMPLE 2

Using the Power Rule for Integration

Determine the following indefinite integrals.

(a) $\displaystyle\int x^8\, dx$ 　　　　(b) $\displaystyle\int \sqrt[4]{x}\, dx$ 　　　　(c) $\displaystyle\int \frac{1}{x^5}\, dx$

SOLUTION

(a) Applying the Power Rule for Integration, we get

$$\int x^8\, dx = \frac{1}{8+1} x^{8+1} + C = \frac{1}{9} x^9 + C = \frac{x^9}{9} + C$$

Notice that the coefficient $\frac{1}{9}$ is the reciprocal of the exponent 9.

(b) Here we use the rules of algebra to write the integrand in the form x^n. Since $\sqrt[4]{x} = x^{1/4}$, we can write

$$\int \sqrt[4]{x}\, dx = \int x^{1/4}\, dx = \left(\dfrac{1}{\frac{1}{4}+1}\right) \cdot x^{(1/4)+1} + C$$

$$= \dfrac{1}{\frac{5}{4}} \cdot x^{(1/4)+(4/4)} + C = \dfrac{4}{5}x^{5/4} + C$$

Notice that the coefficient $\frac{4}{5}$ is the reciprocal of the exponent $\frac{5}{4}$.

(c) Again, the rules of algebra come into play when rewriting the integrand. Here we can write $\dfrac{1}{x^5}$ as x^{-5} and then compute

$$\int \dfrac{1}{x^5}\, dx = \int x^{-5} dx = \dfrac{1}{-5+1}x^{-5+1} + C = \dfrac{1}{-4}x^{-4} + C$$

$$= -\dfrac{1}{4} \cdot \dfrac{1}{x^4} + C = -\dfrac{1}{4x^4} + C$$

Notice that the coefficient $-\dfrac{1}{4}$ is the reciprocal of the exponent -4.

✓ CHECKPOINT 1

Now work Exercise 17.

The next rule involves integration of a constant. To find $\int 5\, dx$, we need a function whose derivative is 5. Since we know that $\dfrac{d}{dx}(5x) = 5$, we can write $\int 5dx = 5x + C$. This suggests the following rule:

> **Constant Rule for Integration**
>
> If k is any real number, then the indefinite integral of k is
>
> $$\int k\, dx = kx + C$$

Interactive Activity

Use differentiation to show that the indefinite integral of $f(x) = e^{2x}$ is $F(x) = \frac{1}{2}e^{2x}$. Do the same for $f(x) = \dfrac{1}{x+3}$ and $F(x) = \ln(x+3)$. Explain why the Power Rule for Integration does not apply in these cases.

Some of the properties of integration are much the same as for limits and for differentiation. Consider the integral $\int (2x - 5)\, dx$. We know that $\int (2x)\, dx = x^2 + C$, and we can show from the Constant Rule for Integration that $\int 5\, dx = 5x + C$. So its seems natural that the integral of the linear function $2x - 5$ is $\int (2x - 5)\, dx = x^2 - 5x + C$, which you should check using differentiation. We could have rewritten the expression as

$$\int (2x - 5)\, dx = \int 2x\, dx - \int 5\, dx = x^2 - 5x + C$$

Some may be concerned that since there are two integrals there must be two constants of integration C. But here we are simply adding the two constants to get a new constant, which we call C.

> **Sum and Difference Rule for Integration**
>
> For integrable functions f and g, we have
>
> $$\int [f(x) \pm g(x)]\, dx = \int f(x)\, dx \pm \int g(x)\, dx$$

NOTE: This rule means that we are allowed to determine indefinite integrals term by term.

EXAMPLE 3 | **Using the Sum and Difference Rule for Integration**

Apply the Sum and Difference Rule for Integration to determine the indefinite integrals.

(a) $\displaystyle\int (x^2 + 3)\, dx$ (b) $\displaystyle\int \left(\sqrt[3]{x} + 5\right) dx$

SOLUTION

(a) The Sum and Difference Rule for Integration yields

$$\int (x^2 + 3)\, dx = \int x^2\, dx + \int 3\, dx = \frac{1}{3}x^3 + 3x + C$$

(b) We begin by writing the first term in the integrand in the form x^n to get

$$\int \left(\sqrt[3]{x} + 5\right) dx = \int (x^{1/3} + 5)\, dx$$

Applying the Sum and Difference Rule gives

$$= \int x^{1/3}\, dx + \int 5\, dx = \frac{1}{\frac{1}{3}+1} x^{(1/3)+1} + 5x + C$$

$$= \frac{1}{\frac{4}{3}} x^{4/3} + 5x + C = \frac{3}{4} x^{4/3} + 5x + C$$

Our next rule addresses functions with coefficients other than 1. In Chapter 2 we learned that

$$\frac{d}{dx}[k \cdot f(x)] = k \cdot \frac{d}{dx}[f(x)]$$

For example,

$$\frac{d}{dx}[3x^2] = 3 \cdot \frac{d}{dx}[x^2] = 3 \cdot 2x = 6x$$

Since integration is the inverse operation of differentiation, we know that

$$\int 6x\, dx = 3x^2 + C$$

But how would we get the general antiderivative if we did not already know it? We could try "pulling out" the constant much as we did with the constant multiple differentiation rule.

$$\int 6x\, dx = 6 \cdot \int x\, dx = 6 \cdot \frac{1}{2}x^2 + C = 3x^2 + C$$

This illustrates that when we are integrating a coefficient times a function we can put the coefficient in front of the integral sign and integrate the function.

Coefficient Rule for Integration

Given any real number coefficient c and integrable function f,

$$\int c \cdot f(x)\, dx = c \cdot \int f(x)\, dx$$

Armed with these integration rules, we can find the indefinite integral of any polynomial function. For example,

$$\int (x^2 - 3x + 4)\, dx = \int x^2\, dx - 3\int x\, dx + \int 4\, dx$$

$$= \frac{x^3}{3} - 3 \cdot \frac{x^2}{2} + 4x + C = \frac{1}{3}x^3 - \frac{3}{2}x^2 + 4x + C$$

Example 4 shows that we do not necessarily have to integrate a function with respect to x.

EXAMPLE 4

Integrating Polynomial Functions

Determine $\int (-2t^3 + 3t + 5)\, dt$.

SOLUTION

Here the dt indicates that we are integrating the cubic $-2t^3 + 3t + 5$ with respect to t. The same rules that we have developed in this section for the independent variable x apply and give us

$$\int (-2t^3 + 3t + 5)\, dt = \int -2t^3\, dt + \int 3t\, dt + \int 5\, dt$$

$$= -2 \int t^3\, dt + 3 \int t\, dt + \int 5\, dt$$

$$= -2 \cdot \frac{t^4}{4} + 3 \cdot \frac{1}{2}t^2 + 5t + C$$

$$= -\frac{1}{2}t^4 + \frac{3}{2}t^2 + 5t + C$$

✓ **CHECKPOINT 2**

Now work Exercise 27.

Recovering Functions from Rate Functions

Through much of our remaining study of the calculus, we will study the properties of rate functions.

Rate Function

A function that gives an instantaneous rate of change is a **rate function.**

Recovering functions from rate functions is very useful in many applications in which the rate of change is given. For example, let's say that we have a derivative $f'(x) = x^2$ and we wish to know the function from which it came, that is, f. By integrating, we know that

$$f(x) = \int x^2\, dx = \frac{1}{3}x^3 + C$$

Each value of C produces a different function, as shown in Figure 6.1.1.

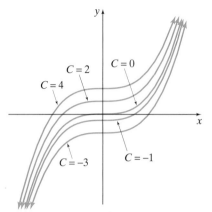

Figure 6.1.1   Graphs of $f(x) = \frac{1}{3}x^3 + C$ for $C = -3, -1, 0, 2, 4.$

Now suppose we know that an ordered pair, say (3,11), is on the graph of f. By replacing $f(x)$ with 11 and x with 3, we can write an equation to find the value of the integration constant C.

$$f(x) = \frac{1}{3}x^3 + C$$

$$11 = \frac{1}{3}(3)^3 + C$$

$$11 = \frac{1}{3} \cdot 27 + C$$

$$11 = 9 + C$$

$$2 = C$$

So, for the derivative $f'(x) = x^2$ and point (3, 11) on the graph of f, the corresponding function f is $f(x) = \frac{1}{3}x^3 + 2$. This is illustrated in Figure 6.1.2.

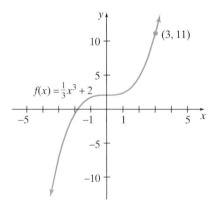

Figure 6.1.2

EXAMPLE 5 | **Recovering a Function from a Rate Function**

The HonCo marketing firm finds the rate of change of total sales of a product featured on an infomercial is given by the function

$$R(t) = 15\sqrt[3]{t}, \qquad 0 \leq t \leq 20$$

where t represents the number of months that the infomercial has aired and $R(t)$ represents the rate of change of total sales of the product measured in $\frac{\text{thousand units}}{\text{month}}$.

(a) Compute $R(10)$ and interpret.

(b) Knowing that there were 500 units sold in trial marketing just as the infomercial began airing, find a total sales function S for the product.

(c) Compute $S(10)$ and interpret.

SOLUTION

(a) First, note that R is a rate function because it measures the rate of change in total sales with respect to time. Evaluating the sales rate function at $t = 10$ yields

$$R(10) = 15\sqrt[3]{10} \approx 32.32 \; \frac{\text{thousand units}}{\text{month}}$$

This means that in the tenth month of airing the infomercial the total sales of the product were increasing at a rate of $32.32 \; \frac{\text{thousand units}}{\text{month}}$, or about $32{,}320$ units per month.

(b) Integrating the sales rate function R with respect to time t recovers the total sales function.

$$S(t) = \int R(t)\, dt = \int 15\sqrt[3]{t}\, dt = \int 15 t^{1/3}\, dt$$

$$= 15 \int t^{1/3}\, dt = 15 \left(\frac{1}{\frac{1}{3}+1} t^{(1/3)+1} + C \right)$$

$$= 15 \cdot \frac{3}{4} \cdot t^{4/3} + C = \frac{45}{4} \sqrt[3]{t^4} + C$$

Thus, the sales function has the general form $S(t) = \frac{45}{4}\sqrt[3]{t^4} + C$. Since we know that 500 units were sold when the infomercial began airing, this means that, when $t = 0$, $S(t) = 0.5$ [that is, the initial value is $(t, S(t)) = (0, 0.5)$. Substituting these values in our general form, we get

$$S(t) = \frac{45}{4}\sqrt[3]{t^4} + C$$

$$0.5 = \frac{45}{4}\sqrt[3]{0^4} + C$$

$$0.5 = 0 + C$$

$$0.5 = C$$

Thus, the total sales function is

$$S(t) = \frac{45}{4}\sqrt[3]{t^4} + 0.5$$

where $S(t)$ represents the total sales in thousands.

(c) Evaluating the sales function at $t = 10$, we get

$$S(10) = \frac{45}{4} \sqrt[3]{10^4} + 0.5 = \frac{45}{4} \cdot \sqrt[3]{10,000} + 0.5$$

$$\approx 242.87 \text{ thousand units}$$

Thus, during the tenth month of airing the infomercial the total sales were 242.87 thousand, or about 242,870 units of the product.

In Example 5, the rate function R is measured in $\frac{\text{thousand units}}{\text{month}}$ and the recovered sales function is measured in thousands of units. *Generally, for a rate function, the units are given in $\frac{\text{dependent units}}{\text{independent unit}}$ and the function recovered is measured with dependent units.*

Recovering Business Functions

We have seen that given a rate function, more commonly called a derivative, we can recover the function from which it came. The rate functions, or derivatives, associated with the cost C, revenue R, and profit P functions are the **marginal functions,** denoted by MC, MR and MP, respectively. Let's review these functions.

From Your Toolbox

Let x be the number of units produced and sold of a certain product in a specific time interval.

- The **Marginal Cost Function** $MC(x) = C'(x)$ is the approximate cost of producing one additional unit at a production level x.
- The **Marginal Revenue Function** $MR(x) = R'(x)$ is the approximate loss or gain in revenue from producing and selling one additional unit at a production level x.
- The **Marginal Profit Function** $MP(x) = P'(x)$ is the approximate loss or gain in profit from producing and selling one additional unit at a production level x.

Recovering business functions involves integrating their corresponding marginal functions and evaluating at some given value. We illustrate this recovery in Example 6.

EXAMPLE 6 | **Recovering a Profit Function**

The Best Dressed Clothing Company finds that its marginal profit MP is linear and has the form $MP(q) = mq + b$, where m and b are constants. The company gets about \$171 additional profit from producing the 101st sport coat and \$169 additional profit from producing the 151st sport coat in each production run.

(a) Determine the marginal profit function MP.

(b) Knowing that the company gets \$11,300 profit from 150 sport coats, find the profit function P.

SOLUTION

(a) Since the marginal profit function is linear, we start by finding the slope m of the linear function. From the definition of marginal profit, we know that $MP(100) = 171$ and $MP(150) = 169$. So the slope, or average rate of change, of the marginal profit function is

$$m = \frac{\text{change in } MP(q)}{\text{change in } q} = \frac{169 - 171}{150 - 100} = \frac{-2}{50} = -\frac{1}{25}$$

Now we can use the CF form with point $(q, MP(q)) = (100, 171)$.

$$MP(q) = -\frac{1}{25}(q - 100) + 171$$

$$= -\frac{1}{25}q + 4 + 171 = -\frac{1}{25}q + 175$$

So the marginal profit function is $MP(q) = -\frac{1}{25}q + 175$.

(b) To recover the profit function P, we must integrate MP with respect to q. Mathematically, this means that

$$P(q) = \int MP(q)\, dq = \int \left(-\frac{1}{25}q + 175 \right) dq$$

$$= -\frac{1}{25} \cdot \frac{1}{2}q^2 + 175q + C = -\frac{1}{50}q^2 + 175q + C$$

To find the value of C, we use the point $(q, P(q)) = (150, 11{,}300)$ [that is, $q = 150$, $P(q) = 11{,}300$].

$$11{,}300 = -\frac{1}{50}(150)^2 + 175(150) + C$$

$$11{,}300 = -\frac{1}{50}(22{,}500) + 26{,}250 + C$$

$$11{,}300 = -450 + 26{,}250 + C$$

$$11{,}300 = 25{,}800 + C$$

$$-14{,}500 = C$$

Thus, the recovered profit function is $P(q) = -\frac{1}{50}q^2 + 175q - 14{,}500$.

SUMMARY

This section began with a discussion of the **antiderivative** and quickly led us to the **indefinite integral.** We examined how to integrate powers of functions, sums and differences of functions, and simple radical functions. Then we discussed how to recover a function when its rate function is given. We used this method to recover the business functions from the marginal business functions.

Important Integration Rules

- **Power Rule:** $\displaystyle \int x^n\, dx = \frac{1}{n+1}x^{n+1} + C, n \neq -1$

- **Constant Rule:** $\displaystyle \int k\, dx = kx + C$

- **Sum and Difference Rule:** $\int [f(x) \pm g(x)]\, dx = \int f(x)\, dx \pm \int g(x)\, dx$

- **Coefficient Rule:** $\int c \cdot f(x)\, dx = c \cdot \int f(x)\, dx$

SECTION 6.1 EXERCISES

In Exercises 1–10, determine if the function F is the general antiderivative of the function f.

1. $F(x) = 6x + C;\ f(x) = 6$

2. $F(x) = \dfrac{1}{2}x + C;\ f(x) = \dfrac{1}{2}$

3. $F(x) = x^8 + C;\ f(x) = x^7$

4. $F(x) = \dfrac{1}{2}x^3 + C;\ f(x) = x^2$

5. $F(t) = 8t + \dfrac{t^2}{2} + C;\ f(t) = 8 + t$

6. $F(t) = 7t + et + C;\ f(t) = 7 + e$

7. $F(x) = x + C;\ f(x) = 0$

8. $F(x) = C;\ f(x) = 0$

9. $F(x) = \dfrac{x^{11}}{11} + C;\ f(x) = x^{10}$

10. $F(x) = -\dfrac{5}{6x^6} + C;\ f(x) = \dfrac{5}{x^7}$

In Exercises 11–20, use the Power Rule for Integration to find the indefinite integrals.

11. $\int x^4\, dx$

12. $\int x^9\, dx$

13. $\int x^{2.31}\, dx$

14. $\int x^{0.27}\, dx$

15. $\int \dfrac{1}{t^3}\, dt$

16. $\int \dfrac{1}{t^{11}}\, dt$

✓ 17. $\int \sqrt[4]{x^5}\, dx$

18. $\int \sqrt{x^5}\, dx$

19. $\int \dfrac{1}{\sqrt[3]{x}}\, dx$

20. $\int \dfrac{1}{\sqrt{x}}\, dx$

For Exercises 21–44, determine the indefinite integral.

21. $\int 0.4x^6\, dx$

22. $\int 0.5x^7\, dx$

23. $\int (2x + 3)\, dx$

24. $\int (5x - 2)\, dx$

25. $\int \left(\dfrac{2}{3}x + 4\right) dx$

26. $\int \left(\dfrac{1}{2}t + 5\right) dt$

✓ 27. $\int (3t^2 + 2t + 10)\, dt$

28. $\int (4x^4 + 5x - 6)\, dx$

29. $\int (1 - 2x^2 + 3x^3)\, dx$

30. $\int (5 - 2x + x^3)\, dx$

31. $\int (6.21x^2 + 0.03x - 4.01)\, dx$

32. $\int (0.03x^2 - 0.21x + 4.02)\, dx$

33. $\int \left(\dfrac{1}{x^2} - \dfrac{3}{x^3}\right) dx$

34. $\int \left(\dfrac{3}{x^4} + 6x^5\right) dx$

35. $\int (3 + 2\sqrt{x})\, dx$

36. $\int (\sqrt{x} - 3x^{3/2})\, dx$

37. $\int \left(\dfrac{3}{x^3} + 2x^{3/2} - 4\right) dx$

38. $\int \left(x^{5/2} - \dfrac{4}{x^5} - \sqrt{x}\right) dx$

39. $\int \dfrac{3t^3 - 2t}{6t}\, dt$

40. $\int \dfrac{4x^4 - 5x^3}{x^2}\, dx$

41. $\int (0.1z^{-3} + 2z^{-2} + z^3)\, dz$

42. $\int (5x^{-5} + 3x^{-4})\, dx$

43. $\int (2t^{0.13} + 5)\, dt$

44. $\int (0.1x^{0.318} + 7x)\, dx$

In Exercises 45–58, solve the given rate function using the given value. Note that $f(a) = b$ corresponds to the ordered pair (a, b).

45. $f'(x) = -2;\ f(0) = 4$

46. $f'(x) = 7;\ f(0) = 1$

47. $f'(x) = 5x;\ f(0) = 0$

48. $f'(x) = 3x;\ f(0) = 0$

49. $f'(x) = 2x - 3;\ f(0) = 4$

50. $g'(x) = 5 - 6x;\ g(0) = 6$

51. $f'(t) = 500 - 0.05t;\ f(0) = 40$

52. $\dfrac{dy}{dx} = 4x^2 - 6x;\ y(1) = 0$

53. $s'(x) = 2x^{-2} + 3x^{-3} - 1;\ s(1) = 2$

54. $M'(x) = \dfrac{2x^4 - x}{x^3};\ M(1) = 10$

55. $y' = \dfrac{5t + 2}{\sqrt[3]{t}};\ (t, y) = (0, 1)$

56. $p'(t) = -\dfrac{10}{t^2} + t$; $p = 20$ when $t = 1$

57. $R'(t) = \dfrac{1 - t^4}{t^3}$; $(t, R) = (1, 4)$

58. $y'(t) = \dfrac{\sqrt{t^3} - t}{\left(\sqrt{t}\right)^3}$; $y = 4$ when $t = 9$

APPLICATIONS

59. Consider the marginal profit for producing q units of a product given by the linear function

$$MP(q) = 40 - 0.05q$$

where $MP(q)$ is in dollars per unit.

(a) Knowing that $P = 0$ when $q = 0$, recover the profit function P.

(b) Use your solution from part (a) to find the total profit realized from selling 200 units of product.

60. Consider the marginal profit for producing x units of a product given by the linear function

$$MP(x) = -0.20x + 40$$

where $MP(x)$ is in dollars per unit.

(a) Knowing that $P = -10$ when $x = 0$, recover the profit function P.

(b) Use your solution from part (a) to find the total profit realized from selling 150 units of product.

61. The marginal revenue function for the FrontRide Bus Company is given by

$$MR(x) = R'(x) = 0.000045x^2 - 0.03x + 3.75,$$
$$0 \le x \le 500$$

(a) Knowing that $R(x) = 0$ when $x = 0$, recover the revenue function R.

(b) Find the price–demand function p for the bus company. Recall that the general formula for the revenue function is

$$R(x) = \left(\begin{array}{c}\text{quantity} \\ \text{produced}\end{array}\right) \cdot \left(\begin{array}{c}\text{price of} \\ \text{each unit}\end{array}\right) = x \cdot p(x)$$

(c) What should the price be when the demand is 100 passengers?

62. The daily marginal revenue function for the BlackDay Sunglasses Company is given by

$$MR(x) = R'(x) = 30 - 0.0003x^2, \qquad 0 \le x \le 540$$

where x represents the number of sunglasses produced and sold.

(a) Knowing that $R(x) = 1487.5$ when $x = 50$, recover the revenue function R.

(b) Find the price–demand function p for the sunglasses. Recall that the general formula for the revenue function is

$$R(x) = \left(\begin{array}{c}\text{quantity} \\ \text{produced}\end{array}\right) \cdot \left(\begin{array}{c}\text{price of} \\ \text{each unit}\end{array}\right) = x \cdot p(x)$$

(c) What should the price be when the demand is 250 sunglasses?

63. The marginal average cost function for producing x promotional banners is given by

$$MAC(x) = -\dfrac{100}{x^2}, \qquad x > 0$$

(a) Knowing that it costs \$2.50 per banner to produce 100 banners, recover the average cost function AC.

(b) Knowing the average cost function AC from part (a), find the cost function C. Recall that, by definition, $AC(x) = \dfrac{C(x)}{x}$.

(c) Using the cost function from part (b), evaluate $C(100)$ and interpret.

64. The marginal average cost function for producing x QuickVid digital cameras is given by

$$MAC(x) = 0.03x^2 - 0.04x + 5, \qquad 0 < x \le 50$$

(a) Knowing that it costs \$422 per camera to produce 20 cameras, recover the average cost function AC.

(b) Knowing the average cost function AC from part (a), find the cost function C. Recall that, by definition, $AC(x) = \dfrac{C(x)}{x}$.

(c) Using the cost function from part (b), evaluate $C(20)$ and interpret.

65. The rate of arrests for drug abuse violations by U.S. adults is given by the function

$$a'(t) = -0.3t^2 + 10.56t - 40.31, \qquad 0 \le t \le 25$$

where t represents the number of years since 1970 and $a'(t)$ represents the arrest rate measured in $\dfrac{\text{thousands of arrests}}{\text{year}}$.

(a) Graph a' in the viewing window $[0, 25]$ by $[-40, 55]$. Use the graph to analyze where the arrests are increasing and decreasing. Interpret.

(b) Knowing that there were 480 thousand adults arrested for drug abuse violations in 1978, recover a, the arrest function for the drug violations.

(c) Graph the model from part (b).

(d) Use the model from part (b) to estimate the number of adults arrested for drug abuse violations in 1991.

66. The sales rate of Lotto tickets sold in the United States is given by the rate function

$$R(x) = -14.58x^2 + 186.84x + 383.93, \qquad 0 \le x \le 16$$

where x represents the number of years since 1980 and $R(x)$ represents the rate of sales measured in $\dfrac{\text{millions of dollars}}{\text{year}}$.

(a) Graph R in the viewing window $[0, 16]$ by $[-400, 1000]$. Use the graph to analyze where the ticket sales are increasing and decreasing. Interpret.

(b) Knowing that there were $52 million in Lotto tickets sold in 1980, recover S, the ticket sales function.

(c) Graph the model from part (b) in the viewing window [0, 16] by [0, 12,000].

(d) Evaluate $S(5)$ and interpret.

SECTION PROJECT

Consider the data in Table 6.1.1 for the Luv-n-Care Shop, which produces handcrafted art. The management tracks its marginal costs for producing x units of art at various production levels, as displayed in the table.

(a) Use the regression capabilities of your calculator to compute a power regression model of the form $y = a \cdot x^b$ for the marginal cost function MC.

(b) Evaluate and interpret $MC(45)$.

TABLE 6.1.1

ITEMS PRODUCED, x	MARGINAL COST, $MC(x)$
10	50
20	56
30	62
40	67
50	73
60	77

(c) Evaluate $MC(40)$ and compare to the marginal cost given in the table. Explain why there is a difference in the marginal cost values.

(d) If the fixed costs are set at 1500 (that is, $C(x) = 1500$ when $x = 0$), recover the cost function C.

(e) Evaluate $C(45)$ and interpret.

SECTION 6.2   AREA AND THE DEFINITE INTEGRAL

In Chapter 2, we learned how to complete a mathematical task that, up until then, was seemingly impossible—to find the slope of a curve at a single point. The calculus tool we used to solve this problem was the *derivative*. Now we must try to solve yet another problem that challenged mathematicians for years. To find the area A under a curve given by a function $y = f(x)$ between two endpoints a and b or, as we also say, on a *closed interval* $[a, b]$. An illustration of our challenge is shown in Figure 6.2.1.

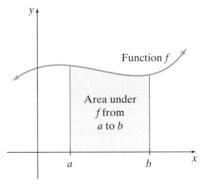

Figure 6.2.1

Finding the Area under a Curve Numerically: The Left Sum Method

To begin to tackle the problem of finding the area under a curve, we rely on area formulas learned in geometry, such as the area of a rectangle. Recall that, for a rectangle with base b and height h, the area of the rectangle is given by

$$A = \text{base} \cdot \text{height} = b \cdot h$$

To convince ourselves that finding a definite area under a curve is possible, we will start with a continuous function $f(x) = x^2 + 1$ on a closed interval $[0, 2]$, as shown in Figure 6.2.2.

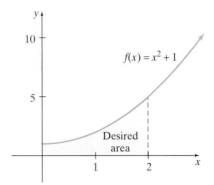

Figure 6.2.2 Area under $f(x) = x^2 + 1$ from $x = 0$ to $x = 2$.

Since we know how to find the area of a rectangle, we can *approximate* the area under the curve by "slicing up" the whole area into thin rectangular strips, computing the area of each strip, and then adding up the areas. Let's say that we want to start with $n = 4$ of these rectangular strips. We start the process by dividing the interval $[a, b]$ into four equal parts. These parts are called **subintervals** and the width of each is called the **step size,** denoted by Δx. To get the stepsize, we simply divide the length of the closed interval $[a, b]$ by the number of subintervals n.

Step Size

For a closed interval $[a, b]$ divided into n equally spaced subintervals, the **step size,** denoted by Δx, is given by

$$\Delta x = \frac{\text{length of interval } [a, b]}{\text{number of subintervals}} = \frac{b - a}{n}$$

EXAMPLE 1 | **Computing a Step Size**

For the interval $[0, 2]$ and $n = 4$:

(a) Calculate the step size Δx.

(b) List the coordinates of the endpoints of the subintervals x_0, x_1, x_2, x_3, x_4.

SOLUTION

(a) Knowing that $a = 0$, $b = 2$, and $n = 4$, we get

$$\Delta x = \frac{b - a}{n} = \frac{2 - 0}{4} = \frac{1}{2}$$

(b) For the interval $[0, 2]$ and step size $\Delta x = \frac{1}{2}$, we get the coordinates shown in Table 6.2.1.

TABLE 6.2.1

x-COORDINATE	x-COORDINATE ON $[0, 2]$ FOR $n = 4$
x_0	0
x_1	$0 + \dfrac{1}{2} = \dfrac{1}{2}$
x_2	$\dfrac{1}{2} + \dfrac{1}{2} = 1$
x_3	$1 + \dfrac{1}{2} = \dfrac{3}{2}$
x_4	$\dfrac{3}{2} + \dfrac{1}{2} = 2$

The coordinates in Table 6.2.1 are the x-coordinates of the endpoints of the bases of the four rectangles. We can either use the left endpoint or the right endpoint to determine the height of the rectangles. Picking the left endpoints to determine the heights is shown in Figure 6.2.3a, and choosing the right endpoints to determine the heights is shown in Figure 6.2.3b.

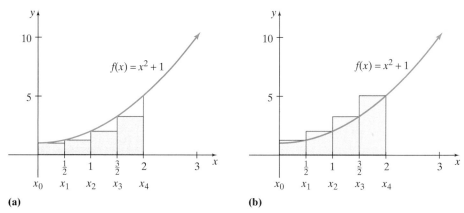

Figure 6.2.3 **(a) Left** endpoints used for the heights of rectangles. **(b) Right** endpoints used for the heights of rectangles.

For our first computation, we will let the *left* endpoint of the rectangles determine the height. The height is found by computing the value of the function $f(x) = x^2 + 1$ at the left endpoint of each of the four rectangles. We now include these values in our table (see Table 6.2.2).

TABLE 6.2.2

x-COORDINATE	x-COORDINATE ON $[0, 2]$ FOR $n = 4$	RECTANGLE HEIGHT $f(x_i)$
x_0	0	$f(x_0) = f(0) = 0^2 + 1 = 1$
x_1	$0 + \dfrac{1}{2} = \dfrac{1}{2}$	$f(x_1) = f\left(\dfrac{1}{2}\right) = \left(\dfrac{1}{2}\right)^2 + 1 = \dfrac{5}{4}$
x_2	$\dfrac{1}{2} + \dfrac{1}{2} = 1$	$f(x_2) = f(1) = 2$
x_3	$1 + \dfrac{1}{2} = \dfrac{3}{2}$	$f(x_3) = f\left(\dfrac{3}{2}\right) = \dfrac{13}{4}$

So it appears that, using these rectangles, the approximate area A under $f(x) = x^2 + 1$ on the interval $[0, 2]$ is

$$A \approx \left(\begin{array}{c}\text{area of first} \\ \text{rectangle}\end{array}\right) + \left(\begin{array}{c}\text{area of second} \\ \text{rectangle}\end{array}\right) + \left(\begin{array}{c}\text{area of third} \\ \text{rectangle}\end{array}\right) + \left(\begin{array}{c}\text{area of fourth} \\ \text{rectangle}\end{array}\right)$$

$$= \Delta x f(x_0) + \Delta x f(x_1) + \Delta x f(x_2) + \Delta x f(x_3)$$

$$= \frac{1}{2}(1) + \frac{1}{2}\left(\frac{5}{4}\right) + \frac{1}{2}(2) + \frac{1}{2}\left(\frac{13}{4}\right)$$

$$= \frac{1}{2} + \frac{5}{8} + 1 + \frac{13}{8}$$

$$= \frac{15}{4} \text{ square units} \quad \text{or} \quad 3.75 \text{ un}^2$$

This area is shown in Figure 6.2.4. We call this way of approximating the area under the curve the **Left Sum Method.**

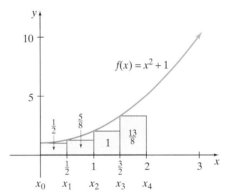

Figure 6.2.4

Left Sum Method

Given a continuous function $y = f(x)$ defined on a closed interval $[a, b]$, we compute the **Left Sum** using n equally spaced subintervals by the following steps:

1. Compute the step size Δx by calculating $\Delta x = \dfrac{b-a}{n}$.
2. Find the x-coordinates $x_0, x_1, x_2, \ldots, x_{n-1}$ of the left endpoints of each of the rectangles by starting with $x_0 = a$ and successively adding step size Δx.
3. Compute the Left Sum:

$$A = \left(\begin{array}{c}\text{area of first} \\ \text{rectangle}\end{array}\right) + \left(\begin{array}{c}\text{area of second} \\ \text{rectangle}\end{array}\right) + \left(\begin{array}{c}\text{area of third} \\ \text{rectangle}\end{array}\right) + \cdots + \left(\begin{array}{c}\text{area of } n\text{th} \\ \text{rectangle}\end{array}\right)$$

$$= \Delta x f(x_0) + \Delta x f(x_1) + \Delta x f(x_2) + \cdots + \Delta x f(x_{n-1})$$

Using **summation notation,** we could write the Left Sum Method formula in step 3 as

$$A = \sum_{i=0}^{n-1} \Delta x f(x_i) \quad \text{or} \quad \sum_{i=0}^{n-1} f(x_i) \Delta x$$

This method is sometimes called **rectangular approximation** and is a straightforward way to estimate the area under a curve.

The Right Sum Method

Another way to approximate area using rectangles is to use the *right* endpoint of each rectangle to determine the rectangles' heights.

EXAMPLE 2 | **Approximating Area Using the Right Sum Method**

For the function $f(x) = x^2 + 1$ on the interval $[0, 2]$, and $n = 4$:

(a) Determine the height of each rectangle using the *right* endpoint of the rectangles.

(b) Use the heights to approximate the area.

SOLUTION

(a) Here we use the step size $\Delta x = \dfrac{1}{2}$ that we determined in Example 1. We get the heights shown in Table 6.2.3 for right endpoints x_1, x_2, x_3, x_4.

	TABLE 6.2.3		
x-COORDINATE	x-COORDINATE ON $[0, 2]$ FOR $n = 4$	RECTANGLE HEIGHT $f(x_i)$	
x_1	$0 + \dfrac{1}{2} = \dfrac{1}{2}$	$f(x_1) = f\left(\dfrac{1}{2}\right) = \left(\dfrac{1}{2}\right)^2 + 1 = \dfrac{5}{4}$	
x_2	$\dfrac{1}{2} + \dfrac{1}{2} = 1$	$f(x_2) = f(1) = 2$	
x_3	$1 + \dfrac{1}{2} = \dfrac{3}{2}$	$f(x_3) = f\left(\dfrac{3}{2}\right) = \dfrac{13}{4}$	
x_4	$\dfrac{3}{2} + \dfrac{1}{2} = 2$	$f(x_4) = f(2) = 5$	

(b) Now, approximating the area using the right endpoint gives us

$$A \approx \left(\begin{array}{c}\text{area of first}\\\text{rectangle}\end{array}\right) + \left(\begin{array}{c}\text{area of second}\\\text{rectangle}\end{array}\right) + \left(\begin{array}{c}\text{area of third}\\\text{rectangle}\end{array}\right) + \left(\begin{array}{c}\text{area of fourth}\\\text{rectangle}\end{array}\right)$$

$$= \Delta x f(x_1) + \Delta x f(x_2) + \Delta x f(x_3) + \Delta x f(x_4)$$

$$= \frac{1}{2}\left(\frac{5}{4}\right) + \frac{1}{2}(2) + \frac{1}{2}\left(\frac{13}{4}\right) + \frac{1}{2}(5)$$

$$= \frac{5}{8} + 1 + \frac{13}{8} + \frac{5}{2}$$

$$= \frac{23}{4} \text{ square units, or } 5.75 \text{ un}^2$$

✓ **CHECKPOINT 1**

Now work Exercise 15.

Right Sum Method

Given a continuous function $y = f(x)$ defined on a closed interval $[a, b]$, we compute the **Right Sum** using n equally spaced subintervals by the following steps:

1. Compute the step size Δx by calculating $\Delta x = \dfrac{b - a}{n}$.

2. Find the x-coordinates $x_1, x_2, \ldots, x_n$ of the right endpoints of each rectangle by starting with $x_1 = a + \Delta x$ and successively adding step size Δx.

3. Compute the Right Sum:

$$A = \begin{pmatrix} \text{area of first} \\ \text{rectangle} \end{pmatrix} + \begin{pmatrix} \text{area of second} \\ \text{rectangle} \end{pmatrix} + \begin{pmatrix} \text{area of third} \\ \text{rectangle} \end{pmatrix} + \cdots + \begin{pmatrix} \text{area of } n\text{th} \\ \text{rectangle} \end{pmatrix}$$

$$= \Delta x f(x_1) + \Delta x f(x_2) + \Delta x f(x_3) + \cdots + \Delta x f(x_n)$$

$$= \sum_{i=1}^{n} \Delta x f(x_i) \text{ or } \sum_{i=1}^{n} f(x_i) \Delta x$$

The approximations using the two methods is compared in Figure 6.2.5 through Figure 6.2.8. Notice that in order to get better approximations for the area under the curve, we could add more and more rectangles and compute their sums.

Left Sum Method

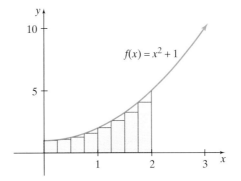

Figure 6.2.5 $n = 8$ rectangles.

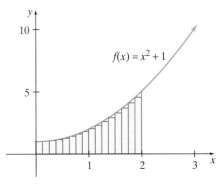

Figure 6.2.6 $n = 16$ rectangles.

Right Sum Method

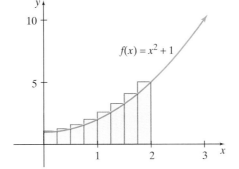

Figure 6.2.7 $n = 8$ rectangles.

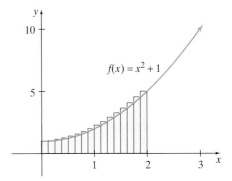

Figure 6.2.8 $n = 16$ rectangles.

The shortcoming of adding more and more rectangles is that computing these approximations by hand can quickly become taxing. Fortunately, our calculators not only serve as graphing devices, but also are programmable handheld computers that complete these computational tasks in seconds. Following are calculator programs for the Left and Right Sum Methods written for the Texas Instruments TI-83. You may need to consult your calculator manual for programming your model of calculator.

Left Sum and Right Sum Method Calculator Programs

These programs approximate the area under the graph of a continuous function f on a closed interval $[a, b]$ for n rectangles.

The function $y = f(x)$ must be entered in Y_1 before the program is executed. When the program executes, the user is prompted to enter a left endpoint A, a right endpoint B, and number of rectangles N.

```
PROGRAM:LEFTSUM                    PROGRAM:RTSUM
Prompt A                           Prompt A
Prompt B                           Prompt B
Prompt N                           Prompt N
(B-A)/N→D                          (B-A)/N→D
0→T                                0→T
For(I,0,N-1)                       For(I,1,N)
A+I*D→X                            A+I*D→X
T+D*Y₁→T                           T+D*Y₁→T
End                                End
Disp "LEFT SUM APPROX",T           Disp "RIGHT SUM APPROX",T
```

TECHNOLOGY NOTE
PROMPT, FOR, END and DISP are all commands found in your calculator's program menu. Consult www.prenhall.com/armstrong for more on this as well as programs for other calculator models.

Using the Left and Right Sum Calculator Programs, we approximate the area under the graph of $f(x) = x^2 + 1$ on the interval $[0, 2]$ in Table 6.2.4.

TABLE 6.2.4

N Rectangles	LEFTSUM Approximation	RTSUM Approximation	|LEFTSUM–RTSUM|
5	3.92	5.52	$\|3.92 - 5.52\| = 1.6$
10	4.28	5.08	0.8
100	4.6268	4.7068	0.08
1,000	4.662668	4.670668	0.008
10,000	4.66626668	4.66706668	0.0008

We can see from the rightmost column in Table 6.2.4 that, *as the number of rectangles increases, the approximations using the Left and Right Sum Programs come closer and closer to one another.* So it appears that a definite area under the graph of $f(x) = x^2 + 1$ on $[0, 2]$ does indeed exist and is a number between 4.66626668 and 4.66706668.

As the number of rectangles gets larger and larger, that is, $n \to \infty$, the approximations approach the actual area. The following formula, which results from letting $n \to \infty$, was derived by Georg Riemann, who was a disciple of the famous German mathematician Karl Friedreich Gauss. Riemann also proved that the end result is the same, even if the rectangles have different widths. This is why we name the width of the rectangle Δx_i and not just Δx.

Area Under a Curve Formula

If f is a continuous function defined on a closed interval $[a, b]$, then the area under the graph of f from a to b is

$$\begin{pmatrix} \text{Area under the graph of } f \\ \text{from } a \text{ to } b \end{pmatrix} = f(x_1)\Delta x_1 + f(x_2)\Delta x_2 + f(x_3)\Delta x_3 + \cdots$$

or, alternatively,

$$= \lim_{n \to \infty} \sum_{i=1}^{n} f(x_i) \cdot \Delta x_i$$

The Definite Integral

In the Area Under a Curve Formula, we saw that an alternative way to represent the area under f from a to b is

$$\lim_{n \to \infty} \sum_{i=1}^{n} f(x_i) \cdot \Delta x_i$$

We can also use the integral that has beginning and ending values, called a **definite integral** to represent this limit.

Definite Integral

For a continuous function f, the **definite integral** of f on the interval $[a, b]$ is

$$\int_{a}^{b} f(x)\, dx = \lim_{n \to \infty} \sum_{i=1}^{n} f(x_i) \cdot \Delta x_i$$

where a and b are the **limits of integration.**

NOTE: The definite integral is often referred to as a **limit of a sum.** Also, a definite integral can be used to represent the area under a curve.

EXAMPLE 3 | **Applying the Definition of Definite Integral**

Write a definite integral to represent the shaded area in Figure 6.2.9.

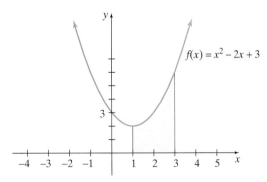

Figure 6.2.9

SOLUTION

The shaded area is below the graph of the function $f(x) = x^2 - 2x + 3$ on the interval $[1, 3]$, which means that $a = 1$ and $b = 3$. We represent the shaded area as

$$\int_{1}^{3} (x^2 - 2x + 3)\, dx$$

In Section 6.1 we saw several properties of the indefinite integral. Many of the properties that we first saw in that section apply here, too.

Properties of the Definite Integral

1. The properties of the indefinite integral also apply to the definite integral.
   - $\int_a^b k \cdot f(x)\, dx = k \int_a^b f(x)\, dx$. This is the *constant rule* for definite integrals.
   - $\int_a^b [f(x) \pm g(x)]\, dx = \int_a^b f(x)\, dx \pm \int_a^b g(x)\, dx$. This is the *sum and difference rule* for definite integrals.

2. If f is continuous on $[a, b]$ and there is a value c between a and b (see Figure 6.2.10), then

$$\int_a^b f(x)\, dx = \int_a^c f(x)\, dx + \int_c^b f(x)\, dx$$

This property, called the **Interval Addition Rule** for definite integrals, means that we can "break up" the limits of integration at a point $x = c$ that is between a and b.

3. $\int_a^a f(x)\, dx = 0$. This property tells us that the area under a single point on a curve is zero.

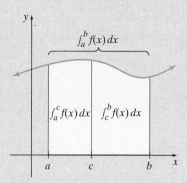

Figure 6.2.10 $\int_a^b f(x)\, dx = \int_a^c f(x)\, dx + \int_c^b f(x)\, dx$.

The Interval Addition Rule is particularly useful when integrating piecewise-defined functions. Its use is demonstrated in Example 4.

| EXAMPLE 4 | **Applying the Interval Addition Rule for Definite Integrals** |

Use the Interval Addition Rule to write a definite integral to represent the shaded region for the piecewise-defined function

$$f(x) = \begin{cases} x + 1, & x \le 2 \\ -2x + 7, & x > 2 \end{cases}$$

A graph of the piecewise-defined function is given in Figure 6.2.11.

SOLUTION

We must break up the integral into two pieces. On the interval $[-1, 2]$ we integrate the expression $x + 1$, and on the interval $\left[2, \frac{7}{2}\right]$ we integrate $-2x + 7$. Since the "switch" in the expressions occurs at $x = 2$, and 2 is between -1 and $\frac{7}{2}$, we can use the Interval Addition Rule to write

$$\int_{-1}^{7/2} f(x)\, dx = \int_{-1}^{2} (x + 1)\, dx + \int_{2}^{7/2} (-2x + 7)\, dx$$

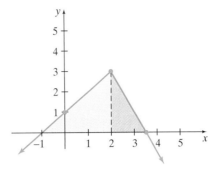

Figure 6.2.11 Graph of $f(x) = \begin{cases} x + 1, & x \le 2 \\ -2x + 7, & x > 2 \end{cases}$

✓ CHECKPOINT 2

Now work Exercise 41.

SUMMARY

In this section we saw that the area under a curve could be approximated numerically by using the Left and Right Sum Methods. We also said that, as the number of rectangles becomes infinitely large, we get the definite integral.

- **Step Size:** $\Delta x = \dfrac{b - a}{n}$

- **Left Sum Method:** $\displaystyle\sum_{i=0}^{n-1} f(x_i)\, \Delta x$

- **Right Sum Method:** $\displaystyle\sum_{i=1}^{n} f(x_i)\, \Delta x$

- **Definite Integral:** $\displaystyle\int_a^b f(x)\, dx = \lim_{n \to \infty} \sum_{i=1}^{n} f(x_i) \cdot \Delta x_i$

- **Constant Rule:** $\displaystyle\int_a^b k \cdot f(x)\, dx = k \int_a^b f(x)\, dx$

- **Sum and Difference Rule:** $\displaystyle\int_a^b [f(x) \pm g(x)]\, dx = \int_a^b f(x)\, dx \pm \int_a^b g(x)\, dx$

- **Interval Addition Rule:** $\displaystyle\int_a^b f(x)\, dx = \int_a^c f(x)\, dx + \int_c^b f(x)\, dx$

SECTION 6.2 EXERCISES

In Exercises 1–10, determine the step size Δx for the given interval $[a, b]$ and number of subintervals n.

1. $[0, 2], n = 4$
2. $[0, 5], n = 5$
3. $[1, 4], n = 6$
4. $[1, 2], n = 4$
5. $[3, 5], n = 4$
6. $[3, 6], n = 6$
7. $[-1, 2], n = 6$
8. $[-2, 2], n = 4$
9. $[-3, 5], n = 28$
10. $[-10, 3], n = 52$

In Exercises 11–20, given f, the closed interval $[a, b]$, and n number of equally spaced subintervals:

(a) Calculate by hand the Left Sum to approximate the area under the graph of f. Do not use your calculator program.

(b) Calculate by hand the Right Sum to approximate the area under the graph of f. Do not use your calculator program.

(c) Determine |Left Sum − Right Sum|.

11. $f(x) = 5, [0, 2], n = 4$ 12. $f(x) = 7, [0, 5], n = 5$

13. $f(x) = 3x, [1, 4], n = 6$ 14. $f(x) = 2x, [1, 2], n = 4$

✓ 15. $f(x) = 2x - 1, [3, 5], n = 4$ 16. $f(x) = 4x - 2, [3, 6], n = 6$

17. $f(x) = 2x^2, [0, 4], n = 4$ 18. $f(x) = x^3, [0, 2], n = 4$

19. $f(x) = 4 - x^2, [-1, 2], n = 6$ 20. $f(x) = x^2 + 2, [-2, 2], n = 4$

Complete the table for each of the functions f on $[a, b]$ in Exercises 21–30 using the appropriate calculator program.

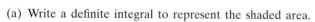

N RECTANGLES	LEFTSUM APPROXIMATION	RTSUM APPROXIMATION	|LEFTSUM − RTSUM|
10			
100			
1000			

21. $f(x) = 3x^2, [-2, 0]$ 22. $f(x) = \dfrac{x^2}{2}, [0, 9]$ 33.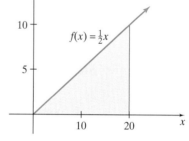

23. $f(x) = 8\sqrt{x}, [1, 4]$ 24. $f(x) = 6x + \sqrt{x}, [1, 2]$

25. $f(x) = x^2 + 2x, [1, 4]$ 26. $f(x) = 4x^2 - x, [0, 5]$

27. $f(x) = x^3 - 3x, [0, 4]$ 28. $f(x) = x^3 + 2x, [0, 2]$

29. $f(x) = 0.2x^2 + 1.3x + 2.3, [0, 5]$

30. $f(x) = 0.3x^3 - 2.7x + 4.1, [0, 3]$

For Exercises 31–42:

(a) Write a definite integral to represent the shaded area.

(b) Use the Left Sum and Right Sum Programs with $n = 100$ to approximate the area.

31. 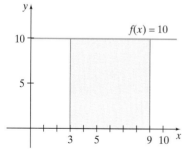   34.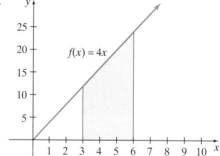

32. 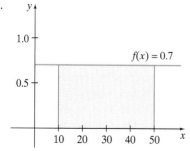   35.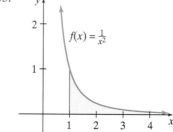

36.

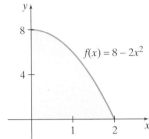

$f(x) = 8 - 2x^2$

37.

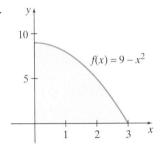

$f(x) = 9 - x^2$

38.

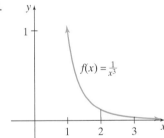

$f(x) = \dfrac{1}{x^3}$

39.

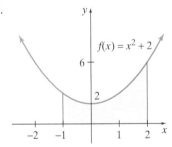

$f(x) = x^2 + 2$

40.

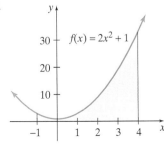

$f(x) = 2x^2 + 1$

✓ 41.

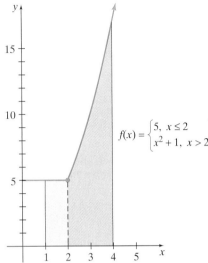

$f(x) = \begin{cases} 5, & x \le 2 \\ x^2 + 1, & x > 2 \end{cases}$

42.

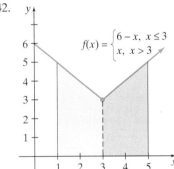

$f(x) = \begin{cases} 6 - x, & x \le 3 \\ x, & x > 3 \end{cases}$

In Exercises 43–50, sketch and shade the region given by the definite integrals.

43. $\displaystyle\int_0^3 x^2 \, dx$ 44. $\displaystyle\int_1^2 x^3 \, dx$

45. $\displaystyle\int_1^4 \frac{1}{x^2} \, dx$ 46. $\displaystyle\int_2^4 \frac{3}{x^4} \, dx$

47. $\displaystyle\int_1^5 \sqrt[3]{x} \, dx$ 48. $\displaystyle\int_3^6 \sqrt{x^3} \, dx$

49. $\displaystyle\int_1^2 \frac{2}{\sqrt{x}} \, dx$ 50. $\displaystyle\int_2^8 \frac{1}{\sqrt[3]{x}} \, dx$

SECTION PROJECT

Consider the function $f(x) = |x| + 1$.

(a) Graph the function in the standard viewing window and classify the function as symmetric to the x-axis, symmetric to the y-axis, symmetric to the origin, or not symmetric. Explain your answer.

(b) Rewrite f as a piecewise-defined function.

(c) Use the Left Sum program to determine the area under the graph of f on the interval $[-6, 6]$ using $n = 50$ equally spaced subintervals.

(d) Use the Left Sum program to determine the area under the graph of f on the interval $[0, 6]$ using $n = 50$ equally spaced subintervals. How does this answer compare to part (c)?

(e) How are the approximations in parts (c) and (d) related? How is this related to the symmetry found in part (a)? Explain.

SECTION 6.3   FUNDAMENTAL THEOREM OF CALCULUS

In the previous section, we saw that the *area under a curve formula* and the *definite integral* are indeed the same. Mathematically, this means that

$$\lim_{n \to \infty} \sum_{i=1}^{n} f(x_i) \Delta x_i = \int_a^b f(x)\, dx$$

In this section, we will learn a powerful tool that is used to quickly evaluate a definite integral—the Fundamental Theorem of Calculus. We will also examine some common applications of the definite integral. In particular, we will see how the integral can be used to compute the **continuous sum of a given rate function**.

Fundamental Theorem of Calculus

As you can imagine, it can be difficult to find the area under a curve by computing the left or right sum for thousands of rectangles. Then why is the Area under a Curve Formula so important? It turns out that the founders of calculus discovered an amazing relationship between the area under a curve and the definite integral. To understand this relationship, let's reconsider an example from Section 6.1.

Flashback

INDEFINITE INTEGRALS REVISITED

 In part (a) of Example 3 in Section 6.1, we computed $\int (x^2 + 3)\, dx$ to be

$$F(x) = \frac{x^3}{3} + 3x + C$$

(a) For $F(x) = \frac{x^3}{3} + 3x + C$, evaluate $F(6) - F(3)$.

(b) Write a definite integral to represent the area under $f(x) = x^2 + 3$ on the interval $[3, 6]$ and use the Left Sum program to approximate the area using $n = 1000$ rectangles.

Solution

(a) Evaluating $F(6) - F(3)$, we get

$$F(6) - F(3) = \left[\frac{6^3}{3} + 3(6) + C \right] - \left[\frac{3^3}{3} + 3(3) + C \right]$$

$$= [72 + 18 + C] - [9 + 9 + C]$$

$$= 72 + 18 + C - 9 - 9 - C$$

$$= 72$$

(b) From the definition of definite integral, we get

$$\int_3^6 (x^2 + 3)\, dx$$

Executing the Left Sum Program with $N = 1000$, we get the output displayed in Figure 6.3.1.

Figure 6.3.1

The Flashback shows that the difference $F(6) - F(3)$ is very close to the approximation for the area under the curve on the interval [3, 6] found by using the Left Sum program. This result is no accident. In fact, we can show that

$$\left(\begin{array}{c} \text{Area under } f \\ \text{from } a \text{ to } b \end{array}\right) = \lim_{n \to \infty} \sum_{i=1}^{n} f(x_i) \Delta x_i = F(b) - F(a)$$

Since $\lim\limits_{n \to \infty} \sum\limits_{i=1}^{n} f(x_i) \Delta x_i = \int_a^b f(x)\, d(x)$ (Definition of Definite Integral), it appears that we have

$$\int_a^b f(x)\, dx = F(b) - F(a)$$

where F is any antiderivative of f. If we can find an antiderivative of f, we can employ one of the most powerful theorems in all of mathematics, the **Fundamental Theorem of Calculus**. A proof of this theorem is in Appendix C.

Fundamental Theorem of Calculus

If f is a continuous function defined on a closed interval $[a, b]$ and F is an *antiderivative* of f, then

$$\int_a^b f(x)\, dx = F(b) - F(a)$$

where

- $\int_a^b f(x)\, dx$ is the **definite integral** of f on the interval $[a, b]$.
- a and b are the **limits of integration**.

Before continuing, let's list a few key points of the Fundamental Theorem of Calculus.

- $\int_a^b f(x)\, dx = F(b) - F(a)$ gives the area under the graph of f from $x = a$ to $x = b$ is a special case of the Fundamental Theorem of Calculus. This is true when $f(x) \geq 0$ everywhere on $[a, b]$. See Figure 6.3.2.
- The difference $F(b) - F(a)$ is also written as $F(x)\big|_a^b$.

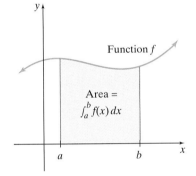

Figure 6.3.2

Function f

Area $= \int_a^b f(x)\, dx$

EXAMPLE 1

Using the Fundamental Theorem of Calculus

Evaluate the definite integrals.

(a) $\displaystyle\int_{-2}^{0} (x^2 - 3x)\, dx$

(b) $\displaystyle\int_{1}^{5} \frac{1}{\sqrt{x}}\, dx$

SOLUTION

(a) $\displaystyle\int_{-2}^{0} (x^2 - 3x)\, dx = \left(\frac{x^3}{3} - \frac{3x^2}{2}\right)\Bigg|_{-2}^{0}$

$$= \left(\frac{0^3}{3} - \frac{3(0)^2}{2}\right) - \left(\frac{(-2)^3}{3} - \frac{3(-2)^2}{2}\right)$$

$$= (0 - 0) - \left(\frac{-8}{3} - 6\right)$$

$$= \frac{8}{3} + \frac{18}{3} = \frac{26}{3}$$

(b) $\displaystyle\int_1^5 \frac{1}{\sqrt{x}}\,dx = \int_1^5 x^{-1/2}\,dx = (2x^{1/2})\Big|_1^5 = 2(\sqrt{x})\Big|_1^5$

$$= 2(\sqrt{5} - \sqrt{1}) = 2\sqrt{5} - 2$$

To see how the Fundamental Theorem of Calculus can be interpreted, let's take another look at an example from Section 6.2.

EXAMPLE 2	**Determining an Exact Area**

Determine the exact area shaded in Figure 6.3.3.

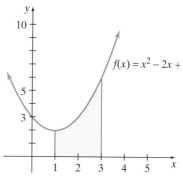

$f(x) = x^2 - 2x + 3$

Figure 6.3.3

SOLUTION

As we stated in Example 3 in Section 6.2, the corresponding definite integral for the shaded area is $\int_1^3 (x^2 - 2x + 3)\,dx$. Evaluating this integral yields

$$\int_1^3 (x^2 - 2x + 3)\,dx = \left(\frac{x^3}{3} - x^2 + 3x\right)\Big|_1^3$$

$$= \left[\frac{3^3}{3} - 3^2 + 3(3)\right] - \left[\frac{1^3}{3} - 1^2 + 3(1)\right]$$

$$= (9 - 9 + 9) - \left(\frac{1}{3} - 1 + 3\right)$$

$$= 9 - \frac{7}{3} = \frac{20}{3}\ \text{un}^2$$

Thus, the area under the graph of the function $f(x) = x^2 - 2x + 3$ on the interval $[1, 3]$ is $\frac{20}{3}$ square units.

The value of the definite integral depends not only on the limits of integration, but also on the orientation of the function relative to the x-axis.

- If a function $f(x) > 0$ on $[a, b]$, then $\int_a^b f(x)\,dx > 0$. Thus, if the graph of f is above the x-axis for all the values $a \le x \le b$, then the definite integral is positive.
- If a function $f(x) < 0$ on $[a, b]$, then $\int_a^b f(x)\,dx < 0$. Thus, if the graph of f is below the x-axis for all the values $a \le x \le b$, then the definite integral is negative.

These properties lead to the notions of net and gross area. Generally, the area between the graph of f and the x-axis on $[a, b]$ is called the **net area.** If we compute the absolute value of the region below the x-axis, we call this the **gross area.** See Figure 6.3.4.

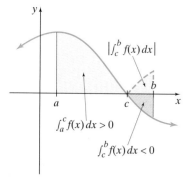

$\left|\int_c^b f(x)\,dx\right|$

$\int_a^c f(x)\,dx > 0$

$\int_c^b f(x)\,dx < 0$

Figure 6.3.4

EXAMPLE 3 | **Determining Net and Gross Area**

Consider the cubic function $f(x) = x^3 - x^2 - 4x + 4$.

(a) Evaluate the net area by computing $\int_{-2}^{2} f(x)\,dx$. Graph the region.

(b) Calculate the gross area between the graph of f and the x-axis.

SOLUTION

(a) Computing the net area, we get

$$\int_{-2}^{2} f(x)\,dx = \int_{-2}^{2} (x^3 - x^2 - 4x + 4)\,dx$$

$$= \left(\frac{x^4}{4} - \frac{x^3}{3} - 2x^2 + 4x \right) \Big|_{-2}^{2}$$

$$= \left[\frac{(2)^4}{4} - \frac{(2)^3}{3} - 2(2)^2 + 4(2) \right]$$

$$\quad - \left[\frac{(-2)^4}{4} - \frac{(-2)^3}{3} - 2(-2)^2 + 4(-2) \right]$$

$$= \frac{4}{3} - \left(-\frac{28}{3} \right) = \frac{32}{3} \ \text{un}^2$$

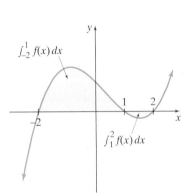

Figure 6.3.5
Net area of $\int_{-2}^{2} (x^3 - x^2 - 4x + 4)\,dx$.

A graph of the integrated region is shown in Figure 6.3.5

(b) To get the gross area, we need to consider the interval $[1, 2]$, where the graph of f lies below the x-axis. Since this integrated region has a negative value, we need to take the absolute value of the integral on $[1, 2]$.

$$\text{Gross area} = \int_{-2}^{1} (x^3 - x^2 - 4x + 4)\,dx + \left| \int_{1}^{2} (x^3 - x^2 - 4x + 4)\,dx \right|$$

$$= \left(\frac{x^4}{4} - \frac{x^3}{3} - 2x^2 + 4x \right) \Big|_{-2}^{1} + \left| \left(\frac{x^4}{4} - \frac{x^3}{3} - 2x^2 + 4x \right) \Big|_{1}^{2} \right|$$

$$= \left[\frac{(1)^4}{4} - \frac{(1)^3}{3} - 2(1)^2 + 4(1) \right] - \left[\frac{(-2)^4}{4} - \frac{(-2)^3}{3} - 2(-2)^2 + 4(-2) \right]$$

$$\quad + \left| \left[\frac{(2)^4}{4} - \frac{(2)^3}{3} - 2(2)^2 + 4(2) \right] - \left[\frac{(1)^4}{4} - \frac{(1)^3}{3} - 2(1)^2 + 4(1) \right] \right|$$

$$= \frac{23}{12} - \left(-\frac{28}{3} \right) + \left| \frac{4}{3} - \frac{23}{12} \right|$$

$$= \frac{45}{4} + \left| -\frac{7}{12} \right| = \frac{71}{6} \ \text{un}^2$$

So the area bounded by the graph of the function f and the x-axis is $\frac{71}{6}\,\text{un}^2$. This is illustrated in Figure 6.3.6. ∎

✓ **CHECKPOINT 1**

Now work Exercise 39.

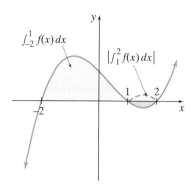

Figure 6.3.6 Gross area of $\int_{-2}^{2}(x^3 - x^2 - 4x + 4)\,dx$.

The Integral as a Continuous Sum

We stated earlier that the definite integral can be thought of as a limit of a sum. The definite integral can also be thought of as a *continuous sum of a rate function over a closed interval*. To see why this observation is important, consider the daily cost function for the TiBike Company, whose daily cost for producing their titanium seat bolts is given by

$$C(x) = 0.02x^2 + 3x + 100, \qquad 0 \le x \le 70$$

The manufacturer has a special order for the first 10 seat bolts for a local bike shop and then receives another order for 50. The cost of producing these 50 is given by

$$\left(\begin{array}{c}\text{cost of producing}\\ 60\,\text{items}\end{array}\right) - \left(\begin{array}{c}\text{cost of producing}\\ 10\,\text{items}\end{array}\right)$$

The increased cost for this production run is

$$C(60) - C(10) = (0.02(60)^2 + 3(60) + 100) - (0.02(10)^2 + 3(10) + 100)$$
$$= 352 - 132 = \$220$$

So the cost of the second order is \$220. Now let's examine this problem using the marginal cost for the product, which is $C'(x) = MC(x) = 0.04x + 3$. Since the integral is a continuous summation, we can compute

$$\int_{10}^{60}(0.04x + 3)\,dx = (0.02x^2 + 3x)\Big|_{10}^{60}$$
$$= [0.02(60)^2 + 3(60)] - [0.02(10)^2 + 3(10)]$$
$$= 352 - 132 = \$220$$

The identical result indicates that we can use the definite integral to find the sum of any quantity if we are given its rate function or, in other words, its derivative. This result is very important since in many applications we are given rate functions, but do not know the functions from which they came. We can state this property in words by saying that ***the integral is a total accumulation of a rate function between two values a and b***. We illustrate some of its uses in Figure 6.3.7.

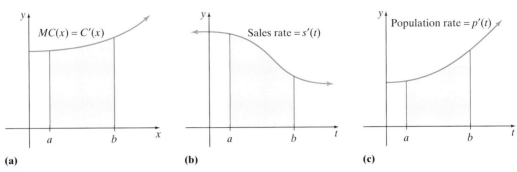

Figure 6.3.7 **(a)** Total increase in cost of producing from a to b units. **(b)** Total increase in sales from a to b units. **(c)** Total increase in population from years a to b.

EXAMPLE 4 **Integrating a Rate Function**

The annual rate of change in the price per pack of cigarettes in the United States from 1960 to 1995 can be modeled by the rate function

$$p'(t) = 0.06t^2 - 1.5t + 8.24, \qquad 0 \le t \le 35$$

where t is the number of years since 1960 and $p'(t)$ is the rate of change in the price per pack measured in $\frac{\text{cents}}{\text{year}}$. See Figure 6.3.8.

(a) Integrate $\int_{23}^{28} p'(t)\, dt$ and interpret.

(b) Integrate $\int_{10}^{14} p'(t)\, dt$ and interpret.

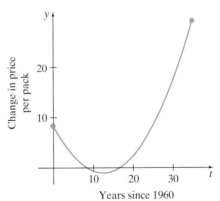

Figure 6.3.8 The annual rate of change in price per pack.

SOLUTION

(a) Integrating, we get

$$\int_{23}^{28} p'(t)\, dt = \int_{23}^{28} (0.06t^2 - 1.5t + 8.24)\, dt$$

$$= (0.02t^3 - 0.75t^2 + 8.24t)\Big|_{23}^{28}$$

$$= [(0.02(28)^3 - 0.75(28)^2 + 8.24(28)]$$
$$\quad -[0.02(23)^3 - 0.75(23)^2 + 8.24(23)]$$

$$= 45.65$$

This means that the total *increase* in the price of a pack of cigarettes from 1983 to 1988 was about 45.65 cents.

(b) Integrating, we get

$$\int_{10}^{14} p'(t)\,dt = \int_{10}^{14} (0.06t^2 - 1.5t + 8.24)\,dt$$
$$= (0.02t^3 - 0.75t^2 + 8.24t)\Big|_{10}^{14}$$
$$= [0.02(14)^3 - 0.75(14)^2 + 8.24(14)]$$
$$\qquad -[0.02(10)^3 - 0.75(10)^2 + 8.24(10)]$$
$$= -4.16$$

This result is illustrated in Figure 6.3.9. This means that between 1970 and 1974, the total *decrease* in the price of a pack of cigarettes was about 4.16 cents.

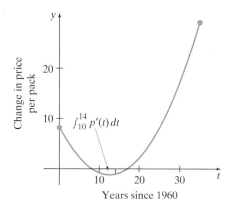

Figure 6.3.9

✓ **CHECKPOINT 2**

Now work Exercise 55.

This notion of the continuous sum is also used in the physical sciences. To see how, let's start with a quick review.

> **From Your Toolbox**
>
> For the position function s, we have the *velocity* of an object at time t given by $v(t) = \frac{d}{dt}[s(t)]$.

Thus, to find the total movement or **displacement** of an object with a given velocity, we can use the definite integral.

EXAMPLE 5 | **Finding Total Displacement Using a Definite Integral**

The velocity of an object can be modeled by

$$v(t) = 8t + 3$$

where t is the time in seconds and $v(t)$ is the velocity measured in $\frac{\text{feet}}{\text{second}}$.

(a) Evaluate $v(10)$ and interpret.

(b) Determine the total movement, or displacement, of the object between 2 and 8 seconds.

SOLUTION

(a) Evaluating, we get

$$v(10) = 8(10) + 3 = 80 + 3 = 83$$

This means that after 10 seconds the velocity of the object is $83 \frac{\text{feet}}{\text{second}}$.

(b) To find the total displacement, we need to compute

$$\int_{2}^{8} v(t) \, dt = \int_{2}^{8} (8t + 3) \, dt = (4t^2 + 3t)\big|_{2}^{8}$$
$$= \left[4\,(8)^2 + 3(8)\right] - \left[4\,(2)^2 + 3(2)\right]$$
$$= 280 - 22 = 258$$

This means that according to the velocity model, the object moved 258 feet between 2 and 8 seconds.

SUMMARY

In this section we were introduced to the Fundamental Theorem of Calculus and saw how we could use this theorem to compute areas under curves. We also saw that we could integrate a rate function to find the total accumulation of a quantity from a beginning point to an ending point.

- **Fundamental Theorem of Calculus:** If f is a continuous function defined on a closed interval $[a, b]$ and F is an antiderivative of f, then $\int_{a}^{b} f(x) \, dx = F(b) - F(a)$.
- If a function $f(x) > 0$ on $[a, b]$, then $\int_{a}^{b} f(x) \, dx > 0$.
- If a function $f(x) < 0$ on $[a, b]$, then $\int_{a}^{b} f(x) \, dx < 0$.

SECTION 6.3 EXERCISES

Exercises 1–24, evaluate the definite integrals.

1. $\int_{1}^{4} 3x \, dx$

2. $\int_{1}^{2} 2x \, dx$

3. $\int_{0}^{2} 5 \, dx$

4. $\int_{0}^{5} 7 \, dx$

5. $\int_{3}^{5} (2x - 1) \, dx$

6. $\int_{3}^{6} (4x - 2) \, dx$

7. $\int_{0}^{4} 2x^2 \, dx$

8. $\int_{0}^{2} x^3 \, dx$

9. $\int_{-1}^{2} (4 - x^2) \, dx$

10. $\int_{-2}^{2} (x^2 + 2) \, dx$

11. $\int_{-2}^{0} 3x^2 \, dx$

12. $\int_{0}^{9} \frac{x^2}{2} \, dx$

13. $\int_{1}^{4} 8\sqrt{x} \, dx$

14. $\int_{1}^{2} \left(6x + \sqrt{x}\right) dx$

15. $\int_{1}^{4} (x^3 - 3x) \, dx$

16. $\int_{0}^{2} (x^3 + 2x) \, dx$

17. $\int_{0}^{5} (0.2x^2 + 1.3x + 2.3) \, dx$

18. $\int_{0}^{3} (0.3x^3 - 2.7x + 4.1) \, dx$

19. $\int_{-2}^{3} (5 + x - 6x^2) \, dx$

20. $\int_{1}^{4} (x^2 - 4x - 3) \, dx$

21. $\int_{0}^{2} (x^4 - 2x^3) \, dx$

22. $\int_{-2}^{3} (8x^3 + 3x - 1) \, dx$

23. $\int_{4}^{9} \frac{x - 3}{\sqrt{x}} \, dx$

24. $\int_{-2}^{-1} \frac{2x - 7}{x^3} \, dx$

25. Consider the absolute value function $f(x) = |x - 3|$.
    (a) Graph f in the standard viewing window.
    (b) Rewrite f as a piecewise-defined function.
    (c) Use the Interval Addition Rule to evaluate the definite integral $\int_{-2}^{5} f(x) \, dx$.

26. Consider the absolute value function $f(x) = |x + 6|$.
    (a) Graph f in the standard viewing window.
    (b) Rewrite f as a piecewise-defined function.

(c) Use the Interval Addition Rule to evaluate the definite integral $\int_{-10}^{0} f(x)\,dx$.

27. Consider the absolute value function $f(x) = |x| - 4$.
(a) Graph f in the standard viewing window.
(b) Rewrite f as a piecewise-defined function.
(c) Determine the x-intercepts of the graph of f.
(d) Use the Interval Addition Rule to determine the net and gross areas of $\int_{-2}^{6} f(x)\,dx$.

28. Consider the absolute value function $f(x) = |x| - 2$.
(a) Graph f in the standard viewing window.
(b) Rewrite f as a piecewise-defined function.
(c) Determine the x-intercepts of the graph of f.
(d) Use the Interval Addition Rule to determine the net and gross areas of $\int_{0}^{9} f(x)\,dx$.

In Exercises 29–34, use the Interval Addition Rule to evaluate the indicated definite integral to find the area between the graph of g and the x-axis.

29. $\displaystyle\int_{-3}^{4} g(x)\,dx$, where $g(x) = \begin{cases} -x, & x < 0 \\ x, & x \geq 0 \end{cases}$

30. $\displaystyle\int_{-3}^{3} g(x)\,dx$, where $g(x) = \begin{cases} -x, & x \leq 1 \\ x - 2, & x > 1 \end{cases}$

31. $\displaystyle\int_{-1}^{2} g(x)\,dx$, where $g(x) = \begin{cases} 1 - x^2, & x < 0 \\ x + 1, & x \geq 0 \end{cases}$

32. $\displaystyle\int_{0}^{5} g(x)\,dx$, where $g(x) = \begin{cases} x^2, & x < 3 \\ x + 6, & x \geq 3 \end{cases}$

33. $\displaystyle\int_{0}^{5} g(x)\,dx$, where $g(x) = \begin{cases} 10 - x^2, & x \leq 2 \\ 6, & x > 2 \end{cases}$

34. $\displaystyle\int_{-4}^{0} g(x)\,dx$, where $g(x) = \begin{cases} 8, & x < -2 \\ x^2 + 4, & x \geq -2 \end{cases}$

For Exercises 35–44:
(a) Determine the x-intercepts of the graph of the function f.
(b) Integrate to determine the net area.
(c) Use the Interval Addition Rule to determine the gross areas for the indicated integral.

35. $f(x) = x + 2;\ \displaystyle\int_{-5}^{5} f(x)\,dx$

36. $f(x) = x - 5;\ \displaystyle\int_{0}^{10} f(x)\,dx$

37. $f(x) = 2\sqrt{x} - 4;\ \displaystyle\int_{0}^{9} f(x)\,dx$

38. $f(x) = \sqrt{x} - 1;\ \displaystyle\int_{0}^{4} f(x)\,dx$

✓ 39. $f(x) = x^2 - 9;\ \displaystyle\int_{-1}^{4} f(x)\,dx$

40. $f(x) = x^2 - 4;\ \displaystyle\int_{-2}^{4} f(x)\,dx$

41. $f(x) = x^2 - x - 2;\ \displaystyle\int_{-1}^{4} f(x)\,dx$

42. $f(x) = x^2 + 4x + 3;\ \displaystyle\int_{-3}^{1} f(x)\,dx$

43. $f(x) = x^3 + x^2 - 2x;\ \displaystyle\int_{-2}^{2} f(x)\,dx$

44. $f(x) = x^3 - 2x^2 - 3x;\ \displaystyle\int_{-2}^{3} f(x)\,dx$

APPLICATIONS

45. The TuffyToe Company determines that the marginal cost function for their new walking shoe is given by

$$MC(x) = C'(x) = 50 - 0.8x, \qquad 0 \leq x \leq 60$$

where x is the number of shoes produced per shift and $MC(x) = C'(x)$ is the marginal cost measured in dollars per shoe.

(a) Evaluate $C'(10)$ and interpret.
(b) Evaluate $\int_{0}^{50} C'(x)\,dx$ and interpret.

46. The TinyTot Toy Company determines that the marginal cost for producing a new action figure is given by

$$MC(x) = C'(x) = 4 - 0.02x, \qquad 0 \leq x \leq 100$$

where x is the number of toys made daily and $MC(x) = C'(x)$ is the marginal cost measured in dollars per toy.

(a) Evaluate $C'(30)$ and interpret.
(b) Evaluate $\int_{0}^{30} C'(x)\,dx$ and interpret.

47. By market analysis, the SatSet Satellite TV Company determines that their marginal revenue function for their best-selling satellite dish is given by

$$MR(x) = R'(x) = 50 + 0.4x^3, \qquad 0 \leq x \leq 40$$

where x represents the number of units produced and sold monthly and $MR(x) = R'(x)$ is the marginal revenue measured in dollars per satellite dish.

(a) Evaluate $\int_{20}^{30} R'(x)\,dx$ and interpret.
(b) Sketch the region determined in part (a).

48. The WiredWorld Audio Company determines that their marginal revenue function for producing and selling a waterproof headset radio is given by

$$MR(x) = R'(x) = 70 + x, \qquad 0 \leq x \leq 70$$

where x is the number of headset radios produced and sold in thousands and $MR(x) = R'(x)$ is the marginal revenue measured in dollars per headset radio.

(a) Evaluate $\int_{10}^{40} R'(x)\,dx$ and interpret.
(b) Sketch the region determined in part (a).

49. The ScandiTrac Company determines that their marginal profit function for producing and selling a new economy model of cross-country ski machine at a mall is given by

$$MP(x) = P'(x) = 0.3x^2 + 0.2x, \qquad 0 \le x \le 30$$

where x is the number of machines produced and sold and $P'(x)$ is the marginal profit function measured in dollars per ski machine.

(a) Knowing that $704 profit is made when 20 ski machines are sold, recover the profit function P.

(b) Evaluate $\int_{10}^{20} P'(x)\, dx$ and interpret.

50. The See-the-Fine-Print Company determines that their marginal profit function for producing and selling over-the-counter reading glasses at a regional store is given by

$$MP(x) = P'(x) = 0.0015x^2 - 0.01x, \qquad 0 \le x \le 60$$

where x is the number of reading glasses sold at the store and $P'(x)$ is the marginal profit function measured in dollars per pair of glasses.

(a) Knowing that $23 in profit is made when 40 pairs of reading glasses are sold, recover the profit function P.

(b) Evaluate $\int_{30}^{40} P'(x)\, dx$ and interpret.

51. The U.S. Department of Energy reports that the import rate of petroleum products into the United States is given by the function

$$P'(t) = 6.79t^2 - 11.44t + 325.32, \qquad 3 \le t \le 25$$

where t is the number of years since 1970 and $P'(t)$ represents the rate of imports per day annually measured in $\dfrac{\text{thousands of barrels imported each day}}{\text{year}}$.

(a) Evaluate $P'(6)$ and interpret the result.

(b) Integrate the rate function to determine the total increase in imports of petroleum into the United States from 1973 to 1978.

52. The rate of U.S. military sales deliveries to Israel can be modeled by

$$s(x) = -57.75x^2 + 630.58x - 1528.43, \qquad 1 \le x \le 9$$

where x represents the number of years since 1986 and $s(x)$ represents the rate of sales measured in $\dfrac{\text{millions of dollars}}{\text{year}}$.

(a) Evaluate and interpret $s(2)$.

(b) Integrate the rate function to determine the total increase in U.S. military sales deliveries to Israel from 1988 to 1992.

53. The rate of military expenditures for NATO countries can be modeled by the function

$$m(x) = -2.08x - 1.85, \qquad 1 \le x \le 9$$

where x represents the number of years since 1985 and $m(x)$ represents the rate of military expenditures for NATO countries, measured in billions of dollars per year.

(a) Evaluate and interpret $m(4)$.

(b) Evaluate and interpret $\int_4^8 m(x)\, dx$.

54. The rate of monthly precipitation in Seattle, Washington, can be modeled by

$$p(x) = 0.32x - 2.02, \qquad 1 \le x \le 12$$

where $x = 1$ corresponds to January, $x = 2$ corresponds to February, and so on, and $p(x)$ is the rate of precipitation each month, measured in inches per month.

(a) Evaluate and interpret $p(12)$.

(b) Evaluate $\int_1^6 p(x)\, dx$.

55. The rate of consumption of jet fuel in the United States each year can be modeled by

$$f(x) = 3.78x^2 + 25.56x - 119.48, \qquad 0 \le x \le 25$$

where x represents the number of years since 1970 and $f(x)$ represents the rate of consumption of jet fuel, measured in millions of gallons per year.

(a) Evaluate and interpret $f(20)$.

(b) Integrate the rate function to determine the total increase in consumption of jet fuel in the United States from 1980 to 1990.

56. The rate of change in the total amount of outstanding automobile loans in the United States can be modeled by

$$c(t) = -1.04t^3 + 15.66t^2 - 61.86t + 56.15, \qquad 0 \le t \le 10$$

where t represents the number of years since 1985 and $c(t)$ represents the rate of change in the total amount of outstanding automobile loans, measured in $\dfrac{\text{billions of dollars}}{\text{year}}$.

(a) Evaluate and interpret $c(10)$.

(b) Integrate the rate function to determine the total increase in the amount of outstanding automobile loans from 1985 to 1995.

57. The rate of change in the number of employees in the U.S. federal government's executive branch can be modeled by

$$f(x) = 0.36x^3 - 10.26x^2 + 78.52x - 123.18, \qquad 0 \le x \le 15$$

where x represents the number of years since 1980 and $f(x)$ represents the rate of change in the number of employees in the U.S. federal government's executive branch measured in $\dfrac{\text{thousands of employees}}{\text{year}}$.

(a) Knowing that there were 2961 thousand employees in the executive branch in 1980, determine F, the model for the number of employees in the U.S. federal government's executive branch annually.

(b) Evaluate and interpret $\int_5^{15} f(x)\, dx$.

In Exercises 58–61, the velocity function v for an object is given, measured in feet per second.

(a) Evaluate $v(t)$ at the point $t = t_1$.

(b) Evaluate and interpret $\int_{t_1}^{t_2} v(t)\, dt$.

58. $v(t) = 4t$; $t_1 = 2$, $t_2 = 8$

59. $v(t) = 0.3t$; $t_1 = 10$, $t_2 = 20$

60. $v(t) = 3t + 6$; $t_1 = 1$, $t_2 = 5$

61. $v(t) = 2t + 10$; $t_1 = 3$, $t_2 = 5$

SECTION PROJECT

The rate of change in the price of natural gas in the United States from 1990 to 1996 is shown in Table 6.3.1.

(a) Use your calculator to determine a quartic regression model in the form

$$f'(x) = ax^4 + bx^3 + cx^2 + dx + e, \qquad 0 \le x \le 6$$

where x represents the number of years since 1990 and $f'(x)$ represents the rate of change in the price of natural gas in the United States measured in $\dfrac{\text{dollars per 1000 cubic feet}}{\text{year}}$. Round all coefficients to the nearest hundredth.

TABLE 6.3.1

YEAR	RATE OF CHANGE IN PRICE
1990	1.71
1991	1.64
1992	1.74
1993	2.04
1994	1.85
1995	1.55
1996	2.25

(b) Evaluate and interpret $f'(3)$ and explain why this answer is different from the rate of change in price found in Table 6.3.1.

(c) Compute the second derivative and determine when the rate of change in the price was increasing most rapidly.

(d) Evaluate $\int_0^6 f'(x)\,dx$ and interpret.

SECTION 6.4   INTEGRATION BY *u*-SUBSTITUTION

The rules of integration we have learned so far apply to functions that are straightforward. We have found that polynomial functions and simple root functions are easy to integrate. But what if we need to integrate a function like $\int x \sqrt[3]{4x^2 + 1}\,dx$? We really have no way to perform the integration with the rules that we have established. In this section we will learn how to rewrite an integral that contains a composite function so that it can be expressed in a simplified form that can easily be integrated. This is called the **method of *u*-substitution.** We will see how this method is used to determine both indefinite and definite integrals and the effect that *u*-substitution has on the limits of integration. We will also examine applications that take advantage of this technique.

u-Substitution with Indefinite Integrals

We begin by reviewing the definition of the differential in the Toolbox to the left.

Using the definition of differential, we see that if we substitute y with u we have $u = f(x)$ then $\dfrac{du}{dx} = f'(x)$, so the **differential in *u*** is $du = f'(x)\,dx$. To see how this can help us with integration, let's consider the indefinite integral $\int 4(4x + 3)^5\,dx$. We could integrate by expanding $4x + 3$, but we would have to raise this binomial to the fifth power! To avoid all this unnecessary algebra, let's see if a substitution in the integrand will make it simpler. If we let $u = 4x + 3$ and compute the differential in u, we get

$$u = 4x + 3$$

$$\frac{du}{dx} = 4$$

$$du = 4\,dx$$

From Your Toolbox

The **differential in *y***, denoted by *dy*, of the dependent variable is given by

$$dy = f'(x)\,dx$$

Notice that we now have an expression in terms of *u* for every part of the integrand. Since $u = 4x + 3$ and $du = 4\,dx$, we can rewrite the integral

$$\int \overbrace{4\underbrace{(4x+3)^5}_{u}dx}^{du} \text{ in terms of } u \text{ as}$$

$$\int u^5\,du$$

Now we can use the Power Rule to find the integral with respect to *u* as

$$\int u^5\,du = \frac{1}{6}u^6 + C$$

Since we are given the integral originally in terms of *x*, and knowing that $u = 4x + 3$, we can resubstitute and write the solution as

$$\int 4(4x+3)^5\,dx = \frac{1}{6}(4x+3)^6 + C$$

We learned in Section 6.1 that *the derivative of the indefinite integral yields the integrand*, hence we can check our answer by differentiating the result via the Chain Rule.

$$\frac{d}{dx}\left[\frac{1}{6}(4x+3)^6 + C\right] = \frac{1}{6} \cdot 6(4x+3)^5 \cdot \frac{d}{dx}(4x+3) + 0$$

$$= (4x+3)^5 \cdot 4 = 4(4x+3)^5$$

Now let's generalize by stating the Power Rule with *u*-substitution.

Power Rule with *u*-Substitution

If *f* is a differentiable function of *x* and $u = f(x)$, then we can write an integral of the form $\int [f(x)]^n f'(x)\,dx$ as $\int u^n\,du$ with

$$\int u^n\,du = \frac{1}{n+1}u^{n+1} + C, \qquad n \neq -1$$

where $du = f'(x)\,dx$.

EXAMPLE 1 | **Using the Power Rule with *u*-Substitution**

Use the method of *u*-substitution to evaluate the given indefinite integrals.

(a) $\displaystyle\int \sqrt{x+2}\,dx$ \qquad (b) $\displaystyle\int 2(2x+1)^3\,dx$

SOLUTION

(a) For this integration, we can let *u* be the radicand and write $u = x + 2$. Differentiating to get an expression for *du*, we find that

$$\frac{du}{dx} = 1$$

$$du = 1 \cdot dx = dx$$

Since $du = dx$, the substitution is

$$\int \sqrt{x+2}\,dx = \int \sqrt{u}\,du = \int u^{1/2}\,du = \frac{2}{3}u^{3/2} + C$$

Since $u = x + 2$, we resubstitute and get

$$\int \sqrt{x + 2} \, dx = \frac{2}{3}(x + 2)^{3/2} + C$$

(b) Here we will let u be the expression in the parentheses and write $u = 2x + 1$. Differentiating with respect to x, we find that

$$\frac{du}{dx} = 2$$

$$du = 2 \, dx$$

Rewriting the integral in terms of u and integrating yields

$$\int 2(2x + 1)^3 \, dx = \int u^3 \, du = \frac{1}{4}u^4 + C$$

Resubstituting gives

$$\int 2(2x + 1)^3 \, dx = \frac{1}{4}(2x + 1)^4 + C$$

✓ **CHECKPOINT 1**

Now work Exercise 3.

In words, the Power Rule with u-substitution states that if an integrand can be expressed as an expression raised to a power, and the derivative of the expression is another factor in the integrand, we can apply the u-substitution. This form of the integral

$$\int (\text{expression})^{\text{power}} \cdot \frac{d}{dx}(\text{expression}) \, dx$$

occurs more frequently than one might think. As we have seen in the first example, we commonly represent u as an expression raised to a power or as a radicand.

EXAMPLE 2 | **Integrating Using a u-Substitution**

Use the Power Rule with u-substitution to integrate $\int \sqrt[3]{(4x^3 + x)} \, (12x^2 + 1) \, dx$.

SOLUTION

To get $\int \sqrt[3]{(4x^3 + x)} \, (12x^2 + 1) \, dx$ in the form $\int u^n \, du$, we can let u be the radicand of the cube root and get

$$u = 4x^3 + x$$

So the differential in u is

$$du = (12x^2 + 1) \, dx$$

Now, writing the integral with respect to u and integrating yields

$$\int \sqrt[3]{(4x^3 + x)} \, (12x^2 + 1) \, dx = \int \sqrt[3]{u} \, du = \int u^{1/3} \, du = \frac{3}{4}u^{4/3} + C$$

Knowing that $u = 4x^3 + x$, we can write the solution in terms of x as

$$\int \sqrt[3]{(4x^3 + x)} \, (12x^2 + 1) \, dx = \frac{3}{4}(4x^3 + x)^{4/3} + C$$

$$= \frac{3}{4}\sqrt[3]{(4x^3 + x)^4} + C$$

$$= \frac{3}{4}(4x^3 + x)\sqrt[3]{(4x^3 + x)} + C \qquad \text{Simplifying the radical}$$

Sometimes, we must alter the differential in u to get it to match the remaining part of the integrand. We can do this by multiplying or dividing both sides of the differential by a constant. Let's say that we wish to integrate $\int x^2 \sqrt{x^3 + 9}\, dx$. If we let u represent the radicand, we have $u = x^3 + 9$. Computing du, we find that

$$\frac{du}{dx} = 3x^2$$

$$du = 3x^2\, dx$$

But we want du to just be $x^2\, dx$. We can get this desired expression by just multiplying both sides of the differential by $\frac{1}{3}$. This gives us

$$du = 3x^2\, dx$$

$$\frac{1}{3}\, du = x^2\, dx$$

Now, writing the integral in terms of u, we get

$$\int x^2 \sqrt{x^3 + 9}\, dx = \int \left(\sqrt{u} \cdot \frac{1}{3} \right) du = \frac{1}{3} \int u^{1/2}\, du$$

$$= \frac{1}{3} \cdot \frac{2}{3} \cdot u^{3/2} + C = \frac{2}{9} u^{3/2} + C$$

In terms of x we get, by substitution,

$$= \frac{2}{9}(x^3 + 9)^{3/2} + C$$

EXAMPLE 3 | Adjusting a Constant with *u*-substitution

Evaluate the indefinite integral $\int \dfrac{1}{(4x - 1)^3}\, dx$.

SOLUTION

Here we let $u = 4x - 1$ so that $du = 4\, dx$. Since we want $du = 1 \cdot dx$, we can divide both sides of the differential by 4 to get

$$\frac{1}{4}\, du = dx$$

So the integral in terms of u becomes

$$\int \frac{1}{(4x - 1)^3}\, dx = \int \left(\frac{1}{4} \cdot \frac{1}{u^3} \right) du$$

$$= \frac{1}{4} \int u^{-3}\, du$$

$$= \frac{1}{4} \cdot \frac{1}{-2} u^{-2} + C$$

$$= -\frac{1}{8} \cdot \frac{1}{u^2} + C$$

$$= -\frac{1}{8(4x - 1)^2} + C \qquad \text{Written in terms of } x$$

Interactive Activity

Verify the solution to Example 3 by differentiating

$$\frac{d}{dx}\left(-\frac{1}{8(4x - 1)^2} + C \right).$$

Let's summarize what we have learned so far and list some tips on integration by u-substitution.

Tips on u-Substitution

- The correct choice of u is usually a radicand, a denominator, or an expression in parentheses.
- Compute the differential in u; $du = f'(x)\,dx$.
- Multiply or divide du by a constant, if needed, to make the remaining part of the integral match the differential in u.
- After the u-substitution, no factors can contain x.
- Integrate and rewrite the integral in terms of the original independent variable (usually x).

Sometimes we must use a "trick" to complete the substitution. If the expression for u is linear, and so is the remainder of the integrand, we have to solve the expression for u in terms of x. We demonstrate this technique in Example 4.

EXAMPLE 4

Solving the u-Expression for x to Complete a Substitution

Determine the indefinite integral $\int x(x-7)^2\,dx$.

SOLUTION

If we assign u to the expression in parentheses, we see that

$$u = x - 7$$
$$du = dx$$

This doesn't help because the desired du should be $x\,dx$. However, since $u = x - 7$, we can write

$$x = u + 7$$

Now, making the substitution, we write

$$\int x(x-7)^2\,dx = \int (u+7)u^2\,du$$

$$= \int (u^3 + 7u)\,du \qquad \text{Distribute } u^2$$

$$= \frac{1}{4}u^4 + \frac{7}{2}u^2 + C \qquad \text{Integrate with respect to } u$$

$$= \frac{1}{4}(x-7)^4 + \frac{7}{2}(x-7)^2 + C \qquad \text{Written in terms of } x$$

✓ CHECKPOINT 2

Now work Exercise 29.

u-Substitution with Definite Integrals

We also can apply the integration technique of u-substitution to definite integrals in one of two ways.

1. Define u, perform the integration with respect to du, and rewrite the answer in terms of x. Evaluate the definite integral using the original limits of integration.

2. Define u and rewrite the integrand *and* the limits of integration in terms of the variable u. Integrate with respect to u and evaluate the integral using the new limits of integration.

We illustrate these two methods in Example 5.

EXAMPLE 5

Evaluating a Definite Integral by *u*-Substitution

Evaluate $\displaystyle\int_0^4 \frac{x}{\sqrt{x^2+9}}\,dx$.

SOLUTION

By letting $u = x^2 + 9$, we have $du = 2x\,dx$, so $\frac{1}{2}du = x\,dx$. We can now proceed in either of the ways outlined prior to Example 5.

1.
$$\int \frac{x}{\sqrt{x^2+9}}\,dx = \int \frac{1}{\sqrt{u}} \cdot \frac{1}{2}\,du$$

$$= \frac{1}{2}\int \frac{1}{\sqrt{u}}\,du = \frac{1}{2}\int u^{-(1/2)}\,du$$

$$= \frac{1}{2}(2)u^{1/2} = \sqrt{u} = (x^2+9)^{1/2}$$

Now we use the Fundamental Theorem of Calculus with the original limits to get

$$\int_0^4 \frac{x}{\sqrt{x^2+9}}\,dx = (x^2+9)^{1/2}\Big|_0^4 = (25)^{1/2} - (9)^{1/2} = 2$$

2. Since we assign $u = x^2 + 9$, we have:

If $x = 0$, then $u = 0^2 + 9 = 9$.

If $x = 4$, then $u = 4^2 + 9 = 25$.

Thus, after we do the substitution, we can rewrite the limits as follows:

$$\int_0^4 \frac{x}{\sqrt{x^2+9}}\,dx = \frac{1}{2}\int_9^{25} \frac{1}{\sqrt{u}}\,du = \frac{1}{2}\int_9^{25} u^{-(1/2)}\,du$$

$$= u^{1/2}\Big|_9^{25} = (25)^{1/2} - (9)^{1/2} = 2$$

Notice in Example 5 that changing the limits of integration saved a few steps in the evaluation process.

✓ CHECKPOINT 3

Now work Exercises 35 and 47.

Applications

The *u*-substitution technique allows us to examine new families of applications. One of these is recovering business functions.

EXAMPLE 6

Using *u*-Substitution to Recover Business Functions

A manager at the Black Box microprocessor manufacturing company finds through data gathered in research that the marginal cost function for a certain type of automobile computer chip made at the facility is given by

$$MC(x) = C'(x) = 6x\sqrt{x^2 + 11}$$

where x represents the number of auto computer chips produced each hour and $MC(x)$ represents the marginal cost. The manager also knows that it costs $1932 to manufacture five chips. Recover the cost function C and find the fixed costs. (To avoid confusion, we will call the arbitrary constant d.)

SOLUTION

Recall from Section 6.2 that $C(x) + d = \int MC(x)\,dx$, where d is a constant that can be found given an initial value. So we must integrate $\int 6x\sqrt{x^2 + 11}\,dx$. It is reasonable to let $u = x^2 + 11$, making

$$du = 2x\,dx$$

However, the remaining part of the integrand is $6x\,dx$. This means we must multiply the differential in u by 3 to get the parts to match.

$$3 \cdot du = 3 \cdot 2x\,dx = 6x\,dx$$

So now our substitution is

$$\int 6x\sqrt{x^2 + 11}\,dx = \int \sqrt{u} \cdot 3\,du = 3\int \sqrt{u}\,du$$

$$= 3\int u^{1/2}\,du = 3 \cdot \left(\frac{2}{3}\right)u^{3/2} + d = 2u^{3/2} + d$$

So, in terms of x, we get the partially recovered cost function as

$$C(x) = 2(x^2 + 11)^{3/2} + d = 2\sqrt{(x^2 + 11)^3} + d$$

where d is a constant. To find d, we can use the initial value that when $x = 5$ $C(x) = 1932$ to get

$$1932 = 2\sqrt{(5^2 + 11)^3} + d$$
$$1932 = 2\sqrt{46{,}656} + d$$
$$1932 = 432 + d$$
$$1500 = d$$

So the recovered cost function is $C(x) = 2\sqrt{(x^2 + 11)^3} + 1500$.

When the cost function was represented by a polynomial function, the fixed cost was always the same as the constant. But here the cost function is a radical function, so we need to use the definition of the cost function in its literal sense and find the cost of producing zero items. So the fixed costs are found by evaluating $C(x)$ at $x = 0$, which yields

$$C(0) = 2\sqrt{(0^2 + 11)^3} + 1500$$
$$= 2\sqrt{1331} + 1500 \approx 1572.97$$

Thus, the fixed costs are about $1572.97.

As in the previous section, we can also compute the continuous sum of a rate function using the u-substitution technique. But now we can integrate composite rate functions, as illustrated in Example 7.

EXAMPLE 7 | **Integrating a Rate Function Using *u*-Substitution**

Media consultants for the new local magazine *Rave!* have projected that the number of subscriptions will grow during the first five years at a rate given by

$$S'(t) = \frac{1000}{(1+0.3t)^{3/2}}, \qquad 0 \le t \le 60$$

where t is the number of months since the magazine's first issue and $S'(t)$ is the rate of change in the number of subscriptions measured in $\frac{\text{subscriptions}}{\text{month}}$.

(a) Evaluate and interpret $S'(12)$.

(b) Evaluate and interpret $\int_0^6 S'(t)\, dt$

SOLUTION

(a) Evaluating, we get

$$S'(12) = \frac{1000}{(1+0.3(12))^{3/2}} = \frac{1000}{(4.6)^{3/2}} \approx 101.36$$

This means that at the end of the first year the subscriptions will be growing at a rate of about $101 \frac{\text{subscriptions}}{\text{month}}$.

(b) Here we need to integrate $\int_0^6 \frac{1000}{(1+0.3t)^{3/2}}\, dt$. If we let $u = 1+0.3t$, then $du = 0.3\, dt$, so $\frac{1}{0.3}\, du = dt$. Changing the limits of integration, we see that:

$$\text{If } t = 0, \text{ then } u = 1 + 0.3(0) = 1.$$
$$\text{If } t = 6, \text{ then } u = 1 + 0.3(6) = 2.8.$$

So, in terms of u, the integral is

$$\int_0^6 \frac{1000}{(1+0.3t)^{3/2}}\, dt = 1000 \int_0^6 \frac{1}{(1+0.3t)^{3/2}}\, dt$$

$$= 1000 \int_1^{2.8} \frac{1}{u^{3/2}} \left(\frac{1}{0.3}\right) du = \frac{1000}{0.3} \int_1^{2.8} \frac{1}{u^{3/2}}\, du$$

$$= \frac{1000}{0.3} \int_1^{2.8} u^{-(3/2)}\, du = \frac{1000}{0.3}(-2)[u^{-(1/2)}]\Big|_1^{2.8}$$

$$= \frac{-2000}{0.3}[(2.8)^{-(1/2)} - (1)^{-(1/2)}] \approx 2682.57$$

Thus, at the end of the first six months the number of subscriptions is estimated to be about 2683. ◼

✓CHECKPOINT **4** | Now work Exercise 63.

SUMMARY |

This section concentrated on the frequently used integration technique called ***u*-substitution.** We saw that the technique was useful in a variety of applications.

• **Power rule with *u*-substitution:** $\displaystyle\int u^n\, du = \frac{1}{n+1}u^{n+1} + C$

- **Tips on *u*-Substitution:**
  - The correct choice of u is usually a radicand, a denominator, or an expression in parentheses.
  - Compute the differential in u; $du = f'(x)\ dx$.
  - Multiply or divide du by a constant if needed to make the remaining part of the integral match the differential in u.
  - After the u-substitution, no factors can contain x.
  - Integrate and rewrite the integral in terms of the original independent variable.

SECTION 6.4 EXERCISES

In Exercises 1–12, use u-substitution to evaluate the indefinite integrals. Check by differentiating the solution.

1. $\int 3(3x+4)^2\, dx$

2. $\int 5(5x+3)^2\, dx$

✓ 3. $\int 8x(4x^2+1)^3\, dx$

4. $\int 6x(3x^2-5)^3\, dx$

5. $\int 2t\sqrt{t^2-1}\, dt$

6. $\int 10t\sqrt[3]{5t^2-11}\, dt$

7. $\int (3x^2-2)(x^3-2x)\, dx$

8. $\int (4x-1)(2x^2-x)\, dx$

9. $\int \dfrac{3x^2}{\sqrt[3]{x^3-5}}\, dx$

10. $\int \dfrac{8}{\sqrt{8x+9}}\, dx$

11. $\int \dfrac{4x}{(2x^2+3)^3}\, dx$

12. $\int \dfrac{3x^2-2}{(x^3-2x)^2}\, dx$

In Exercises 13–28, use u-substitution to evaluate the indefinite integrals.

13. $\int (4x+7)^4\, dx$

14. $\int (12x-1)^9\, dx$

15. $\int t(t^2-3)^5\, dt$

16. $\int t^2(t^3-1)^3\, dt$

17. $\int \dfrac{x^2}{\sqrt{x^3-5}}\, dx$

18. $\int \dfrac{x^3}{\sqrt{x^4-4}}\, dx$

19. $\int \dfrac{5x}{(10x^2-4)^2}\, dx$

20. $\int \dfrac{x^2}{(3x^3-8)^5}\, dx$

21. $\int \dfrac{1}{x^2}\sqrt{1-x^{-1}}\, dx$

22. $\int \dfrac{1}{\sqrt{x}}\sqrt[3]{1+\sqrt{x}}\, dx$

23. $\int (t-7)^{10}\, dt$

24. $\int (6t-10)^5\, dt$

25. $\int \dfrac{x+1}{(x^2+2x+5)^5}\, dx$

26. $\int \dfrac{2x+3}{(2x^2+6x-10)^4}\, dx$

27. $\int (6x+9)(x^2+3x-10)\, dx$

28. $\int (20x+20)(x^2+2x-9)^2\, dx$

In Exercises 29–34, determine the integrals by solving the u-expression to complete the substitution. See Example 4.

✓ 29. $\int x\sqrt{x+1}\, dx$

30. $\int x\sqrt[3]{x-2}\, dx$

31. $\int t(t-1)^5\, dt$

32. $\int 2t(t+1)^3\, dt$

33. $\int \dfrac{3x}{\sqrt{x-1}}\, dx$

34. $\int \dfrac{x}{\sqrt[3]{x+2}}\, dx$

In Exercises 35–46, evaluate the definite integrals. Do not change the limits of integration.

✓ 35. $\displaystyle\int_0^4 \dfrac{2x}{\sqrt{x^2+9}}\, dx$

36. $\displaystyle\int_{-2}^0 \dfrac{1}{(x-4)^2}\, dx$

37. $\displaystyle\int_0^2 x(x^2-2)^3\, dx$

38. $\displaystyle\int_0^1 3x(x^2-1)^2\, dx$

39. $\displaystyle\int_0^4 (3t-5)^2\, dt$

40. $\displaystyle\int_0^2 (2t+3)^3\, dt$

41. $\displaystyle\int_{-2}^1 \sqrt{1-x}\, dx$

42. $\displaystyle\int_{-5}^0 \sqrt[3]{2-x}\, dx$

43. $\displaystyle\int_0^1 \dfrac{x}{(1+3x^2)^2}\, dx$

44. $\displaystyle\int_{-1}^1 \dfrac{2x}{(1+x^2)}\, 3dx$

45. $\displaystyle\int_2^{10} \dfrac{1}{\sqrt{t-1}}\, dt$

46. $\displaystyle\int_0^7 \dfrac{1}{\sqrt[3]{t+1}}\, dt$

In Exercises 47–58, evaluate the definite integrals. Change the limits of integration to u-limits.

✓ 47. $\displaystyle\int_0^3 \dfrac{x}{(x^2+1)^2}\, dx$

48. $\displaystyle\int_0^1 \dfrac{8x}{(2x^2+1)^3}\, dx$

49. $\displaystyle\int_1^4 \frac{2x+1}{(x^2+x-1)^2}\,dx$ 50. $\displaystyle\int_1^3 \frac{2x+3}{(x^2+3x)^3}\,dx$

51. $\displaystyle\int_0^2 \frac{3t^2}{(1+t^3)^5}\,dt$ 52. $\displaystyle\int_{-1}^0 \frac{t^3}{(2-t^4)^4}\,dt$

53. $\displaystyle\int_0^2 3x^2\sqrt{x^3+1}\,dx$ 54. $\displaystyle\int_0^1 2x^3\sqrt{x^4+4}\,dx$

55. $\displaystyle\int_{-1}^1 t(t^2-1)^3\,dt$ 56. $\displaystyle\int_{-2}^0 t^2(t^3-2)\,dt$

57. $\displaystyle\int_2^{10} \frac{3}{\sqrt{5x-1}}\,dx$ 58. $\displaystyle\int_0^4 \frac{x}{\sqrt{x^2+9}}\,dx$

APPLICATIONS

In Exercises 59–62, determine the exact area of the shaded region.

59.

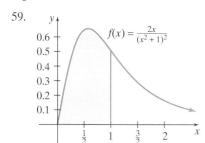

60.

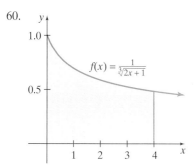

61.

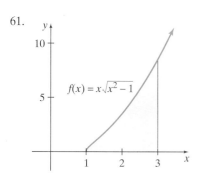

62.

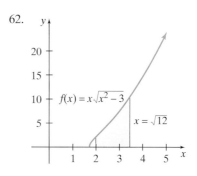

✓ 63. The CustomKey Company determines that the marginal cost of producing sterling silver key fobs is given by

$$MC(x) = \frac{10x}{\sqrt{x^2+10{,}000}}, \qquad 0 \le x \le 100$$

where x represents the number of fobs produced daily and $MC(x)$ is the marginal cost in dollars.

(a) Evaluate $MC(75)$ and interpret.

(b) Evaluate $\int_0^{75} MC(x)\,dx$ and interpret.

(c) Knowing that the cost to produce 75 fobs is \$3250, recover the cost function C.

(d) Determine the fixed costs.

64. The TightNut Company determines that the marginal cost of producing center punches is given by

$$MC(x) = 0.003x\sqrt{x+1}, \qquad 0 \le x \le 50$$

where x is the number of center punches produced in a work shift and $MC(x)$ is the marginal cost in dollars.

(a) Evaluate and interpret $MC(48)$.

(b) Evaluate $\int_{35}^{48} MC(x)\,dx$ and interpret.

(c) Knowing that it costs \$23.40 to produce 24 center punches, recover the cost function C.

(d) Determine the fixed costs.

65. The marginal profit function for seasonal flags produced by the WaveFree Company is given by

$$MP(x) = \frac{2x}{\sqrt{2x^2-400}}, \qquad 15 \le x \le 105$$

where x is the number of flags produced and sold monthly and $MP(x)$ is the marginal profit in hundreds of dollars.

(a) Evaluate and interpret $MP(70)$.

(b) Evaluate $\int_{20}^{100} MP(x)\,dx$ and interpret.

(c) Knowing that the WaveFree Company breaks even when 20 units are sold [that is, $P(x) = 0$ when $x = 20$], recover the profit function P.

66. The FemTouch Accessory Manufacturer determines their marginal profit function for decorative scarves is

$$MP(x) = \frac{4}{\sqrt[3]{12x-200}}, \qquad 20 \le x \le 150$$

where x is the number of scarves produced and sold daily and $MP(x)$ is the marginal profit in hundreds of dollars.

(a) Evaluate $MP(100)$ and interpret.

(b) Evaluate $\int_{20}^{100} MP(x)\,dx$ and interpret.

(c) Knowing that the FemTouch accessory manufacturer breaks even when 100 units are sold [that is, $P(x) = 0$ when $x = 100$], recover the profit function P.

67. Suppose that the velocity of an object is given by the function

$$v(t) = \frac{t}{\sqrt{t^2 + 9}}$$

where t is the time in seconds and $v(t)$ is the velocity in $\frac{\text{feet}}{\text{second}}$.

(a) Determine the total distance moved between 3 and 5 seconds.

(b) Knowing that when $t = 4$ seconds $s(t) = 8$ feet, determine the position function s.

68. Suppose that the velocity of an object is given by the function

$$v(t) = \frac{2\sqrt{t + 1} + 1}{\sqrt{t + 1}}$$

where t is the time in seconds and $v(t)$ is the velocity in $\frac{\text{feet}}{\text{second}}$.

(a) Determine the total distance moved between 4 and 16 seconds.

(b) Knowing that when $t = 3$ seconds, $s(t) = 16$ feet, determine the position function s.

69. The number of homes being built in the computer-designed town of EastWorld is increasing at an annual rate of

$$h'(t) = \frac{15t^2}{\sqrt{t^3 + 4}}, \qquad 0 \le t \le 10$$

where t is the number of years since the town was first incor-

porated and $h'(t)$ represents the rate of homes being built measured in $\frac{\text{homes}}{\text{year}}$.

(a) Evaluate and interpret $h'(5)$. Round your answer to the nearest whole number.

(b) Evaluate and interpret $\int_0^5 h'(t)\,dt$. Round your answer to the nearest whole number.

(c) Knowing that there were no homes completed when the town was first incorporated, determine h, the number of homes in the town after t years.

(d) Evaluate and interpret $h(5)$. Round your answer to the nearest whole number.

SECTION PROJECT

Conservationists find that the rate of growth of carp in the first year in a stocked pond can be modeled by

$$f'(x) = \frac{10x}{\sqrt{x + 1}}, \qquad 0 \le x \le 12$$

where x is the number of months since the pond was stocked and $f'(x)$ is the rate of growth in the number of carp in the pond, measured in $\frac{\text{carp}}{\text{month}}$.

(a) Evaluate and interpret $f'(6)$. Round your answer to the nearest whole number.

(b) Evaluate and interpret $\int_0^{12} f'(x)\,dx$. Round your answer to the nearest whole number.

(c) Knowing that the pond was stocked with 100 carp, determine $f(x)$, the number of carp in the pond after x months.

(d) Evaluate and interpret $f(5)$. Round your answer to the nearest whole number.

(e) Determine an equation of the line tangent to the graph of f when $x = 6$. Evaluate the tangent line equation at the value of $x = 12$ and interpret.

SECTION 6.5   INTEGRALS THAT YIELD LOGARITHMIC AND EXPONENTIAL FUNCTIONS

In Chapter 4, we discussed the derivatives of the logarithmic and exponential functions, along with their associated applications. Now we turn our attention to **integrals** that involve **exponential and logarithmic functions.** Just as we need special rules to differentiate this family of functions, we also need an additional set of integration rules. We will see how to use the u-substitution technique with these functions and check out applications that lend themselves to these functions.

Integrals That Yield Logarithmic Functions

Before examining the role that the logarithm plays in integration, let's review the applicable differentiation rules.

From Your Toolbox

- $\dfrac{d}{dx}[\ln x] = \dfrac{1}{x}$.
- If f is a differentiable function of x, then, by the Chain Rule,

$$\frac{d}{dx}\ln[f(x)] = \frac{1}{f(x)} \cdot f'(x)$$

- The derivative of the indefinite integral yields the integrand. Mathematically, this means that $\int f(x)\,dx = F(x) + C$ if and only if

$$\frac{d}{dx}[F(x) + C] = f(x).$$

Remember that the Power Rule for Integration stated that $\int x^n\,dx = \dfrac{1}{n+1}x^{n+1} + C$, provided that $n \neq -1$. To find the integral of x^n when $n = -1$, we must integrate $\int x^{-1}\,dx = \int \dfrac{1}{x}dx$. Since $\dfrac{d}{dx}[\ln x] = \dfrac{1}{x}$, we can conclude that $\int \dfrac{1}{x}dx = \ln x + C$. But we need to be careful here, because the natural logarithm function is only defined for positive numbers. Let's say that x is negative, in which case $(-x)$ is positive. Then, by the Chain Rule for the logarithm function,

$$\frac{d}{dx}\ln(-x) = \frac{1}{-x}\frac{d}{dx}(-x) = \frac{1}{-x}(-1) = \frac{1}{x}$$

So we see that $\int \dfrac{1}{x}dx = \ln(-x) + C$ if x is negative. Instead of writing two integration rules for positive and negative integrands, we can use the absolute value to write one neat integration formula.

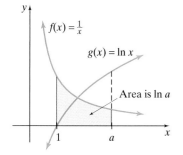

$f(x) = \frac{1}{x}$

$g(x) = \ln x$

Area is $\ln a$

Figure 6.5.1 $\displaystyle\int_1^a \frac{1}{x}\,dx = \ln a.$

Integration Formula for $\frac{1}{x}$

For the function $f(x) = \dfrac{1}{x}$, the indefinite integral is given by

$$\int \frac{1}{x}\,dx = \ln|x| + C$$

This relationship between the functions $f(x) = \dfrac{1}{x}$ and $g(x) = \ln x$ is shown in Figure 6.5.1.

NOTE: For $a > 1, \int_1^a \frac{1}{x}dx = \ln x\big|_1^a = \ln a - \ln 1 = \ln a - 0 = \ln a$. This means that the area under the graph of $f(x) = \frac{1}{x}$ on the interval $[1, a]$ is the same as the value of $\ln a$. Some books define the natural logarithm function exactly this way.

EXAMPLE 1 | **Integrating Functions of the Form $f(x) = \dfrac{k}{x}$**

Evaluate the following definite integrals.

(a) $\displaystyle\int_{-6}^{-2} \frac{1}{x}\,dx$ (b) $\displaystyle\int_1^5 \frac{6}{x}\,dx$

SOLUTION

(a) Using the Integration Formula for $\dfrac{1}{x}$, we get

$$\int_{-6}^{-2} \frac{1}{x}\,dx = \ln|x|\,\bigg|_{-6}^{-2} = \ln|-2| - \ln|-6| = \ln(2) - \ln(6) \approx -1.10$$

(b) We can begin by factoring out the constant 6.

$$\int_1^5 \frac{6}{x}\,dx = 6 \int_1^5 \frac{1}{x}\,dx = 6 \cdot \ln|x| \Big|_1^5$$
$$= 6[\ln|5| - \ln|1|] = 6[\ln(5) - \ln(1)]$$
$$= 6[\ln(5) - 0] = 6\ln(5) \approx 9.66$$

Since we rarely must integrate $f(x) = \frac{1}{x}$, the usefulness of the Integration Formula for $\frac{1}{x}$ is limited. However, the u-substitution form of the formula is quite useful.

Integration Formula for $\frac{1}{u}$

If u is a differentiable function of x, then

$$\int \frac{1}{u}\,du = \ln|u| + C$$

EXAMPLE 2

Integrating using u-Substitution

Determine the following indefinite integrals.

(a) $\displaystyle\int \frac{1}{t+2}\,dt$ (b) $\displaystyle\int \frac{4x}{x^2+5}\,dx$

SOLUTION

(a) We can use a u-substitution here with $u = t + 2$. Differentiating both sides of the equation gives

$$du = dt$$

So, in terms of u, the integral becomes

$$\int \frac{1}{t+2}\,dt = \int \frac{1}{u}\,du = \ln|u| + C$$

In terms of t, the solution is

$$= \ln|t+2| + C$$

(b) To use the Integration Formula for $\frac{1}{u}$, we let $u = x^2 + 5$; then $du = 2x\,dx$. Since we want the expression containing dx to include $4x$, we can multiply by 2 to get $2\,du = 4x\,dx$. In terms of u, the integral becomes

$$\int \frac{4x}{x^2+5}\,dx = \int \frac{1}{u} \cdot 2\,du = 2 \int \frac{1}{u}\,du = 2\ln|u| + C$$

So, in terms of x, the solution is

$$= 2\ln|x^2+5| + C = 2\ln(x^2+5) + C$$

Notice that, since the expression $x^2 + 5$ is positive for all values of x, we can remove the absolute value signs.

✓**CHECKPOINT 1**

Now work Exercise 13.

Many applications that involve integrating a rate function use our new integration rule.

EXAMPLE 3 | **Integrating a Rate Function That Yields a Logarithmic Function**

Since running a series of first-come, first-served promotions, the FineHomes Furniture Store has found that its sales rate during its three-month sales drive is given by the function

$$s(t) = \frac{10}{t} + 2, \qquad 1 \le t \le 12$$

where t represents the number of weeks that the promotion has been running and $s(t)$ is the sales rate measured in $\frac{\text{thousands of dollars}}{\text{week}}$.

(a) Determine the total increase in sales generated from the first to the fifth week.

(b) Knowing that $6000 was made during the first week, recover $R(t)$, the revenue generated after t weeks.

SOLUTION

(a) Since we are given a rate function, we need to compute $\int_1^5 s(t)\,dt$ to obtain the total increase in sales generated from the first to the fifth week. Evaluating the definite integral, we get

$$\int_1^5 s(t)\,dt = \int_1^5 \left(\frac{10}{t} + 2\right) dt = \int_1^5 \frac{10}{t}\,dt + \int_1^5 2\,dt$$

$$= 10\int_1^5 \frac{1}{t}\,dt + 2\int_1^5 dt = (10\ \ln|t| + 2t)\Big|_1^5$$

$$= (10\ \ln|5| + 2(5)) - (10\ \ln|1| + 2(1))$$

$$= 10\ \ln 5 + 10 - 0 - 2 = 10\ \ln 5 + 8$$

$$\approx 24.09\ \text{thousand dollars}$$

So, from weeks 1 to 5 of the promotion, the total increase in sales was about $24,090.

(b) To recover the revenue $R(t)$, we begin by evaluating the indefinite integral $\int s(t)\,dt$. This gives us

$$R(t) = \int s(t)\,dt = \int \left(\frac{10}{t} + 2\right) dt$$

$$= 10\int \frac{1}{t}\,dt + \int 2\,dt$$

$$= 10\ \ln|t| + 2t + C$$

Since the time is represented by positive numbers only, we can omit the absolute value and write the partial solution $R(t) = 10\ln t + 2t + C$. To find the constant C, we use the fact that, when $t = 1$, $R(t) = 6$, and get

$$R(t) = 10\ \ln t + 2t + C$$
$$6 = 10\ \ln(1) + 2(1) + C$$
$$6 = 10(0) + 2 + C$$
$$6 = 2 + C$$
$$4 = C$$

So the revenue function for the store is $R(t) = 10\ \ln t + 2t + 4$.

Integration of Exponential Functions

Now we turn our attention from integrals yielding logarithmic functions to those involving exponential functions. Let's consult our Toolbox to the left to recall some facts on exponential functions.

From the first bulleted fact, we know that $\dfrac{d}{dx}[e^x] = e^x$. This, along with the fact that differentiation and integration are inverse operations, gives the following integration rule.

> **Integration Formula for e^x**
>
> For the exponential function $f(x) = e^x$,
>
> $$\int e^x \, dx = e^x + C$$

EXAMPLE 4 | **Integration of Functions Containing e^x**

Determine the definite integral $\int_0^4 (2 - e^x)\, dx$.

SOLUTION

Integrating, we get

$$\int_0^4 (2 - e^x)\, dx = \int_0^4 2\, dx - \int_0^4 e^x\, dx$$
$$= (2x - e^x)\big|_0^4 = [2(4) - e^4] - [2(0) - e^0]$$
$$= 8 - e^4 - 0 + 1 = 9 - e^4 \approx -45.60$$

In practice, few integrands are simply e^x. We can easily extend the methods of u-substitution to the exponential function, as is shown in the integration formula for e^u.

> **Integration Formula for e^u**
>
> If u is a differentiable function of x, then,
>
> $$\int e^u \, du = e^u + C$$

EXAMPLE 5 | **Integration of Functions Containing e^u**

Determine the given indefinite integrals.

(a) $\displaystyle\int 10xe^{x^2}\, dx$ (b) $\displaystyle\int (2x+1)e^{2x^2+2x}\, dx$

SOLUTION

(a) Since the integration formula for e^u suggests letting u be the expression in the exponent, we get $u = x^2$, so $du = 2x\, dx$. Multiplying by 5, we get

$$5\, du = 10x\, dx$$

So the indefinite integral is

$$\int 10xe^{x^2}\, dx = \int e^u \cdot 5\, du = 5\int e^u\, du = 5e^u + C$$

In terms of x, this gives

$$\int 10xe^{x^2}\, dx = 5e^{x^2} + C$$

(b) Again, if we let $u = 2x^2 + 2x$, then $du = (4x + 2)\, dx$. Since we want the du expression to contain $2x + 1$, we multiply by $\frac{1}{2}$ to get

$$\frac{1}{2}\, du = \frac{1}{2}(4x + 2)\, dx = (2x + 1)\, dx$$

This gives us

$$\int (2x + 1)e^{2x^2+2x}\, dx = \int e^u \cdot \frac{1}{2}\, du = \frac{1}{2}\int e^u\, du = \frac{1}{2}e^u + C$$

In terms of x, this is

$$\int (2x + 1)e^{2x^2+2x}\, dx = \frac{1}{2}e^{2x^2+2x} + C$$

■

✓ CHECKPOINT 2

Now work Exercise 43.

In many applications, we are asked to integrate functions of the form $f(x) = e^{kx}$, where k is a real number constant. To determine this integral, we use a u-substitution with $u = kx$. Then we see that $du = k\, dx$ and that $\frac{1}{k}\, du = dx$. So, in terms of u, the integral becomes

$$\int e^{kx}\, dx = \frac{1}{k}\int e^u\, du = \frac{1}{k}e^u + C = \frac{1}{k}e^{kx} + C$$

Integration Formula for e^{kx}

If k is a nonzero real number coefficient of x, then, for the exponential function $f(x) = e^{kx}$,

$$\int e^{kx}\, dx = \frac{1}{k}e^{kx} + C$$

EXAMPLE 6

Applying the Integration Formula for e^{kx}

The growth rate of the number of web sites on the Internet from 1995 to 1997 on a monthly basis can be modeled by

$$g(t) = 24.53e^{0.17t}, \qquad 1 \le t \le 20$$

where t represents the number of months starting in 1995 (January 1995 corresponds to $t = 1$) and $g(t)$ represents the growth rate of web sites measured in $\frac{\text{hundreds of new sites}}{\text{month}}$. The graph of g is displayed in Figure 6.5.2.

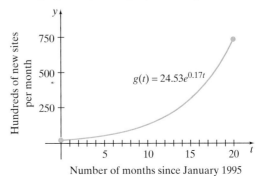

Figure 6.5.2

(a) Evaluate $g(6)$ and interpret.

(b) Use the model to estimate the total increase in the number of new web sites added to the Internet from June 1995 to June 1996.

SOLUTION

(a) Evaluating $g(t)$ at $t = 6$, we get

$$g(6) = 24.53e^{0.17(6)} = 24.53e^{1.02} \approx 68.03$$

So in June 1995 (when $t = 6$) the number of web sites was growing at a rate of about $68.03 \, \frac{\text{hundreds of new sites}}{\text{month}}$, or about 6803 new sites per month.

(b) In June 1995, the t-value is 6, and in June 1996, the value is 18, so the total number of new web sites added during this period is given by the integral

$$\int_6^{18} 24.53e^{0.17t} \, dt = 24.53 \int_6^{18} e^{0.17t} \, dt$$

From the integration formula for e^{kx}, we get

$$= 24.53 \cdot \frac{1}{0.17}(e^{0.17t}) \Big|_6^{18} = \frac{24.53}{0.17}(e^{0.17(18)} - e^{0.17(6)})$$

$$= \frac{24.53}{0.17}(e^{3.06} - e^{1.02}) \approx 2677.29$$

So, according to the model, the total increase in the number of new web sites between June 1995 and June 1996 was about 2677.29 hundred, or about 267,729. ■

✓**CHECKPOINT 3**

Now work Exercise 61.

Integrating General Exponential Functions

Thus far we have discussed how to integrate functions of the type $f(x) = e^{kx}$. But what about general exponential functions such as $f(x) = 2^x$ or $g(x) = 3^{0.6x}$? Surely they can be integrated, too. Some calculus courses derive a formula specifically for these functions, but we will rely on techniques that we have already learned. Refer to the Toolbox to the left to recall key ideas from Chapters 1 and 4 on general exponential functions.

From Your Toolbox

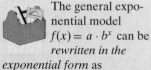
The general exponential model $f(x) = a \cdot b^x$ can be *rewritten in the exponential form* as $g(x) = a \cdot e^{kx}$, where $k = \ln b$.

To determine the indefinite integral whose integrand has the form b^x, such as $\int 2^x \, dx$, we rewrite the integrand in the exponential form e^{kx}, where $k = \ln b$. Thus, we can rewrite $\int 2^x \, dx$ as $\int e^{\ln 2 \cdot x} \, dx$ and then integrate using the integration formula for e^{kx}.

EXAMPLE 7

Integrating a General Exponential Function

Evaluate the definite integral $\int_{-2}^{1} \left(\frac{3}{4}\right)^x \, dx$.

SOLUTION

We begin by rewriting the integrand as $e^{\ln(3/4) \cdot x}$ and use the Integration Formula for e^{kx}.

$$\int_{-2}^{1} \left(\frac{3}{4}\right)^x \, dx = \int_{-2}^{1} e^{\ln(3/4) \cdot x} \, dx = \frac{1}{\ln\left(\frac{3}{4}\right)}(e^{\ln(3/4) \cdot x}) \Big|_{-2}^{1}$$

TECHNOLOGY NOTE
To see how to use the $\int f(x)\,dx$
or **FNINT** command on your
calculator, consult the online
calculator manual at
www.prenhall.com/armstrong

Rewriting $e^{\ln(3/4)\cdot x}$ as $\left(\dfrac{3}{4}\right)^{x}$ yields

$$= \frac{1}{\ln\left(\dfrac{3}{4}\right)}\left[\left(\dfrac{3}{4}\right)^{x}\right]\Big|_{-2}^{1} = \frac{1}{\ln\left(\dfrac{3}{4}\right)}\left[\left(\dfrac{3}{4}\right)^{1} - \left(\dfrac{3}{4}\right)^{-2}\right]$$

$$= \frac{1}{\ln\left(\dfrac{3}{4}\right)}\left(\dfrac{3}{4} - \dfrac{16}{9}\right) = \frac{1}{\ln\left(\dfrac{3}{4}\right)}\left(-\dfrac{37}{36}\right)$$

$$= -\frac{37}{36\,\ln\left(\dfrac{3}{4}\right)} \approx 3.57$$

Many times we are given data that are always increasing or always decreasing on an interval. This kind of data is said to be **monotonic.** We can model these
data with the general exponential mathematical model $y = a \cdot b^{x}$, where a is any
nonzero real number and with $b > 0$, $b \neq 1$. To integrate functions of the type
$\int a \cdot b^{x}\,dx$, we *pull out* the constant a and then we can determine $\int b^{x}\,dx$.

EXAMPLE 8

Integrating a General Exponential Function Application

The rate of increase in consumption of bottled water in the United States
between 1975 and 1995 can be modeled by

$$f'(t) = 18.82 \cdot (1.13)^{t}, \qquad 5 \leq t \leq 25$$

where t represents the number of years since 1970 and $f'(t)$ is the rate of change
in consumption measured in $\dfrac{\text{millions of gallons}}{\text{year}}$.

(a) Use the model to find the total increase in gallons of bottled water consumed from 1991 to 1995.

(b) Knowing that, in 1990, 2 billion (or 2000 million) gallons of bottled water
were consumed, recover the consumption function f for bottled water.

(c) Use your solution from part (b) to estimate the number of gallons consumed in 1983.

SOLUTION

(a) Since the years 1991 and 1995 correspond to the independent variable values $t = 21$ and $t = 25$, respectively, we must integrate

$$\int_{21}^{25} 18.82 \cdot (1.13)^{t}\,dt = 18.82 \int_{21}^{25} (1.13)^{t}\,dt$$

$$= 18.82 \int_{21}^{25} e^{\ln(1.13)\cdot t}\,dt = \frac{18.82}{\ln(1.13)}\left[e^{\ln(1.13)\cdot t}\right]\Big|_{21}^{25}$$

$$= \frac{18.82}{\ln(1.13)}(1.13^{t})\Big|_{21}^{25} = \frac{18.82}{\ln(1.13)}(1.13^{25} - 1.13^{21})$$

$$\approx 1264.15 \text{ million gallons}$$

This means that from 1991 to 1995 the total increase in bottled water consumption in the United States was about 1264.15 million gallons.

(b) To recover the consumption function f, we start by evaluating the indefinite integral $\int f'(t)\, dt$.

$$\int 18.82 \cdot (1.13)^t \, dt = 18.82 \int (1.13)^t \, dt$$

$$= 18.82 \int e^{\ln(1.13)t}$$

$$= \frac{18.82}{\ln(1.13)} e^{\ln(1.13)t} + C$$

$$= \frac{18.82}{\ln(1.13)} (1.13)^t + C$$

Since 2 billion gallons of water was consumed in 1990, we know that when $t = 20$, $f(t) = 2000$ million. This gives us

$$f(t) = \frac{18.82}{\ln(1.13)} (1.13)^t + C$$

$$2000 = \frac{18.82}{\ln(1.13)} (1.13)^{20} + C$$

$$2000 - \frac{18.82}{\ln(1.13)} (1.13)^{20} = C$$

$$225.59 \approx C$$

So the consumption of bottled water since 1970 can be modeled by

$$f(t) = \frac{18.82}{\ln(1.13)} (1.13)^t + 225.59, \qquad 5 \leq t \leq 25$$

(c) To estimate the number of gallons consumed in 1983, we need to evaluate $f(t)$ at $t = 13$. Doing so, we get

$$f(13) = \frac{18.82}{\ln(1.13)} (1.13)^{13} + 225.59 \approx 979.82 \text{ million gallons}$$

Thus, according to the model, in 1983 about 979.82 million gallons of bottled water were consumed. ◼

SUMMARY

In this section we analyzed and applied integration formulas for the exponential and logarithmic functions.

- **Integration formula for $\dfrac{1}{x}$:** $\displaystyle\int \frac{1}{x}\, dx = \ln|x| + C$

- **Integration formula for $\dfrac{1}{u}$:** $\displaystyle\int \frac{1}{u}\, du = \ln|u| + C$

- **Integration formula for e^x:** $\displaystyle\int e^x \, dx = e^x + C$

- **Integration formula for e^u:** $\displaystyle\int e^u \, du = e^u + C$

- **Integration formula for e^{kx}:** $\displaystyle\int e^{kx} \, dx = \frac{1}{k} e^{kx} + C$

SECTION 6.5 EXERCISES

In Exercises 1–10, evaluate the indefinite and definite integrals.

1. $\displaystyle\int \frac{2}{x}\,dx$

2. $\displaystyle\int \frac{5}{x}\,dx$

3. $\displaystyle\int \frac{-2}{3t}\,dt$

4. $\displaystyle\int \frac{-7}{2t}\,dt$

5. $\displaystyle\int \frac{1+e}{x}\,dx$

6. $\displaystyle\int \frac{e^2-1}{x}\,dx$

7. $\displaystyle\int_1^5 \frac{3}{x}\,dx$

8. $\displaystyle\int_1^8 \frac{5}{x}\,dx$

9. $\displaystyle\int_2^5 \frac{1}{2}x^{-1}\,dx$

10. $\displaystyle\int_5^8 \frac{2}{5}x^{-1}\,dx$

In Exercises 11–24, evaluate the integrals by u-substitution.

11. $\displaystyle\int \frac{1}{2x+3}\,dx$

12. $\displaystyle\int \frac{1}{7-5x}\,dx$

✓ 13. $\displaystyle\int \frac{t}{t^2+1}\,dt$

14. $\displaystyle\int \frac{2t}{t^2+3}\,dt$

15. $\displaystyle\int \frac{x-2}{x^2-4x+9}\,dx$

16. $\displaystyle\int \frac{3x^2-3}{x^3-3x-7}\,dx$

17. $\displaystyle\int \frac{x^3}{x^4+10}\,dx$

18. $\displaystyle\int \frac{3x^2}{x^3-4}\,dx$

19. $\displaystyle\int \frac{1}{x\cdot\ln 2x}\,dx$

20. $\displaystyle\int \frac{3}{x\cdot\ln 6x}\,dx$

21. $\displaystyle\int_1^2 \frac{2}{\sqrt{x}(\sqrt{x}+4)}\,dx$

22. $\displaystyle\int_1^2 \frac{\ln x}{x}\,dx$

23. $\displaystyle\int_0^3 \frac{2}{3x-2}\,dx$

24. $\displaystyle\int_{-1}^0 \frac{1}{4-5x}\,dx$

APPLICATIONS

25. The Top-2-Bottom Dress Store has a clearance sale to remove old inventory. Past records show that the rate of sales is given by

$$S(t) = \frac{40}{t}, \qquad 1 \le t \le 8$$

where t represents the number days that the sale has been running and $S(t)$ represents the sales rate, measured in $\frac{\text{dresses}}{\text{day}}$.

(a) Evaluate $S(2)$ and interpret.

(b) Determine the total increase in the number of dresses sold between days 2 and 4.

26. The Silver Spur Gun Shop runs a last chance sale based on the sales rate model

$$S(t) = \frac{82}{t}, \qquad 1 \le t \le 10$$

where t represents the number days that the sale has been running and $S(t)$ represents the sales rate, measured in $\frac{\text{guns}}{\text{day}}$.

(a) Evaluate $S(4)$ and interpret.

(b) If the store owner must sell 150 guns in the first five days to break even, will the owner do so according to the model?

27. The rate of change function for foreign student enrollment in U.S. colleges can be modeled by

$$g(x) = \frac{92.43}{x}, \qquad 1 \le x \le 21$$

where x represents the number of years since 1975 and $g(x)$ represents the rate of change in the foreign student enrollment in U.S. colleges, measured in $\frac{\text{thousands of students}}{\text{year}}$.

(a) Evaluate and interpret $g(6)$.

(b) Evaluate and interpret $\int_1^6 g(x)\,dx$.

28. The rate of change in the average dollar amount awarded in the Pell Grant system can be modeled by

$$g(x) = \frac{237.73}{x+1}, \qquad 0 \le x \le 16$$

where x represents the number of years since 1979 and $g(x)$ represents the rate of change in the average dollar amount awarded in the Pell Grant system, measured in $\frac{\text{dollars}}{\text{year}}$.

(a) Evaluate and interpret $g(2)$.

(b) Evaluate and interpret $\int_0^2 g(x)\,dx$.

29. The rate of change function for the number of employees in the solid waste management industry is modeled by

$$g(x) = \frac{53.44}{x+1}, \qquad 0 \le x \le 16$$

where x represents the number of years since 1980 and $g(x)$ represents the rate of change in employees, measured in $\frac{\text{thousands of employees}}{\text{year}}$.

(a) Evaluate and interpret $g(16)$.

(b) Knowing that there were 209.5 thousand employees in the solid waste management industry in 1990, recover the model f that represents the number of employees in the solid waste management industry. [*Hint: f* has the form $f(x) = a + b \ln x$.]

30. The rate of change in the number of public schools with interactive video disks can be modeled by

$$g(x) = \frac{15.1}{x}, \qquad 1 \le x \le 7$$

where x represents the number of years since 1991 and $g(x)$ represents the rate of change in the number of public schools with interactive video disks, measured in $\dfrac{\text{thousands of schools}}{\text{year}}$.

(a) Evaluate and interpret $g(2)$.

(b) Knowing that in 1992 there were 6.5 thousand schools with the video disks, recover the model f, where $f(x)$ represents the number of public schools with interactive video disks and f has the form $f(x) = a + b \ln x$.

In Exercises 31–40, evaluate the indefinite and definite integrals.

31. $\displaystyle\int 2e^x \, dx$

32. $\displaystyle\int 6e^x \, dx$

33. $\displaystyle\int (e^x - 1) \, dx$

34. $\displaystyle\int (e^x + 3) \, dx$

35. $\displaystyle\int (4e^t + t - 1) \, dt$

36. $\displaystyle\int (t^3 - 3e^t + t) \, dt$

37. $\displaystyle\int_0^1 (e^x - 1) \, dx$

38. $\displaystyle\int_0^1 (6 - e^x) \, dx$

39. $\displaystyle\int_1^2 \dfrac{e^x + 4}{2} \, dx$

40. $\displaystyle\int_1^2 \dfrac{3 - e^x}{6} \, dx$

In Exercises 41–58, evaluate the integrals by u-substitution.

41. $\displaystyle\int e^{4x+1} \, dx$

42. $\displaystyle\int e^{8x} \, dx$

✓ 43. $\displaystyle\int 2xe^{x^2+1} \, dx$

44. $\displaystyle\int xe^{x^2} \, dx$

45. $\displaystyle\int \dfrac{e^x}{e^x + 2} \, dx$

46. $\displaystyle\int \dfrac{e^x - e^{-x}}{e^x + e^{-x}} \, dx$

47. $\displaystyle\int \dfrac{3}{e^x(1 - e^{-x})} \, dx$

48. $\displaystyle\int x^2 e^{x^3} \, dx$

49. $\displaystyle\int \dfrac{e^{\sqrt{x}}}{\sqrt{x}} \, dx$

50. $\displaystyle\int \dfrac{(e^x + 1)^2}{e^x} \, dx$

51. $\displaystyle\int_1^3 e^{-4x} \, dx$

52. $\displaystyle\int_0^1 e^{2x+3} \, dx$

53. $\displaystyle\int_0^1 \dfrac{1 + e^x}{e^x} \, dx$

54. $\displaystyle\int_1^2 \left(e^{2x} - \dfrac{2}{x} \right) \, dx$

55. $\displaystyle\int xe^{x^2} \, dx$

56. $\displaystyle\int 3x^5 e^{x^6} \, dx$

57. $\displaystyle\int e^{4x}(e^{4x} + 4) \, dx$

58. $\displaystyle\int e^x(e^x + 10) \, dx$

APPLICATIONS

59. The FlowStop Company determines that the marginal cost for their new line of faucet parts is given by

$$MC(x) = 1.50 + 0.04e^{0.02x}$$

(a) Evaluate $MC(20)$ and interpret.

(b) If the number of units of the part produced increases from 100 to 150, what is the total increase in cost?

60. The population of a small town is growing at a rate given by

$$R(t) = 250e^{0.04t}$$

where t represents the number of years from the present and $R(t)$ represents the town's population growth rate, measured in $\dfrac{\text{people}}{\text{year}}$.

(a) Evaluate and interpret $\int_0^3 R(t) \, dt$.

(b) A civil planner estimates that 3000 more people can move into the town without negatively affecting the quality of life. How many more years will it take for the population to reach that number?

✓ 61. The rate of change in the number of infant deaths per 1000 live births in Guam is given by

$$g(x) = -1.38e^{-0.45x}, \qquad 1 \le x \le 31$$

where x represents the number of years since 1964 and $g(x)$ represents the rate of change in the number of infant deaths per 1000 live births in Guam, measured in $\dfrac{\text{infant deaths per 1000 live births}}{\text{year}}$.

(a) Evaluate and interpret $g(11)$.

(b) Evaluate and interpret $\int_1^{11} g(x) \, dx$.

62. The rate of change in the number of infant deaths per 1000 live births in Puerto Rico is given by

$$g(x) = -1.7e^{-0.4x}, \qquad 0 \le x \le 36$$

where x represents the number of years since 1960 and $g(x)$ represents the rate of change in the number of infant deaths per 1000 live births in Puerto Rico, measured in $\dfrac{\text{infant deaths per 1000 live births}}{\text{year}}$.

(a) Evaluate and interpret $g(20)$.

(b) Evaluate and interpret $\int_0^{20} g(x) \, dx$.

63. The rate of change in the U.S. population can be modeled by

$$g(x) = 1.03e^{0.013x}, \qquad 0 \le x \le 100$$

where x represents the number of years since 1900 and $g(x)$ represents the rate of change in the U.S. population, measured in $\dfrac{\text{millions}}{\text{year}}$.

(a) Evaluate and interpret $g(50)$.

(b) Determine the total increase in the U.S. population from 1900 to 1950.

64. The rate of change in the tuition and fees for in-state students at four-year colleges in the United States can be modeled by

$$g(x) = 98.15e^{0.08x}, \qquad 1 \le x \le 12$$

where x represents the number of years since 1984 and $g(x)$ represents the rate of change in the tuition and fees for in-state students at four year colleges, measured in $\dfrac{\text{dollars}}{\text{year}}$.

(a) Graph g in the viewing window [0, 12] by [95, 220].

(b) Evaluate and interpret $g(5)$.

(c) Determine the increase in the total paid in tuition and fees for in-state students at four-year colleges in the United States from 1985 to 1989.

In Exercises 65–74, evaluate the indefinite and definite integrals.

65. $\displaystyle\int 5^x \, dx$

66. $\displaystyle\int 10^x \, dx$

67. $\displaystyle\int 5 \cdot \left(\frac{1}{2}\right)^x \, dx$

68. $\displaystyle\int 2 \cdot \left(\frac{5}{8}\right)^x \, dx$

69. $\displaystyle\int 7^{-x} \, dx$

70. $\displaystyle\int 10^{-x} \, dx$

71. $\displaystyle\int_1^2 10^{3x} \, dx$

72. $\displaystyle\int_1^5 5^{-2x} \, dx$

73. $\displaystyle\int_{-1}^1 2^{3x-1} \, dx$

74. $\displaystyle\int_1^{\sqrt{2}} x \cdot 2^{x^2} \, dx$

APPLICATIONS

75. The rate of change function for the number of people confined in state or correctional facilities for more than one year can be modeled by

$$P(x) = 0.2612(1.086)^x, \qquad 0 \le x \le 15$$

where x represents the number of years since 1980 and $P(x)$ represents the rate of change in the U.S. prison population, measured by $\dfrac{100{,}000 \text{ inmates}}{\text{year}}$.

(a) Evaluate $P(10)$ and interpret the result.

(b) Find the total change in the prison population from 1985 to 1990.

76. The rate of change in dormitory charges for in-state students at two-year colleges in the United States can be modeled by

$$f(x) = 32.67(1.04)^x, \qquad 1 \le x \le 7$$

where x represents the number of years since 1983 and $f(x)$ represents the rate of change in dormitory charges in $\dfrac{\text{dollars}}{\text{year}}$.

(a) Evaluate $f(6)$ and interpret.

(b) Evaluate $\int_1^6 f(x)\, dx$ and interpret.

SECTION PROJECT

The rate of change in dormitory charges for in-state students at four-year colleges in the United States is shown in Table 6.5.1. The rate is measured in $\dfrac{\text{dollars}}{\text{year}}$.

TABLE **6.5.1**	
YEAR	**RATE OF CHANGE IN DORM PRICES**
1984, $t = 1$	57.14
1985	60.0
1986	63.0
1987	66.15
1988	69.45
1989	72.93
1990	76.57

(a) Use the regression capabilities of your calculator to determine a general exponential growth model in the form $f'(t) = a \cdot b^t$ for $1 \le t \le 7$.

(b) Write the units used for t and $f(t)$.

(c) Determine $\int_1^6 f'(t)\, dt$ and interpret the solution.

(d) Use the regression capabilities of your calculator to determine a linear model in the form $g'(t) = at + b$ for $1 \le t \le 7$.

(e) Determine $\int_1^6 g'(t)\, dt$ and interpret the solution. Why is the answer different from that found in part (c)?

SECTION 6.6   DIFFERENTIAL EQUATIONS: SEPARATION OF VARIABLES

So far in this chapter, we have seen equations of the form

$$P'(x) = \frac{x}{15} + 20 \quad \text{and} \quad \frac{dy}{dx} = 4x^2 - 3x$$

These are examples of differential equations. Quite simply, an equation that involves a function and one (or more) of its derivatives is called a **differential equation.** Entire courses and textbooks are dedicated to the study of these equations.

In this section and the next, we will study **first-order differential** equations. These are differential equations that involve a first derivative but no higher derivatives. The technique we will use to solve this kind of differential equation is called **separation of variables.**

The Differential Equation $y = f'(x)$

Since the beginning of Chapter 6, we have really been solving differential equations that have the form $y = f'(x)$. When we see

$$\frac{dy}{dx} = 3x^2$$

we are saying that the derivative of some function is $3x^2$. This differential equation is solved by integrating

$$y = \int 3x^2\, dx = x^3 + C$$

EXAMPLE 1 | **Solving a Differential Equation**

Solve the differential equation $y' = x^4$.

SOLUTION

If we replace y' with $\frac{dy}{dx}$, we have

$$\frac{dy}{dx} = x^4$$

Since $\frac{dy}{dx} = x^4$, we know that y is an antiderivative of x^4; hence,

$$y = \int x^4\, dx = \frac{1}{5}x^5 + C \qquad\qquad \blacksquare$$

Example 1 indicates that:

> In general, a differential equation of the form $y = f'(x)$ or $\frac{dy}{dx} = f(x)$ is solved by integrating
>
> $$y = \int f(x)\, dx$$

In Example 1 we saw that the solution to $y' = x^4$ is $y = \frac{1}{5}x^5 + C$. We call $y = \frac{1}{5}x^5 + C$ the **general solution** of the differential equation $y' = x^4$ because it represents all possible solutions, one for each choice of the constant C. When the constant C is replaced with a particular value, we then have a **particular solution.** A few solutions to the differential equation $y' = x^4$ are

$$y = \frac{1}{5}x^5 + 1 \qquad C = 1$$

$$y = \frac{1}{5}x^5 \qquad C = 0$$

$$y = \frac{1}{5}x^5 - 1 \qquad C = -1$$

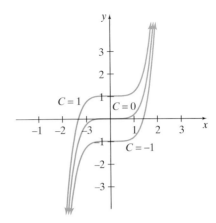

Figure 6.6.1 Graphs of $y = \frac{1}{5}x^5 + C$ for
$C = -1, C = 0,$ and $C = 1.$

Notice in Figure 6.6.1 that each curve has the same shape. This is because each curve has the same derivative $y' = x^4$. For this reason, we say that the different values for C in a general solution of a differential equation produces a *family of curves*, where the general solution $y = f(x) + C$ represents the entire family.

Separation of Variables

One important type of differential equation is called a separable differential equation. We say that a differential equation is *separable* if we can move everything involving x's to one side of the equation and everything involving y's to the other side. Example 2 illustrates the process.

EXAMPLE 2

Finding a General Solution by Separation of Variables

Determine the general solution to $\frac{dy}{dx} = \frac{x^2}{y^3}.$

SOLUTION

The method known as separation of variables requires us to separate the x's from the y's algebraically.

$$\frac{dy}{dx} = \frac{x^2}{y^3}$$

$$dy = \frac{x^2}{y^3}\, dx \qquad \text{Multiply both sides by } dx$$

$$y^3\, dy = x^2\, dx \qquad \text{Multiply both sides by } y^3$$

We now integrate both sides of the equation and get

$$\int y^3 dy = \int x^2 dx \qquad \text{Integrate both sides}$$

$$\frac{1}{4}y^4 + C_0 = \frac{1}{3}x^3 + C_1 \qquad \text{Utilize the Power Rule for integration}$$

$$\frac{1}{4}y^4 = \frac{1}{3}x^3 + C \qquad \text{Combine both constants and rewrite with one constant}$$

Solving for y yields

$$y^4 = \frac{4}{3}x^3 + 4C$$

$$y^4 = \frac{4}{3}x^3 + C \qquad \text{Four times a constant is just another constant}$$

$$y = \sqrt[4]{\frac{4}{3}x^3 + C} \qquad \text{The general solution using principal fourth root} \qquad \blacksquare$$

Notice that we can check our solution for Example 2 as follows:

$$y = \sqrt[4]{\frac{4}{3}x^3 + C} = \left(\frac{4}{3}x^3 + C\right)^{1/4}$$

$$\frac{dy}{dx} = \frac{1}{4}\left(\frac{4}{3}x^3 + C\right)^{-(3/4)} \cdot 4x^2$$

$$= \frac{x^2}{\left(\frac{4}{3}x^3 + C\right)^{3/4}} = \frac{x^2}{\left[\left(\frac{4}{3}x^3 + C\right)^{1/4}\right]^3}$$

$$= \frac{x^2}{y^3} \qquad \text{Substitute } y \text{ for } \left(\frac{4}{3}x^3 + C\right)^{1/4}$$

So we have $\frac{dy}{dx} = \frac{x^2}{y^3}$ as desired. Example 2 outlines the following process for separation of variables.

Separation of Variables

A separable first-order differential equation has the form

$$\frac{dy}{dx} = f(x) \cdot g(y) \quad \text{or} \quad \frac{dy}{dx} = \frac{f(x)}{g(y)}, \qquad g(y) \neq 0$$

It is solved by separating the variables and integrating

$$\int g(y)\,dy = \int f(x)\,dx$$

In Example 2 we found a general solution. Many times, we are given some additional information that allows us to determine the **particular solution**. We call the additional information an **initial condition**. Example 3 illustrates how to determine a particular solution.

EXAMPLE 3 | **Determining a Particular Solution**

Solve the differential equation $\frac{dy}{dx} = \frac{x^2}{y^3}$ with the condition that $x = 0, y = 1$.

SOLUTION

From Example 2, we know that the general solution is

$$y = \sqrt[4]{\frac{4}{3}x^3 + C}$$

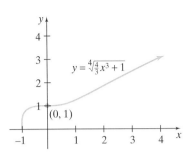

Figure 6.6.2

Given that $y = 1$, when $x = 0$, we can substitute this into the general solution and determine C.

$$y = \sqrt[4]{\frac{4}{3}x^3 + C}$$

$$1 = \sqrt[4]{\frac{4}{3}(0)^3 + C}$$

$$1 = \sqrt[4]{C}$$

$$1 = C$$

So the particular solution is $y = \sqrt[4]{\frac{4}{3}x^3 + 1}$ (see Figure 6.6.2).

✓ **CHECKPOINT 1**

Now work Exercise 33.

Example 3 illustrates that, when solving a separable differential equation with an initial condition, we use the following steps:

1. Separate the variables.
2. Determine the general solution.
3. Use the initial condition to determine C.

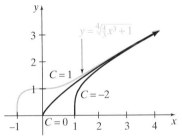

Figure 6.6.3

Graphs of $y = \sqrt[4]{\frac{4}{3}x^3 + C}$ of the solutions to $\frac{dy}{dx} = \frac{x^2}{y^3}$.

Several solutions to $\frac{dy}{dx} = \frac{x^2}{y^3}$ are shown in Figure 6.6.3, with our particular solution from Example 3 shown in blue.

The computing capabilities of a graphing calculator can help us to visualize the solutions of a differential equation. For example, the differential equation $\frac{dy}{dx} = \frac{x^2}{y^3}$ in Example 3 tells us that the slope at any point (x, y) on a graph is given by the expression $\frac{x^2}{y^3}$. A graph of these slopes is called a **slope field.** Calculator programs can draw slope fields for us. A slope field program for the TI-83 is included in Appendix B.

EXAMPLE 4 | **Finding a Particular Solution When a Logarithm Is Involved**

Determine the particular solution for $\frac{dy}{dx} = 3xy$ with the condition $x = 0$, $y = 3$.

SOLUTION

We start by separating the variables.

$$\frac{dy}{dx} = 3xy$$

$$\frac{1}{y}\frac{dy}{dx} = 3x$$

$$\frac{1}{y}dy = 3x\,dx$$

Integrating both sides yields

$$\int \frac{1}{y}dy = \int 3x\,dx$$

$$\ln y = \frac{3}{2}x^2 + C, \qquad y > 0$$

From Your Toolbox

Recall, the definition of a logarithm, $y - \ln x$ if, and only if, $e^y = x$.

Using the Toolbox to the left, we can rewrite this equation as

$$y = e^{(3/2)x^2 + C}$$
$$y = e^{(3/2)x^2} \cdot e^C$$

Since e^C represents an arbitrary constant, we can simply write C for e^C and write the general solution as

$$y = C \cdot e^{(3/2)x^2}$$

Now, applying the initial condition $x = 0$, $y = 3$,

$$3 = C \cdot e^{(3/2)(0)^2}$$
$$3 = C \cdot e^0 = C \cdot 1 = C$$

So the particular solution is $y = 3e^{(3/2)x^2}$.

✓ **CHECKPOINT 2**

Now work Exercise 63.

Sometimes we will find that there can be more than one general solution to a differential equation. This is illustrated in Example 5.

EXAMPLE 5

Finding All General Solutions to a Differential Equation

Find the general solution to $y \cdot \dfrac{dy}{dx} = 2x$.

SOLUTION

Since the x's and y's are already separated, we can simply multiply by dx and then integrate both sides.

$$y \cdot \frac{dy}{dx} = 2x$$
$$y\, dy = 2x\, dx$$
$$\int y\, dy = \int 2x\, dx$$
$$\frac{1}{2}y^2 = x^2 + C$$
$$y^2 = 2x^2 + 2C \qquad \text{Multiply by 2}$$
$$y^2 = 2x^2 + C \qquad \text{Since } 2C \text{ is simply another constant } C$$
$$y = \pm\sqrt{2x^2 + C} \qquad \text{Take the square root of both sides}$$

So we have two square roots, one positive and the other negative. Thus, these two solutions

$$y = \sqrt{2x^2 + C} \quad \text{and} \quad y = -\sqrt{2x^2 + C}$$

together are the general solution to the differential equation.

EXAMPLE 6

Solving a Differential Equation by u-Substitution

Determine the general solution for $\dfrac{dy}{dx} = xy - 3x$.

SOLUTION

To separate the variables for this differential equation, we need to factor on the right side.

$$\frac{dy}{dx} = xy - 3x$$

$$\frac{dy}{dx} = x(y - 3) \qquad\qquad \text{Factor } x \text{ from each term}$$

$$\frac{dy}{y - 3} = x\, dx \qquad\qquad \text{Multiply by } dx \text{ and divide by } y - 3$$

$$\int \frac{1}{y - 3} dy = \int x\, dx \qquad\qquad \text{Integrate both sides}$$

$$\int \frac{1}{u} du = \int x\, dx \qquad\qquad \text{Substitute } u = y - 3 \text{ so that } du = dy$$

$$\ln u = \frac{1}{2}x^2 + C \qquad\qquad \text{Integrate and assume that } u > 0$$

$$\ln(y - 3) = \frac{1}{2}x^2 + C \qquad\qquad \text{Resubstitute } y - 3 \text{ for } u$$

$$y - 3 = e^{(1/2)x^2 + C} \qquad\qquad \text{Definition of logarithm}$$

$$y - 3 = e^{(1/2)x^2} \cdot e^C = Ce^{(1/2)x^2} \qquad \text{Rule of exponents and replace } e^C \text{ by } C$$

$$y = Ce^{(1/2)x^2} + 3$$

Applications

Many of the applications of separable differential equations involve differential rate equations, where a rate of change function is given. Usually, we use the separation of variables technique to recover an "original" function.

EXAMPLE 7 | **Application of Separation of Variables**

The annual rate of increase in employees between 1950 and 1985 at the NewMedia Company can be modeled by the differential rate equation

$$\frac{dy}{dt} = 0.3t^{0.6}, \qquad 0 \le t \le 35$$

where t represents the number of years since 1950 and y represents the number of employees in hundreds.

(a) Knowing that the company had 1800 employees in 1965, find the employee growth function y for the company.

(b) Use the result from part (a) to estimate the number of employees at the company in 1983.

SOLUTION

(a) We start by determining the general solution to the differential rate equation.

$$\frac{dy}{dt} = 0.3t^{0.6}$$

$$dy = (0.3t^{0.6})\,dt$$

$$\int dy = \int (0.3t^{0.6})\,dt$$

$$y = 0.3 \cdot \frac{1}{1.6}t^{1.6} + C$$

$$y = \frac{3}{16}t^{1.6} + C$$

Since the company had 1800 employees in 1965, we know that $y = 18$ when $t = 15$. This allows us to solve for C.

$$18 = \frac{3}{16}(15)^{1.6} + C$$

$$18 \approx 14.28 + C$$

$$18 - 14.28 \approx C$$

$$3.72 \approx C$$

So the employee growth of the company can be modeled by

$$y = \frac{3}{16}t^{1.6} + 3.72$$

(b) To estimate the number of employees at the company in 1983, we need to use the model found in part (a) and evaluate when $t = 33$.

$$y = \frac{3}{16}(33)^{1.6} + 3.72$$

$$y \approx 54.14$$

So, according to the model, the New Media Company had about 54.14 hundred, or about 5414, employees in 1983.

SUMMARY

- A differential equation of the form $y = f'(x)$ or $\frac{dy}{dx} = f(x)$ is solved by integrating $y = \int f(x)\,dx$.
- A separable first-order differential equation $\frac{dy}{dx} = f(x) \cdot g(y)$, $g(y) \neq 0$, is solved by separating the variables and integrating $\int g(y)\,dy = \int f(x)\,dx$.

SECTION 6.6 EXERCISES

In Exercises 1–14, determine if the given function f is the solution to the given differential equation by determining the derivative f'.

1. $f(x) = 5x + 3$, $y' = 5$

2. $f(x) = x^2$, $f'(x) = 2x - 7$

3. $f(x) = x^5 - 4$, $y' = x^4$

4. $f(x) = x^6 + x - 4$, $y' = 6x^5 - 1$

5. $f(x) = \dfrac{1}{x}$, $y' = \dfrac{-1}{x^2}$

6. $f(x) = \dfrac{1}{x}$, $xy' = -y$

7. $f(x) = 2x^3 + 10$, $y' = 3x^2$

8. $f(x) = x^2 + 1$, $3xy' = 6y - 6$

9. $f(x) = \ln x$, $xy' = 1$

10. $f(x) = \dfrac{1}{2}\ln x$, $2xy' = 1$

11. $f(x) = \dfrac{3}{e^{4x}}$, $y' + 4y = 0$

12. $f(x) = xe^x$, $xy' = y(x+1)$

13. $f(x) = e^{3x} + 5$, $y' + 15 = 3y$

14. $f(x) = \dfrac{e^{2x} + x^2}{2}$, $y' - 3 = e^{2x} + x$

In Exercises 15–24, determine the general solution to the differential equation.

15. $\dfrac{dy}{dx} = -2$

16. $\dfrac{dy}{dx} = e$

17. $\dfrac{dy}{dx} = 1 - 3x$

18. $\dfrac{dy}{dx} = 2x$

19. $\dfrac{dy}{dx} = 5x^3 + x - 2$

20. $\dfrac{dy}{dx} = 10x^4 - x^2 + 3$

21. $\dfrac{dy}{dx} = \sqrt{x} + 2$

22. $\dfrac{dy}{dx} = \sqrt[4]{x} + x$

23. $\dfrac{dy}{dx} = \dfrac{5}{x^2}$

24. $\dfrac{dy}{dx} = x^2 - \dfrac{1}{x^2}$

In Exercises 25–30, graph a family of curves for the general solutions of the given differential equations using the values $C = 0$, $C = -2$, and $C = 3$.

25. $\dfrac{dy}{dx} = 2$

26. $\dfrac{dy}{dx} = -5$

27. $\dfrac{dy}{dx} = 3x + 1$

28. $\dfrac{dy}{dx} = 1 - 2x$

29. $\dfrac{dy}{dx} = \sqrt{x} + 3$

30. $\dfrac{dy}{dx} = \sqrt{x+2}$

For Exercises 31–44, determine the particular solution to the differential equation using the given condition.

31. $\dfrac{dy}{dx} = 12x$, $x = 1$, $y = 8$

32. $\dfrac{dy}{dx} = 12x$, $x = 1$, $y = -1$

✓ 33. $\dfrac{dy}{dx} = 3x^2 + 4x$, $x = 1$, $y = 6$

34. $\dfrac{dy}{dx} = xe^{x^2}$, $x = 0$, $y = 2$

35. $\dfrac{dy}{dt} = 7 - 4t$, $t = 0$, $y = 3$

36. $\dfrac{dy}{dt} = 7 - 4t$, $t = -1$, $y = 3$

37. $\dfrac{dy}{dx} = x^3 - 2x$, $x = 0$, $y = -2$

38. $\dfrac{dy}{dx} = -x^2 - x$, $x = 1$, $y = 1$

39. $\dfrac{dy}{dt} = t^2 + 2t + 3$, $t = 0$, $y = 4$

40. $\dfrac{dy}{dx} = \sqrt[3]{x^2 - x}$, $x = 0$, $y = 6$

41. $\dfrac{dy}{dx} = 2x^{0.3}$, $x = 0$, $y = 4$

42. $\dfrac{dy}{dx} = 4x^{1.7}$, $x = 0$, $y = 1$

43. $\dfrac{dy}{dx} = 1 - \dfrac{2}{x}$, $x = 1$, $y = 6$

44. $\dfrac{dy}{dx} = 3 - \dfrac{1}{x}$, $x = 1$, $y = 8$

APPLICATIONS

45. The Strike Now Company determines that the marginal cost for their MaxiPak of lighters is given by the differential rate equation

$$\frac{dC}{dx} = -0.02x + 6, \qquad 0 \le x \le 300$$

where x is the number of MaxiPaks produced and C is the cost in dollars.

(a) Determine the general solution of the cost function C.

(b) Find the particular cost function C, knowing that the cost of producing 10 MaxiPaks of lighters is $400.

46. The Sleep-4-Ever Company determines that the marginal revenue for their queen-sized mattress is given by the differential rate equation

$$\frac{dR}{dx} = 157$$

where x is the number of queen-sized mattresses sold and R is the revenue in dollars.

(a) Determine the general solution of the revenue function R.

(b) Find the particular revenue function R, knowing that there is zero revenue for zero sales.

47. Suppose that the ChocoCola Company has changed their formula for their Chocolate Cherry Cola. The public is slow in accepting this new formula, but the company is winning over Chocolate Cherry Cola drinkers over time. The company's research department determines that the rate of acceptance of the new cola can be modeled by the differential equation

$$\frac{dA}{dt} = t + 1.1, \qquad 0 \le t \le 12$$

where t is the time that the new formula for Chocolate Cherry Cola has been on the market, measured in months, and A is the percentage of the Chocolate Cherry Cola drinkers who have accepted the new formula.

(a) Evaluate $\frac{dA}{dt}$ when $t = 10$ and interpret.

(b) Determine the general solution for the differential rate equation and interpret.

(c) Find the particular solution, knowing that 5% of the drinkers had accepted the new formula as it was released. [That is, the initial condition is $(0, 5)$.]

48. The increase in leasable area in shopping centers in the United States can be modeled by the differential rate equation

$$\frac{dA}{dt} = 0.12, \qquad 1 \le t \le 10$$

where t represents the number of years since 1989 and A represents the leasable area in shopping centers in billions of square feet.

(a) Determine the general solution for the differential rate equation.

(b) Knowing that there was 4.87 billion square feet of leasable area in shopping centers in 1994, determine the particular solution for the differential equation.

(c) Evaluate $A(8)$ and interpret.

49. The change in the death rate from heart disease in the United States can be modeled by the differential rate equation

$$\frac{dD}{dt} = \frac{-20.17}{t^{1.06}}, \qquad 1 \le t \le 20$$

where t represents the number of years since 1979 and D is the number of deaths per 100,000 Americans caused by heart disease.

(a) Evaluate $\frac{dD}{dt}$ when $t = 10$ and interpret.

(b) Determine the particular solution $D(t)$ to the differential equation, knowing that the number of deaths in 1980 was found to be 336.18 deaths per 100,000 Americans.

(c) Evaluate $D(15)$ and interpret.

50. A conservationist determines that the rate of population growth of the rare blue-spotted frog on a wildlife reserve during a year long study can be modeled by the differential rate equation

$$\frac{dp}{dt} = \frac{15}{t^{0.6}}, \qquad 1 \le t \le 12$$

where t is the number of months since the start of the study and p represents the number of frogs.

(a) Evaluate $\frac{dp}{dt}$ when $t = 5$ and interpret.

(b) Determine the particular solution $p(t)$ of the differential equation knowing that the number of frogs observed after $t = 9$ months of the study was 90.

(c) Evaluate $p(5)$ and interpret.

For Exercises 51–61, determine the general solution to the separable differential equations.

51. $\dfrac{dy}{dx} = \dfrac{x}{y}$

52. $\dfrac{dy}{dx} = \dfrac{3y}{x}$

53. $\dfrac{dy}{dx} = 3x^2 y$

54. $\dfrac{dy}{dx} = 4x^3 y$

55. $\dfrac{dy}{dx} = e^{x+y}$ (*Hint:* Rewrite the right-hand side as two factors.)

56. $\dfrac{dy}{dx} = e^{x-y}$ (*Hint:* Rewrite the right-hand side as two factors.)

57. $\dfrac{dy}{dx} = (2x + 2)y$

58. $\dfrac{dy}{dx} = 4xy + 7y$

59. $\dfrac{dy}{dx} = \sqrt{xy}$ (*Hint:* Rewrite the right-hand side as two factors.)

60. $\dfrac{dy}{dx} = \dfrac{y^2}{x^2}$

61. $\dfrac{dy}{dx} = ye^x$

For Exercises 62–72, determine the particular solution to the separable differential equations using the given condition.

62. $\dfrac{dy}{dx} = xy, x = 0, y = -1$ ✓ 63. $\dfrac{dy}{dx} = 2xy, x = 0, y = 1$

64. $\dfrac{dy}{dx} = \dfrac{y}{x}, x = 1, y = 3$

65. $\dfrac{dy}{dx} = 2xy^4, x = 0, y = 1$

66. $\dfrac{dy}{dx} = \dfrac{x+1}{xy}$, $x = 1$, $y = 2$

67. $\dfrac{dy}{dx} = e^{x-y}$, $x = 0$, $y = 0$

68. $x \cdot \dfrac{dy}{dx} = xy$, $x = 0$, $y = 1$

69. $x \cdot \dfrac{dy}{dx} = x^4 y$, $x = 3$, $y = 10$

70. $\dfrac{dy}{dx} = xye^{x^2}$, $x = 0$, $y = e$

71. $\dfrac{dy}{dx} = \dfrac{xy}{1+x^2}$, $x = -1$, $y = 2$

72. $\dfrac{dy}{dx} = \dfrac{1+x^2}{xy}$, $x = 2$, $y = 4$

SECTION PROJECT

Consider the New Skin Company, which tracks the marginal revenue (in hundreds of dollars), at various weekly sales levels, of its designer face cream. The data are summarized in Table 6.6.1.

TABLE 6.6.1

SALES LEVEL, x	MARGINAL REVENUE, $\frac{dR}{dx}$
1	3.97
5	3.91
10	3.77
15	3.68
20	3.60
25	3.48
30	3.43

(a) Use the regression capabilities of your calculator to find a linear regression model for the marginal revenue in the form $\dfrac{dR}{dx} = ax + b$ for $1 \le xc\ 30$.

(b) Knowing that there is no revenue when sales are zero, determine the revenue function R.

(c) Find the sales level x that will maximize the revenue.

(d) Determine the price–demand function p.

SECTION 6.7   DIFFERENTIAL EQUATIONS: GROWTH AND DECAY

Quite often in the business, life, and social sciences, we wish to study the behavior of a quantity as it increases or decreases over time. In this section, we will study **exponential growth** and **exponential decay** phenomena that are modeled by a specific type of *differential equation*. These growth and decay models have the properties of exponential functions that we have studied so far. We will also examine how exponential growth and decay models affect both the business and life sciences. And we will examine **limited** and **logistic models** in which the growth is limited by some *y*-value. In Chapter 2 we called these *y*-values **horizontal asymptotes.**

Unlimited Growth

There are many applications in which *the rate of change increases proportionally to the amount present.* With compound interest, for example, we know that as interest in the account accumulates, the faster the account balance grows. In the social sciences, we know that as a population increases, more individuals are added to the population, and the population increases faster. These are all examples of **unlimited growth models.** If we call *y* the quantity at any given time *t*, the *exponential growth model* satisfies the equation

$$\left(\begin{array}{c} \text{rate of change} \\ \text{in quantity} \end{array} \right) = \left(\begin{array}{c} \text{is proportional} \\ \text{to} \end{array} \right) \left(\begin{array}{c} \text{the amount} \\ \text{present} \end{array} \right)$$

or, as a differential equation with constant k, we write

$$y' = ky$$

To solve this differential equation, we rely on our separation of variables technique. Since the independent variable for this kind of model is usually time t, we can write

$$\frac{dy}{dt} = ky \qquad \text{Replace } y' \text{ with } \frac{dy}{dt}$$

$$\int \frac{1}{y} dy = \int k\, dt \qquad \text{Divide by } y \text{ and integrate each side}$$

$$\ln y = kt + C \qquad \text{Perform the integration, assume } y > 0$$

$$y = e^{kt+C} \qquad \text{Definition of logarithm}$$

$$y = e^{kt} \cdot e^{C} = Ce^{kt} \qquad \text{Rule of exponents and replace } e^{C} \text{ with } C$$

Unlimited Growth Model

The solution to the differential equation $y' = ky$ is the **unlimited growth model**

$$y = Ce^{kt}$$

with initial value when $t = 0$, $y = C$ (see Figure 6.7.1).

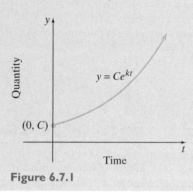

Figure 6.7.1

NOTES:
1. When the rate of growth is proportional to the present size, we have a condition called **unlimited growth** (unlimited means that y grows without limit).
2. Whenever we encounter this situation, the differential equation of the form $y' = ky$, we can immediately write the solution as $y = Ce^{kt}$. Remember that C is the initial value.
3. k is called the **rate of growth**.

EXAMPLE 1 | **Writing an Exponential Growth Model with a Given Rate**

A biological researcher initially starts a culture of 200 bacteria and knows from past experience that the number of bacteria grows proportionally to the amount present. Past studies have shown that the rate of growth is 11% each hour.

(a) Write an exponential growth model for the number of bacteria present after t hours.

(b) Evaluate y and y', when $t = 6$, and interpret.

SOLUTION

(a) We know that when the time starts the number of bacteria is 200. Thus, when $t = 0$, $y = 200$, meaning that the initial amount $C = 200$. This gives us the exponential growth model

$$y = 200\, e^{kt}$$

Since past studies have shown the rate of growth to be 11% each hour, we have $k = 0.11$. So the function for the number of bacteria present after t hours is given by

$$y = 200\,e^{0.11t}$$

(b) Evaluating y at $t = 6$ gives

$$y = 200\,e^{0.11(6)} = 200\,e^{0.66} \approx 386.96$$

This means that, according to the model, there will be about 387 bacteria present after 6 hours. The derivative of $y = 200\,e^{0.11t}$ is

$$y' = \frac{d}{dt}(200\,e^{0.11t}) = 200\,e^{0.11t}\,0.11 = 22\,e^{0.11t}$$

Evaluating y' at $t = 6$ yields

$$y' = 22\,e^{0.11(6)} = 22\,e^{0.66} \approx 42.57\,\frac{\text{bacteria}}{\text{hour}}$$

Thus, after 6 hours the number of bacteria in the culture is growing at a rate of about 42.57 bacteria per hour.

Interactive Activity

 Using the model in Example 1, graph y and y' in the same viewing window. Is y' positive or negative for $t \geq 0$? Is y' increasing or decreasing? What does this imply about the properties of y?

✓ **CHECKPOINT 1**

Now work Exercise 3.

In practice, we are usually not given the growth rate for a differential equation as we were in Example 1. Instead, we are usually given an initial amount C and then given some other ordered pair, which we will call (t_1, y_1). We can use this ordered pair to determine the rate of growth. (This technique will be demonstrated in Example 2.)

Exponential Decay

Another family of applications relating to exponential models is called **exponential decay.** This type of model is used to exhibit behavior in which the rate of change decreases proportionally to the amount present. In this case, the differential equation for such behavior is

$$\left(\begin{array}{c}\text{rate of change} \\ \text{in quantity}\end{array}\right) = \left(\begin{array}{c}\text{is proportional} \\ \text{to}\end{array}\right)\left(\begin{array}{c}\text{the amount} \\ \text{present}\end{array}\right)$$

$$f'(t) = -k \cdot f(t)$$

The exponential model associated with this differential equation is $f(t) = a \cdot e^{-kt}$, where k is the decay constant.

Rule for Exponential Decay

The general exponential model $y = a \cdot e^{-kt}$ satisfies the differential equation

$$y' = -k \cdot f(t)$$

where the rate of change decreases proportionally to the amount present. a is called the **initial amount** and k is the **rate of decay.**

Just as with growth models, exponential decay models are usually presented with an initial amount, $f(0) = a$, and some other condition, given by an ordered pair (t_1, y_1). One common application of exponential decay models is the *half-life of a radioactive substance.* Half-life is the amount of time that a radioactive substance will be reduced by half to an inert substance.

EXAMPLE 2

Determining Radioactive Decay

A scientist at the Los Altos Laboratory has a 100-milligram sample of the radioactive substance cobalt 60. The half-life of cobalt 60 is known to be 5.3 years.

(a) Determine an exponential decay model for the number of milligrams of cobalt 60 present after t years.

(b) How many years will it take for the sample of cobalt 60 to decay to 10 milligrams?

SOLUTION

(a) We know that the initial amount is 100 milligrams, so this gives us

$$f(t) = 100\,e^{-kt}$$

where t represents the number of years and $f(t)$ represents the amount of cobalt 60 present. Knowing that the half-life of cobalt 60 is 5.3 years tells us that, when $t_1 = 5.3$, $y_1 = \frac{1}{2}(100) = 50$ milligrams. So we arrive at the equation

$$50 = 100\,e^{-k(5.3)}$$

$$\frac{1}{2} = e^{-k(5.3)}$$

$$\ln\left(\frac{1}{2}\right) = \ln\,e^{-k(5.3)}$$

$$\ln\left(\frac{1}{2}\right) = -k(5.3)$$

$$-\frac{\ln\left(\frac{1}{2}\right)}{5.3} = k$$

$$0.13 \approx k$$

So the exponential decay model for the cobalt 60 sample is

$$f(t) = 100\,e^{-0.13t}$$

(b) Here we need the t-value so that $f(t) = 10$. So we need to solve the following equation:

$$10 = 100\,e^{-0.13t}$$

$$\frac{1}{10} = e^{-0.13t}$$

$$\ln\left(\frac{1}{10}\right) = \ln\,e^{-0.13t}$$

$$\ln\left(\frac{1}{10}\right) = -0.13t$$

$$-\frac{\ln\left(\frac{1}{10}\right)}{0.13} = t$$

$$17.71 \approx t$$

Interactive Activity

Notice that for radioactive decay models the decay rate is independent of the initial amount. Solve the exponential equation $\frac{1}{2}a = a \cdot e^{-kT}$ to find a formula for the decay rate k of a radioactive substance with a half-life of T years.

This means that it will take about 17.71 years for the 100-milligram cobalt 60 sample to decay to 10 milligrams. ◼

Limited Growth Models

There are some pitfalls to the unlimited growth model. For example, a town of 4 thousand that is currently growing at a rate of 18% per year has an unlimited growth model of

$$y = 4000e^{0.18t}$$

If we use the model to project the population 50 years from now, we get

$$y = 4000e^{0.18(50)} = 4000e^9 \approx 32{,}412{,}336 \text{ people}$$

It seems rather unrealistic to believe that a town with a population of 4 thousand will swell to over 32 million in 50 years. There must be a reasonable limit, or **upper bound,** for almost every such quantity. For these kind of situations, we use the **limited growth model.** Here the rate of change is directly proportional to the distance from a limiting point, or upper bound. If we denote L as the upper bound and y as the value of the function at time t, then this distance from the upper bound is given by the difference $L - y$. So the differential equation for this situation is expressed as

$$\begin{pmatrix} \text{rate of} \\ \text{growth} \end{pmatrix} = \begin{pmatrix} \text{constant of} \\ \text{variation} \end{pmatrix} \cdot \begin{pmatrix} \text{distance from} \\ \text{upper bound } L \end{pmatrix}$$

or, mathematically,

$$y' = k(L - y)$$

For these applications, we assume that the initial condition is $y = 0$ when $t = 0$. Let's now use the separation of variables technique to solve this differential equation. Since time t is usually our independent variable, we can write

$$\frac{dy}{dt} = k(L - y)$$

Now, separating the variables t and y, we get

$$\frac{1}{L - y} \, dy = k \, dt$$

$$\int \frac{1}{L - y} \, dy = \int k \, dt \qquad \text{Integrate both sides}$$

We need to use a u-substitution on the left-hand side. Let $u = L - y$ so that $du = -dy$ to get

$$\int \frac{1}{u} (-du) = \int k \, dt$$

$$-\int \frac{1}{u} \, du = \int k \, dt$$

$$-\ln u = kt + C$$

In terms of y this is

$$-\ln (L - y) = kt + C$$

$$\ln (L - y) = -kt - C$$

Now, using the definition of a logarithm gives us,

$$L - y = e^{-kt-C}$$
$$L - y = e^{-kt} \cdot e^{-C}$$
$$-y = -L + e^{-kt} \cdot e^{-C}$$
$$y = L - e^{-kt} \cdot e^{-C}$$

Using the initial condition that $(t, y) = (0, 0)$, we can find the value of the constant e^{-C}.

$$0 = L - e^{-k(0)} \cdot e^{-C}$$
$$0 = L - e^{-C}$$
$$e^{-C} = L$$

Thus, we get the solution

$$y = L - Le^{-kt}$$

Factoring L from the right side and writing the model as a function of t, we get the limited growth model as a function of t to be

$$y = L(1 - e^{-kt})$$

Limited Growth Model

The solution to the differential equation $y' = k(L - y)$ is the **limited growth model** given by the function

$$y = L(1 - e^{-kt})$$

where L is the **limit of growth** and k is the **rate of growth.**

NOTES: 1. $\lim_{t \to \infty} f(t) = L$. That is, the limit of growth is L.
2. y is increasing and the graph is concave down for all $t \geq 0$.
3. Applications of the limited growth model include learning curves, company growth, depreciation of equipment, and diffusion of mass media information.

EXAMPLE 3

Deriving a Limited Growth Model

In a study of the memorization of unrelated shapes, a person can memorize no more than 150 of them in three hours. In a psychology class study, it is found that, on average, a subject can memorize 23 of the shapes after a 10-minute period.

(a) Write a limited growth model where t is the time used to memorize the shapes and y is the number of shapes memorized after t minutes.

(b) Estimate the number of shapes that could be memorized in 45 minutes.

SOLUTION

(a) To determine the particular solution, we need to find the value of k. Given that $L = 150$ (the upper bound) and $y = 23$ when $t = 10$, we can get k

by solving

$$y = L(1 - e^{-kt})$$

$$23 = 150(1 - e^{-k \cdot 10})$$

$$\frac{23}{150} = 1 - e^{-k \cdot 10}$$

$$e^{-k \cdot 10} = 1 - \frac{23}{150}$$

$$e^{-k \cdot 10} = \frac{127}{150}$$

$$\ln(e^{-k \cdot 10}) = \ln\left(\frac{127}{150}\right)$$

$$-k \cdot 10 = \ln\left(\frac{127}{150}\right)$$

$$k = -\frac{1}{10}\ln\left(\frac{127}{150}\right) \approx 0.017$$

This gives us the learning curve model $y = 150(1 - e^{-0.017t})$. The graph is shown in Figure 6.7.2.

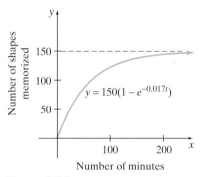

Figure 6.7.2

(b) To estimate the number of shapes memorized in 45 minutes, we need to evaluate y when $t = 45$ to get

$$y = 150(1 - e^{-0.017(45)}) = 150(1 - e^{-0.765}) \approx 80.20$$

Thus, according to the model, about 80 shapes can be memorized in a 45-minute period. Keep in mind that the valid range for this model is the set of whole numbers from 0 to 150 since the units are measured in number of shapes.

✓ **CHECKPOINT 2**

Now work Exercise 31.

Logistic Growth Model

Many phenomena in the business, life, and social sciences follow a combination of initial rapid growth, only to flatten out for large values of t. The population of many developing countries follows this pattern of a rapid increase in population early on and then a leveling off as the country develops (see Figure 6.7.3). This

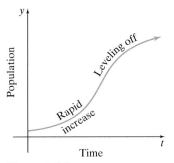

Figure 6.7.3

results in a differential equation that exhibits a combination of both unlimited growth and limited growth. In this type of differential equation, the rate of change y' is proportional to the amount currently present times the distance from an upper bound L.

$$\left(\begin{array}{c}\text{rate of}\\\text{growth}\end{array}\right) = \left(\begin{array}{c}\text{constant of}\\\text{variation}\end{array}\right) \cdot \left(\begin{array}{c}\text{amount currently}\\\text{present}\end{array}\right) \cdot \left(\begin{array}{c}\text{distance from}\\\text{upper bound } L\end{array}\right)$$

or, mathematically,

$$y' = ky(L - y)$$

By using the separation of variables technique with independent variable t, we get

$$\frac{dy}{dt} = y' = ky(L - y)$$

$$\int \frac{1}{y(L - y)}\, dy = \int k\, dt$$

Computing the general solution to this differential equation is done in the exercises. The result is the **logistic growth model.**

Logistic Growth Model

The solution to the differential equation $y' = ky(L - y)$ is the **logistic growth model** given by the function

$$y = \frac{L}{1 + ae^{-kLt}}$$

where L is the limit of growth and a and k are real number constants.

NOTES:
1. To determine a logistic model, we are often given three pieces of information:
   - An upper limit L
   - An initial value that allows us to find the constant a
   - A boundary condition to find the exponent kL in the denominator of the model
2. Applications of the logistic growth model include new product sales, dispersion of mass media information and rumors, spread of disease, and long-term population growth.
3. Notice that when $t = 0$

$$y = \frac{L}{1 + ae^{-kL(0)}} = \frac{L}{1 + a}$$

so $\left(0, \dfrac{L}{1 + a}\right)$ is the initial value. Many of the properties of the logistic growth model are verified in the exercises.

One common application of logistic growth models occurs in the *harvest models* of trees. The model applies in this situation because, as the population of trees in a forest or tree farm increases, the room for roots to spread becomes limited, the amount of sunlight hitting the trees can become restricted, and the number of seedlings that have room and light to grow in becomes less and less. This results in a tapering of the number of trees that can thrive in a fixed area. Consequently, conservationists use a logistic model for the population size for trees.

EXAMPLE 4 | **Deriving a Logistic Growth Model**

Suppose that commercial tree growers estimate that the maximum number of pine trees that can be supported at a large nursery is 15,000. The nursery was started with 1200 harvestable trees on the land. After 14 years, they find 6000 of the trees are ready to be harvested. Write a differential equation for the rate of growth in population of the nursery trees; then determine a logistic model for the number of trees that are harvestable after t years.

SOLUTION

In the differential equation $y' = ky(L - y)$, we know that the upper bound is $L = 15{,}000$, so we get $y' = ky(15{,}000 - y)$. To get the harvest function, we know that its general solution has the form

$$y = \frac{15{,}000}{1 + ae^{-k \cdot 15{,}000 t}}$$

To simplify our computations, we replace $k \cdot 15{,}000$ with another constant b.

$$y = \frac{15{,}000}{1 + ae^{-bt}}$$

Since the nursery started with 1200 trees, we know that $y = 1200$ when $t = 0$.

$$1200 = \frac{15{,}000}{1 + ae^0}$$

$$1200 = \frac{15{,}000}{1 + a}$$

$$1 + a = \frac{15{,}000}{1200}$$

$$a = \frac{15{,}000}{1200} - 1 = 11.5$$

This gives us the general solution

$$y = \frac{15{,}000}{1 + 11.5\,e^{-bt}}$$

To get the proper value of b, we use the condition that $y = 6000$ when $t = 14$. This gives us

$$6000 = \frac{15{,}000}{1 + 11.5\,e^{-b(14)}}$$

$$1 + 11.5\,e^{-b(14)} = \frac{15{,}000}{6000}$$

$$11.5\,e^{-b(14)} = \frac{15{,}000}{6000} - 1$$

$$e^{-b(14)} = \frac{1}{11.5}\left(\frac{15{,}000}{6000} - 1\right)$$

$$-b(14) = \ln\left[\frac{1}{11.5}\left(\frac{15{,}000}{6000} - 1\right)\right]$$

$$b = -\frac{1}{14}\left\{\ln\left[\frac{1}{11.5}\left(\frac{15{,}000}{6000} - 1\right)\right]\right\} \approx 0.15$$

Thus, the logistic model for the number of trees that are harvestable after t years is

$$y = \frac{15,000}{1 + 11.5 e^{-0.15t}}$$

The graph of the model for the first five decades is displayed in Figure 6.7.4.

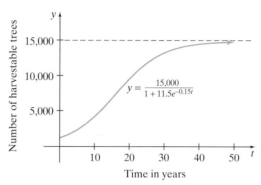

Figure 6.7.4

Many times, we are not given information such as the upper bound or growth rate for an application. As a matter of fact, these are the pieces of information that are expected to be discovered. In practice, logistic model applications come from collected data and are generated by regression techniques. Example 5 is one such application.

EXAMPLE 5

Modeling the Logistic Curve

According to the U.S. Bureau of the Census data, the number of deaths occurring annually from acquired immune deficiency syndrome (AIDS) from 1981 to 1994 can be modeled by the logistic function

$$y = \frac{48,171}{1 + 68.53 e^{-0.49t}}, \qquad 1 \le t \le 14$$

where t represents the number of years since 1980 and y represents the number of deaths occurring annually from AIDS.

(a) Graph the function and determine $\lim\limits_{t \to \infty} \dfrac{48,171}{1 + 68.53 e^{-0.49t}}$ and interpret.

(b) Compute y when $t = 10$ and interpret.

SOLUTION

(a) The graph of the model is shown in Figure 6.7.5. Computing the limit, we get

$$\lim_{t \to \infty} \frac{48,171}{1 + 68.53 e^{-0.49t}} = \frac{48,171}{1 + 68.53 (0)} = 48,171$$

Thus, according to the logistic model, the upper bound for the number of deaths occurring from AIDS is about 48,171 each year.

(b) Evaluating the model at $t = 10$ yields

$$y = \frac{48,171}{1 + 68.53 e^{-0.49(10)}} \approx 31,894.68$$

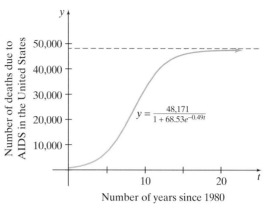

Figure 6.7.5 Logistic model of U.S. AIDS deaths.

This means that according to the model, in 1990 approximately 31,895 people died due to AIDS.

SUMMARY

In this section, we focused on four different types of growth and decay models. We found that each differential equation from which the model came could be found by separation of variables.

- The **exponential growth model** $y = Ce^{kt}$ is the solution to the differential equation $y' = ky$.
- The **exponential decay model** $y = Ce^{-kt}$ is the solution to the differential equation $y' = -ky$.
- The **limited growth model** $y = L(1 - e^{-kt})$ is the solution to the differential equation $y' = k(L - y)$.
- The **logistic growth model** $y = \dfrac{L}{1 + a \cdot e^{-kLt}}$ is the solution to the differential equation $y' = ky(L - y)$.

SECTION 6.7 EXERCISES

APPLICATIONS

1. The population of Sensenville is increasing at a rate of 5% per year. The current population of the town is 50,000. Assume that the rate of change in the population is increasing proportionally to the population size.

(a) Write a differential equation and an initial condition for the population of Sensenville.

(b) Write an exponential growth model for the population of Sensenville in t years.

(c) Use the model to estimate the population in 10 years.

2. The population of Burgentown is increasing at a rate of 2% per year. The current population of the city is 250,000. Assume that the rate of change in the population is increasing proportionally to the population size.

(a) Write a differential equation and an initial condition for the population of Burgentown.

(b) Write an exponential growth model for the population of Burgentown in t years.

(c) Use the model to estimate the population in 10 years.

✓ 3. The number of bacteria in a culture starts at 5000 and increases at a rate of 12% per hour. Assume that the rate of change in the number of bacteria is increasing proportionally to the number of bacteria present.

(a) Write a differential equation for the number of bacteria present after t hours.

(b) Write an exponential growth model of the bacteria present after t hours.

(c) Find the number of bacteria after $t = 3, 12,$ and 24 hours.

4. The number of bacteria in a culture starts at 500 and increases at a rate of 7% per hour. Assume that the rate of change in the number of bacteria is increasing proportionally to the number of bacteria present.

(a) Write a differential equation for the number of bacteria present after t hours.

(b) Write an exponential growth model of the bacteria present after t hours.

(c) Find the number of bacteria after $t = 0, 3, 6,$ and 12 hours.

5. In 1970 the consumption of timber in the United States was 39.9 million board feet, and in 1994 the consumption rose to 60.4 million board feet.

(a) Write a natural resource function $N(t) = a \cdot e^{kt}$, where t represents the number of years since 1970 and $N(t)$ represents the amount of timber consumed in millions of board feet.

(b) Compute and graph $\int_0^{20} N(t)\, dt$ to determine the total amount of timber consumed between 1970 and 1990.

6. In 1970 the human consumption of fish in the United States was 11.3 billion pounds, and in 1995 the consumption increased to 16.6 billion pounds.

(a) Write a natural resource function $N(t) = a \cdot e^{kt}$, where t represents the number of years since 1970 and $N(t)$ represents the amount of fish consumed in billions of pounds.

(b) Compute and graph $\int_0^{25} N(t)\, dt$ to determine the total amount of fish consumed between 1970 and 1995.

7. The percentage of wood waste generated in the United States can be modeled by the exponential growth model

$$y = 6.0\, e^{0.034t}, \qquad 0 \leq t \leq 9$$

where t represents the number of years since 1990 and y represents the percentage of wood waste generated.

(a) Write a differential equation and an initial condition for the percentage of wood waste generated in the United States.

(b) Evaluate y and y' when $t = 3$ and interpret.

8. The percentage of paper waste generated in the United States can be modeled by the exponential growth model

$$f(t) = 35.29\, e^{0.022t}, \qquad 0 \leq t \leq 9$$

where t represents the number of years since 1990 and $f(t)$ represents the percentage of paper waste generated.

(a) Write a differential equation and an initial condition for the percentage of paper waste generated in the United States.

(b) Evaluate y and y' when $t = 6$ and interpret.

9. Suppose a scientist has a 3000-milligram sample of the radioactive substance phosphorus 32. The half-life of phosphorus 32 is known to be 14.2 days.

(a) Determine an exponential decay model for the number of milligrams of phosphorus 32 present after t days.

(b) How many days will it take for the sample of phosphorus 32 to decay to 2000 milligrams?

10. Suppose that a scientist has a 300-gram sample of the radioactive substance radon. The half-life of radon is known to be 3.82 days.

(a) Determine an exponential decay model for the number of grams of radon present after t days.

(b) How many days will it take for the sample of radon to decay to 50 grams?

11. The population of Devonburg was 920,000 in 1980. Because of changing economic conditions in the community, the population had decreased to 700,000 in 1995. Assume that the rate of change in the population is decreasing proportionally to the population size.

(a) Write a differential equation and an initial condition for the population of Devonburg.

(b) Write an exponential decay model for the population of Devonburg in t years.

(c) Use the model to estimate the population in 1998.

12. The population of Grandland was 604,000 in 1985. Due to urban flight, the population decreased to 500,000 in 1990. Assume that the rate of change in the population is decreasing proportionally to the population size.

(a) Write a differential equation and an initial condition for the population of Grandland.

(b) Write an exponential decay model for the population of Grandland in t years.

(c) Use the model to estimate the population in 2000.

13. The percentage of Caucasians in the U.S. labor force who are unemployed can be modeled by

$$y = 6.54\, e^{-0.09t}, \qquad 0 \leq t \leq 8$$

where t represents the number of years since 1990 and $f(t)$ represents the percentage of Caucasians in the U.S. labor force who are unemployed.

(a) Write a differential equation and an initial condition for the percentage of Caucasians in the U.S. labor force who are unemployed.

(b) Evaluate y and y' when $t = 3$ and interpret.

14. The number of new car sales that were imports can be modeled by

$$y = 2.54\, e^{-0.09t}, \qquad 0 \leq t \leq 9$$

where t represents the number of years since 1990 and y represents the number of new car sales that were imports, measured in millions.

(a) Write a differential equation and an initial condition for the number of new car sales which were imports.

(b) Evaluate y and y' when $t = 1$ and interpret.

For Exercises 15–26:

(a) Write the corresponding differential equation $y' = k(L - y)$ using the definition of the limited growth model.

(b) Evaluate y when $t = 0$ to determine the initial value and identify the limit of growth L. Use the definition of the limited growth model.

15. $y = 10(1 - e^{-t})$

16. $y = 5(1 - e^{-t})$

17. $y = 8(1 - e^{-2t})$

18. $y = 8(1 - e^{-3t})$

19. $y = 20(1 - e^{-0.5t})$

20. $y = 16(1 - e^{-0.3t})$

21. $y = 12 - 12e^{-0.1t}$

22. $y = 24 - 24e^{-0.2t}$

23. $y = 1.5(1 - e^{-0.2t})$

24. $y = 0.01(1 - e^{-0.3t})$

25. $y = 15.5 - 15.5e^{-0.1t}$

26. $y = 62 - 62e^{-0.02t}$

APPLICATIONS

27. Suppose that the Wesson and Selverstone Investment Group has introduced a new computer system for its 3000 employees. The developer of the software has determined that the employees will learn the new system according to the model

$$y = 3000(1 - e^{-0.3t}), \qquad t \geq 0$$

where t represents the number of weeks since the system has been installed and y represents the number of employees that have learned the new system.

(a) Evaluate y and y' when $t = 2$ and interpret.

(b) According to the model, how many weeks will it take for at least 2500 employees to learn the new system?

28. Consider a college dormitory of 600 students where a flu epidemic is infecting students according to the model

$$y = 600(1 - e^{-0.22t}), \qquad t \geq 0$$

where t represents the number of days since the flu was first diagnosed and y represents the number of dorm residents infected.

(a) Evaluate y and y' when $t = 4$ and interpret.
(b) According to the model, how many days will it take so that at least 400 students have been infected by the flu?

29. The MidWest Fabricating Company installs a new metal press for its 300 employees. The company that makes the press has determined through experience that employees learn to use the new press in a way that follows the limited growth model

$$y = L(1 - e^{-kt}), \qquad t \geq 0$$

where t is the number of weeks since the new press was installed and y represents the number of employees who have learned to use the new equipment. They find that after the first week 40 employees have learned to use the new metal press.

(a) Identify the limit of growth L.
(b) Solve the equation $40 = 300(1 - e^{-k(1)})$ to determine the rate of growth k.
(c) Graph the model for the first half-year (that is, 26 weeks) after installation.

(d) If the company has set a goal of having half of the employees skilled at using the machine in the first month (that is, in the first 4 weeks), will the company reach its goal?

30. A municipal park is being overrun by 200 deer, so park officials decide to open the park to hunting for a short time in an attempt to curb the deer population. The officials believe that the number of deer killed by the hunters will follow the limited growth model

$$y = L(1 - e^{-kt}), \qquad t \geq 0$$

where t is the number of days since the hunting season began and y represents the number of deer killed. After the first two days of hunting, 24 deer were killed.

(a) Identify the limit of growth L.
(b) Solve the equation $24 = 200(1 - e^{-k(2)})$ to determine the rate of growth k.
(c) Graph y to display the number of deer killed if the hunting season lasted 30 days.
(d) If the officials plan to call an end to the hunting season when the deer population is cut in half, how many days will this take?

✓ 31. The Newportage Company conducts an aptitude test in which a maximum of 75 manufacturing procedures are learned. Studies show that, on the average, 15 of the procedures can be memorized in a 1.5-day period. Assume that the rate of learning is proportional to the distance from the maximum number of learning procedures (that is, the number of procedures learned can be modeled by a limited growth model).

(a) Identify the limit of growth L and determine the rate of growth k.
(b) Write a limited growth model in the form

$$y = L(1 - e^{-kt}), \qquad t \geq 0$$

where t is the number of days that an employee has been memorizing the procedures and y represents the number of procedures memorized. Graph the model for $0 \leq t \leq 30$.
(c) Evaluate y' when $x = 5$ and compare to y' when $x = 25$.

For Exercises 32–41:

(a) Write the corresponding differential equation $y' = ky(L - y)$ for the model. Use the definition of the logistic growth model.
(b) Evaluate y when $t = 0$ to determine the initial value and identify the limit of growth L.

32. $y = \dfrac{30}{1 + e^{-t}}$

33. $y = \dfrac{15}{1 + e^{-t}}$

34. $y = \dfrac{50}{1 + 2e^{-t}}$

35. $y = \dfrac{10}{1 + 5e^{-t}}$

36. $y = \dfrac{150}{1 + e^{-2t}}$

37. $y = \dfrac{90}{1 + e^{-3t}}$

38. $y = \dfrac{400}{1 + 2e^{-3t}}$

39. $y = \dfrac{2500}{1 + 5e^{-4t}}$

40. $y = \dfrac{1000}{1 + 4.5e^{-0.9t}}$

41. $y = \dfrac{4000}{1 + 6e^{-1.2t}}$

APPLICATIONS

42. According to estimates from the Whitehead Research Group, the percentage of households who own video disk recorders in a certain county is given by the logistic model

$$y = \frac{90}{1 + 22e^{-0.7t}}, \qquad 0 \le t \le 10$$

where t represents the number of years since 1990 and $f(t)$ represents the percentage of households who own video disk recorders.

(a) According to the model, what percentage of the households in the county owned video disk recorders in 1999?

(b) Evaluate y' when $t = 6$ and interpret.

(c) According to the model, what is the maximum number of households in the county that are expected to own a video disk recorder?

43. On a certain university campus, the 12,000 students who attend home basketball games are quickly swept by the new fad of tossing toilet paper onto the basketball court after the first basket is made for the home team. Research students in the sociology department think that the fad started with about 100 fans and that the number of students at the home games who throw toilet paper follow the logistic model

$$y = \frac{12{,}000}{1 + 119e^{-0.8x}}, \qquad 0 \le x \le 12$$

where x is the number of home games completed and y represents the number of students expected to toss toilet paper onto the basketball court after the first basket is made for the home team.

(a) Evaluate y when $x = 8$ and interpret.

(b) If the university's athletic director threatens to search all students coming to the game for toilet paper when at least half of the student fans are engaging in the activity, how many home games will this take?

44. The Top Cove Fishery, established a harvesting area that is designed to hold a maximum of 7500 catfish. The fishery was initially stocked with nearly 1000 catfish. After two years, the number of catfish in the fishery had grown to nearly 2500.

(a) What is the limit of growth L?

(b) Solve the equation $2500 = \dfrac{7500}{1 + 1000e^{-b(2)}}$ to determine the rate of growth where $b = kL$.

(c) Write the logistic function in the form $y = \dfrac{L}{1 + ae^{-bt}}$ for the first 5 years, since the harvesting area was established where $b = kL$.

(d) Conservationists define the maximum sustained yield (MSY) as the largest number of the population that can be

removed while sustaining the population. For logistic functions, the MSY is found at the inflection point of the curve that is the point $\left(\dfrac{\ln a}{b}, \dfrac{L}{2}\right)$, where $b = kL$. Determine the MSY for the harvesting area.

45. Consider a town with a population of 30,000 whose city council president has just resigned. He announces the resignation at a closed session of the 10-member city council. It takes only 6 hours for 27,300 citizens to learn about the resignation.

(a) What is the limit of growth L?

(b) Solve the equation $27{,}300 = \dfrac{30{,}000}{1 + 3000e^{-b(6)}}$ to determine the rate of growth, where $b = kL$.

(c) Write the logistic function in the form $y = \dfrac{L}{1 + ae^{-bt}}$ for the first 7 hours after the mayor's announcement.

46. The percentage of African Americans 25 years old and over who completed 4 years or more of high school can be modeled by

$$y = \frac{94.3}{1 + 4e^{-0.07t}}, \qquad 1 \le x \le 40$$

where t represents the number of years since 1959 and y represents the percentage of African Americans 25 years old and over who completed 4 years or more of high school.

(a) Evaluate and interpret y when $x = 16$.

(b) Write the corresponding differential equation $y' = ky(L - y)$ for the model. (*Hint: $b = kL = 0.07$.*)

(c) Evaluate y' when $x = 16$ and interpret.

(d) Determine $\displaystyle\lim_{t \to \infty} \left(\frac{94.3}{1 + 4e^{-0.07t}}\right)$ and interpret.

47. Consider the differential equation corresponding to the logistic growth model $y' = ky(L - y)$.

(a) Use the separation of variables technique to write an equation with expressions involving y on the left side and the constant k on the right.

(b) Verify that the derivative of the expression $\dfrac{1}{L} \ln\left(\dfrac{y}{L - y}\right)$ is $\dfrac{1}{y(L - y)}$ using the differentiation rules. Keep in mind that L is a constant.

(c) Show that when we solve the equation $\dfrac{1}{L} \ln\left(\dfrac{y}{L - y}\right) = kt + C$ for y we get $y = \dfrac{L}{1 + ae^{-kLt}}$, where $a = e^{-CL}$. (*Hint: Isolate y and divide the numerator and denominator of the resulting fraction by ae^{kLt}.*)

(d) Verify that $y = \dfrac{L}{1 + ae^{-kLt}}$ satisfies the differential equation $y' = ky(L - y)$, and explain why $y' > 0$ for any $t > 0$. This shows that the logistic function is increasing over its domain.

(e) Use the properties of limits and exponential functions to explain why $\displaystyle\lim_{t \to \infty} \left(\frac{L}{1 + ae^{-kLt}}\right) = L$.

SECTION PROJECT

Consider the U.S. population figures from 1790 to 1980 given in Table 6.7.1.

(a) Make a scatterplot of the data, where t is the number of years past 1800 and $-10 \leq t \leq 180$.

(b) Use your calculator to get a logistic regression model y that represents the U.S. population and t is the number of years past 1800. Now graph the scatterplot along with the logistic model y for $-10 \leq t \leq 300$ and then for $0 \leq t \leq 370$.

(c) Determine the year in which the rate of growth in the U.S. population started to decrease. (*Hint:* Find the inflection point.) What do you notice about this answer and the answer to part (b)?

(d) According to the model, what is the largest that the U.S. population can reach?

TABLE 6.7.1

YEAR	POPULATION (IN MILLIONS)	YEAR	POPULATION (IN MILLIONS)
1790	3.929	1890	62.980
1800	5.308	1900	76.212
1810	7.240	1910	92.228
1820	9.638	1920	106.022
1830	12.861	1930	123.203
1840	17.063	1940	132.165
1850	23.192	1950	151.326
1860	31.443	1960	179.324
1870	50.189	1970	203.302
1880	50.189	1980	226.549

CHAPTER REVIEW EXERCISES

In Exercises 1–4, determine if the function F is an antiderivative of the function f.

1. $F(x) = x^2 + C$; $f(x) = x$

2. $F(x) = x^3 - 5x + C$; $f(x) = 3x^2 - 5$

3. $F(t) = \frac{1}{5}t^5 + \frac{1}{4}t^4 + \frac{1}{3}t^3 + C$; $f(t) = t^4 + t^3 + t^2$

4. $F(x) = \frac{2}{x^3} + 4x + C$; $f(x) = \frac{6}{x^4} + 4$

In Exercises 5–8, use the Power Rule for Integration to find the indefinite integrals.

5. $\int x^7 \, dx$

6. $\int x^{5.13} \, dx$

7. $\int \frac{1}{t^5} \, dt$

8. $\int \frac{1}{\sqrt[4]{x^3}} \, dx$

For Exercises 9–12, determine the indefinite integral.

9. $\int (0.5x^7 - 6x) \, dx$

10. $\int (6t^5 + 4t^4 - 3) \, dt$

11. $\int (x^2 + 0.4x - \sqrt{x^5}) \, dx$

12. $\int \frac{5x^2 + 3x}{x} \, dx$

In Exercises 13 and 14, the rate function f' is shown. Use the graph to determine the intervals where its corresponding function f is increasing or decreasing and where the graph of f is concave up or down and where the relative maximum and minimum values occur.

13.

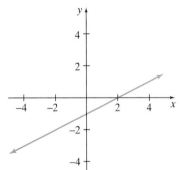

14.

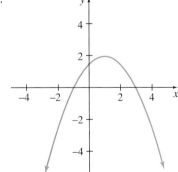

In Exercises 15–18, solve the given problem given the rate function and the given value. Note that $f(a) = b$ corresponds to the ordered pair (a, b).

15. $f'(x) = 3x - 4$; $f(0) = 11$

16. $\frac{dy}{dx} = x^2 - 5x$; $y = 10$ when $x = 6$

17. $y' = \sqrt{x}; (x, y) = (9, -5)$

18. $P'(t) = 0.5 + 0.15\sqrt[3]{t^4}; P(27) = 3.1$

19. Consider the marginal profit (in dollars per unit) for producing q units of a product given by the linear function

$$MP(q) = 48 - 0.03q$$

(a) Knowing that $P = 1600$ when $q = 100$, recover the profit function P.

(b) Use your solution from part (a) to find the total profit realized from selling 500 units of product.

20. The monthly marginal revenue function for the Byrne Rubber Tire Company is given by

$$MR(x) = R'(x) = -0.3x^2 - 1.4x + 35.2$$

where x is the number of thousands of tires produced and sold and $MR(x)$ is the marginal revenue in thousands of dollars per thousand tires.

(a) Knowing that $R(x) = 0$ when $x = 0$, recover the revenue function R.

(b) Find the price–demand function p for the tires. Recall that the general formula for the revenue function is

$$R(x) = (\text{quantity produced}) \cdot (\text{price of each unit})$$

$$= x \cdot p(x)$$

(c) What should the price be when the demand is 6000 tires (that is, $x = 6$)?

In Exercises 21–24, determine the step size Δx for the given interval $[a, b]$ and number of subintervals n.

21. $[0, 9], n = 27$ 22. $[-5, 5], n = 10$

23. $[7, 11], n = 32$ 24. $[-4, 9], n = 39$

In Exercises 25–28, calculate (a) the Left Sum and (b) the Right Sum to approximate the area under the graph of f on the closed interval $[a, b]$ using n equally spaced subintervals. Do the computations by hand and do not use your calculator program.

25. $f(x) = 5x, [0, 5], n = 5$

26. $f(x) = 3x - 8, [6, 8], n = 4$

27. $f(x) = x^2, [0, 3], n = 6$

28. $f(x) = 3x^2 + 4, [-1, 1], n = 4$

Complete the table for each of the functions f on $[a, b]$ in Exercises 29–32.

29. $f(x) = 5x^2 - 3, [0, 8]$

30. $f(x) = \sqrt{x - 2}, [3, 5]$

31. $f(x) = 3x^4 - 5x^3, [0, 4]$

32. $f(x) = 0.6x^2 - 1.3x + 7.3, [0, 10]$

In Exercises 33–38, evaluate the definite integrals.

33. $\displaystyle\int_3^7 6x\, dx$ 34. $\displaystyle\int_0^3 (3x - 4)\, dx$

35. $\displaystyle\int_{-2}^2 (10 - x^2)\, dx$ 36. $\displaystyle\int_{-2}^8 (x^5 + 9x^2 - 2)\, dx$

37. $\displaystyle\int_0^5 (2.7x^2 - 1.4x + 0.8)\, dx$ 38. $\displaystyle\int_1^{64} x\left(\sqrt{x} - \sqrt[3]{x}\right)\, dx$

In Exercises 39–41, determine the area of the shaded region.

39.

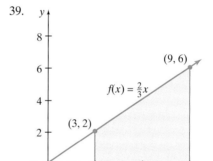

40.

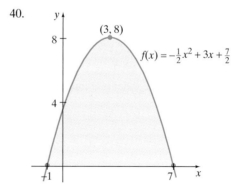

N RECTANGLES	LEFTSUM APPROXIMATION	RTSUM APPROXIMATION	|LEFTSUM−RTSUM|
10			
100			
1000			

41.

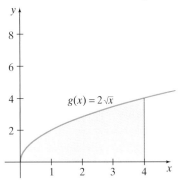

$f(x) = \frac{8}{x^2} + 2$

$(2, 4)$

$\left(8, \frac{17}{8}\right)$

42. Consider the areas under the graphs of f and g.

(a) Determine the area of the shaded region in the figure.

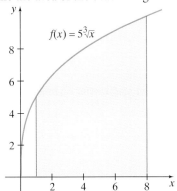

$f(x) = 5\sqrt[3]{x}$

(b) Determine the area of the shaded region in the figure.

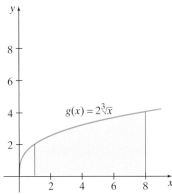

$g(x) = 2\sqrt[3]{x}$

43. Consider the areas under the graphs of f and g.

(a) Determine the area of the shaded region in the figure.

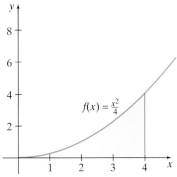

$f(x) = \frac{x^2}{4}$

(b) Determine the area of the shaded region in the figure.

$g(x) = 2\sqrt{x}$

44. Consider the absolute value function $f(x) = |x| - 3$.

(a) Graph f in the standard viewing window.

(b) Rewrite f as a piecewise-defined function.

(c) Determine the x-intercepts of the graph of f.

(d) Use the interval addition rule to determine the net and gross areas of $\int_{-2}^{5} f(x)\,dx$.

For Exercises 45–48:

(a) Graph the piecewise-defined function g in the standard viewing window.

(b) Use the interval addition rule to evaluate the indicated definite integral to find the area between the graph of g and the x-axis.

45. $g(x) = \begin{cases} -2x, & x < 0 \\ 3x, & x \geq 0 \end{cases}$; $\int_{-5}^{2} g(x)\,dx$

46. $g(x) = \begin{cases} 6 - 3x, & x < 1 \\ 3x, & x \geq 1 \end{cases}$; $\int_{0}^{6} g(x)\,dx$

47. $g(x) = \begin{cases} x^2 - 5, & x \leq -3 \\ x + 7, & x > -3 \end{cases}$; $\int_{-5}^{2} g(x)\,dx$

48. $g(x) = \begin{cases} 5, & x < 4 \\ x^2 - 11, & x \geq 4 \end{cases}$; $\int_{0}^{10} g(x)\,dx$

For Exercises 49–52:

(a) Determine the x-intercepts of the graph of the function f.

(b) Use the interval addition rule to determine the net and gross areas of the indicated integral.

49. $f(x) = x - 3$; $\int_{0}^{10} f(x)\,dx$

50. $f(x) = 2 - \sqrt{x}$; $\int_{0}^{10} f(x)\,dx$

51. $f(x) = x^2 - x - 2$; $\int_{-1}^{5} f(x)\,dx$

52. $f(x) = x^3 + 2x^2 - 3x$; $\int_{-5}^{5} f(x)\,dx$

53. The AddEmUp Calculator Company determines that the marginal cost function for a new calculator is given by

$$MC(x) = C'(x) = \frac{1}{20}x + 12, \qquad 0 \le x \le 300$$

where x is the number of calculators produced per day and $MC(x) = C'(x)$ is the marginal cost in dollars per calculator.
(a) Evaluate $C'(160)$ and interpret.
(b) Evaluate $\int_0^{200} C'(x)\, dx$ and interpret.

54. The Voll-E Ball Company determines that the marginal profit function for selling a new kind of football is

$$MP(x) = P'(x) = 0.0002x^2 - 0.01x + 2.85, \qquad 0 \le x \le 100$$

where x is the number of footballs made and sold and $P'(x)$ is the marginal profit function in dollars per football.
(a) Knowing that the company breaks even if 20 footballs are sold (that is, the profit for making and selling 20 footballs is $0), recover the profit function P.
(b) Evaluate $\int_{50}^{100} P'(x)\, dx$ and interpret.

55. The rate of U.S. government expenditures on defense- and space-related research and development is given by

$$r(t) = -0.0167t^3 + 0.448t^2 - 0.85t + 12.5, \qquad 1 \le t \le 21$$

where t represents the number of years since 1974 and $r(t)$ represents the rate of expenditures, measured in $\frac{\text{billions of dollars}}{\text{year}}$.
(a) Evaluate and interpret $r(7)$.
(b) Determine the total expenditures on defense- and space-related research and development expenditures from 1979 to 1985.

56. The velocity function for an object, in feet per second, is
$v(t) = 6t - 9$
(a) Evaluate $v(t)$ at the point $t = 2$.
(b) Evaluate and interpret $\int_{50}^{100} v(t)\, dt$.

In Exercises 57–60, use the u-substitution method to evaluate the indefinite integrals. Check by differentiating the solution.

57. $\int 5(5x - 7)^3\, dx$

58. $\int 3t^2(t^3 + 5)^4\, dt$

59. $\int (2x - 5)(x^2 - 5x + 3)\, dx$

60. $\int \frac{2x}{\sqrt{x^2 - 5}}\, dx$

In Exercises 61–64, use the u-substitution method to evaluate the indefinite integrals.

61. $\int (3x + 6)^8\, dx$

62. $\int t^4(2t^5 - 7)^3\, dt$

63. $\int \frac{x + 2}{x^2 + 4x + 6}\, dx$

64. $\int (8x - 28)(x^2 - 7x + 9)^3\, dx$

In Exercises 65–68, determine the integrals by solving the u-expression to complete the substitution.

65. $\int 3t(t + 4)^2\, dt$

66. $\int x(x - 1)^6\, dx$

67. $\int \frac{x}{\sqrt{x + 3}}\, dx$

68. $\int x\sqrt{x - 5}\, dx$

In Exercises 69–72, evaluate the definite integrals, and evaluate the limits of integration in terms of x.

69. $\int_{10}^{17} 2x\sqrt{x^2 - 64}\, dx$

70. $\int_0^{12} (x - 4)^3\, dx$

71. $\int_2^{11} \frac{3x^2}{\sqrt{x^3 - 4}}\, dx$

72. $\int_5^{20} \sqrt[4]{21 - t}\, dt$

In Exercises 73–76, evaluate the definite integrals. Change the limits of integration to u-limits.

73. $\int_1^4 \frac{2x - 5}{x^2 - 5x + 1}\, dx$

74. $\int_0^2 (3t^2 - 4)(t^3 - 4t + 7)^4\, dt$

75. $\int_4^8 3x^2\sqrt{x^3 + 17}\, dx$

76. $\int_5^9 \frac{2x}{\sqrt{x^2 + 144}}\, dx$

77. Determine the exact area of the shaded region.

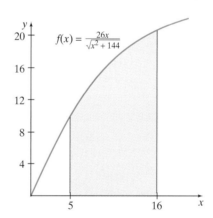

$f(x) = \frac{26x}{\sqrt{x^2 + 144}}$

78. The marginal profit function for hard disks produced by the Elephant Media Company is given by

$$MP(x) = \frac{80x}{\sqrt{x^2 + 576}}, \qquad 0 \le x \le 100$$

where x is the number of hard disks produced in a work shift and $MP(x)$ is the marginal profit.
(a) Evaluate and interpret $MP(7)$.
(b) Evaluate $\int_7^{45} MP(x)$ and interpret.
(c) Knowing that the Elephant Media Company breaks even when 32 hard disks are sold [that is, $P(x) = 0$ when $x = 32$], recover the profit function P.

79. Suppose that the velocity of an object is given by

$$v(t) = (t^3 - 5t)^3(3t^2 - 5)$$

where t is the time in seconds and $v(t)$ is the velocity in $\dfrac{\text{meters}}{\text{second}}$.

(a) Determine the total distance moved between $t = 3$ and 6 seconds.

(b) Knowing that when $t = 3$ seconds $s(t) = 736$ meters, determine the position function s.

In Exercises 80–87, evaluate the indefinite and definite integrals.

80. $\displaystyle\int \frac{3}{x}\, dx$

81. $\displaystyle\int \frac{3}{4} t^{-1}\, dt$

82. $\displaystyle\int 3e^x\, dx$

83. $\displaystyle\int \left(5e^x - 3x\right) dx$

84. $\displaystyle\int \left(\frac{1}{2}\right)^t dt$

85. $\displaystyle\int_3^5 \frac{6}{x}\, dx$

86. $\displaystyle\int_0^5 \frac{e^t - 5}{3}\, dt$

87. $\displaystyle\int_0^2 5^x\, dx$

In Exercises 88–95, evaluate the integrals using u-substitution.

88. $\displaystyle\int \frac{2x + 5}{x^2 + 5x - 3}\, dx$

89. $\displaystyle\int \frac{5}{\sqrt{x}\left(\sqrt{x} + 4\right)}\, dx$

90. $\displaystyle\int e^{3t-5}\, dt$

91. $\displaystyle\int e^x (5 - e^x)^3\, dx$

92. $\displaystyle\int 3x^2 \cdot 5^{x^3}\, dx$

93. $\displaystyle\int_0^3 \frac{2x}{x^2 + 16}\, dx$

94. $\displaystyle\int_3^7 e^{2t-6}\, dt$

95. $\displaystyle\int_{-2}^2 2^{3x}\, dx$

96. The PaperCut Discount Stationery Store is having a sale on pen and pencil sets. The rate of sales is given by

$$S(t) = \frac{60}{t}, \qquad 1 \le t \le 10$$

where t represents the number of days that the sale has been running and $S(t)$ represents the sales rate, measured in $\dfrac{\text{pen and pencil sets}}{\text{day}}$.

(a) Evaluate $S(4)$ and interpret.

(b) Use an integral to determine the total increase in the number of pen and pencil sets sold between days 4 and 8.

🌐 97. The rate of change in the number of U.S. billed international telephone calls can be modeled by

$$r(t) = 63.8e^{0.798t}, \qquad 1 \le t \le 11$$

where t represents the number of years since 1984 and $r(t)$ represents the rate of change in the number of U.S. billed international telephone calls, measured in $\dfrac{\text{millions of calls}}{\text{year}}$.

(a) Evaluate and interpret $r(6)$.

(b) Evaluate and interpret $\int_1^{11} r(t)\, dt$.

In Exercises 98–101, write an exponential growth model in the form $f(t) = a \cdot e^{kt}$ that satisfies the given differential equation and initial condition.

98. $f'(t) = 0.7 \cdot f(t);\ f(0) = 300$

99. $f'(t) = 1.37 \cdot f(t);\ f(0) = 8.2$

100. $f'(t) = -0.07 \cdot f(t);\ f(t) = 5$ when $t = 0$

101. $f'(t) = -0.87 \cdot f(t);\ f(0) = 384$

102. The population of Popuville is increasing at a rate of 11% per year. The current population of the town is 42,000. Assume that the rate of change in the population is increasing proportionally to the population size.

(a) Write a differential equation and an initial condition for the population of Popuville.

(b) Write an exponential growth model for the population of Popuville after t years.

(c) Use the model to estimate the population after 15 years.

103. The number of bacteria in a culture starts at 300 and increases at a rate of 8% per hour. Assume that the rate of change in the number of bacteria is increasing proportionally to the number of bacteria present.

(a) Write a differential equation for the number of bacteria present after t hours.

(b) Write an exponential growth model for the number of bacteria present after t hours.

(c) Complete the table.

NUMBER OF HOURS, t	NUMBER OF BACTERIA, $f(t)$
0	
2	
4	
10	
20	

(d) Graph the exponential growth model for the values $0 \le t \le 20$.

🌐 104. In 1970, 8.6 million tons of solid waste was recovered for recycling in the United States. By 1995, this number had increased to 56.2 million tons.

(a) Considering recycled waste as a "natural resource," write a natural resource function

$$N(t) = a \cdot e^{kt}$$

where t represents the number of years since 1970 and $N(t)$ represents the number of millions of tons of solid waste recovered each year for recycling.

(b) Evaluate and interpret $\displaystyle\int_0^{25} N(t)\, dt$.

105. The amount of chlorofluorocarbon gases (CFCs) released into the environment each year in the United States can be modeled by

$$f(t) = 342e^{-0.17t}, \qquad 1 \le t \le 7$$

where t represents the number of years since 1988 and $f(t)$ represents the amount of CFCs in thousands of metric tons.

(a) Write a differential equation and an initial condition for the number of thousands of metric tons of CFCs released each year in the United States.

(b) Graph f in the viewing window [0, 12] by [0, 300].

(c) Evaluate and interpret $f(5)$.

(d) Evaluate and interpret $f'(5)$.

Applications of Integral Calculus

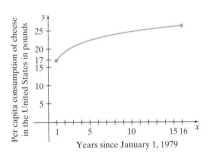

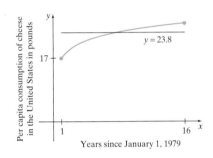

The per capita consumption of cheese in the United States, in pounds, from January 1, 1980 to January 1, 1995 can be modeled by

$$y = 17.13 + 3.4 \ln x.$$

Average yearly per capita consumption of cheese in the United States from January 1, 1980 to January 1, 1995 is about 23.8 pounds

$$23.8 \approx \frac{1}{15} \int_1^{16} (17.13 + 3.4 \ln x) \, dx.$$

What We Know

In Chapter 6 we saw how the definite integral can be used to give a total accumulation. We also saw how integration and differentiation were related through the Fundamental Theorem of Calculus.

Where Do We Go

In this chapter we will see even more applications of integration. Included applications will be average value, perpetuities, the Gini coefficient, consumers' and producers' surplus, and many others. We continue to strengthen the concepts set forth in Chapter 6, especially an accumulation.

SECTION 7.1

AVERAGE VALUE OF A FUNCTION AND THE DEFINITE INTEGRAL IN FINANCE

In this section we examine the **average value** of a function. Compound interest will be revisited as we examine what is known as an *income stream*. The process of computing an average value for a function over an interval will be exploited throughout Chapter 7, and the concept of an income stream will also reappear in later sections of this chapter. We encourage you to work to understand these concepts as they arise, to insure your continued success in these later sections.

Average Value Function

Before we analyze how to determine an average value of a function, let's first recall that the average (or mean) of n numbers, $y_1, y_2, y_3, \ldots, y_n$, is simply the sum of the numbers divided by n. That is,

$$\bar{y} = \frac{1}{n} \sum_{i=1}^{n} y = \frac{y_1 + y_2 + y_3 + \cdots + y_n}{n}$$

In the following Flashback, we review how to determine the average of a set of numbers. The Flashback also provides the necessary foundation for the definition of the **average value of a function.**

Flashback

CARBON DIOXIDE (CO_2) EMISSIONS REVISITED

In Section 5.3 we analyzed the carbon content of CO_2 emissions, measured in millions of metric tons, into the atmosphere by the United States from 1989 to 1995. The data are given in Table 7.1.1. We can represent the data graphically by plotting the data points and then making a bar graph, as shown in Figure 7.1.1.

(a) Using the data values, determine the average value of the carbon content of CO_2 emissions over the 7-year period. Round to the nearest tenth.

(b) Determine the total area of the seven rectangles in Figure 7.1.1.

Flashback Solution

(a) The average value of the carbon content of the CO_2 emissions into the atmosphere by the United States over the 7-year period is given by

$$\bar{y} = \frac{1}{n} \sum_{i=1}^{n} y = \frac{\begin{matrix}1384.3 + 1372.0 + 1358.8 + 1379.5 \\ + 1404.8 + 1431.4 + 1442.3\end{matrix}}{7}$$

$$= \frac{9773.1}{7} \approx 1396.2 \text{ million metric tons per year}$$

TABLE 7.1.1	
YEAR	**CARBON CONTENT**
1989	1384.3
1990	1372.0
1991	1358.8
1992	1379.5
1993	1404.8
1994	1431.4
1995	1442.3

Source: U.S. Census Bureau web site.

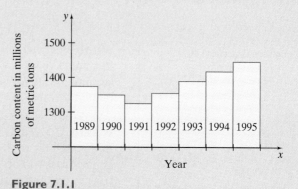

Figure 7.1.1

(b) The base of the first rectangle is 1, and the height is the carbon content, 1384.3. So the area of this rectangle is $1 \cdot 1384.3 = 1384.3$, which is just the carbon content for that year. Repeating this process for the other rectangles, we see that the total area of the seven rectangles is the sum of the values of the carbon content, 9773.1, which we determined in part (a).

If we divide the area of the seven rectangles by the number of rectangles, 7, and round to the nearest tenth, we get

$$\frac{9773.1}{7} \approx 1396.2$$

which is the average value we found in the Flashback, part (a). Thus, we can interpret this result to mean

$$\frac{\text{area under the bar graph}}{\text{width of the bar graph}} = \text{average value}$$

Now, if a continuous function f models the data, we can find the average value of the function over the interval $[a, b]$. Using the preceding result for bar graphs, it can be shown that

$$\frac{\text{area under the curve}}{\text{width of the interval}} = \text{average value}$$

Since we know that the area under a curve is given by $\int_a^b f(x) \, dx$, these results suggest the following definition for the average value of a continuous function f on the interval $[a, b]$.

Average Value of a Continuous Function on $[a, b]$

The **average value** of a continuous function, f, on the interval $[a, b]$ is given by

$$\frac{1}{b-a} \int_a^b f(x) \, dx$$

EXAMPLE 1

Determining an Average Value

The carbon content, in million metric tons, of CO_2 emissions into the atmosphere by the United States can be modeled by

$$y = f(x) = -1.32x^3 + 19.66x^2 - 72.76x + 1441.47, \qquad 0 \le x \le 7$$

where $x = 0$ corresponds to January 1, 1989, and y represents the carbon content, in million metric tons, of CO_2 emissions into the atmosphere by the United States. Use the model to calculate the average value of CO_2 emissions for the period January 1, 1989, to January 1, 1996.

SOLUTION

We first observe that $x = 0$ corresponds to January 1, 1989, and $x = 7$ corresponds to January 1, 1996. Thus, we need to determine the average value of $y = f(x) = -1.32x^3 + 19.66x^2 - 72.76x + 1441.47$ on the interval $[0, 7]$. So we compute

$$\frac{1}{7-0} \int_0^7 (-1.32x^3 + 19.66x^2 - 72.76x + 1441.47)\, dx$$

$$= \frac{1}{7}\left[-\frac{1.32}{4}x^4 + \frac{19.66}{3}x^3 - \frac{72.76}{2}x^2 + 1441.47x \right]\Bigg|_0^7$$

$$\approx 1394.7 \text{ million metric tons per year (rounded to the nearest tenth)}$$

■

Notice that the average value determined in Example 1 was fairly close to the actual average value that we determined in the Flashback. Some discrepancy is expected, since the function in Example 1 is a model that *approximates* the data.

EXAMPLE 2 | **Determining an Average Value**

Determine the average value of $f(x) = x^3 - 2x^2 - 5x + 11$ on the interval $[-2, 4]$.

SOLUTION

Here we compute the average value to be

$$\frac{1}{b-a}\int_a^b f(x)\, dx = \frac{1}{4-(-2)} \int_{-2}^4 (x^3 - 2x^2 - 5x + 11)\, dx$$

$$= \frac{1}{6}\left[\frac{x^4}{4} - \frac{2x^3}{3} - \frac{5x^2}{2} + 11x \right]\Bigg|_{-2}^4$$

$$= \frac{1}{6}\left[\left(64 - \frac{128}{3} - 40 + 44\right) - \left(4 + \frac{16}{3} - 10 - 22\right) \right]$$

$$= \frac{1}{6}\left[\frac{76}{3} - \left(\frac{-68}{3}\right) \right] = \frac{1}{6}(48) = 8$$

So the average value of $f(x) = x^3 - 2x^2 - 5x + 11$ on $[-2, 4]$ is 8. ■

✓CHECKPOINT 1

Now work Exercise 5.

The average value can be thought of as giving the typical function value for a function on the closed interval $[a, b]$. At this time, we offer the following analogy and comparison between a discrete average value and a continuous average value.

TYPES OF VALUES	DISCRETE	CONTINUOUS
Sum of values	$\sum_{i=1}^n y$	$\int_a^b f(x)\, dx$
Number of "points"	n	$b - a$
Average	$\frac{1}{n}\sum_{i=1}^n y$	$\frac{1}{b-a}\int_a^b f(x)\, dx$

EXAMPLE 3

Determining an Average Value

The annual U.S. per capita consumption (adult population) of alcoholic beverages can be modeled by

$$f(x) = 43.75e^{-0.012x}, \qquad 1 \leq x \leq 11$$

where x is measured in years ($x = 1$ corresponds to January 1, 1980), and $f(x)$ is the annual per capita consumption of alcoholic beverages in gallons. Use the model to determine the average per capita consumption of alcoholic beverages, rounded to the nearest tenth, during the 1980s.

SOLUTION

When we say "during the 1980s", we mean from January 1, 1980, to January 1, 1990 (or January 1, 1980, through December 31, 1989). This means that we are to determine the average value of f on the interval $[1, 11]$. Computing this integral yields

$$\frac{1}{b-a}\int_a^b f(x)\,dx = \frac{1}{11-1}\int_1^{11}(43.75e^{-0.012x})\,dx = \frac{1}{10}\left[\frac{43.75}{-0.012}e^{-0.012x}\right]\Big|_1^{11}$$

$$= \frac{1}{10}\left[\frac{43.75}{-0.012}e^{-0.012(11)}\right] - \frac{1}{10}\left[\frac{43.75}{-0.012}e^{-0.012(1)}\right]$$

$$\approx \frac{1}{10}[-3194.993 - (-3602.345)]$$

$$= \frac{1}{10}[407.352] = 40.7 \text{ rounded to the nearest tenth.}$$

Our answer is that during the 1980s the annual average alcoholic beverage consumption per capita was about 40.7 gallons.

✓CHECKPOINT **2**

Now work Exercise 15.

Compound Interest

So far we have examined how the average value of a function is used in the life and social sciences. Now we turn our attention to economics, finance, and the managerial sciences. To do this, we need to recall a formula first encountered in Chapter 1.

> **CONTINUOUS COMPOUND**
> **INTEREST FORMULA**
>
> **From Your Toolbox**
>
> If a principal of P dollars is invested into an account earning annual interest rate r (in decimal form) compounded continuously, then the amount A in the account at the end of t years is given by
>
> $$A(t) = Pe^{rt}$$

EXAMPLE 4

Determining an Average Balance and Computing a Bonus

(a) Wilma deposits $5000 into a money market account earning 6.5% interest compounded continuously. Determine Wilma's average balance, to the nearest cent, over one year.

(b) Suppose that the institution where Wilma has deposited her money also pays a bonus at the end of the year of 1.25% of the average balance in the account during the year. Compute the size of Wilma's bonus.

SOLUTION

(a) From the formula for continuous compound interest, we have

$$A(t) = 5000e^{0.065t}$$

We are asked to compute the average value of $A(t) = 5000e^{0.065t}$ on $[0, 1]$. So

$$\frac{1}{b-a} \int_a^b A(t)\, dt = \frac{1}{1-0} \int_0^1 5000e^{0.065t}\, dt$$

$$= \frac{5000}{0.065} e^{0.065t} \Big|_0^1 = \frac{5000}{0.065} e^{0.065(1)} - \frac{5000}{0.065} e^{0.065(0)} \approx 5166.08$$

So Wilma's average balance over one year was $5166.08.

(b) Computing 1.25% of the $5166.08 average balance from part (a) yields a bonus of

$$0.0125(5166.08) \approx \$64.58 \qquad \blacksquare$$

In Example 5, we review how to determine a total amount when given a rate of change function, as well as computing an average value.

EXAMPLE 5

Determining an Account Change and an Average Balance

Stella deposits $7000 into an account where the rate of change of the amount in the account t years after the initial deposit is given by $A'(t) = 490e^{0.07t}$.

(a) Determine by how much the account changes from the end of the third year to the end of the fifth year. Round to the nearest cent.

(b) Determine the average balance, to the nearest cent, in the account over the first 5 years.

SOLUTION

(a) Notice that we are given a **rate of change function,** $A'(t) = 490e^{0.07t}$. Recall from Chapter 6 that the total change in the account from the end of the third year to the end of the fifth year can be found by the definite integral. That is,

$$A(5) - A(3) = \int_3^5 490e^{0.07t}\, dt = \frac{490}{0.07} e^{0.07t} \Big|_3^5$$

$$= 7000e^{0.07(5)} - 7000e^{0.07(3)} \approx 1297.73$$

Stella's account changed by about $1297.73 from the end of the third year to the end of the fifth year.

(b) Here we need to determine the average value of A on $[0, 5]$. The first order of business is to determine $A(t)$. To do this, we integrate

$$A(t) = \int A'(t)\, dt = \int 490 e^{0.07t}\, dt$$

$$= \frac{490}{0.07} e^{0.07t} + C = 7000 e^{0.07t} + C$$

So $A(t) = 7000 e^{0.07t} + C$. To determine the constant C, we note that, at $t = 0$, $A(t) = 7000$. (An initial deposit corresponds to $t = 0$.) Substituting gives

$$A(0) = 7000 e^{0.07(0)} + C = 7000$$

$$7000 + C = 7000$$

$$C = 0$$

Since $A(t) = 7000 e^{0.07t}$, we compute the average value of $A(t)$ on $[0, 5]$ as follows:

$$\frac{1}{b-a} \int_0^5 7000 e^{0.07t}\, dt = \frac{1}{5-0} \left[\frac{7000}{0.07} e^{0.07t} \right]\Big|_0^5$$

$$= \frac{1}{5} [100{,}000 e^{(0.07 \cdot 5)} - 100{,}000 e^{(0.07 \cdot 0)}] \approx 8381.35$$

The average amount in Stella's account during the first 5 years was \$8381.35. ∎

Continuous Stream of Income

The prior two examples reviewed the concept of continuous compounding. Continuous compounding does not mean that the bank or other institution where the money is invested is continuously placing money (the interest earned) into our account. However, we compute what is in our account as if it were. We can think of our account as being a **continuous income stream.** In fact, this is very similar to the electric company, the natural gas company, or the water company, which has a meter on our house that continuously monitors and totals the amount of electricity, natural gas, or water that we are consuming. Example 6 describes a continuous stream of income.

EXAMPLE 6 | **Determining Total Income from an Income Stream**

Suppose that the rate of change of income, in thousands of dollars per year, for an oil field is projected to be $f(t) = 350 e^{-0.09t}$, where t is the number of years from when oil extraction began. See Figure 7.1.2. Determine the total amount of money generated from this oil field during the first two years of operation.

SOLUTION

Once again we recognize that we have a rate of change function. To determine the total amount of money for the first 2 years, we need to determine the area under the curve on the interval $[0, 2]$. Integrating yields

$$\int_0^2 350 e^{-0.09t}\, dt = \left(\frac{350}{-0.09} e^{-0.09t} \right)\Big|_0^2$$

$$= \frac{350}{-0.09} e^{-0.09(2)} - \frac{350}{-0.09} e^{-0.09(0)} \approx 640.61584$$

So the first 2 years of operation will generate about \$640,615.84 in income.

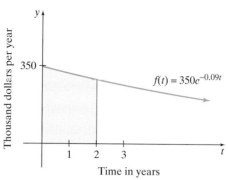

Figure 7.1.2 The area of the shaded region gives the total money generated in the first 2 years.

In reality, income from the oil field in Example 6 is not received in one lump sum payment at the end of 2 years. The income is probably collected on a regular basis, possibly every month or quarter. In situations like Example 6, however, it is convenient to assume that the income is actually received in a *continuous stream;* that is, we assume that the income is a continuous function of time. The rate of change of income, $f(t) = 350e^{-0.09t}$ in Example 6, is called a **rate of flow function.**

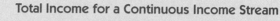

Total Income for a Continuous Income Stream

If f is the rate of flow function of a continuous income stream, then the **total income** produced from time $t = a$ to $t = b$ is found by

$$\int_a^b f(t)\, dt$$

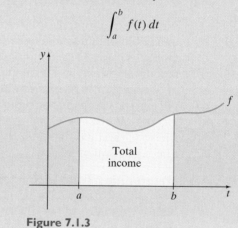

Figure 7.1.3

✓ **CHECKPOINT 3**

Now work Exercise 39.

Annuities

We conclude this section with a look at what is known as an annuity. An **annuity** is defined as a sequence of equal payments made at equal time intervals. For example, if money is borrowed to finance the purchase of a home, a car, or even an education, the loan usually is repaid in equal monthly payments for a specified length of time. This sequence of equal monthly payments forms an annuity. Also, payments that one receives at retirement from a pension plan, a 401-k, a 403-b, or an IRA are other examples of annuities.

Assume that Maria deposits $2000 at the beginning of each year into an IRA and that her investment earns an annual interest rate of 6% compounded continuously. To compute the amount in her account after 4 years requires that we compute what happens to the initial deposit, the second deposit, the third deposit, and so on, for the 4 years. For example, using $A(t) = Pe^{rt}$:

- The initial deposit $(t = 0)$ of \$2000 is in the account for 4 years and grows to $2000e^{(0.06)(4)}$.

- The second deposit $(t = 1)$ is in the account for 3 years $(4 - 1)$ and grows to $2000e^{(0.06)(4-1)} = 2000e^{(0.06)(3)}$.

Continuing in this manner, the total amount S in the account at the end of 4 years can be represented by

$$S = \left(\begin{array}{c} \text{value of initial payment} \\ t = 0 \end{array} \right) + \left(\begin{array}{c} \text{value of second payment} \\ t = 1 \end{array} \right)$$
$$+ \left(\begin{array}{c} \text{value of third payment} \\ t = 2 \end{array} \right) + \left(\begin{array}{c} \text{value of final payment} \\ t = 3 \end{array} \right)$$
$$S = 2000e^{0.06(4)} + 2000e^{0.06(4-1)} + 2000e^{0.06(4-2)} + 2000e^{0.06(4-3)}$$
$$S = 2000e^{0.06(4)} + 2000e^{0.06(3)} + 2000e^{0.06(2)} + 2000e^{0.06(1)}$$
$$S \approx 2542.50 + 2394.43 + 2254.99 + 2123.67 = \$9315.59$$

To see how we can approximate the sum, S, with a definite integral, we plot each term of the sum as shown in Figure 7.1.4.

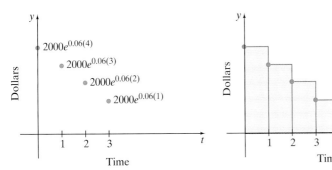

Figure 7.1.4

Figure 7.1.5

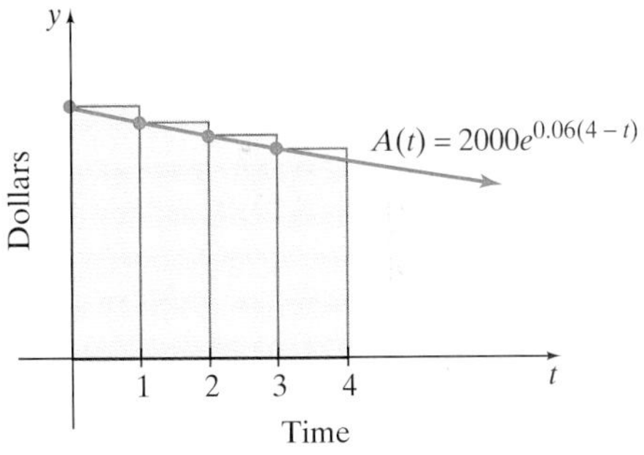

Figure 7.1.6

The sum of the amount in the IRA can be thought of as being equal to the sum of the four rectangles shown in Figure 7.1.5, since the base (width) of each rectangle is 1. Figure 7.1.6 shows the graph of $A(t) = 2000e^{0.06(4-t)}$ with the four rectangles. What we are about to do should look quite familiar by now. The area under A on the interval [0, 4] is approximately equal to the area of the four rectangles. Hence, the area under the curve can be used to approximate the amount or sum in the IRA at the end of 4 years. We illustrate in Example 7.

EXAMPLE 7   **Applying the Definite Integral**

Write and evaluate a definite integral to find the area under A in Figure 7.1.6 that approximates the sum in the IRA at the end of 4 years.

SOLUTION

The definite integral that determines the area under A is

$$\int_0^4 2000e^{0.06(4-t)} \, dt$$

We evaluate the definite integral using u-substitution as follows:

Let $u = 4 - t$; then $du = -dt$ or $-du = dt$.

By u-substitution, the integral may be written as

$$-\int 2000e^{0.06u}\, du = -\left(\frac{2000}{0.06}e^{0.06u}\right) = -\left(\frac{2000}{0.06}e^{0.06(4-t)}\right)\Big|_0^4$$

$$= -\left(\frac{2000}{0.06}e^{0.06(4-4)} - \frac{2000}{0.06}e^{0.06(4-0)}\right)$$

$$\approx -(33,333.33 - 42,374.97) = \$9041.64 \qquad \blacksquare$$

Example 7 indicates that the definite integral gives an *approximation* to the actual sum. With this in mind, the explanation preceding Example 7, as well as Example 7, outlines a procedure that we can now generalize to give an approximation for S, the **future value of an annuity**.

Future Value of an Annuity

If n equal payments of P dollars is deposited into an annuity at an annual interest rate r (in decimal form), then the **future value, S,** for a duration of the annuity T is given by

$$S \approx \int_0^T Pe^{r(T-t)}\, dt$$

NOTE: It is very important to keep the units of time for T, r, and P consistent when using the formula to approximate the future value of an annuity, S. Example 8 illustrates.

EXAMPLE 8 | **Determining a Future Value**

Irene has just started a new job teaching. At the beginning of each month she deposits \$25 into her 403-b plan, which earns 6.6% annual interest compounded continuously. Approximate how much Irene will have in her account after 20 years.

SOLUTION

We approximate the amount in Irene's 403-b plan after 20 years by using

$$S \approx \int_0^T Pe^{r(T-t)}\, dt$$

If we have $P = \$25$ per month, then r and T must also be in terms of months. Thus,

$$T = 12(20) = 240 \text{ months} \quad \text{and} \quad r = \frac{0.066}{12} = 0.0055 \text{ per month}$$

So the approximation for S is

$$S \approx \int_0^T Pe^{r(T-t)}\, dt = \int_0^{240} 25e^{0.0055(240-t)}\, dt$$

Using a u-substitution yields $u = 240 - t$ and $du = -dt$ or $-du = dt$, and thus

$$\int_0^{240} 25e^{0.0055(240-t)} \, dt = -\int 25e^{0.0055u} \, du$$

$$= -\left(\frac{25}{0.0055}e^{0.0055u}\right) = -\left(\frac{25}{0.0055}e^{0.0055(240-t)}\right)\bigg|_0^{240}$$

$$= -\left(\frac{25}{0.0055}e^{0.0055(240-240)} - \frac{25}{0.0055}e^{0.0055(240-0)}\right)$$

$$\approx -(4545.45 - 17{,}015.55) = \$12{,}470.10$$

So the total amount in the account after 20 years is approximately \$12,470.10.

✓ **CHECKPOINT 4**

Now work Exercise 45.

SUMMARY

In this introductory section on applications of the integral, we examined how recognizing a definite integral as a limit of a sum and recognizing the area under a curve as a definite integral were very important. The former was utilized to determine a formula for the average value, and the latter was used in determining a formula to approximate the future value of an annuity. We also saw how integrating a rate of change function gives a total accumulation. In this section, it was seen again in the context of **income streams.**

- The **average value** of a continuous function, f, on the interval $[a, b]$ is given by

$$\frac{1}{b-a}\int_a^b f(x) \, dx$$

- If n equal payments of P dollars are deposited into an annuity at an annual interest rate r (in decimal form), then the **future value, S,** for a duration of the annuity T is given by

$$S \approx \int_0^T Pe^{r(T-t)} \, dt$$

- If f is the **rate of flow function** of a continuous income stream, then the **total income** produced from time $t = a$ to $t = b$ is found by

$$\int_a^b f(t) \, dt$$

SECTION 7.1 EXERCISES

In Exercises 1–10, determine the average value of the given function on the stated interval.

1. $f(x) = 2x + 5$ $[0, 7]$

2. $f(x) = 3x + 1$ $[1, 4]$

3. $f(x) = 3x^2 - 2x$ $[-1, 2]$

4. $g(x) = 4x - 2x^2$ $[0, 3]$

✓ 5. $y = x^3 + 2x^2 - x + 1$ $[0, 2]$

6. $y = -2x^3 + x + 2$ $[-2, 1]$

7. $g(x) = x^{2/3}$ $[1, 8]$

8. $g(x) = \sqrt{x+1}$ $[3, 7]$

9. $y = 4e^{0.2x}$ $[0, 10]$

10. $y = 2e^{-0.15x}$ $[0, 5]$

For Exercises 11–14:

(a) Use your **RTSUM** program, with $n = 100$, to approximate $\int_a^b f(x)\, dx$.

(b) Use your result from part (a) to approximate an average value of f on $[a, b]$.

11. $f(x) = \dfrac{2x + 1}{x^2 + 1}$ $[-1, 1]$

12. $f(x) = \dfrac{x}{x + 1}$ $[0, 4]$

13. $f(x) = \ln(1 + x^2)$ $[0, 2]$

14. $f(x) = \dfrac{1}{x^2 + 4}$ $[0, 5]$

APPLICATIONS

✓ 15. The amount of money spent on home health care in the United States can be approximated by

$$f(x) = 0.13x^2 - 0.39x + 2.79, \qquad 1 \le x \le 16$$

where x is measured in years ($x = 1$ corresponds to January 1, 1980) and $f(x)$ is the amount of money, in billions of dollars. Use the model to compute the average amount spent on home health care in the United States during the 1980s.

16. The number of Medicare enrollees in the United States can be modeled by

$$f(x) = 0.61x + 27.68, \qquad 1 \le x \le 16$$

where x is measured in years ($x = 1$ corresponds to January 1, 1980) and $f(x)$ is the number of Medicare enrollees, in millions, in the United States. Use the model to compute the average number of Medicare enrollees during the 1980s.

17. The number of Medicaid recipients in the United States can be modeled by

$$f(x) = 0.11x^2 - 0.91x + 22.59, \qquad 1 \le x \le 16$$

where x is measured in years ($x = 1$ corresponds to January 1, 1980) and $f(x)$ is the number of Medicaid recipients, in millions, in the United States. Use the model to compute the average number of Medicaid recipients during the 1980s.

18. The annual U.S. per capita consumption, in pounds, of beef can be modeled by

$$B(t) = 0.03t^3 - 0.76t^2 + 4.54t + 68.34, \qquad 1 \le t \le 16$$

where t is measured in years ($t = 1$ corresponds to January 1, 1980) and $B(t)$ is the per capita consumption, in pounds, of beef. Determine the average per capita consumption of beef during the 1980s.

19. The revenue, in billions of dollars, in the resource recovery industry in the United States can be modeled by

$$f(x) = -0.02x^2 + 1.091x + 3.318, \qquad 1 \le x \le 16$$

where x is measured in years ($x = 1$ corresponds to January 1,

1980). Determine the average revenue in the resource recovery industry during the 1980s.

20. The cost, in cents per gallon, of unleaded regular gasoline in the United States, can be modeled by

$$f(x) = 0.01x^4 - 0.25x^3 + 2.77x^2 - 11.31x + 133.29, \qquad 1 \le x \le 16$$

where x is measured in years ($x = 1$ corresponds to January 1, 1980).

(a) Determine the average price, in cents per gallon, for unleaded regular gasoline in the United States during the 1980s.

(b) Determine the average price, in cents per gallon, for unleaded regular gasoline in the United States during the first half of the 1990s.

21. The 1998 prospectus of a certain mutual fund states that during 1997 the amount of money that was kept in cash reserves was approximately linear and can be represented by $CR(x) = 0.8x + 3$, where x is the number of months after the first of the year and $CR(x)$ is the cash reserves in millions of dollars.

(a) Compute the average cash reserves on the interval $[0, 3]$, the first quarter of 1997.

(b) Compute the average cash reserves on the interval $[9, 12]$, the last quarter of 1997.

22. The 1998 prospectus of a certain mutual fund states that during 1997 the amount of money that was kept in cash reserves was approximately linear and can be represented by

$$CR(x) = -0.9x + 12.5$$

where x is the number of months after the first of the year and $CR(x)$ is the cash reserves in millions of dollars.

(a) Compute the average cash reserves on the interval $[0, 3]$, the first quarter of 1997.

(b) Compute the average cash reserves on the interval $[9, 12]$, the last quarter of 1997.

23. The price–demand function for Linguini's Pizza Palace all-you-can-eat pizza buffet is given by

$$p(x) = -0.02x + 8.3$$

where $p(x)$ is the price in dollars and x is the quantity demanded. Determine the average price on the demand interval $[120, 200]$.

24. The price–demand function for Linguini's Pizza Palace specialty pizzas is given by

$$p(x) = 15.22e^{-0.015x}$$

where $p(x)$ is the price in dollars and x is the quantity demanded. Determine the average price on the demand interval $[40, 80]$.

25. Researchers have determined through experimentation that the percent concentration of a certain medication

during the first 20 hours after it has been administered can be approximated by

$$p(t) = \frac{200t}{2t^2 + 5}, \qquad 0 \le t \le 20$$

where t is the time in hours after the medication is taken and $p(t)$ is the percent concentration. Determine the average percent concentration on the interval $[0, 20]$.

26. Researchers have determined through experimentation that the percent concentration of a certain medication during the first 20 hours after it has been administered can be approximated by

$$p(t) = \frac{300t}{6t^2 + 5}, \qquad 0 \le t \le 20$$

where t is the time in hours after the medication is taken and $p(t)$ is the percent concentration.

(a) Determine the average percent concentration on the interval $[0, 10]$.

(b) Determine the average percent concentration on the interval $[10, 20]$.

(c) Is the average percent concentration greater during the first 10 hours after it has been taken or during the second 10 hours after it has been taken? Explain.

27. Dixco Engines has determined that the cost for producing x diesel engines is given by

$$C(x) = 60{,}000 + 300x$$

where $C(x)$ is the cost in dollars.

(a) Determine the average value of the cost function over the interval $[0, 500]$.

(b) Determine the average cost function, AC. $\left[\text{Recall that } AC(x) = \frac{C(x)}{x}.\right]$ Evaluate $AC(500)$.

(c) Explain the differences between parts (a) and (b).

28. Digital Pet has determined that the cost for producing x virtual pets in one day is given by

$$C(x) = 150 + 3x + \frac{2x^2}{30}$$

where $C(x)$ is the cost in dollars.

(a) Determine the average value of the cost function over the interval $[0, 200]$.

(b) Determine the average cost function, AC. $\left[\text{Recall that } AC(x) = \frac{C(x)}{x}.\right]$ Evaluate $AC(200)$.

(c) Explain the differences between parts (a) and (b).

29. Jan deposits $3000 into a money market account earning 5.85% interest compounded continuously. Determine Jan's average balance, to the nearest cent, over 5 years.

30. Eugene deposits $12,000 into a money market account earning 6.5% interest compounded continuously. Determine Eugene's average balance, to the nearest cent, over 3 years.

31. Lisa deposits $6000 into a bank account that earns 4.5% interest compounded continuously. Determine Lisa's average balance, to the nearest cent, over 1 year.

32. Wilhelm deposits $9000 into a money market account earning 7% interest compounded continuously. The institution where Wilhelm has deposited his money also pays a bonus at the end of the year of 1% of the average balance in the account during the year.

(a) Compute Wilhelm's average balance, to the nearest cent, over 1 year.

(b) Compute the size of Wilhelm's bonus.

33. Elmer deposits $3500 into a bank account that earns 3.8% interest compounded continuously. The bank also pays a bonus at the end of the year of 0.75% of the average balance in the account during the year.

(a) Compute Elmer's average balance, to the nearest cent, over 1 year.

(b) Compute the size of Elmer's bonus.

34. Rusty has $4000 to invest in a money market account. At Cold Cash Company, Rusty is offered an account that earns 5% interest compounded continuously with a 1% bonus at the end of the year on his average balance in the account during the year. At Money Time Company, Rusty is offered an account that earns 4.9% interest compounded continuously with a 1.2% bonus at the end of the year on his average balance in the account during the year. If Rusty's goal is to achieve the largest bonus at the end of the year, into which account should he place his money? Explain.

35. Shannon deposits $8000 into an account where the rate of change of the amount in the account is given by $A'(t) = 480e^{0.06t}$, t years after the initial deposit.

(a) Determine by how much the account changes from the end of the second year to the end of the fifth year. Round to the nearest cent.

(b) Determine the average balance, to the nearest cent, in the account over the first 5 years.

36. Kevan deposits $2000 into an account where the rate of change of the amount in the account is given by $A'(t) = 110e^{0.055t}$, t years after the initial deposit.

(a) Determine by how much the account changed from the end of the first year to the end of the third year. Round to the nearest cent.

(b) Determine the average balance, to the nearest cent, in the account over the first 3 years.

37. Muggy is a company that produces coffee mugs. The research department at Muggy determined the following marginal cost function:

$$C'(x) = 1 + \frac{x}{20}$$

Here $C'(x)$ is in dollars per mug and x is the number of coffee mugs produced per day.

(a) Compute the increase in cost going from a production level of 25 mugs per day to 75 mugs per day.

(b) If daily fixed costs are $50, compute the average value of $C(x)$ over the interval [25, 75] and interpret.

38. Binky Inc. is a company that produces pacifiers. The research department at Binky Inc. has determined the following marginal cost function:

$$C'(x) = 0.06 + \frac{x}{300}$$

Here $C'(x)$ is in dollars per pacifier and x is the number of pacifiers produced per day.

(a) Compute the increase in cost going from a production level of 200 pacifiers per day to 300 pacifiers per day.

(b) If daily fixed costs are $150, compute the average value of $C(x)$ over the interval [200, 300] and interpret.

✓ 39. Geologists estimate that an oil field will produce oil at a rate given by $f(t) = 600e^{-0.1t}$ thousand barrels per month, t months into production.

(a) Graph f in the viewing window [0, 12] by [0, 600].

(b) Write a definite integral to estimate the total production for the first year of production.

(c) Evaluate the integral from part (b) to estimate the total production for the first year of operation. Round to the nearest whole number.

40. The rate of change of income, in thousands of dollars per year, for the oil field in Exercise 39, is projected to be $f(t) = 8400e^{-0.1t}$, where t is the number of months into production.

(a) Graph f in the viewing window [0, 12] by [0, 8400].

(b) Write a definite integral to estimate the total amount of money generated from this oil field for the first year of operation.

(c) Evaluate the integral from part (b) to estimate the total amount of money generated from this oil field for the first year of operation. Round to the nearest cent.

41. Smart Ones Car Wash generates income in the first 3 years of operation at a rate given by

$$f(t) = 2, \qquad 0 \le t \le 10$$

where t is the number of years in operation and $f(t)$ is in millions of dollars per year. Determine the total income produced in the first 3 years of operation.

42. The rate of change of income produced by a vending machine located in a college dorm is given by

$$f(t) = 2500e^{0.05t}, \qquad 0 \le t \le 10$$

where t is time in years since the installation of the vending machine and $f(t)$ is in dollars per year. Determine the total income generated by the vending machine during the first 3 years of operation.

43. The rate of change of income produced by a vending machine located in a busy airport is given by

$$f(t) = 6000e^{0.08t}, \qquad 0 \le t \le 10$$

where t is time in years since the installation of the vending machine and $f(t)$ is in dollars per year. Determine the total income generated by the vending machine during the first 3 years of operation.

44. Rich's Lawn Service has determined that his new landscaping company has a projected rate of change of income given by

$$f(t) = 30e^{0.15t}, \qquad 0 \le t \le 10$$

where t is time in years since starting the business and $f(t)$ is in thousands of dollars per year. Determine the total income generated by Rich's Lawn Service during the first 2 years of operation.

✓ 45. At the beginning of each month, Antonio deposits $250 into his 403-b plan, which earns 6.5% annual interest compounded continuously. Determine how much Antonio will have in his account after 25 years. Round to the nearest cent.

46. At the beginning of each month, Susan deposits $300 into her 403-b plan, which earns 7.25% annual interest compounded continuously. Determine how much Susan will have in her account after 30 years. Round to the nearest cent.

47. At the age of 22, Vicki opened an IRA. At the beginning of each year, she deposits $1800 into her IRA, which earns 8% annual interest compounded continuously. Determine how much Vicki has in her account in 43 years when she retires at age 65.

48. (Continuation of Exercise 47.) Determine how much of Vicki's final amount is interest.

49. At the age of 25, Jeff opened an IRA. At the beginning of each year, he deposits $2000 into his IRA, which earns 8% annual interest compounded continuously. Determine how much Jeff has in his account in 40 years when he retires at age 65.

50. (Continuation of Exercise 49.) Determine how much of Jeff's final amount is interest.

51. Comparing Exercises 47 and 49:

(a) How much did Vicki contribute to her IRA?

(b) How much did Jeff contribute to his IRA?

(c) By how much did Vicki's final amount surpass Jeff's final amount? Explain how Vicki's final amount could surpass Jeff's final amount if Jeff contributed more to his IRA than did Vicki.

52. On the birth of their new son Dylan, Bill and Lisa opened a college savings account. They deposit $50 at the beginning of each month into an account earning 5.5% annual interest compounded continuously. Determine how much is in the account in 18 years when Dylan starts college.

53. On the birth of their new son Randon, Donald and Melissa opened a college savings account. They deposit $75 at the beginning of each month into an account earning 5.25% annual interest compounded continuously. Determine how much is in the account in 18 years when Randon starts college.

54. On the birth of their new daughter Hannah, Steve and Felicia opened a college savings account. They deposit $100 at the beginning of each month into an account earning 5% annual interest compounded continuously. Determine how much is in the account in 18 years when Hannah starts college.

SECTION PROJECT

Maria has decided to open an IRA for retirement purposes. She has found an account that earns 7% annual interest compounded continuously. She has decided to invest $1800 each year into the IRA. She can invest this $1800 in one of two ways: either $1800 at the beginning of each year or $150 at the beginning of each month. Maria plans on investing for 30 years, at which time she will retire. To help Maria decide between the two investment options:

(a) Determine how much is in Maria's account in 30 years if she invests $1800 at the beginning of each year.

(b) Determine how much is in Maria's account in 30 years if she invests $150 at the beginning of each month.

(c) Based on parts (a) and (b), which plan would you advise Maria to take? Explain.

SECTION 7.2   AREA BETWEEN CURVES AND APPLICATIONS

In Chapter 6, as well as in Section 7.1, we saw that the area under a curve on a closed bounded interval can be determined by the definite integral. In this section we extend this concept to compute the area between two curves on a closed interval. We will also see how to determine the closed interval over which the area trapped between two curves can be calculated. The latter will require us to find points of intersection of two curves. We can do this algebraically or graphically utilizing the calculator's INTERSECT command.

Area between Curves on a Closed Interval [a, b]

We begin this section utilizing what was learned in Chapter 6, specifically, that the definite integral gives the area under a curve. Our first example reviews this important concept.

EXAMPLE 1 | **Representing Area by a Definite Integral**

(a) Write a definite integral to represent the shaded area in Figure 7.2.1.

(b) Write a definite integral to represent the shaded area in Figure 7.2.2.

(c) Use the results of parts (a) and (b) to write a definite integral to represent the shaded area in Figure 7.2.3.

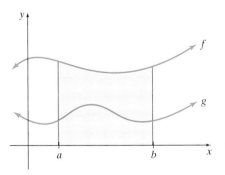

Figure 7.2.1

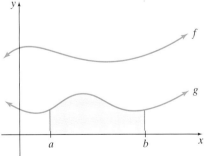

Figure 7.2.2

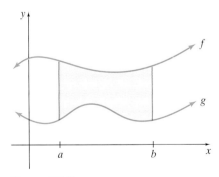

Figure 7.2.3

SOLUTION

(a) Since the shaded area is below the graph of f on the interval $[a, b]$, the definite integral that represents the shaded area is $\int_a^b f(x)\, dx$.

(b) Since the shaded area is below the graph of g on the interval $[a, b]$, the definite integral that represents the shaded area is $\int_a^b g(x)\,dx$.

(c) Here it appears that the shaded area is simply

Area under the graph of f $-$ area under the graph of g

$$\int_a^b f(x)\,dx - \int_a^b g(x)\,dx$$

We can now use a property from Section 6.1 and write these integrals as

$$\int_a^b [f(x) - g(x)]\,dx \qquad \blacksquare$$

In Example 1c, what we really found was the area between the graphs of f and g on the interval $[a, b]$. In fact, the process outlined in Example 1 can be summarized as follows.

Area between Two Curves

On a closed interval $[a, b]$, the area between two continuous functions f and g is given by:

1. If $f(x) \geq g(x)$, then the area between the graphs of f and g is $\int_a^b [f(x) - g(x)]\,dx$. See Figure 7.2.4a.

2. If $g(x) \geq f(x)$, then the area between the graphs of f and g is $\int_a^b [g(x) - f(x)]\,dx$. See Figure 7.2.4b.

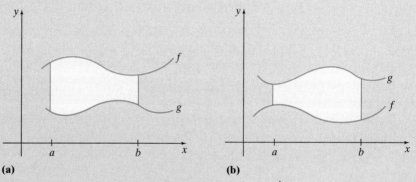

(a)   **(b)**

Figure 7.2.4   **(a)** Area of shaded region is given by $\int_a^b [f(x) - g(x)]\,dx$.
(b) Area of shaded region is given by $\int_a^b [g(x) - f(x)]\,dx$.

NOTE: An easy way to handle *both* cases listed is to remember the following: To determine the area between two curves on $[a, b]$, integrate

$$\int_a^b [(\text{top curve} - \text{bottom curve})]\,dx$$

EXAMPLE 2 | **Determining the Area between Two Curves**

Determine the area between the x-axis and $f(x) = x^2 - x - 6$ on the interval $[1, 3]$.

SOLUTION

We first graph the function and shade the area as an aid in setting up the correct integral. See Figure 7.2.5.

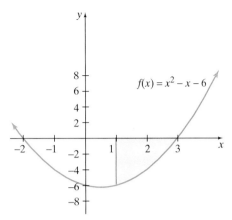

Figure 7.2.5

If we recall that the x-axis can be represented by the function $y = 0$, we can apply the techniques of this section. Thus,

$$\int_1^3 [\text{top curve} - \text{bottom curve}] \, dx = \int_1^3 [0 - (x^2 - x - 6) \, dx$$

$$= \int_1^3 (-x^2 + x + 6) \, dx$$

$$= \left(-\frac{1}{3}x^3 + \frac{1}{2}x^2 + 6x\right)\bigg|_1^3$$

$$= \left[\left(-\frac{1}{3}(3)^3 + \frac{1}{2}(3)^2 + 6(3)\right) - \left(-\frac{1}{3}(1)^3 + \frac{1}{2}(1)^2 + 6(1)\right)\right]$$

$$= 7\frac{1}{3} \text{ square units}$$

EXAMPLE 3 | **Determining Area between Two Curves**

Determine the area between $f(x) = -x^2 + 4$ and $g(x) = 2x + 7$ on the interval $[-3, 2]$.

SOLUTION

As in Example 2, we begin with a graph of f and g and shade the area. See Figure 7.2.6. Since $g(x) \geq f(x)$ for all x on the interval $[-3, 2]$, the graph of g is the top curve and the graph of f is the bottom curve.

$$\int_{-3}^2 [\text{top curve} - \text{bottom curve}] \, dx = \int_{-3}^2 [(2x + 7) - (-x^2 + 4)] \, dx$$

$$= \int_{-3}^2 (x^2 + 2x + 3) \, dx$$

$$= \left(\frac{1}{3}x^3 + x^2 + 3x\right)\bigg|_{-3}^2$$

$$= \left(\frac{8}{3} + 4 + 6\right) - (-9 + 9 - 9)$$

$$= \frac{65}{3} \text{ square units}$$

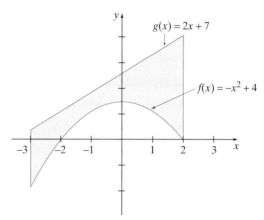

Figure 7.2.6

Now work Exercise 15.

In Example 4, we extend the concept set forth in Chapter 6 that integrating a rate of change function gives an accumulation.

EXAMPLE 4 | **Determining Plant Growth**

From past records, a botanist knows that a certain species of tree grows at a rate of $\frac{3}{2}x^{-1/2}$ feet per year, where x is the age of the tree in years. However, if a special nitrogen-rich nutrient is given to the tree, it grows at a rate of $2x^{-1/2}$ feet per year. On the interval [1, 4], how many more feet in growth would result from the special nitrogen-rich nutrient?

SOLUTION

In Figure 7.2.7 we have graphed the two curves representing the two different growth rates. To determine how many more feet of growth results from the special nitrogen-rich nutrient, we need to integrate the difference of the rates from $x = 1$ to $x = 4$. In other words, we need to determine the area between the curves.

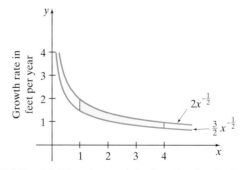

Figure 7.2.7   Additional growth is given by the shaded region.

The area between the curves is found by

$$\int_1^4 \left(2x^{-1/2} - \frac{3}{2}x^{-1/2} \right) dx = \int_1^4 \frac{1}{2} x^{-1/2}\, dx$$

$$= x^{1/2}\Big|_1^4 = (4)^{1/2} - (1)^{1/2} = 1$$

So on the interval [1, 4], that is, from year 1 to year 4, the nitrogen-rich nutrient would result in an additional 1 foot of growth. ◆

✓**CHECKPOINT 2**

Now work Exercise 49.

Area Bounded by Two Curves

So far in this section we have presented examples in which the interval of integration has been given. We now consider how to determine the area of a region that is trapped or enclosed by two curves and when no interval is given. In these situations, where the curves completely enclose an area, we simply determine where the curves intersect.

EXAMPLE 5 | **Determining the Area Enclosed by Two Curves**

Determine the area bounded by the curves $y = x^2 + x - 5$ and $y = 2x + 1$.

SOLUTION

As always, our first step is to graph the curves so that we can see the region whose area we want. See Figure 7.2.8. For this region, the top curve is $y = 2x + 1$ and the bottom curve is $y = x^2 + x - 5$.

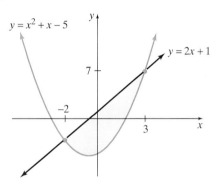

Figure 7.2.8

  To determine the limits of integration, we must find the points where the curves intersect. Algebraically, this is handled by setting the two equations equal to each other and solving for x.

$$2x + 1 = x^2 + x - 5$$
$$x^2 - x - 6 = 0$$
$$(x - 3)(x + 2) = 0$$
$$x = 3 \quad \text{or} \quad x = -2$$

So the curves intersect at x-values of $x = 3$ and $x = -2$. The area enclosed by the two curves is then found by integrating

Interactive Activity

 Graph $y_1 = 2x + 1$ and $y_2 = x^2 + x - 5$, and use the INTERSECT command to verify the points of intersection found algebraically in Example 5.

$$\int_{-2}^{3} [(2x + 1) - (x^2 + x - 5)]\, dx = \int_{-2}^{3} (-x^2 + x + 6)\, dx$$

$$= \left(-\frac{1}{3}x^3 + \frac{1}{2}x^2 + 6x\right)\Bigg|_{-2}^{3} = 13.5 - \left(-7\frac{1}{3}\right)$$

$$= 20\frac{5}{6}, \quad \text{or approximately 20.83 square units.} ◆$$

In our next example, we revisit marginal revenue and marginal cost.

EXAMPLE 6 | **Computing a Total Profit**

The FrigAir Company knows that its marginal cost to produce x refrigerators is given by $C'(x) = 4x + 23$ and the marginal revenue is given by $R'(x) = -0.6x + 460$, where both marginals are in dollars per unit. Compute the total profit from $x = 0$ to $x = 95$, the production level where profit is maximized.

SOLUTION

Figure 7.2.9 has a graph of R' and C'. In Chapter 6 we saw that to determine total profit we needed to integrate P', the marginal profit function. But, since $P(x) = R(x) - C(x)$, we have $P'(x) = R'(x) - C'(x)$. Hence, to find total profit, we need to integrate

$$\int_0^{95} P'(x)\, dx = \int_0^{95} [R'(x) - C'(x)]\, dx$$

From Figure 7.2.9 we notice that the total profit is simply the area between the graphs of R' and C'. Continuing, we have

$$\int_0^{95} [R'(x) - C'(x)]\, dx = \int_0^{95} [(-0.6x + 460) - (4x + 23)]\, dx$$

$$= \int_0^{95} (-4.6x + 437)\, dx$$

$$= \left(\frac{-4.6}{2} x^2 + 437x \right) \Big|_0^{95}$$

$$= \left[\frac{-4.6}{2}(95)^2 + 437(95) \right] - \left[\frac{-4.6}{2}(0)^2 + 437(0) \right]$$

$$= 20{,}757.5$$

So the total profit from a production level of $x = 0$ to $x = 95$ is \$20,757.50.

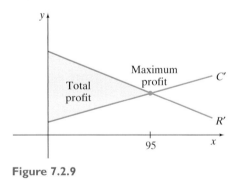

Figure 7.2.9

NOTE: In Example 6 we stated that profit is maximized at $x = 95$. As shown in Figure 7.2.9, that is where $R'(x) = C'(x)$. In general, **profit is maximized when $R'(x) = C'(x)$.**

✓ CHECKPOINT 3

Now work Exercise 53.

Area between Curves That Cross on [a, b]

The final scenario we consider in this section is what happens if two curves cross each other on a given interval. When this occurs, what was a top curve may become a bottom curve, as we illustrate in Example 7.

EXAMPLE 7 | **Determining the Area between Curves That Cross on an Interval**

Determine the area between the curves $y = x + 1$ and $y = \frac{2}{x}$ on the interval $\left[\frac{1}{2}, 3\right]$.

SOLUTION

The first thing we do is graph the two curves on the stated interval to see which is the top curve and which is the bottom curve. We will also be able to see if the two curves cross each other in the interval. Figure 7.2.10 shows the graph, and we have shaded the area that we want to determine.

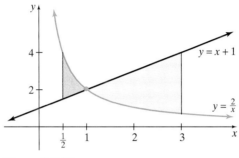

Figure 7.2.10

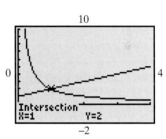

Figure 7.2.11

To the left of the point of intersection, $y = \frac{2}{x}$ is the top curve, while to the right it is the bottom curve. At this time we need to determine where the curves cross, or intersect if you prefer. Employing the INTERSECT command, we determine the point of intersection. See Figure 7.2.11.

Since the curves cross each other at $x = 1$, we integrate over the intervals $\left[\frac{1}{2}, 1\right]$ and $[1, 3]$ separately and then add the results to determine the area between the curves. This gives

$$\int_{1/2}^{1} \left[\frac{2}{x} - (x + 1)\right] dx + \int_{1}^{3} \left(x + 1 - \frac{2}{x}\right) dx$$

$$= \int_{1/2}^{1} \left(\frac{2}{x} - x - 1\right) dx + \int_{1}^{3} \left(x + 1 - \frac{2}{x}\right) dx$$

$$= \left(2 \ln|x| - \frac{1}{2}x^2 - x\right)\Bigg|_{1/2}^{1} + \left(\frac{1}{2}x^2 + x - 2 \ln|x|\right)\Bigg|_{1}^{3} \qquad \text{Recall, } \int \frac{1}{x} dx = \ln|x| + c$$

$$= \left[\left(2 \ln 1 - \frac{1}{2}(1)^2 - 1\right) - \left(2 \ln \frac{1}{2} - \frac{1}{2}\left(\frac{1}{2}\right)^2 - \frac{1}{2}\right)\right] + \left[\left(\frac{1}{2}(3)^2 + 3 - 2 \ln 3\right) - \left(\frac{1}{2}(1)^2 + 1 - 2 \ln 1\right)\right]$$

$$\approx 4.314$$

So the area between the curves $y = x + 1$ and $y = \frac{2}{x}$ on the interval $\left[\frac{1}{2}, 3\right]$ is about 4.31 square units. ∎

✓ CHECKPOINT 4

Now work Exercise 25.

Interactive Activity

Algebraically determine the point of intersection for the functions given in Example 7.

Our final example in this section analyzes U.S. trade with Mexico. In this example, we will determine when the United States had a **trade deficit**. A trade deficit occurs when imports from Mexico exceed exports to Mexico. We will also determine when the United States had a **trade surplus**. A trade surplus occurs when exports to Mexico exceed imports from Mexico. Finally, we will determine the U.S. **trade balance** with Mexico.

EXAMPLE 8

Determining a Trade Balance

The following represent the U.S. imports and exports with Mexico from 1987 to 1994.

$$\text{Imports:} \quad y = 0.16x^3 - 1.82x^2 + 8.9x + 12.51, \quad 1 \le x \le 8$$
$$\text{Exports:} \quad y = 0.06x^3 - 0.74x^2 + 7.22x + 8.21, \quad 1 \le x \le 8$$

Here x represents the number of years since 1986 and y represents the value of imports or exports in billions of dollars.

(a) Determine for what years the United States had a trade deficit with Mexico and determine for what years the United States had a trade surplus with Mexico.

(b) For the years that the United States had a trade deficit with Mexico, write and evaluate an integral to determine the total accumulated trade deficit.

(c) For the years that the United States had a trade surplus with Mexico, write and evaluate an integral to determine the total accumulated trade surplus.

(d) From 1987 to 1994, determine the U.S. trade balance with Mexico.

SOLUTION

(a) Figure 7.2.12 shows a graph of the two models in the viewing window [1, 9] by [0, 60]. It appears that there are two values for x where the curves intersect and cross each other. We determine these points of intersection by using the INTERSECT command on our calculator. See Figures 7.2.13 and 7.2.14.

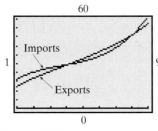

Figure 7.2.12

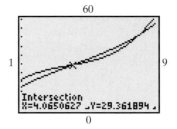

Figure 7.2.13

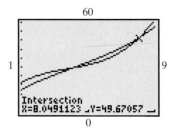

Figure 7.2.14

We round the x-value of the left-hand intersection point to 4 and the x-value of the right-hand intersection point to 8. Note that on the interval [1, 4] the imports curve is above the exports curve. This means that the United States had a trade deficit with Mexico from 1987 to 1990. On the interval [4, 8], the exports curve is above the imports curve. This means that the United States had a trade surplus with Mexico from 1990 to 1994.

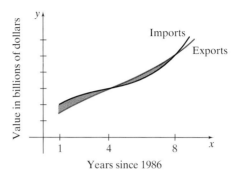

Figure 7.2.15

(b) To determine the total accumulated trade deficit, we need to compute the red shaded area between the curves shown in Figure 7.2.15. The total accumulated trade deficit from 1987 to 1990 is found by integrating

$$\int_1^4 (\text{imports} - \text{exports})\, dx$$

$$= \int_1^4 [(0.16x^3 - 1.82x^2 + 8.9x + 12.51) - (0.06x^3 - 0.74x^2 + 7.22x + 8.21)]\, dx$$

$$= \int_1^4 (0.1x^3 - 1.08x^2 + 1.68x + 4.3)\, dx$$

$$= \left(\frac{0.1}{4}x^4 - \frac{1.08}{3}x^3 + \frac{1.68}{2}x^2 + 4.3x\right)\Big|_1^4$$

$$= 14 - 4.805 = 9.195$$

So the U.S. total accumulated trade deficit with Mexico from 1987 to 1990 was approximately 9.2 billion dollars.

(c) To determine the total accumulated trade surplus from 1990 to 1994, we need to compute the black shaded area between the two curves shown in Figure 7.2.15. The total accumulated trade surplus is found by integrating

$$\int_4^8 (\text{exports} - \text{imports})\, dx$$

$$= \int_4^8 [(0.06x^3 - 0.74x^2 + 7.22x + 8.21) - (0.16x^3 - 1.82x^2 + 8.9x + 12.51)]\, dx$$

$$= 7.76 \text{ (The details of this computation are left as an Exercise).}$$

So the U.S. total accumulated trade surplus with Mexico from 1990 to 1994 was approximately 7.76 billion dollars.

(d) The result from part (b) gives the accumulated trade deficit, while the result from part (c) gives the accumulated trade surplus. Notice that the result from part (b) is larger than the result from part (c). This implies that from 1987 to 1994 the U.S. trade balance with Mexico is a negative number, indicating an overall deficit. This can be found by

$$\int_4^8 (\text{exports} - \text{imports})\, dx - \int_1^4 (\text{imports} - \text{exports})\, dx = 7.76 - 9.2$$

$$= -1.44$$

So from 1987 to 1994 the United States had an accumulated trade deficit with Mexico of 1.44 billion dollars.

Interactive Activity

Determine the Mexican trade balance with the United States from 1987 to 1994.

SUMMARY

In this section we saw how to compute the area between two curves and how to apply this concept. To determine the area between two curves, we offer the following guidelines:

1. Graph the two curves to see the region whose area we want.
2. If an interval is not given, or if the curves cross each other on a given interval, determine the appropriate limits of integration by determining where the curves intersect each other. This can be done algebraically or graphically using the INTERSECT command on your calculator.
3. Integrate top curve minus bottom curve on each interval.

Area between Two Curves

On a closed interval $[a, b]$, the area between the graphs of two continuous functions f and g is given by:

- If $f(x) \geq g(x)$, the area is found by integrating $\int_a^b [f(x) - g(x)]\, dx$.
- If $g(x) \geq f(x)$, the area is found by integrating $\int_a^b [g(x) - f(x)]\, dx$.

SECTION 7.2 EXERCISES

In Exercises 1–8, write a definite integral to represent the area of the shaded region.

1.

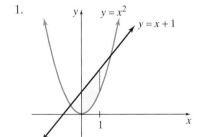

2.

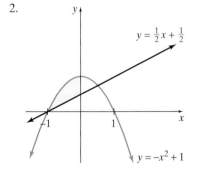

3.

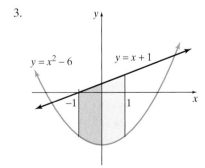

4.

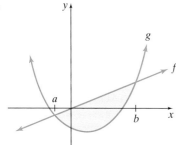

5.

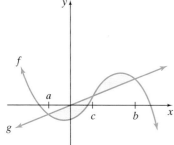

6.

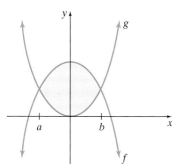

7.

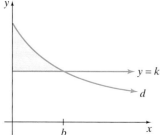

8.

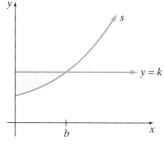

In Exercises 9–26, determine the area of the region bounded by the given conditions.

9. $f(x) = 3x + 1$ and the x-axis on $[0, 4]$

10. $f(x) = 2x + 3$ and the x-axis on $[1, 3]$

11. $f(x) = 2x^2 + x + 1$ and the x-axis on $[-2, 1]$

12. $f(x) = x^2 + 2x + 3$ and the x-axis on $[-3, 2]$

13. $f(x) = -x^2 + x + 2$ and the x-axis on $[1, 4]$

14. $f(x) = -2x^2 - x + 1$ and the x-axis on $[-2, 3]$

✓ 15. $f(x) = 8 - x^2$ and $y = 4$ on $[-2, 2]$

16. $f(x) = x^2 - 2$ and $y = 2$ on $[-2, 2]$

17. $f(x) = -x + 4$ and $g(x) = -x^2 + 1$ on $[-2, 2]$

18. $f(x) = 2x + 3$ and $g(x) = -x^2 + 1$ on $[-1, 2]$

19. $f(x) = x - 3$ and $g(x) = x^2 + x + 1$ on $[-2, 1]$

20. $f(x) = \frac{1}{2}x - 1$ and $g(x) = x^2 + 2x - 1$ on $[1, 5]$

21. $f(x) = 16 - \sqrt{x}$ and $g(x) = 8 - \sqrt[3]{x}$ on $[1, 5]$

22. $f(x) = x^2$ and $g(x) = \sqrt{x}$ on $[0, 1]$

23. $f(x) = 5e^{-0.15x}$ and $g(x) = 10e^{-0.08x}$ on $[0, 5]$

24. $f(x) = 7.2e^{-0.07x}$ and $g(x) = 2.34e^{-0.11x}$ on $[1, 4]$

✓ 25. $f(x) = \frac{3}{x}$ and $g(x) = \frac{4}{x^2}$ on $[2, 8]$

26. $f(x) = \frac{3}{x}$ and $g(x) = \frac{4}{x^2}$ on $[1, 8]$

In Exercises 27–34, determine the area bounded by the given curves.

27. $f(x) = 4 - x^2$ and $g(x) = -x + 2$

28. $f(x) = -x - 3$ and $g(x) = 9 - x^2$

29. $f(x) = -2x^2 - 5x + 3$ and $g(x) = 2x^2 + 3x - 2$

30. $f(x) = x^2 - 6$ and $g(x) = 12 - x^2$

31. $f(x) = 3x^2 - 2x$ and $g(x) = x^3$

32. $f(x) = x^2$ and $g(x) = x^3$

33. $f(x) = x^3$ and $g(x) = x$

34. $f(x) = 2x^3$ and $g(x) = -x^3 + x^2 + 2x$

APPLICATIONS

🌐 35. The average wage, $W_c(t)$, in dollars per hour, for individuals in construction in the United States can be approximated by

$$W_c(t) = 0.28t + 13.4, \qquad 1 \le t \le 6$$

where t is the number of years since 1989. The average wage, $W_m(t)$, in dollars per hour, for individuals in manufacturing in the United States can be approximated by

$$W_m(t) = 0.32t + 10.51, \qquad 1 \le t \le 6$$

where t is the number of years since 1989. During the period 1990 through 1995, how much more money has a worker in construction earned than a worker in manufacturing? Assume that each worked 40 hours per week and earned a paycheck 52 weeks each year.

36. The average wage, $W_{mi}(t)$, in dollars per hour, for individuals in mining in the United States can be approximated by

$$W_{mi}(t) = 0.3t + 13.5, \qquad 1 \le t \le 6$$

where t is the number of years since 1989. The average wage, $W_{ma}(t)$, in dollars per hour, for an individual in manufacturing in the United States can be approximated by

$$W_{ma}(t) = 0.32t + 10.51, \qquad 1 \le t \le 6$$

where t is the number of years since 1989. During the period 1990 through 1995, how much more money has a worker in mining earned than a worker in manufacturing? Assume that each worked 40 hours per week and earned a paycheck 52 weeks each year.

37. The receipts, in billions of dollars, in the National Income and Products Accounts for State and Local Governments can be modeled by

$$f(x) = 1.13x^2 + 23.51x + 337.88, \qquad 1 \le x \le 16$$

where x is the number of years since 1979. The expenditures, in billions of dollars, in the National Income and Products Accounts for State and Local Governments can be modeled by

$$g(x) = 1.28x^2 + 18.77x + 283.09, \qquad 1 \le x \le 16$$

where x is the number of years since 1979.

(a) Graph f and g in the viewing window [0, 17] by [300, 1000].

(b) Write an integral to represent the area between the two curves on the interval [1, 16].

(c) Evaluate the integral from part (b) and interpret. (When receipts exceed expenditures, a surplus exists, whereas when expenditures exceed receipts, a deficit exists.)

38. The size of a certain bacteria culture grows at a rate of $4t^{5/2}$ milligrams per minute. If a special nutrient is introduced into the culture, it grows at a rate of $4t^{18/5}$ milligrams per minute.

(a) Set up an integral to determine the area between the two curves on [0, 1].

(b) Evaluate the integral from part (a) and interpret.

39. (Continuation of Exercise 38.) The size of a certain bacteria culture grows at a rate of $4t^{5/2}$ milligrams per minute. If a special nutrient is introduced into the culture, it grows at a rate of $4t^{18/5}$ milligrams per minute.

(a) Set up an integral to determine the area between the two curves on [1, 2].

(b) Evaluate the integral from part (a) and interpret.

(c) On the interval [0, 2], how many more bacteria result from introducing the special nutrient?

40. The imports and exports for the United States with Japan can be modeled by

Imports: $\quad I(x) = -0.61x^3 + 8.31x^2 - 27.42x + 114.37$

Exports: $\quad E(x) = -0.12x^4 + 2.28x^3 - 13.93x^2$
$$+ 32.78x + 23.22$$

where x is the number of years since 1988 and imports and exports are in billions of dollars. Determine the area between the two curves on the interval [1, 8] and interpret.

41. The imports and exports for the United States with Canada can be modeled by

Imports: $\quad I(x) = -0.24x^3 + 4.6x^2 - 14.49x + 99.96$

Exports: $\quad E(x) = -0.17x^3 + 3.14x^2 - 8.14x + 85.58$

where x is the number of years since 1988 and imports and exports are in billions of dollars. Determine the area between the two curves on the interval [1, 8] and interpret.

42. Verify that the total accumulated trade surplus found in Example 8 is 7.76 billion dollars.

43. The rate at which the number of public schools in the United States with interactive videodisk units has grown according to $\dfrac{15.1}{x}$ thousand schools per year, where x is the number of years since 1991. From 1992 through 1996, how many schools received these interactive videodisk units?

44. The rate at which the number of public schools in the United States with computer networks has grown according to $\dfrac{17.58}{x}$ thousand schools per year, where x is the number of years since 1991. From 1992 through 1996, how many schools received computer networks?

45. Double D cola is planning to release a new Double Dose Caffeine Cola on the market. In the past, Double D cola ran a 20-week ad campaign for any new product that it released on the market. For example, when Double D ran a 20-week ad campaign for its new caffeine-free cola, weekly sales $s(t)$ in thousands of dollars per week t weeks after the campaign began were modeled by

$$s(t) = -0.51t^2 + 20.20t + 49.49$$

An agent for a TV personality has approached Double D cola and claims that, if Double D uses his client in the 20-week ad campaign, the weekly sales $w(t)$ in thousands of dollars per week t weeks after the campaign begins will be approximated by

$$w(t) = -0.71t^2 + 28.14t + 57.22$$

(a) On the interval [0, 20], determine the area between s and w and interpret.

(b) If the endorsement of the TV personality costs Double D cola $1,000,000, will this endorsement be paid off in the first 20 weeks of the campaign? Explain.

46. A market analyst for Chocolate Time estimates that with no promotion the annual sales of its candy bar, Chocoloco Dream, can be modeled by

$$s_1(t) = 0.76t + 8.42$$

million dollars per year, t years from now. This same analyst estimates that, with a modest promotional campaign, annual sales can be modeled by

$$s_2(t) = 11.4e^{0.05t}$$

During the first 5 years, what will the total increase in sales be in response to the promotion?

47. A market analyst for Chocolate Time estimates that with no promotion the annual sales of its candy bar, Chocoholic Dream, can be modeled by

$$s_1(t) = 0.76t + 8.42$$

million dollars per year, t years from now. This same analyst estimates that, with a modest promotional campaign, annual sales can be modeled by

$$s_2(t) = 12.15e^{0.04t}$$

During the first 5 years, what will the total increase in sales be in response to the promotion?

48. A psychologist has determined that people can memorize digits at the rate of $4.4e^{-0.2x}$ digits per minute, where x is the time in minutes that the individuals have been memorizing. She discovered that if a scent of peppermint oil is present while people are attempting to memorize and when asked to recall the digits memorized, the rate of learning appeared to be $5.3e^{-0.15x}$ digits per minute.

(a) Graph both rates in the viewing window $[0, 5]$ by $[0, 6]$.

(b) On the interval $[1, 5]$, determine the area between the two curves and interpret.

✓ **49.** A psychologist has determined that the rate at which a child learns to recognize new words is given by $25t + 50\sqrt{t}$ words per year. If the child listens to special vocabulary-building tapes, the rate at which the child recognizes new words is given by $35t + 70\sqrt{t}$. In both cases, t represents the child's age in years.

(a) Graph both rates in the viewing window $[0, 10]$ by $[0, 600]$.

(b) On the interval $[4, 8]$, determine the area between the two curves and interpret.

50. Dynatronics has determined that its costs, in millions of dollars, have been increasing, mainly due to inflation, at a rate given by $1.25e^{0.12x}$, where x is time measured in years since its new telephone equipment began being produced. At the beginning of the third year of production, a breakthrough in the production process resulted in costs increasing at a rate given by $0.85\sqrt{x}$. Determine the total savings in costs on the interval $[3, 6]$ due to the breakthrough.

51. Quality Chips has determined that its costs, in millions of dollars, have been increasing, mainly due to inflation, at a rate given by $2.67e^{0.09x}$, where x is time measured in years since its snack food started in production. At the beginning of the third year, a breakthrough in the production process resulted in costs increasing at a rate given by $0.68x^{2/3}$. Determine the total savings in costs on the interval $[3, 8]$ due to the breakthrough.

52. See It Inc. determines that the marginal revenue to produce x graphing calculators is given by $R'(x) = -0.3x + 100$, while the marginal cost is given by $C'(x) = 0.5x + \dfrac{750}{x+1}$, where both marginals are in dollars per unit. Determine the total profit from $x = 10$ to $x = 100$.

✓ **53.** Monkey Works determines that the marginal revenue to produce x squeegees is given by $R'(x) = -0.2x + 40$, while the marginal cost is given by $C'(x) = \dfrac{300}{x+1}$, where both marginals are in dollars per unit. Determine the total profit from $x = 10$ to $x = 100$.

54. Music Time determines that the marginal revenue to produce x integrated stereo amplifiers is given by $R'(x) = 216 - 0.24x^2$, while marginal cost is given by $C'(x) = 20$, where both marginals are in dollars per unit. Determine the total profit from $x = 10$ to $x = 30$.

55. Java Buddy determines that the marginal revenue to produce x coffee mugs is given by $R'(x) = 5 - 0.05x$, while marginal cost is given by $C'(x) = 1 + \dfrac{x}{40}$, where both marginals are in dollars per unit. Determine the total profit from $x = 5$ to $x = 40$.

56. A factory worker can assemble units at the rate of $-3x^2 + 13x + 11$ units per hour during the first 4 hours on the job, where x is the number of hours on the job. Determine how many units can be assembled by this worker during the first 4 hours.

57. (Continuation of Exercise 56.) If the factory worker in Exercise 56 has breakfast along with two cups of coffee, then she can assemble units at the rate of $-2x^2 + 11x + 13$ units per hour during the first 4 hours on the job, where x is the number of hours on the job.

(a) In this case, how many units can she assemble during the first 4 hours?

(b) How many more units can she assemble in the first 4 hours if she has breakfast and 2 cups of coffee?

58. A botanist knows from past records that a certain species of oak tree grows at a rate of $\dfrac{4x^2 + 16x + 9}{2x + 4}$ feet per year, where x is the age of the tree in years. Determine how much growth takes place on the interval $[3, 8]$.

59. (Continuation of Exercise 58.) The botanist in Exercise 58 knows that restricting the amount of light that the oak tree receives inhibits its growth. When restricting the light, the oak tree grows at a rate of $\dfrac{2x^2 + 8x + 9}{2x + 4}$ feet per year, where x is the age of the tree in years. On the interval $[3, 8]$, how many fewer feet in growth will result from restricting the amount of light that the tree receives?

SECTION PROJECT

(a) The average net income for a General and Family practice physician in the United States is modeled by

$$f(x) = -0.09x^3 + 1.52x^2 - 1.54x + 77.95, \qquad 1 \le x \le 10$$

thousands of dollars per year, where x is the number of years since 1984. Determine the area between the graph of f and the x-axis on the interval $[1, 9]$ and interpret.

(b) The average net income for a surgeon in the United States is modeled by

$$s(x) = -0.04x^3 - 0.24x^2 + 18.97x + 136.5, \qquad 1 \le x \le 10$$

thousands of dollars per year, where x is the number of years since 1984. Determine the area between the graph of s and the x-axis on the interval $[1, 9]$ and interpret.

(c) The average net income for a pediatrician in the United States is modeled by

$$k(x) = -0.19x^3 + 3x^2 - 5.9x + 79.52, \qquad 1 \le x \le 10$$

thousands of dollars per year, where x is the number of years since 1984. Determine the area between the graph of k and the x-axis on the interval $[1, 9]$ and interpret.

Assume from 1985 to 1993 that Dylan is a family practice physician earning the average net income each year, Lisa is a surgeon earning the average net income each year, and Austin is a pediatrician earning the average net income each year.

(d) From 1985 to 1993, how much more money does Lisa earn than Dylan?

(e) From 1985 to 1993, how much more money does Lisa earn than Austin?

(f) From 1985 to 1993, how much more money does Austin earn than Dylan?

SECTION 7.3   ECONOMIC APPLICATIONS OF AREA BETWEEN TWO CURVES

In Section 7.2 we introduced how to determine the area between two curves and considered many applications of this concept. In this section we focus mainly on two specific applications of area between two curves: **consumers' and producers' surplus** and **income distributions.** Our discussion of consumers' and producers' surplus will also include a look at **market price, total consumer expenditure, revenue, market demand,** and **equilibrium point.** Our discussion of income distribution will include a look at **Lorenz curves** and the **Gini Index.**

Price–Demand Function Revisited

In Chapter 1, we were first introduced to what is known as the price–demand function.

From Your Toolbox   **PRICE–DEMAND FUNCTION**

    $p(x)$ gives the price at which people buy exactly x units of a product.

A common way to represent the price–demand function in economics is with the letter d. Quite simply, $d(x)$ yields the price at which exactly x units of a product will be sold.

NOTE: This section contains applications that are also taught in many economics courses. Since many of you will see the letter d for the price–demand function in your economics course, we will use this notation in this section. In this section, we will use d to represent the price–demand function, which we will simply call the *demand function.*

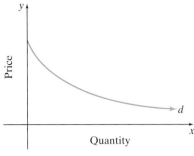

Figure 7.3.1

As shown in Figure 7.3.1, the demand function is a decreasing function. Intuitively, this seems reasonable. As the price decreases, more people will buy the product. Economists refer to the curve in Figure 7.3.1 simply as a *demand curve*. This is because it expresses the price per unit as a function of the quantity, x, in demand.

EXAMPLE 1 | **Evaluating a Demand Function**

Skinner Bikes, a bicycle retailer, has determined that a demand function for a certain brand of mountain bike is given by $d(x) = -0.3x + 330$, where x is the quantity demanded each month and $d(x)$ is the price per bicycle in dollars. Compute $d(200)$, $d(250)$, and $d(300)$ and interpret each.

SOLUTION

Since $d(x) = -0.3x + 330$, we have

$$d(200) = -0.3(200) + 330 = \$270 \text{ per bicycle}$$
$$d(250) = -0.3(250) + 330 = \$255 \text{ per bicycle}$$
$$d(300) = -0.3(300) + 330 = \$240 \text{ per bicycle}$$

The interpretation of these results are, respectively,

200 bicycles will be sold each month at a price of $270 per bicycle.

250 bicycles will be sold each month at a price of $255 per bicycle.

300 bicycles will be sold each month at a price of $240 per bicycle.

Consumers' Surplus

For the sake of argument, let's assume that you really like the bicycle discussed in Example 1 and were willing to pay $275 for it. If it currently costs, say, $200, then in a way you "saved" $75; the $275 you were willing to pay minus the $200 price. Now consider *any* price above $200 that some consumer is willing to pay minus the actual price of $200. The sum of these savings over all possible prices above $200 is known as the *consumers' surplus* for this bicycle. In general, consumers' surplus measures the benefit that consumers get from an economy where competition keeps prices down.

The actual price at which any item is sold, called the **market price,** is influenced by many factors. However, when a market price is determined, we can compute the quantity that consumers demand using the demand function. Returning to our bicycle retailer in Example 1, we note that at a price of $255 per bicycle 250 bicycles are demanded each month.

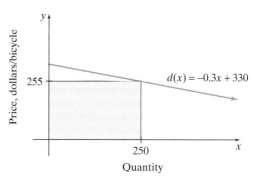

Figure 7.3.2 Actual consumer expenditure.

In Figure 7.3.2 is a shaded rectangle whose area represents what economists call the *actual consumer expenditure*. In general, the actual consumer expenditure is the amount of money that consumers spend for x bicycles at price $d(x)$ dollars per bicycle. In this case, the actual consumer expenditure is computed to be

$$(\$255)(250) = \$63{,}750$$

The area of the rectangle in Figure 7.3.2 is the **actual consumer expenditure,** while the area below the demand curve $d(x) = -0.3x + 330$ and above the rectangle gives a *total savings* that consumers realize by buying at $255 per bicycle. See Figure 7.3.3. This total savings is what is known as **consumers' surplus.**

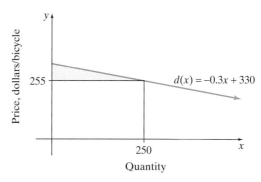

Figure 7.3.3 Shaded area represents consumers' surplus.

Consumers' Surplus

The **consumers' surplus** is the total amount saved by consumers who are willing to pay more than market price, d_{mp}, for a product, yet are able to purchase the product for d_{mp}.

EXAMPLE 2 | **Using a Definite Integral to Determine Consumers' Surplus**

(a) Set up a definite integral to determine the area of the shaded region in Figure 7.3.3, which represents the consumers' surplus for the bicycle from Example 1.
(b) Evaluate the integral from part (a) to determine the area of the shaded region, as well as determining the consumers' surplus.

Interactive Activity

 There is another way to determine the area representing consumers' surplus in Example 2. We can take the area under the graph of d from $x = 0$ to $x = 250$ and subtract the area of the rectangle that represents actual consumer expenditure. This corresponds to integrating $\int_0^{250}(-0.3x + 330)\,dx$ and subtracting \$63,750 (the actual consumer expenditure). Perform this integration and subtraction to verify the consumers' surplus determined in Example 2.

SOLUTION

(a) Here we need to determine the area between two curves. We have

top curve is $d(x) = -0.3x + 330$ and bottom curve is $y = 255$

From our work in Section 7.2, we represent the area of the shaded region as

$$\int_0^{250} [(-0.3x + 330) - 255]\,dx = \int_0^{250} (-0.3x + 75)\,dx$$

(b) We evaluate the integral from part (a) to be

$$\int_0^{250} (-0.3x + 75)\,dx = \left(\frac{-0.3}{2}x^2 + 75x\right)\Big|_0^{250} = \$9375$$

So the consumers' surplus is \$9375 each month. ◼

Example 2 illustrates how we can determine consumers' surplus for any situation.

Computing Consumers' Surplus

For a demand function d and a point (x_m, d_{mp}) on the graph of d, where x_m is called the **market demand** and d_{mp} is called the **market price,** we determine the consumers' surplus by integrating

$$\int_0^{x_m} [d(x) - d_{mp}]\,dx$$

EXAMPLE 3 | **Computing Consumers' Surplus**

Assume that the market price of the bicycle in Example 1 is \$210 per bicycle. Determine the consumers' surplus.

SOLUTION

Recall that $d(x) = -0.3x + 330$; its graph is shown in Figure 7.3.4. We know that the market price is \$210, but we do not have the market demand. Since we need the market demand in our definite integral (it is the upper limit of integration), we must solve

$$210 = -0.3x + 330$$
$$-120 = -0.3x$$
$$400 = x$$

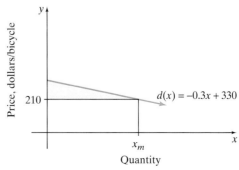

Figure 7.3.4

To find the consumers' surplus, we need to integrate

$$\int_0^{400} [(-0.3x + 330) - 210]\, dx = \int_0^{400} (-0.3x + 120)\, dx$$

$$= \left(\frac{-0.3}{2}x^2 + 120x \right) \Big|_0^{400}$$

$$= \$24{,}000$$

So the consumers' surplus is $24,000 each month.   ■

✔**CHECKPOINT 1**

Now work Exercise 15.

Supply Function

Now let's take a look at things from the producers' perspective. We know that when prices go up consumers usually demand less. However, manufacturers tend to respond to higher prices by *supplying more*. So a typical supply curve, denoted *s*, that relates the price per unit $s(x)$ as a function of quantity, *x*, supplied should be an increasing function, as shown in Figure 7.3.5. Notice that as the quantity *x* increases so does the price.

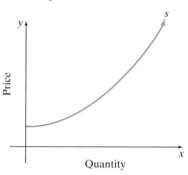

Figure 7.3.5

The supply function is defined as follows:

Supply Function

The **supply function**, *s*, for any product gives the price $s(x)$ at which exactly *x* units of the product will be supplied.

EXAMPLE 4   **Evaluating a Supply Function**

Suppose that the supplier of the bicycle that we have discussed in this section has a supply function given by $s(x) = 0.2x + 75$, where *x* is the quantity supplied each month and $s(x)$ is the price per bicycle in dollars. Compute $s(100)$, $s(200)$, and $s(300)$ and interpret each.

SOLUTION

Since $s(x) = 0.2x + 75$, we have

$$s(100) = 0.2(100) + 75 = \$95 \text{ per bicycle}$$
$$s(200) = 0.2(200) + 75 = \$115 \text{ per bicycle}$$
$$s(300) = 0.2(300) + 75 = \$135 \text{ per bicycle}$$

The interpretations of these results are, respectively,

The supplier will supply 100 bicycles each month at a price of $95 per bicycle.

The supplier will supply 200 bicycles each month at a price of $115 per bicycle.

The supplier will supply 300 bicycles each month at a price of $135 per bicycle.

Producers' Surplus

Let's return to the scenario where the market price for a bicycle is $200 per bicycle. Suppose that the bicycle supplier is willing to stay in business if the price per bicycle dropped to $150. The fact that bicycles sell for $200 gives a "gain" of $50 per bicycle to the supplier. Now consider any price below $200 at which some supplier is willing to supply these bicycles, subtracted from the actual price of $200. The sum of these gains over all possible prices below $200 is known as the *producers' surplus*.

However, once a market price, d_{mp}, is determined for an item, we can compute the quantity that producers will supply by the supply function. Our bicycle supplier in Example 4 supplies 100 bicycles each month when the price is $95 per bicycle. The shaded region in Figure 7.3.6 gives the actual amount that the supplier received or, if you prefer, the *revenue*. (From the consumers' perspective, this amount is called *consumer expenditure*.) We can quickly compute this amount to be $95(100) = $9500.

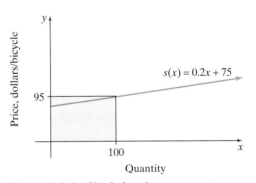

Figure 7.3.6   Shaded region represents revenue.

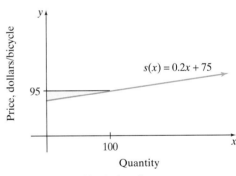

Figure 7.3.7   Shaded region represents producer's surplus.

Figure 7.3.6 shows the actual amount (or revenue) received, while Figure 7.3.7 shows the area above the supply curve, $s(x) = 0.2x + 75$, and below the line $y = 95$. This area in Figure 7.3.7 is the extra amount that suppliers receive over what they are willing to receive. This extra amount is called the **producers' surplus.**

> **Producer's Surplus**
>
> **Producers' surplus** is the total amount gained by producers who are willing to receive less than market price, d_{mp}, for a product, yet are able to receive d_{mp} for the product.

To determine the area of the shaded region in Figure 7.3.7, and therefore the producers' surplus, we need to determine the area between the two curves $y = 95$ and $s(x) = 0.2x + 75$. This is found by integrating

$$\int_0^{100} [95 - (0.2x + 75)]\, dx = \int_0^{100} (-0.2x + 20)\, dx$$

Notice that, just like computing consumers' surplus, determining producers' surplus is analogous to determining the area between two curves. The following shows how to determine producers' surplus in any situation.

Computing Producers' Surplus

For a supply function s and a point (x_m, d_{mp}) on the graph of s, where x_m is called the **market demand** and d_{mp} is called the **market price**, we determine the producers' surplus by integrating

$$\int_0^{x_m} [d_{mp} - s(x)]\, dx$$

EXAMPLE 5 | **Determining Producers' Surplus**

Assume the supply function from Example 4 for the bicycle to be $s(x) = 0.2x + 75$, where $s(x)$ is the price per bicycle in dollars and x is the number of bicycles. Determine the producers' surplus if the market price is $210 per bicycle.

SOLUTION

Since $s(x) = 0.2x + 75$ and the market price is $210 per bicycle, to determine producers' surplus we must first determine the quantity supplied, or the market demand. To determine the market demand, we need to solve

$$210 = 0.2x + 75$$
$$135 = 0.2x$$
$$x = 675$$

To find producers' surplus, we integrate

$$\int_0^{675} [210 - (0.2x + 75)]\, dx = \int_0^{675} (135 - 0.2x)\, dx = \left(135x - \frac{0.2x^2}{2}\right) \Big|_0^{675}$$
$$= \$45{,}562.50 \text{ each month.}$$

So the producers' surplus is $45,562.50 each month. ◼

✓**CHECKPOINT 2**

Now work Exercise 29.

Interactive Activity

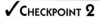 There is another way to determine the area representing producers' surplus in Example 5. We can take the area under the graph of s from $x = 0$ to $x = 675$, and subtract it from the area of the rectangle that represents the actual amount that producers' receive to supply 675 bicycles. This corresponds to integrating $\int_0^{675} (0.2x + 75)\, dx$ and subtracting the result from $675(210) = \$141{,}750$. Perform the integration and subtraction to verify the producers' surplus determined in Example 5.

Equilibrium Point

If we graph a demand function, d, and a supply function, s, together on the same set of axes, we have the situation shown in Figure 7.3.8. The demand, x, at which the demand and supply curves intersect is called the *equilibrium demand*, while the price is called the *equilibrium price*. The point of intersection for d and s is called the **equilibrium point**. Also notice in Figure 7.3.8 that both the consumers' surplus and producers' surplus can be shown together.

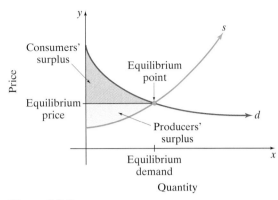

Figure 7.3.8

EXAMPLE 6 | **Locating an Equilibrium Point**

Continuing with our bicycle scenario, if the demand function is given by $d(x) = -0.3x + 330$ and the supply function is given by $s(x) = 0.2x + 75$:

(a) Determine the equilibrium point.

(b) Determine the consumers' surplus at the equilibrium demand.

(c) Determine the producers' surplus at the equilibrium demand.

SOLUTION

(a) Figure 7.3.9 has a graph of d and s. We will determine the equilibrium point algebraically and ask that you check it by using the INTERSECT command on your calculator.

$$-0.3x + 330 = 0.2x + 75$$
$$255 = 0.5x$$
$$510 = x$$

We have the equilibrium demand to be 510 bicycles each month. The equilibrium price is found by using either $d(x)$ or $s(x)$. Using $s(x)$, we get

$$s(510) = 0.2(510) + 75 = 177$$

The equilibrium price is $177, which means that the equilibrium point is $(510, \$177)$.

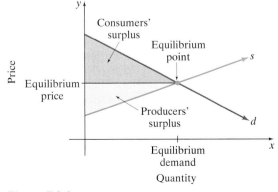

Figure 7.3.9

(b) As shown in Figure 7.3.9, the consumers' surplus is found by integrating

$$\int_0^{510} [(-0.3x + 330) - 177]\,dx$$

$$= \int_0^{510} (-0.3x + 153)\,dx$$

$$= \left(\frac{-0.3}{2}x^2 + 153x \right)\Big|_0^{510}$$

$$= \left[\frac{-0.3}{2}(510)^2 + 153(510) \right] - \left[\frac{-0.3}{2}(0)^2 + 153(0) \right]$$

$$= \$39{,}015 \text{ each month.}$$

So the consumers' surplus is $39,015 each month.

(c) As shown in Figure 7.3.9, the producers' surplus is determined by integrating

$$\int_0^{510} [177 - (0.2x + 75)]\,dx = \int_0^{510} (102 - 0.2x)\,dx = \left(102x - \frac{0.2}{2}x^2 \right)\Big|_0^{510}$$

$$= \left[102(510) - \frac{0.2}{2}(510)^2 \right] - \left[102(0) - \frac{0.2}{2}(0)^2 \right]$$

$$= \$26{,}010 \text{ each month}$$

So the producers' surplus is $26,010 each month. ◆

✓ **CHECKPOINT 3**

Now work Exercise 41.

Lorenz Curves and Gini Index

It is a fact that in our society some individuals make more money than others. Economists measure the gap between the rich and the poor by examining the proportion of the total income that is earned by the lowest 20% of the population, and then the proportion earned by the lowest 40% of the population, and so on. For example, data for income distribution in the United States in 1990 is shown in Table 7.3.1.

The table tells us that in 1990 the lowest 40% of the population earns only $0.135 = 13.5\%$ of the total income. The graph of $y = L(x)$ through the data points in Figure 7.3.10 is called a **Lorenz curve.** A Lorenz curve gives us the proportion of total income earned by the lowest proportion, x, of the population. The domain for L is [0, 1] and the range is [0, 1].

TABLE 7.3.1

PROPORTION OF POPULATION	PROPORTION OF INCOME
0.20	0.039
0.40	0.135
0.60	0.294
0.80	0.534
1.00	1.00

Source: U.S. Census Bureau web site.

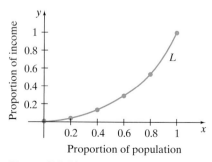

Figure 7.3.10

There are two extreme cases of income distribution, which means that there are two extreme cases for Lorenz curves. They are:

1. Absolute equality of income distribution
2. Absolute inequality of income distribution

Absolute equality of income distribution means that everyone earns the same income. The lowest 20% earns 20% of the income, the lowest 40% earns 40% of the income, and so on. The Lorenz curve for absolute equality is $L(x) = x$, as shown in Figure 7.3.11. Absolute inequality of income distribution means that no one earns any income except one person who earns it all. This would yield the Lorenz curve shown in Figure 7.3.12.

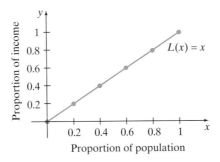

Figure 7.3.11 Lorenz curve for absolute equality.

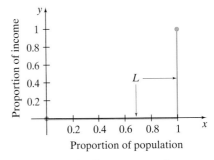

Figure 7.3.12 Lorenz curve for absolute inequality.

All Lorenz curves that we will analyze lie between these two extremes. See Figure 7.3.13. To measure how the actual income distribution differs from absolute equality, we will calculate the area between $L(x) = x$ (absolute equality) and the Lorenz curve modeling our income distribution.

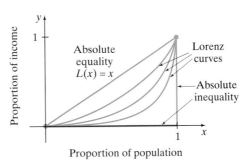

Figure 7.3.13

Since the domain and range of any Lorenz curve are $[0, 1]$, the area between absolute equality and absolute inequality is $\frac{1}{2}$. (The region between these two extremes is $\frac{1}{2}$ of a square whose area is 1 square unit.) It follows that the area between any Lorenz curve and absolute equality is at most $\frac{1}{2}$. However, economists multiply this area by 2 to get a number between 0 and 1, with 0 being absolute equality and 1 being absolute inequality.

The area between a Lorenz curve and absolute equality gives a measure that economists call the **Gini Index** or the **Gini coefficient.** A larger Gini Index means more area between the Lorenz curve and absolute equality, which then means a greater inequality in income distribution.

EXAMPLE 7 | **Computing a Gini Index**

In Utopia Land, the Lorenz curve for income distribution is modeled by $L(x) = x^2$.

(a) Graph $L(x) = x$, absolute equality, and $L(x) = x^2$, the Lorenz curve for Utopia Land on the same set of axes.

(b) Compute the Gini Index for Utopia Land.

SOLUTION

(a) Figure 7.3.14 gives the required graphs.

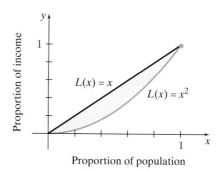

Figure 7.3.14

(b) The shaded region shown in Figure 7.3.14 is the area between two curves. We compute it to be

$$\int_0^1 (x - x^2)\, dx = \left(\frac{1}{2}x^2 - \frac{1}{3}x^3\right)\Big|_0^1 = \left[\frac{1}{2}(1)^2 - \frac{1}{3}(1)^3\right] - \left[\frac{1}{2}(0)^2 - \frac{1}{3}(0)^3\right]$$
$$= \frac{1}{6}$$

We now multiply by 2 to get the Gini Index of

$$2\left(\frac{1}{6}\right) = \frac{1}{3}$$

So the Gini Index for Utopia Land is $\frac{1}{3}$ or approximately 0.33.

We can generalize the work done in Example 7 as follows:

> **Computing a Gini Index**
>
> For a Lorenz curve, L, the **Gini Index** is found by
> $$2 \cdot \int_0^1 [x - L(x)]\, dx$$

EXAMPLE 8 | **Computing a Gini Index**

The Lorenz curve for the distribution of income in the United States in 1990 can be modeled by

$$L(x) = 2.71x^4 - 4.56x^3 + 3.27x^2 - 0.41x$$

Compute the Gini Index for the United States in 1990.

SOLUTION

The Gini Index is found by

$$2 \cdot \int_0^1 [x - L(x)] \, dx = 2 \cdot \int_0^1 [x - (2.71x^4 - 4.56x^3 + 3.27x^2 - 0.41x)] \, dx$$

$$= 2 \cdot \int_0^1 (-2.71x^4 + 4.56x^3 - 3.27x^2 + 1.41x) \, dx$$

$$= 2 \cdot \left[\left(\frac{-2.71}{5}x^5 + \frac{4.56}{4}x^4 - \frac{3.27}{3}x^3 + \frac{1.41}{2}x^2 \right) \Big|_0^1 \right]$$

$$= 2 \cdot [0.213]$$

$$= 0.426$$

So in 1990 the Gini Index in the United States was about 0.426.

✓ **CHECKPOINT 4**

Now work Exercise 65.

SUMMARY

In this section we studied some applications for area between two curves. **Consumers' surplus** and **producers' surplus** are simply areas between two curves, and we stress that they should be viewed that way, as opposed to simply memorizing a formula. We also discussed where to locate an **equilibrium point** and the relationship between **consumer expenditure** and revenue. The final topic in this section dealt with income distributions and a measure of income distribution called the **Gini Index.** Again, this is an application of area between two curves.

- **Computing consumers' surplus:** For demand function d and point (x_m, d_{mp}) on the graph of d, where x_m is the **market demand** and d_{mp} is the **market price,** we determine the consumers' surplus by integrating $\int_0^{x_m} [d(x) - d_{mp}] \, dx$.
- **Computing producers' surplus:** For supply function s and a point (x_m, d_{mp}) on the graph of s, where x_m is the **market demand** and d_{mp} is the **market price,** we determine the producers' surplus by integrating $\int_0^{x_m} [d_{mp} - s(x)] \, dx$.
- **Computing a Gini Index:** For a Lorenz curve, $y = L(x)$, the Gini Index is found by $2 \cdot \int_0^1 [x - L(x)] \, dx$.

SECTION 7.3 EXERCISES

Exercises 1–4 refer to Figure 7.3.15.

1. Shade the region that corresponds to the consumers' surplus. Write an integral to determine consumers' surplus.

2. Shade the region that corresponds to the producers' surplus. Write an integral to determine producers' surplus.

3. Shade the region that corresponds to the actual consumer expenditure. Determine the actual consumer expenditure.

4. Shade the region that corresponds to the revenue. Determine the actual revenue.

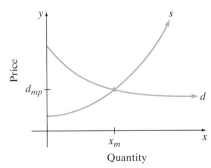

Figure 7.3.15

Exercises 5 and 6 refer to Figure 7.3.16.

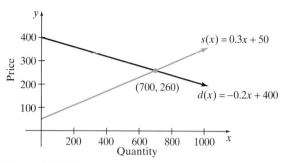

Figure 7.3.16

5. Shade the region that corresponds to the producers' surplus. Write an integral to determine producers' surplus.

6. Shade the region that corresponds to the consumers' surplus. Write an integral to determine consumers' surplus.

Exercises 7 and 8 refer to Figure 7.3.17.

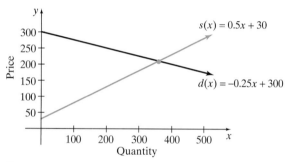

Figure 7.3.17

7. Shade the region that corresponds to the producers' surplus. Write an integral to determine producers' surplus.

8. Shade the region that corresponds to the consumers' surplus. Write an integral to determine consumers' surplus.

Exercises 9 and 10 refer to Figure 7.3.18.

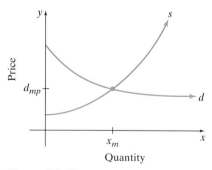

Figure 7.3.18

9. Shade the region that corresponds to the consumers' surplus. Write an integral to determine consumers' surplus.

10. Shade the region that corresponds to the producers' surplus. Write an integral to determine producers' surplus.

In Exercises 11–24, for each demand function given and demand level x, determine the consumers' surplus.

11. $d(x) = 3500 - 3x$, $x = 200$

12. $d(x) = 3500 - 3x$, $x = 500$

13. $d(x) = 220 - \frac{1}{3}x$, $x = 100$

14. $d(x) = 330 - \frac{1}{5}x$, $x = 200$

✓ 15. $d(x) = 600 - 0.6x$, $x = 150$

16. $d(x) = 540 - 0.4x$, $x = 300$

17. $d(x) = 1610 - 0.08x^2$, $x = 100$

18. $d(x) = 2720 - 0.05x^2$, $x = 200$

19. $d(x) = 400 - \frac{1}{3}x^2$, $x = 30$

20. $d(x) = 300 - \frac{1}{4}x^2$, $x = 30$

21. $d(x) = 600e^{-0.01x}$, $x = 300$

22. $d(x) = 550e^{-0.02x}$, $x = 150$

23. $d(x) = 230e^{-0.02x}$, $x = 175$

24. $d(x) = 320e^{-0.01x}$, $x = 100$

In Exercises 25–38, for each supply function given and demand level x, determine the producers' surplus.

25. $s(x) = 0.2x + 50$, $x = 75$

26. $s(x) = 0.3x + 66$, $x = 90$

27. $s(x) = 0.15x + 80$, $x = 100$

28. $s(x) = 0.27x + 60$, $x = 80$

✓ 29. $s(x) = \frac{1}{3}x + 55$, $x = 60$

30. $s(x) = \frac{1}{2}x + 40$, $x = 70$

31. $s(x) = 0.05x^2 + 20$, $x = 100$

32. $s(x) = 0.01x^2 + 25$, $x = 200$

33. $s(x) = 0.1x^2 + 10$, $x = 150$

34. $s(x) = 0.2x^2 + 15$, $x = 110$

35. $s(x) = 10e^{0.02x}$, $x = 90$

36. $s(x) = 20e^{0.01x}$, $x = 120$

37. $s(x) = 13e^{0.01x}$, $x = 110$

38. $s(x) = 25e^{0.03x}$, $x = 55$

For Exercises 39–48:

(a) Determine the equilibrium point. In Exercises 47 and 48, round equilibrium demand and equilibrium price to the nearest whole number.

(b) Determine the consumers' surplus at equilibrium demand.

(c) Determine the producers' surplus at equilibrium demand.

39. $d(x) = 3553 - 13x;\ s(x) = 5.70x$

40. $d(x) = 329 - \dfrac{1}{5}x;\ s(x) = 0.27x$

✓ 41. $d(x) = 398 - 0.4x;\ s(x) = 0.5x + 11$

42. $d(x) = 89 - 0.25x;\ s(x) = 0.85x + 19.7$

43. $d(x) = 2743 - 0.04x^2;\ s(x) = 0.06x^2 + 20.5$

44. $d(x) = 2528 - 0.01x^2;\ s(x) = 0.02x^2 + 5$

45. $d(x) = 500 - 0.2x^2;\ s(x) = 0.2x^2 + 10$

46. $d(x) = 1426 - 0.02x^2;\ s(x) = 0.01x^2 + 13.33$

47. $d(x) = 600e^{-0.01x};\ s(x) = 1.125e^{0.02x}$

48. $d(x) = 550e^{-0.02x};\ s(x) = 20e^{0.01x}$

APPLICATIONS

49. Lindsay's Department Store has determined that the demand function for a new type of nonsticking frying pan is given by

$$d(x) = -1.4x + 25$$

where x is the number of pans demanded each day and $d(x)$ is the price per pan in dollars.

(a) Assuming that the equilibrium price is $11 per pan, determine the equilibrium demand.

(b) Determine the consumers' surplus at equilibrium demand.

50. The demand function for a new crock pot is given by

$$d(x) = 98 - 7\sqrt{x}$$

where x is the number of crock pots demanded each day and $d(x)$ is the price per crock pot in dollars.

(a) Assuming that the equilibrium price is $28 per crock pot, determine the equilibrium demand.

(b) Determine the consumers' surplus at equilibrium demand.

51. Frickel's Department Store has determined that the demand function for an extra-wide toaster is given by

$$d(x) = 27.3e^{-0.09x}$$

where x is the number of toasters demanded each day and $d(x)$ is the price per toaster in dollars. Assuming that the equilibrium price is $12.14, determine the consumers' surplus at equilibrium demand. Round equilibrium demand to the nearest whole number.

52. The Hacker's Delight has determined that the demand function for the new Doppler Don Driving Iron is given by

$$d(x) = -x^2 - x + 650$$

where x is the number of driving irons demanded each

month and $d(x)$ is the price per driving iron in dollars. Assuming that the equilibrium price is $230, determine the consumers' surplus.

53. Balata Inc., a producer of golf balls, has determined that the supply function for the new Xtrah golf ball is given by

$$s(x) = 0.24x + 3.70$$

where x is the number of dozens supplied each month and $s(x)$ is the price per dozen.

(a) If the equilibrium price is $19.30 per dozen, determine the equilibrium demand.

(b) Determine the producers' surplus at equilibrium demand.

54. Linguini's Pizza Palace has determined that the supply function for a large pizza is given by

$$s(x) = 0.02x^2 + 0.78$$

where x is the number of pizzas supplied each day and $s(x)$ is the price per pizza.

(a) If the equilibrium price of a pizza is $8 per pizza, determine the equilibrium demand.

(b) Determine producers' surplus at equilibrium demand.

55. Balata Inc., a producer of golf balls, has determined that the supply function for the new Equalizer golf ball is given by

$$s(x) = 6.7e^{0.02x}$$

where x is the number of dozens supplied each month and $d(x)$ is the price per dozen. Assuming that the equilibrium price is $19.75 per dozen, determine the producers' surplus. Round equilibrium demand to the nearest whole number.

56. Baker's Bake Shoppe, a supplier of specialty baking pans, has determined that the supply function for a certain birthday cake baking pan is given by

$$s(x) = 4e^{0.05x}$$

where x is the number of these baking pans supplied each month and $s(x)$ is the price per pan in dollars. Assuming that the equilibrium price is $8 per pan, determine the producers' surplus. Round equilibrium demand to the nearest whole number.

57. BuildIt, a local home improvement store, has determined that the demand function for a 30-gallon trash can is given by

$$d(x) = 13 - 0.01x^2$$

while the related producer supply function is given by

$$s(x) = 0.1x + 1$$

where x is the daily quantity and $d(x)$ and $s(x)$ are in dollars per can.

(a) Determine the equilibrium point.

(b) Determine the consumers' surplus at equilibrium.

(c) Determine the producers' surplus at equilibrium.

58. Fresh Paint, a local paint supply and paint accessories store, has determined that the demand function for a step

ladder is given by

$$d(x) = -0.2x^2 + 100$$

while the related producer supply function is given by

$$s(x) = x + 1$$

where x is the weekly quantity and $d(x)$ and $s(x)$ are in dollars per ladder.

(a) Determine the equilibrium point. Round the equilibrium demand to the nearest whole number and round the equilibrium price to the nearest dollar.

(b) Determine the consumers' surplus at equilibrium.

(c) Determine the producers' surplus at equilibrium.

59. Office House, an office supply store, has determined that the demand function for a floppy disk storage case is given by

$$d(x) = 30 - x$$

while the related producer supply function is given by

$$s(x) = \sqrt{x}$$

where x is the weekly quantity and $d(x)$ and $s(x)$ are in dollars per storage case.

(a) Determine the equilibrium point.

(b) Determine the consumers' surplus at equilibrium.

(c) Determine the producers' surplus at equilibrium.

60. Just Smell It, a scented candle shop, has determined that the demand function for a large vanilla-scented candle is given by

$$d(x) = 20 - \frac{1}{2}x$$

while the related producer supply function is given by

$$s(x) = \frac{1}{2}\sqrt{x}$$

where x is the monthly quantity and $d(x)$ and $s(x)$ are in dollars per candle.

(a) Determine the equilibrium point. Round the equilibrium demand to the nearest whole number and round the equilibrium price to the nearest dollar.

(b) Determine the consumers' surplus at equilibrium.

(c) Determine the producers' surplus at equilibrium.

In Exercises 61–66, the techniques of regression were used on data from the U.S. Census Bureau Web site to obtain the Lorenz curves for selected years. Recall that x represents the proportion of the population and $L(x)$ represents the proportion of income. Determine the Gini Index for each year. Round to the nearest ten-thousandth.

61. In 1980, $L(x) = 1.98x^4 - 3.31x^3 + 2.61x^2 - 0.28x$.

62. In 1982, $L(x) = 2.16x^4 - 3.61x^3 + 2.76x^2 - 0.31x$.

63. In 1984, $L(x) = 2.16x^4 - 3.57x^3 + 2.71x^2 - 0.3x$.

64. In 1986, $L(x) = 2.37x^4 - 3.95x^3 + 2.93x^2 - 0.35x$.

65. In 1988, $L(x) = 2.53x^4 - 4.22x^3 + 3.07x^2 - 0.38x$.

66. In 1990, $L(x) = 2.71x^4 - 4.56x^3 + 3.27x^2 - 0.41x$.

67. The Gini Index is similar to a snapshot in time. The index does not depend on time. However, a collection of indices over some period of time produces a "moving picture" of income distribution. Exercises 61–66 are a collection over a 10-year period of time, specifically, 1980 to 1990. Analyze the Gini Indices over this 10-year period of time. Does the Gini Index appear to be increasing or decreasing over this time period? What does this mean about income distribution in the United States over this time period?

68. Use the data in Table 7.3.2 for U.S. income distribution in 1995.

TABLE 7.3.2

PROPORTION OF POPULATION	PROPORTION OF INCOME
0	0
0.4	0.145
0.6	0.303
0.8	0.535
1	1

Source: U.S. Census Bureau web site.

(a) Let x represent the proportion of the population and let y represent the proportion of income. Enter the data into your calculator and determine a quartic regression equation to model the data. Round all coefficients to the nearest hundredth.

(b) Compute the Gini Index for 1995. Round to the nearest ten-thousandth.

(c) Compare the Gini Index for 1995 with the Gini Index for 1982. See Exercise 62. Interpret the results.

SECTION PROJECT

An office supply store has collected the following data for a certain brand of briefcase.

NUMBER OF BRIEFCASES DEMANDED PER WEEK	PRICE ($ PER BRIEFCASE)
2	79.99
5	61.99
7	51.99
9	43.99
12	33.99
15	25.99

Meanwhile, the supplier of the briefcases has collected the following data:

NUMBER SUPPLIED EACH WEEK	PRICE ($ PER BRIEFCASE)
1	23.99
3	29.99
6	39.99
8	47.49
11	61.99
13	70.99
16	89.99

(a) Let x represent the quantity demanded per week. Enter the data into your calculator, plot the data, and determine a quadratic regression model for d. Round coefficients to the nearest thousandth.

(b) Now let x represent the quantity supplied each week. Enter the data into your calculator, plot the data, and determine a quadratic regression model for s. Round coefficients to the nearest thousandth.

(c) Graph d and s in the viewing window [0, 20] by [0, 70].

(d) Determine the equilibrium point by using the INTERSECT command. Round the equilibrium demand to the nearest whole number and round the equilibrium price to the nearest dollar.

(e) Determine the consumers' surplus at equilibrium.

(f) Determine the producers' surplus at equilibrium.

SECTION 7.4   INTEGRATION BY PARTS

In this section we present another technique of integration called **integration by parts.** Integration by parts is somewhat analogous to the Product Rule for differentiating. Applications involving this new technique will include total production from oil fields and a return to continuous income streams, first presented in Section 7.1. Here we will analyze the **present value** of a continuous income stream.

Integration by Parts

In many real-life situations, managers encounter integrals that cannot be evaluated with the integration techniques that we have learned so far. Other integration techniques are needed to solve these problems. For example, the manager of a taxicab fleet knows that the variable cost per mile for maintaining a cab that has been in service for x years since 1997 is modeled by

$$C(x) = 8.3 + 0.88 \ln x$$

To find the average variable cost per mile for a cab that has been in service for 2 years since 1997, the manager would need to evaluate the integral

$$\frac{1}{2} \int_0^2 (8.3 + 0.88 \ln x)\, dx$$

We cannot evaluate this integral because we have not yet determined $\int \ln x\, dx$. (We know how to *differentiate* $\ln x$.) A new method of integration, called **integration by parts,** can help to solve this problem.

To see where the formula for integration by parts comes from, let's assume that we have two differentiable functions, u and v. We need to recall the definition of the differential from Section 3.1. (Consult the Toolbox to the left.) So in differential notation, for our two functions u and v, we have

$$du = u'\, dx \quad \text{and} \quad dv = v'\, dx$$

The derivative of $u \cdot v$ is determined by the Product Rule to be

$$\frac{d}{dx}(u \cdot v) = u'v + uv'$$

From Your Toolbox

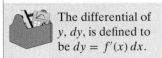

The differential of y, dy, is defined to be $dy = f'(x)\, dx$.

Now, if we integrate both sides of this equation with respect to x, we get

$$\int \frac{d}{dx}(u \cdot v)\,dx = \int (u'v + uv')\,dx$$

$$u \cdot v = \int u'v\,dx + \int uv'\,dx$$

Since $du = u'\,dx$ and $dv = v'\,dx$, we rewrite this as

$$u \cdot v = \int v\,du + \int u\,dv$$

We now solve this equation for $\int u\,dv$ and get

$$\int u\,dv = u \cdot v - \int v\,du$$

This last equation is the formula for the integration technique know as **integration by parts.**

Integration by Parts

For two differentiable functions u and v,

$$\int u\,dv = u \cdot v - \int v\,du$$

When applying the integration by parts formula, we are really performing a *double substitution.* Example 1 shows how this works. After Example 1, we will supply some general guidelines for using integration by parts.

EXAMPLE 1 | **Using the Integration by Parts Formula**

Determine $\int xe^x\,dx$.

SOLUTION

Our integration by parts formula has four parts that must be determined: u, du, v, and dv. Our goal is to turn our original problem, $\int xe^x\,dx$, into an integral of the form $\int u\,dv$. This is where the double substitution comes in. We will let u equal some part of $xe^x\,dx$, and then dv will equal the rest of $xe^x\,dx$. We decide to select $u = x$, since $du = 1 \cdot dx$, and $dv = e^x\,dx$, since $\int e^x\,dx = e^x$. So we have

$$u = x \qquad\qquad dv = e^x\,dx$$

$$du = 1 \cdot dx = dx \qquad v = \int e^x\,dx = e^x$$

Applying the integration by parts formula yields

$$\int u\,dv = u \cdot v - \int v\,du$$

$$\int xe^x\,dx = xe^x - \int e^x\,dx$$

$$= xe^x - e^x + C$$

NOTE: 1. Notice that the differentials du and dv include the dx.
2. When we integrate dv to get v, we can omit the constant of integration C. One constant C at the very end is enough.

The key step in using integration by parts is selecting the *two* substitutions *u* and *dv*. We rewrite the original integral in the form $\int u\,dv$ by substituting *u* for part of the original integrand and *dv* for the rest. We choose *u* and *dv* so that the resulting expression is simpler than the original. In Example 1 we chose *u* and *dv* so that the resulting expression has $\int e^x\,dx$, which we can easily integrate. We offer the following guidelines for choosing *u* and *dv*.

> ### Guidelines for the Selection of *u* and *dv*
> - Either select *dv* to be the most complicated part of the integrand that is easily integrated. Then *u* is the rest.
> - Or select *u* so that its derivative, *du*, is a simpler function than *u*. Then *dv* is the rest.

EXAMPLE 2

Using the Integration by Parts Formula

Determine $\int x \ln x\,dx$.

SOLUTION

In the integrand $x \ln x$, we can easily integrate *x*, but not $\ln x$. With this in mind, we select $u = \ln x$ and $dv = x\,dx$, which means that the four required pieces are

$$u = \ln x \qquad dv = x\,dx$$
$$du = \frac{1}{x}\,dx \qquad v = \int x\,dx = \frac{1}{2}x^2$$

Applying the integration by parts formula yields

$$\int u\,dv = u \cdot v - \int v\,du$$
$$\int x \ln x\,dx = (\ln x) \cdot \frac{1}{2}x^2 - \int \frac{1}{2}x^2 \cdot \frac{1}{x}\,dx$$
$$= \frac{1}{2}x^2 \ln x - \frac{1}{2}\int x\,dx$$
$$= \frac{1}{2}x^2 \ln x - \frac{1}{2} \cdot \frac{1}{2}x^2 + C = \frac{1}{2}x^2 \ln x - \frac{1}{4}x^2 + C$$

Interactive Activity

 Check the result of Example 2 by differentiating the answer, that is,

$$\frac{d}{dx}\left[\frac{1}{2}x^2 \ln x - \frac{1}{4}x^2 + C \right]$$

✓ CHECKPOINT 1

Now work Exercise 3.

EXAMPLE 3

Using the Integration by Parts Formula

Determine $\int \ln x\,dx$.

SOLUTION

This one looks a bit unusual in that there does not appear to be anything multiplied by $\ln x$. If we imagine $\ln x$ to be $1 \cdot \ln x$, then we can employ our guidelines. We urge caution here. It may be tempting to let $u = 1$, but this would force $dv = \ln x\,dx$. We do not know the integral of $\ln x$—that is what we are trying to determine! Basically, we have no choice but to select $u = \ln x$ and $dv = 1 \cdot dx$. This gives

$$u = \ln x \qquad dv = 1 \cdot dx$$
$$du = \frac{1}{x}\,dx \qquad v = \int 1\,dx = x.$$

The integration by parts formula gives

$$\int u \, dv = u \cdot v - \int v \, du$$

$$\int \ln x \, dx = x \ln x - \int x \cdot \frac{1}{x} \, dx$$

$$= x \ln x - \int 1 \, dx$$

$$= x \ln x - x + C$$

Interactive Activity

 Verify the result of Example 3 by differentiating the answer.

EXAMPLE 4

Using the Integration by Parts Formula

Determine $\int x e^{0.2x} \, dx$.

SOLUTION

From the guidelines we select $u = x$ and $dv = e^{0.2x} dx$. This gives

$$u = x \qquad dv = e^{0.2x} \, dx$$

$$du = dx \qquad v = \int e^{0.2x} \, dx = \frac{1}{0.2} e^{0.2x} = 5 e^{0.2x}$$

So, via the integration by parts formula, we have

$$\int u \, dv = u \cdot v - \int v \, du$$

$$\int x e^{0.2x} dx = 5 x e^{0.2x} - \int 5 e^{0.2x} dx$$

$$= 5 x e^{0.2x} - 5 \int e^{0.2x} dx$$

$$= 5 x e^{0.2x} - 5 \left(\frac{1}{0.2} e^{0.2x} \right) + C$$

$$= 5 x e^{0.2x} - 25 e^{0.2x} + C$$

NOTE: Many times, when evaluating an integral using integration by parts and part of the integrand involves e^{ax}, choose $dv = e^{ax} dx$. We then immediately have $v = \frac{1}{a} e^{ax}$.

✓ **CHECKPOINT 2**

Now work Exercise 21.

EXAMPLE 5

Evaluating an Indefinite Integral

Determine $\int 2 x e^{x^2} dx$.

SOLUTION

Do not be fooled! This is nothing more than a u-substitution first encountered in Section 6.4. To integrate $\int 2 x e^{x^2} dx$, we let $u = x^2$; then $du = 2x \, dx$ and by u-substitution we have

$$\int 2 x e^{x^2} dx = \int e^{u} \, du = e^{u} + C = e^{x^2} + C$$

Example 5 serves as a reminder that, even though we are currently learning a new integration technique in integration by parts, we should not forget our previously learned techniques. Our final example, before we consider some applications, requires two uses of the integration by parts formula.

EXAMPLE 6

Utilizing the Integration by Parts Formula Twice

Determine $\int x^2 e^{2x}\,dx$.

SOLUTION

From the guidelines and the Note following Example 4, we select $u = x^2$ and $dv = e^{2x}\,dx$. This gives

$$u = x^2 \qquad dv = e^{2x}\,dx$$
$$du = 2x\,dx \qquad v = \frac{1}{2}e^{2x}$$

The integration by parts formula gives

$$\int x^2 e^{2x}\,dx = \frac{1}{2}x^2 e^{2x} - \int \frac{1}{2}e^{2x} \cdot 2x\,dx$$
$$= \frac{1}{2}x^2 e^{2x} - \int x e^{2x}\,dx$$

This time, the integration by parts formula did not produce an integral that we could quickly evaluate. However, the new integral is simpler than the original, and it also looks similar to the one that we did in Example 4. So let's use integration by parts again. For the integral $\int x e^{2x}\,dx$, we select $u = x$ and $dv = e^{2x}\,dx$. This gives

$$u = x \qquad dv = e^{2x}\,dx$$
$$du = dx \qquad v = \frac{1}{2}e^{2x}$$

Integration by parts yields

$$\int x e^{2x}\,dx = \frac{1}{2}x e^{2x} - \int \frac{1}{2}e^{2x}\,dx$$
$$= \frac{1}{2}x e^{2x} - \frac{1}{2}\int e^{2x}\,dx$$
$$= \frac{1}{2}x e^{2x} - \frac{1}{2}\left(\frac{1}{2}e^{2x}\right) + C$$
$$= \frac{1}{2}x e^{2x} - \frac{1}{4}e^{2x} + C$$

Putting it all together results in

$$\int x^2 e^{2x}\,dx = \frac{1}{2}x^2 e^{2x} - \int x e^{2x}\,dx$$
$$= \frac{1}{2}x^2 e^{2x} - \left(\frac{1}{2}x e^{2x} - \frac{1}{4}e^{2x} + C\right)$$
$$= \frac{1}{2}x^2 e^{2x} - \frac{1}{2}x e^{2x} + \frac{1}{4}e^{2x} + C$$

✓ **CHECKPOINT 3**

Now work Exercise 33.

Applications

Most of the applications that we consider in this section require us to evaluate a definite integral. We will disregard whatever limits of integration we have until after we have found an indefinite integral, using whatever technique is required.

EXAMPLE 7 | **Estimating Oil Field Production**

Geologists have estimated that an oil field will produce oil at a rate given by

$$B(t) = 5te^{-0.2t}$$

where $B(t)$ is thousands of barrels per month t months from the start of operation. Estimate the total production in the first year of operation.

SOLUTION

In Figure 7.4.1 we have a graph of B. We notice that the total production in the first year of operation is equivalent to the area under the curve from $t = 0$ to $t = 12$. So we need to evaluate

$$\int_0^{12} 5te^{-0.2t}\, dt$$

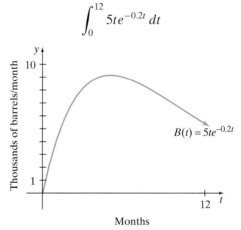

Figure 7.4.1

As mentioned prior to Example 7, we will ignore the limits of integration until after we have found an antiderivative. Using the guidelines for integration by parts, we select $u = 5t$ and $dv = e^{-0.2t}\, dt$. This gives

$$u = 5t \qquad dv = e^{-0.2t}\, dt$$

$$du = 5\, dt \qquad v = \frac{1}{-0.2}e^{-0.2t} = -5e^{-0.2t}$$

So, by integration by parts, we have

$$\int 5te^{-0.2t}\, dt = 5t(-5e^{-0.2t}) - \int (-5e^{-0.2t}) \cdot 5\, dt$$

$$= -25te^{-0.2t} + 25 \int e^{-0.2t}\, dt$$

$$= -25te^{-0.2t} + 25\left(\frac{1}{-0.2}e^{-0.2t}\right) + C$$

$$= -25te^{-0.2t} - 125e^{-0.2t} + C$$

Now to evaluate the *definite* integral and answer the question, we evaluate from $t = 0$ to $t = 12$ and get

$$(-25te^{-0.2t} - 125e^{-0.2t})\Big|_0^{12} = [(-25(12)e^{-0.2(12)} - 125e^{-0.2(12)})$$
$$- (-25(0)e^{-0.2(0)} - 125e^{-0.2(0)})]$$
$$\approx 86.44$$

So, in the first year of operation, the total production is estimated to be 86.44 thousand barrels.

✓ **CHECKPOINT 4**

Now work Exercise 49.

Flashback

CONTINUOUS INCOME STREAM REVISITED

In Section 7.1 we introduced the concept of a continuous income stream. Recall that a continuous income stream is very similar to the electric company, natural gas company, or water company that has a meter on our house that continuously monitors and totals the amount of electricity, natural gas, or water that we are consuming. Suppose that the rate of change of income in thousands of dollars per year for the oil field in Example 7 is

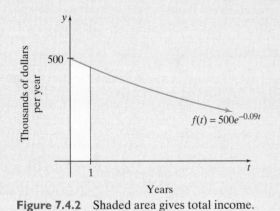

Figure 7.4.2 Shaded area gives total income.

projected to be

$$f(t) = 500e^{-0.09t}$$

where t is the number of years from when oil extraction began. See Figure 7.4.2. Find the total income from this field during the first year of operation.

Flashback Solution

Again, since we have a rate of change function, determining the total income generated during the first year of operation is equivalent to determining the area under the curve from $t = 0$ to $t = 1$. Integrating yields

$$\int_0^1 500e^{-0.09t}\, dt = \left(\frac{500}{-0.09}e^{-0.09t}\right)\Big|_0^1$$
$$= \frac{500}{-0.09}e^{-0.09(1)} - \frac{500}{-0.09}e^{-0.09(0)}$$
$$\approx 478.1600818$$

So in the first year of operation this oil field will generate approximately \$478,160.08 in income.

Recall from Section 7.1 that the rate of change of income in the Flashback is called the *rate of flow function.* Also in Section 7.1 we stated that in general the *total income* produced from time $t = a$ to $t = b$ is found by $\int_a^b f(t)\, dt$.

Now that we have properly reviewed continuous income streams, we turn our attention to the **present value of a continuous income stream.**

Present Value of a Continuous Income Stream

Suppose that we have a continuous income stream from an oil field and we know the rate of flow function. The ability to determine its *present value* is very important if we wish to sell (or buy) this oil field. We need to recall that, if P dollars are invested at an annual interest rate r compounded continuously, then the amount

A after t years is given by $A = Pe^{rt}$. Solving this equation for P yields

$$P = Ae^{-rt}$$

which we say is the **present value** of A. This means that P is the amount that needs to be invested *now* at interest rate r compounded continuously so that we have A dollars t years from now.

We now generalize the present value concept to continuous income streams by revisiting rectangles and the definite integral. First, we assume that the present money can be invested at the given rate, r, compounded continuously over the given time interval. Even though this assumption may sound unrealistic, managers make these assumptions regularly based on historical data.

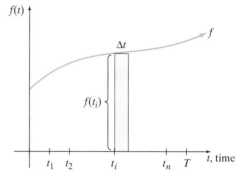

Figure 7.4.3

Suppose that f is the rate of flow function for a continuous income stream. Divide the interval $[0, T]$ into n equal subintervals of length Δt, as shown in Figure 7.4.3. In a typical subinterval, we can pick any point t_i. Let's say that it is the left-hand endpoint of the subinterval. *The total income produced over this subinterval is approximately equal to the area of the rectangle shaded in Figure 7.4.3.* This area is given by

$$f(t_i)\,\Delta t$$

Using the present value formula, $P = Ae^{-rt}$, with $A = f(t_i)\,\Delta t$ and $t = t_i$, the present value of the income received over this subinterval, P_i, is approximately equal to

$$P_i \approx f(t_i)\,\Delta t \cdot e^{-rt_i} = f(t_i)e^{-rt_i}\,\Delta t$$

To get the total present value on the interval $[0, T]$ requires us to sum *all* these present values. We know from our work in Chapter 6 that this is given by the definite integral $\int_0^T f(t)e^{-rt}\,dt$.

Present Value of a Continuous Income Stream

If f is the rate of flow function for a continuous income stream, then the **present value, P**, at annual interest rate, r, compounded continuously for T years is given by

$$P = \int_0^T f(t)e^{-rt}\,dt$$

EXAMPLE 8 | **Determining the Present Value of a Continuous Income Stream**

A window washing business generates income at the rate of $3t$ thousand dollars per year, where t is the number of years from now.

(a) Determine the present value of this continuous income stream for the next 7 years at 8% compounded continuously.

(b) Determine the total amount (income plus interest) produced by the window washing business over this 7-year period.

SOLUTION

(a) We know that $f(t) = 3t$, so the present value for the next 7 years at 8% compounded continuously is

$$P = \int_0^T f(t)e^{-rt}\, dt = \int_0^7 3te^{-0.08t}\, dt$$

Since this definite integral requires integration by parts, we will ignore the limits of integration until we have determined an antiderivative. Here we select $u = 3t$ and $dv = e^{-0.08t}\, dt$. This yields

$$u = 3t \qquad dv = e^{-0.08t}\, dt$$

$$du = 3\, dt \qquad v = \frac{1}{-0.08}e^{-0.08t} = -12.5e^{-0.08t}$$

Using integration by parts, we have

$$\int 3te^{-0.08t}\, dt = 3t(-12.5e^{-0.08t}) - \int (-12.5e^{-0.08t}) \cdot 3\, dt$$

$$= -37.5te^{-0.08t} + 37.5\int e^{-0.08t}\, dt$$

$$= -37.5te^{-0.08t} + 37.5\left(\frac{1}{-0.08}e^{-0.08t}\right) + C$$

$$= -37.5te^{-0.08t} - 468.75e^{-0.08t} + C$$

We now evaluate this from $t = 0$ to $t = 7$ and get

$$(-37.5te^{-0.08t} - 468.75e^{-0.08t})\big|_0^7 \approx 51.053372$$

So the present value of this continuous income stream over the next 7 years is about \$51,053.37.

(b) The total amount (income plus interest) earned over the next 7 years is equivalent to the amount earned by investing the present value \$51,053.37 in an account earning 8% annual interest compounded continuously for 7 years. So we use the compound interest formula and get

$$A = Pe^{rt} = 51,053.37e^{0.08(7)} \approx \$89,377.73 \qquad \blacksquare$$

✓ CHECKPOINT 5

Now work Exercise 59.

Example 8 illustrates how to determine the present value of a continuous income stream and also how to determine the total amount (income plus interest) generated by the continuous income stream. If we wish to compute just the total income generated by the continuous income stream, we would evaluate $\int_0^7 3t\, dt$.

SUMMARY

In this section we presented an integration technique known as **integration by parts.** We suggest becoming familiar with the guidelines for selecting u and dv, as well as *when* we need to use integration by parts. A mathematical application of integration by parts was the determination of an antiderivative for ln x to be $\int \ln x \, dx = x \ln x - x + C$. The section concluded with some applications involving integration by parts. We also determined a formula for the present value of a continuous income stream.

- **Integration by Parts Formula:** $\int u \, dv = uv - \int v \, du$.
- **Guidelines for the Selection of u and dv:**   (1) Select dv to be the most complicated part of the integral that is easily integrated. Then u is the rest. (2) Or select u so that its derivative, du, is a simpler function than u. Then dv is the rest.
- **Present Value of a Continuous Income Stream Formula:** $P = \int_0^T f(t)e^{-rt} \, dt$.

SECTION 7.4 EXERCISES

Evaluate the given integrals in Exercises 1–26. If integration by parts is required, consult the guidelines for selecting u and dv given in this section.

1. $\displaystyle\int 2xe^{3x} \, dx$

2. $\displaystyle\int 8xe^{5x} \, dx$

✓ 3. $\displaystyle\int xe^{4x} \, dx$

4. $\displaystyle\int xe^{6x} \, dx$

5. $\displaystyle\int xe^{-x} \, dx$

6. $\displaystyle\int xe^{-3x} \, dx$

7. $\displaystyle\int xe^{-0.03x} \, dx$

8. $\displaystyle\int xe^{-0.07x} \, dx$

9. $\displaystyle\int_0^2 xe^{-x} \, dx$

10. $\displaystyle\int_0^1 5xe^{-2x} \, dx$

11. $\displaystyle\int_0^5 xe^{-0.04x} \, dx$

12. $\displaystyle\int_0^6 xe^{-0.03x} \, dx$

13. $\displaystyle\int_1^e x \ln x \, dx$

14. $\displaystyle\int_1^3 x \ln x \, dx$

15. $\displaystyle\int x^2 \ln x \, dx$

16. $\displaystyle\int x^3 \ln x \, dx$

17. $\displaystyle\int \ln 2x \, dx$

18. $\displaystyle\int \ln 3x \, dx$

19. $\displaystyle\int (x + 3)e^x \, dx$ (*Hint:* Let $u = x + 3$)

20. $\displaystyle\int (x - 5)e^x \, dx$ (*Hint:* Let $u = x - 5$)

✓ 21. $\displaystyle\int 6te^{-0.1t} \, dt$

22. $\displaystyle\int_0^3 6te^{-0.1t} \, dt$

23. $\displaystyle\int_0^5 5te^{-0.06t} \, dt$

24. $\displaystyle\int 5te^{-0.06t} \, dt$

25. $\displaystyle\int 1.2te^{-0.08t} \, dt$

26. $\displaystyle\int 2.3te^{-0.05t} \, dt$

In Exercises 27–30, evaluate the integral of the form $\int (x + c)(x + d)^n \, dx$ by letting $u = x + c$ and $dv = (x + d)^n$ and applying the integration by parts formula.

27. $\displaystyle\int (x + 2)(x + 1)^5 \, dx$

28. $\displaystyle\int (x + 2)(x - 3)^6 \, dx$

29. $\displaystyle\int (x - 2)(x + 3)^6 \, dx$

30. $\displaystyle\int (x - 1)(x - 2)^4 \, dx$

Exercises 31–40 require two (or more) applications of integration by parts.

31. $\displaystyle\int x^2 e^{3x} \, dx$

32. $\displaystyle\int x^2 e^x \, dx$

✓ 33. $\displaystyle\int x^2 e^{-x} \, dx$

34. $\displaystyle\int x^2 e^{4x} \, dx$

35. $\displaystyle\int x^2 e^{5x} \, dx$

36. $\displaystyle\int (x + 3)^2 e^x \, dx$

37. $\displaystyle\int (\ln x)^2 \, dx$

38. $\displaystyle\int x(\ln x)^2 \, dx$

39. $\displaystyle\int x(\ln x)^3 \, dx$

40. $\displaystyle\int (\ln x)^3 \, dx$

In Exercises 41–48, determine the method needed to evaluate the given integral. If the method selected is integration by parts, state the choice for u and dv. If the method selected is u-substitution, simply state the choice for u.

41. $\displaystyle\int xe^{-x} \, dx$

42. $\displaystyle\int xe^{-x^2} \, dx$

43. $\displaystyle\int \frac{1}{x \ln x} \, dx$

44. $\displaystyle\int \frac{\ln x}{x} \, dx$

45. $\displaystyle\int (x + 6)e^x \, dx$

46. $\displaystyle\int (x - 5)e^x \, dx$

47. $\displaystyle\int 3xe^{2x} \, dx$

48. $\displaystyle\int 3x^2 e^{x^3} \, dx$

APPLICATIONS

✓ 49. It is estimated that an oil field will produce oil at a rate given by $B(t) = 6te^{-0.15t}$ thousand barrels per month, t months into production.

(a) Write a definite integral to estimate the total production for the first year of operation.

(b) Evaluate the integral from part (a) to estimate the total production for the first year of operation.

50. It is estimated that an oil field will produce oil at a rate given by $B(t) = 7.5te^{-0.13t}$ thousand barrels per month, t months into production.

(a) Write a definite integral to estimate the total production for the first year of operation.

(b) Evaluate the integral from part (a) to estimate the total production for the first year of operation.

51. The BrenKev Corporation has determined that its marginal profit function is given by

$$P'(t) = 2te^{0.2t}$$

where t is time in years and $P'(t)$ is in millions of dollars per year. Assuming that $P(0) = 0$, determine $P(t)$.

52. The VivaMix Corporation has determined that its marginal profit function is given by

$$P'(t) = 3te^{0.15t}$$

where t is time in years and $P'(t)$ is in millions of dollars per year. Assuming that $P(0) = 0$, determine $P(t)$.

53. The BrenKev Corporation has determined that its marginal cost function is given by

$$C'(t) = te^{-0.15t}$$

where t is time in years and $C'(t)$ is in millions of dollars per year. Assuming that $C(0) = 2$, determine $C(t)$.

54. The VivaMix Corporation has determined that its marginal cost function is given by

$$C'(t) = 0.3te^{-0.1t}$$

where t is time in years and $C'(t)$ is in millions of dollars per year. Assuming that $C(0) = 1.5$, determine $C(t)$.

55. Medical researchers have determined that the rate of absorption of a certain medication is given by $4te^{-0.51t}$ milligrams per hour, where t is the number of hours since the medication was taken. Determine the total amount of the medication absorbed during the first 6 hours.

56. Medical researchers have determined that the rate of absorption of a certain medication is given by $6te^{-0.42t}$ milligrams per hour, where t is the number of hours since the medication was taken. Determine the total amount of the medication absorbed during the first 8 hours.

57. Dena's Car Wash generates income at the rate of $2t$ million dollars per year, where t is the number of years that the car wash has been in business. Determine the present value of this continuous income stream over the first 5 years at a continuous compound interest rate of 6%.

58. SterilizeIt is a medical instrument company that generates income at the rate of $500{,}000 + 30{,}000t$ dollars per year, where t is the number of years that the company has been in operation. Determine the present value of this continuous income stream over the first 7 years at a continuous compound interest rate of 6.5%.

✓ 59. Suppose that an oil well produces income at the rate of $200e^{-0.1t}$ thousand dollars per year for t years. Determine the present value of this continuous income stream over the first 4 years of operation assuming a continuous compound interest rate of 5.75%.

60. Suppose that an oil well produces income at the rate of $150e^{-0.15t}$ thousand dollars per year for t years. Determine the present value of this continuous income stream over the first 3 years of operation assuming a continuous compound interest rate of 5.5%.

61. Reggie has a choice of two investments. Each choice requires the same initial investment and each produces a continuous income stream at an interest rate of 8% compounded continuously. The first investment is an oil well that generates income at a rate of $50t$ thousand dollars per year, where t is the number of years from now. The second investment is a natural gas well that generates income at a rate of $30 + 40t$ thousand dollars per year, where t is the number of years from now. Compare the present value of each investment to determine which is the better choice over the next 5 years.

62. Repeat Exercise 61 except here determine which is the better investment over the next 10 years.

🌐 63. The college enrollment for males, in millions, in the United States can be approximated by

$$f(x) = 2.18 + 1.1 \ln x, \qquad 1 \le x \le 31$$

where x is in years ($x = 1$ corresponds to January 1, 1960) and $f(x)$ is the male enrollment in millions.

(a) Determine the average value of f on the interval $[1, 11]$ and interpret.

(b) Determine the average value of f on the interval $[11, 21]$ and interpret.

(c) Determine the average value of f on the interval $[21, 31]$ and interpret.

🌐 64. The college enrollment for females, in millions, in the United States can be approximated by

$$g(x) = 0.22x + 0.94, \qquad 1 \le x \le 31$$

where x is in years ($x = 1$ corresponds to January 1, 1960) and $g(x)$ is the female enrollment in millions.

(a) Determine the average value of g on the interval $[1, 11]$ and interpret.

(b) Determine the average value of g on the interval $[11, 21]$ and interpret.

(c) Determine the average value of g on the interval $[21, 31]$ and interpret.

🌐 65. The percent of solid waste generated in the United States that is paper and paperboard can be modeled by

$$f(x) = 35.8 + 1.96 \ln x, \qquad 1 \le x \le 6$$

where x is in years ($x = 1$ corresponds to January 1, 1991) and $f(x)$ is the percent of solid waste that is paper and paperboard. Determine the average value of f on $[1, 6]$ and interpret.

🌐 66. The revenue generated in the solid waste management industry in the United States is approximated by

$$R(x) = 8.13 + 8.5 \ln x, \qquad 1 \le x \le 16$$

where x is in years ($x = 1$ corresponds to January 1, 1980) and $R(x)$ is the revenue in billions of dollars.

(a) Determine the average value of R on the interval $[1, 11]$ and interpret.

(b) Determine the average value of R on the interval $[11, 16]$ and interpret.

67. The U.S. imports of crude oil can be modeled by

$$g(x) = 2096 + 264 \ln x, \qquad 1 \le x \le 7$$

where x is in years ($x = 1$ corresponds to January 1, 1990) and $g(x)$ is the crude oil imports in millions of barrels. Determine the average value of g on the interval $[1, 7]$ and interpret.

68. The annual per capita consumption of light and skim milk can be modeled by

$$m(x) = 10.12 + 2 \ln x, \qquad 1 \le x \le 16$$

where x is in years ($x = 1$ corresponds to January 1, 1980) and $m(x)$ is the annual per capita consumption in gallons.

(a) Determine the average value of m on the interval $[1, 11]$ and interpret.

(b) Determine the average value of m on the interval $[11, 16]$ and interpret.

69. The annual per capita consumption of cheese (excluding cottage cheese) is approximated by

$$h(x) = 17.13 + 3.4 \ln x, \qquad 1 \le x \le 16$$

where x is in years ($x = 1$ corresponds to January 1, 1980) and $h(x)$ is the annual per capita consumption in pounds.

(a) Determine the average value of h on the interval $[1, 11]$ and interpret.

(b) Determine the average value of h on the interval $[11, 16]$ and interpret.

70. The amount of money spent annually for admission to spectator sports in the United States is modeled by

$$s(x) = 2.1 + 1.3 \ln x, \qquad 1 \le x \le 12$$

where x is in years ($x = 1$ corresponds to January 1, 1985) and $s(x)$ is the amount spent in billions of dollars.

(a) Determine the average value of s on the interval $[1, 12]$ and interpret.

(b) Evaluate $\int_1^{12} s(x)\,dx$ and interpret.

71. The variable cost (gas, oil, maintenance, and the like) per mile of owning and operating an automobile in the United States is approximated by

$$v(x) = 8.3 + 0.88 \ln x, \qquad 1 \le x \le 6$$

where x is in years ($x = 1$ corresponds to January 1, 1990) and $v(x)$ is the variable cost in cents per mile. Determine the average value on the interval $[1, 6]$ and interpret.

72. The average annual salary for a high school principal in the United States can be modeled by

$$f(x) = 40.9 + 10.3 \ln x, \qquad 1 \le x \le 12$$

where x is in years ($x = 1$ corresponds to January 1, 1985)

and $f(x)$ is the average salary in thousands of dollars per year. Suppose that Fred is a high school principal who earned the average salary from January 1, 1985, to January 1, 1997.

(a) Write an integral to determine Fred's total salary from January 1, 1985, to January 1, 1997.

(b) Evaluate the integral from part (a) and interpret.

SECTION PROJECT

The data in Table 7.4.1 give the average annual salaries, in thousands, for certain employees in the U.S. public school systems. The year column corresponds to January 1 of the given year.

TABLE 7.4.1

YEAR	SUPERINTENDENT	BUSINESS ADMINISTRATOR	TEACHER (K–12)
1985	56.954	40.344	23.587
1990	75.425	52.354	31.278
1992	83.342	57.036	34.565
1993	85.120	57.864	35.291
1994	87.717	59.997	36.531
1995	90.198	61.323	37.264
1996	94.229	63.840	38.706

Source: U.S. Census Bureau web site.

(a) Let x represent years ($x = 1$ correspond to January 1, 1985) and y_1 represent the average salary in thousands of a superintendent. Determine a linear regression model for the data. Round coefficients to the nearest hundredth.

(b) Let x represent years ($x = 1$ correspond to January 1, 1985) and y_2 represent the average salary in thousands of a business administrator. Determine a linear regression model for the data. Round coefficients to the nearest hundredth.

(c) Let x represent years ($x = 1$ correspond to January 1, 1985) and y_3 represent the average salary in thousands of a teacher. Determine a natural logarithmic regression model for the data. Round coefficients to the nearest hundredth.

(d) Suppose that Christine is a superintendent who earned the average salary from January 1, 1985, to January 1, 1997. Write an integral to determine Christine's total salary from January 1, 1985, to January 1, 1997.

(e) Evaluate the integral from part (d) and interpret.

(f) Suppose that Joe is a business administrator who earned the average salary from January 1, 1985, to January 1, 1997. Write an integral to determine Joe's total salary from January 1, 1985, to January 1, 1997.

(g) Evaluate the integral from part (f) and interpret.

(h) Suppose that Rich is a teacher who earned the average salary from January 1, 1985, to January 1, 1997. Write an integral to determine Rich's total salary from January 1, 1985, to January 1, 1997.

(i) Evaluate the integral from part (h) and interpret.

SECTION 7.5   NUMERICAL INTEGRATION

In Chapters 6 and 7 we presented several techniques of integration. In spite of all the varied techniques of integration, we still encounter integrals that cannot be antidifferentiated by any of our methods. For example, $\int e^{-x^2}\, dx$ has no elementary antiderivative and cannot be evaluated using the Fundamental Theorem of Calculus. (Incidentally, $\int e^{-x^2}\, dx$ is related to the bell-shaped curve seen in probability and statistics courses.)

So how do we evaluate $\int_0^1 e^{-x^2}\, dx$? We approximate it numerically by interpreting it as an area under a curve. We have already seen how to approximate a definite integral using rectangles, and we have used the calculator's FNINT or $\int f(x)\, dx$ (or something similar) command by now. Hence, we are familiar with some numerical methods. In this section, we explore the Trapezoidal Rule as another numerical method. The main use, and beauty, of the Trapezoidal Rule is that it is ideally suited for discrete data.

Trapezoidal Rule

To gain an understanding of the Trapezoidal Rule, we need to review the process behind rectangular approximations.

EXAMPLE 1

Reviewing Rectangular Approximations

Consider $f(x) = \frac{1}{3}x^3 - 2.5x^2 + 4x + 4$. Use four left-endpoint rectangles to approximate $\int_2^6 \left(\frac{1}{3}x^3 - 2.5x^2 + 4x + 4 \right) dx$, that is, the area under the curve from $x = 2$ to $x = 6$.

SOLUTION

Figure 7.5.1 shows the graph of f along with the four rectangles. First, recall that:

- a is the left endpoint of the interval.
- b is the right endpoint of the interval.
- n is the number of rectangles.
- Δx is the width, or base, of each rectangle.

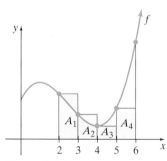

Figure 7.5.1

So here we have $a = 2$, $b = 6$, and $n = 4$. Then Δx is

$$\Delta x = \frac{b - a}{n} = \frac{6 - 2}{4} = 1$$

Each rectangle has a base $= \Delta x = 1$, and the height is given by f(left endpoint). With this in mind, we find the area of rectangle A_1 to be

$$A_1 = \text{base} \cdot \text{height}$$
$$= \Delta x \cdot f(\text{left endpoint})$$
$$= 1 \cdot f(2) = 1 \cdot \frac{14}{3} = \frac{14}{3}$$

For rectangle A_2 we get

$$A_2 = \Delta x \cdot f(\text{left endpoint})$$
$$= 1 \cdot f(3) = 1 \cdot \frac{5}{2} = \frac{5}{2}$$

In a similar fashion we compute the area of rectangles A_3 and A_4 to be

$$A_3 = \frac{4}{3} \quad \text{and} \quad A_4 = \frac{19}{6}$$

The sum of the four rectangles gives the rectangular approximation. This yields

$$\int_2^6 \left(\frac{1}{3}x^3 - 2.5x^2 + 4x + 4\right) dx \approx A_1 + A_2 + A_3 + A_4$$

$$= \frac{14}{3} + \frac{5}{2} + \frac{4}{3} + \frac{19}{6} = \frac{35}{3}$$

Interactive Activity

Evaluate $\int_2^6 \left(\frac{1}{3}x^3 - 2.5x^2 + 4x + 4\right) dx$ by determining an antiderivative and using the Fundamental Theorem of Calculus.

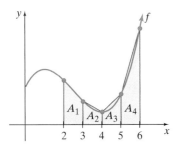

Figure 7.5.2

Recall that in Chapter 6 we improved on this approximation by placing smaller and smaller rectangles under the curve. Now we will investigate another numerical technique that may yield a better approximation with the same value for n. To do this we will use *trapezoids* rather than rectangles.

Let's look at Figure 7.5.2. Here we have redrawn $f(x) = \frac{1}{3}x^3 - 2.5x^2 + 4x + 4$, but instead of four rectangles we have drawn four trapezoids. It appears that the four trapezoids will yield a better approximation than the four rectangles did. All we need is to recall the formula for the area of a trapezoid:

$$A_{\text{trap}} = \frac{1}{2}(h_1 + h_2)b$$

where h_1 and h_2 represents the two heights of a trapezoid and b represents the base. *Remember that h_1 is parallel to h_2.* Let's rework the function in Example 1 with four trapezoids.

EXAMPLE 2

Approximating an Integral with Trapezoids

Consider $f(x) = \frac{1}{3}x^3 - 2.5x^2 + 4x + 4$ from Example 1. Use four trapezoids to approximate $\int_2^6 f(x)\, dx$, that is, the area under the graph of f from $x = 2$ to $x = 6$.

SOLUTION

Just as with rectangles we have $a = 2$, $b = 6$, $n = 4$, and $\Delta x = \frac{b-a}{n} = \frac{6-2}{2} = 1$. So each trapezoid has a base $= \Delta x = 1$. Notice that h_1 and h_2 are found to be f (left endpoint) and f (right endpoint), respectively. The areas of the four trapezoids are

$$A_1 = \frac{1}{2}(h_1 + h_2)b$$

$$= \frac{1}{2}[f(\text{left endpoint}) + f(\text{right endpoint})] \cdot b$$

$$= \frac{1}{2}[f(2) + f(3)] \cdot \Delta x$$

$$= \frac{1}{2}\left[\frac{14}{3} + \frac{5}{2}\right] \cdot 1 = \frac{43}{12}$$

$$A_2 = \frac{1}{2}(h_1 + h_2)b$$

$$= \frac{1}{2}[f(\text{left endpoint}) + f(\text{right endpoint})] \cdot b$$

$$= \frac{1}{2}[f(3) + f(4)] \cdot \Delta x$$

$$= \frac{1}{2}\left[\frac{5}{2} + \frac{4}{3}\right] \cdot 1 = \frac{23}{12}$$

$$A_3 = \frac{1}{2}(h_1 + h_2)b$$

$$= \frac{1}{2}[f(\text{left endpoint}) + f(\text{right endpoint})] \cdot b$$

$$= \frac{1}{2}[f(4) + f(5)] \cdot \Delta x$$

$$= \frac{1}{2}\left[\frac{4}{3} + \frac{19}{6}\right] \cdot 1 = \frac{9}{4}$$

$$A_4 = \frac{1}{2}(h_1 + h_2)b$$

$$= \frac{1}{2}[f(\text{left endpoint}) + f(\text{right endpoint})] \cdot b$$

$$= \frac{1}{2}[f(5) + f(6)] \cdot \Delta x$$

$$= \frac{1}{2}\left[\frac{19}{6} + 10\right] \cdot 1 = \frac{79}{12}$$

The sum of the four trapezoids gives the approximation

$$\int_2^6 \left(\frac{1}{3}x^3 - 2.5x^2 + 4x + 4\right) dx \approx A_1 + A_2 + A_3 + A_4$$

$$= \frac{43}{12} + \frac{23}{12} + \frac{9}{4} + \frac{79}{12} = \frac{43}{3} \qquad \blacksquare$$

As you discovered in the Interactive Activity following Example 1, the exact answer to $\int_2^6 \left(\frac{1}{3}x^3 - 2.5x^2 + 4x + 4\right) dx$ is $\frac{40}{3}$. We can see that four trapezoids is more accurate than four rectangles. Also notice that, just as we did with rectangular approximations, our subintervals are of equal length when using trapezoids.

We want to look back at our work in Example 2 and refer to Figure 7.5.3 in order to generalize the trapezoid process. To find the area of the trapezoids, we calculated

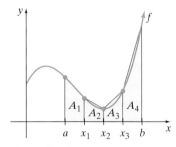

Figure 7.5.3

$$A_1 = \frac{1}{2}[f(a) + f(x_1)] \cdot \Delta x$$

$$A_2 = \frac{1}{2}[f(x_1) + f(x_2)] \cdot \Delta x$$

$$A_3 = \frac{1}{2}[f(x_2) + f(x_3)] \cdot \Delta x$$

$$A_4 = \frac{1}{2}[f(x_3) + f(b)] \cdot \Delta x$$

The total area of the four trapezoids is the sum

$A_1 + A_2 + A_3 + A_4$

$= \frac{1}{2}[f(a) + f(x_1)] \cdot \Delta x + \frac{1}{2}[f(x_1) + f(x_2)] \cdot \Delta x + \frac{1}{2}[f(x_2) + f(x_3)] \cdot \Delta x + \frac{1}{2}[f(x_3) + f(b)] \cdot \Delta x$

Factoring out a $\frac{1}{2}\Delta x$ yields

$$= \frac{\Delta x}{2}[f(a) + f(x_1) + f(x_1) + f(x_2) + f(x_2) + f(x_3) + f(x_3) + f(b)]$$

$$= \frac{\Delta x}{2}[f(a) + 2f(x_1) + 2f(x_2) + 2f(x_3) + f(b)]$$

The last equation is the Trapezoidal Rule. In general, if n trapezoids are used, then the approximate value of the definite integral is given by the trapezoidal rule, as follows:

Trapezoidal Rule

If f is continuous on $[a, b]$, then

$$\int_a^b f(x)\, dx \approx \frac{\Delta x}{2}[f(a) + 2f(x_1) + 2f(x_2) + \cdots + 2f(x_{n-1}) + f(b)]$$

where a is the left endpoint, b is the right endpoint, n is the number of trapezoids, and $\Delta x = \dfrac{b-a}{n}$. Notice that $x_1 = a + \Delta x$, $x_2 = x_1 + \Delta x$, and so on.

As with rectangular approximations, the larger the value of n the more trapezoids we will use and the smaller Δx will be. Also, the result of using larger values for n is a better approximation to the definite integral.

EXAMPLE 3	**Using the Trapezoidal Rule**

Use the trapezoidal rule with four trapezoids to approximate $\displaystyle\int_0^2 \frac{1}{x^2+16}\, dx$.

SOLUTION

We have $a = 0$, $b = 2$, $n = 4$, and $\Delta x = \frac{2-0}{4} = \frac{1}{2}$. So by the trapezoid rule we have

$$\int_0^2 \frac{1}{x^2+16}\, dx \approx \frac{\frac{1}{2}}{2}\left[f(0) + 2f\left(\frac{1}{2}\right) + 2f(1) + 2f\left(\frac{3}{2}\right) + f(2)\right]$$

$$\approx \frac{1}{4}[0.0625 + 0.1230 + 0.1176 + 0.1096 + 0.05]$$

$$\approx 0.1157$$

See Figure 7.5.4.

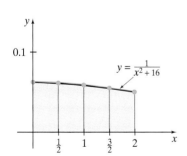

Figure 7.5.4

✓ **CHECKPOINT 1**

Now work Exercise 3.

Because of the structure of the Trapezoid Rule, it can be easily programmed into our calculator (or computer). We supply the following program for the TI-83. As always, if you do not have a TI-83, your code may be different. For this reason, we also supply a line-by-line explanation of what the code is performing. Also, consult www.prenhall.com/armstrong for programs.

TI-83 Program for Trapezoid Rule When Given $f(x)$	
Disp"ENTER FUNCTION IN Y_1"	Stores function in Y_1.
Prompt A, B, N	Asks for A, B, and N, the left endpoint, right endpoint, and number of trapezoids.
$(B - A)/N \rightarrow D$	Computes the base of each trapezoid.
$Y_1(A) + Y_1(B) \rightarrow S$	Evaluates function in Y_1 at A and B; adds the results and stores sum in S.
$A + D \rightarrow X$	Left endpoint plus base stored in X.
Lbl 1	Beginning of loop.
$2^*Y_1(X) + S \rightarrow S$	Two times function evaluated at X; result added to sum in S; result stored in S.
$X + D \rightarrow X$	Current X increased by base; result in X.
If $X < B$	If current $X <$ right endpoint, execute next line.
Then	Prior line true; execute next line.
Goto 1	Return to beginning of loop.
Else	If $X \geq B$, execute next line.
$S^*(D/2) \rightarrow S$	Current sum in S multiplied by D/2; result in S.
Disp"TRAP APPROX IS", S	Display trapezoid approximation.

To use the trapezoidal rule program to approximate $\int_{-1}^{1} e^{-x^2} dx$ using $n = 10, 50$, and 100 trapezoids, we first enter $y_1 = e^{-x^2}$ into our calculator. Next we execute the program. Results are listed in Table 7.5.1. Note that $a = -1$ and $b = 1$. We conclude that $\int_{-1}^{1} e^{-x^2} dx \approx 1.49$.

TABLE 7.5.1	
n	**TRAPEZOIDAL APPROXIMATION TO $\int_{-1}^{1} e^{-x^2} dx$**
10	1.48873668
50	1.493452053
100	1.493599214

NOTE: There is some error involved in the trapezoidal approximation. The formula to compute the error is beyond the scope of this text. However, it is worth noting that doubling the number of trapezoids reduces the maximum error by a factor of 4.

Trapezoidal Rule for Discrete Data

We have now seen how to use the Trapezoidal Rule to approximate $\int_{a}^{b} f(x) dx$. Many times, however, we have a set of data and we cannot determine a model for the data. In other words, we have no $f(x)$. In these circumstances, the trapezoidal rule is extremely powerful, as shown in Example 4.

EXAMPLE 4   **Using the Trapezoidal Rule for Discrete Data**

Table 7.5.2 shows the rate of change of revenue in millions of dollars per year for a bowling alley, where x represents the number of years since 1989.

TABLE 7.5.2

x	RATE OF CHANGE OF REVENUE
1	3.3
2	2
3	2.5
4	2.8
5	2.9
6	2.2
7	2

(a) Let $x = 1$ correspond to January 1, 1990, and y represent the rate of change of revenue in millions of dollars per year. Plot the data.

(b) Approximate the bowling alley's revenue from January 1, 1990, to January 1, 1996, by using the trapezoid rule with the given data.

SOLUTION

(a) Figure 7.5.5 gives a plot of the data.

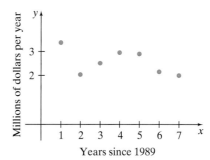

Figure 7.5.5

(b) Figure 7.5.6 is a plot of the data with the trapezoids sketched in as well. Here is where the power of the trapezoid rule comes through. Without a model for the data, we can use the data as is and approximate the revenue. To use the trapezoid rule, note that $a = 1$, $b = 7$, $n = 6$, and $\triangle x = 1$. So by the trapezoid rule we have

$$\frac{1}{2}[f(1) + 2f(2) + 2f(3) + 2f(4) + 2f(5) + 2f(6) + f(7)]$$

Now, instead of a formula for our function, we use the functional values from the data in our table! For example, from the table we know that $f(1) = 3.3$, $f(2) = 2$, and so on. Proceeding, we get the approximation to be

$$\frac{1}{2}[3.3 + 2(2) + 2(2.5) + 2(2.8) + 2(2.9) + 2(2.2) + 2] = 15.05$$

So the bowling alley's revenue from January 1, 1990, to January 1, 1996, was approximately 15.05 million dollars.

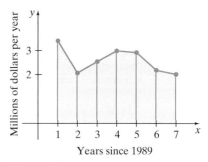

Figure 7.5.6

✓ **CHECKPOINT 2**

Now work Exercise 23.

Applying the trapezoidal rule to a set of data can be a bit tedious, especially if there are many data points. We supply the following program that performs the

TECHNOLOGY NOTE
Many calculators utilize lists to store data. Usually, they can be found under the STATS command. Consult the on-line calculator manual at www.prenhall.com/armstrong

trapezoidal rule on a set of data. Again, the code is for a TI-83 and the program makes use of Lists. In L_1 (List 1) we shall place the values for the independent variable, while in L_2 (List 2) we place the values of the dependent variable.

TI-83 Program for the Trapezoidal Rule When Given Data Points

Disp"NUMBER DATA POINTS"	Asks for the number of data points.
Prompt P	Enter number of data points.
$(L_1(P) - L_1(1))/(P - 1) \to D$	Computes the base of each trapezoid; stores in D.
$L_2(1) + L_2(P) \to S$	Sums first and last dependent variable values; stores sum in S.
$2 \to I$	2 stored in I.
Lbl 1	Beginning of loop.
$2*L_2(I) + S \to S$	Two times dependent variable value in current I added to value in S; result stored in S.
$I + 1 \to I$	Increment size of I.
If $I < P$	If updated I < P, execute next line.
Then	If previous line true, execute next line.
Goto 1	Return to beginning of loop.
Else	If $I \geq P$, execute next line.
$S*(D/2) \to S$	Current value in S times (D/2) stored in S.
Disp"TRAP APPROX IS",S	Display trapezoidal approximation.

TABLE 7.5.3

x	RATE OF CHANGE OF REVENUE
1	2.5
2	2.8
3	1.2
4	0.9
5	2.7
6	2.1
7	3.1

Figure 7.5.7

Table 7.5.3 shows the rate of change of revenue in millions of dollars per year for a restaurant, where x represents the number of years since 1989. To approximate the restaurant's revenue from 1990 to 1996 using the trapezoidal rule program for data points, we first enter the data in the x column into L_1 and the rate of change of revenue column into L_2. Figure 7.5.7 shows our screen after doing this. We now run our trapezoidal rule program for data points. We are first asked how many data points and we enter a 7. The program returns 12.5. So we conclude that from 1990 to 1996 the restaurant's revenue was approximately 12.5 million dollars.

EXAMPLE 5

Using the Trapezoidal Rule to Analyze Sulfur Dioxide Emissions

Table 7.5.4 shows the U.S. sulfur dioxide emissions in thousands of tons from 1940 to 1990. Figure 7.5.8 gives a plot of the data, where $x = 1$ corresponds to the beginning of 1940 and y represents the sulfur dioxide emissions in thousands of tons.

TABLE 7.5.4

X	SULFUR DIOXIDE EMISSIONS
1	19,953
11	22,358
21	22,227
31	31,161
41	25,905
51	22,433

Source: U.S. Census Bureau web site.

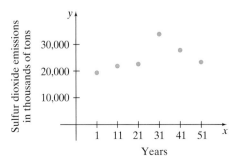

Figure 7.5.8

(a) Approximate the total amount of sulfur dioxide emissions by the United States from 1940 to 1990.

(b) Approximate the average amount of sulfur dioxide emissions on a yearly basis from 1940 to 1990.

SOLUTION

(a) We enter the data into our calculator, just as we did to estimate the restaurant revenue, and run the Trapezoidal Rule Program for data points. Note here that we have six data points. The result we get from the program is 1,228,440. Since this is in thousands of tons, we conclude that from 1940 to 1990 the amount of U.S. sulfur dioxide emissions was approximately 1,228,440,000 tons.

(b) To compute this average value, we take the result from part (a) and divide by $b - a$, just as we learned in Section 7.1. This yields

$$\frac{1,228,440,000}{51 - 1} = \frac{1,228,440,000}{50} = 24,568,800$$

We conclude that from 1940 to 1990 the average value of the U.S. sulfur dioxide emissions was approximately 24,568,800 tons. ◼

✓ **CHECKPOINT 3**

Now work Exercise 35.

NOTE: Both trapezoidal rule programs are designed for subintervals of equal length. When using the trapezoidal rule program for data points, the independent variable values should be equally spaced.

Simpson's Rule

Another popular numerical integration technique is called **Simpson's Rule.** Simpson's Rule utilizes adjacent parabolic segments over a closed interval $[a, b]$. The sum of the areas below these parabolic segments approximates $\int_a^b f(x)\, dx$. Approximations using Simpson's Rule are very accurate, but it requires an *even* number n of equal subintervals. This could be limiting if our data do not have an even number of data points. For the curious, Simpson's Rule can be found in almost any engineering calculus textbook. We will not pursue this numerical technique in this text.

SUMMARY

In this section we introduced our final numerical technique for integration, the **trapezoidal rule.** At this time the numerical techniques that we have discussed include:

• Rectangular approximations
• Built-in numerical integration on the calculator
• Trapezoidal approximations

We supplied two programs for the trapezoidal rule: one if we are given a function and another if we do not have a function, or we cannot determine one, for a set of data. Applications utilizing the trapezoidal rule program for data points were discussed. We want to stress that if some data are *not* easily modeled, the trapezoidal rule is a powerful weapon to analyze the data. The Trapezoidal

Rule assumes that the independent variable values of the data points are equally spaced.

SECTION 7.5 EXERCISES

In Exercises 1–8, use the Trapezoidal Rule to approximate the given definite integral using $n = 4$ trapezoids. Do not use a calculator program for these exercises.

1. $\int_{-1}^{3} x^3 \, dx$

2. $\int_{0}^{2} x^4 \, dx$

✓ 3. $\int_{1}^{3} \frac{5x}{5 + 2x^2} \, dx$

4. $\int_{0}^{2} \frac{1}{(1 + 2x)^2} \, dx$

5. $\int_{0}^{1} x^2 e^x \, dx$

6. $\int_{0}^{1} x^5 e^x \, dx$

7. $\int_{3}^{5} (\ln x)^2 \, dx$

8. $\int_{2}^{5} (\ln x)^3 \, dx$

In Exercises 9–22, use the TI-83 program for the Trapezoidal Rule when given $f(x)$ to approximate the given definite integral using $n = 10$, $n = 50$, and $n = 100$ trapezoids.

9. $\int_{0}^{2} x^4 \, dx$

10. $\int_{-1}^{3} x^3 \, dx$

11. $\int_{0}^{2} \frac{1}{(1 + 2x)^2} \, dx$

12. $\int_{1}^{3} \frac{5x}{5 + 2x^2} \, dx$

13. $\int_{1}^{4} \frac{e^{2x}}{1 + e^{4x}} \, dx$

14. $\int_{1}^{2} \frac{e^x - e^{-x}}{2} \, dx$

15. $\int_{1}^{2} \frac{e^x + e^{-x}}{2} \, dx$

16. $\int_{0}^{1} x^2 e^x \, dx$

17. $\int_{0}^{1} x^5 e^x \, dx$

18. $\int_{3}^{5} (\ln x)^2 \, dx$

19. $\int_{2}^{5} (\ln x)^3 \, dx$

20. $\int_{0}^{2} \frac{2x + 3}{\sqrt{x^2 + 2x + 5}} \, dx$

21. $\int_{1}^{3} \frac{x + 2}{\sqrt[3]{(6x - x^2)}} \, dx$

22. $\int_{0}^{3} e^{x^2} \, dx$

APPLICATIONS

In Exercises 23–28, use the Trapezoidal Rule. For these exercises do not use any calculator program.

✓ 23. The following data give the rate of change of revenue in thousands of dollars per week for Sparkle Time Car Wash. Here x represents the number of weeks after April 1. Approximate Sparkle Time's revenue for this period of time.

x	RATE OF CHANGE OF REVENUE
1	2.35
2	3.75
3	1.25
4	2.5
5	1.75

24. The following data give the rate of change of sales, in millions of dollars per month, for Spring Brook Mall. Here x corresponds to the month of the year. Approximate Spring Brook Mall's total sales for this period of time.

x	RATE OF CHANGE OF SALES
1	3.35
2	2.69
3	1.72
4	1.11
5	2.35
6	2.51

25. The following data give the marginal cost for different levels of production at Muggy Inc. Here x represents the number of coffee mugs produced and $C'(x)$ is in dollars per mug. Approximate the total cost in going from a production level of 10 mugs to 50 mugs.

x	10	20	30	40	50
$C'(x)$	1.5	2	2.4	3.1	3.5

26. The following data give the marginal cost for different levels of production at Binky Inc. Here x represents the number of pacifiers produced and $C'(x)$ is in dollars per pacifier. Approximate the total cost in going from a production level of 220 pacifiers to 300 pacifiers.

x	220	240	260	280	300
$C'(x)$	1.18	1.25	1.36	1.48	1.55

27. The Beef Place keeps a monthly record of the rate of increase in profits, $P'(x)$, during the first 5 months of the year.

The following are the data where x is the month of the year and $P'(x)$ is measured in thousands of dollars per month. Approximate the total profit during this period of time.

x	1	2	3	4	5
$P'(x)$	0.62	0.77	0.78	0.85	1.03

28. The amount of energy produced, in quadrillion Btu, by the United States from the beginning of 1985 to the beginning of 1993 is listed in the following table. Here $x = 1$ corresponds to the beginning of 1985.

x	BTU PRODUCED
1	84.55
3	84.73
5	86.85
7	89.31
9	88.52

Source: U.S. Census Bureau web site.

Approximate the total amount of energy produced by the United States from the beginning of 1985 to the beginning of 1993.

For the remainder of the exercises, use the TI-83 program for the Trapezoidal Rule when given data points, where appropriate.

29. The following data give the rate of change of revenue, in thousands of dollars per week, for Timber!, a local tree trimming company. Here x represents the number of weeks after May 1. Approximate the company's revenue for this time period.

x	RATE OF CHANGE OF REVENUE
1	2.25
2	3.75
3	1.35
4	2.55
5	1.85
6	2.12
7	3.11
8	2.11
9	1.50
10	3.12

30. Anthony has been keeping a monthly record of a company's rate of increase in profits, $P'(x)$, over a 1-year period. The following are the data that Anthony collected, where x is the month of the year and $P'(x)$ is in thousands of dollars per month. Approximate the total profit for this time period.

x	$P'(x)$
1	6.2
2	7.7
3	7.8
4	8.5
5	10.3
6	10.2
7	10.7
8	11.1
9	12.1
10	12.3
11	12.1
12	12.5

31. Tina has been asked to determine the marginal cost for different levels of production. She collected the following data, where x represents the number of items produced and $C'(x)$ is in hundreds of dollars per item. Use the data that Tina collected to determine the total cost of producing 20 items.

x	0	2	4	6	8	10	12	14	16	18	20
$C'(x)$	7	11	13	16	18	21	24	28	31	35	39

32. Vanessa is the leader of a sales management team for a company, and her team has been asked to prepare a year-end report. Vanessa's team gathered data showing the rate of change of sales, $s'(t)$, in millions of dollars per month, for the prior year, where t is measured in months. The following are the data that were collected.

t	$s'(t)$
0	0.8
1	0.7
2	1.13
3	1.44
4	0.68
5	0.22
6	1.01
7	1.51
8	2.13
9	1.78
10	1.42
11	0.77
12	0.96

Use the data that Vanessa's team gathered to approximate the total sales for the year.

33. The following data give the number of females, single and married, in the U.S. labor force, in thousands, from 1960

to 1995. Here x is in years where $x = 1$ corresponds to the beginning of 1960.

x	SINGLE	MARRIED
1	5,410	12,893
6	5,976	14,829
11	7,265	18,475
16	9,125	21,484
21	11,865	24,980
26	13,163	27,894
31	14,612	30,901
36	15,467	33,359

Source: U.S. Census Bureau web site.

(a) Approximate the average number of single females, on a yearly basis, in the U.S. labor force from the beginning of 1960 to the beginning of 1995.

(b) Approximate the average number of married females, on a yearly basis, in the U.S. labor force from the beginning of 1960 to the beginning of 1995.

In Exercises 34, 35, and 36, use the following data for U.S. air pollution emissions from 1940 to 1990 measured in thousands of tons per year. Here x is in years where $x = 1$ corresponds to the beginning of 1940.

x	NITROGEN DIOXIDES	VOLATILE ORGANIC COMPOUNDS	CARBON MONOXIDE
1	7,374	17,161	93,615
11	10,093	20,936	102,609
21	14,140	24,459	109,745
31	20,625	30,646	128,079
41	23,281	25,893	115,625
51	23,038	23,599	100,650

Source: U.S. Census Bureau web site.

34. (a) Approximate the total amount of nitrogen dioxide emissions in the United States from the beginning of 1940 to the beginning of 1990.

(b) Approximate the average amount of nitrogen oxide emissions on a yearly basis from the beginning of 1940 to the beginning of 1990.

✓ 35. (a) Approximate the total amount of volatile organic compounds emissions in the United States from the beginning of 1940 to the beginning of 1990.

(b) Approximate the average amount of volatile organic compounds on a yearly basis from the beginning of 1940 to the beginning of 1990.

36. (a) Approximate the total amount of carbon monoxide emissions in the United States from the beginning of 1940 to the beginning of 1990.

(b) Approximate the average amount of carbon monoxide emissions on a yearly basis from the beginning of 1940 to the beginning of 1990.

In Exercises 37 and 38, use the following data on work stoppages in the United States from 1960 to 1995. Here x is in years where $x = 1$ corresponds to the beginning of 1960, and the number of workers involved are in thousands.

x	NUMBER OF WORK STOPPAGES	NUMBER OF WORKERS INVOLVED
1	222	896
6	268	999
11	381	2468
16	235	965
21	187	795
26	54	324
31	44	185
36	31	192

Source: U.S. Census Bureau web site.

37. (a) Approximate the total number of work stoppages in the United States from the beginning of 1960 to the beginning of 1995.

(b) Approximate the average number of work stoppages, on a yearly basis, in the United States from the beginning of 1960 to the beginning of 1995.

38. (a) Approximate the total number of workers involved in work stoppages in the United States from the beginning of 1960 to the beginning of 1995.

(b) Approximate the average number of workers involved in work stoppages, on a yearly basis, in the United States from the beginning of 1960 to the beginning of 1995.

In Exercises 39 and 40, use the following data for the number of high school graduates, male and female, in thousands from 1960 to 1992. Here x is in years where $x = 1$ corresponds to the beginning of 1960.

x	MALES	FEMALES
1	756	923
5	997	1148
9	1184	1422
13	1420	1541
17	1450	1537
21	1500	1589
25	1429	1583
29	1334	1339
33	1216	1182

Source: U.S. Census Bureau web site.

39. (a) Approximate the total number of males who graduated from high school from the beginning of 1960 to the beginning of 1992.

(b) Approximate the average number of males who graduated from high school, on a yearly basis, from the beginning of 1960 to the beginning of 1992.

40. (a) Approximate the total number of females who graduated from high school from the beginning of 1960 to the beginning of 1992.

(b) Approximate the average number of females who graduated from high school, on a yearly basis, from the beginning of 1960 to the beginning of 1992.

41. The following data give the cost in constant 1987 dollars for the regulation and monitoring of pollution abatement in the United States. Here x is in years where $x = 1$ corresponds to the beginning of 1985, and the regulation and monitoring costs are in billions of constant 1987 dollars.

x	1	3	5	7	9
Costs	1.361	1.519	1.657	1.654	1.656

Source: U.S. Census Bureau web site.

Approximate the total cost in constant 1987 dollars spent on regulation and monitoring from the beginning of 1985 to the beginning of 1993.

42. Crude oil imports, in millions of barrels per year, into the United States from 1970 to 1995 are given in the following table. Here x is in years where $x = 1$ corresponds to the beginning of 1970.

x	Number of Barrels (in millions)
1	483
6	1498
11	1926
16	1168
21	2151
26	2608

Source: U. S. Census Bureau web site.

(a) Approximate the total number of barrels of oil imported into the United States from the beginning of 1970 to the beginning of 1995.

(b) Approximate the average number of barrels, on a yearly basis, of oil imported into the United States from the beginning of 1970 to the beginning of 1995.

(c) A barrel contains 42 gallons. How many total gallons of oil were imported into the United States from the beginning of 1970 to the beginning of 1995?

43. The following data give the number of pregnancies in the United States from 1976 to 1992. Here x is in years where $x = 1$ corresponds to the beginning of 1976, and the number of pregnancies is in millions.

x	Number
1	5.002
3	5.433
5	5.912
7	6.024
9	6.019
11	6.129
13	6.341
15	6.668
17	6.484

Source: U.S. Census Bureau web site.

(a) Approximate the total number of pregnancies in the United States from the beginning of 1976 to the beginning of 1992.

(b) Approximate the average number of pregnancies, on a yearly basis, in the United States from the beginning of 1976 to the beginning of 1992.

In Exercises 44 and 45, use the following data for the first-year enrollment of students pursuing a registered nursing degree, in thousands, where the degree is either a 4-year baccalaureate or a 2-year associate degree. Here x is in years where $x = 1$ corresponds to the beginning of 1985.

x	4-Year Degree	2-Year Degree
1	39.573	63.776
3	28.026	54.330
5	29.042	63.973
7	33.437	69.869
9	41.290	75.382
11	43.451	76.016

Source: U.S. Census Bureau web site.

44. Approximate the average number of first-year registered nursing students in a 4-year baccalaureate degree program from the beginning of 1985 to the beginning of 1995.

45. Approximate the average number of first-year registered nursing students in a 2-year associate degree program from the beginning of 1985 to the beginning of 1995.

Section Project

The following data give the average tuition and fees as well as room and board charges, in dollars, from 1985 to 1995 for public and private 2- and 4-year colleges. Here x is in years where $x = 1$ corresponds to the beginning of 1985.

PUBLIC				
	TUITION AND FEES		ROOM AND BOARD	
x	2-YEAR COLLEGES	4-YEAR COLLEGES	2-YEAR COLLEGES	4-YEAR COLLEGES
1	584	1386	2223	2513
3	660	1651	2328	2819
5	730	1846	2453	3059
7	824	2159	2644	3425
9	1025	2604	2774	3838
11	1194	2982	2955	4100

Source: U.S. Census Bureau web site.

PRIVATE				
	TUITION AND FEES		ROOM AND BOARD	
x	2-YEAR COLLEGES	4-YEAR COLLEGES	2-YEAR COLLEGES	4-YEAR COLLEGES
1	3485	6843	2718	3400
3	3684	8118	2700	4160
5	4817	9451	3149	4622
7	5570	11,379	3733	5124
9	6059	13,055	3845	5843
11	6865	14,510	4194	6500

Source: U.S. Census Bureau web site.

(a) Approximate the average tuition and fees on a yearly basis from 1985 to 1995 at public 2-year colleges.

(b) Approximate the average tuition and fees on a yearly basis from 1985 to 1995 at private 2-year colleges.

(c) Approximate the average room and board on a yearly basis from 1985 to 1995 at public 2-year colleges.

(d) Approximate the average room and board on a yearly basis from 1985 to 1995 at private 2-year colleges.

(e) Approximate the average tuition and fees on a yearly basis from 1985 to 1995 at public 4-year colleges.

(f) Approximate the average tuition and fees on a yearly basis from 1985 to 1995 at private 4-year colleges.

(g) Approximate the average room and board on a yearly basis from 1985 to 1995 at public 4-year colleges.

(h) Approximate the average room and board on a yearly basis from 1985 to 1995 at private 4-year colleges.

(i) Assume the average yearly amount determined in parts (e) and (f) for a 4-year degree. How much more does it cost in tuition and fees for 4 years at a private 4-year college than at a public 4-year college?

(j) Assume the average yearly amount determined in parts (a) and (b) for a 2-year degree. How much more does it cost in tuition and fees for 2 years at a private 2-year college than at a public 2-year college?

SECTION 7.6 IMPROPER INTEGRALS

So far all the definite integrals we have considered have been on a closed bounded interval $[a, b]$. In this section, we relax this condition and define integrals over an unbounded interval such as $(-\infty, b]$, $[a, \infty)$, or even $(-\infty, \infty)$. Integrals defined on unbounded intervals are called **improper integrals** and have many applications. To understand what is happening with improper integrals, we need a brief review of limits at infinity.

Limits at Infinity

In Chapter 2 we discussed the limit concept. We saw that the definitions of the derivative and the definite integral demonstrate how central the limit concept is to calculus. Let's revisit limits at infinity and recall some concepts from our work in Chapter 2. Consult the Toolbox to the left.

Let's quickly review some limits. Remembering that the limit of a constant is that constant yields

$$\lim_{b \to \infty} \left(3 - \frac{1}{b^2} \right) = \lim_{b \to \infty} 3 - \lim_{b \to \infty} \frac{1}{b^2} = 3 - 0 = 3$$

Employing other results in the Toolbox gives

$$\lim_{b \to \infty} (e^{-2b} - 3) = \lim_{b \to \infty} e^{-2b} - \lim_{b \to \infty} 3 = 0 - 3 = -3$$

From Your Toolbox

- For $f(x) = b^x$, b a real number where $b > 1$,

(a) $\lim_{x \to \infty} b^x = \infty$ and
$\lim_{x \to -\infty} b^x = 0$.

(b) $\lim_{x \to \infty} b^{-x} = 0$ and
$\lim_{x \to -\infty} b^{-x} = \infty$.

- $\lim_{x \to \infty} \ln x = \infty$.

- $\lim_{x \to \infty} \frac{k}{x^n} = 0$ and
$\lim_{x \to -\infty} \frac{k}{x^n} = 0$, where k is any real constant and n is a positive real number.

Improper Integrals

At this time one may be wondering, "Why are we discussing limits at this point in the book?" The answer becomes apparent as we consider Examples 1 and 2.

EXAMPLE 1

Evaluating Definite Integrals Using the Fundamental Theorem of Calculus

Consider $f(x) = \frac{1}{x^3}$. Determine the following.

(a) $\int_1^5 \frac{1}{x^3}\, dx$ (b) $\int_1^{10} \frac{1}{x^3}\, dx$ (c) $\int_1^b \frac{1}{x^3}\, dx$

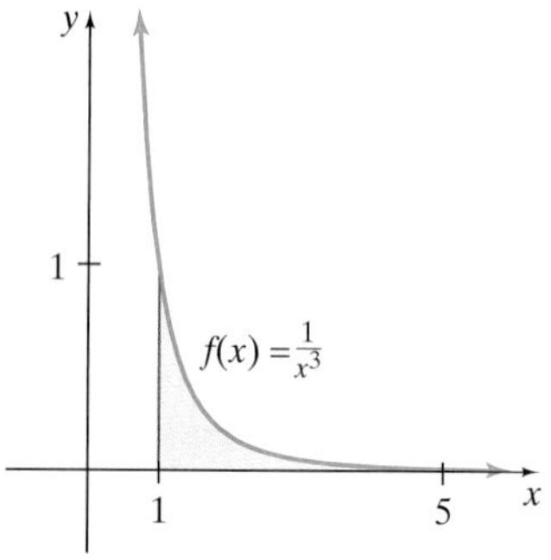

SOLUTION

(a) Evaluation of $\int_1^5 \frac{1}{x^3}\, dx$ will give us the area of the shaded region shown in Figure 7.6.1.

$$\int_1^5 \frac{1}{x^3}\, dx = \int_1^5 x^{-3}\, dx = -\frac{1}{2}x^{-2}\Big|_1^5 = -\frac{1}{2x^2}\Big|_1^5$$

$$= -\frac{1}{2(5)^2} - -\frac{1}{2(1)^2} = \frac{24}{50}, \quad \text{or } 0.48$$

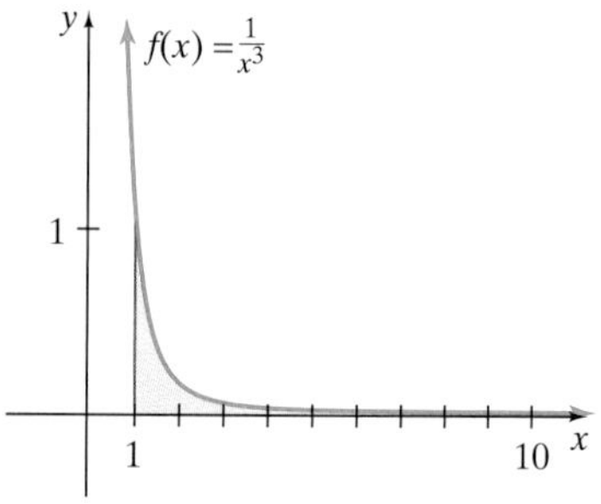

Figure 7.6.1

(b) Evaluation of $\int_1^{10} \frac{1}{x^3}\, dx$ will give us the area shown shaded in Figure 7.6.2.

$$\int_1^{10} \frac{1}{x^3}\, dx = \int_1^{10} x^{-3}\, dx = -\frac{1}{2}x^{-2}\Big|_1^{10} = -\frac{1}{2x^2}\Big|_1^{10}$$

$$= -\frac{1}{2(10)^2} - -\frac{1}{2(1)^2} = \frac{99}{200}, \quad \text{or } 0.495$$

Figure 7.6.2

(c) Here we are asked to determine the area under $f(x) = \frac{1}{x^3}$ from $x = 1$ to some arbitrary point $x = b$, as shown in Figure 7.6.3. The process employed in parts (a) and (b) does not change.

$$\int_1^b \frac{1}{x^3}\, dx = \int_1^b x^{-3}\, dx = -\frac{1}{2}x^{-2}\Big|_1^b = -\frac{1}{2x^2}\Big|_1^b$$

$$= -\frac{1}{2(b)^2} - -\frac{1}{2(1)^2} = -\frac{1}{2b^2} + \frac{1}{2}$$

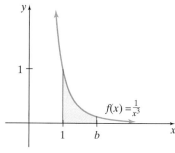

Figure 7.6.3

We are now ready to make sense out of integrating over an unbounded region.

EXAMPLE 2

Evaluating an Improper Integral

Consider the graph of $f(x) = \frac{1}{x^3}$ and the shaded region shown in Figure 7.6.4. The shaded region is unbounded; that is, it continues indefinitely to the right. We represent the area of this unbounded region by

$$\int_1^\infty \frac{1}{x^3}\, dx$$

which is an example of an *improper integral*. Evaluate this integral.

SOLUTION

To do this, we employ a strategy almost identical to Example 1c. That is, we pick an arbitrary point to the right of $x = 1$, call it b, and integrate $\int_1^b \frac{1}{x^3}\,dx$. We know from Example 1c that

$$\int_1^b \frac{1}{x^3}\,dx = -\frac{1}{2b^2} + \frac{1}{2}$$

Now we take the limit as $b \to \infty$ of this result and get

$$\lim_{b\to\infty}\left(-\frac{1}{2b^2} + \frac{1}{2}\right) = \lim_{b\to\infty}\left(-\frac{1}{2b^2}\right) + \lim_{b\to\infty}\frac{1}{2}$$

$$= 0 + \frac{1}{2} = \frac{1}{2}$$

What we really did in this example is compute the following:

$$\lim_{b\to\infty}\left[\int_1^b \frac{1}{x^3}\,dx\right]$$

Notice that this evaluation took place in two stages: first we integrated using the Fundamental Theorem of Calculus; then we did the limit.

It may contradict our intuition that an unbounded region, as in Figure 7.6.4, has a finite area of $\frac{1}{2}$ square unit, but our intuition is somewhat clouded when we encounter the concept of infinity. Do not despair. Some of the greatest minds in history have struggled with the concept of infinity.

At this time we generalize what we did in Example 2.

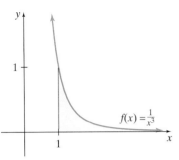

Figure 7.6.4

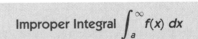

Improper Integral $\int_a^\infty f(x)\,dx$

Let f be continuous and nonnegative for any $x \geq a$. Then

$$\int_a^\infty f(x)\,dx = \lim_{b\to\infty}\int_a^b f(x)\,dx$$

provided the limit exists. If the limit exists, we say the improper integral $\int_a^\infty f(x)\,dx$ **converges.** If the limit does not exist, we say the improper integral $\int_a^\infty f(x)\,dx$ **diverges.**

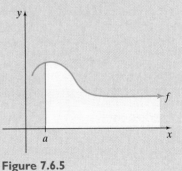

Figure 7.6.5

EXAMPLE 3 | **Evaluating an Improper Integral**

Evaluate

(a) $\int_1^\infty \frac{1}{x^2}\,dx$

(b) $\int_1^\infty \frac{1}{x}\,dx$

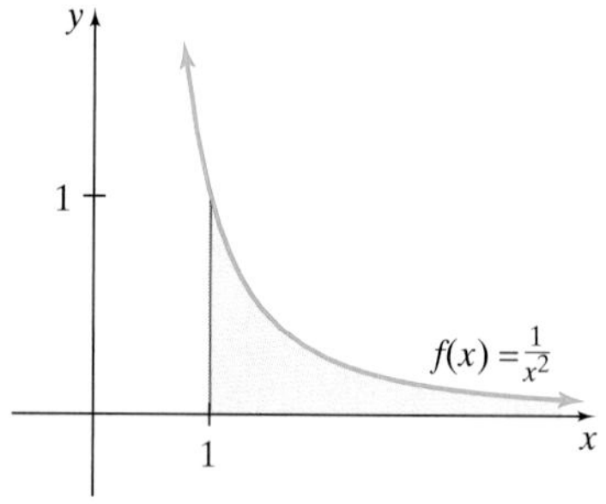

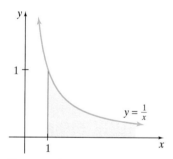

Figure 7.6.6

SOLUTION

(a) Evaluating $\int_1^\infty \frac{1}{x^2}\,dx$ will give us the area of the shaded region in Figure 7.6.6. We begin by rewriting the integral as

$$\int_1^\infty \frac{1}{x^2} = \lim_{b\to\infty}\left[\int_1^b \frac{1}{x^2}\,dx\right]$$

We perform the integration as follows:

$$\int_1^b \frac{1}{x^2}\,dx = \int_1^b x^{-2}\,dx = -x^{-1}\Big|_1^b$$

$$= -\frac{1}{x}\Big|_1^b = -\frac{1}{b} + 1$$

Next we evaluate the limit to get

$$\lim_{b\to\infty}\left(-\frac{1}{b} + 1\right) = \lim_{b\to\infty}\left(-\frac{1}{b}\right) + \lim_{b\to\infty} 1$$

$$= 0 + 1 = 1$$

So the improper integral, $\int_1^\infty \frac{1}{x^2}\,dx$, converges and has a value of 1.

(b) Evaluating $\int_1^\infty \frac{1}{x}\,dx$ will give us the area shaded in Figure 7.6.7. Proceeding as in part (a) gives

$$\int_1^\infty \frac{1}{x}\,dx = \lim_{b\to\infty}\left[\int_1^b \frac{1}{x}\,dx\right]$$

Integrating gives

$$\int_1^b \frac{1}{x}\,dx = \ln x\Big|_1^b$$

$$= \ln b - \ln 1 = \ln b$$

Evaluating the limit of this result gives

$$\lim_{b\to\infty}(\ln b) = \infty$$

This means the improper integral, $\int_1^\infty \frac{1}{x}\,dx$, diverges.

Figure 7.6.7

✓ **CHECKPOINT 1**

Now work Exercise 5.

We will discuss two other types of improper integrals in this section. Consider the shaded areas in Figures 7.6.8 and 7.6.9. To determine the area under $f(x) = \frac{1}{x^2}$ on

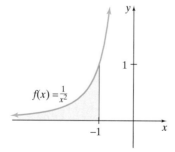

Figure 7.6.8

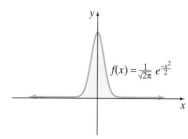

Figure 7.6.9

the interval $(-\infty, -1]$, as shown in Figure 7.6.8, requires that we evaluate the **improper integral** $\int_{-\infty}^{-1} \frac{1}{x^2} \, dx$. To determine the area under $f(x) = \frac{1}{\sqrt{2\pi}} e^{-(x^2/2)}$ on the interval $(-\infty, \infty)$, as shown in Figure 7.6.9, requires that we evaluate the **improper integral** $\int_{-\infty}^{\infty} \frac{1}{\sqrt{2\pi}} e^{-(x^2/2)} \, dx$. (The graph in Figure 7.6.9 is called the *normal curve,* which is seen frequently in statistics and probability.) We can evaluate these improper integrals in a manner that is similar to evaluating $\int_{a}^{\infty} f(x) \, dx$.

Improper Integrals $\int_{-\infty}^{b} f(x) \, dx$ **and** $\int_{-\infty}^{\infty} f(x) \, dx$

Let f be continuous and nonnegative on the indicated interval. Then we have:

1. $\displaystyle \int_{-\infty}^{b} f(x) \, dx = \lim_{a \to -\infty} \int_{a}^{b} f(x) \, dx$

2. $\displaystyle \int_{-\infty}^{\infty} f(x) \, dx = \int_{-\infty}^{c} f(x) \, dx + \int_{c}^{\infty} f(x) \, dx$

In 2, c is any point in $(-\infty, \infty)$, provided that both improper integrals on the right exist. Again, if the limits exist we say that the improper integral **converges**. If the limit does not exist, we say the improper integral **diverges**.

NOTE: If either integral on the right in the second formula in the box diverges, then the improper integral diverges.

EXAMPLE 4

Evaluating an Improper Integral

Evaluate $\int_{-\infty}^{0} e^x \, dx$.

SOLUTION

First we rewrite the improper integral as

$$\int_{-\infty}^{0} e^x \, dx = \lim_{a \to -\infty} \left[\int_{a}^{0} e^x \, dx \right]$$

Just as we did in the previous example, we will integrate first.

$$\int_{a}^{0} e^x \, dx = e^x \Big|_{a}^{0} = e^0 - e^a = 1 - e^a$$

Next we take the limit of this result and get

$$\lim_{a \to -\infty} (1 - e^a) = \lim_{a \to -\infty} 1 - \lim_{a \to -\infty} e^a$$
$$= 1 - 0 = 1$$

So the improper integral $\int_{-\infty}^{0} e^x \, dx$ converges and has a value of 1.

✓ CHECKPOINT **2**

Now work Exercise 17.

EXAMPLE 5

Evaluating an Improper Integral

Evaluate $\int_{-\infty}^{\infty} \frac{4x^3}{(1+x^4)^2} \, dx$.

SOLUTION

We begin by rewriting the improper integral as

$$\int_{-\infty}^{\infty} \frac{4x^3}{(1+x^4)^2}\, dx = \int_{-\infty}^{0} \frac{4x^3}{(1+x^4)^2}\, dx + \int_{0}^{\infty} \frac{4x^3}{(1+x^4)^2}\, dx$$

$$= \lim_{a\to-\infty} \int_{a}^{0} \frac{4x^3}{(1+x^4)^2}\, dx + \lim_{b\to\infty} \int_{0}^{b} \frac{4x^3}{(1+x^4)^2}\, dx$$

To integrate $\int_{a}^{0} \frac{4x^3}{(1+x^4)^2}\, dx$, we use u-substitution. Let $u = 1 + x^4$; then $du = 4x^3\, dx$. (We opt to leave off limits until we have an antiderivative.) Proceeding, we have

$$\int \frac{4x^3}{(1+x^4)^2}\, dx = \int \frac{1}{u^2}\, du = \int u^{-2}\, du$$

$$= -u^{-1} = -\frac{1}{u}$$

$$= -\frac{1}{(1+x^4)}\Bigg|_{a}^{0} = -1 + \frac{1}{1+a^4}$$

We now take the limit and get

$$\lim_{a\to-\infty} \left(-1 + \frac{1}{1+a^4}\right) = \lim_{a\to-\infty}(-1) + \lim_{a\to-\infty}\left(\frac{1}{1+a^4}\right)$$

$$= -1 + 0 = -1$$

Integrating $\int_{0}^{b} \frac{4x^3}{(1+x^4)^2}\, dx$ will result in the same antiderivative as just computed, since it has the same integrand, just different limits. Hence, we can immediately write

$$\int_{0}^{b} \frac{4x^3}{(1+x^4)^2}\, dx = -\frac{1}{(1+x^4)}\Bigg|_{0}^{b} = -\frac{1}{1+b^4} + 1$$

Taking the limit of this result yields

$$\lim_{b\to\infty}\left(-\frac{1}{1+b^4} + 1\right) = \lim_{b\to\infty}\left(-\frac{1}{1+b^4}\right) + \lim_{b\to\infty} 1$$

$$= 0 + 1 = 1$$

Putting this all together results in

$$\int_{-\infty}^{\infty} \frac{4x^3}{(1+x^4)^2}\, dx = \int_{-\infty}^{0} \frac{4x^3}{(1+x^4)^2}\, dx + \int_{0}^{\infty} \frac{4x^3}{(1+x^4)^2}\, dx$$

$$= \lim_{a\to-\infty} \int_{a}^{0} \frac{4x^3}{(1+x^4)^2}\, dx + \lim_{b\to\infty} \int_{0}^{b} \frac{4x^3}{(1+x^4)^2}\, dx$$

$$= -1 + 1 = 0$$

So the improper integral converges.

✓ **CHECKPOINT 3**

Now work Exercise 31.

Applications

One family of applications of improper integrals is in estimating the total amount of oil or natural gas that will be produced by a well, given its production rate. Example 6 illustrates this.

EXAMPLE 6 | **Analyzing Total Oil Production**

Petroleum engineers estimate that an oil field will produce oil at a rate given by $f(t) = 600e^{-0.1t}$ thousand barrels per month, t months into production. Estimate the total amount of oil produced by this well.

SOLUTION

The total amount of oil produced by the oil well in T months of production is given by

$$\int_0^T 600e^{-0.1t} \, dt$$

Since we want to know the potential output of the well, we assume that the well will operate indefinitely. Thus, the total amount of oil produced is given by

$$\int_0^\infty 600e^{-0.1t} \, dt = \lim_{T \to \infty} \left[\int_0^T 600e^{-0.1t} \, dt \right]$$

As we have seen in this section, we integrate first and then apply the limit concept. Integration yields

$$\int_0^T 600e^{-0.1t} \, dt = \frac{600}{-0.1} e^{-0.1t} \Big|_0^T = -6000e^{-0.1t} \Big|_0^T$$

$$= -6000e^{-0.1T} + 6000e^{-0.1(0)}$$

$$= -6000e^{-0.1T} + 6000$$

Applying the limit results in

$$\lim_{T \to \infty} (-6000e^{-0.1T} + 6000) = 0 + 6000 = 6000$$

So the total production of the oil well is estimated to be 6000 thousand, or 6,000,000 barrels.

Capital Value

We begin our study of capital value by reconsidering the idea of the present value of a continuous income stream.

Flashback | **CONTINUOUS INCOME STREAM REVISITED**

In Section 7.4 we analyzed the present value of a continuous income stream. Recall that if $f(t)$ represents the rate of flow of a continuous income stream then the present value, P, at annual interest rate r compounded continuously for T years is given by

$$P = \int_0^T f(t)e^{-rt} \, dt$$

The **capital value** of a continuous income stream is simply the present value on the interval $[0, \infty)$;

that is,

$$\text{capital value} = \int_0^\infty f(t)e^{-rt} \, dt$$

In other words, capital value gives the worth of an investment that generates income forever. B. K. O'Neill has just discovered that some land he inherited has a huge oil deposit on it. He decides to lease the oil rights to Shannon Oil for an indefinite annual payment of $50,000. Determine the capital value of this lease at an annual interest rate of 8% compounded continuously.

(Continued)

Flashback Solution

We first note that the annual payments create a continuous income stream given by

$$f(t) = 50,000$$

that lasts indefinitely, or forever. So the capital value is given by

$$\int_0^\infty f(t)e^{-rt}\,dt = \int_0^\infty 50,000e^{-0.08t}\,dt$$

$$= \lim_{T \to \infty}\left[\int_0^T 50,000e^{-0.08t}\,dt\right]$$

First, the integration gives

$$\int_0^T 50,000e^{-0.08t}\,dt = \frac{50,000}{-0.08}e^{-0.08t}\Big|_0^T = -625,000e^{-0.08t}\Big|_0^T$$

$$= -625,000e^{-0.08T} + 625,000$$

Evaluating the limit of this result gives

$$\lim_{T \to \infty}(-625,000e^{-0.08T} + 625,000) = 625,000$$

So the capital value is $625,000.

Endowments and Perpetuities

A fund that creates a steady income forever (think of a continuous income stream) may also be called a *permanent endowment* or a **perpetuity.** Some scholarship funds, such as the one in Example 7, and endowed chairs at universities are examples of perpetuities.

| EXAMPLE 7 | **Funding a Scholarship** |

Maria Lopez, a wealthy alumna of Old State University, wants to establish a nursing student scholarship in her name at her alma mater. If the annual proceeds of the scholarship are to be $10,000 and the annual rate of interest is 6% compounded continuously, determine the amount, to the nearest cent, that Maria needs to fund this scholarship.

SOLUTION

We first note that the annual proceeds determine a continuous income stream given by

$$f(t) = 10,000$$

that is to last forever. So the amount necessary to fund this scholarship is given by

$$\int_0^\infty 10,000e^{-0.06t}\,dt = \lim_{T \to \infty}\left[\int_0^T 10,000e^{-0.06t}\,dt\right]$$

Integrating first produces

$$\int_0^T 10,000e^{-0.06t}\,dt = \frac{10,000}{-0.06}e^{-0.06t}\Big|_0^T$$

$$= -166,666.67e^{-0.06T} + 166,666.67$$

Evaluating the limit yields

$$\lim_{T \to \infty}(-166,666.67e^{-0.06T} + 166,666.67) = 166,666.67$$

We conclude that the amount Maria needs to fund this scholarship indefinitely is $166,666.67. ∎

✓**CHECKPOINT 4**

Now work Exercise 53.

Probability Density Functions

We conclude this section with a brief, intuitive, and informal look at how to use improper integrals with **probability density functions.** Consider an experiment that is designed so that any real number x in the interval $[a, b]$ is a possible outcome. An example of such an experiment might be to list the height or weight of a person selected at random.

Under some circumstances it is possible to determine a function f that will determine the probability that x assumes a value on a stated subinterval of $(-\infty, \infty)$. Functions that do this are called **probability density functions.** A probability density function must satisfy three conditions (see Figure 7.6.10):

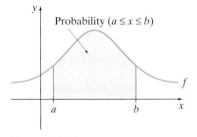

Probability $(a \leq x \leq b)$

Figure 7.6.10

1. For all x in $(-\infty, \infty)$, $f(x) \geq 0$.
2. The area under the graph of f is 1. That is, $\int_{-\infty}^{\infty} f(x)\, dx = 1$.
3. If $[a, b]$ is a subinterval of $(-\infty, \infty)$ then

$$\text{Probability } (a \leq x \leq b) = \int_a^b f(x)\, dx$$

EXAMPLE 8

Calculating a Probability

The length of telephone calls in minutes from a pay phone in a college dormitory has a probability density function of

$$f(x) = \begin{cases} 0.5e^{-0.5x}, & x \geq 0 \\ 0, & x < 0 \end{cases}$$

Calculate the probability that a call selected at random will last between 1 and 2 minutes.

SOLUTION

The first two conditions of a probability density function are left to you to verify in the Interactive Activity to the left of this example. The probability that a call selected at random will last between 1 and 2 minutes is

$$\text{Probability } (1 \leq x \leq 2) = \int_1^2 0.5e^{-0.5x}\, dx = \left. \frac{0.5}{-0.5} e^{-0.5x} \right|_1^2$$

$$= \left. -e^{-0.5x} \right|_1^2 = -e^{-0.5(2)} + e^{-0.5(1)} \approx 0.2387$$

Interactive Activity

Graph f from Example 8 to verify that it satisfies condition 1 of probability density functions. Also, verify that

$$\int_{-\infty}^{\infty} 0.5e^{-0.5x}\, dx = 1,$$

showing that condition 2 is satisfied. [*Hint:* Since $f(x) = 0$ for $x < 0$, $\int_{-\infty}^0 f(x) = 0$.]

EXAMPLE 9

Calculating a Probability

For the probability density function given in Example 8, determine the probability that a telephone call selected at random is 4 or more minutes in length.

SOLUTION

The probability that a call selected at random is 4 or more minutes in length is determined by integrating

$$\text{Probability}(x \geq 4) = \int_4^{\infty} 0.5e^{-0.5x}\, dx = \lim_{b \to \infty} \left[\int_4^b 0.5e^{-0.5x}\, dx \right]$$

As always, we do the integration first. This yields

$$\int_4^b 0.5e^{-0.5x}\, dx = \left. -e^{-0.5x} \right|_4^b = -e^{-0.5b} + e^{-2}$$

Now we determine the limit as follows:

$$\lim_{b \to \infty} (-e^{-0.5b} + e^{-2}) = e^{-2} \approx 0.1353$$

So we have Probability$(x \geq 4) \approx 0.1353$.

✓ **CHECKPOINT 5**

Now work Exercise 63.

SUMMARY

In this section we saw that an integral in which a limit of integration was ∞ (or $-\infty$) is called an **improper integral.** We saw how to use the limit concept to evaluate improper integrals. In particular, we determined that if f is continuous and nonnegative we have:

- $$\int_a^\infty f(x)\,dx = \lim_{b \to \infty} \int_a^b f(x)\,dx$$

- $$\int_{-\infty}^b f(x)\,dx = \lim_{a \to -\infty} \int_a^b f(x)\,dx$$

- $$\int_{-\infty}^\infty f(x)\,dx = \int_{-\infty}^c f(x)\,dx + \int_c^\infty f(x)\,dx$$

If the limit exists, we said that the improper integral **converges,** and if the limit does not exist, we said that the improper integral **diverges.** We applied improper integrals to determine the total output of an oil well, to assess a capital value, to determine the size of a **perpetuity,** and to briefly explore **probability density functions.** Recall that the **capital value** is determined by

$$\text{capital value} = \int_0^\infty f(t)e^{-rt}\,dt$$

where $f(t)$ is a continuous income stream.

SECTION 7.6 EXERCISES

In Exercises 1–16, evaluate each improper integral or state that it diverges.

1. $$\int_2^\infty \frac{1}{x^4}\,dx$$

2. $$\int_4^\infty \frac{1}{x^3}\,dx$$

3. $$\int_3^\infty \frac{1}{\sqrt{x}}\,dx$$

4. $$\int_5^\infty \frac{3}{\sqrt{x}}\,dx$$

✓ 5. $$\int_2^\infty \frac{1}{x^{1.5}}\,dx$$

6. $$\int_3^\infty \frac{1}{x^{2.5}}\,dx$$

7. $$\int_1^\infty \frac{1}{x^{2/3}}\,dx$$

8. $$\int_1^\infty \frac{1}{x^{3/4}}\,dx$$

9. $$\int_1^\infty e^{-x}\,dx$$

10. $$\int_0^\infty e^{-x}\,dx$$

11. $$\int_1^\infty e^{-0.03x}\,dx$$

12. $$\int_1^\infty e^{0.03x}\,dx$$

13. $$\int_1^\infty \frac{\ln x}{x}\,dx$$

14. $$\int_{10}^\infty \ln x\,dx$$

15. $$\int_1^\infty \frac{2x}{x^2+3}\,dx$$

16. $$\int_0^\infty \frac{x^4}{(x^5+2)^2}\,dx$$

In Exercises 17–28, evaluate each improper integral or state that it diverges.

✓ 17. $$\int_{-\infty}^0 e^{2x}\,dx$$

18. $$\int_{-\infty}^0 e^{-2x}\,dx$$

19. $$\int_{-\infty}^0 e^{-x}\,dx$$

20. $$\int_{-\infty}^0 e^{2.3x}\,dx$$

21. $$\int_{-\infty}^0 e^{0.2x}\,dx$$

22. $$\int_{-\infty}^0 e^{x+1}\,dx$$

23. $$\int_{-\infty}^{-4} x^{-2}\,dx$$

24. $$\int_{-\infty}^{-2} x^{-4}\,dx$$

25. $$\int_{-\infty}^{-2} \frac{x^3}{(x^4-1)^2}\,dx$$

26. $$\int_{-\infty}^0 \frac{3x}{(x^2+5)^4}\,dx$$

27. $$\int_0^\infty 2xe^{-x^2}\,dx$$

28. $$\int_{-\infty}^0 2xe^{-x^2}\,dx$$

In Exercises 29–36, evaluate the given improper integral or state that it diverges.

29. $\int_{-\infty}^{\infty} \dfrac{4x^3}{(x^4+3)^2}\, dx$

30. $\int_{-\infty}^{\infty} \dfrac{4x^3}{(x^4+1)^2}\, dx$

✓ 31. $\int_{-\infty}^{\infty} 2xe^{-x^2}\, dx$

32. $\int_{-\infty}^{\infty} 2xe^{x^2}\, dx$

33. $\int_{-\infty}^{\infty} e^{-x}\, dx$

34. $\int_{-\infty}^{\infty} e^{x}\, dx$

35. $\int_{-\infty}^{\infty} \dfrac{2x}{x^2+1}\, dx$

36. $\int_{-\infty}^{\infty} \dfrac{1}{\sqrt{(x^2+1)}}\, dx$

APPLICATIONS

37. Petrohas Engineering estimates that an oil well produces oil at a rate of $B(t) = 60e^{-0.05t} - 60e^{-0.1t}$ thousand barrels per month, where t is the number of months from now. Estimate the total amount of oil produced by this well.

38. A geologist estimates that an oil well produces oil at a rate of $B(t) = 85e^{-0.02t} - 85e^{-0.1t}$ thousand barrels per month, where t is the number of months from now. Estimate the total amount of oil produced by this well.

39. GeoSurveys Inc. estimates that a natural gas well has a rate of production given by $CF(t) = te^{-0.25t}$, where $CF(t)$ is millions of cubic feet per month, t months from now. Estimate the total amount of natural gas produced.

40. EarthWorks Inc. estimates that a natural gas well has a rate of production given by $CF(t) = 2te^{-0.37t}$, where $CF(t)$ is millions of cubic feet per month, t months from now. Estimate the total amount of natural gas produced.

41. The rate at which a certain hazardous chemical is being released into a lake from an abandoned dump is given by $350e^{-0.3t}$ tons per year, t years from now. Determine the total amount of the hazardous chemical that will be released into the lake if the leak continues indefinitely.

42. The rate at which a certain toxic chemical is being released into a river from an abandoned dump is given by $300e^{-0.25t}$ tons per year, t years from now. Determine the total amount of the toxic chemical that will be released into the river if the leak continues indefinitely.

43. It is a fact that when people take medication the human body does not absorb all the medication. Researchers at a pharmaceutical company know that one way to determine the amount of medication absorbed by the body is to determine the rate at which the body eliminates the medication. For a certain arthritis medication, researchers have determined that the rate at which the body eliminates the medication is given by $5e^{-0.02t} - 5e^{-0.035t}$ milliliters per minute, t minutes after the medication was administered. Determine the amount of the medication that was eliminated from the body.

44. For a certain asthma medication, the rate at which the body eliminates the medication is given by $6.5e^{-0.025t} - 6.5e^{-0.05t}$ milliliters per minute, t minutes after the medication

was administered. Determine the amount of the medication that was eliminated from the body.

45. Elle owns a rental property that generates an indefinite annual rent of $10,000. Determine the capital value of this property at an annual interest rate of 6% compounded continuously.

46. Ginger owns a rental property that generates an indefinite annual rent of $6000. Determine the capital value of this property at an annual interest rate of 8.5% compounded continuously.

47. Austin has created a new computer game. He decides to lease the rights to his computer game to Macrohard for an indefinite annual payment of $30,000. Determine the capital value of this lease at an annual interest rate of 7.25% compounded continuously.

48. Nikita has created a new computer game. She decides to lease the rights to her computer game to Star Systems for an indefinite annual payment of $15,000. Determine the capital value of this lease at an annual interest rate of 9.75% compounded continuously.

49. Dylan has inherited some land that has a huge gold deposit on it. He decides to lease the rights to the gold to Hoofer Mining Company for an indefinite annual payment of $70,000. Determine the capital value of this lease at an annual interest rate of 6.5% compounded continuously.

50. Emily owns a property that has a huge silver deposit on it. She decides to lease the rights to the silver to Hoofer Mining Company for an indefinite annual payment of $25,000. Determine the capital value of this lease at an annual interest rate of 8.8% compounded continuously.

51. Lisa owns a rental property that generates an indefinite annual rent of $12,000. Determine the capital value of this property at an annual interest rate of 5.5% compounded continuously.

52. Gene owns a rental property that generates an indefinite annual rent of $2000. Determine the capital value of this property at an annual interest rate of 7% compounded continuously.

✓ 53. If the annual proceeds from the Emma Lou Smith scholarship fund will be $8000 indefinitely and the annual interest rate is 7.5% compounded continuously, how much should be invested to fund this scholarship?

54. If the annual proceeds from the Count Moncheech de Squeeg cinematography scholarship fund will be $5000 indefinitely and the annual interest rate is 7% compounded continuously, how much should be invested to fund this scholarship?

55. The Shannon Oil Company has decided to provide an annual scholarship to individuals studying environmental science in the amount of $3000. If the annual rate of interest

is 6.6% compounded continuously, how much should be invested to fund this scholarship perpetually?

56. The Beagle Works Company has decided to provide an annual scholarship to individuals studying veterinary medicine in the amount of $10,000. If the annual rate of interest is 7.75% compounded continuously, how much should be invested to fund this scholarship perpetually?

57. You wish to leave a legacy at your alma mater in the form of a scholarship that bears your name. You want the scholarship to be an annual award of $5000. Assume that the annual interest rate is 6.25% compounded continuously.

(a) How much should you invest to fund this annual $5000 scholarship for the next 75 years?

(b) How much should you invest to fund this annual $5000 scholarship indefinitely?

58. You wish to leave a legacy at your alma mater in the form of a scholarship that bears your name. You want the scholarship to be an annual award of $8000. Assume that the annual interest rate is 6.25% compounded continuously.

(a) How much should you invest to fund this annual $8000 scholarship for the next 75 years?

(b) How much should you invest to fund this annual $8000 scholarship indefinitely?

59. The length of telephone calls in minutes from a pay phone in a college dormitory has a probability density function of

$$f(x) = \begin{cases} 0.3e^{-0.3x}, & x \geq 0 \\ 0, & x < 0 \end{cases}$$

(a) Verify that $\int_{-\infty}^{\infty} f(x)\, dx = 1$.

(b) Calculate the probability that a call selected at random lasts between 3 and 4 minutes.

(c) Calculate the probability that a call selected at random lasts 3 minutes or longer.

60. The length of telephone calls in minutes from a pay phone in a college cafeteria has a probability density function of

$$f(x) = \begin{cases} 0.6e^{-0.6x}, & x \geq 0 \\ 0, & x < 0 \end{cases}$$

(a) Verify that $\int_{-\infty}^{\infty} f(x)\, dx = 1$.

(b) Calculate the probability that a call selected at random lasts between 1 and 2 minutes.

(c) Calculate the probability that a call selected at random lasts 3 minutes or longer.

61. The length of time for which a calculus book on a 4-hour reserve at a college library is checked out has a probability density function of

$$f(x) = \begin{cases} \dfrac{1}{8}x, & 0 \leq x \leq 4 \\ 0, & \text{otherwise} \end{cases}$$

(a) Verify that $\int_{-\infty}^{\infty} f(x)\, dx = 1$.

(b) Compute the probability that a student selected at random has the book checked out from 1 to 2 hours.

(c) Compute the probability that a student selected at random has the book checked out from 30 minutes to 1 hour.

62. The length of time for which a solutions manual for a certain calculus book on a 6-hour reserve at a college library is checked out has a probability density function of

$$f(x) = \begin{cases} \dfrac{1}{18}x, & 0 \leq x \leq 6 \\ 0, & \text{otherwise} \end{cases}$$

(a) Verify that $\int_{-\infty}^{\infty} f(x)\, dx = 1$.

(b) Compute the probability that a student selected at random has the solutions manual checked out from 2 to 4 hours.

(c) Compute the probability that a student selected at random has the solutions manual checked out from 1 to 2 hours.

✓ **63.** The length of a prison term for convicted felons has a probability density function of

$$f(x) = \begin{cases} 0.1e^{-0.1x}, & x \geq 0 \\ 0, & x < 0 \end{cases}$$

(a) Verify that $\int_{-\infty}^{\infty} f(x)\, dx = 1$.

(b) Determine the probability that a randomly selected felon has a prison term of 5 to 10 years.

(c) Determine the probability that a randomly selected felon has a prison term of 1 to 5 years.

(d) Determine the probability that a randomly selected felon has a prison term of 8 years or more.

64. The number of months until failure for a certain part on a copying machine has a probability density function of

$$f(x) = \begin{cases} 0.02e^{-0.02x}, & x \geq 0 \\ 0, & x < 0 \end{cases}$$

(a) Verify that $\int_{-\infty}^{\infty} f(x)\, dx = 1$.

(b) Compute the probability that the part will fail in the first year.

(c) Compute the probability that the part will fail in the second year.

65. The number of months until failure for a certain computer chip has a probability density function of

$$f(x) = \begin{cases} 0.03e^{-0.03x}, & x \geq 0 \\ 0, & x < 0 \end{cases}$$

(a) Verify that $\int_{-\infty}^{\infty} f(x)\, dx = 1$.

(b) Compute the probability that the part will fail in the first year.

(c) Compute the probability that the part will fail in the second year.

66. At Robotics Inc. the number of minutes that a randomly selected telephone customer is on hold has a probability density function of

$$f(x) = \begin{cases} 0.6e^{-0.6x}, & x \geq 0 \\ 0, & x < 0 \end{cases}$$

(a) Verify that $\int_{-\infty}^{\infty} f(x)\, dx = 1$.

(b) Compute the probability that a customer is on hold from 1 to 4 minutes.

(c) Compute the probability that a customer is on hold for 2 minutes or longer.

67. For a certain variation of the flu, the amount of time in days for a complete recovery has a probability density function of

$$f(x) = \begin{cases} 0.025e^{-0.025x}, & x \geq 0 \\ 0, & x < 0 \end{cases}$$

(a) Verify that $\int_{-\infty}^{\infty} f(x)\, dx = 1$.

(b) Determine the probability that an infected individual recovers between 3 and 6 days.

(c) Determine the probability that an infected individual recovers between 7 and 10 days.

(d) Determine the probability that an infected individual takes 10 days or more to recover.

SECTION PROJECT

Peggy never finishes her calculus lecture before the end of the period, but always finishes her lectures within 5 minutes

of the end of the period. The length of time that elapses between the end of the period and the end of her lecture has a probability density function of

$$f(x) = \begin{cases} \dfrac{3}{125}x^2, & 0 \leq x \leq 5 \\ 0, & \text{otherwise} \end{cases}$$

(a) Verify that $\int_{-\infty}^{\infty} f(x)\, dx = 1$.

(b) Determine the probability that the lecture continues for 1 to 2 minutes beyond the end of the period.

(c) Compute the probability that the lecture ends within 1 minute of the end of the period.

Donald never starts his calculus lecture at the beginning of a period, but always starts his lecture within 2 to 9 minutes from the beginning of the period. The length of time that elapses from 2 minutes after the beginning of the period and the beginning of his lecture is given by the probability density function of

$$f(x) = \begin{cases} \dfrac{3}{721}x^2, & 2 \leq x \leq 9 \\ 0, & \text{otherwise} \end{cases}$$

(d) Verify that $\int_{-\infty}^{\infty} f(x)\, dx = 1$.

(e) Compute the probability that his lecture begins within 3 to 5 minutes from the beginning of the period.

(f) Compute the probability that his lecture begins within 7 to 9 minutes from the beginning of the period.

CHAPTER REVIEW EXERCISES

In Exercises 1–4, determine the average value of the given function on the stated interval.

1. $f(x) = 3x + 4$ $[0, 6]$

2. $f(x) = x^2 - 2x$ $[1, 5]$

3. $g(x) = 4x^3 - 3x^2 + 2$ $[0, 5]$

4. $y = 3e^{-0.2x}$ $[0, 2]$

5. For $f(x) = \ln(1 + x)$:

(a) Use your RTSUM program to approximate $\int_a^b f(x)\, dx$ on $[0, 8]$. Use $n = 100$

(b) Use your approximation from part (a) to determine an (approximate) average value for $f(x)$ on $[0, 8]$.

6. The price–demand function for the 5-minute special lunch at the Reggie's Veggies Restaurant is given by

$$p(x) = 9.10 - 0.07x$$

where $p(x)$ is the price in dollars and x is the quantity demanded. Determine the average price in the demand interval $[24, 60]$.

7. The Tinnitus Phone Factory has determined that the cost for producing x telephones is given by

$$C(x) = 1300 + 18x$$

where $C(x)$ is the cost in dollars.

(a) Determine the average value of the cost function over the interval $[0, 400]$.

(b) Determine the average cost function, $AC(x) = \dfrac{C(x)}{x}$. Evaluate $AC(400)$.

(c) Explain the differences between parts (a) and (b).

8. Frederick deposits $1400 into a money market account earning 5.25% interest compounded continuously. Determine Frederick's average balance, to the nearest cent, over 1 year.

9. Penultimate Savings and Loan offers a money market account that earns 4.7% interest compounded continuously. At the end of the year, the account also pays a bonus of 0.85% of the average balance in the account during the year. Carlene deposits $2400 into a Penultimate money market account.

(a) Compute Carlene's average balance, to the nearest cent, over 1 year.

(b) Compute the size of Carlene's bonus.

10. Sylvia deposits $800 into an account where the rate of change of the amount in the account is given by $A'(t) = 56e^{0.07t}$, t years after the initial deposit.

(a) Determine by how much the account will change from the end of the first year to the end of the fourth year. Round to the nearest cent.

(b) Determine the average balance, to the nearest cent, in the account over the first 4 years.

11. The rate of change of the total income produced by a pinball machine located in a college dorm is given by

$$f(t) = 4000e^{0.09t}$$

where t is the time in years since the installation of the pinball machine and $f(t)$ is in dollars per year. Determine the total income generated by the pinball machine during its first 5 years of operation.

12. Calvin contributes $2000 to his IRA account at the beginning of each year for 30 years. The account earns 6.2% annual interest compounded continuously. Determine how much money Calvin has in his account after 30 years.

13. Upon adopting their daughter Felicia, Chris and Pat opened a college savings account. They deposit $65 at the beginning of each month into an account earning 4.8% annual interest, compounded continuously. Determine how much is in the account in 16 years, when Felicia starts college.

14. The percentage of wage and salary workers in the United States who are members of unions can be modeled by

$$f(x) = -0.0069x^3 + 0.18x^2 - 1.68x + 21.61, \quad 1 \le x \le 14$$

where $x = 1$ corresponds to January 1, 1983, and $f(x)$ is the percentage of wage and salary workers in the United States who were members of unions at time x. Use the model to compute the average percentage of salary workers in the United States who were members of unions between January 1, 1983, and January 1, 1996.

For Exercises 15 and 16, write a definite integral to represent the area of the shaded region.

15.

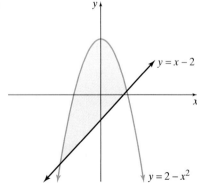

16.

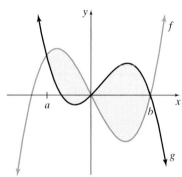

In Exercises 17–24, determine the area of the region bounded by the given conditions.

17. $f(x) = 9 - x^2$ and the x-axis on $[-3, 3]$

18. $f(x) = x^3 - 5x^2 - 2x + 3$ and the x-axis on $[1, 5]$

19. $f(x) = x^2 - 4$ and $g(x) = x + 2$ on $[-2, 3]$

20. $f(x) = x^3 + 2x^2 - 20$ and $g(x) = -3x^2 + 4x$ on $[-5, 2]$

21. $f(x) = 4x + 6$ and $g(x) = x^2 - 6$ on $[-2, 6]$

22. $f(x) = 4 + \sqrt{x}$ and $g(x) = \sqrt[3]{x}$ on $[0, 6]$

23. $f(x) = 5e^x$ and $g(x) = 3.7^x$ on $[-2, 5]$

24. $f(x) = \dfrac{8}{x^3}$ and $g(x) = \dfrac{2}{x}$ on $[1, 4]$

In Exercises 25–28, determine the area of the region bounded by the given curves.

25. $f(x) = x^2$ and $g(x) = x + 6$

26. $f(x) = x^3 + 2x^2 - 5x - 6$ and the x-axis

27. $f(x) = x^3$ and $g(x) = 2x^2 + 8x$

28. $f(x) = \sqrt[3]{x}$ and $g(x) = \dfrac{1}{4}x$

29. The Krazy Dog Company is planning to sell a new tricycle for dogs. It estimates that its monthly sales, $s(t)$, in units per month, t months after the product is released, will be given by

$$s(t) = 1.7t^2 - 16.1t + 160$$

An agent for Lucky, the star of the famous movie *Rover's Revenge*, claims that if Lucky is featured in an ad campaign the monthly sales, $m(t)$, in units per month, t months after the product is released, will be

$$m(t) = 164e^{0.09t}$$

On the interval $[0, 15]$, determine the area between the graphs of s and m and interpret.

30. The size of bacteria culture A grows at a rate of $20x^{1.5}$ bacteria per minute and the size of bacteria culture B grows at a rate of $10x^{2.5}$ bacteria per minute.

(a) Set up an integral to determine the area between the two curves on $[0, 2]$.

(b) Evaluate the integral from part (a) and interpret.

(c) Set up an integral to determine the area between the two curves on [2, 4].

(d) Evaluate the integral from part (c) and interpret.

(e) On the interval [0, 4], determine which culture has more bacterial growth and by how many bacteria.

31. The Loose Nail Furniture Company has determined that its costs, in thousands of dollars, have been increasing mainly due to inflation at a rate given by $215e^{0.14x}$, where x is time measured in years since a certain item started being produced. At the beginning of the fourth year, a breakthrough in the production process resulted in costs increasing at a rate given by $80x^{3/4}$. Determine the total savings in the interval [4, 8] due to the breakthrough.

32. The Lost Sole Shoe Company determines that the marginal revenue to produce x pairs of shoes is given by $R'(x) = 85 - 0.005x^2$, while the marginal cost is given by $C'(x) = 32$, where both marginals are in dollars per pair of shoes. Determine the total profit from $x = 45$ to $x = 84$.

33. Fred and Ginger are assembly workers at the Clobini Factory. Fred assembles units at the rate of $f(x) = -5x^2 + 23x + 31$ units per hour and Ginger assembles units at the rate of $g(x) = -4x^2 + 25x + 32$, where x is the number of hours on the job, during the first 4 hours on the job.

(a) How many units does Fred assemble during his first 4 hours?

(b) How many more units does Ginger assemble than Fred during the same time period?

🌐 34. The imports and exports for the United States with Belgium can be modeled by

$$\text{Imports: } I(x) = 0.55x + 4.11, \qquad 1 \le x \le 5$$

$$\text{Exports: } E(x) = -0.37x^3 + 3.42x^2 - 8.38x + 15.08,$$
$$1 \le x \le 5$$

where x is the number of years since 1991 and imports and exports are in billions of dollars. Determine the area between the two curves on [1, 5] and interpret.

Exercises 35–38 refer to the accompanying figure.

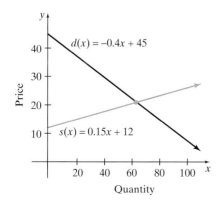

35. Shade the region that corresponds to the consumers' surplus. Write an integral to determine the consumers' surplus.

36. Shade the region that corresponds to the producers' surplus. Write an integral to determine the producers' surplus.

37. Shade the region that corresponds to the actual consumer expenditure. Determine the actual consumer expenditure.

38. Shade the region that corresponds to the actual revenue. Determine the actual revenue.

In Exercises 39–42, for each demand function given and demand level x, determine the consumers' surplus.

39. $d(x) = 450 - 0.8x, x = 100$

40. $d(x) = 3600 - 0.02x^2, x = 300$

41. $d(x) = 100 - \sqrt{x}, x = 36$

42. $d(x) = 225e^{-0.01x}, x = 250$

In Exercises 43–46, for each supply function given and demand level x, determine the producers' surplus.

43. $s(x) = 0.5x + 1200, x = 400$

44. $s(x) = \dfrac{1}{15}x + 10, x = 150$

45. $s(x) = 0.02x^2 + 25, x = 50$

46. $s(x) = 84e^{0.02x}, x = 200$

For Exercises 47–50:

(a) Determine the equilibrium point. Round equilibrium demand and equilibrium price to the nearest whole number.

(b) Determine the consumers' surplus at equilibrium demand.

(c) Determine the producers' surplus at equilibrium demand.

47. $d(x) = 250 - 0.3x; s(x) = 60 + 0.2x$

48. $d(x) = 1280 - 0.8x; s(x) = 272 + 1.6x$

49. $d(x) = 4500 - 0.04x^2; s(x) = 0.03x^2 + 1700$

50. $d(x) = 700e^{-0.05x}; s(x) = 50e^{0.06x}$

51. Knutsen Bolts Hardware has determined that the demand function for a new type of electric drill is given by

$$d(x) = 22 - 2\sqrt{x}$$

where x is the number of drills demanded each day and $d(x)$ is the price per drill in dollars.

(a) Assuming that the equilibrium price is $14 per drill, determine the equilibrium demand. Round to the nearest whole number.

(b) Determine the consumers' surplus at equilibrium demand.

52. Carpets Galore has determined that the demand function for a certain kind of carpet is given by

$$d(x) = -0.5x + 400$$

where x is the number of square yards sold per week and $d(x)$ is the price per square yard in dollars. Assuming that the equilibrium price is $18 per square yard, determine the consumers' surplus.

53. The Heavy Stuff Sporting Goods company has determined that the supply function for its bowling ball is given by

$$s(x) = 18e^{0.007x}$$

where x is the number of bowling balls supplied each month and $s(x)$ is the price per bowling ball in dollars. Assuming that the equilibrium price is $47 per bowling ball, determine the producers' surplus.

54. A cushion manufacturer has determined that the supply function for a certain kind of cushion is given by

$$s(x) = 0.04x^2 + 11$$

where x is the number of these cushions supplied each week and $s(x)$ is the price per cushion in dollars. Assuming that the equilibrium price is $23.96, determine the producers' surplus.

55. Springs 'n' Things has determined that the demand function for its deluxe clipboard is given by

$$d(x) = 12.48 - 0.005x^2$$

while the related producer supply function is given by

$$s(x) = \sqrt{x}$$

where x is the weekly quantity and $d(x)$ and $s(x)$ are in dollars per clipboard.

(a) Determine the equilibrium point.

(b) Determine the consumers' surplus at equilibrium.

(c) Determine the producers' surplus at equilibrium.

56. Using the techniques of regression and data from the U.S. Bureau of the Census web site, the following Lorenz curve models income distribution in the United States in 1992. Recall that x represents the proportion of the population and $L(x)$ represents the proportion of income.

$$L(x) = 2.65x^4 - 4.32x^3 + 2.92x^2 - 0.25x$$

Determine the Gini Index for 1992. Round to the nearest ten-thousandth.

Evaluate the given integrals in Exercises 57–62. If integration by parts is required, consult the guidelines for selecting u and dv given in Section 7.4.

57. $\int 4xe^{2x}\, dx$

58. $\int xe^{-0.05x}\, dx$

59. $\int x^4 \ln x\, dx$

60. $\int (x-2)e^x\, dx$

61. $\int 8te^{-0.2t}\, dt$

62. $\int (x+5)(x-2)^7\, dx$

Exercises 63–66 require two (or more) applications of integration by parts.

63. $\int x^2 e^{-x}\, dx$

64. $\int (x+1)^2\, e^x\, dx$

65. $\int (\ln x)^4\, dx$

66. $\int x(\ln x)^4\, dx$

In Exercises 67–70, determine the method needed to evaluate the given integral. If the method selected is integration by parts, state the choice for u and dv. If the method selected is u-substitution, simply state the choice for u.

67. $\int x^2 e^{-x^3}\, dx$

68. $\int 2x \ln x\, dx$

69. $\int xe^{5x}\, dx$

70. $\int \frac{(\ln x)^5}{x}\, dx$

71. It is estimated that an oil field will produce oil at a rate given by $B(t) = 8te^{-0.17t}$ thousand barrels per month, t months into production.

(a) Write a definite integral to estimate the total production for the first year of operation.

(b) Evaluate the integral from part (a) to estimate the total production for the first year of operation.

72. Hoppin Pepper Inc. has determined that its marginal profit function is given by $P'(t) = 42te^{0.12t}$, where t is time in years and $P'(t)$ is in thousands of dollars per year. Assuming that $P(0) = 0$, determine $P(t)$.

73. Kevin's business generates income at the rate of $90t$ thousand dollars per year, where t is the number of years since the business started. Determine the present value of this continuous income stream over the first 4 years at a continuous compound interest rate of 7%.

74. Suppose that a gold mine produces income at the rate of $140e^{-0.08t}$ thousand dollars per year, where t is the number of years since mining began. Determine the present value of this continuous income stream over the first 6 years of operation, assuming a continuous compound interest rate of 5%.

75. The average sales price for a home in the United States can be approximated by

$$h(x) = 5.63x + 39.25, \qquad 1 \le x \le 21$$

where $x = 1$ corresponds to October 1, 1976, and $h(x)$ is the average selling price, in thousands of dollars.

(a) Determine the average value of h on the interval $[1, 6]$ and interpret.

(b) Determine the average value of h on the interval $[8, 12]$ and interpret.

(c) Determine the average value of h on the interval $[16, 21]$ and interpret.

76. The average annual expenditures on food by U.S. households can be modeled by

$$f(x) = 4.14 + 0.15 \ln x, \qquad 1 \le x \le 7$$

where $x = 1$ corresponds to January 1, 1989, and $f(x)$ is the amount spent on food by the average household in a year, in thousands of dollars.

(a) Determine the average value of f on the interval $[1, 4]$ and interpret.

(b) Determine the average value of f on the interval $[4, 7]$ and interpret.

In Exercises 77–80, use the Trapezoidal Rule to approximate the given definite integral using $n = 4$ trapezoids. Do not use a calculator program for these exercises.

77. $\displaystyle\int_{-3}^{1} x^5 \, dx$

78. $\displaystyle\int_{4}^{8} (\ln x)^4 \, dx$

79. $\displaystyle\int_{0}^{1} (x^2 - 2x) \, dx$

80. $\displaystyle\int_{0}^{8} \frac{1}{x+1} \, dx$

In Exercises 81–84, use the Trapezoidal Rule to approximate the given definite integral using $n = 10$, $n = 50$, and $n = 100$ trapezoids. You may use a calculator program for these exercises.

81. $\displaystyle\int_{1}^{11} \sqrt{x^2 + 3x + 2} \, dx$

82. $\displaystyle\int_{3}^{8} x^2 e^{0.5x} \, dx$

83. $\displaystyle\int_{1}^{5} \frac{(\ln x)^3}{x} \, dx$

84. $\displaystyle\int_{0}^{10} \frac{3x+4}{x^2+2} \, dx$

85. The following data give the rate of change of the year-to-date revenues in thousands of dollars per month for Jack's Handcrafted Birdhouses. Here x represents the number of months after May 1.

x	RATE OF CHANGE OF YEAR-TO-DATE REVENUE
1	2047
2	3124
3	2387
4	1937
5	2183
6	2706

Use the Trapezoidal Rule to approximate the revenues for Jack's Handcrafted Birdhouses for this period of time. (Do not use a calculator program for this exercise.)

86. The following data give the marginal revenue for different levels of production at David's DVD Division. Here x represents the number of DVDs produced and $R'(x)$ is in dollars per DVD.

x	$R'(x)$
100	30
200	20
300	15
400	10
500	8

Use the Trapezoidal Rule to approximate the total revenue increase that would be obtained by going from a production level of 100 to 500 DVDs. (Do not use a calculator program for this exercise.)

87. The GHI company has determined the marginal cost for different levels of production. The following data have been collected, where x represents the number of items produced and $C'(x)$ is in thousands of dollars per item.

x	$C'(x)$	x	$C'(x)$
0	3.8	14	5.8
2	3.6	16	6.9
4	3.2	18	7.8
6	3.4	20	9.4
8	3.9	22	12.7
10	4.2	24	18.5
12	4.7		

Assume that $C(0) = 0$ and use the Trapezoidal Rule to approximate the total cost of producing 24 items. (You may use a calculator program.)

88. The following data give the annual amount spent by the U.S. federal government on human resources, in billions of dollars, from 1973 through 1997. Here, $x = 1$ corresponds to the beginning of 1973.

x	AMOUNT SPENT (BILLIONS OF DOLLARS)
1	119.5
5	221.9
9	362.0
13	471.8
17	568.7
21	827.5
25	1019.4 (est.)

Source: U.S. Census Bureau.

Use the Trapezoidal Rule to approximate the total amount spent from the beginning of 1973 to the beginning of 1997. (You may use a calculator program.)

In Exercises 89 and 90, use the following information and the Trapezoidal Rule to answer the questions. You may use a calculator program for these exercises.

The following data give the number of new incorporations (in thousands) and the number of business failures (in thousands) in the United States from 1985 through 1995. Here $x = 1$ represents the beginning of 1985.

x	NUMBER OF NEW INCORPORATIONS (THOUSANDS)	BUSINESS FAILURES (THOUSANDS)
1	664	57
3	686	61
5	677	50
7	629	88
9	707	86
11	768	71

Source: U.S. Census Bureau.

89. (a) Approximate the total number of businesses that incorporated from the beginning of 1985 to the beginning of 1995.

(b) Approximate the average number of businesses that incorporated, on a yearly basis, from the beginning of 1985 to the beginning of 1995.

90. (a) Approximate the total number of business failures that occurred from the beginning of 1985 to the beginning of 1995.

(b) Approximate the average number of business failures that occurred, on a yearly basis, from the beginning of 1985 to the beginning of 1995.

In Exercises 91–100, evaluate each improper integral or state that it is divergent.

91. $\int_5^\infty \frac{1}{x^2}\, dx$

92. $\int_1^\infty \frac{4}{\sqrt[3]{x}}\, dx$

93. $\int_{10}^\infty \frac{3}{x}\, dx$

94. $\int_0^\infty (8x - 2)\, dx$

95. $\int_0^\infty e^{0.1x}\, dx$

96. $\int_5^\infty \frac{4x}{2x^2 + 3}\, dx$

97. $\int_{-\infty}^0 e^{3x}\, dx$

98. $\int_{-\infty}^0 xe^{-0.2x^2}\, dx$

99. $\int_{-\infty}^\infty e^{0.05x}\, dx$

100. $\int_{-\infty}^\infty \frac{2x}{\sqrt{x^2 + 5}}\, dx$

101. An oil well has been estimated to produce oil at a rate of $B(t) = 43e^{-0.03t} - 43e^{-0.08t}$ thousand barrels per month, where t is the number of months from now. Estimate the total amount of oil that will be produced by this well.

102. Kevin owns a rental property that generates an indefinite annual rent of $14,000. Determine the capital value of this property at an annual interest rate of 5.25% compounded continuously.

103. A scholarship fund is to provide an annual scholarship in the amount of $8000. If the annual rate of interest is 9%, compounded continuously, how much should be invested to fund this scholarship perpetually?

104. The rate at which a certain toxic chemical is being released into an ocean from an abandoned dump is given by $260e^{-0.2t}$ tons per year t years from now. Determine the total amount of the toxic chemical that will be released into the ocean if the leak continues indefinitely.

105. The number of days for which a book is checked out of the Springfield Public Library has a probability density function of

$$f(x) = \begin{cases} \dfrac{2}{441}x, & 0 \le x \le 21 \\ 0, & \text{otherwise} \end{cases}$$

(a) Verify that $\int_{-\infty}^\infty f(x)\, dx = 1$.
(b) Compute the probability that the length of time for which a book is checked out is from 7 to 14 days.
(c) Compute the probability that the length of time for which a book is checked out is from 14 to 21 days.

106. In a certain country, the number of days for a mailed letter to reach its destination has a probability function of

$$f(x) = \begin{cases} 0.08e^{-0.08x}, & x \ge 0 \\ 0, & x < 0 \end{cases}$$

(a) Verify that $\int_{-\infty}^\infty f(x)\, dx = 1$.
(b) Compute the probability that a letter takes 2 to 5 days to reach its destination.
(c) Compute the probability that a letter takes 10 or more days to reach its destination.

CHAPTER 8

Calculus of Several Variables

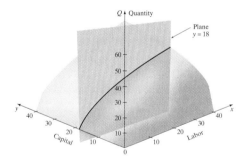

Surface of a production function $Q = f(x, y) = 1.64x^{0.6}y^{0.4}$, where x is the number of units of labor, y is the number of units of capital, and Q is the quantity produced, sliced by the plane $y = 18$.

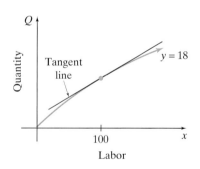

When the surface of a production function is sliced with plane $y = 18$, the result produces a curve. Line is tangent to curve at $x = 100$.

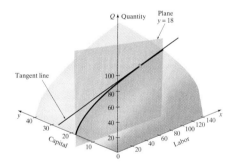

Tangent line to the curve that results from slicing surface with plane $y = 18$ is also tangent to the surface.

What We Know

In the first seven chapters of this text, we saw how the derivative and the integral can be used in a variety of settings. In all these settings the functions that we applied calculus to were functions of a single independent variable.

Where Do We Go

In this chapter we extend our calculus knowledge to functions that involve two independent variables. We will still see that rates of change are central to the derivative concept in the many varied applied settings.

SECTION 8.1   FUNCTIONS OF SEVERAL INDEPENDENT VARIABLES

In this section we introduce the concept of a function of two or more independent variables and the coordinate system used to analyze the graphs of such functions. Some calculators can graph a function of two independent variables. At this time we suggest that you consult the owner's manual of your calculator to determine if your calculator does this.

Functions of Two or More Independent Variables

Many quantities in the world depend on several independent variables. For example, the wind chill factor and the heat index, two numbers that we may hear on the local weather report, each depend on two variables. The wind chill depends on the air temperature and the wind speed, while the heat index depends on the air temperature and the relative humidity. In mathematical terms, if WCI represents the wind chill index, then we may write

$$\text{WCI} = f(v, t)$$

where v is the wind speed in miles per hour and t is the air temperature in degrees Fahrenheit. We can evaluate a function that has more than one independent variable. Before we show how to do this, we first present a definition for a function of two independent variables.

Function of Two Independent Variables

An equation of the form $z = f(x, y)$ represents a function of two independent variables if, for each ordered pair of real numbers (x, y), the equation determines a unique real number z.

NOTE: The variables x and y are the **independent variables**, and z is the **dependent variable**.

The **domain** of a function of two variables is the set of all ordered pairs (x, y) for which z is defined. Example 1 illustrates how to evaluate a function of two independent variables.

EXAMPLE 1   **Evaluating a Function of Two Independent Variables**

Consider $f(x, y) = 2x^2 - 3y^3$. Determine the following:

(a) $f(2, 1)$ (b) $f(-1, 2)$ (c) $f(1, -2)$

SOLUTION

(a) Evaluating a function of two independent variables is just like evaluating a function with only one independent variable. That is, we substitute 2 in for x and 1 in for y. This produces

$$f(x, y) = 2x^2 - 3y^3$$
$$f(2, 1) = 2(2)^2 - 3(1)^3 = 8 - 3 = 5$$

So we have $f(2, 1) = 5$.

(b) Evaluating as we did in part (a) yields

$$f(x, y) = 2x^2 - 3y^3$$
$$f(-1, 2) = 2(-1)^2 - 3(2)^3 = 2 - 24 = -22$$

So we have $f(-1, 2) = -22$.

(c) Evaluating as we did in parts (a) and (b) gives

$$f(x, y) = 2x^2 - 3y^3$$
$$f(1, -2) = 2(1)^2 - 3(-2)^3 = 2 + 24 = 26$$

So we have $f(1, -2) = 26$. ∎

✓ **CHECKPOINT 1**

Now work Exercise 7.

In Example 2, we illustrate how a function of two variables can be constructed to determine a cost function.

EXAMPLE 2 **Determining a Cost Function**

DynaBall Corporation manufactures two types of golf balls. One has a balata cover favored by professionals and other low-handicap players. The other has a durable cover favored by high-handicap players. The cost to make a balata ball is $0.97, whereas the cost to make a durable-cover ball is $0.89.

(a) Determine a cost function, C as a function of x, for producing x balata balls.

(b) Determine a cost function, C as a function of y, for producing y durable-cover balls.

(c) Determine a cost function, C as a function of x and y, for the balata and durable-cover balls.

SOLUTION

(a) The cost function for the balata balls is simply

$$C(x) = (\text{unit cost}) \cdot (\text{quantity}) = 0.97x$$

(b) For the durable-cover balls, the cost function is

$$C(y) = (\text{unit cost}) \cdot (\text{quantity}) = 0.89y$$

(c) Notice here, in the language of functions, that we are asked to determine cost as a function of x *and* y. Intuitively, this should be the sum of the cost functions from parts (a) and (b). This gives us

$$C(x, y) = 0.97x + 0.89y$$ ∎

The cost function in Example 2c, $C(x, y) = 0.97x + 0.89y$, gives us the cost of producing x balata balls and y durable-cover balls. This function is an example of a **function of two independent variables x and y**. If this company decides to expand its golf ball line by producing and selling a special 80 compression ball for senior citizens, we would introduce a third independent variable and rewrite our cost function as a *function of three independent variables*. For the most part, we will be discussing functions of two independent variables in this chapter. For many of the problems that we encounter, we may have to place restrictions on the domain to determine the realistic domain. Example 3 illustrates this idea.

EXAMPLE 3 **Determining a Domain and Evaluating a Function of Two Variables**

In Example 2 we determined that the cost function to produce x balata balls and y durable-cover balls is given by $C(x, y) = 0.97x + 0.89y$.

(a) Determine the realistic domain for $C(x, y)$.

(b) Evaluate $C(250, 700)$ and interpret.

SOLUTION

(a) At first glance, we notice that $C(x, y)$ is defined for all values of x and y. In other words, any ordered pair (x, y) is in the domain. But since $C(x, y)$ represents a *cost function,* we claim that it is unrealistic for x or y to be negative. (How does one produce -3 golf balls?) It is possible for x or y to be 0, so we could write that the domain is all ordered pairs (x, y) with real numbers x and y such that $x \geq 0$ and $y \geq 0$.

(b) We evaluate $C(250, 700)$ by simply substituting as follows:

$$C(x, y) = 0.97x + 0.89y$$
$$C(250, 700) = 0.97(250) + 0.89(700) = 865.50$$

This means that it costs \$865.50 to produce 250 balata balls and 700 durable-cover balls.

Cartesian Coordinates in 3-Space

So far in this text we have considered graphs of functions where the function has one independent variable and one dependent variable. The graph of this type of function is determined by points in the xy-plane that we called **ordered pairs.** To graph a function of two independent variables and one dependent variable, we simply extend this concept. Since we now have a total of three variables, two independent and one dependent, we need three coordinate axes and we plot points determined by **ordered triples** (x, y, z). In Figure 8.1.1 we have three coordinate axes that meet, at right angles to each other, at a point called the **origin** $(0, 0, 0)$. Think of our familiar xy-plane as being horizontal, like the floor of a room, while the z-axis extends vertically above and below the plane. See Figure 8.1.2.

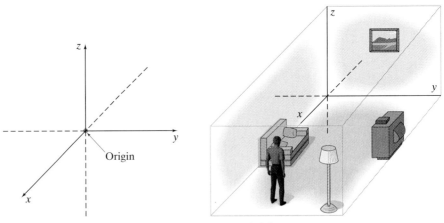

Figure 8.1.1

Figure 8.1.2

In Figure 8.1.1 we have labeled the x-, y-, and z-axes. Consider the solid portion of each axis to be the positive portion of the axis, while the dashed part corresponds to the negative portion of each axis. These three axes may be used to determine the location of any point in three-dimensional space.

Any point in three-dimensional space, or **3-space,** can be represented with an ordered triple (x, y, z). When considering points in space, imagine the first two coordinates of an ordered triple as indicating where to go in the xy-plane. The third coordinate, the z-coordinate, tells us whether to go up or down or to stay put. Example 4 shows how to plot points in space.

EXAMPLE 4 | **Plotting Points in Space**

Plot the following points:

(a) $(2, 3, 1)$ (b) $(-2, -4, -2)$

SOLUTION

(a) As shown in Figure 8.1.3a, we first locate the point $(2, 3)$ in the xy-plane. We start at the origin and move 2 units along the positive x-axis and then 3 units to the right (in the positive direction), taking care to remain parallel to the y-axis. We use a "✕" to locate $(2, 3)$ in the xy-plane. Since our z-coordinate is a *positive* 1, we move *up* 1 unit directly above this location, taking care to remain parallel to the z-axis, and plot the point $(2, 3, 1)$ as shown in Figure 8.1.3b.

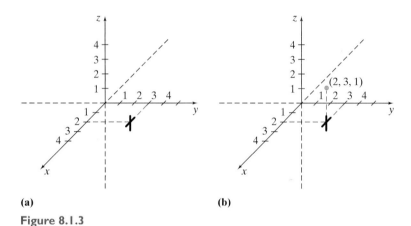

(a)   **(b)**

Figure 8.1.3

(b) In a manner similar to part (a), we first locate $(-2, -4)$ in the xy-plane. See Figure 8.1.4a. Here our z-coordinate is a *negative* 2, so we move *down* 2 units from this location and plot the point $(-2, -4, -2)$, as shown in Figure 8.1.4b.

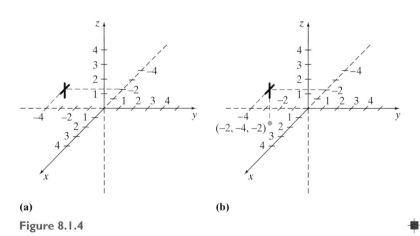

(a)   **(b)**

Figure 8.1.4

✓CHECKPOINT **2** | Now work Exercise 31.

Graphing in 3-Space

Recall that for a function f of one independent variable the graph of $y = f(x)$ is a curve that is above or below the x-axis. This curve is a collection of all points (x, y) such that $y = f(x)$. Also note that the curve lies in the xy-plane.

For a function of two independent variables, say $z = f(x, y)$, the graph of $f(x, y)$ is a collection of all points (x, y, z) such that $z = f(x, y)$. The graph of $z = f(x, y)$ is called a **surface** and is either above or below the xy-plane. Recall that an equation of the form $z = f(x, y)$ represents a function of two independent variables if for each ordered pair of real numbers (x, y) the equation determines a unique real number z. We can think of the domain of $z = f(x, y)$ as being any ordered pair (x, y) for which z is defined.

As we can imagine, graphing functions of two variables is difficult since it involves drawing three-dimensional graphs. Computers and some calculators can generate a 3D graph fairly quickly. Figure 8.1.5 shows several graphs of surfaces.

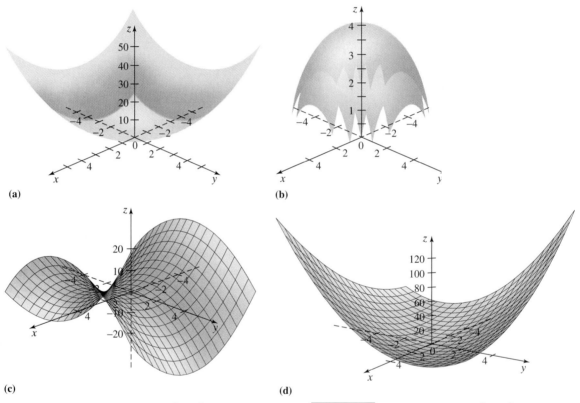

Figure 8.1.5 **(a)** Graph of $z = x^2 + y^2$. **(b)** Graph of $z = \sqrt{16 - x^2 - y^2}$. **(c)** Graph of $z = x^2 - y^2$. **(d)** Graph of $z = x^2 - 2xy + 2y^2$.

In the next section we will use two-dimensional techniques to aid us in visualizing the three-dimensional (or 3D) graphs. In this section we direct our focus on the domains of functions of two variables and some very basic 3D graphs. The most elementary 3D graphs are **planes.** Some planes are illustrated in Example 5.

EXAMPLE 5 | **Graphing Planes**

Discuss the domain of each of the following and sketch a graph.

(a) $z = 2$ (b) $x = 2$ (c) $y = 3$

SOLUTION

(a) The equation $z = 2$ is satisfied by all ordered triples having the form $(x, y, 2)$. To find a point on this graph, we locate any point (x, y) in the xy-plane and then move up 2 units. The result is a plane 2 units above the xy-plane and parallel to the xy-plane. The surface is shown in Figure 8.1.6.

(b) The equation $x = 2$ is satisfied by all ordered triples of the form $(2, y, z)$. To find any point on this graph, we locate any point of the form $(2, y)$ in the xy-plane or, equivalently, any point on the line $x = 2$ in the xy-plane and plot all points above, on, or below this line. The graph of $x = 2$ is a plane parallel to the yz-plane, as shown in Figure 8.1.7.

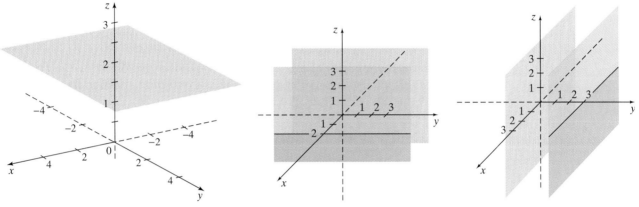

Figure 8.1.6 Graph of $z = 2$.   **Figure 8.1.7** Graph of $x = 2$.   **Figure 8.1.8** Graph of $y = 3$.

(c) The equation $y = 3$ is satisfied by all ordered triples of the form $(x, 3, z)$. To find any point on this graph, we locate any point of the form $(x, 3)$ in the xy-plane or, equivalently, any point on the line $y = 3$ in the xy-plane and plot all points above, on, or below this line. The graph of $y = 3$ is a plane parallel to the xz-plane, as shown in Figure 8.1.8. ◼

In Example 5, all the planes that we graphed had only one variable in the equation. We notice that when this occurs the resulting graph is a plane that is parallel to one of the **coordinate planes.** The three coordinate axes determine three coordinate planes, which are the xy-, the xz-, and the yz-plane. See Figure 8.1.9.

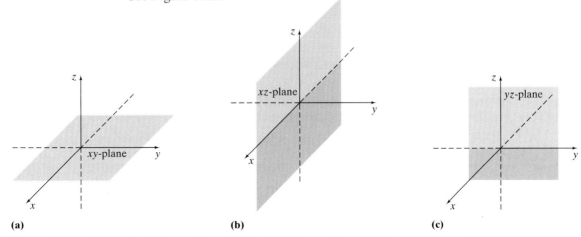

(a)   (b)   (c)

Figure 8.1.9   **(a)** Graph of the xy-plane. **(b)** Graph of the xz-plane. **(c)** Graph of the yz-plane.

Planes parallel to the coordinate planes are used in the next section when we look at *level curves* and *cross sections* as part of our two-dimensional analysis of a 3D surface. Since this will be used in future work, we offer the following:

> For any real constant c, we have the following:
> - The graph of $z = c$ is parallel to the xy-plane.
> - The graph of $x = c$ is parallel to the yz-plane.
> - The graph of $y = c$ is parallel to the xz-plane.

✓ **CHECKPOINT 3**

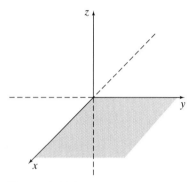

Figure 8.1.10

Now work Exercise 47.

Most applications that we encounter in this chapter will have the domain restricted to quadrant 1 in the xy-plane, that is, $x \geq 0$ and $y \geq 0$. Figure 8.1.10 shows the domain for most applications that we will examine. The surface of a 3D graph is either entirely above, entirely below, or has portions above and portions below this quadrant of the xy-plane. The last part of this section focuses on one such application.

Cobb–Douglas Production Function

When producing any item, many inputs are used in the production process. Mathematically, if we let Q represent the output, then we have

$$Q = f(x_1, x_2, x_3, \ldots, x_k)$$

where each $x_1, x_2, \ldots$ is a different input. Clearly, Q is a function of several variables. However, economists tend to reduce all of these different inputs in this **production function** to two, L and K, and use

$$Q = f(L, K)$$

where Q is the number of units of output, L is the number of units of **labor,** and K is the number of units of **capital.** Capital includes many items, such as buildings, equipment, and insurance. This production function was developed and made popular by the mathematician Charles Cobb and the economist Paul Douglas and is called the Cobb–Douglas production function. The most general form of a Cobb–Douglas production function is the following:

> **Cobb–Douglas Production Function**
>
> If L represents the units of labor and K represents the units of capital, then the total production Q is given by the Cobb–Douglas production function
>
> $$Q = a L^b K^c$$
>
> where $a, b,$ and c are constants.

EXAMPLE 6 | **Analyzing a Cobb–Douglas Production Function**

The Cobb–Douglas production function was introduced in 1928. It was originally constructed for all the manufacturing output in the United States for the years 1899 to 1922. The production function for all U.S. manufacturing output from 1899 to 1922 is given by

$$Q = 1.01 L^{0.75} K^{0.25}$$

where Q is the total yearly production, K is the capital investment, and L is the total labor force. Using x for labor and y for capital, we can rewrite this production function as

$$Q = f(x, y) = 1.01x^{0.75}y^{0.25}$$

(a) Determine the realistic domain of $Q = f(x, y)$.

(b) Determine if the graph of Q is above or below the xy-plane.

(c) Evaluate Q when $x = 50$ and $y = 60$ and interpret.

SOLUTION

(a) Since x represents the total labor force and y represents the total capital investment, we conclude that realistically neither can be negative. So the domain is all ordered pairs (x, y) such that $x \geq 0$ and $y \geq 0$. Graphically, it is the region shaded in Figure 8.1.11.

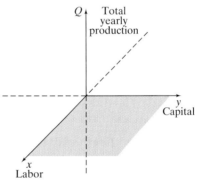

Figure 8.1.11

(b) From part (a) we know that $x \geq 0$ and $y \geq 0$. If $x = 0$ or if $y = 0$, we have $Q = 0$. If $x \neq 0$ and $y \neq 0$, then $Q > 0$. So we conclude that the graph of Q, the surface, lies above the xy-plane. (It intersects the x-axis and the y-axis.) Figure 8.1.12 gives two different perspectives of the graph of Q. In Figure 8.1.12a the axes are in a standard position. In Figure 8.1.12b the axes are rotated 180°.

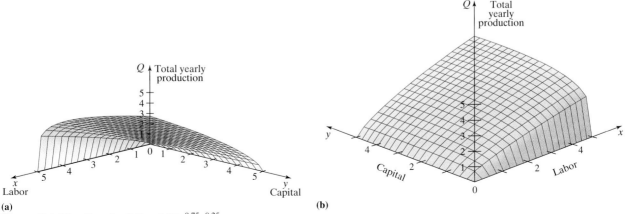

(a)

(b)

Figure 8.1.12   Graph of $Q = 1.01x^{0.75}y^{0.25}$.

(c) To evaluate Q when $x = 50$ and $y = 60$, we simply substitute these values for x and y into the production function, as follows:

$$Q = 1.01x^{0.75}y^{0.25} = 1.01(50)^{0.75}(60)^{0.25} \approx 52.86$$

Thus, when 50 units of labor and 60 units of capital are used, the number of units produced is approximately 52.86.

NOTE: For future graphs of production functions we will use the perspective in Figure 8.1.12b unless stated otherwise. This perspective shows the important features of the surface of a production function quite nicely.

When Cobb and Douglas first introduced their production function, they gave the more specific form

$$Q = aL^b K^{1-b}$$

They used this form because they restricted the results to what economists call **constant return to scales**. Return to scales is discussed in economics theory, but for our purposes we just want to notice that, to have constant return to scales, the sum of the exponents is equal to 1. In fact, this is the only difference between Cobb and Douglas's specific form and the general form presented prior to Example 6.

SUMMARY

In this section we introduced the concept of a function of two independent variables.

- **Function of two independent variables:** An equation of the form $z = f(x, y)$ represents a function of two independent variables if for each ordered pair of real numbers (x, y) the equation determines a unique real number z.

    We saw that the domain of a function of two variables is in the xy-plane and the graph is in **3-space.** We plotted points in 3-space, looked at graphs of some surfaces in space, and graphed some very basic surfaces in space called **planes.**

- For any real constant c, we have the following:

    The graph of $z = c$ is parallel to the xy-plane. For example, see Figure 8.1.6.
    The graph of $x = c$ is parallel to the yz-plane. For example, see Figure 8.1.7.
    The graph of $y = c$ is parallel to the xz-plane. For example, see Figure 8.1.8.

SECTION 8.1 EXERCISES

In Exercises 1–4, evaluate $f(x, y) = x^2 - 2xy + 3y^2$ for the indicated values.

1. $f(2, 3)$ 2. $f(-1, 4)$ 3. $f(-2, -1)$ 4. $f(3, -2)$

In Exercises 5–8, evaluate $f(x, y) = \dfrac{x^2 + y^2}{x + y}$ for the indicated values.

5. $f(3, 1)$ 6. $f(-1, 2)$ ✓ 7. $f(-2, -3)$ 8. $f(1, -2)$

In Exercises 9–12, evaluate $g(x, y) = \dfrac{3x - y + 1}{x^2 - y^2}$ for the indicated values.

9. $g(1, 4)$ 10. $g(-2, -5)$ 11. $g(-2, 2)$ 12. $g(2, 0)$

In Exercises 13–16, evaluate $f(x, y) = y + x \ln x + xe^y$ for the indicated values.

13. $f(1, 2)$ 14. $f(1, 0)$ 15. $f(e, 0)$ 16. $f(e^2, 1)$

For Exercises 17–22, determine the domain of the given function. Recall that here the domain is the set of all ordered pairs (x, y) in the xy-plane.

17. $f(x, y) = x + 2y$

18. $g(x, y) = 2x - y$

19. $g(x, y) = \dfrac{1}{x + y}$

20. $f(x, y) = \dfrac{3x}{x - y}$

21. $f(x, y) = \dfrac{3}{x - 4y}$

22. $g(x, y) = e^y + x \ln x$

For Exercises 23–36, plot the given points.

23. $(4, 4, 2)$ 24. $(-3, 2, -1)$ 25. $(-2, 1, 2)$

26. $(-3, 4, 1)$ 27. $(-3, -1, 1)$ 28. $(2, -2, 3)$

29. $(1, -3, -2)$ 30. $(2, -3, -2)$ ✓ 31. $(4, 2, -1)$

32. $(2, 5, -3)$ 33. $(-1, -3, -2)$ 34. $(-2, -4, -1)$

35. $(2, -4, 0)$ 36. $(-3, 1, 0)$

For Exercises 37–44, without plotting the point, determine if the given point is above or below the xy-plane. Explain.

37. $(2, 2, 3)$ 38. $(3, 1, 2)$ 39. $(-3, -2, 5)$

40. $(-1, -2, 4)$ 41. $(4, -2, -1)$ 42. $(2, 0, -5)$

43. $(0, 3, -0.5)$ 44. $(1, -1, 0.1)$

In Exercises 45–50, sketch the graph of the given plane in 3-space.

45. $x = 3$ 46. $x = -2$ ✓ 47. $z = 3$

48. $z = -2$ 49. $y = -1$ 50. $y = 2$

Applications

51. Seamount Boats spends x thousand dollars each week on newspaper advertising and y thousand dollars each week on radio advertising. The company has weekly sales, in tens of thousands of dollars, given by

$$S(x, y) = 2x^2 + y$$

(a) Determine $S(5, 3)$ and interpret.
(b) Determine $S(3, 5)$ and interpret.

52. Lakeway Boats spends x thousand dollars each week on radio advertising and y thousand dollars each week on television advertising. The company has weekly sales, in tens of thousands of dollars, given by

$$S(x, y) = 3x + 2y^3$$

(a) Determine $S(2, 6)$ and interpret.
(b) Determine $S(6, 2)$ and interpret.

53. Tube Town, a recently opened water park, spends x thousand dollars on radio advertising and y thousand dollars on television advertising. The park has weekly ticket sales, in tens of thousands of dollars, given by

$$TS(x, y) = 1.5x^2 + 3.2y^3$$

(a) Determine $TS(1, 0.5)$ and interpret.
(b) Determine $TS(0.5, 1)$ and interpret.

54. In their study of human groupings, anthropologists often use an index that indicates the shape of the head, the *cephalic index*. The cephalic index is given by

$$C(W, L) = 100 \cdot \frac{W}{L}$$

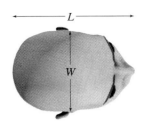

where W is the width and L is the length of an individual's head. Both measurements are made across the top of the head and are in inches. Determine $C(6, 8.2)$ and $C(8.8, 9.7)$.

55. Poiseuille's law states that the resistance, R, for blood flowing in a blood vessel is given by

$$R(L, r) = K \cdot \frac{L}{r^4}$$

where K is a constant, L is the length of the blood vessel, and r is the radius of the blood vessel. Determine $R(36, 1)$ and $R(36, 2)$.

56. An individual's body surface area is approximated by

$$BSA(w, h) = 0.007184 w^{0.425} h^{0.725}$$

where BSA is in square meters, w is weight in kilograms, and h is height in centimeters. Determine $BSA(70, 160)$ and interpret.

57. The *Doyle log rule* is one method used to determine the yield of a log, measured in board-feet. In English, the rule states

> Deduct 4 inches from the diameter of the log as an allowance for slab; square one-quarter of the remainder and multiply the result by the length of the log in feet." (*Source:* http://www.forestry.uga.edu.docs/950-measurementscaling.html.)

Mathematically, this is translated as

$$f(d, L) = \left(\frac{d - 4}{4}\right)^2 \cdot L$$

where d is the diameter in inches, L is the length in feet, and $f(d, L)$ is the number of board-feet. Determine $f(30, 12)$ and interpret.

58. The wind chill index is given by

$$f(v, T) = 91.4 - \left(0.474677 - 0.020425v + 0.303107\sqrt{v}\right) \times (91.4 - T)$$

where v represents the velocity of the wind in miles per hour, T represents the actual temperature in degrees Fahrenheit, and $f(v, T)$ is the wind chill index in degrees Fahrenheit. (Wind chill is the temperature that it actually feels like.) Evaluate $f(25, 10)$ and interpret. Round to the nearest integer.

59. The Leaf Eater Company manufactures and sells leaf blowers and a special 10-foot blower attachment to clean

gutters. The monthly cost function, in dollars, for the company is given by

$$C(x, y) = 1000 + 35x + 1.5y$$

where x is the number of leaf blowers produced each month and y is the number of 10-foot blower attachments produced each month. Determine $C(50, 30)$ and interpret.

60. The Leaf Eater Company from Exercise 59 has determined the following:

$p = 120 - 0.8x - 0.1y$, the price in dollars for a leaf blower

$q = 52.3 - 0.15x + 0.015y$, the price in dollars for a 10-foot attachment

where x is the number of leaf blowers sold each month and y is the number of 10-foot blower attachments sold each month.

(a) Determine the revenue function $R(x, y)$.

(b) Determine $R(50, 30)$ and interpret.

61. Refer to Exercises 59 and 60.

(a) Using the cost function from Exercise 59 and the revenue function from Exercise 60, determine the profit function $P(x, y)$.

(b) Determine $P(50, 30)$ and interpret.

62. The Chalet Bicycle Company manufactures 21-speed racing bicycles and 21-speed mountain bicycles. Let x represent the weekly demand for a 21-speed racing bicycle and y represent the weekly demand for a 21-speed mountain bicycle. The weekly price–demand equations are given by

$p = 350 - 4x + y$, the price in dollars for a 21-speed racing bicycle

$q = 450 + 2x - 3y$, the price in dollars for a 21-speed mountain bicycle

The cost function is given by

$$C(x, y) = 390 + 95x + 100y$$

(a) Determine the weekly revenue function $R(x, y)$.

(b) Evaluate $R(15, 20)$ and interpret.

(c) Determine the weekly profit function $P(x, y)$.

(d) Evaluate $P(15, 20)$ and interpret.

63. A T-shirt maker produces two types of tie-dyed T-shirts. The full-rainbow T-shirt costs $5 each to produce and the partial-rainbow T-shirt costs $4 each to produce. She sells the full-rainbow T-shirts for $12 each and the partial-rainbow T-shirts for $9.50 each.

(a) Determine the cost function $C(x, y)$ for making x full-rainbow T-shirts and y partial-rainbow T-shirts. Assume that the fixed costs are $50.

(b) Determine the revenue function $R(x, y)$.

(c) Determine the profit function $P(x, y)$.

64. If $5000 is invested at an annual interest rate of r (in decimal form) compounded quarterly for t years, the total amount accumulated is given by

$$f(r, t) = 5000\left(1 + \frac{r}{4}\right)^{4t}$$

Determine $f(0.075, 15)$ and interpret.

65. If an amount P dollars is invested at an annual interest rate of 6.125% compounded monthly for t years, the total amount accumulated is given by

$$f(P, t) = P\left(1 + \frac{0.06125}{12}\right)^{12t}$$

Determine $f(2000, 20)$ and interpret.

66. If an amount P dollars is invested at an annual interest rate of 6% compounded continuously for t years, the total amount accumulated is given by

$$f(P, t) = Pe^{0.06t}$$

Determine $f(3000, 10)$ and interpret.

67. If $2000 is invested at an annual interest rate of r (in decimal form) compounded continuously for t years, the total amount accumulated is given by

$$f(r, t) = 2000e^{r \cdot t}$$

Determine $f(0.0725, 25)$ and interpret.

68. A golf club manufacturer has a Cobb–Douglas production function given by

$$Q = f(x, y) = 21x^{0.3}y^{0.75}$$

where x is utilization of labor and y is utilization of capital. Determine the number of units of golf clubs produced when 200 units of labor and 75 units of capital are used.

69. A sports shoe manufacturer has a Cobb–Douglas production function given by

$$Q = f(x, y) = 42x^{0.37}y^{0.66}$$

where x is utilization of labor and y is utilization of capital. Determine the number of units of sports shoe produced when 300 units of labor and 100 units of capital are used.

70. The volume of a cylinder, such as a soup can, is given by $\pi r^2 h$, where r is the radius and h is the height. Since the volume is a function of the radius and the height, we can write

$$V(r, h) = \pi r^2 h$$

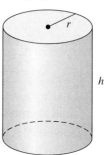

(a) Determine $V(2, 6)$ and interpret. (Each is measured in inches.)

(b) If the radius and height are equal, we can express the volume of the cylinder as a function of one variable, either r or h. Determine $V(r)$ and $V(h)$.

71. An individual's IQ is defined to be

$$\text{IQ} = f(a, m) = \frac{100m}{a}$$

where m is the individual's mental age (as determined by a test) in years and a is the individual's actual age in years.

(a) Evaluate $f(9, 12)$ and interpret.

(b) Evaluate $f(12, 9)$ and interpret.

(c) Evaluate $f(12, 12)$ and interpret.

(d) Determine IQ if $a = m$. In your own words, what does this tell us about an IQ of 100?

72. A bicycle seat manufacturer has a Cobb–Douglas production function given by

$$Q = f(x, y) = 22x^{0.75}y^{0.25}$$

where x is utilization of labor and y is utilization of capital. Determine the number of units of bicycle seats produced when 100 units of labor and 50 units of capital are used.

73. (Continuation of Exercise 72.) Suppose that we wish to know what combinations of labor and capital would result in 2000 units of bicycle seats being produced. In other words, we want to know what values of x and y satisfy

$$2000 = 22x^{0.75}y^{0.25}$$

If we solve this equation for y, we get

$$\frac{2000}{22x^{0.75}} = y^{0.25}$$

$$(90.91x^{-0.75})^4 = y$$

(a) In the viewing window [0, 200] by [0, 1000], graph $y = (90.91x^{-0.75})^4$. The curve that you see is called an *isoquant*. We will discuss isoquants in Section 8.2.

(b) Use the TRACE command to approximate the number of units of capital needed if 70 units of labor is used to produce 2000 units of bicycle seats.

(c) Use the TRACE command to approximate the number of units of capital needed if 100 units of labor is used to produce 2000 units of bicycle seats.

(d) Determine three combinations of labor and capital that will produce 2000 units of bicycle seats. (Answers may vary.)

74. The O'Neill Corporation has 10 soft drink bottling plants located in the United States. In a recent year the data for each plant gave the number of labor hours (in thousands), capital (total net assets, in millions), and the total quantity produced (in thousands of gallons). The data are shown in Table 8.1.1. The plants all use the same technology, so a production function can be determined. A Cobb–Douglas production function modeling these data is given by

$$Q = f(L, K) = 1.64L^{0.623}K^{0.357}$$

TABLE 8.1.1

LABOR	CAPITAL	QUANTITY
100	11	68
100	13	72
110	14	79
125	16	89
133	17	95
140	20	105
151	23	114
152	23	115
160	24	120
166	26	127

(a) Determine $f(125, 18)$ and interpret.

(b) Determine $f(70, 30)$ and interpret.

SECTION PROJECT

The program in Appendix B that performs multiple regression is needed to do this Section Project.

The Basich Company has eight plants in the United States producing handheld calculators. In a recent year the data for each plant gave the number of labor hours (in thousands), capital (in millions), and total quantity produced. The data are shown in Table 8.1.2. The plants all use the same technology, so a production function can be determined.

TABLE 8.1.2

LABOR	CAPITAL	QUANTITY
96	15	12,234
103	20	13,907
104	21	14,187
109	22	14,903
111	27	15,915
121	30	17,507
122	31	17,768
127	36	19,047

(a) Enter the data into your calculator and execute the program to determine a Cobb–Douglas production function to model the data. Round all values to the nearest hundredth.

(b) Suppose that Richard, CEO of the Basich Company, wants to know all combinations of labor and capital that produce 15,000 calculators. To do this, use the production function from part (a) and rename L with an x and K with a y. Substitute 15,000 for Q and solve the resulting equation for y.

(c) Graph the equation found in part (b) in an appropriate viewing window. Use the TRACE command to determine three realistic combinations of labor and capital that will produce 15,000 calculators.

SECTION 8.2

LEVEL CURVES, CONTOUR MAPS, AND CROSS-SECTIONAL ANALYSIS

In Section 8.1 we were introduced to functions of two independent variables and saw that the graphs of these functions are surfaces in 3-space. Figure 8.2.1 shows a graph of a production function.

As mentioned in Section 8.1, graphing surfaces in 3-space is very difficult without the aid of a computer. For this reason **cross-sectional analysis** can aid in understanding the behavior of a surface by analyzing the surface using graphs in two dimensions. For example, consider Figure 8.2.2, which shows the surface in Figure 8.2.1 sliced with a horizontal plane $z = 4$.

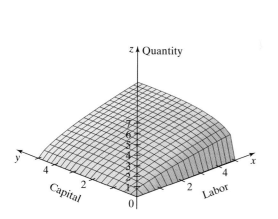

Figure 8.2.1   Graph of $z = 1.3x^{0.75}y^{0.3}$.

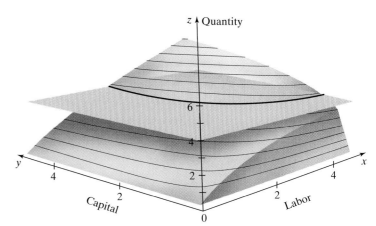

Figure 8.2.2   Graph of $z = 1.3x^{0.75}y^{0.3}$ and the horizontal plane $z = 4$.

The horizontal plane $z = 4$ intersects the surface to form a curve that is parallel to the xy-plane, which means that we can graph the curve in the xy-plane. Slicing of surfaces with horizontal plane(s) and graphing the resulting curve(s) in a single xy-plane is where we begin our analysis of surfaces in 3-space.

Horizontal Cross Sections, Level Curves, and Isoquants

To have an idea of what we will be doing, let's consider the **topographical map** shown in Figure 8.2.3. A *topographical map* is simply a two-dimensional graph of a three-dimensional surface, such as a mountain. The lines we see on the topographical map connect points with the same elevation. In mathematics, we call lines that connect points of equal elevation **contour lines** or **level curves.** Also, we call a collection of contour lines a **contour map.** Assuming that the elevation between the contour lines on the topographical map changes by a constant amount, we see that the more closely packed the lines are, the steeper the terrain. The more spread apart the contour lines are, the flatter the terrain. A topographical map gives a good overall picture of the terrain, indicating where hills and flat areas are located. In Example 1 we employ the concept illustrated in the topographical map to a surface in 3-space.

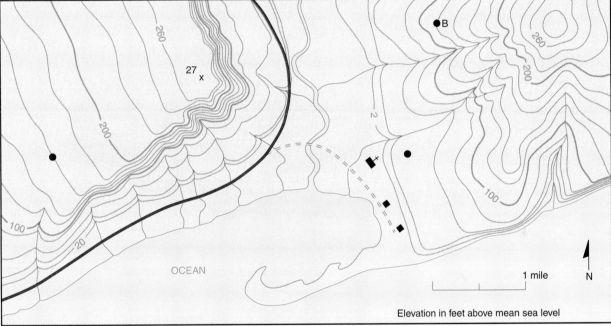

Figure 8.2.3

EXAMPLE 1 | **Determining a Level Curve**

An economist for the Linger Golf Cart Corporation has computed a production function for the manufacture of their golf carts to be

$$f(x, y) = Q = 1.3x^{0.75}y^{0.25}$$

where Q is the number of golf carts produced each week, x is the number of labor hours each day, and y is the daily usage of capital investment. What combinations of labor hours each day and daily usage of capital investment will result in 13 golf carts being produced each week?

SOLUTION

We are being asked to determine *all* possible combinations of labor hours (x) and daily usage of capital investment (y) such that $Q = 13$. In other words, we want to know for what values of x and y does

$$13 = 1.3x^{0.75}y^{0.25}$$

Solving this equation for y gives

$$10 = x^{0.75}y^{0.25}$$
$$10x^{-0.75} = y^{0.25}$$
$$y = (10x^{-0.75})^4 = 10,000x^{-3}$$

To determine all combination of x and y, we can simply graph $y = (10x^{-0.75})^4 = 10,000x^{-3}$ in the xy-plane. Figure 8.2.4 shows a graph of $f(x, y) = Q = 1.3x^{0.75}y^{0.25}$, and Figure 8.2.5 shows a graph of $y = 10,000x^{-3}$. The graph in Figure 8.2.5 answers the question.

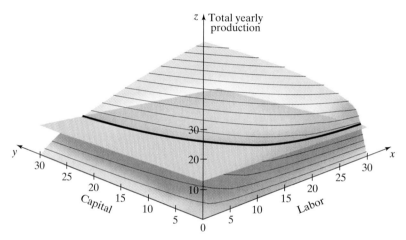

Figure 8.2.4 Graph of $f(x, y) = Q = 1.3x^{0.75}y^{0.25}$ and the horizontal plane $Q = 13$.

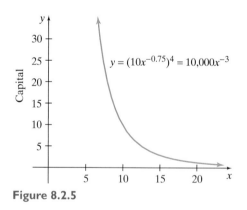

Figure 8.2.5

Technology Note
To see the isoquant in Figure 8.2.5 on your calculator, simply graph $y_1 = 10,000 \, x^{-3}$ in the viewing window [0, 20] by [0, 30].

Economists call the graph in Figure 8.2.5 an **isoquant** (*iso* means same and *quant* is short for quantity). It gives all combinations of x and y that yield a production of 13 golf carts each week. Graphically, it is the result of slicing the surface given by $f(x, y) = Q = 1.3x^{0.75}y^{0.25}$ with the horizontal plane $Q = 13$. (We can think of the variable Q as behaving like the variable z in our xyz-coordinate system.)

If we take the curve in Figure 8.2.5 and lift it 13 units above the xy-plane, we get a picture of the behavior of the *surface* 13 units above the xy-plane. Figure 8.2.6 illustrates this.

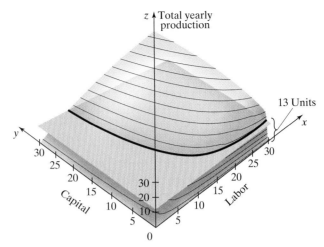

Figure 8.2.6

As we saw with the topographical map, the general mathematical word for an isoquant is *level curve*. It has this name since every point on the curve has the same dependent variable value. Also, recall that another name for level curve is a *contour line*. In general, we have the following:

Level Curve or Contour Line

A **level curve**, or **contour line**, is obtained from a surface $z = f(x, y)$ by slicing it with a horizontal plane $z = c$. The equation for the level curve at height c is given by

$$c = f(x, y)$$

A collection of level curves is called a **contour map** or **contour diagram.** Example 2 illustrates how to construct a contour map.

EXAMPLE 2 | **Constructing a Contour Map**

Let $z = f(x, y) = x^2 + y^2$. Construct a contour map using $c = 2, 4,$ and 6. Compare the contour map to the graph of the surface.

SOLUTION

The level curve at any height c is given by

$$c = x^2 + y^2$$

This is simply the equation of a circle, centered at the origin, with a radius of $\sqrt{c}$. So the level curve at $c = 2$ is found by graphing

$$2 = x^2 + y^2$$

For $c = 4$, we graph $4 = x^2 + y^2$, and for $c = 6$, we graph $6 = x^2 + y^2$. The contour map is shown in Figure 8.2.7, and the graph of $z = x^2 + y^2$ is shown in Figure 8.2.8. Notice that the graph of $z = x^2 + y^2$ gets steeper as we move farther away from the origin. This can be observed on the contour map in Figure 8.2.7, since the level curves become more tightly packed together as we move away from the origin.

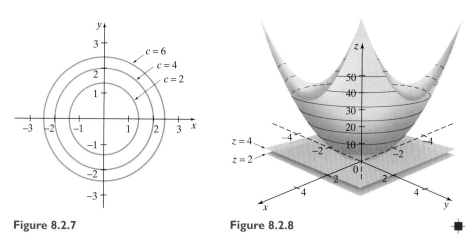

Figure 8.2.7   **Figure 8.2.8**

✓ **CHECKPOINT 1**

Now work Exercise 5.

EXAMPLE 3

Constructing a Contour Map in an Application

Recall that the golf cart manufacturer in Example 1 has a production function of $Q = f(x, y) = 1.3x^{0.75}y^{0.25}$, where Q is the number of golf carts produced each week, x is the number of labor hours each day, and y is the daily usage of capital investment. Construct a contour map (here it is a collection of isoquants) using $c = 10, 20, 30$, and 40.

SOLUTION

Recall that we can think of Q as being the same as z. The isoquant (level curve) at any production level (height) is given by

$$c = 1.3x^{0.75}y^{0.25}$$

Solving this equation for y gives

$$\frac{c}{1.3} = x^{0.75}y^{0.25}$$

$$\frac{c}{1.3}x^{-0.75} = y^{0.25}$$

$$y = \left(\frac{c}{1.3}x^{-0.75}\right)^4 = \frac{c^4}{2.8561}x^{-3}$$

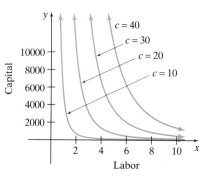

Figure 8.2.9

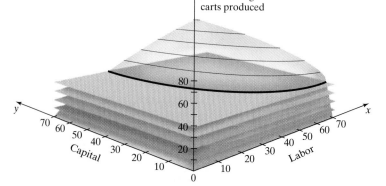

Figure 8.2.10

We now substitute the given values of c into this equation and graph the resulting equations, as shown in Figure 8.2.9. The graph of the surface is in Figure 8.2.10.

Notice in Examples 2 and 3 that the contour map is constructed on one xy-plane. Also notice that the values of c in both examples are equally spaced. Having equally spaced c-values and constructing a contour map on a single xy-plane are necessary if we want the contour map to give us an idea of the behavior of the surface.

Indifference Curves

So far we have seen that the level curves of a production function are called iso-quants. We now look at another level curve from economics that is called an **indifference curve.** First we need a brief discussion on what economists call a **utility function.** Utility functions and indifference curves are the basis of the modern theory of consumer behavior.

We say that an individual derives *satisfaction* or *utility* from commodities consumed during a given time period. In the time period, the individual will consume a large variety of different commodities. Economists refer to this collection of different commodities as a **commodity bundle.** For different commodity bundles, economists assume that each individual compares alternative commodity bundles and states a preference. Example 4 illustrates this idea.

EXAMPLE 4 | **Analyzing Commodity Bundles**

Tom enjoys eating cheeseburgers and drinking colas. He was asked to rank the following commodity bundles (of cheeseburgers and colas) for a typical week, with a preferred bundle assigned a higher number. His rankings are given in Table 8.2.1.

TABLE **8.2.1**

BUNDLE	COLAS (x)	CHEESEBURGERS (y)	RANK
A	3	5	3
B	4	3	3
C	5	2	3
D	1	4	1
E	2	2	1
F	3	1	1

(a) Which bundle(s) is(are) most preferred by Tom? Least preferred?

(b) Make a plot of the data, placing quantity of colas on the x-axis and quantity of cheeseburgers on the y-axis. Connect with a smooth curve those points that have the same rank.

SOLUTION

(a) It appears that Tom prefers bundles A, B, and C the most since he assigned each a rank of 3. He is said to be *indifferent* among these three bundles. Tom prefers bundles D, E, and F the least since he assigned each a rank of 1. Tom is said to be *indifferent* among these three bundles.

(b) Figure 8.2.11 gives a plot of the data.

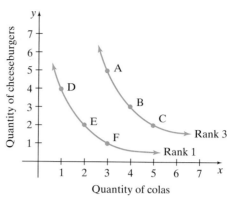

Figure 8.2.11

The two curves seen in Figure 8.2.11 are called **indifference curves,** since the consumer in question, Tom, is indifferent to any bundle on each curve. In other words, on the rank 3 indifference curve, Tom would receive the same utility from bundle A as from bundle B or C.

Since the indifference curves in Figure 8.2.11 are labeled with rank 1 and rank 3, we can easily imagine that the *indifference curves are level curves* of some function. The function in question is called a **utility function** and is denoted

$$U(x, y) = f(x, y)$$

The utility function assigns a numerical value (or utility level) to commodity bundles for goods x and y. Without going too far into economics theory, all that we require of the utility function is that it reflect the same rankings that a consumer assigns to alternative commodity bundles. So if a consumer prefers bundle A to bundle D, the utility function has to assign a *larger* numerical value to bundle A than to bundle D. Example 5 shows how this is done.

EXAMPLE 5

Sketching an Indifference Map

Suppose that the utility from consuming x colas and y cheeseburgers is given by $U(x, y) = \sqrt{xy}$. Draw a contour map for $c = 1, 2, 3$, and 4. Here we will have four indifference curves. A collection of indifference curves is also known as an **indifference map.**

SOLUTION

As we have seen so far in this section, we can algebraically set this up as

$$U(x, y) = \sqrt{xy}$$
$$c = \sqrt{xy}$$
$$c^2 = xy$$
$$y = \frac{c^2}{x}$$

Now we substitute $1, 2, 3$, and 4 in for c and graph the following in the xy-plane.

$$y = \frac{1}{x}, \quad y = \frac{4}{x}, \quad y = \frac{9}{x} \quad \text{and} \quad y = \frac{16}{x}$$

Figure 8.2.12 is the contour map and Figure 8.2.13 is a graph of the utility function. Again notice that the curves in Figure 8.2.12 are the result of slicing the surface in Figure 8.2.13 with horizontal planes 1, 2, 3 and 4 units, respectively, above the xy-plane.

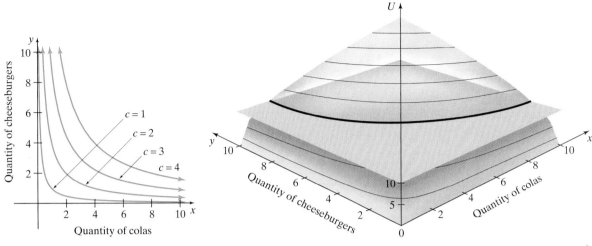

Figure 8.2.12   **Figure 8.2.13**

✓ CHECKPOINT **2**   Now work Exercise 17.

Vertical Cross Sections

So far we have seen how slicing a surface with a horizontal plane produces a level curve that can be drawn in the xy-plane. A collection of several slices at different heights produces a collection of level curves that gives us an idea of the behavior of the surface in 3-space. However, if we look at **vertical cross sections** of the form $x = c$ and $y = c$, we can sometimes improve on our 3-space visualization.

In Example 2, we sliced the surface with planes of the form $z = c$. See Figure 8.2.8. Notice that these planes are perpendicular to the z-axis and that the collection of level curves, that is, the contour map, was graphed on the xy-plane. The procedure for vertical cross-sections is basically the same. If we slice a surface with planes of the form $x = c$ (these will be vertical planes), the planes are perpendicular to the x-axis, and the resulting cross sections are graphed on the yz-plane. Similarly, if we slice a surface with planes of the form $y = c$ (again, vertical planes), the planes are perpendicular to the y-axis, and the resulting cross sections are graphed on the xz-plane. We demonstrate this process in Example 6.

EXAMPLE 6 | **Constructing Vertical Cross Sections**

Consider $z = y^2 - x^2$.

(a) Sketch the cross sections on the xz-plane with y fixed at 0, ± 1, and ± 2.

(b) Sketch the cross sections on the yz-plane with x fixed at 0, ± 1, and ± 2.

SOLUTION

(a) Here y is fixed, which means that we are slicing the surface with vertical planes. The vertical planes are perpendicular to the y-axis and are of the form $y = 0$, $y = \pm 1$, and $y = \pm 2$. So we need to graph the following curves in the xz-plane.

$$z = -x^2 \qquad (y = 0)$$
$$z = 1 - x^2 \qquad (y = \pm 1)$$
$$z = 4 - x^2 \qquad (y = \pm 2)$$

To graph these curves, we simply need to treat z as the dependent variable and x as the independent variable. The graphs of these curves are given in Figure 8.2.14, and the graph of the surface with a vertical cross section is shown in Figure 8.2.15.

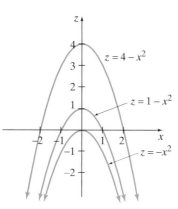

Figure 8.2.14

Figure 8.2.15 Vertical plane $y = -2$ intersects the surface in a curve.

(b) Here x is fixed, which means that we are slicing the surface with vertical planes. The vertical planes are perpendicular to the x-axis and are of the form $x = 0$, $x = \pm 1$, and $x = \pm 2$. Hence, we need to graph the following curves in the yz-plane.

$$\begin{aligned} z &= y^2 & (x = 0) \\ z &= y^2 - 1 & (x = \pm 1) \\ z &= y^2 - 4 & (x = \pm 2) \end{aligned}$$

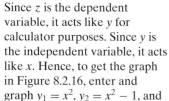

Notice that here z is the dependent variable and y is the independent variable. The graphs of these cross sections are given in Figure 8.2.16, and the graph of the surface with a vertical cross section is shown in Figure 8.2.17.

Notice the upward opening parabolas in the y-direction, which matches our work in part (b). Also, notice the downward opening parabolas in the x-direction, which matches our work in part (a).

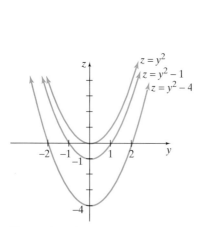

Figure 8.2.16

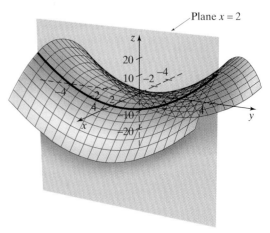

Figure 8.2.17 Vertical plane $x = 2$ intersects the surface in a curve.

Interactive Activity

Construct a contour map for the function in Example 6 for c-values of $c = 0$, ± 1, and ± 2. Using these horizontal cross sections along with the vertical cross section from parts (a) and (b) in Example 6, do you "see" the surface as shown in Figure 8.2.17?

Before we look at our concluding example, we want to mention that our work with vertical cross sections will be very important in Section 8.3 when we tackle derivatives of functions of two variables. These derivatives, called *partial derivatives*, are related to vertical cross sections.

EXAMPLE 7 | **Applying Cross-sectional Analysis to a Production Function**

Consider the production function for the Linger Golf Cart Corporation in Example 1. It is

$$f(x, y) = Q = 1.3x^{0.75}y^{0.25}$$

where Q is the number of golf carts produced per week, x is the number of labor hours each day, and y is the daily usage of capital investment. Construct a cross section on an xz-plane with y fixed at $y = 20$. Use the graph to approximate how many labor hours are required each day to produce 10 golf carts per week.

SOLUTION

We begin by replacing the dependent variable Q with z so that we have

$$z = 1.3x^{0.75}y^{0.25}$$

Again, since y is fixed, we are slicing the surface with a vertical plane perpendicular to the y-axis of the form $y = 20$. So we need to graph in an xz-plane the following:

$$z = 1.3x^{0.75}(20)^{0.25} \approx 2.75x^{0.75}$$

The graph of the cross section is given in Figure 8.2.18, and the graph of the surface is given in Figure 8.2.19.

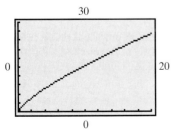

Figure 8.2.18
Graph of $y = 2.75x^{0.75}$, representing the vertical cross section.

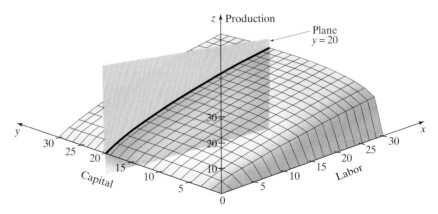

Figure 8.2.19

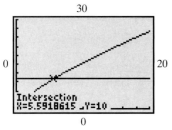

Figure 8.2.20

To use the graph in Figure 8.2.18 to answer this question, we must remember that on the calculator, for this graph, y is acting like the variable z, the quantity produced. We still have x representing labor hours each day. So to answer this question we graph $y = 10$ and use the INTERSECT command. See Figure 8.2.20.

From Figure 8.2.20 we conclude that approximately 5.6 labor hours each day will produce a quantity of 10 golf carts per week when the capital investment is fixed at $y = 20$ units each day.

✓ **CHECKPOINT 3**

Now work Exercise 37.

SUMMARY

We began this section by analyzing a surface using horizontal cross sections, called **level curves.** We then analyzed surfaces in space using vertical cross sections. We sliced a surface with planes of the form $x = c$ and $y = c$ and graphed the resulting cross sections on the yz-plane and the xz-plane, respectively. Figures 8.2.4 through 8.2.10 and 8.2.14 through 8.2.17 show some surfaces sliced by planes and the corresponding cross sections. Notice that each type of cross section tells us something about the shape of the surface.

• If the surface has an equation of the form $z = f(x, y)$, a level curve is obtained by slicing the surface with a horizontal plane $z = c$. The equation for the level curve at height c is given by $c = f(x, y)$. See Figure 8.2.7.

SECTION 8.2 EXERCISES

In Exercises 1–8, make a contour map of the given function for the specified c-values,

1. $f(x, y) = x + y + 1$ $c = 0, 1, 2, 3$

2. $f(x, y) = 1 - x - y$ $c = 0, 3, 6, 9$

3. $f(x, y) = 2x + 3y + 6$ $c = 0, 2, 4, 6$

4. $f(x, y) = 2y - 3x + 1$ $c = 0, 2, 4, 6$

✓ 5. $f(x, y) = \sqrt{16 - x^2 - y^2}$ $c = 0, 2, 4$

6. $f(x, y) = \sqrt{25 - x^2 - y^2}$ $c = 1, 3, 5$

7. $f(x, y) = e^x + y$ $c = 0, 2, 4, 6$

8. $f(x, y) = \ln x + y - 1$ $c = 0, 2, 4, 6$

In Exercises 9–14, make a contour map of the given production functions for the specified c-values.

9. $Q = 24x^{0.5}y^{0.5}$ $c = 10, 20, 30, 40$

10. $Q = 36.5x^{0.5}y^{0.5}$ $c = 100, 200, 300, 400$

11. $Q = 5.6x^{0.65}y^{0.41}$ $c = 10, 20, 30, 40$

12. $Q = 6.3x^{0.7}y^{0.35}$ $c = 20, 40, 60, 80$

13. $Q = 180x^{0.7}y^{0.3}$ $c = 100, 200, 300, 400$

14. $Q = 120x^{0.75}y^{0.25}$ $c = 100, 200, 300, 400$

In Exercises 15–20, make a contour map of the given utility function for the specified c-values.

15. $U(x, y) = \sqrt{xy}$ $c = 2, 4, 6, 8$

16. $U(x, y) = x^{0.25}y^{0.75}$ $c = 1, 2, 3, 4$

✓ 17. $U(x, y) = x^{0.75}y^{0.25}$ $c = 1, 2, 3, 4$

18. $U(x, y) = x^{0.65}y^{0.35}$ $c = 1, 2, 3, 4$

19. $U(x, y) = x^{0.55}y^{0.45}$ $c = 1, 2, 3, 4$

20. $U(x, y) = x^{0.55}y^{0.45}$ $c = 2, 4, 6, 8$

In Exercises 21–28, sketch on the appropriate plane, either the xz-plane or yz-plane, cross sections for the given function for the specified vertical plane values. See Example 6.

21. $z = 15x^{0.5}y^{0.5}$ $x = 2, 4, 6$

22. $z = 15x^{0.5}y^{0.5}$ $x = 3, 6, 9$

23. $z = 15x^{0.5}y^{0.5}$ $y = 2, 4, 6$

24. $z = 15x^{0.5}y^{0.5}$ $y = 3, 6, 9$

25. $z = 120x^{0.75}y^{0.25}$ $x = 10, 20, 30$

26. $z = 120x^{0.75} y^{0.25}$ $x = 20, 40, 60$

27. $z = 120x^{0.75} y^{0.25}$ $y = 20, 40, 60$

28. $z = 120x^{0.75} y^{0.25}$ $y = 10, 20, 30$

APPLICATIONS

29. Cranky Corporation has three crankshaft producing plants in the United States and a Cobb–Douglas production function of $Q = f(x, y) = 1.2x^{0.7}y^{0.3}$, where x represents the number of labor hours (in thousands), y represents the capital (total net assets in dollars), and Q represents the quantity produced (in thousands). Sketch isoquants for $Q = 20, 40,$ and 60.

30. Ling Incorporated has four pretzel making plants in the United States and a Cobb–Douglas production function of $Q = f(x, y) = 2.3x^{0.2} y^{0.8}$, where x represents the number of labor hours (in hundreds), y represents the capital (total net assets in dollars), and Q represents the quantity produced (in hundreds of pounds). Sketch isoquants for $Q = 30, \ 60,$ and 90.

31. Suppose that the utility from consuming x colas and y slices of pizza in a typical week is given by $U(x, \ y) = x^{2/3}y^{1/3}$. Make an indifference map for $c = 1, 2,$ and 3.

32. Suppose that the utility from consuming x chocolate milkshakes and y cheeseburgers in a typical week is given by $U(x, \ y) = x^{1/3}y^{2/3}$. Make an indifference map for $c = 1, 2,$ and 3.

33. In their study of human groupings, anthropologists often use an index called the **cephalic index.** The cephalic index is given by $z = C(x, \ y) = 100 \cdot \dfrac{x}{y}$, where x is the width and y is the length of an individual's head. Both measurements are made across the top of the head and both are in inches. Construct a contour map for $c = 75, 80, 85,$ and 90.

34. The **Doyle log rule** is one method used to determine the yield of a log, measured in board feet. It is given by

$$f(x, \ y) = \left(\frac{x - 4}{4} \right)^2 \cdot y^*$$

where x is the diameter of the log in inches, y is the length of the log in feet, and $f(x, \ y)$ is the number of board-feet.

(a) Construct a contour map for $c = 400, 500, 600,$ and 700.

(b) In your own words, describe what the level curve for $c = 500$ represents.

35. The O'Neill Corporation has 10 soft drink bottling plants in the United States and a Cobb–Douglas production function of $Q = f(x, y) = 1.64x^{0.6}y^{0.4}$, where x is the number of labor hours (in thousands), y is the capital (total net assets

Based on data gathered at http://www.forestry.uga.edu/docs/ 950-measurementscaling.html.

in millions), and Q is the quantity produced (in thousands of gallons). Sketch isoquants for $Q = 80, 100, 120,$ and 140.

36. (Continuation of Exercise 35.) James, the CEO of the O'Neill Corporation, wants to keep the number of labor hours at each plant fixed at $x = 110$.

(a) By fixing labor at $x = 110$, is James looking at a vertical or horizontal cross section?

(b) Graph the cross section when $x = 110$.

(c) Use the graph to approximate how much capital is required to produce 100.

(d) In your own words, describe what the graph in part (b) represents.

✓ 37. (Continuation of Exercise 35.) James, the CEO of the O'Neill Corporation, wants to keep capital at each plant fixed at $y = 18$.

(a) By fixing capital at $y = 18$, is James looking at a vertical or horizontal cross section?

(b) Graph the cross section when $y = 18$.

(c) Use the graph to approximate how many labor hours would be required to produce 100.

(d) In your own words, describe what the graph in part (b) represents.

38. A golf club manufacturer has a Cobb–Douglas production function given by

$$Q = f(x, \ y) = 21x^{0.3} y^{0.75}$$

where x is the utilization of labor, y is the utilization of capital, and Q is the number of units of golf clubs produced. Sketch the isoquants for $Q = 500, 1000, 1500,$ and 2000.

39. (Continuation of Exercise 38.) The CEO of the golf club manufacturer wants to keep labor fixed at $x = 100$ units.

(a) By fixing labor at $x = 100$ units, is the CEO looking at a vertical or horizontal cross section?

(b) Graph the cross section when $x = 100$.

(c) Use the graph to approximate how many units of capital would be required to produce 1500 units of golf clubs.

(d) In your own words, describe what the graph in part (b) represents.

40. (Continuation of Exercise 38.) The CEO of the golf club manufacturer wants to keep capital fixed at $y = 100$ units.

(a) By fixing capital at $y = 100$ units, is the CEO looking at a vertical or horizontal cross section?

(b) Graph the cross section when $y = 100$.

(c) Use the graph to approximate how many units of labor would be required to produce 1500 units of golf clubs.

(d) In your own words, describe what the graph in part (b) represents.

41. The Brody Corporation has 11 plants worldwide and a Cobb–Douglas production function given by

$$Q = f(x, \ y) = 1.7x^{0.8} y^{0.2}$$

where x is the number of units of labor (in thousands), y is the number of units of capital (in millions), and Q is the number of units of quantity produced. Construct isoquants for $c = 300, 400, 500,$ and 600.

42. (Continuation of Exercise 41.) Cheryl, the CEO of the Brody Corporation, wants to keep labor fixed at $x = 350$ units.

(a) Graph the vertical cross section when $x = 350$.

(b) Use the graph to approximate how much capital is required to produce 560 units.

43. (Continuation of Exercise 41.) Cheryl, the CEO of the Brody Corporation, wants to keep capital fixed at $y = 50$ units.

(a) Graph the vertical cross section when $y = 50$.

(b) Use the graph to approximate how many units of labor are required to produce 560 units.

44. The volume of a cylinder with radius r and height h is given by $V(r, h) = \pi r^2 h$. Substituting x for r and y for h gives $V(x, y) = \pi x^2 y$. Make a contour map for $c = 10, 20, 30,$ and 40.

45. Suppose that the height of the cylinder in Exercise 44 is fixed at 4 inches, that is, $y = 4$.

(a) Graph the vertical cross section when $y = 4$.

(b) Use the graph to approximate the radius necessary to produce a cylinder with a volume of approximately 113 cubic inches.

(c) Use the graph to approximate the volume if the radius is 6 inches.

(d) In your own words, what does the graph in part (a) represent.

46. IQ is defined to be $IQ = f(x, y) = \dfrac{100y}{x}$, where y is the individual's mental age in years and x is the individual's actual age in years.

(a) Make a contour map for $c = 90, 100, 110,$ and 120. Keep in mind that $x > 0$ and $y > 0$.

(b) Give three combinations of actual age and mental age that gives an IQ of 120.

47. For the IQ function in Exercise 46, consider a vertical cross section by keeping mental age fixed at 20, that is, $y = 20$. Our function then becomes

$$IQ = \frac{100 \cdot 20}{x} = \frac{2000}{x}$$

(a) To accommodate our calculator, substitute y for IQ. This yields $y = \dfrac{2000}{x}$. Graph this in the window $[0, 60]$ by $[0, 300]$.

(b) The point $\left(15, 133\frac{1}{3}\right)$ is on the graph. Interpret what this means.

(c) The point $(16, 125)$ is on the graph. Interpret what this means.

48. For the IQ function in Exercise 46, consider a vertical cross section by keeping actual age fixed at 20, that is, $x = 20$.

Our function then becomes

$$IQ = \frac{100y}{20} = 5y$$

(a) To accommodate our calculator, substitute y for IQ and x for y. This gives $y = 5x$. Graph this in the viewing window $[0, 60]$ by $[0, 300]$.

(b) The point $(30, 150)$ is on the graph. Interpret what this means.

(c) The point $(31, 155)$ is on the graph. Interpret what this means.

(d) The point $(29, 145)$ is on the graph. Interpret what this means.

(e) Fill in the blanks.

> When actual age is fixed at 20, each 1-year increase in mental age _____ (increases/decreases) IQ by _____ units.

Section Project

The program in Appendix B that performs multiple regression is needed to do this section project.

The Riblet Engine Company has six plants around the United States to produce small engines. In a recent year the data for each plant gave the number of labor hours (in thousands), capital (in millions), and total quantity produced, as follows:

Labor	Capital	Quantity
42	5	27,825
44	6	31,058
47	7	34,633
49	9	39,762
50	10	42,201
52	13	48,621

(a) The plants all operate with the same technology, so a production function can be determined. Enter the data into your calculator and execute the program to determine a Cobb–Douglas production function to model the data. Round all values to the nearest hundredth.

(b) Graph isoquants for $q = 20,000, 30,000, 40,000,$ and $50,000$.

(c) The CEO of the company, Gene, wants to keep labor fixed at 48. Is Gene asking for a vertical or horizontal cross section?

(d) Graph the cross section when labor is fixed at 48.

(e) Use the graph to approximate how much capital is required to produce a quantity of 35,000.

SECTION 8.3

PARTIAL DERIVATIVES AND SECOND-ORDER PARTIAL DERIVATIVES

In this section we discuss how to measure the rate of change of a function of two independent variables. We will observe how the vertical cross sections discussed in Section 8.2 are central to the study of this rate of change, which is known as **partial derivatives.** Partial derivatives simply measure the rate of change of a function of two independent variables, and they will be compared to the derivative that was discussed in Section 2.3. We also will discuss *second-order* partial derivatives. Second-order partial derivatives are analogous to second derivatives, which we discussed in Section 5.2.

From Your Toolbox

- Recall that for a given function f the derivative of f at x is

$$f'(x) = \lim_{h \to 0} \frac{f(x+h) - f(x)}{h}$$

- f' gives the rate of change of f.

Partial Derivatives

We have encountered many interpretations and applications of the derivative of a function of one independent variable, f.

Naturally, we would like to extend this rate of change concept to a function of two variables, $f(x, y)$. To do this, we consider how the function values change as x changes, that is, when y is kept constant, and we consider how the function values change as y changes, that is, when x is kept constant. Geometrically, we are talking about the **vertical cross sections** introduced in Section 8.2.

Flashback

O'NEILL CORPORATION REVISITED

In Section 8.2 we were introduced to the O'Neill Corporation. Recall that the O'Neill Corporation has 10 soft drink bottling plants and a production function of

$$Q = f(x, y) = z = 1.64x^{0.6}y^{0.4}$$

where x is the number of labor hours (in thousands), y is the capital (total net assets in millions), and Q is the quantity produced (in thousands of gallons). James, the CEO of the O'Neill Corporation, wants to keep the number of labor hours at each plant fixed at $x = 110$. Graph the cross section when $x = 110$ and describe what the graph represents.

Flashback Solution

Figure 8.3.1 gives the graph of the surface and the plane $x = 110$, and Figure 8.3.2 is a graph of the cross section. Notice that the cross section is graphed on the yz-plane.

The graph of the cross section shows us that quantity is a function of capital when labor is held constant at $x = 110$. Specifically, we have

$$Q = f(x, y) = 1.64x^{0.6}y^{0.4}$$
$$= f(110, y) = 1.64(110)^{0.6}y^{0.4}$$

So, by holding labor constant at $x = 110$, we now have quantity as a function of one variable, capital or y.

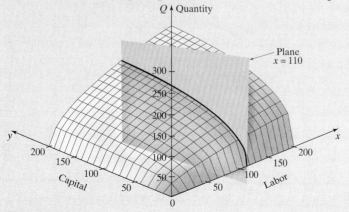

Figure 8.3.1

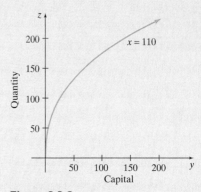

Figure 8.3.2

Let's take another look at Figure 8.3.2. Notice that when labor is fixed at $x = 110$ the vertical cross section is a function of *one* variable. (All along the curve in Figure 8.3.2, x is held constant at $x = 110$.) As we saw in the Flashback, this function of one variable has the form

$$f(110, y) = 1.64(110)^{0.6} y^{0.4}$$

From our work with derivatives, we know that we can find the rate at which $f(110, y) = 1.64(110)^{0.6} y^{0.4}$ changes, as y changes, by taking the derivative. What exactly does all this tell us? Example 1 provides the answer.

EXAMPLE 1 | **Determining a Rate of Change**

For the O'Neill Corporation production function, we determined in the Flashback that when labor is held constant at $x = 110$ we get a function of one variable

$$f(110, y) = 1.64(110)^{0.6} y^{0.4}$$

(a) Determine $\frac{d}{dy}[1.64(110)^{0.6} y^{0.4}]$.

(b) If the number of labor hours is fixed at $x = 110$, how fast is the number of gallons produced increasing as the capital increases from a level of 20?

SOLUTION

(a) The derivative is

$$\frac{d}{dy}[1.64(110)^{0.6} y^{0.4}] = 1.64(110)^{0.6}(0.4) y^{-0.6}$$

(b) Evaluating the derivative in part (a) at $y = 20$ gives

$$1.64(110)^{0.6}(0.4)(20)^{-0.6} \approx 1.8244$$

Evaluating the derivative at $y = 20$ gives the *slope of the line tangent to the curve at $y = 20$*. See Figure 8.3.3. The curve in Figure 8.3.3 is the graph of the cross section when $x = 110$, so it is the graph of quantity as a function of capital when labor is fixed at 110. This tells us that, when the number of labor hours is held constant at $x = 110$ and capital is at 20 units, as capital increases 1 unit, production is increasing at a rate of about $1.8 \frac{\text{thousand gallons}}{\text{unit of capital}}$!

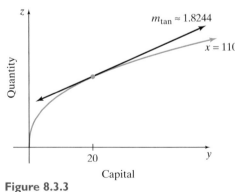

Figure 8.3.3

Let's summarize what was done in the Flashback and Example 1.

- When x is held constant at $x = 110$, we produce a **vertical cross section** of the surface.

- The graph of this cross section in the yz-plane shows production as a function of *one* variable.
- We can differentiate the function of one variable to give us a rate of change.

Definition of Partial Derivatives

The two-dimensional analysis that we did in the Flashback and in Example 1 tells us something remarkable about the three-dimensional surface. Since x is constant at $x = 110$ and we looked specifically at the scenario when $y = 20$, this means that we were at the point $(110, 20, 91.22)$ on the graph of the surface (z-coordinate rounded to nearest hundredth) of our function. If we are at the point $(110, 20, 91.22)$ on the surface and move *in the y-direction,* that is, parallel to the y-axis, the surface is changing at a rate of approximately $1.8\frac{\text{units of } z}{\text{unit of } y}$. This is exactly the same as the slope of the tangent line to the curve that resulted from slicing the surface with the plane $x = 110$! See Figure 8.3.4.

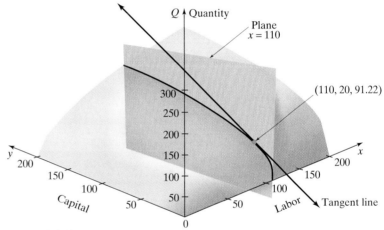

Figure 8.3.4

The process of holding x constant at any value $x = a$, moving in the y-direction (that is, parallel to the y-axis), and determining the rate of change of the surface is one interpretation of the **partial derivative of $f(x, y)$ with respect to y.** In a completely analogous manner, the process of holding y constant at any value $y = b$, moving in the x-direction (that is, parallel to the x-axis), and determining the rate of change of the surface is one interpretation of the **partial derivative of $f(x, y)$ with respect to x.** See Figures 8.3.5a–c.

The previous explanation serves as a basis for the following formal definition:

Partial Derivative

Let $f(x, y)$ be a function of two variables. The **partial derivative of $f(x, y)$ with respect to x is**

$$f_x(x, y) = \lim_{h \to 0} \frac{f(x + h, y) - f(x, y)}{h}$$

The **partial derivative of $f(x, y)$ with respect to y is**

$$f_y(x, y) = \lim_{h \to 0} \frac{f(x, y + h) - f(x, y)}{h}$$

Of course, each of these partial derivatives exists only if the appropriate limit exists.

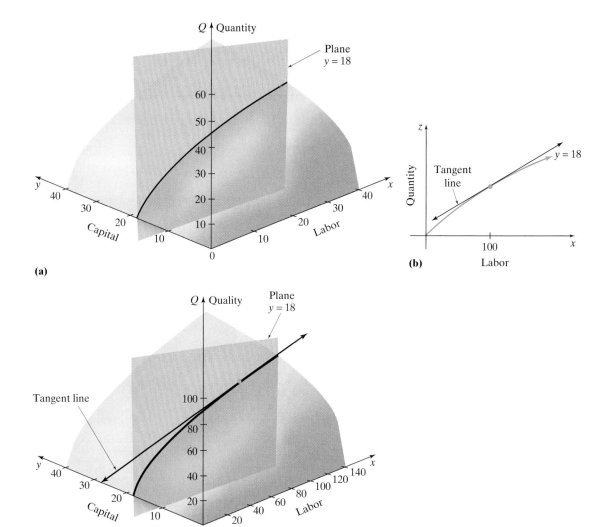

(a)

(b)

(c)

Figure 8.3.5

The important practical point on how to compute these partial derivatives is given in the following note.

NOTE:
1. The **partial derivative of $f(x, y)$ with respect to x** is found by treating y as a constant and performing our ordinary differentiation techniques. The notation for this partial derivative is $f_x(x, y)$ or $\dfrac{\partial f}{\partial x}$. The units of this partial derivative are units of f per unit of x.
2. The **partial derivative of $f(x, y)$ with respect to y** is found by treating x as a constant and performing our ordinary differentiation techniques. The notation for this partial derivative is $f_y(x, y)$ or $\dfrac{\partial f}{\partial y}$. The units of this partial derivative are units of f per unit of y.

EXAMPLE 2 | **Computing a Partial Derivative**

For $f(x, y) = x^2 y - y^2$, determine $f_x(x, y)$ and $f_y(x, y)$.

SOLUTION

To determine $f_x(x, y)$, we treat y as a constant and take the derivative with respect to x. This gives

$$f(x, y) = x^2y - y^2$$

$$f_x(x, y) = \frac{\partial}{\partial x}(x^2y) - \frac{\partial}{\partial x}(y^2)$$

$$f_x(x, y) = 2x \cdot y - 0 = 2xy$$

Observe that $\frac{\partial}{\partial x}(y^2) = 0$ because y is treated as a constant.

To compute $\frac{\partial}{\partial y}$, we treat x as a constant and take the derivative with respect to y. This yields

$$f(x, y) = x^2y - y^2$$

$$f_y(x, y) = \frac{\partial}{\partial y}(x^2y) - \frac{\partial}{\partial y}(y^2)$$

$$f_y(x, y) = x^2 \cdot 1 - 2y = x^2 - 2y$$

✓ CHECKPOINT 1

Now work Exercise 3.

EXAMPLE 3

Computing Partial Derivatives

For $f(x, y) = x^2y + y^3x - 2xy + y$, determine $f_x(x, y)$ and $f_y(x, y)$.

SOLUTION

To determine $f_x(x, y)$, we treat y as a constant and differentiate with respect to x.

$$f(x, y) = x^2y + y^3x - 2xy + y$$
$$f_x(x, y) = (2 \cdot x)y + y^3 \cdot 1 - 2 \cdot 1 \cdot y + 0$$
$$= 2xy + y^3 - 2y$$

To find $f_y(x, y)$, we treat x as a constant and differentiate with respect to y. This gives

$$f(x, y) = x^2y + y^3x - 2xy + y$$
$$f_y(x, y) = x^2 \cdot 1 + (3y^2) \cdot x - 2x \cdot 1 + 1$$
$$= x^2 + 3y^2x - 2x + 1$$

✓ CHECKPOINT 2

Now work Exercise 7.

Interactive Activity

In your own words, explain why there was no usage of the Product Rule for derivatives when computing the partial derivatives in Example 3.

EXAMPLE 4

Evaluating and Interpreting a Partial Derivative

Let $f(x, y) = 2x^3y^4 + \frac{1}{3}y^3 + e^{3x} - e^{2y}$.

(a) Determine $f_x(x, y)$.   (b) Evaluate $f_x(x, y)$ at $\left(1, \frac{3}{2}\right)$ and interpret.

SOLUTION

(a) To determine $f_x(x, y)$, we treat y as a constant and differentiate with respect to x.

$$f(x, y) = 2x^3y^4 + \frac{1}{3}y^3 + e^{3x} - e^{2y}$$

$$f_x(x, y) = (6x^2)y^4 + 0 + 3e^{3x} - 0 = 6x^2y^4 + 3e^{3x}$$

(b) Evaluating $f_x(x, y)$ at $\left(1, \frac{3}{2}\right)$ yields

$$f_x\left(1, \frac{3}{2}\right) = 6(1)^2\left(\frac{3}{2}\right)^4 + 3e^{3(1)} = 30.375 + 3e^3 \approx 90.63$$

This means that if we are at the point $\left(1, \frac{3}{2}, 11.25\right)$ on the surface and move along the surface in the x-direction, parallel to the x-axis, the function is changing at a rate of approximately $90.63 \frac{\text{units of } f}{\text{unit of } x}$. Equivalently, we say that at the point $\left(1, \frac{3}{2}, 11.25\right)$ the slope of the surface in the x-direction is approximately 90.63. See Figure 8.3.6.

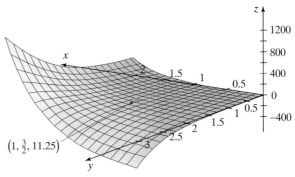

Figure 8.3.6

✓CHECKPOINT 3   Now work Exercise 33.

Applications

In the next example, we analyze the relationship between a partial derivative and **marginal analysis.**

EXAMPLE 5 | **Analyzing Marginal Revenue**

A company sells leaf blowers and a special 10-foot attachment to the leaf blower to clean gutters. Let x and y be the number of leaf blowers sold per month and the number of 10-foot attachments sold per month, respectively. Suppose that

$$p = 1200 - 8x - y, \qquad \text{the price in dollars for a leaf blower}$$
$$q = 523 - 1.5x - 0.15y, \quad \text{the price in dollars for an attachment}$$

(a) Determine the revenue function $R(x, y)$.

(b) Determine $R_x(x, y)$ and $R_y(x, y)$. Interpret each.

(c) Evaluate and interpret $R_x(50, 25)$.

SOLUTION

(a) Since revenue is price times quantity, we have

$$R(x, y) = x(1200 - 8x - y) + y(523 - 1.5x - 0.15y)$$
$$= 1200x - 8x^2 - xy + 523y - 1.5xy - 0.15y^2$$
$$= 1200x - 8x^2 - 2.5xy + 523y - 0.15y^2$$

(b) To determine $R_x(x, y)$, we treat y as a constant and differentiate with respect to x.

$$R_x(x, y) = 1200 \cdot 1 - 8 \cdot 2x - 2.5(1)(y) + 0 - 0$$
$$= 1200 - 16x - 2.5y$$

$R_x(x, y)$ gives the amount that each additional leaf blower adds to the total revenue. In other words, it gives the *marginal revenue* for leaf blowers.

To determine $R_y(x, y)$, we treat x as a constant and differentiate with respect to y.

$$R_y(x, y) = 0 - 0 - 2.5(x)(1) + 523(1) - 0.15(2y)$$
$$= -2.5x + 523 - 0.3y$$

$R_y(x, y)$ gives the amount that each additional 10-foot attachment adds to the total revenue. In other words, it is the *marginal revenue* for the 10-foot attachments.

(c) Since $x = 50$ and $y = 25$, we have

$$R_x(50, 25) = 1200 - 16(50) - 2.5(25) = 337.5$$

This means that when 50 leaf blowers and 25 of the 10-foot attachments have been sold, the company receives $337.50 for each *additional* leaf blower sold. ◈

Interactive Activity

Evaluate and interpret $R_y(50, 25)$ for the revenue function in Example 5.

EXAMPLE 6

Computing Partial Derivatives of a Volume Formula

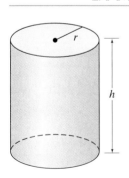

Recall that the formula for the volume of a cylinder is given by $V(r, h) = \pi r^2 h$, where r is the radius and h is the height.

(a) Determine $V_r(r, h)$ and $V_h(r, h)$.

(b) Evaluate each partial derivative when $r = 2$ centimeters (cm) and $h = 3$ cm and interpret.

SOLUTION

(a) Keep in mind that r and h are the independent variables. Thus, $V_r(r, h)$ is found by keeping h constant. This gives

$$V_r(r, h) = 2\pi r h$$

To find $V_h(r, h)$, we keep r constant, which gives

$$V_h(r, h) = \pi r^2$$

(b) Evaluating $V_r(r, h)$ when $r = 2$ cm and $h = 3$ cm yields

$$V_r(2, 3) = 2\pi(2)(3) = 12\pi$$

This means that when $r = 2$ cm and $h = 3$ cm, if the radius increases by 1 cm while the height remains constant at $h = 3$ cm, the volume of the cylinder increases by approximately 12π cm^3.

Evaluating $V_h(r, h)$ when $r = 2$ cm and $h = 3$ cm yields

$$V_h(2, 3) = \pi(2)^2 = 4\pi$$

Thus, when $r = 2$ cm and $h = 3$ cm, if the height increases by 1 cm while the radius remains constant at $r = 2$ cm, the volume of the cylinder increases by approximately 4π cm^3. ◼

Marginal Productivity of Labor and Capital

Example 1 foreshadowed the concept known as **marginal productivity of labor** and **marginal productivity of capital.** These two partial derivatives are very important in economics.

Marginal Productivity of Labor and Capital

For any production function of the form $Q = f(x, y) = ax^m y^n$, where a, m, and n are positive constants and x represents units of labor and y represents units of capital, we have the following:

- $f_x(x, y)$ gives the approximate change in productivity per unit change in labor and is called **marginal productivity of labor.**
- $f_y(x, y)$ gives the approximate change in productivity per unit change in capital and is called **marginal productivity of capital.**

EXAMPLE 7 | **Computing Marginal Productivity of Labor and Capital**

Recall that the production function for the O'Neill Corporation is given by

$$Q = f(x, y) = 1.64x^{0.6}y^{0.4}$$

where x is the number of labor hours in thousands, y is capital in millions, and Q is the quantity in thousands of gallons.

(a) Compute the marginal productivity of labor and the marginal productivity of capital.

(b) Evaluate the marginal productivity of labor at $x = 100$ and $y = 18$ and interpret.

(c) Evaluate the marginal productivity of capital at $x = 110$ and $y = 20$ and interpret.

SOLUTION

(a) Since marginal productivity of labor is $f_x(x, y)$, we have

$$f_x(x, y) = 1.64(0.6)x^{-0.4}y^{0.4} = 0.984x^{-0.4}y^{0.4}$$

Marginal productivity of capital is $f_y(x, y)$, which yields

$$f_y(x, y) = 1.64x^{0.6}(0.4)y^{-0.6} = 0.656x^{0.6}y^{-0.6}$$

(b) Since $x = 100$ and $y = 18$, we evaluate $f_x(100, 18)$, which gives

$$f_x(100, 18) = 0.984(100)^{-0.4}(18)^{0.4} \approx 0.5$$

This means that when the O'Neill Corporation uses 100 units of labor and 18 units of capital and keeps capital fixed at 18 units, production is increasing at a rate of approximately $0.5 \frac{\text{thousand gallons}}{\text{unit of labor}}$.

(c) Since $x = 110$ and $y = 20$, we need to evaluate $f_y(110, 20)$, which yields

$$f_y(110, 20) = 0.656(110)^{0.6}(20)^{-0.6} \approx 1.8$$

Thus, when the O'Neill Corporation uses 110 units of labor and 20 units of capital and keeps labor fixed at 110 units, production is increasing at the rate of approximately $1.8 \, \frac{\text{thousand gallons}}{\text{unit of capital}}$.

Interactive Activity

Compare the result of Example 7c with the result from Example 1.

For some amounts currently invested in labor and in capital, a natural question to ask is whether production would increase more rapidly if additional resources were invested in labor or in capital. We can approximate the change in productivity by simply doing the following:

- The change in productivity with respect to labor is the product of

$$f_x(x, y) \cdot (\text{change in } x)$$

- The change in productivity with respect to capital is the product of

$$f_y(x, y) \cdot (\text{change in } y)$$

EXAMPLE 8

Allocating Additional Resources

Suppose that a production function is given by

$$f(x, y) = 4.23x^{0.37}y^{0.66}$$

where x represents dollars in millions spent on labor and y represents dollars in millions spent on capital equipment. Currently, $x = 5$ and $y = 1$. Would production increase more by spending an additional $1 million on labor or $500,000 on capital equipment?

SOLUTION

We need to compute and compare $f_x(5, 1) \cdot (1)$ and $f_y(5, 1) \cdot (0.5)$. Notice that, since capital, y, is in millions, $500,000 is the same as $0.5 million. Proceeding with the partial derivatives, we have

$$f_x(x, y) = 4.23(0.37)x^{-0.63}y^{0.66} = 1.5651x^{-0.63}y^{0.66}$$
$$f_x(5, 1) = 1.5651(5)^{-0.63}(1)^{0.66} \approx 0.568$$

The change in productivity with respect to labor is $f_x(5, 1) \cdot (1) \approx 0.568$. Continuing, we get for capital

$$f_y(x, y) = 4.23(0.66)x^{0.37}y^{-0.34} = 2.7918x^{0.37}y^{-0.34}$$
$$f_y(5, 1) = 2.7918(5)^{0.37}(1)^{-0.34} \approx 5.06$$

The change in productivity with respect to capital is $f_y(5, 1) \cdot (0.5) \approx 2.53$. Since the change in productivity with respect to capital is greater than the change in productivity with respect to labor, production increases more by spending $500,000 on capital equipment than by spending $1 million on labor.

Second-order Partial Derivatives

Just like the second derivatives for a function of one variable, there are **second-order partial derivatives.** These second-order partial derivatives are very important for Section 8.4, when we discuss locating relative extrema on the surface of $z = f(x, y)$.

> ### Second-order Partial Derivatives
>
> If $z = f(x, y)$, then the four possible second-order partial derivatives are as follows:
>
> $$f_{xx}(x, y) = \frac{\partial}{\partial x}\left(\frac{\partial f}{\partial x}\right) \qquad f_{yy}(x, y) = \frac{\partial}{\partial y}\left(\frac{\partial f}{\partial y}\right)$$
>
> $$f_{xy}(x, y) = \frac{\partial}{\partial y}\left(\frac{\partial f}{\partial x}\right) \qquad f_{yx}(x, y) = \frac{\partial}{\partial x}\left(\frac{\partial f}{\partial y}\right)$$

NOTE: In subscript notation, we differentiate from left to right. That is, to compute f_{yx}, we first compute f_y and then differentiate this result with respect to x.

EXAMPLE 9

Computing a Second-order Partial Derivative

Determine all four second-order partial derivatives of

$$f(x, y) = x^2 y + y^3 x - 2xy + y.$$

SOLUTION

First we calculate $f_x(x, y)$. It is given by

$$f_x(x, y) = 2xy + y^3 - 2y$$

We now can compute $f_{xx}(x, y)$ from this result by finding the derivative of $f_x(x, y)$ with respect to x.

$$f_{xx}(x, y) = \frac{\partial}{\partial x}[f_x(x, y)] = 2y + 0 - 0 = 2y$$

Also, we can compute $f_{xy}(x, y)$ from $f_x(x, y)$ by finding the derivative of $f_x(x, y)$ with respect to y.

$$f_{xy}(x, y) = \frac{\partial}{\partial y}[f_x(x, y)] = 2x + 3y^2 - 2$$

To determine the other two second-order partial derivatives, we return to our original function, $f(x, y) = x^2 y + y^3 x - 2xy + y$, and compute $f_y(x, y)$. It is

$$f_y(x, y) = x^2 + 3y^2 x - 2x + 1$$

From this result, we have

$$f_{yy}(x, y) = \frac{\partial}{\partial y}[f_y(x, y)] = 6yx$$

$$f_{yx}(x, y) = \frac{\partial}{\partial x}[f_y(x, y)] = 2x + 3y^2 - 2$$

Notice in Example 9 that $f_{xy}(x, y) = f_{yx}(x, y)$. This is true for many functions that we encounter, but not all functions.

✓ CHECKPOINT 4

Now work Exercise 45.

SUMMARY

In this section we saw how the vertical cross sections introduced in Section 8.2 are key to our understanding of a partial derivative. We supplied the following note on computing partial derivatives:

- The **partial derivative of $f(x, y)$ with respect to x** is found by treating y as a constant and performing our ordinary differentiation techniques. The notation for this partial derivative is $f_x(x, y)$ or $\dfrac{\partial f}{\partial x}$.

- The **partial derivative of $f(x, y)$ with respect to y** is found by treating x as a constant and performing our ordinary differentiation techniques. The notation for this partial derivative is $f_y(x, y)$ or $\dfrac{\partial f}{\partial y}$.

We encountered several applications of partial derivatives, including marginal productivity of labor and marginal productivity of capital. These were defined as follows:

- For any production function of the form $Q = f(x, y) = ax^m y^n$:

  $f_x(x, y)$ gives the approximate change in productivity per unit change in labor and is called **marginal productivity of labor.**

  $f_y(x, y)$ gives the approximate change in productivity per unit change in capital and is called **marginal productivity of capital.**

The section concluded with a look at second-order partial derivatives.

- If $z = f(x, y)$, then the four possible **second-order partial derivatives** are as follows:

$$f_{xx}(x, y) = \frac{\partial}{\partial x}\left(\frac{\partial f}{\partial x}\right) \quad f_{yy}(x, y) = \frac{\partial}{\partial y}\left(\frac{\partial f}{\partial y}\right)$$

$$f_{xy}(x, y) = \frac{\partial}{\partial y}\left(\frac{\partial f}{\partial x}\right) \quad f_{yx}(x, y) = \frac{\partial}{\partial x}\left(\frac{\partial f}{\partial y}\right)$$

SECTION 8.3 EXERCISES

In Exercises 1–30, determine $f_x(x, y)$ and $f_y(x, y)$.

1. $f(x, y) = 5x^2 - 6y^3$
2. $f(x, y) = 3x + 2y + 10$
3. $f(x, y) = 2xy - y^2 + 1$
4. $f(x, y) = 3x^4 + 3x + 5y^2$
5. $f(x, y) = x^3 y^2$
6. $f(x, y) = x^2 y^4$
7. $f(x, y) = x^2 + 3x^2 y^3 - 2y^3 - xy$
8. $f(x, y) = 2x^4 + x^2 y^2 - 3y^2 - y$
9. $f(x, y) = 2x^3 - y^2 + 2x - 3$
10. $f(x, y) = 3x^4 - 2y^3 + 3x^2 - 5xy$
11. $f(x, y) = (2x + 3y)^3$
12. $f(x, y) = (3x - 5y)^4$
13. $f(x, y) = 37.21x^{0.15} y^{0.87}$
14. $f(x, y) = 2.41x^{0.27} y^{0.71}$
15. $f(x, y) = e^x \ln y$
16. $f(x, y) = e^y \ln x$
17. $f(x, y) = \dfrac{y}{x}$
18. $f(x, y) = \dfrac{x}{y}$
19. $f(x, y) = 3x^2 - 2x^3 y^4 + 7$
20. $f(x, y) = 4y^3 - 5x^2 y + 2$

21. $f(x, y) = (x^2 + y^3)^2$
22. $f(x, y) = (x^3 - y^2)^4$
23. $f(x, y) = \ln(x^2 + y^3)$
24. $f(x, y) = \ln(y^2 - x)$
25. $f(x, y) = \dfrac{x^2 y}{y + x}$
26. $f(x, y) = \dfrac{xy^2}{y - x}$
27. $f(x, y) = 3x^2 y^3 + e^{x+y}$
28. $f(x, y) = -2x^3 y - e^{x-y}$
29. $f(x, y) = e^{xy} + \ln y$
30. $f(x, y) = y^2 e^{xy} + \ln x$

In Exercises 31–40, for the given function, evaluate the stated partial derivative and interpret. See Example 4.

31. $f(x, y) = 4y^3 - 5x^2 y + 2$ — Determine $f_x(1, 2)$.
32. $f(x, y) = 3x^4 - 2y^3 + 3x^2 - 5xy$ — Determine $f_x(2, 1)$.
33. $f(x, y) = 2x^4 + x^2 y^2 - 3y^2 - y$ — Determine $f_y(2, 3)$.
34. $f(x, y) = \ln(y^2 - x)$ — Determine $f_y(2, 5)$.
35. $f(x, y) = \ln(y^2 - x)$ — Determine $f_x(2, 5)$.
36. $f(x, y) = 37.21x^{0.15} y^{0.87}$ — Determine $f_y(5, 2)$.
37. $f(x, y) = 37.21x^{0.15} y^{0.87}$ — Determine $f_x(5, 2)$.
38. $f(x, y) = 3x^2 y^3 + e^{x+y}$ — Determine $f_y(1, 2)$.

39. $f(x, y) = -2x^3 y - e^{x-y}$ Determine $f_x(2, 4)$.

40. $f(x, y) = \dfrac{\ln x}{x^2 - y^2}$ Determine $f_x(4, 3)$.

In Exercises 41–56, determine $f_{xx}(x, y)$, $f_{xy}(x, y)$, and $f_{yy}(x, y)$.

41. $f(x, y) = 4y^3 - 5x^2 y + 2$

42. $f(x, y) = 2x^3 - x^2 y^3 + 2y^4$

43. $f(x, y) = 3x^2 - 2x^3 y^2 + 2x^3$

44. $f(x, y) = 3x^4 + 2x^3 y^2 - y$

✓ 45. $f(x, y) = 5xy - 6x^3 + 7y$

46. $f(x, y) = 5xy^3 - 7x^3 y$

47. $f(x, y) = y^5 - 2x^3 y^2 + 7x^2$

48. $f(x, y) = y^3 + 3x^2 y - 8x^3$

49. $f(x, y) = y^2 e^{xy} + \ln x$ 50. $f(x, y) = x^3 e^{xy} + \ln y$

51. $f(x, y) = 5.2x^{0.65} y^{0.4}$ 52. $f(x, y) = 2.41x^{0.27} y^{0.71}$

53. $f(x, y) = \dfrac{2y}{x}$ 54. $f(x, y) = \dfrac{y^2}{x}$

55. $f(x, y) = ye^{xy}$ 56. $f(x, y) = y^2 e^{xy}$

57. For $f(x, y) = x^2 + y^2 - xy + 3y$, determine values for x and y such that $f_x(x, y) = 0$ and $f_y(x, y) = 0$ simultaneously.

58. For $f(x, y) = x^2 + y^2 - 8x + 2y + 7$, determine values for x and y such that $f_x(x, y) = 0$ and $f_y(x, y) = 0$ simultaneously.

APPLICATIONS

59. Seamego Boats spends x thousand dollars each week on newspaper advertising and y thousand dollars each week on radio advertising. The company has weekly sales, in tens of thousands of dollars, given by

$$S(x, y) = 2x^2 + y$$

(a) Determine $S_x(x, y)$ and $S_y(x, y)$.
(b) Determine $S_x(3, 5)$ and $S_y(3, 5)$ and interpret each.

60. Hilltop Boats spends x thousand dollars each week on radio advertising and y thousand dollars each week on television advertising. The company has weekly sales, in tens of thousands of dollars, given by

$$S(x, y) = 3x + 2y^3$$

(a) Determine $S_x(x, y)$ and $S_y(x, y)$.
(b) Determine $S_x(2, 6)$ and $S_y(2, 6)$ and interpret each.

61. Tube Town, a recently opened water park, spends x thousand dollars on radio advertising and y thousand dollars on television advertising. The park has weekly ticket sales, in tens of thousands of dollars, of

$$TS(x, y) = 1.5x^2 + 3.2y^2$$

(a) Determine $TS_x(x, y)$ and $TS_y(x, y)$.
(b) Determine $TS_x(1, 0.5)$ and $TS_y(1, 0.5)$ and interpret each.

62. In Section 8.1 we saw that the wind chill index is given by

$$f(v, T) = 91.4 - (0.4474677 - 0.020425v + 0.303107\sqrt{v}\,)$$
$$\times (91.4 - T)$$

where v is the wind speed in miles per hour and T is the actual air temperature in degrees Fahrenheit.
(a) Determine $f_T(v, T)$ and $f_v(v, T)$.
(b) Evaluate $f(25, 5)$, $f_T(25, 5)$, and $f_v(25, 5)$ and interpret each.

63. If an amount P dollars is invested at an annual interest rate of 6% compounded continuously for t years, the total amount accumulated is given by $f(P, t) = Pe^{0.06t}$. Determine $f_P(P, t)$ and $f_t(P, t)$ and interpret each.

64. In human cells, sodium ions are transported from inside the cell to outside the cell; at the same time, potassium ions are transported from outside the cell to inside. When this takes place across a nerve fiber in a nerve cell, a slight negative electric charge is left behind in the nerve fiber. The amount of voltage that will develop across a membrane is given by the Nernst equation, $f(x, y) = -61 \log \dfrac{x}{y}$, where x is the concentration of potassium ions inside the nerve fiber, y is the concentration of potassium ions outside the nerve fiber, and $f(x, y)$ is measured in millivolts. Determine $f_x(x, y)$ and $f_y(x, y)$ and interpret each.

65. The formula for a cone is given by

$$V(r, h) = \frac{1}{3}\pi r^2 h$$

where r is the radius and h is the height.
(a) Determine $V_r(r, h)$ and $V_h(r, h)$.
(b) Compute $V(3, 5)$, $V_r(3, 5)$, and $V_h(3, 5)$ and interpret each.

66. Poiseuille's law states that the resistance, R, for blood flowing in a blood vessel is given by

$$R(L, r) = k \cdot \frac{L}{r^4}$$

where k is a constant, L is the length of the blood vessel, and r is the radius of the blood vessel.
(a) Determine $R_L(L, r)$ and $R_r(L, r)$.
(b) Evaluate $R_L(6, 0.3)$ and $R_r(6, 0.3)$ and interpret each.

67. In their study of human groupings, anthropologists often use an index called the cephalic index. The cephalic index is given by

$$C(W, L) = 100 \cdot \frac{W}{L}$$

where W is the width and L is the length of an individual's head. Both measurements are made across the top of the head and are in inches. Determine $C_W(6, 8.2)$ and $C_L(6, 8.2)$ and interpret each.

68. An individual's body surface area is approximated by

$$BSA(w, h) = 0.007184w^{0.425}h^{0.725}$$

where BSA is in square meters, w is weight in kilograms, and h is height in centimeters.

(a) Determine $BSA_w(w, h)$ and $BSA_h(w, h)$.

(b) Evaluate $BSA(70, 160)$, $BSA_w(70, 160)$, and $BSA_h(70, 160)$ and interpret each.

69. The Chalet Bicycle Company manufactures 21-speed racing bicycles and 21-speed mountain bicycles. Let x represent the weekly demand for a 21-speed racing bicycle and y represent the weekly demand for a 21-speed mountain bicycle. The weekly price–demand equations are given by

$p = 350 - 4x + y$, the price in dollars for a 21-speed racing bicycle

$q = 450 + 2x - 3y$, the price in dollars for a 21-speed mountain bicycle.

(a) Determine the revenue function $R(x, y)$.

(b) Determine $R_x(x, y)$ and $R_y(x, y)$. Interpret each.

(c) Evaluate $R(15, 20)$ and $R_x(15, 20)$ and interpret each.

70. (Continuation of Exercise 69.) The cost function for Chalet Bicycle Company is given by

$$C(x, y) = 390 + 95x + 100y$$

(a) Determine the weekly profit function $P(x, y)$.

(b) Determine $P_x(x, y)$ and $P_y(x, y)$ and interpret each.

(c) Evaluate $P(15, 20)$ and $P_x(15, 20)$ and interpret each.

71. The Kerr Company produces two types of graphing calculators, x units of a 2D graphing calculator and y units of a 3D graphing calculator each month. The monthly revenue, in dollars, is given by

$$R(x, y) = 70x + 95y + 0.5xy - 0.04x^2 - 0.04y^2$$

(a) Determine the marginal revenue function for the 2D graphing calculator.

(b) Determine the marginal revenue function for the 3D graphing calculator.

72. (Continuation of Exercise 71.) The Kerr Company knows that the cost function for producing the two types of calculators is

$$C(x, y) = 4x + 5y + 3200$$

(a) Determine the marginal cost function for the 2D graphing calculator.

(b) Determine the marginal cost function for the 3D graphing calculator.

73. The Smolki Engine Company produces two types of small engines, x thousand two-stroke engines and y thousand

four-stroke engines. The vice president of the company, Jamie, has determined that the revenue and cost functions for a year are, in thousands of dollars,

$$R(x, y) = 42x + 39.5y$$
$$C(x, y) = 0.5x^2 + 0.8xy + y^2 + 5x + 4y$$

(a) Determine the profit function $P(x, y)$.

(b) Determine $P_x(x, y)$ and $P_y(x, y)$ and interpret each.

(c) Compute $P(18, 10)$ and $P_x(18, 10)$ and interpret each.

(d) Compute $P_y(18, 10)$ and interpret.

74. A golf club manufacturer has a Cobb–Douglas production function given by

$$Q = f(x, y) = 21x^{0.3}y^{0.75}$$

where x is the utilization of labor (in millions), y is the utilization of capital (in millions), and Q is the number of units of golf clubs produced.

(a) Compute $f_x(x, y)$ and $f_y(x, y)$.

(b) If the golf club manufacturer is currently using 150 units of labor and 100 units of capital, determine the marginal productivity of labor and the marginal productivity of capital.

(c) Would production increase more by spending an additional \$1 million on labor or \$500,000 on capital? Explain.

75. The Kevshan Company has a production function of

$$Q = f(x, y) = 161x^{0.8}y^{0.25}$$

where x is the number of labor hours in thousands, y is the capital equipment in millions, and Q is the quantity of calculators produced.

(a) Compute $f_x(x, y)$ and $f_y(x, y)$.

(b) If the Kevshan Company is now using 111 units of labor and 25 units of capital, determine the marginal productivity of labor and the marginal productivity of capital and interpret each.

76. The Baker Compact Disc Company has eight plants around the United States that manufacture compact disc players. The Cobb–Douglas production function for the eight plants is given by

$$Q = f(x, y) = 3.14x^{0.6}y^{0.45}$$

where x is the dollars (in millions) spent on labor, y is the dollars (in millions) spent on capital, and Q is the quantity produced (in thousands).

(a) Compute $f_x(x, y)$ and $f_y(x, y)$.

(b) If the Baker Compact Disc Company is currently using 1.7 units of labor and 5 units of capital, determine the marginal productivity of labor and the marginal productivity of capital and interpret each.

(c) Would production increase more by spending an additional \$1 million on labor or \$500,000 on capital? Explain.

77. The Toth Engine Company has a production function given by

$$Q = f(x, y) = 1434.33x^{0.6}y^{0.45}$$

where x is the number of labor hours in thousands, y is the capital in millions, and Q is the total quantity produced.

(a) Compute $f_x(x, y)$ and $f_y(x, y)$.

(b) If the Toth Engine Company is currently using 47 units of labor and 8 units of capital, determine the marginal productivity of labor and the marginal productivity of capital and interpret each.

78. Anastasia's, a manufacturer of skis, has a monthly production function given by

$$Q = f(x, y) = 210x^{0.6}y^{0.4}$$

where Q is the number of pairs of skis produced, x is the amount of labor used, and y is the amount of capital used.

(a) Compute $f_x(x, y)$ and $f_y(x, y)$.

(b) If Anastasia's is currently using 15 units of labor and 23 units of capital, determine the marginal productivity of labor and the marginal productivity of capital and interpret each.

79. The Jendrag Motorcycle Company has a production function given by

$$Q = f(x, y) = 60x^{0.55}y^{0.5}$$

where x is the dollars (in millions) spent on labor, y is the dollars (in millions) spent on capital, and Q is the quantity produced (in thousands).

(a) Compute $f_x(x, y)$ and $f_y(x, y)$.

(b) If the Jendrag Company is currently using 220 units of labor and 140 units of capital, determine the marginal productivity of labor and the marginal productivity of capital and interpret each.

(c) Would production increase more by spending an additional $1 million on labor or $1.5 million on capital? Explain.

SECTION PROJECT

The program in Appendix B that performs multiple regression is needed to do the section project.

The IronWorks Company has six plants in the U.S. Midwest that make fireplace accessories. Data for each plant give the dollars (in millions) spent on labor, the dollars (in millions) spent on capital, and the total quantity produced (in thousands) for 1999. The plants all operate at the same level of technology, so a production function can be determined.

LABOR	CAPITAL	QUANTITY
2.2	1.7	10.7
2.3	1.6	10.8
2.5	1.75	11.8
2.9	1.8	13.3
3	1.85	13.7
3.2	2	14.7

(a) Enter the data into your calculator and execute the program to determine a Cobb–Douglas production function. Round all values to the nearest hundredth.

(b) On average, the IronWorks is currently investing 2.7 units of labor and 1.8 units of capital. Find the marginal productivity of labor and marginal productivity of capital at these levels of labor and capital investment and interpret each.

(c) At its current levels of capital and labor investment, would production increase more by spending, on average, an additional $800,000 on labor or $400,000 on capital?

SECTION 8.4   MAXIMA AND MINIMA

We are now ready for a brief analysis of extreme values for a function of two variables. If we can locate the high points (**relative maxima**) and the low points (**relative minima**) on a surface, then we can optimize the quantity represented by the surface. For example, we could determine a maximum for profit or a minimum for cost. This should sound familiar, because we optimized functions of one variable in Sections 5.4 and 5.5. There we learned how the **Second Derivative Test** aided us in locating the relative extrema. We shall see in this section that there is a Second Derivative Test for relative extrema of a function of two variables. Just as we did with a function of one variable, we must first discuss the concept of **critical points**.

Relative Extrema, Critical Points, and the Second Derivative Test

We begin our analysis by making an assumption. Given $z = f(x, y)$, we assume that all second-order partial derivatives exist for $f(x, y)$ in some circular region in the xy-plane. This assumption guarantees that we are dealing with what is known

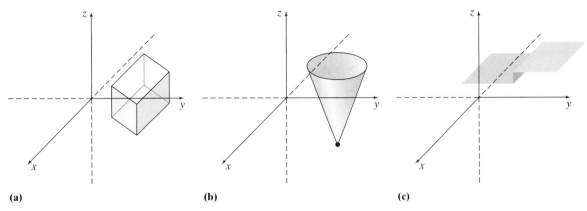

(a)   **(b)**   **(c)**

Figure 8.4.1   **(a)** Surface has an edge. **(b)** Surface has a sharp point. **(c)** Surface has a break.

as a **smooth surface. A smooth surface** is a one that has no edges (like a shoe box) or sharp points (like the point of a nail) or breaks (like the San Andreas fault line). These surfaces are illustrated in Figure 8.4.1.

Finally, we are not going to concern ourselves with boundary points. Thus, we are not going to concern ourselves with absolute extrema.

In Section 5.1 we defined relative extrema for a function of one variable as follows:

From Your Toolbox   **DEFINITION OF RELATIVE EXTREMA**

 For some open interval containing c:

1. f has a **relative maximum** at c if $f(c) \geq f(x)$ for all x in the interval.
2. f has a **relative minimum** at c if $f(c) \leq f(x)$ for all x in the interval.

We extend this definition to functions of two variables as follows:

Definition of Relative Extrema in 3-Space

Let $f(x, y)$ be a function of two variables. The value $f(a, b)$ is:

1. **A relative maximum** if $f(a, b) \geq f(x, y)$ for all points (x, y) in some circular region in the xy-plane around (a, b), where (a, b) is the center of the circular region.
2. **A relative minimum** if $f(a, b) \leq f(x, y)$ for all points (x, y) in some circular region in the xy-plane around (a, b), where (a, b) is the center of the circular region.

NOTE: The circular region described in these definitions should be small. If it is too large, it may contain some values of $f(x, y)$ that are larger or smaller than our relative extremum.

Figure 8.4.2 illustrates a relative maximum and Figure 8.4.3 illustrates a relative minimum.

Figure 8.4.4 illustrates a **saddle point,** which is neither a relative maximum nor a relative minimum. No matter how small a circular region we draw with (a, b) as the center, there are always values of $f(x, y)$ such that $f(x, y) > f(a, b)$ and $f(x, y) < f(a, b)$.

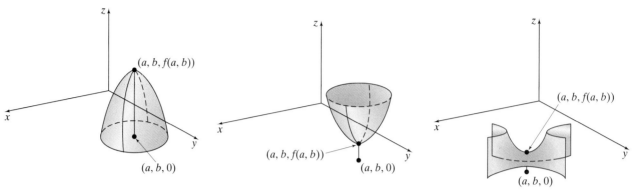

Figure 8.4.2 Relative maximum.   **Figure 8.4.3** Relative minimum.   **Figure 8.4.4** Saddle point.

Look carefully at Figures 8.4.2 and 8.4.3. Recall that a vertical cross section of the form $x = c$ is parallel to the y-axis and provides the geometric meaning of $f_y(x, y)$. Likewise, a vertical cross section of the form $y = c$ is parallel to the x-axis and provides the geometric meaning of $f_x(x, y)$. Now in Figures 8.4.2 and 8.4.3 it appears that *any* vertical cross section will have a horizontal tangent at the relative extreme, in particular, both $f_x(x, y) = 0$ and $f_y(x, y) = 0$.

Recall our definition of the **critical value** of a function of one variable:

From Your Toolbox   **DEFINITION OF CRITICAL VALUES**

A **critical value** for f is an x-value in the domain of f for which
(1) $f'(x) = 0$ or   (2) $f'(x)$ is undefined.

The possibilities for a function of two variables are beyond the scope of this text. We shall restrict our discussion to **critical points** for which $f_x(x, y) = 0$ and $f_y(x, y) = 0$.

Critical Point

The point (a, b) is a **critical point** for $f(x, y)$ if $f_x(a, b) = 0$ **and** $f_y(a, b) = 0$.

NOTE: We call it a critical *point* because the domain of $f(x, y)$ is a region in the xy-plane.

EXAMPLE 1 | **Determining Critical Points**

Determine the critical points for $f(x, y) = x^2 + y^2 - 8x + 2y + 7$.

SOLUTION

Critical points are determined by setting the partial derivatives $f_x(x, y) = 0$ and $f_y(x, y) = 0$ and solving this system of equations for x and y. The partial derivatives are

$$f_x(x, y) = 2x - 8 \quad \text{and} \quad f_y(x, y) = 2y + 2$$

Setting each equal to zero and solving yields

$$0 = 2x - 8 \quad \text{and} \quad 0 = 2y + 2$$
$$4 = x \quad \quad \text{and} \quad -1 = y$$

So the only critical point is the point $(4, -1)$. Notice that this point is in the xy-plane. If we wanted to know the point on the *surface,* we would find the z-coordinate by evaluating $f(4, -1)$. This yields a z-coordinate of -10. So the point on the surface would be $(4, -1, -10)$. See Figure 8.4.5.

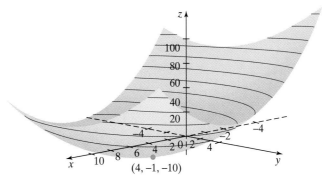

Figure 8.4.5

✓ **CHECKPOINT 1**

Now work Exercise 3.

EXAMPLE 2 | **Determining Critical Points**

Determine the critical points for $f(x, y) = x^2 + y^2 - xy + 3y$.

SOLUTION

The partial derivatives are

$$f_x(x, y) = 2x - y \quad \text{and} \quad f_y(x, y) = 2y - x + 3$$

Setting each to zero gives

$$0 = 2x - y \quad \text{and} \quad 0 = 2y - x + 3$$

Here we have a system of two equations. We need to determine values of x and y that satisfy *both* equations. There are several methods that we could use to solve this system. We use an algebraic technique known as *substitution.* A review of the substitution method for solving a system of equations is given in Appendix A. This method yields

$$0 = 2x - y \quad \text{and} \quad 0 = 2y - x + 3$$
$$y = 2x$$

We substitute $y = 2x$ into the second equation for y and get

$$0 = 2(2x) - x + 3$$
$$0 = 4x - x + 3$$
$$0 = 3x + 3$$
$$-1 = x$$

We simply substitute $x = -1$ into either of the original equations and solve for y. We choose the first equation and this gives

$$0 = 2x - y$$
$$0 = 2(-1) - y$$
$$-2 = y$$

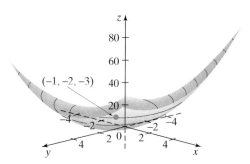

Figure 8.4.6

So $x = -1$ and $y = -2$ are the solutions to both equations. Thus, our critical point is $(-1, -2)$. See Figure 8.4.6. ∎

✓CHECKPOINT 2

Now work Exercise 7.

Now that we know how to find critical points, our next task is to determine whether we have a relative extrema at a given critical point. Just as not all critical values of a function of one variable give relative extrema, there is no guarantee that we will find a relative extrema at critical points of a function of two variables. For example, the saddle point in Figure 8.4.4 is a critical point that does not produce a relative extremum. The following **Second Derivative Test** enables us to determine the shape of the surface in the vicinity of critical points, which in turn tells us if we have a relative extremum or not.

Second Derivative Test

Let $z = f(x, y)$ be a function of two variables such that $f_{xx}(x, y), f_{yy}(x, y)$, and $f_{xy}(x, y)$ all exist. If $f_x(a, b) = 0$ and $f_y(a, b) = 0$, that is, (a, b) is a critical point, then we define a number D to be

$$D = f_{xx}(a, b) \cdot f_{yy}(a, b) - [f_{xy}(a, b)]^2$$

The Second Derivative Test has the following form:

1. If $D > 0$ and $f_{xx}(a, b) < 0$, then $f(a, b)$ is a **relative maximum.**
2. If $D > 0$ and $f_{xx}(a, b) > 0$, then $f(a, b)$ is a **relative minimum.**
3. If $D < 0$, then $f(a, b)$ is a **saddle point.**
4. If $D = 0$, the test gives no information about $f(a, b)$.

Although this test appears complicated, it is fairly easy to apply, as Example 3 illustrates.

EXAMPLE 3

Determining Relative Extrema

Determine any relative extrema for $f(x, y) = x^2 + y^2 - xy + 3y$.

SOLUTION

This is the function from Example 2. There we determined the partial derivatives to be

$$f_x(x, y) = 2x - y \quad \text{and} \quad f_y(x, y) = 2y - x + 3$$

and the only critical point occurred at $x = -1$ and $y = -2$, that is, at $(-1, -2)$. To use the Second Derivative Test, we need the following second-order partial derivatives:

$$f_{xx}(x, y) = 2, \quad f_{yy}(x, y) = 2, \quad \text{and} \quad f_{xy}(x, y) = -1$$

We now evaluate our critical point $(-1, 2)$ at each second-order partial derivative.

$$f_{xx}(-1, -2) = 2, \quad f_{yy}(-1, -2) = 2, \quad \text{and} \quad f_{xy}(-1, -2) = -1$$

We then define D to be

$$D = f_{xx}(-1, -2) \cdot f_{yy}(-1, -2) - [f_{xy}(-1, -2)]^2$$
$$= 2 \cdot 2 - [-1]^2 = 4 - 1 = 3$$

Since $D > 0$ and $f_{xx}(-1, -2) > 0$, the Second Derivative Test asserts that there is a *relative minimum* at $(-1, -2)$. Specifically, on the *surface* there is a relative minimum at $(-1, -2, -3)$. See Figure 8.4.6.　　■

We offer the following guidelines on applying the Second Derivative Test:

1. Determine all critical points (a, b).
2. Determine $f_{xx}(x, y), f_{yy}(x, y)$, and $f_{xy}(x, y)$.
3. Evaluate the second-order partial derivatives at each critical point.
4. Determine D.
5. Apply the Second Derivative Test.

    Before we address some applications, let's do Example 4 to ensure that the process is understood.

EXAMPLE 4 | **Determining Relative Extrema**

Determine the relative extrema for $f(x, y) = x^3 + 3x^2 - y^2 + 2y + 4$.

SOLUTION

We begin by locating the critical points. First, the partial derivatives are

$$f_x(x, y) = 3x^2 + 6x \quad \text{and} \quad f_y(x, y) = -2y + 2$$

Setting each equal to zero and solving yields

$$0 = 3x^2 + 6x \quad \text{and} \quad 0 = -2y + 2$$
$$0 = 3x(x + 2) \qquad\qquad 1 = y$$
$$x = 0, \quad x = -2$$

So here we have two critical points, when $x = 0$ and $y = 1$ as well as when $x = -2$ and $y = 1$. This gives the points $(0, 1)$ and $(-2, 1)$ as the critical points.

    Next we determine the second-order partial derivatives to be

$$f_{xx}(x, y) = 6x + 6, \quad f_{yy}(x, y) = -2, \quad \text{and} \quad f_{xy}(x, y) = 0$$

Evaluating the second-order partial derivatives at the critical point $(0, 1)$ gives

$$f_{xx}(0, 1) = 6(0) + 6 = 6, \quad f_{yy}(0, 1) = -2, \quad \text{and} \quad f_{xy}(0, 1) = 0$$

For the critical point $(0, 1)$, we determine D to be

$$D = f_{xx}(0, 1) \cdot f_{yy}(0, 1) - [f_{xy}(0, 1)]^2$$
$$= 6(-2) - [0]^2 = -12$$

Since $D < 0$ at the critical point $(0, 1)$, the Second Derivative Test asserts that we have a *saddle point* at $(0, 1)$. Specifically, we have a saddle point at $(0, 1, 5)$.

Evaluating the second-order partial derivatives at the critical point $(-2, -1)$ gives

$$f_{xx}(-2, -1) = 6(-2) + 6 = -6, \quad f_{yy}(-2, -1) = -2, \quad \text{and} \quad f_{xy}(-2, -1) = 0$$

For the critical point $(-2, -1)$, we now determine D to be

$$D = f_{xx}(-2, -1) \cdot f_{yy}(-2, -1) - [f_{xy}(-2, -1)]^2$$
$$= (-6)(-2) - [0]^2 = 12$$

Since $D > 0$ at the critical point $(-2, -1)$ and $f_{xx}(-2, -1) < 0$, the Second Derivative Test tells us that we have a *relative maximum* at $x = -2$, $y = -1$. Specifically, we have a *relative maximum* at $(-2, -1, 9)$. Figure 8.4.7 shows the surface with the saddle point and the relative maximum.

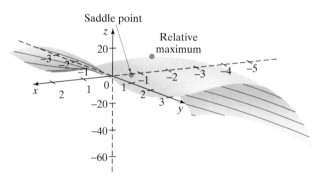

Figure 8.4.7

✓ **CHECKPOINT 3**

Now work Exercise 21.

Applications

The final two examples of this section illustrate how, through applications, the Second Derivative Test is a very powerful method in locating relative extrema.

EXAMPLE 5 | **Maximizing Revenue**

The All Clear Company sells two types of car windshield wiper fluid: *regular,* which can be used at temperatures above 10°F, and a *no-freeze,* which can be used at temperatures above −30°F. Let x represent the number of gallons of no-freeze sold each year, in millions, and let y represent the number of gallons of regular sold each year, also in millions. Suppose that

$$p = 3.9 - 0.5x - 0.1y, \quad \text{the price in dollars of a gallon of no-freeze}$$
$$q = 3.9 - 0.1x - 0.8y, \quad \text{the price in dollars of a gallon of regular}$$

(a) Determine the revenue function $R(x, y)$.

(b) Determine x and y such that revenue is maximized and find the maximum revenue.

SOLUTION

(a) As always, revenue equals price times quantity, which gives

$$R(x, y) = x(3.9 - 0.5x - 0.1y) + y(3.9 - 0.1x - 0.8y)$$
$$= 3.9x - 0.5x^2 - 0.1xy + 3.9y - 0.1xy - 0.8y^2$$
$$= -0.5x^2 - 0.8y^2 + 3.9x + 3.9y - 0.2xy$$

(b) Since we want to maximize revenue, first locate any critical points of $R(x, y)$. Computing $R_x(x, y)$ and $R_y(x, y)$ yields

$$R_x(x, y) = -x + 3.9 - 0.2y \quad \text{and} \quad R_y(x, y) = -1.6y + 3.9 - 0.2x$$

Setting each equal to zero and solving gives

$$0 = -x + 3.9 - 0.2y \quad \text{and} \quad 0 = -1.6y + 3.9 - 0.2x$$

We opt to solve by using the substitution method, as we did in Example 2. From the first equation we have

$$x = 3.9 - 0.2y$$

Substituting this into the second equation yields

$$0 = -1.6y + 3.9 - 0.2(3.9 - 0.2y)$$
$$0 = -1.6y + 3.9 - 0.78 + 0.04y$$
$$0 = -1.56y + 3.12$$
$$y = 2$$

Substituting $y = 2$ into the first equation gives us

$$0 = -x + 3.9 - 0.2(2)$$
$$x = 3.5$$

So we have a critical point at $x = 3.5$ and $y = 2$, or more precisely at $(3.5, 2)$. We now proceed to determine if the critical point yields a relative extremum. First we compute the second-order partial derivatives to be

$$R_{xx}(x, y) = -1, \quad R_{yy}(x, y) = -1.6, \quad \text{and} \quad R_{xy}(x, y) = -0.2$$

Evaluating our critical point at these second-order partial derivatives gives

$$R_{xx}(3.5, 2) = -1, \quad R_{yy}(3.5, 2) = -1.6, \quad \text{and} \quad R_{xy}(3.5, 2) = -0.2$$

Now, for the critical point $(3.5, 2)$ we define D as

$$D = R_{xx}(3.5, 2) \cdot R_{yy}(3.5, 2) - [R_{xy}(3.5, 2)]^2$$
$$= (-1)(-1.6) - [-0.2]^2 = 1.56$$

Since $D > 0$ for the critical point $(3.5, 2)$ and $R_{xx}(3.5, 2) < 0$, we have a *relative maximum* at $x = 3.5$, $y = 2$. So All Clear needs to sell 3.5 million gallons of the no-freeze and 2 million gallons of the regular to maximize revenue for the year. The maximum revenue is

$$R(3.5, 2) = -0.5(3.5)^2 - 0.8(2)^2 + 3.9(3.5) + 3.9(2) - 0.2(3.5)(2)$$
$$= 10.725$$

So the maximum revenue is $10.725 million dollars. ◼

Interactive Activity

Determine the critical point in Example 5 by setting each partial derivative equal to zero, solving for y, graphing, and using the INTERSECTION command.

EXAMPLE 6 | **Maximizing Volume**

North-South Airlines allows each passenger to carry up to three suitcases as long as the sum of the width, length, and height of each suitcase is less than or equal to 60 inches. Determine the dimensions of a suitcase of maximum volume that a passenger may carry with this restriction.

SOLUTION

First we sketch a picture of a generic suitcase. See Figure 8.4.8. We represent length with an x, width by y, and height by z. The airline's restriction is that

$$x + y + z \leq 60$$

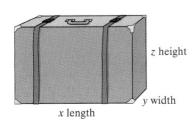

z height

y width

x length

Figure 8.4.8

The volume of our suitcase is given by

$$V = x \cdot y \cdot z$$

Notice that the volume is a function of three independent variables. We need to somehow rewrite this as a function of two variables so that we may use the techniques learned in this section. Since we believe that the largest volume for the suitcase results when $x + y + z = 60$, we solve this equation for z and get

$$z = 60 - x - y$$

We substitute this into our equation for volume and get a function of two variables,

$$V(x, y) = xy(60 - x - y) = 60xy - x^2y - xy^2$$

To maximize the volume, we proceed by determining the critical points and applying the Second Derivative Test. First, we determine $V_x(x, y)$ and $V_y(x, y)$.

$$V_x(x, y) = 60y - 2xy - y^2 \quad \text{and} \quad V_y(x, y) = 60x - x^2 - 2xy$$

Setting each equal to zero yields

$$0 = 60y - 2xy - y^2 \quad \text{and} \quad 0 = 60x - x^2 - 2xy$$
$$0 = y(60 - 2x - y) \quad \text{and} \quad 0 = x(60 - x - 2y)$$

Thus,

$$0 = y \quad \text{or} \quad 0 = 60 - 2x - y \quad \text{and} \quad 0 = x \quad \text{or} \quad 0 = 60 - x - 2y$$

We can immediately reject the situation when $x = 0$ and $y = 0$ since this would produce a suitcase with a volume of 0. We are then left with solving

$$0 = 60 - 2x - y \quad \text{and} \quad 0 = 60 - x - 2y$$

Using the substitution method, we solve the first equation for y and get

$$y = 60 - 2x$$

Substituting this into our second equation gives

$$0 = 60 - x - 2(60 - 2x)$$
$$0 = 60 - x - 120 + 4x$$
$$0 = -60 + 3x$$
$$x = 20$$

Substituting $x = 20$ into the first equation to determine y gives

$$y = 60 - 2(20) = 20$$

So our critical point is $x = 20$ and $y = 20$. Next we need the second-order partial derivatives.

$$V_{xx}(x, y) = -2y, \quad V_{yy}(x, y) = -2x, \quad \text{and} \quad V_{xy}(x, y) = 60 - 2x - 2y$$

Evaluating each second-order partial derivative at the critical point $(20, 20)$ yields

$$V_{xx}(20, 20) = -40$$
$$V_{yy}(20, 20) = -40$$
$$V_{xy}(20, 20) = 60 - 2(20) - 2(20) = 60 - 40 - 40 = -20$$

Interactive Activity

How was the height of the suitcase in Example 6 computed to be 20 inches?

For the critical point $x = 20$, $y = 20$, we determine D to be

$$D = V_{xx}(20, 20) \cdot V_{yy}(20, 20) - [V_{xy}(20, 20)]^2$$
$$= (-40)(-40) - [-20]^2 = 1200$$

Since $D > 0$ for the critical point $(20, 20)$ and $V_{xx}(20, 20) < 0$, we have a relative maximum at $x = 20$, $y = 20$. So the dimensions that would maximize volume would be 20 by 20 by 20 inches.

SUMMARY

In this section we defined the **relative extrema** of a function of two variables.

- Let $f(x, y)$ be a function of two variables. The value $f(a, b)$ is:

  A **relative maximum** if $f(a, b) \geq f(x, y)$ for all points (x, y) in some circular region in the xy-plane around (a, b), where (a, b) is the center of the circular region.

  A **relative minimum** if $f(a, b) \leq f(x, y)$ for all points (x, y) in some circular region in the xy-plane around (a, b), where (a, b) is the center of the circular region.

  We also defined critical points for a function of two variables as follows:

- The point (a, b) is a **critical point** for $f(x, y)$ if $f_x(a, b) = 0$ and $f_y(a, b) = 0$.

  We presented the **Second Derivative Test,** which allows us to determine if the critical point produces a relative maximum, a relative minimum, a saddle point, or none of these.

- If (a, b) is a critical point, then we define a number D to be

$$D = f_{xx}(a, b) \cdot f_{yy}(a, b) - [f_{xy}(a, b)]^2$$

The Second Derivative Test has the following form:

1. If $D > 0$ and $f_{xx}(a, b) < 0$, the $f(a, b)$ is a **relative maximum**.
2. If $D > 0$ and $f_{xx}(a, b) > 0$, then $f(a, b)$ is a **relative minimum**.
3. If $D < 0$, then $f(a, b)$ is a **saddle point**.
4. If $D = 0$, the test gives no information about $f(a, b)$.

SECTION 8.4 EXERCISES

In Exercises 1–10, determine the critical points for the given function.

1. $f(x, y) = x^2 + y^2 - 4x + 6y - 4$

2. $f(x, y) = x^2 + y^2 - 6x - 4y + 3$

✓ 3. $f(x, y) = x^2 + y^2 + 2x - 4y - 3$

4. $f(x, y) = x^2 + y^2 - 8x + 2y - 2$

5. $f(x, y) = 2x^2 + 3y^2 - 8x + 6y + 5$

6. $f(x, y) = x^2 + y^2 + xy - 6y + 1$

✓ 7. $f(x, y) = x^2 + y^2 - xy - 3y + 5$

8. $f(x, y) = x^2 + y^2 - xy - 3x - 3y + 2$

9. $f(x, y) = x^2 - y^2 + xy + 2x - 9y - 5$

10. $f(x, y) = 2x^2 + 3y^2 + 2xy + 4x - 8y + 3$

In Exercises 11–30, use the Second Derivative Test to locate any relative extrema and saddle points.

11. $f(x, y) = x^2 + y^2 - 6x - 4y + 3$

12. $f(x, y) = x^2 + y^2 - 4x + 6y - 4$

13. $f(x, y) = x^2 + y^2 - 6x + 4y + 2$

14. $f(x, y) = 3x^2 - y^2 - 4x + 6y + 1$

15. $f(x, y) = -x^2 - 3y^2 - 4x - 4y - 1$

16. $f(x, y) = x^2 + y^2 - xy - 3y + 5$

17. $f(x, y) = x^2 + y^2 + xy - 6y + 1$

18. $f(x, y) = x^2 - y^2 + xy + 2x - 9y - 5$

19. $f(x, y) = x^2 + y^2 - xy - 3x - 3y + 2$

20. $f(x, y) = 3x^2 + y^2 + xy + 3y + 4$

✓ 21. $f(x, y) = \dfrac{4}{3}x^3 - y^2 - 4x^2 + 2y - 1$

22. $f(x, y) = 4x^3 - 6x^2 - 24x + 2y^2 - 4y + 6$

23. $f(x, y) = 2y^3 - 6y^2 - 18y + 2x^2 - 4x + 12$

24. $f(x, y) = x^3 + 3xy - y^3$

25. $f(x, y) = x^3 - 3xy + y^3$

26. $f(x, y) = 2x^3y - 2x + 16y - 5$

27. $f(x, y) = e^{-x^2 - y^2}$

28. $f(x, y) = e^{xy}$

29. $f(x, y) = 6xy + \dfrac{12}{x} - \dfrac{3}{y}$

30. $f(x, y) = 4xy + \dfrac{8}{x} - \dfrac{2}{y}$

APPLICATIONS

31. The annual cost for labor and specialized robotics equipment for an automobile manufacturer, in millions of dollars, is given by

$$C(x, y) = 5x^2 + 5xy + 7.5y^2 - 40x - 45y + 135$$

where x is the amount, in millions of dollars, spent each year on labor and y is the amount, in millions of dollars, spent each year on robotics equipment.

(a) Determine how much should be spent on each, per year, to minimize cost.

(b) Determine the minimum cost.

32. New Jeans, a specialty blue jeans manufacturer, produces two types of blue jeans each day, x pairs of straight leg and y pairs of wide leg. The daily profit function, in dollars, is given by

$$P(x, y) = 78x + 3xy - 3x^2 - y^2 - 2y$$

(a) How many straight-leg jeans and how many wide-leg jeans should be produced and sold each day to maximize profit?

(b) What is the maximum profit?

33. The Wokon Company produces two types of woks each day, x number of regular sized and y number of jumbo sized. The daily profit from the sale of x number of regular-sized woks and y number of jumbo-sized woks is given by

$$P(x, y) = -0.3y^3 - 0.2x^2 + 6xy - 2$$

where $P(x, y)$ is measured in hundreds of dollars.

(a) How many regular-sized woks and how many jumbo-sized woks should be sold each day to maximize profit?

(b) What is the maximum profit?

34. The BigBang Engine Company produces two types of engines, x thousand two-stroke engines and y thousand four-stroke engines. Vice-president Kathy determined the revenue and cost functions for a year to be (in millions of dollars)

$$R(x, y) = 42x + 39.5y$$
$$C(x, y) = 0.5x^2 + 0.8xy + y^2 + 5x + 4y$$

(a) Determine how many two-stroke engines and how many four-stroke engines should be produced each year to maximize profit.

(b) What is the maximum profit?

35. Crispy Chips produces two types of potato chips each year, x 18-ounce bags (in millions) of sour cream and chives and y 18-ounce bags (in millions) of barbecue. The revenue and cost functions for the year, in millions of dollars, are

$$R(x, y) = 1.5x + 2y$$
$$C(x, y) = x^2 - xy + 2y^2 + 2.5x - 9y + 2$$

(a) Determine the profit function $P(x, y)$.

(b) How many 18-ounce bags of each type of chip should be produced each year to maximize profit?

(c) What is the maximum profit?

36. The Palton Company produces two types of graphing calculators each month, x units of a 2D graphing calculator and y units of a 3D graphing calculator. The monthly revenue and cost functions, in dollars, are

$$R(x, y) = 75x + 101y + 0.05xy - 0.04x^2 - 0.04y^2$$
$$C(x, y) = 13.98x + 15.02y + 32{,}000$$

(a) Determine how many 2D graphing calculators and how many 3D graphing calculators should be sold each month to maximize profit.

(b) What is the maximum profit?

37. Sammy's Cycles manufactures 21-speed touring bicycles and 21-speed mountain bicycles. Let x represent the weekly demand for a 21-speed touring bicycle, and let y represent the weekly demand for a 21-speed mountain bicycle. The weekly price–demand equations are given by

$p = 349 - 4x + y$,   the price in dollars for a 21-speed touring bicycle

$q = 446 + 2x - 3y$,   the price in dollars for a 21-speed mountain bicycle

(a) Determine the revenue function $R(x, y)$.

(b) How many of each type of bicycle should be produced each week to maximize revenue?

(c) What is the maximum revenue?

38. (Continuation of Exercise 37.) Sammy's Cycles has a cost function given by

$$C(x, y) = 390 + 95x + 96y$$

(a) Determine the profit function $P(x, y)$.

(b) How many of each type of bicycle should be produced to maximize profit?

(c) What is the maximum profit?

39. The Leaf Chewer Company manufactures and sells leaf blowers and a special 8-foot blower attachment to clean gutters. Let x represent the number of leaf blowers produced and sold each month, and let y represent the number of special 8-foot attachments produced and sold each month. The monthly price–demand equations are given by

$p = 228 - 8x - y$,   the price in dollars for a leaf blower

$q = 31 - x - 0.15y$,   the price in dollars for a 8-foot attachment

(a) Determine the revenue function $R(x, y)$.

(b) Determine how many leaf blowers and how many 8-foot attachments should be produced and sold each month to maximize revenue.

40. (Continuation of Exercise 39.) The monthly cost function for the Leaf Chewer Company is given by

$$C(x, y) = 1000 + 28x + 5y$$

(a) Determine the monthly profit function $P(x, y)$.

(b) Determine how many leaf blowers and how many 8-foot attachments should be sold each month to maximize profit.

(c) What is the maximum profit?

41. You have been challenged to design a rectangular box with no top and two parallel partitions that holds 125 cubic

inches. Determine the dimensions that will require the least amount of material to do this.

SECTION PROJECT

(a) *Federal Express* states that the maximum length plus girth for its FedEx Overnight Freight and its FedEx 2Day Freight shipments is 300 inches (data gathered at the Federal Express web site). See Figure 8.4.9. Determine the dimensions of the largest-volume package that can be sent by FedEx Overnight Freight and FedEx 2Day Freight. Assume that the length is y.

(b) To avoid U.S. Domestic Freight Service charges, any package sent by *Federal Express* must have a length plus girth not exceeding 165 inches (data gathered at the Federal Express web site). See Figure 8.4.9. Determine the dimensions of the largest-volume package that can be sent by *Federal Express* without incurring Domestic Freight Service charges. Assume that the length is y.

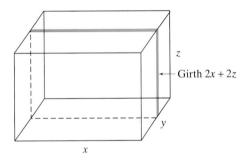

Girth $2x + 2z$

Figure 8.4.9

SECTION 8.5   LAGRANGE MULTIPLIERS

Many optimization problems that we encounter are actually *constrained* by some external circumstances. For example, North-South Airlines in Example 6 from Section 8.4 had a *constraint* that length plus width plus height of a suitcase cannot exceed 60 inches. We maximized the volume of a suitcase, $V = x \cdot y \cdot z$, subject to the constraint $x + y + z \leq 60$. Other examples include maximizing an area of an enclosure given a finite amount of fence or maximizing production given a budget constraint. Each of these problems may be solved by using the **Method of Lagrange Multipliers.**

Method of Lagrange Multipliers

The method that we are about to study is for solving maximum or minimum problems when dealing with restrictions on the independent variables. The method is named after the French mathematician Joseph Louis Lagrange (1736–1813), who discovered the method at the age of 19. We introduce the method through a Flashback.

THE BELOVED BEAGLE REVISITED

In Section 5.4, we saw that one of the authors needs to build an enclosure for his beloved beagle. He has 200 feet of fence and wants to make a rectangular enclosure along a straight wall of his house (see Figure 8.5.1).

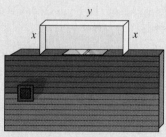

Figure 8.5.1

(a) Determine a function of two independent variables for the area of the enclosure.

(b) Determine any restrictions on x and y, assuming that we want to use all 200 feet of fence to maximize the area of the enclosure.

Flashback Solution

(a) We know that area is equal to length times width. For the enclosure shown in Figure 8.5.1, this gives

$$A(x, y) = xy$$

(b) We only have 200 feet of fence and we wish to use all of it to make this enclosure. From Figure 8.5.1 we determine that this means that x and y are such that

$$2x + y = 200$$

This equation represents the restrictions placed on the independent variables x and y. These restrictions placed on x and y are called **constraints.**

In short, what we are asked to do in situations like the Flashback is to

Maximize $\quad A(x, y) = xy$

Subject to $\quad 2x + y = 200 \quad$ or $\quad 2x + y - 200 = 0$

This is a specific example of the more general class of maximum/minimum problems of the form

Maximize/Minimize	$z = f(x, y)$
Subject to	$g(x, y) = 0$

In Section 5.4 we determined the dimensions that maximized the area of the enclosure in the Flashback by solving the *constraint equation* for y and substituted the result into the area equation. This gave us a function of one variable, which we solved using methods developed in Section 5.4. So why do we need a new method to solve this problem? There are several reasons.

- Sometimes our constraint equation is complicated and cannot be solved for one variable.
- The *Method of Lagrange Multipliers* generalizes to functions of several variables subject to one or more constraints.

Theorem of Lagrange

The relative extrema of the function $z = f(x, y)$ subject to the constraint equation $g(x, y) = 0$ occur among those points (x, y) for which there exists a value of λ (*lambda*),

$$F(x, y, \lambda) = f(x, y) + \lambda \cdot g(x, y)$$

where we have the following:

$$F_x(x, y, \lambda) = 0, \quad F_y(x, y, \lambda) = 0, \quad \text{and} \quad F_\lambda(x, y, \lambda) = 0$$

provided all partial derivatives exist.

NOTE: The λ (Greek letter *lambda*) in the definition of $F(x, y, \lambda)$ is called the *Lagrange multiplier*. To organize our work when using the method of Lagrange, we use the following steps.

How to Use the Method of Lagrange

1. Write the situation in the form

$$\text{Maximize (or minimize)} \quad z = f(x, y)$$
$$\text{Subject to} \quad g(x, y) = 0$$

2. Define $F(x, y, \lambda) = f(x, y) + \lambda \cdot g(x, y)$.
3. Determine the partial derivatives

$$F_x(x, y, \lambda), \quad F_y(x, y, \lambda), \quad \text{and} \quad F_\lambda(x, y, \lambda)$$

4. Solve the system of equations

$$F_x(x, y, \lambda) = 0, \quad F_y(x, y, \lambda) = 0, \quad F_\lambda(x, y, \lambda) = 0$$

5. The maximum (or minimum) of $f(x, y)$ is among the values in step 4. Simply evaluate $z = f(x, y)$ at each.

EXAMPLE 1

Utilizing the Method of Lagrange

Use the Method of Lagrange to determine the maximum area of an enclosure for the beloved beagle in the Flashback.

SOLUTION

We will break down the method of Lagrange into the five steps just given.

Step 1: Write the situation in the appropriate form.

$$\text{Maximize} \quad A(x, y) = xy$$
$$\text{Subject to} \quad 2x - y - 200 = 0$$

Step 2: Define function F using the Lagrange multiplier λ.

$$F(x, y, \lambda) = A(x, y) + \lambda \cdot g(x, y) = xy + \lambda \cdot (2x + y - 200)$$

Step 3: Determine the partial derivatives.

$$F_x(x, y, \lambda) = y + \lambda(2 + 0 - 0) = y + 2\lambda$$
$$F_y(x, y, \lambda) = x + \lambda(0 + 1 - 0) = x + \lambda$$
$$F_\lambda(x, y, \lambda) = 0 + 1 \cdot (2x + y - 200) = 2x + y - 200$$

Step 4: Solve the system. (Solutions to the system are critical points for F.)

$$y + 2\lambda = 0$$
$$x + \lambda = 0$$
$$2x + y - 200 = 0$$

To solve this system, notice from the first two equations that we have

$$y = -2\lambda \quad \text{and} \quad x = -\lambda$$

If we substitute these values for x and y into the third equation and solve for λ, we have

$$2(-\lambda) + (-2\lambda) - 200 = 0$$
$$-4\lambda - 200 = 0$$
$$-4\lambda = 200$$
$$\lambda = -50$$

Substituting $\lambda = -50$ into the first two equations yields

$$y = -2\lambda \qquad \text{and} \quad x = -\lambda$$
$$y = -2(-50) \quad \text{and} \quad x = -(-\lambda)$$
$$y = 100 \qquad \text{and} \quad x = 50$$

Step 5: Here it is important to realize exactly what the Theorem of Lagrange states. Quite simply, the method of Lagrange only finds critical points. It does not tell whether the function is maximized, minimized, or neither at the critical point. In this problem, as in each problem that we do, we must know that the maximum (or minimum) does exist. It then follows that the maximum (or minimum) must occur at a critical point as determined by Lagrange. For this situation, there is certainly an optimal area enclosure for the beagle. Since there is only one possibility from step 4, it follows that the dimensions are $x = 50$ feet by $y = 100$ feet, and the maximum enclosure is $A(50, 100) = 5000$ square feet. ◆

Interactive Activity

Refer to Section 5.4, Exercise 32, to verify that we achieved the same result as well as to compare the two different methods.

EXAMPLE 2 | **Applying the Method of Lagrange**

$$\text{Minimize} \qquad f(x, y) = x^2 + 3y^2$$
$$\text{Subject to} \qquad x + y = 2$$

SOLUTION

Step 1: Write in an appropriate form by writing the constraint equation as $g(x, y) = 0$.

$$\text{Minimize} \qquad f(x, y) = x^2 + 3y^2$$
$$\text{Subject to} \qquad g(x, y) = x + y - 2 = 0$$

Step 2: Define function F using the Lagrange multiplier λ.

$$F(x, y, \lambda) = f(x, y) + \lambda \cdot g(x, y) = x^2 + 3y^2 + \lambda(x + y - 2)$$

Step 3: Determine the partial derivatives.

$$F_x(x, y, \lambda) = 2x + \lambda$$
$$F_y(x, y, \lambda) = 6y + \lambda$$
$$F_\lambda(x, y, \lambda) = x + y - 2$$

Step 4: Solve the system

$$2x + \lambda = 0$$
$$6y + \lambda = 0$$
$$x + y - 2 = 0$$

From the first two equations, we have

$$2x = -\lambda \quad \text{and} \quad 6y = -\lambda$$

Setting these equal to each other yields

$$2x = 6y$$

$$y = \frac{1}{3}x$$

Since we have y in terms of x, we can return to our constraint equation, $x + y - 2 = 0$ and substitute for y, which yields

$$x + \frac{1}{3}x - 2 = 0$$

$$\frac{4}{3}x = 2$$

$$x = \frac{3}{2}$$

From $y = \frac{1}{3}x$ we also have

$$y = \frac{1}{3}\left(\frac{3}{2}\right) = \frac{1}{2}$$

Step 5: There is only one solution from step 4, $x = \frac{3}{2}$ and $y = \frac{1}{2}$. So the function is minimized at $f\left(\frac{3}{2}, \frac{1}{2}\right)$. This gives

$$f\left(\frac{3}{2}, \frac{1}{2}\right) = \left(\frac{3}{2}\right)^2 + 3\left(\frac{1}{2}\right)^2 = 3$$

The minimum value of $f(x, y) = x^2 + 3y^2$ subject to $x + y = 2$ is 3. ◼

✓**CHECKPOINT 1**

Now work Exercise 7.

Geometric Analysis

Now let's look at what is happening geometrically. If we add a *constraint,* $g(x, y) = 0$, in the xy-plane, it gives the sketch in Figure 8.5.2. Now we only look at that part of the surface that is directly above our constraint equation. This yields our *constrained maximum,* (x_0, y_0, z_0), which we notice in Figure 8.5.2 is different from our *unconstrained (or free) maximum,* (x, y, z).

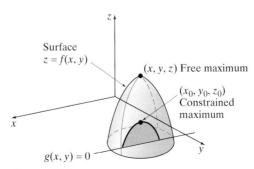

Figure 8.5.2

To gain another perspective, let's return to our two-dimensional setting and look at the level curves that we discussed in Section 8.2. Recall that a level curve is the result of slicing the surface with a horizontal plane.

EXAMPLE 3 | **Analyzing Level Curves and a Constraint Equation**

In Example 1 we

Maximized $\quad A(x, y) = xy$

Subject to $\quad g(x, y) = 2x + y - 200 = 0$

Sketch level curves for $A(x, y)$ for c-values of 3000, 5000, and 7000, and graph the constraint equation in the xy-plane.

SOLUTION

Recall that to sketch level curves for $A(x, y)$, we graph equations of the form

$$c = xy \quad \text{or} \quad y = \frac{c}{x}$$

So for $c = 3000$, $c = 5000$, and $c = 7000$, we need to graph

$$y = \frac{3000}{x}, \quad y = \frac{5000}{x}, \quad \text{and} \quad y = \frac{7000}{x}$$

We also need to graph the constraint equation

$$g(x, y) = 2x + y - 200$$

When $g(x, y) = 0$, the graph is a line in the xy-plane. Thus, we need to graph

$$2x + y - 200 = 0 \quad \text{or} \quad y = 200 - 2x$$

In Figure 8.5.3 we show the graphs of $y = \frac{3000}{x}$, $y = \frac{5000}{x}$, $y = \frac{7000}{x}$, and $y = 200 - 2x$.

Interactive Activity

In the viewing window [0, 200] by [0, 250], graph $y = \frac{3000}{x}$, $y = \frac{5000}{x}$, $y = \frac{7000}{x}$, and $y = 200 - 2x$. Use the INTERSECT command to confirm the results of Figure 8.5.3.

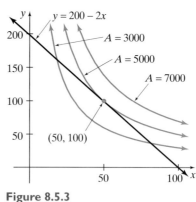

Figure 8.5.3

From Figure 8.5.3, we can make the following observations.

1. The constraint line shows all combinations of x and y that we could use to exhaust our 200 feet of fence.
2. The level curve corresponding to $A = 7000$ is not possible since it does not intersect our constraint equation.
3. The maximum value for A on the constraint equation occurs where the constraint is tangent to the level curve $A = 5000$.

Example 3 demonstrates that the critical points of our function $F(x, y, \lambda)$ in the Lagrange multipliers method occur at points where a level curve of $f(x, y)$ is tangent to the constraint curve $g(x, y)$.

EXAMPLE 4

Maximizing a Production Function

Recall that the O'Neill Corporation has a production function of

$$Q = f(x, y) = 1.64x^{0.6}y^{0.4}$$

where x is labor hours, y is capital, and $f(x, y)$ is thousands of gallons produced. Each unit of labor is $15,000 and each unit of capital is $7000, and the total expense for both (a budget constraint) is limited to $1.8 million. (These figures are for each of the O'Neill Corporation's 10 plants.) Determine the number of units of labor and capital needed to maximize production and give the maximum production.

SOLUTION

This appears to be a maximization problem requiring the Method of Lagrange multipliers.

Step 1: We want to maximize $f(x, y) = 1.64x^{0.6}y^{0.4}$. Since each unit of labor is $15,000 and each unit of capital is $7000, we have the constraint equation $15,000x + 7000y = 1,800,000$. So our problem has the form

Maximize $f(x, y) = 1.64x^{0.6}y^{0.4}$

Subject to $g(x, y) = 15,000x + 7000y - 1,800,000 = 0$

Step 2: Define $F(x, y, \lambda) = f(x, y) + \lambda \cdot g(x, y)$ using the Lagrange multiplier λ.

$$F(x, y, \lambda) = 1.64x^{0.6}y^{0.4} + \lambda(15,000x + 7000y - 1,800,000)$$

Step 3: Determine the partial derivatives.

$$F_x(x, y, \lambda) = 0.984x^{-0.4}y^{0.4} + 15,000\lambda$$

$$F_y(x, y, \lambda) = 0.656x^{0.6}y^{-0.6} + 7000\lambda$$

$$F_\lambda(x, y, \lambda) = 15,000x + 7000y - 1,800,000$$

Step 4: Solve the system

$$0.984x^{-0.4}y^{0.4} + 15,000\lambda = 0$$

$$0.656x^{0.6}y^{-0.6} + 7000\lambda = 0$$

$$15,000x + 7000y - 1,800,000 = 0$$

To solve this nasty looking system, let's solve the first two equations for $-\lambda$. This gives

$$0.984x^{-0.4}y^{0.4} = -15,000\lambda \quad \text{and} \quad 0.656x^{0.6}y^{-0.6} = -7000\lambda$$

$$\frac{0.984x^{-0.4}y^{0.4}}{15,000} = -\lambda \qquad \text{and} \qquad \frac{0.656x^{0.6}y^{-0.6}}{7000} = -\lambda$$

Now we can set these equal to each other to give

$$\frac{0.984x^{-0.4}y^{0.4}}{15,000} = \frac{0.656x^{0.6}y^{-0.6}}{7000}$$

Now for the "trick." Multiply both sides of the equation by $x^{0.4}y^{0.6}$.

$$x^{0.4}y^{0.6}\left(\frac{0.984x^{-0.4}y^{0.4}}{15,000}\right) = x^{0.4}y^{0.6}\left(\frac{0.656x^{0.6}y^{-0.6}}{7000}\right)$$

$$\frac{0.984x^{-0.4+0.4}y^{0.4+0.6}}{15,000} = \frac{0.656x^{0.6+0.4}y^{-0.6+0.6}}{7000}$$

$$\frac{0.984y}{15,000} = \frac{0.656x}{7000}$$

$$y = \frac{15,000}{0.984}\left(\frac{0.656x}{7000}\right) \approx 1.43x$$

Substituting this back into the budget constraint equation for y gives

$$15,000x + 7000(1.43x) - 1,800,000 = 0$$
$$25,010x = 1,800,000$$
$$x \approx 71.97$$

From $y \approx 1.43x$, we also determine that

$$y \approx 1.43(71.97) \approx 102.92$$

Step 5: Since there is only one solution from step 4, $x \approx 71.97$ and $y \approx 102.92$, we conclude that $x \approx 71.97$ units of labor and $y \approx 102.92$ units of capital maximize production. The maximum production is approximately

$$f(71.97, 102.92) = 1.64(71.97)^{0.6}(102.92)^{0.4} \approx 136.19 \text{ thousand gallons}$$

✓ **CHECKPOINT 2**

Now work Exercise 25.

Marginal Productivity of Money

In Example 4 we did not determine the value of $-\lambda$. In economics, $-\lambda$ has a practical meaning. To determine $-\lambda$ in Example 4, we could use either

$$\frac{0.984x^{-0.4}y^{0.4}}{15,000} = -\lambda \quad \text{or} \quad \frac{0.656x^{0.6}y^{-0.6}}{7000} = -\lambda$$

from step 4 and substitute in our values for x and y to give

$$-\lambda \approx 7.57 \times 10^{-5} = 0.0000757$$

The first thing we observe is that $-\lambda$ is positive. Now let's investigate the meaning of $-\lambda$ by looking at what happens when the budget for the O'Neill Corporation increases from \$1.8 million to \$1.9 million. How does this change x, y, and Q? Example 5 shows us.

EXAMPLE 5 | **Maximizing a Production Function**

If the budget for the O'Neill Corporation has been increased from \$1.8 million to \$1.9 million, what combinations of x and y maximize Q?

SOLUTION

Maximize $\quad Q = f(x, y) = 1.64x^{0.6}y^{0.4}$

Subject to $\quad g(x, y) = 15,000x + 7000y - 1,900,000 = 0$

The process outlined in steps 2 through 4 of Example 4 is the same here, except 1,800,000 is replaced with 1,900,000. We save the details for you to work out in the Interactive Activity to the left. The solution is $x \approx 75.97$ and $y \approx 108.64$. The maximum production with $x \approx 75.97$ units of labor and $y \approx 108.64$ units of capital is about $Q \approx 143.76$ thousand gallons.

Notice that our value for Q in Example 5 is approximately equal to our value for Q from Example 4 plus $10^5 \cdot (-\lambda)$. In other words,

$$Q \text{ from Example 5} \approx Q \text{ from Example } 4 + 10^5 \cdot (-\lambda)$$
$$\approx 136.19 + 10^5(7.57 \times 10^{-5}) \approx 143.76$$

Thus, in a production function setting:

- $-\lambda$ gives the extra production achieved by increasing the budget by *one* unit. Since the budget increased by $\$100,000 = 10^5$ in Example 5, we multiplied $-\lambda$ by 10^5.
- The value of $-\lambda$ approximates the increase in the optimal value of Q when the budget is increased by *one* unit.

In the language of calculus this means that $-\lambda$ **gives the rate of change of the optimal value of Q as the budget increases**.

Marginal Productivity of Money

Let $Q = f(x, y)$ be a production function and $g(x, y) = 0$ be a budget constraint. For

$$F(x, y, \lambda) = f(x, y) + \lambda \cdot g(x, y)$$

$-\lambda$ is the **marginal productivity of money** at (x, y), which is the extra production achieved by increasing the budget one unit.

NOTE: $-\lambda$ in this setting is always a positive number.

Now, from our work in Example 5 and the discussion immediately following Example 5, we can observe where the following formula from economics arises.

Formula for Optimal Increase in Production

Assume that $-\lambda$ is the marginal productivity of money and P additional dollars is available to the budget. The **optimal increase in production** is given by $-\lambda P$.

Example 6 ties together elements from Examples 4 and 5 and the marginal productivity of money concept.

EXAMPLE 6   **Maximizing Production and Determining Optimal Increase**

RollyOn Tires has a Cobb–Douglas production function of

$$Q = f(x, y) = 250x^{0.25}y^{0.75}$$

where x represents units of labor and y represents units of capital. Suppose that each unit of labor is $\$100$ and each unit of capital is $\$250$. Assume that the budget is limited to $\$100,000$.

(a) Determine x and y that maximize production and determine the maximum production.

(b) Determine the marginal productivity of money for the division of labor and capital determined in part (a) and interpret. Find the optimal increase production if an additional $\$15,000$ is budgeted.

SOLUTION

(a)

Step 1: We want to

$$\text{Maximize} \quad f(x, y) = 250x^{0.25}y^{0.75}$$
$$\text{Subject to} \quad g(x, y) = 100x + 250y - 100{,}000 = 0$$

Step 2: Define $F(x, y, \lambda) = f(x, y) + \lambda g(x, y)$ using the Lagrange multiplier λ.

$$F(x, y, \lambda) = 250x^{0.25}y^{0.75} + \lambda(100x + 250y - 100{,}000)$$

Step 3: Determine the partial derivatives.

$$F_x(x, y, \lambda) = 62.5x^{-0.75}y^{0.75} + 100\lambda$$
$$F_y(x, y, \lambda) = 187.5x^{0.25}y^{-0.25} + 250\lambda$$
$$F_\lambda(x, y, \lambda) = 100x + 250y - 100{,}000$$

Step 4: As we did in Example 4, we solve the first two equations for $-\lambda$.

$$62.5x^{-0.75}y^{0.75} = -100\lambda \quad \text{and} \quad 187.5x^{0.25}y^{-0.25} = -250\lambda$$
$$\frac{62.5x^{-0.75}y^{0.75}}{100} = -\lambda \quad \text{and} \quad \frac{187.5x^{0.25}y^{-0.25}}{250} = -\lambda$$

Setting these equal to each other and multiplying both sides of the equation by $x^{0.75}y^{0.25}$ gives

$$x^{0.75}y^{0.25}\left(\frac{62.5x^{-0.75}y^{0.75}}{100}\right) = x^{0.75}y^{0.25}\left(\frac{187.5x^{0.25}y^{-0.25}}{250}\right)$$

$$\frac{62.5y}{100} = \frac{187.5x}{250}$$

$$y = \frac{100}{62.5}\left(\frac{187.5x}{250}\right) = 1.2x$$

Substituting this into the budget constraint equation for y gives

$$100x + 250(1.2x) - 100{,}000 = 0$$
$$400x = 100{,}000$$
$$x = 250$$

From $y = 1.2x$ we also determine that $y = 300$.

Step 5: So $x = 250$ units of labor and $y = 300$ units of capital maximize production at $f(250, 300) = 71{,}658.2$ units. We round this to 71,658 units.

(b) The marginal productivity of money at $x = 250$ and $y = 300$ is

$$-\lambda = \frac{62.5x^{-0.75}y^{0.75}}{100} = \frac{62.5(250)^{-0.75}(300)^{0.75}}{100} \approx 0.717$$

Thus, for each additional dollar available to the budget, production increases by about 0.717 unit. The optimal increase in production is $-\lambda P$. Since $P = \$15{,}000$, $-\lambda P \approx 0.717(15{,}000) = 10{,}755$.
Hence, we now have

71,658	(maximum production with a budget of \$100,000)
+10,755	(additional production from extra \$15,000 added to the budget)
82,413	

This 82,413 is the optimal production possible with \$115,000.

✓ **CHECKPOINT 3**   Now work Exercise 27.

SUMMARY

In this section we have seen how to maximize or minimize problems that arise when we have **constraints** on our independent variables. The **Method of Lagrange Multipliers** was used to solve these problems.

- **How to Use the Method of Lagrange**

    1. Write the situation in the form

    $$\text{Maximize (or minimize)} \quad z = f(x, y)$$
    $$\text{Subject to} \quad g(x, y) = 0$$

    2. Define $F(x, y, \lambda) = f(x, y) + \lambda \cdot g(x, y)$.
    3. Determine the partial derivatives $F_x(x, y, \lambda)$, $F_y(x, y, \lambda)$, and $F_\lambda(x, y, \lambda)$.
    4. Solve the system of equations

    $$F_x(x, y, \lambda) = 0, \quad F_y(x, y, \lambda) = 0, \quad F_\lambda(x, y, \lambda) = 0$$

    5. The maximum (or minimum) of $f(x, y)$ is among the values in step 4. Simply evaluate $z = f(x, y)$ at each.

    The section concluded with problems from the world of economics, specifically marginal productivity of money.

- $-\lambda$ is the **marginal productivity of money** at (x, y), which is the extra production achieved by increasing the budget one unit.
- Assume that $-\lambda$ is the marginal productivity of money and P additional dollars is available to the budget. The **optimal increase in production** is given by $-\lambda P$.

SECTION 8.5 EXERCISES

In Exercises 1–18, use the method of Lagrange multipliers to solve each problem.

1. Maximize $f(x, y) = 3xy$ Subject to $x + y = 1$

2. Maximize $f(x, y) = 2xy$ Subject to $2x + y = 20$

3. Maximize $f(x, y) = xy$ Subject to $2x + y = 12$

4. Maximize $f(x, y) = 2xy - 5$ Subject to $x + y = 12$

5. Minimize $f(x, y) = xy$ Subject to $x - y = -8$

6. Maximize $f(x, y) = 3xy + x$ Subject to $x + y = 1$

✓ 7. Minimize $f(x, y) = x^2 + 2y^2$ Subject to $2x + y = 8$

8. Minimize $f(x, y) = x^2 + y^2 - 7$ Subject to $x + 2y = 10$

9. Maximize $f(x, y) = 2xy - 4x$ Subject to $x + y = 12$

10. Maximize $f(x, y) = 25 - x^2 - y^2$ Subject to $2x + y = 10$

11. Maximize $f(x, y) = 1.64x^{0.6} y^{0.4}$ Subject to $12{,}000x + 5000y = 1{,}100{,}000$

12. Maximize $f(x, y) = 125x^{0.75} y^{0.25}$ Subject to $100x + 250y = 100{,}000$

13. Minimize $f(x, y) = x^2 + y^2 - xy - 4$ Subject to $x + y - 6 = 0$

14. Maximize $f(x, y) = 16 - x^2 - y^2$ Subject to $x + 2y - 6 = 0$

15. Maximize $f(x, y) = xy$ Subject to $x^2 + y^2 = 8$

16. Minimize $f(x, y) = xy$ Subject to $x^2 + y^2 = 8$

17. Maximize $f(x, y) = e^{xy}$ Subject to $x^2 + y^2 = 8$

18. Minimize $f(x, y) = e^{x^2 + y^2}$ Subject to $x + y = 5$

APPLICATIONS

19. On-the-Go Music produces two types of walkman's each day, x units of its regular model and y units of its joggers' model. The cost, in dollars, is given by

$$C(x, y) = 0.1x^2 + 0.2y^2$$

Due to limitations on the supply of parts, it is necessary that

$$x + y = 180$$

(a) Determine how many of each type of walkman should be produced to minimize cost.

(b) Determine the minimum cost.

20. DeeDee's produces two types of diapers each week, x packages of its regular absorbent and y packages of its extra absorbent. The cost, in dollars, is given by

$$C(x, y) = 0.1x^2 + 0.5y^2$$

Due to limited materials, it is necessary that

$$x + y = 1200$$

(a) Determine how many of each type of diaper should be produced to minimize cost.

(b) Determine the minimum cost.

21. SalPal produces two types of personal security devices each day, x units of its flashlight and pepper spray device and y units of its flashlight and siren device. The cost, in dollars, is given by

$$C(x, y) = 0.1x^2 + 0.2y^2$$

Due to shipping restrictions, it is necessary that

$$x + y = 120$$

(a) Determine how many of each type of security device should be produced to minimize cost.

(b) Determine the minimum cost.

22. Rikki's Bicycle Company makes two models of bicycles, 10-speeds and 21-speeds. Each week, the cost to make x 10-speeds and y 21-speeds is given by

$$C(x, y) = 20,000 + 100x + 140y - xy$$

where $C(x, y)$ is in dollars. Determine the number of 10-speeds and the number of 21-speeds that should be produced each week to minimize costs if the total number produced each week is 200 bicycles.

23. The Riblet Engine Company has installed a newer, more efficient production line, which is alongside an older production line that is still in use. Let x represent the number of units produced on the older line, and let y represent the number of units produced on the new line. The cost of a production run is given by

$$C(x, y) = 2x^2 - xy + y^2 + 250$$

where $C(x, y)$ is in dollars. A production run has been scheduled for which both lines will be used to complete an order of 104 units.

(a) Determine the number of units produced by each line in order to minimize the cost.

(b) Determine the minimum production cost.

24. Sumption Sonic Incorporated has a production function for the manufacture of solar cells of

$$Q = f(x, y) = 2.5x^{0.7}y^{0.3}$$

where Q is the number of solar cells produced per week, x is the number of units of labor each week, and y is the number of units of capital. A unit of labor costs $400, a unit of capital costs $500, and the total budget for both is $180,000.

(a) Determine the number of units of labor and the number of units of capital required to maximize production.

(b) Determine the maximum production.

✓ 25. The Linger Golf Cart Corporation has a production function for the manufacture of golf carts given by

$$Q = f(x, y) = 13x^{0.75}y^{0.3}$$

where Q is the number of golf carts produced per week, x is the number of labor hours per week, and y is the weekly capital investment. A unit of labor is $15 and a unit of capital is $50, and the total budget for both is limited to $18,000.

(a) Determine the number of units of labor and the number of units of capital required to maximize production.

(b) Determine the maximum production.

26. The $(RB)^2$ Company has a production function of

$$Q = f(x, y) = 1610x^{0.8}y^{0.25}$$

where Q is the quantity of handheld calculators produced per year, x is the number of labor hours in thousands, and y is the capital equipment. A unit of labor is $15,000 and a unit of capital is $4000, and the total budget for both is limited to $1.5 million.

(a) Determine the number of units of labor and the number of units of capital required to maximize production.

(b) Determine the maximum production.

✓ 27. The Arnold Engine Company has a production function of

$$Q = f(x, y) = x^{0.6}y^{0.45}$$

where x is the number of units of labor, y is the number of units of capital, and Q is the total quantity produced. Each unit of labor is $16,000 and each unit of capital is $4000, and the total budget for both is limited to $1.7 million.

(a) Determine the number of units of labor and capital needed to maximize production.

(b) Determine the marginal productivity of money for the labor and capital found in part (a) and interpret.

(c) Determine the optimal increase in production if an additional $200,000 is budgeted.

28. Hogrefe's Incorporated has a production function given by

$$Q = f(x, y) = x^{0.7}y^{0.3}$$

where x is the number of units of labor, y is the number of units of capital, and Q is the total quantity produced. Each unit of labor is $1200, each unit of capital is $800, and the total budget for both is limited to $480,000.

(a) Determine the number of units of labor and capital needed to maximize production.

(b) Determine the marginal productivity of money for the labor and capital found in part (a) and interpret.

(c) Determine the optimal increase in production if an additional $20,000 is budgeted.

29. Hoffman Laboratories has a production function given by

$$Q = f(x, y) = 5.6x^{0.65}y^{0.41}$$

where Q is the quantity of pain-killing tablets produced (in thousands) per week, x is the weekly units of labor, and y is the weekly units of capital. Each unit of labor is $2000, each unit of capital is $1500, and the budget for both is limited to $800,000.

(a) Determine the number of units of labor and capital needed to maximize production.

(b) Determine the marginal productivity of money for the labor and capital found in part (a) and interpret.

(c) Determine the optimal increase in production if an additional $50,000 is budgeted.

30. Blair's Inc. has a production function given by

$$Q = f(x, y) = 25x^{0.8}y^{0.2}$$

where Q is the total quantity produced, x is the number of units of labor used, and y is the number of units of capital used. Each unit of labor costs $30, each unit of capital costs $60, and the budget for both is $300,000.

(a) Determine the number of units of labor and capital needed to maximize production.

(b) Determine the marginal productivity of money for the labor and capital found in part (a) and interpret.

(c) Determine the optimal increase in production if an additional $40,000 is budgeted.

31. New River Outfitters has a production function given by

$$Q = f(x, y) = 100x^{0.6}y^{0.4}$$

where x is the number of units of labor, y is the number of units of capital, and Q is the quantity produced. Each unit of labor costs $100, each unit of capital is $150, and the budget for both is $60,000.

(a) Determine the number of units of labor and capital needed to maximize production.

(b) Determine the marginal productivity of money for the labor and capital found in part (a) and interpret.

(c) Determine the optimal increase in production if an additional $10,000 is budgeted.

32. The Bowen Company has a production function given by

$$Q = f(x, y) = 25x^{0.65}y^{0.35}$$

where x is the number of units of labor, y is the number of units of capital, and Q is the quantity produced. Each unit of labor costs $200, each unit of capital costs $250, and the budget for both is $130,000.

(a) Determine the number of units of labor and capital needed to maximize production.

(b) Determine the marginal productivity of money for the labor and capital found in part (a) and interpret.

(c) Determine the optimal increase in production if an additional $20,000 is budgeted.

33. A company wishes to enclose a rectangular parking lot using an existing building as one side of the boundary and adding fencing for the other boundaries. If 625 feet of fencing is available, determine the dimensions of the largest parking lot that can be enclosed.

34. Determine the dimensions of a rectangular garden with an area of 5000 square feet that minimizes the cost of fencing if one side costs three times as much as the other three sides.

35. *Federal Express* states that the maximum length plus girth for its FedEx Overnight Freight and its FedEx 2Day Freight shipments is 300 inches (data gathered at Federal Express web site). See Figure 8.5.4. Use the method of Lagrange multipliers to determine the dimensions of the largest-volume package that can be sent by FedEx Overnight Freight and FedEx 2Day Freight. Assume that the length is y.

36. To avoid U.S. Domestic Freight Services charges, any packages sent by *Federal Express* must have the length plus girth not exceed 165 inches (data gathered at Federal Express web site). See Figure 8.5.4. Use the method of Lagrange multipliers to determine the dimensions of the largest-volume package that can be sent by *Federal Express* without incurring Domestic Freight Service charges. Assume that the length is y.

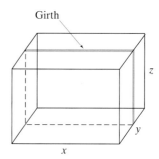

Figure 8.5.4

SECTION PROJECT

This section project synthesizes information studied in Chapter 8 and uses the following data. Also, the program in Appendix B that performs multiple regression is needed to do the section project.

The Gantzler Corporation has six plants around the United States that produce plastic tubing. In a recent year the following data for each plant gave the number of labor hours (in thousands), capital (in millions), and quantity produced.

LABOR	CAPITAL	QUANTITY
38	3.1	3440
39	3.2	3533
41	3.7	3800
42	3.8	3894
42	3.81	3896
43	3.9	3987

(a) The plants have the same technology, so a production function can be determined. Enter the data into your calculator and execute the program to determine a Cobb–Douglas production function to model the data. Use x for labor and y for capital and round all values to the nearest hundredth.

(b) Compute $f_x(x, y)$ and $f_y(x, y)$.

(c) If, on average, the Gantzler Corporation is now using 41 units of labor and 3.6 units of capital, determine the marginal productivity of labor and the marginal productivity of capital and interpret each.

(d) Suppose that each unit of labor costs $12,000 and each unit of capital costs $2000. The budget for both is $1.1 million. Determine the number of units of labor and capital needed to maximize production.

(e) Determine the marginal productivity of money for the division of labor and capital found in part (d).

(f) Determine the optimal increase in production if an additional $100,000 is budgeted.

SECTION 8.6   DOUBLE INTEGRALS

In this chapter we have seen how to generalize the derivative concept to functions of two variables. A natural question to ask is whether we can do the same with integration. The answer is yes, and we give a brief description of the process in this section. To fully understand the process, we begin the section by looking at **partial antidifferentiation.**

Partial Antidifferentiation and Iterated Integrals

In Chapter 6 we saw that antidifferentiation is the reverse operation of differentiation. In other words, antidifferentiation can "undo" differentiation. Remember that at that time we were dealing with functions of a single variable. To antidifferentiate a function of two variables with respect to one variable, we simply treat all other variables as if they are constants. This means that when we see

$$\int f(x, y)\, dx \quad \text{(Notice the } dx \text{ in this integral)}$$

we antidifferentiate $f(x, y)$ with respect to x and treat y as a constant. Likewise, when we see

$$\int f(x, y)\, dy \quad \text{(Notice the } dy \text{ in this integral)}$$

we antidifferentiate $f(x, y)$ with respect to y and treat x as a constant. Example 1 illustrates the process.

EXAMPLE 1 | **Antidifferentiating Functions of Two Variables**

Determine the following indefinite integrals:

(a) $\int (x^2 y + 3y^2 x)\, dx$ (b) $\int (x^2 y + 3y^2 x)\, dy$

SOLUTION

(a) Here we are antidifferentiating with respect to x, so we treat y as a constant. We simply apply some of the following properties of antidifferentiation from Chapter 6.

From Your Toolbox

- $\int [f(x) + g(x)] \, dx = \int f(x) \, dx + \int g(x) \, dx$
- $\int k \cdot f(x) \, dx = k \cdot \int f(x) \, dx$, where k is a constant

---

$$\int (x^2 y + 3y^2 x) \, dx = \int x^2 y \, dx + \int 3y^2 x \, dx$$
$$= y \int x^2 \, dx + 3y^2 \int x \, dx \qquad \text{\footnotesize y is constant}$$
$$= y \left(\frac{1}{3}x^3\right) + 3y^2 \left(\frac{1}{2}x^2\right) + C(y)$$
$$= \frac{1}{3}x^3 y + \frac{3}{2}x^2 y^2 + C(y)$$

Notice that the "constant" is not just a C. It can be *any function of y*, since for any function of y we know

$$\frac{\partial}{\partial x} C(y) = 0$$

(b) Here we antidifferentiate with respect to y, so we treat x as a constant. This gives

$$\int (x^2 y + 3y^2 x) \, dy = \int x^2 y \, dy + \int 3y^2 x \, dy$$
$$= x^2 \int y \, dy + x \int 3y^2 \, dy \qquad \text{\footnotesize x is constant}$$
$$= x^2 \left(\frac{1}{2}y^2\right) + x(y^3) + C(x)$$
$$= \frac{1}{2}x^2 y^2 + xy^3 + C(x)$$

Notice here that our antiderivative has as its arbitrary "constant" $C(x)$.

In Chapter 6 we encouraged checking our antidifferentiation by differentiating the answer. We still encourage this type of checking. In Example 1 our result can be checked by partial differentiation. To check the result in Example 1a, we compute the partial derivative

$$\frac{\partial}{\partial x} \left[\frac{1}{3}x^3 y + \frac{3}{2}x^2 y^2 + C(y)\right] = \left(\frac{1}{3} \cdot 3x^2\right) y + \left(\frac{3}{2} \cdot 2x\right) y^2 + 0$$
$$= x^2 y + 3xy^2$$

Interactive Activity

Check the result in Example 1b by computing the partial derivative

$$\frac{\partial}{\partial y} \left[\frac{1}{2}x^2 y^2 + xy^3 + C(x)\right]$$

Example 1 shows us that we can easily antidifferentiate functions of two variables as long as we remember which variable is being treated as a constant. In Example 2 we extend the process to evaluating definite integrals.

EXAMPLE 2 | **Evaluating Partial Antiderivatives**

Evaluate the following definite integrals.

(a) $\displaystyle\int_0^1 (x^2y + 3y^2x)\,dx$ 　　　　(b) $\displaystyle\int_0^1 (x^2y + 3y^2x)\,dy$

SOLUTION

(a) Using the result from Example 1a, we have

$$\int_0^1 (x^2y + 3y^2x)\,dx = \left(\frac{1}{3}x^3y + \frac{3}{2}x^2y^2\right)\Big|_{x=0}^{x=1}$$

$$= \left(\frac{1}{3}(1)^3y + \frac{3}{2}(1)^2y^2\right) - \left(\frac{1}{3}(0)^3y + \frac{3}{2}(0)^2y^2\right)$$

$$= \frac{1}{3}y + \frac{3}{2}y^2$$

Notice that this definite integral produces a function of y.

(b) Using the result from Example 1b, we have

$$\int_0^1 (x^2y + 3y^2x)\,dy = \left(\frac{1}{2}x^2y^2 + xy^3\right)\Big|_{y=0}^{y=1}$$

$$= \left(\frac{1}{2}x^2(1)^2 + x(1)^3\right) - \left(\frac{1}{2}x^2(0)^2 + x(0)^3\right)$$

$$= \frac{1}{2}x^2 + x$$

Notice that this definite integral produces a function of x.

Interactive Activity

In Example 2a, what did we choose $C(y)$ to equal?
In Example 2b, what did we choose $C(x)$ to equal?

✓ **CHECKPOINT 1**

Now work Exercise 9.

In Example 2, we noticed the following:

1. $\displaystyle\int_0^1 (x^2y + 3y^2x)\,dx = \frac{1}{3}y + \frac{3}{2}y^2 = f(y)$

2. $\displaystyle\int_0^1 (x^2y + 3y^2x)\,dy = \frac{1}{2}x^2 + x = f(x)$

So it appears that when we integrate and evaluate a definite integral of the form

$$\int_a^b f(x, y)\,dx$$

the result is a function of a single variable, y. (It could also produce a constant.) Likewise, when we integrate and evaluate a definite integral of the form

$$\int_c^d f(x, y)\,dy$$

the result is a function of a single variable, x. (It could also produce a constant.) Since each of these produces a function of a single variable, they in turn could be an integrand of a second integral. In Example 3, we evaluate what are called **iterated integrals.** The word iterated simply means repeated.

EXAMPLE 3 | **Evaluating Iterated Integrals**

Evaluate the following:

(a) $\displaystyle\int_0^1 \left[\int_0^1 (x^2 y + 3y^2 x) \, dx \right] dy$

(b) $\displaystyle\int_0^1 \left[\int_0^1 (x^2 y + 3y^2 x) \, dy \right] dx$

SOLUTION

(a) The brackets indicate the order in which the definite integrals are to be evaluated. Hence, we first evaluate the inside integral. From Example 2a we have

$$\int_0^1 (x^2 y + 3y^2 x) \, dx = \frac{1}{3}y + \frac{3}{2}y^2$$

This result is now the integrand for the outer integral. This gives

$$\int_0^1 \left[\int_0^1 (x^2 y + 3y^2 x) \, dx \right] dy = \int_0^1 \left(\frac{1}{3}y + \frac{3}{2}y^2 \right) dy$$

$$= \left(\frac{1}{6}y^2 + \frac{1}{2}y^3 \right) \Big|_0^1$$

$$= \left(\frac{1}{6} + \frac{1}{2} \right) - (0 + 0) = \frac{2}{3}$$

(b) From Example 2b the inside integral is

$$\int_0^1 (x^2 y + 3y^2 x) \, dy = \frac{1}{2}x^2 + x$$

This result is now the integrand for the outer integral. This gives

$$\int_0^1 \left[\int_0^1 (x^2 y + 3y^2 x) \, dy \right] dx = \int_0^1 \left(\frac{1}{2}x^2 + x \right) dx$$

$$= \left(\frac{1}{6}x^3 + \frac{1}{2}x^2 \right) \Big|_0^1$$

$$= \left(\frac{1}{6} + \frac{1}{2} \right) - (0 + 0) = \frac{2}{3}$$

Double Integrals

It is no accident that the results in Examples 3a and b are identical. Examples 1 through 3 suggest the following definition of a **double integral.**

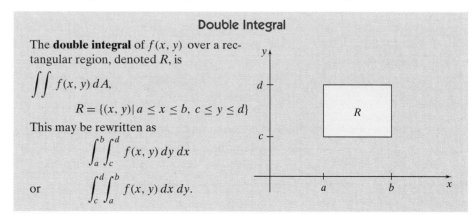

Double Integral

The **double integral** of $f(x, y)$ over a rectangular region, denoted R, is

$$\iint f(x, y) \, dA,$$

$$R = \{(x, y) \mid a \le x \le b, \ c \le y \le d\}$$

This may be rewritten as

$$\int_a^b \int_c^d f(x, y) \, dy \, dx$$

or

$$\int_c^d \int_a^b f(x, y) \, dx \, dy.$$

NOTE: In the double integral, $\iint f(x, y)\, dA$, R is called the **region of integration**. Notice that this region is in the xy-plane. Also, dA indicates that either order of integration, $dy\, dx$ or $dx\, dy$, may be used.

We must mention that a more general definition of double integrals exists. The one that we supply is more applicable to the functions that we will study.

EXAMPLE 4

Writing and Evaluating a Double Integral

For $\iint (2x + y)\, dA$, where $R = \{(x, y) \mid 1 \le x \le 3, 0 \le y \le 2\}$:

(a) Sketch the region of integration R in the xy-plane.

(b) Write and evaluate a double integral with $dy\, dx$ order of integration.

(c) Write and evaluate a double integral with $dx\, dy$ order of integration.

SOLUTION

(a) The region of integration is shown in Figure 8.6.1.

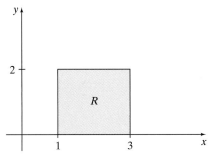

Figure 8.6.1

(b) A $dy\, dx$ order of integration means that the innermost integral has limits determined by the bounds for y and the outermost integral has limits determined by the bounds for x. This gives

$$\int_1^3 \int_0^2 (2x + y)\, dy\, dx$$

We evaluate by performing the innermost integration first. This gives

$$\int_1^3 \left[\int_0^2 (2x + y)\, dy \right] dx = \int_1^3 \left[\left(2xy + \frac{1}{2} y^2 \right) \Big|_{y=0}^{y=2} \right] dx$$

$$= \int_1^3 \left[\left(2x(2) + \frac{1}{2} (2)^2 \right) - \left(2x(0) + \frac{1}{2} (0)^2 \right) \right] dx$$

$$= \int_1^3 (4x + 2)\, dx = (2x^2 + 2x) \Big|_1^3$$

$$= (18 + 6) - (2 + 2) = 20$$

(c) A $dx\, dy$ order of integration means that the innermost integral has limits determined by the bounds for x and the outermost integral has limits determined by the bounds for y. This gives

$$\int_0^2 \int_1^3 (2x + y)\, dx\, dy$$

Evaluating, starting with the innermost integral, yields

$$\int_0^2 \left[\int_1^3 (2x + y)\, dx \right] dy = \int_0^2 \left[(x^2 + xy)\big|_{x=1}^{x=3} \right] dy$$

$$= \int_0^2 \left[(9 + 3y) - (1 + y) \right] dy$$

$$= \int_0^2 (8 + 2y)\, dy = (8y + y^2)\big|_0^2$$

$$= (16 + 4) - (0 + 0) = 20$$

Again Example 4 illustrates that, as long as our function is continuous, either order of integration (*dy dx* or *dx dy*) yields the same result.

✓ CHECKPOINT 2

Now work Exercise 15.

EXAMPLE 5 | **Evaluating a Double Integral**

Given $\int_{-1}^1 \int_0^2 x^2 e^{-y}\, dx\, dy$, sketch the region of integration and evaluate the double integral.

SOLUTION

From the limits of integration and the *dx dy* order of integration, we have

$$-1 \le y \le 1 \quad \text{and} \quad 0 \le x \le 2$$

The region is shown in Figure 8.6.2.

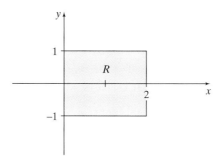

Figure 8.6.2

Interactive Activity

 Evaluate $\int_0^2 \int_{-1}^1 x^2 e^{-y}\, dy\, dx$ to verify the result of Example 5.

Evaluating, starting with the innermost integral, gives

$$\int_{-1}^1 \left[\int_0^2 x^2 e^{-y}\, dx \right] dy = \int_{-1}^1 \left[\left(\frac{1}{3} x^3 e^{-y} \right)\Big|_{x=0}^{x=2} \right] dy = \int_{-1}^1 \left[\frac{1}{3}(2)^3 e^{-y} - \frac{1}{3}(0)^3 e^{-y} \right] dy$$

$$= \int_{-1}^1 \frac{8}{3} e^{-y}\, dy = \left(-\frac{8}{3} e^{-y} \right)\Big|_{-1}^1 = -\frac{8}{3} e^{-1} + \frac{8}{3} e$$

✓ CHECKPOINT 3

Now work Exercise 25.

Average Value over Rectangular Regions

In Section 7.1 we defined the average value of a function, *f*, over an interval [*a*, *b*]. We can extend the average value concept to functions of two variables.

From Your Toolbox

The *average value* of a continuous function, f, on the interval $[a, b]$ is given by

$$\frac{1}{b-a}\int_a^b f(x)\,dx$$

Average Value over Rectangular Regions

The **average value** of $f(x, y)$ over rectangular region R is defined to be

$$\frac{1}{(b-a)(c-d)}\iint f(x, y)\,dA$$

or

$$\frac{1}{\text{area of } R}\iint f(x, y)\,dA$$

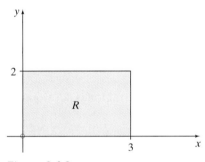

NOTE: Observe that $(b - a)(c - d)$ is simply the area of rectangular region R.

EXAMPLE 6

Computing an Average Value

Determine the average value of $f(x, y) = x + y$ over the region R shown in Figure 8.6.3.

Figure 8.6.3

SOLUTION

From Figure 8.6.3 we see that R is simply

$$0 \le x \le 3 \quad \text{and} \quad 0 \le y \le 2$$

Hence, the area of R is 6 $((3 - 0)(2 - 0))$. Using a $dy\,dx$ order of integration, we compute the average value to be

$$\frac{1}{6}\int_0^3\int_0^2 (x + y)\,dy\,dx = \frac{1}{6}\int_0^3\left[\int_0^2 (x + y)\,dy\right]dx = \frac{1}{6}\int_0^3\left[\left(xy + \frac{1}{2}y^2\right)\Big|_{y=0}^{y=2}\right]dx$$

$$= \frac{1}{6}\int_0^3 [(2x + 2) - (0 + 0)]\,dx$$

$$= \frac{1}{6}\int_0^3 (2x + 2)\,dx = \frac{1}{6}\left[(x^2 + 2x)\Big|_0^3\right]$$

$$= \frac{1}{6}[(9 + 6) - (0 + 0)] = \frac{1}{6}(15) = \frac{15}{6} = \frac{5}{2} = 2.5$$

Thus, 2.5 is simply the average of the z-values for $z = f(x, y) = x + y$ over region R. See Figure 8.6.4.

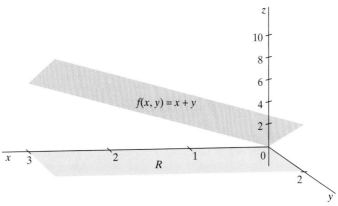

Figure 8.6.4

EXAMPLE 7 | **Computing an Average Value of a Production Function**

The Boroff Engine Company has a production function given by

$$Q = f(x, y) = x^{0.6}y^{0.45}$$

where Q is the number of engines produced per week, x is the number of employees at the company, and y is the weekly operating budget in thousands. Because the company uses a temporary labor agency, it uses anywhere from 40 to 50 employees each week, and its operating budget is anywhere from $10,000 to $15,000 each week. Determine the average number of engines that the company can produce each week.

SOLUTION

Since the Boroff Engine Company uses anywhere from 40 to 50 employees each week, we know that

$$40 \leq x \leq 50$$

Also, since its operating budget is anywhere from $10,000 to $15,000 each week,

$$10 \leq y \leq 15 \qquad \text{Recall that } y \text{ is in thousands}$$

Thus, we need to determine the average value of $Q = f(x, y) = x^{0.6}y^{0.45}$ over the rectangular region shown in Figure 8.6.5. The area of the region is found to be

$$(50 - 40)(15 - 10) = 50$$

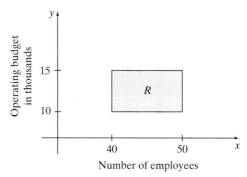

Figure 8.6.5

The average value is

$$\frac{1}{50} \int_{40}^{50} \int_{10}^{15} x^{0.6} y^{0.45} \, dy \, dx = \frac{1}{50} \int_{40}^{50} \left[\int_{10}^{15} x^{0.6} y^{0.45} \, dy \right] dx$$

$$= \frac{1}{50} \int_{40}^{50} \left[\left(\frac{1}{1.45} x^{0.6} y^{1.45} \right) \Big|_{10}^{15} \right] dx$$

$$= \frac{1}{50} \int_{40}^{50} \left(\frac{1}{1.45} x^{0.6} (15)^{1.45} - \frac{1}{1.45} x^{0.6} (10)^{1.45} \right) dx$$

$$\approx \frac{1}{50} \int_{40}^{50} (34.99 x^{0.6} - 19.44 x^{0.6}) \, dx$$

$$= \frac{1}{50} \int_{40}^{50} 15.55 x^{0.6} \, dx$$

$$= \frac{15.55}{50} \int_{40}^{50} x^{0.6} \, dx = \frac{15.55}{50} \left[\frac{1}{1.6} x^{1.6} \Big|_{40}^{50} \right]$$

$$= \frac{15.55}{50} \left[\frac{1}{1.6} (50)^{1.6} - \frac{1}{1.6} (40)^{1.6} \right] \approx 30.51$$

To the nearest engine, the Boroff Engine Company can produce an average of 31 engines each week. Figure 8.6.6 gives a graph of $Q = f(x, y) = x^{0.6} y^{0.45}$ over the region R. Again, the result of 31 is simply the average of the Q-values over R.

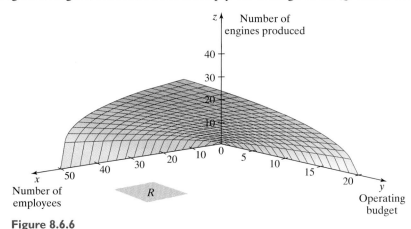

Figure 8.6.6

Volume

We conclude this section with a discussion on volume. In Chapter 6 we saw that the area of the region under a curve given by $y = f(x)$ on the interval $[a, b]$ is determined by $\int_a^b f(x) \, dx$, as long as $f(x)$ is nonnegative on $[a, b]$. See Figure 8.6.7. Here, we can determine the *volume* of the *solid* under a surface given by $f(x, y)$ over rectangular region R using double integrals. See Figure 8.6.8.

Volume under a Surface

If $z = f(x, y)$ is nonnegative and continuous ($f(x, y) \geq 0$) over rectangular region R, $R = \{(x, y) \mid a \leq x \leq b, \, c \leq y \leq d\}$, then the volume of the solid under $f(x, y)$ over R is given by

$$V = \iint f(x, y) \, dA$$

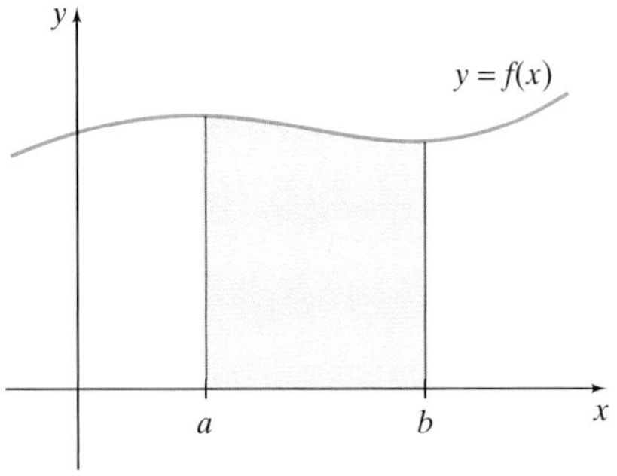

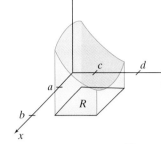

Figure 8.6.7

Figure 8.6.8 Volume $= \iint f(x, y) \, dA.$

EXAMPLE 8 | **Computing a Volume**

Determine the volume under $f(x, y) = 4 - x^2 - y^2$ above the region

$$R = \{(x, y) \mid 0 \le x \le 1, \, 0 \le y \le 1\}.$$

SOLUTION

Figure 8.6.9 shows the region, and Figure 8.6.10 shows the volume that we are try-ing to determine.

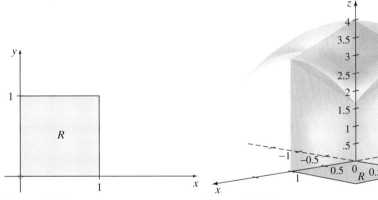

Figure 8.6.9

Figure 8.6.10

The volume is

$$V = \iint (4 - x^2 - y^2) \, dA$$

We select a $dy \, dx$ order of integration, which gives

$$V = \int_0^1 \int_0^1 (4 - x^2 - y^2) \, dy \, dx = \int_0^1 \left[\left(4y - x^2 y - \frac{1}{3} y^3 \right) \Big|_{y=0}^{y=1} \right] dx$$

$$= \int_0^1 \left(4 - x^2 - \frac{1}{3} \right) dx = \int_0^1 \left(\frac{11}{3} - x^2 \right) dx$$

$$= \left(\frac{11}{3} x - \frac{1}{3} x^3 \right) \Big|_0^1 = \left(\frac{11}{3} - \frac{1}{3} \right) - (0 - 0)$$

$$= \frac{10}{3} \text{ cubic units}$$

✓ CHECKPOINT 4 | Now work Exercise 39.

SUMMARY

In this section we have seen how to extend the concept of antidifferentiation to functions of two variables. Partial antidifferentiation led to iterated integrals, which in turn led us to double integrals. We then extended the concept of average value to functions of two variables. We concluded the section with a brief look at how a double integral can be used to determine the volume of a solid under a surface over a rectangular region.

• The **double integral** of $f(x, y)$ over a rectangular region, denoted R, is

$$\iint f(x, y)\, dA, \qquad R = \{(x, y)\,|\, a \le x \le b,\ c \le y \le d\}$$

This may be rewritten as $\int_a^b \int_c^d f(x, y)\, dy\, dx$ or $\int_c^d \int_a^b f(x, y)\, dx\, dy$.

• The **average value** of $f(x, y)$ over rectangular region R is defined to be

$$\frac{1}{(b-a)(c-d)} \iint f(x, y)\, dA \quad \text{or} \quad \frac{1}{\text{area of } R} \iint f(x, y)\, dA$$

• If $z = f(x, y)$ is nonnegative and continuous ($f(x, y) \ge 0$) over rectangular region R, $R = \{(x, y)\,|\, a \le x \le b,\ c \le y \le d\}$, then the volume of the solid under $f(x, y)$ over R is given by $V = \iint f(x, y)\, dA$.

SECTION 8.6 EXERCISES

In Exercises 1–12, evaluate the given integral.

1. $\displaystyle \int_1^3 3x^2 y^2\, dy$

2. $\displaystyle \int_1^4 2x^3 y^3\, dy$

3. $\displaystyle \int_1^3 3x^2 y^2\, dx$

4. $\displaystyle \int_1^4 2x^3 y^3\, dx$

5. $\displaystyle \int_0^2 (2x + 3y)\, dy$

6. $\displaystyle \int_0^3 (3x - 2y)\, dy$

7. $\displaystyle \int_0^2 (2x + 3y)\, dx$

8. $\displaystyle \int_0^3 (3x - 2y)\, dx$

✓ 9. $\displaystyle \int_1^2 (x^3 y^2 - 2xy)\, dy$

10. $\displaystyle \int_2^4 (x^2 y^3 - 3xy)\, dy$

11. $\displaystyle \int_1^2 (x^3 y^2 - 2xy)\, dx$

12. $\displaystyle \int_2^4 (x^2 y^3 - 3xy)\, dx$

For Exercises 13–22:

(a) Sketch the region of integration.

(b) Write and evaluate a double integral with a $dy\, dx$ order of integration.

(c) Write and evaluate a double integral with a $dx\, dy$ order of integration.

13. $\displaystyle \iint 2xy\, dA \qquad R = \{(x, y)\,|\, 0 \le x \le 2,\ 0 \le y \le 3\}$

14. $\displaystyle \iint 3xy\, dA \qquad R = \{(x, y)\,|\, 0 \le x \le 2,\ 0 \le y \le 3\}$

✓ 15. $\displaystyle \iint (3x + y)\, dA \qquad R = \{(x, y)\,|\, 0 \le x \le 1,\ 0 \le y \le 2\}$

16. $\displaystyle \iint (4x - y)\, dA \qquad R = \{(x, y)\,|\, 2 \le x \le 4,\ 1 \le y \le 2\}$

17. $\displaystyle \iint \sqrt{xy}\, dA \qquad R = \{(x, y)\,|\, 1 \le x \le 16,\ 1 \le y \le 4\}$

18. $\displaystyle \iint \sqrt{xy}\, dA \qquad R = \{(x, y)\,|\, 1 \le x \le 4,\ 1 \le y \le 16\}$

19. $\displaystyle \iint x^2 y^2\, dA \qquad R = \{(x, y)\,|\, 1 \le x \le 2,\ 0 \le y \le 1\}$

20. $\displaystyle \iint x^2 y^2\, dA \qquad R = \{(x, y)\,|\, 0 \le x \le 1,\ 1 \le y \le 2\}$

21. $\displaystyle \iint \left(2 - \frac{1}{2}x^2 + y^2\right) dA$
$R = \{(x, y)\,|\, 0 \le x \le 1,\ 0 \le y \le 1\}$

22. $\displaystyle \iint \left(3 + x^2 - \frac{1}{2}y^2\right) dA$
$R = \{(x, y)\,|\, 0 \le x \le 1,\ 0 \le y \le 1\}$

In Exercises 23–28, sketch the region of integration and evaluate the given double integral.

23. $\displaystyle \int_0^1 \int_0^2 xy\, dy\, dx$

24. $\displaystyle \int_1^3 \int_1^2 5xy\, dy\, dx$

✓ 25. $\displaystyle \int_1^2 \int_1^3 (x^2 y + y)\, dx\, dy$

26. $\displaystyle \int_1^3 \int_1^2 (x^2 y - y)\, dx\, dy$

27. $\displaystyle \int_{-1}^1 \int_0^2 x^2 e^y\, dx\, dy$

28. $\displaystyle \int_{-1}^1 \int_0^2 x^2 e^y\, dy\, dx$

In Exercises 29–36, determine the average value of the function over the region R.

29. $f(x, y) = x^2 + y^2$ $R = \{(x, y)| 0 \le x \le 1, 0 \le y \le 1\}$

30. $f(x, y) = x^2 + y^2$ $R = \{(x, y)| 0 \le x \le 2, 0 \le y \le 2\}$

31. $f(x, y) = 9 - x^2 - y^2$
    $R = \{(x, y)| 0 \le x \le 1, 0 \le y \le 1\}$

32. $f(x, y) = 9 - x^2 - y^2$
    $R = \{(x, y)| 0 \le x \le 2, 0 \le y \le 2\}$

33. $f(x, y) = 2xy$ $R = \{(x, y)| 0 \le x \le 2, 0 \le y \le 2\}$

34. $f(x, y) = 2xy$ $R = \{(x, y)| 0 \le x \le 1, 0 \le y \le 1\}$

35. $f(x, y) = x^2 e^y$ $R = \{(x, y)| 0 \le x \le 2, 0 \le y \le 1\}$

36. $f(x, y) = x^2 e^y$ $R = \{(x, y)| 0 \le x \le 1, 0 \le y \le 2\}$

In Exercises 37–44, determine the volume under $f(x, y)$ above the given region R.

37. $f(x, y) = 9 - x^2 - y^2$
    $R = \{(x, y)| 0 \le x \le 1, 0 \le y \le 2\}$

38. $f(x, y) = 12 - x^2 - y^2$
    $R = \{(x, y)| 0 \le x \le 1, 0 \le y \le 2\}$

✓ 39. $f(x, y) = x^2 + 2y^2$ $R = \{(x, y)| 0 \le x \le 1, 0 \le y \le 2\}$

40. $f(x, y) = 2x^2 + y^2$ $R = \{(x, y)| 0 \le x \le 2, 0 \le y \le 2\}$

41. $f(x, y) = 3xy$ $R = \{(x, y)| 0 \le x \le 2, 0 \le y \le 1\}$

42. $f(x, y) = 2xy$ $R = \{(x, y)| 0 \le x \le 1, 0 \le y \le 1\}$

43. $f(x, y) = x^2 e^y$ $R = \{(x, y)| 0 \le x \le 2, 0 \le y \le 1\}$

44. $f(x, y) = x^2 e^y$ $R = \{(x, y)| 0 \le x \le 1, 0 \le y \le 2\}$

Applications

45. Sumption Sonic Inc. has a production function given by

$$Q = f(x, y) = 250x^{0.7}y^{0.3}$$

where Q is the number of solar cells produced each week, x is the number of employees at the company, and y is the weekly operating budget in thousands. Because the company uses a temporary labor agency, it uses anywhere from 30 to 40 employees each week, and its operating budget is anywhere from $150,000 to $180,000 each week. Determine the average number of solar cells that the company can produce each week.

46. The Linger Golf Cart Corporation has a production function given by

$$Q = f(x, y) = 1.3x^{0.75}y^{0.3}$$

where Q is the number of golf carts produced each week, x is the number of employees at the company, and y is the weekly operating budget in thousands. Because the company uses a temporary labor agency, it uses anywhere from 10 to 15 employees each week, and its operating budget is anywhere from $12,000 to $16,000 each week. Determine the average number of golf carts that the company can produce each week.

47. The Basich Company has a production function given by

$$Q = f(x, y) = 161x^{0.8}y^{0.25}$$

where Q is the number of hand-held calculators produced each week, x is the number of employees at the company, and y is the weekly operating budget in thousands. Because the company use a temporary labor agency, it uses anywhere from 70 to 80 employees each week, and its operating budget is anywhere from $20,000 to $30,000 each week. Determine the average number of hand-held calculators that the company can produce each week.

48. Hoffman Laboratories has a production function given by

$$Q = f(x, y) = 5.6x^{0.65}y^{0.41}$$

where Q is the number of pain-killing tablets produced (in thousands) each week, x is the number of employees at the company, and y is the weekly operating budget in thousands. Because the company use a temporary labor agency, it uses anywhere from 25 to 35 employees each week, and its operating budget is anywhere from $40,000 to $50,000 each week. Determine the average number of pain-killing tablets that the company can produce each week.

49. The temperature, in degrees Fahrenheit, x miles east and y miles north of a reporting weather station is given by

$$f(x, y) = 36 + 2x - 3y$$

Determine the average temperature over the region shown in Figure 8.6.11.

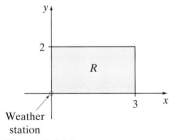

Figure 8.6.11

50. The temperature, in degrees Fahrenheit, x miles east and y miles north of a reporting weather station is given by

$$f(x, y) = 45 + x - 2y$$

Determine the average temperature over the region shown in Figure 8.6.12.

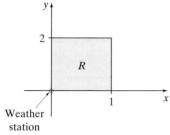

Figure 8.6.12

51. Cedar Falls is a town roughly shaped like a square, as shown in Figure 8.6.13. The **population density**, in people per square mile, x miles east and y miles north of the southwest corner of the town is given by

$$h(x, y) = 30,000e^{-y}$$

Determine the average population density for Cedar Falls.

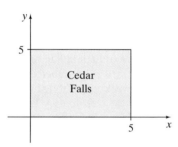

Figure 8.6.13

52. Bucksville is a town situated along a fairly straight river and is fairly rectangular, as shown in Figure 8.6.14. The population density, in people per square mile, x miles east and y miles north of the southwest corner of the town is given by

$$h(x, y) = 15,000(5 - y^2)$$

Determine the average population density for Bucksville.

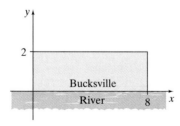

Figure 8.6.14

53. Cedar Falls is a town roughly shaped like a square, as shown in Figure 8.6.13. The **population density,** in people per square mile, x miles east and y miles north of the southwest corner of the town is given by

$$h(x, y) = 30,000e^{-y}$$

Determine the *total population* of Cedar Falls. (To do this, integrate the population density over region R. This illustrates that a double integral can determine a *continuous sum*.)

54. Bucksville is a town situated along a fairly straight river and is fairly rectangular, as shown in Figure 8.6.14. The population density, in people per square mile, x miles east and y miles north of the southwest corner of the town is given by

$$h(x, y) = 15,000(5 - y^2)$$

Determine the total population of Bucksville.

55. A heavy industrial complex in Bakerstown emits pollution into the atmosphere. Due to the prevailing winds, the air pollution (in parts per million) x miles east and y miles north of the complex is given by

$$f(x, y) = 125 - 10x^2 - 10y^2$$

Determine the average concentration of air pollution in the northeast part of Bakerstown. See Figure 8.6.15.

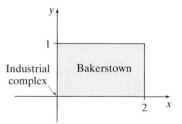

Figure 8.6.15

56. Repeat Exercise 55 for the portion of Bakerstown shown in Figure 8.6.16.

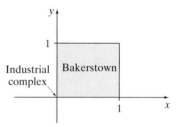

Figure 8.6.16

57. An individual's IQ is defined to be

$$\text{IQ} = f(a, m) = \frac{100m}{a}$$

where m is the individual's mental age in years and a is the individual's actual age in years. In a calculus class, the mental age varies from 23 to 32 years, and the actual age varies from 19 to 25 years. Determine the average IQ for this calculus class.

58. Repeat Exercise 57, except here the mental age varies from 23 to 45 years and the actual age varies from 22 to 32 years.

SECTION PROJECT

In their study of human groupings, anthropologists often use an index called the *cephalic index*. The cephalic index is given by

$$C(W, L) = 100 \cdot \frac{W}{L}$$

where W is the width and L is the length of an individual's head. Both measurements are made across the top of the

head and are in inches. In a calculus class the width of heads varies from 6 to 8.5 inches and the length of heads varies from 8 to 10 inches.

(a) Determine the average value of the cephalic index for this class.

(b) Repeat part (a), except here the width of heads varies from 6.5 to 8.2 inches and the length of heads varies from 7.6 to 9.8 inches.

CHAPTER REVIEW EXERCISES

In Exercises 1–4, evaluate $f(x, y) = \frac{x^2 + 3}{x - y^2}$ for the indicated values.

1. $f(3, 5)$
2. $f(-2, 4)$
3. $f(10, 3)$
4. $f(20, -4)$

In Exercises 5 and 6, determine the domain of the given function. Recall that here the domain is the set of all ordered pairs (x, y) in the xy-plane.

5. $f(x, y) = \sqrt{x + y}$
6. $f(x, y) = \frac{5x}{x - 2y}$

In Exercises 7–10, plot the given points.

7. $(5, 0, 0)$
8. $(3, 8, -5)$
9. $(0, -4, 3)$
10. $(-2, -6, 5)$

In Exercises 11–14, without plotting the point, determine if the given point is above or below the xy-plane. Explain.

11. $(-3, -4, 2)$
12. $(4, 3, -8)$
13. $(6, -2, -1)$
14. $(-5, 9, 4)$

In Exercises 15 and 16, sketch the graph of the given plane in 3-space.

15. $x = -5$
16. $z = 2$

17. The Moondoe Coffee Company has determined that its monthly profit, in dollars, is given by

$$P(x, y) = 0.8x + 1.2y - 6000$$

where x is the number of regular cups of coffee sold in a month and y is the number of cups of cappuccino sold in a month. Determine $P(4000, 3000)$ and interpret.

18. The Stellar Stereo Store spends x thousand dollars each month on television advertising and y thousand dollars each month on direct mail advertising. The company has monthly sales given by

$$S(x, y) = x^2 + 3y$$

(a) Determine $S(4, 7)$ and interpret.
(b) Determine $S(7, 4)$ and interpret.

19. The WriteNow Company sells a regular calligraphy set for $12 and a deluxe calligraphy set for $21. The cost to produce

these items is $7 for the regular set and $13 for the deluxe set. The company also has fixed monthly expenses totaling $5000.

(a) Determine the cost $C(x, y)$ of producing x regular sets and y deluxe sets in a month.
(b) Determine the revenue function $R(x, y)$.
(c) Determine the profit function $P(x, y)$.

20. The Digital Calculator Company manufactures scientific calculators and graphing calculators. Let x represent the weekly demand for scientific calculators, and let y represent the demand for graphing calculators. The weekly price–demand equations are given by

$$p = 60 - 0.3x + 0.1y, \quad \text{the price in dollars for a scientific calculator}$$

$$q = 100 + 0.2x - 0.5y, \quad \text{the price in dollars for a graphing calculator}$$

The cost function is given by

$$C(x, y) = 3000 + 25x + 35y$$

(a) Determine the weekly revenue function $R(x, y)$.
(b) Evaluate $R(40, 60)$ and interpret.
(c) Determine the weekly profit function $P(x, y)$.
(d) Evaluate $P(40, 60)$ and interpret.

21. If an amount P dollars is invested at an annual interest rate of r (in decimal form) compounded continuously for 10 years, the total amount accumulated is given by

$$f(P, r) = Pe^{10r}$$

determine $f(4500, 0.045)$ and interpret.

22. The volume of a pyramid with a square base is given by

$$V(x, h) = \frac{1}{3}x^2 h$$

where x is the length of one side of the square base and h is the height of the pyramid. Assume that x and h are measured in inches.

(a) Determine $V(8, 12)$ and interpret.
(b) If the height is equal to the side length of the square base, the volume can be expressed as a function of one variable, either x or h. Determine $V(x)$ and $V(h)$.

In Exercises 23–26, make a contour map of the given function for the specified c-values.

23. $f(x, y) = 2x - y + 3$ $c = 0, 2, 4, 6$

24. $f(x, y) = x^2 - y$ $c = -2, 0, 2, 4$

25. $f(x, y) = 16 - x^2 - y^2$ $c = 0, 4, 8, 12$

26. $f(x, y) = x - e^y$ $c = 0, 1, 2, 3$

In Exercises 27–30, make a contour map of the given production functions for the specified c-values.

27. $Q = 10x^{0.5}y^{0.5}$ $c = 10, 20, 30, 40$

28. $Q = 2.7x^{0.3}y^{0.75}$ $c = 20, 40, 60, 80$

29. $Q = 8.9x^{0.62}y^{0.42}$ $c = 100, 200, 300, 400$

30. $Q = 80x^{0.21}y^{0.79}$ $c = 50, 100, 150, 200$

In Exercises 31–34, make a contour map of the given utility function for the specified c-values.

31. $U = \sqrt[3]{xy^2}$ $c = 2, 4, 6, 8$

32. $U = x^{0.4}y^{0.6}$ $c = 1, 2, 3, 4$

33. $U = x^{0.15}y^{0.85}$ $c = 2, 4, 6, 8$

34. $U = x^{0.3}y^{0.7}$ $c = 1, 2, 3, 4$

In Exercises 35–38, sketch on the appropriate plane, either the xz-plane or yz-plane, cross sections for the given function for the specified vertical plane values. See Example 6 in Section 8.2.

35. $z = 20x^{0.3}y^{0.7}$ $x = 2, 4, 6$

36. $z = 20x^{0.3}y^{0.7}$ $y = 2, 4, 6$

37. $z = x^2 - y$ $y = 0, 1, 2, 3$

38. $z = x^2 - y$ $x = 0, 1, 2, 3$

39. The McGillicutty Corporation has 12 manufacturing plants and has a Cobb–Douglas production function of $Q = f(x, y) = 1.87x^{0.7}y^{0.3}$, where x is the number of labor-hours (in thousands), y is the capital (total net assets in millions), and Q is the quantity produced (in thousands of units). Sketch isoquants for $Q = 60, 80, 100$, and 120.

40. Refer to Exercise 39. Lucille, the CEO of the McGillicutty Corporation, wants to keep the capital at each plant fixed at $y = 4.1$.

(a) By fixing capital at $y = 4.1$, is Lucille looking at a vertical or horizontal cross section?

(b) Graph the cross section when $y = 4.1$.

(c) Use the graph to approximate how many labor-hours would be required to produce a quantity of 150,000 units (that is, $Q = 150$).

41. Refer to Exercises 39 and 40. Instead of keeping the capital fixed, suppose that Lucille, the CEO of the McGillicutty Corporation, wants to keep the number of labor-hours at each plant fixed at $x = 240$.

(a) By fixing labor at $x = 240$, is Lucille looking at a vertical or horizontal cross section?

(b) Graph the cross section when $x = 240$.

(c) Use the graph to approximate how much capital is required to produce 150,000 units (that is, $Q = 150$).

42. The MaxiSound Company has seven plants around the United States that produce speakers for computers. In a recent year, the data for each plant gave the number of labor-hours (in thousands), capital (in millions), and total quantity of speakers produced, as follows:

LABOR	CAPITAL	QUANTITY
24	4.3	13,016
26	4.9	14,565
29	5.6	16,500
31	6.1	17,852
38	5.8	18,600
45	6.4	21,052
46	7.3	23,084

(a) The plants all use the same technology, so a production function can be determined. Enter the data into your calculator and execute the multiple regression program in Appendix B to determine a Cobb–Douglas production function to model the data. Round all values to the nearest hundredth.

(b) Construct isoquants for $Q = 12,000, 16,000, 20,000$, and $24,000$.

In Exercises 43–50, determine $f_x(x, y)$ and $f_y(x, y)$.

43. $f(x, y) = 5x - 3y^2$

44. $f(x, y) = 2x^3 - 5xy + y^8$

45. $f(x, y) = (8y^2 - 3x)^5$

46. $f(x, y) = 5.83x^{0.38}y^{0.61}$

47. $f(x, y) = \dfrac{x + y}{3x}$

48. $f(x, y) = \ln(5x - y^3)$

49. $f(x, y) = e^{x-y} + 5x^3y$

50. $f(x, y) = \ln xy - e^{x/y}$

In Exercises 51–54, for the given function, evaluate the stated partial derivative and interpret. See Example 4 in Section 8.3.

51. $f(x, y) = 5x^2 + 3xy^4 - 2y^5$; determine $f_y(3, 5)$.

52. $f(x, y) = \ln(x^2 + 5y)$; determine $f_x(2, 7)$.

53. $f(x, y) = 12.85x^{0.38}y^{0.62}$; determine $f_y(3, 2)$.

54. $f(x, y) = \dfrac{\ln 2y}{x^2 + y}$; determine $f_x(-2, 5)$.

In Exercises 55–58, determine $f_{xx}(x, y), f_{xy}(x, y)$, and $f_{yy}(x, y)$.

55. $f(x, y) = -4x^3 + 10xy + xy^5$

56. $f(x, y) = 8x^2y^7 - x^3y^5 + 12x^2$

57. $f(x, y) = (x + 2y)^2$ 58. $f(x, y) = \ln(3x - y)$

59. For $f(x, y) = 6x^2 - 4xy + 5y^2 + x - 3y + 17$, determine values for x and y such that $f_x(x, y) = 0$ and $f_y(x, y) = 0$ simultaneously.

60. The Heavy Stuff Athletic Club spends x thousand dollars each year on television advertising and y thousand dollars each year on newspaper advertising. The club has weekly revenues (in thousands of dollars) given by

$$R(x, y) = 1.5x + 0.6x^2$$

(a) Determine $R_x(x, y)$ and $R_y(x, y)$ and interpret.

(b) Determine $R_x(12, 8)$ and interpret.

(c) Determine $R_x(12, 8)$ and interpret.

61. Pixel Power, Inc., manufactures 15-inch computer monitors and 17-inch computer monitors. Let x represent the weekly demand for a 15-inch monitor, and let y represent the weekly demand for a 17-inch monitor. The weekly price–demand equations are given by

$p = 480 - 0.6x + 0.2y$, the price in dollars for a 15-inch monitor

$q = 720 + 0.3x - 0.7y$, the price in dollars for a 17-inch monitor

(a) Determine the weekly revenue function $R(x, y)$.

(b) Determine $R_x(x, y)$ and $R_y(x, y)$ and interpret each.

(c) Determine $R_x(640, 850)$ and $R_y(640, 850)$ and interpret each.

62. (Continuation of Exercise 61.) The weekly cost function for Pixel Power, Inc., is given by

$$C(x, y) = 12,000 + 128x + 163y$$

(a) Determine the weekly profit function $P(x, y)$.

(b) Determine $P_x(x, y)$ and $P_y(x, y)$ and interpret each.

(c) Determine $P_x(640, 850)$ and $P_y(640, 850)$ and interpret each.

63. The WatchIt Company produces two types of videocassette recorders, x hundred regular VCRs and y hundred stereo VCRs each month. The monthly revenue and cost functions are, in hundreds of dollars,

$$R(x, y) = -3.5x^2 - 3.5xy + 1.2y^2 + 175x + 215y$$
$$C(x, y) = 15xy + 65x + 85y + 600$$

(a) Determine the monthly profit function $P(x, y)$.

(b) Determine $P_x(x, y)$ and $P_y(x, y)$ and interpret each.

(c) Determine $P_x(8, 11)$ and $P_y(8, 11)$ and interpret each.

(d) Determine $P(8, 11)$ and interpret.

64. The Superior Gizmo Company has a production function given by

$$Q = f(x, y) = 5.64x^{0.25}y^{0.8}$$

where x is the number of labor-hours in thousands, y is the capital in millions, and Q is the total quantity produced in thousands.

(a) Compute $f_x(x, y)$ and $f_y(x, y)$.

(b) If the Superior Gizmo Company is currently using 38 units of labor and 1.6 units of capital, determine the marginal productivity of labor and the marginal productivity of capital and interpret each.

(c) Would production increase more by increasing the labor by 1200 labor-hours or by increasing the capital by $40,000?

In Exercises 65–68, determine the critical points for the given function.

65. $f(x, y) = x^2 + y^2 + 8x - 6y + 8$

66. $f(x, y) = x^2 - y^2 + 4x + 10y - 3$

67. $f(x, y) = x^2 + y^2 - 5xy + 6x - 8y + 7$

68. $f(x, y) = 3x^2 + 5y^2 - 8xy + 2x + 6y + 3$

In Exercises 69–76, use the Second Derivative Test to locate any relative extrema and saddle points.

69. $f(x, y) = x^2 + y^2 - 2x + 8y - 7$

70. $f(x, y) = 2x^2 + 5y^2 + 6x - 2y + 12$

71. $f(x, y) = x^2 - 2y^2 + xy + 3x - 3y$

72. $f(x, y) = \dfrac{4}{3}y^3 + x^2 - 4x - 8y - 17$

73. $f(x, y) = 2x^3 + -2x^2 + 4xy + y^2$

74. $f(x, y) = \dfrac{1}{5}x^5 - y^2 - 16x + 6y - 8$

75. $f(x, y) = e^{x^2 - y^2 + 4y}$

76. $f(x, y) = 4xy + \dfrac{1}{x} + \dfrac{2}{y}$

77. Roberta's Hair Salon offers a basic haircut and a deluxe haircut, which includes shampoo and styling. Let x represent the daily demand for basic haircuts, and let y represent the daily demand for deluxe haircuts. The daily price–demand equations are given by

$p = 12 - 0.3x + 0.1y$, the price in dollars for a basic haircut

$q = 20 + 0.1x - 0.2y$, the price in dollars for a deluxe haircut

(a) Determine the revenue function $R(x, y)$.

(b) How many basic haircuts and how many deluxe haircuts should be given per day in order to maximize revenue?

(c) What is the maximum daily revenue?

78. (Continuation of Exercise 77.) Roberta's Hair Salon has a daily cost function given by

$$C(x, y) = 200 + 4x + 5y$$

(a) Determine the profit function $P(x, y)$.

(b) How many basic haircuts and how many deluxe haircuts should be given per day in order to maximize profit?

(c) What is the maximum daily profit?

79. The WriteNow Company makes marking pens with either a fine tip or a large tip. Let x represent the number of thousands of fine-tip pens, and let y represent the number of thousands of large-tip pens produced each day. The daily profit function is given by

$$P(x, y) = -3x^2 - 2y^2 + 2xy + 800x + 1200y - 400$$

where $P(x, y)$ is the profit in dollars for producing x thousand fine-tip pens and y thousand large-tip pens.

(a) How many fine-tip pens and how many large-tip pens should be sold each day to maximize profit?

(b) What is the maximum profit?

80. You have been challenged to design a rectangular box with no top and three parallel partitions that holds 80 cubic inches. Determine the dimensions that will require the least amount of material to do this.

In Exercises 81–86, use the method of Lagrange multipliers to solve each problem.

81. Maximize $f(x, y) = 5xy$ Subject to $x + y = 10$

82. Minimize $f(x, y) = 3xy$ Subject to $x - 2y = 6$

83. Maximize $f(x, y) = 16 - 2x^2 + 5xy - 18y^2$
       Subject to $x + 3y = 6$

84. Minimize $f(x, y) = 3 + x^2 + y^2$
       Subject to $2x - y = 10$

85. Maximize $f(x, y) = 3.8x^{0.35}y^{0.6}$
       Subject to $120x + 260y = 2500$

86. Minimize $f(x, y) = xy$ Subject to $x^2 + 4y^2 = 8$

87. The Liberty Bell Telephone Company has a production function for the manufacture of portable telephones of

$$Q = f(x, y) = 28x^{0.6}y^{0.4}$$

where Q is the number of thousands of telephones produced per week, x is the number of units of labor each week, and y is the number of units of capital. A unit of labor is \$500 and a unit of capital is \$800. The total expense for both is limited to \$4000.

(a) Determine the number of units of labor and the number of units of capital required to maximize production.

(b) Determine the maximum production.

88. The Dizzy Dog Company has a production function for the manufacture of dog toys given by

$$Q = f(x, y) = 4.3x^{0.55}y^{0.5}$$

where Q is the quantity of dog toys produced (in thousands) per week, x is the number of units of labor, and y is the number of units of capital. Each unit of labor is \$600 and each unit of capital is \$400. The budget for both is limited to \$24,000.

(a) Determine the number of units of labor and the number of units of capital required to maximize production.

(b) Determine the marginal productivity of money for the division of labor and capital found in part (a) and interpret.

(c) Determine the optimal increase in production if an additional \$2000 is allotted toward the budget.

89. A family plans to plant a garden with an area of 300 square feet. Three sides of the garden will have fencing, and the fourth side will have a wall that costs five times as much (per linear foot) as the fencing. Use the method of Lagrange multipliers to determine the dimensions of the garden that will minimize the cost.

90. Packages weighing less than 15 pounds sent by Priority Mail are charged at the standard rate (based on the weight of the package), provided that the length plus girth is no more than 84 inches. Use the method of Lagrange multipliers to determine the dimensions of the largest volume package that can be sent via Priority Mail at the standard rate. Assume that the length is y. (*Source:* United States Postal Service web site.)

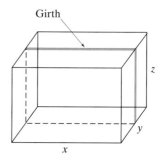

In Exercises 91–94, evaluate the given integral.

91. $\displaystyle\int_5^8 2x^2y^3\,dx$ 92. $\displaystyle\int_0^5 (3x - 4y)\,dy$

93. $\displaystyle\int_2^6 (x^2y - 4y^3)\,dy$ 94. $\displaystyle\int_3^9 (x^2y^3 + 4xy)\,dx$

For Exercises 95 and 96:

(a) Sketch the region of integration.

(b) Write and evaluate a double integral with a $dy\,dx$ order of integration.

(c) Write and evaluate a double integral with a $dx\,dy$ order of integration.

95. $\displaystyle\int (3x - 4xy)\,dA$ $R = \{(x, y) | 0 \le x \le 5, 0 \le y \le 2\}$

96. $\displaystyle\int \sqrt[3]{xy}\,dA$ $R = \{(x, y) | 0 \le x \le 8, 1 \le y \le 27\}$

For Exercises 97 and 98:

(a) Sketch the region of integration.

(b) Evaluate the given double integral.

97. $\displaystyle\int_0^4\int_1^3 3xy^2\,dy\,dx$ 98. $\displaystyle\int_{-2}^6\int_0^3 (2x + ye^x)\,dx\,dy$

In Exercises 99 and 100, determine the volume under $f(x, y)$ above the given region R.

99. $f(x, y) = 25 - x^2 - y^2$
$$R = \{(x, y) | 0 \leq x \leq 3, \ 0 \leq y \leq 4\}$$

100. $f(x, y) = \dfrac{e^x}{y}$ $R = \{(x, y) | 0 \leq x \leq 4, \ 1 \leq y \leq e\}$

In Exercises 101–104, determine the average value of the function over the region R.

101. $f(x, y) = x^2 + 3xy - y^2$
$$R = \{(x, y) | 0 \leq x \leq 3, \ 0 \leq y \leq 4\}$$

102. $f(x, y) = 3x + 4y^3$ $R = \{(x, y) | 0 \leq x \leq 5, \ 0 \leq y \leq 2\}$

103. $f(x, y) = x^3 y$ $R = \{(x, y) | 1 \leq x \leq 5, \ 3 \leq y \leq 7\}$

104. $f(x, y) = e^{x + y/2}$ $R = \{(x, y) | 0 \leq x \leq 1, \ 0 \leq y \leq 2\}$

105. The Cripton Company has a production function given by

$$Q = f(x, y) = 11.8x^{0.68} y^{0.45}$$

where Q is the number of radios produced (in thousands) each week, x is the number of employees at the company, and y is the weekly operating budget in thousands. Because the company uses a temporary labor agency, it has anywhere from 18 to 30 employees each week, and its operating budget is anywhere from \$16,000 to \$24,000 each week. Determine the average number of radios that the company produces each week.

106. The temperature, in degrees Fahrenheit, x miles east and y miles north of a reporting weather station is given by

$$f(x, y) = 68 + 2x - 3y$$

Determine the average temperature over the region shown in the figure.

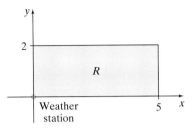

107. Brandonville is a town whose shape is fairly rectangular, as shown. A heavy industrial complex located at the southwest corner of the town emits pollution into the atmosphere. The air pollution (in parts per million) x miles east and y miles north of the complex is given by

$$f(x, y) = 150 - x^2 - 2y^2$$

Determine the average concentration of air pollution in Brandonville.

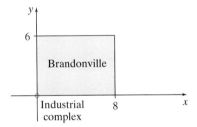

108. Refer to Exercise 107. The population density of Brandonville, in people per square mile, x miles east and y miles north of the southwest corner of the town is given by

$$h(x, y) = 8000(3 + x - y^2)$$

Determine the average population of Brandonville.

Appendices

603

APPENDIX A ESSENTIALS OF ALGEBRA

Exponents and Radicals

If a is a real number and n is a positive integer, then a^n *is the product of n factors of a.* That is, $a^n = a \cdot a \cdot a \cdot \cdots \cdot a$. We can use this definition to give us the Properties of Exponents.

Properties of Exponents

Let m and n be integers and a and b be real numbers. Then

1. $a^0 = 1,\ a \neq 0$

2. $a^1 = a$

3. $a^m \cdot a^n = a^{m+n}$

4. $\dfrac{a^m}{a^n} = a^{m-n},\ a \neq 0$

5. $(ab)^n = a^n \cdot b^n$

6. $\left(\dfrac{a}{b}\right)^n = \dfrac{a^n}{b^n},\ b \neq 0$

7. $a^{-n} = \dfrac{1}{a^n},\ a \neq 0$

8. $a^{m/n} = \sqrt[n]{a^m} = \left(\sqrt[n]{a}\right)^m$

Property 8 is a special type of exponent called a **rational exponent.** An expression containing a rational exponent can be written as a radical expression, and vice versa. Many times in calculus we wish to write expressions in the exponential form $ax^{m/n}$.

EXAMPLE 1 | **Writing Expressions in Exponential Form**

Write the following expressions in the form $ax^{m/n}$.

(a) $\dfrac{3}{x^5}$ (b) $4\sqrt[3]{x}$ (c) $\dfrac{1}{\sqrt{x}}$

SOLUTION

(a) We use exponent property 7 to write $\dfrac{3}{x^5} = 3x^{-5}$.

(b) From exponent properties 2 and 8, we see that $4\sqrt[3]{x} = 4\sqrt[3]{(x)^1} = 4x^{1/3}$.

(c) From exponent properties 8 and 7, we see that $\dfrac{1}{\sqrt{x}} = \dfrac{1}{x^{1/2}} = x^{-1/2}$. ◼

The Properties of Exponents can also be applied to write expressions given in rational exponent form in radical form.

EXAMPLE 2 | **Writing Expressions in Radical Form**

Write the following expressions in radical form.

(a) $x^{3/2}$ (b) $7x^{-1/4}$

SOLUTION

(a) Using exponent property 8, we see that $x^{3/2} = \sqrt{x^3}$.

(b) We start by applying exponent property 7 to get $7x^{-1/4} = \dfrac{7}{x^{1/4}}$. Exponent property 8 then gives us $7x^{-1/4} = \dfrac{7}{x^{1/4}} = \dfrac{7}{\sqrt[4]{x}}$. ◼

Systems of Equations

A **system of equations** is a collection of two or more equations with each having one or more variables. The following are examples of systems of equations.

$$\begin{cases} 2x + y = 6 \\ x - y = 6 \end{cases} \qquad \begin{cases} 2x + y - z = 2 \\ x + 3y + 2z = 1 \\ x + y + z = 2 \end{cases}$$

A **solution** to a system of equations are the values of the variables that make each equation in the system true. For example, $x = 4, y = -2$ is a solution to the system

$$\begin{cases} 2x + y = 6 \\ x - y = 6 \end{cases} \quad \text{since} \quad \begin{cases} 2(4) + (-2) = 6 & \text{is true} \\ 4 - (-2) = 6 & \text{is true.} \end{cases}$$

The algebraic method used to solve many of the systems in this textbook is the **Substitution Method.**

Substitution Method

1. Choose one of the equations and solve for one of the variables.
2. Substitute the result in the remaining equation(s).
3. If one equation with one variable results, solve the equation. Otherwise, repeat step 1.
4. Find the values of the remaining variables by back-substitution.

EXAMPLE 3 | **Solving a System of Equations by Substitution**

Solve the system of equations $\begin{cases} 3x + 2y = 11 \\ -x + y = 3 \end{cases}$ by the substitution method.

SOLUTION

We begin by solving the second equation for y to get $y = x + 3$. Now we substitute this result into the first equation to get

$$3x + 2(x + 3) = 11$$
$$3x + 2x + 6 = 11$$
$$5x + 6 = 11$$
$$5x = 5$$
$$x = 1$$

Now we back-substitute $x = 1$ in the second equation to determine y.

$$-1 + y = 3$$
$$y = 4$$

So the solution to the system is $x = 1, y = 4$.

Many of the systems we solve have one solution, just as in Example 3. We call these systems **consistent.** If a system has no solutions, it is called **inconsistent.** Some systems have an infinite number of solutions. These systems are called **dependent.**

Another technique used to solve systems of equations is the **Graphical Method.** This method is based on the following observation.

For a consistent system, the coordinates of the points of intersection on the graphs of the equations are the solutions of the system.

EXAMPLE 4 | **Solving a System of Equations by Graphing**

Verify the solution of the system $\begin{cases} 3x + 2y = 11 \\ -x + y = 3 \end{cases}$ by the graphical method.

SOLUTION

To graph the system, we need to solve each equation for y to get

$$\begin{cases} y = \dfrac{11 - 3x}{2} \\ y = x + 3 \end{cases}$$

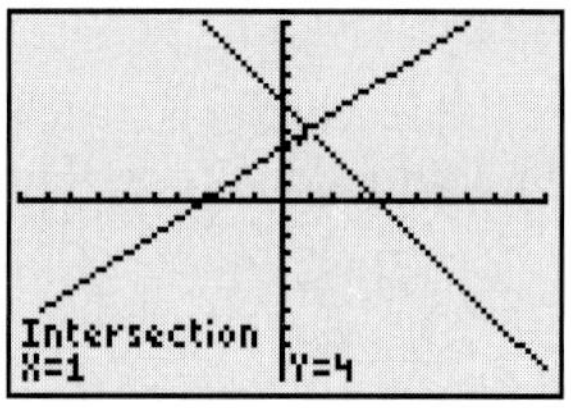

The graph of the equations is shown in Figure A.1 with the INTERSECT command used to determine the intersection point. We see that the result is the same as in Example 3.

Figure A.1

Sometimes we must solve a system of three equations with three variables. The graphical method is of little use in this case, so we use the substitution method.

EXAMPLE 5 | **Solving a System of Three Equations**

Solve the system of equations $\begin{cases} x + z = 0 \\ y + 2z = 0 \\ 2x + y = 200 \end{cases}$ using the substitution method.

SOLUTION

We begin by solving the first equation for x and the second equation for y to get $x = -z$ and $y = -2z$, respectively. Now we substitute these expressions in the third equation to get

$$2(-z) + (-2z) = 200$$
$$-2z - 2z = 200$$
$$-4z = 200$$
$$z = -50$$

Since we know that $z = -50$, we can substitute this value into the first equation in the system to get $x - 50 = 0$; so we see that $x = 50$. Using $z = -50$ again, we back-substitute in the second equation to get

$$y + 2(-50) = 0$$
$$y - 100 = 0$$
$$y = 100$$

So the solution to this system is $x = 50$, $y = 100$, and $z = -50$.

Properties of Logarithms

We can define a logarithm base b of x by saying that $y = \log_b x$ **if and only if** $b^y = x$, **where** $b > 0, b \neq 1, x > 0$. We can use this definition to prove the Properties of Logarithms. Some of these properties are proved in Appendix C.

Properties of Logarithms
Let $b > 0, b \neq 1$, where m and n are positive real numbers and r is any real number. Then

1. $\log_b 1 = 0$
2. $\log_b b = 1$
3. $\log_b (m \cdot n) = \log_b m + \log_b n$
4. $\log_b \left(\dfrac{m}{n}\right) = \log_b m - \log_b n$
5. $\log_b m^r = r \log_b m$
6. $\log_b b^r = r$

Some logarithms have special bases. If $b = 10$, then $\log_{10} x = \log x$ is a **common logarithm.** Many applications in this textbook use the number e as the base. This is called a **natural logarithm** and is denoted by $\log_e x = \ln x$.

EXAMPLE 6

Using the Properties of Logarithms

Use the properties of logarithms to write the following as a sum, difference, or product of logs.

(a) $\log_b \dfrac{(x-3)y}{x}$

(b) $\log_b [(x+7)y]^4$

SOLUTION

(a) $\log_b \dfrac{(x-3)y}{x} = \log_b [(x-3)y] - \log_b x$ Logarithm property 4

$= \log_b(x+3) + \log_b y - \log_b x$ Logarithm property 3

(b) $\log_b [(x+7)y]^4 = 4\log_b [(x+7)y]$ Logarithm property 5

$= 4[\log_b(x+7) + \log_b y]$ Logarithm property 3   ◼

Working with Logarithms

There are times when we need to solve **logarithmic equations.** The method used to solve these equations is based on the following property:

> **If $\log_b m = \log_b n$, then $m = n$.**

Here m and n are called the **arguments** of the logarithmic expressions. The next example illustrates the use of this property.

EXAMPLE 7

Solving Logarithmic Equations

Solve the equation for x: $\log_b (x+6) - \log_b (x+2) = \log_b x$

SOLUTION

We start by rewriting the left side as a single logarithmic expression to get

$$\log_b \left(\frac{x+6}{x+2} \right) = \log_b x$$

Now we use the property for solving logarithmic equations by setting the arguments of the logarithmic expression equal to each other.

$$\frac{x+6}{x+2} = x$$

$$x + 6 = x(x+2)$$

$$x + 6 = x^2 + 2x$$

$$0 = x^2 + x - 6$$

$$0 = (x-2)(x+3)$$

$$x = 2 \quad \text{or} \quad x = -3$$

Since we cannot take the logarithm of a negative number, we see that $x = -3$ is not a solution. So the only solution to the equation is $x = 2$.   ◼

Another way we use logarithms is to solve **exponential equations.** If we are given an equation of the form $b^x = k$, we can solve for x by taking the logarithm of each side of the equation and using logarithm property 5. Since it does not matter which base of logarithm we use, we usually take the natural logarithm of each side.

EXAMPLE 8 | **Solving Exponential Equations**

Solve the equation $3 \cdot 5^x = 21$ for x.

SOLUTION

$$3 \cdot 5^x = 21 \qquad \text{Given equation}$$
$$5^x = 7 \qquad \text{Divide by 3}$$
$$\ln(5^x) = \ln 7 \qquad \text{Take the natural logarithm of both sides}$$
$$x \ln 5 = \ln 7 \qquad \text{By logarithm property 5}$$
$$x = \frac{\ln 7}{\ln 5} \approx 1.21 \qquad \text{Divide by ln 5}$$

APPENDIX A EXERCISES

In Exercises 1–6, write the expression in exponential form.

1. $\dfrac{-4}{x^6}$

2. $\dfrac{17}{x^2}$

3. $1.6\sqrt{x}$

4. $7.2\sqrt[4]{x}$

5. $\dfrac{8}{\sqrt[3]{x^2}}$

6. $\dfrac{-1}{\sqrt[4]{x^5}}$

In Exercises 7–12, write the expressions in radical form.

7. $x^{3/7}$

8. $x^{1/8}$

9. $6.3x^{4/5}$

10. $2.1x^{2/5}$

11. $2x^{-2/3}$

12. $5x^{-1/2}$

In Exercises 13–18, solve the system of equations using the substitution method.

13. $\begin{cases} y = x + 2 \\ x + 2y = 16 \end{cases}$

14. $\begin{cases} x - y = 2 \\ 2x + y = 13 \end{cases}$

15. $\begin{cases} x + 2y = 8 \\ 3x - 4y = 9 \end{cases}$

16. $\begin{cases} 2x - y = -21 \\ 4x + 5y = 7 \end{cases}$

17. $\begin{cases} 4x + 3z = 4 \\ 2y - 6z = -1 \\ 8x + 4y + 3z = 9 \end{cases}$

18. $\begin{cases} x + y + z = 2 \\ 2x + y - z = 5 \\ x - y + z = -2 \end{cases}$

For Exercises 19 and 20, solve the system of equations using the graphical method.

19. $\begin{cases} 2x + 5y = 21 \\ 3x - 2y = -16 \end{cases}$

20. $\begin{cases} 3x - 4y = 1 \\ 2x + 3y = 12 \end{cases}$

In Exercises 21–26, use the properties of logarithms to write the following as a sum, difference, or product of logs.

21. $\log_b \left(\dfrac{2x}{5y} \right)^3$

22. $\log_b \left(\dfrac{7y}{3x} \right)^2$

23. $\log \left(\dfrac{xy^2}{y + 4} \right)$

24. $\log \left(\dfrac{x^5 y^3}{x - 2} \right)$

25. $\ln \sqrt{\dfrac{5x^3}{y^9}}$

26. $\ln \sqrt[3]{\dfrac{y^4}{x^2}}$

In Exercises 27–30, solve the logarithmic equation for x.

27. $\log_b (x - 2) = \log_b (x - 7) + \log_b 4$

28. $\log_b x - \log_b (x + 1) = \log_b 5$

29. $\ln (4x + 5) - \ln (x + 3) = \ln 3$

30. $\ln (4x - 2) = \ln 4 - \ln (x - 2)$

In Exercises 31–34, solve the exponential equation for x.

31. $3^x = 6$

32. $4^x = 12$

33. $2^{5x+2} = 8$

34. $10^{3x-7} = 5$

APPENDIX B   CALCULATOR PROGRAMS

> All the programs in this section are written for the Texas Instruments TI-83. To convert the programs for your model of calculator, consult your owner's manual or consult the online calculator manual at www.prenhall.com/armstrong

Quadratic Formula Program

This program approximates the real roots of the quadratic equation $ax^2 + bx + c = 0$ or reports that there are no real roots. The values of the coefficients a, b, and c are entered after the program is executed.

```
PROGRAM:QUADFORM
Prompt A,B,C
(B^2-4*A*C)→D
If D<0
Then
Disp"NO REAL SOLUTIONS"
Goto 1
Else
(-B+√(D))/(2*A)→M
(-B-√(D))/(2*A)→N
Disp"SOLUTIONS",M,N
Lbl 1
```

EXAMPLE 1

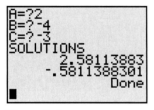

Figure B.1

Use the Quadratic Formula Program to approximate the roots of the quadratic equation $2x^2 - 4x = 3$.

SOLUTION

First we need to write the equation in the standard form $2x^2 - 4x - 3 = 0$. When we enter $a = 2$, $b = -4$, and $c = -3$ after the program is executed, we get the output shown in Figure B.1.

Newton's Method Program

We can extend the ideas of tangent lines and linear approximations discussed in Section 3.1 to derive what is known as Newton's Method, named after one of the founding fathers of the calculus, Isaac Newton. This method can be used to numerically determine zeros (or roots) of differentiable functions. For the method, we let f be a differentiable function and suppose that r is a real number zero. If x_n is an approximation to r, then the next approximation is given by $x_{n+1} = x_n - \frac{f(x_n)}{f'(x_n)}$, provided that $f'(x_n) \neq 0$, where $n = 0, 1, 2, \ldots$. Before running the program, $f(x)$ must be stored in Y_1 and its derivative $f'(x)$ stored in Y_2. Upon execution of the program, an initial guess $X = x_0$ is prompted.

```
PROGRAM:NEWTON
Prompt X
Lbl A
X Y₁(X)/Y₂(X)→X
Disp X
Pause
Goto A
```

EXAMPLE 2

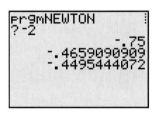

Figure B.2

For the function $f(x) = x^2 - 4x - 2$, compute the first three iterations of an approximation to a real root of f using the Newton's Method program. Use the initial guess $x = -2$.

SOLUTION

First we enter $f(x)$ in Y_1 and $\dfrac{d}{dx}(x^2 - 4x - 2) = 2x - 4$ in Y_2 and then execute the program. The first three iterations are displayed in Figure B.2. The ENTER key must be pressed to get the next approximation. ∎

Left/Right Sum Program

This program combines the left sum and right sum methods, which were discussed in Section 6.2. To run the program, the integrand of $\int_a^b f(x)\,dx$ must be entered in Y_1. When executed, the program prompts the user to enter values for the limits of integration a and b and the desired number of rectangles n.

```
PROGRAM:LFTRTSUM
Prompt A,B,N
(B-A)/N→D
0→T
0→Z
For(I,0,N-1)
A+I*D→X
T+D*Y₁→T
End
Disp"LEFT SUM APPROX",T
For(J,1,N)
A+J*D→X
Z+D*Y₁→Z
End
Disp"RIGHT SUM APPROX",Z
Disp"DIFFERENCE",abs(T-Z)
```

EXAMPLE 3

Use the Left/Right Sum Program to approximate $\displaystyle\int_0^3 \frac{1}{1+x}\,dx$ using $n = 20$ rectangles.

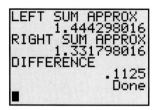

Figure B.3

SOLUTION

Here we enter $\frac{1}{1+x}$ in Y_1. After the program is executed, we enter $a = 0$, $b = 3$, and $n = 20$. The output is shown in Figure B.3.

Midsum Program

This program is another way to approximate the area under a curve based on the rectangular approximation methods discussed in Section 6.2. It uses the midpoint rule, which states that the area under the graph of f on the closed interval $[a, b]$ can be approximated by

$$\text{Area} \approx \sum_{i=1}^{n} f\left(\frac{x_i + x_{i-1}}{2}\right) \Delta x$$

where the height of each rectangle is determined by the middle of its base. The integrand of $\int_a^b f(x) \; dx$ must be entered in Y_1. When executed, the program prompts the user to enter values for the limits of integration a and b and the number of desired rectangles n.

```
PROGRAM:MIDSUM
Prompt A,B,N
0→T
(B-A)/N→H
For(I,1,N,1)
A+I*H→F
A+(I-1)*H→G
(F+G)/2→X
T+Y₁*H→T
End
Disp"APPROX AREA",T
```

EXAMPLE 4

Use the Midsum Program to approximate $\int_0^3 \frac{1}{1+x} \; dx$ using $n = 20$ rectangles.

SOLUTION

Here we enter $\frac{1}{1+x}$ in Y_1. When we enter $a = 0$, $b = 3$, and $n = 20$ after the program is executed, we get the output shown in Figure B.4.

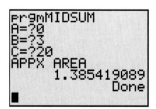

Figure B.4

Slope Field Program

This program sketches the slope field, or family of solution curves, for the differential equation $\frac{dy}{dx} = f(x,y)$. Here $\frac{dy}{dx}$ is entered in Y_1.

```
PROGRAM:SLPFILD
Clr Draw Fnoff
7*(Xmax-Xmin)/83→H
7*(Ymax-Ymin)/55→K
6.25/H^2→A
6.25/K^2→B
Xmin+0.5*H→X
Ymin+0.5*K→V
1→L
Lbl 1
V→Y
1→J
Lbl 2
Y₁→T
1/√(A+B*T^2)→M
T*M→N
Line(X-M,Y-N,X+M,Y+N)
Y+K→Y
IS>(J,8)
Goto 2
X+H→X
IS>(L,12)
Goto 1
```

EXAMPLE 5

Graph the slope field for the differential equation $\frac{dy}{dx} = \frac{x}{2y}$ in the viewing window $[-3, 3]$ by $[-3, 3]$.

SOLUTION

We enter the differential equation as shown in Figure B.5. When the program is executed, we get the output shown in Figure B.6.

Figure B.5

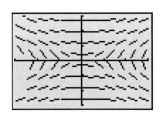

Figure B.6

Multiple Regression Program

This program models the Cobb–Douglas production function $Q(L, K) = kL^aK^b$, where the labor data (L) is entered in list L_1, the capital data in L_2, and the quantity data in L_3. Upon execution, the program prompts the user to enter the number of data points used.

```
PROGRAM:MULTIREG
Disp "NUMBER OF DATA POINTS"
Prompt N
log(L₁)→L₁
log(L₂)→L₂
log(L₃)→L₃
(sum(L₁))/N→P
(sum(L₂))/N→Q
(sum(L₃))/N→R
sum(L₁*L₃)-N*P*R→A
sum(L₂*L₃)-N*Q*R→B
sum((L₁)^2)-N*P^2→C
sum(L₁*L₂)-N*P*Q→D
sum((L₂)^2)-N*Q^2→E
(A*E-B*D)/(C*E-D^2)→F
(B*C-A*D)/(C*E-D^2)→G
R-F*P-G*Q→H
Disp"EXPONENT FOR L IS",F
Disp"EXPONENT FOR K IS",G
Disp"CONSTANT IS",10^H
```

EXAMPLE 6

The Brandy Corporation has 11 plants worldwide. In a recent year, the data for each plant gave the number of labor hours (in thousands), capital (total net assets, in millions) and total quantity produced, as follows:

LABOR	250	270	300	320	350	400	440	440	450	460	460
CAPITAL	30	34	44	50	70	76	84	86	104	110	116
QUANTITY	245	240	300	320	390	440	520	520	580	600	600

The plants all use the same technology, so a production function can be determined. Enter the data and execute the program to determine a Cobb–Douglas production function to model the data.

SOLUTION

We enter the labor, capital, and quantity data in lists L_1, L_2, and L_3, respectively. The output is displayed in Figure B.7.

So, based on the data, the Cobb–Douglas production function can be modeled by $Q(L, K) = 0.73L^{0.83}K^{0.35}$.

Figure B.7

APPENDIX C   SELECTED PROOFS

Quadratic Formula

The solution of the quadratic equation $ax^2 + bx + c = 0$ is given by $x = \dfrac{-b \pm \sqrt{b^2 - 4ac}}{2a}$.

Proof

We will solve the quadratic equation $ax^2 + bx + c = 0$ by completing the square.

$$ax^2 + bx = -c \qquad \text{Subtract } c$$

$$x^2 + \frac{b}{a}x = -\frac{c}{a} \qquad \text{Divide by } a$$

$$x^2 + \frac{b}{a}x + \left(\frac{b}{2a}\right)^2 = \left(\frac{b}{2a}\right)^2 - \frac{c}{a} \qquad \text{Complete the square}$$

$$x^2 + \frac{b}{a}x + \left(\frac{b}{2a}\right)^2 = \frac{b^2}{4a^2} - \frac{c}{a} \qquad \text{Simplify the right side}$$

$$\left(x + \frac{b}{2a}\right)^2 = \frac{b^2}{4a^2} - \frac{c}{a} \qquad \text{Write the left side as a squared binomial}$$

$$x + \frac{b}{2a} = \pm\sqrt{\frac{b^2}{4a^2} - \frac{c}{a}} \qquad \text{Take the square root of both sides}$$

$$x + \frac{b}{2a} = \pm\sqrt{\frac{b^2 - 4ac}{4a^2}} \qquad \text{Write the radicand as a single fraction}$$

$$x + \frac{b}{2a} = \pm\frac{\sqrt{b^2 - 4ac}}{2a} \qquad \text{Take the square root of the denominator on the right side}$$

$$x = -\frac{b}{2a} \pm \frac{\sqrt{b^2 - 4ac}}{2a} \qquad \text{Subtract } -\frac{b}{2a}$$

$$x = \frac{-b \pm \sqrt{b^2 - 4ac}}{2a} \qquad \text{Write the right side as a single fraction} \quad ◆$$

Properties of Logarithms

Here we will prove three of the properties of logarithms listed in Appendix A.

Property: $\log_b (m \cdot n) = \log_b m + \log_b n$

Proof

Let $x = \log_b m$ and $y = \log_b n$. Then, by the definition of logarithm, we may write $b^x = m$ and $b^y = n$. Multiplying these equations gives

$$b^x \cdot b^y = m \cdot n = b^{x+y}$$

Using the definition of logarithm, we may write the equation $m \cdot n = b^{x+y}$ as

$$\log_b (m \cdot n) = x + y$$

Since $x = \log_b m$ and $y = \log_b n$, we get

$$\log_b (m \cdot n) = \log_b m + \log_b n \qquad\qquad ◆$$

Property: $\log_b \left(\dfrac{m}{n} \right) = \log_b m - \log_b n$

Proof

Let $x = \log_b m$ and $y = \log_b n$. Then, by the definition of logarithm, we may write $b^x = m$ and $b^y = n$. Dividing these equations gives

$$\frac{b^x}{b^y} = \frac{m}{n} = b^{x-y}$$

Using the definition of logarithm, we may write the equation $\dfrac{m}{n} = b^{x-y}$ as

$$\log_b \left(\frac{m}{n} \right) = x - y$$

Since $x = \log_b m$ and $y = \log_b n$, we get

$$\log_b \left(\frac{m}{n} \right) = \log_b m - \log_b n$$

Property: $\log_b m^r = r \log_b m$

Proof

Let $x = \log_b m$. Then, by the definition of logarithm, we may write the equation in exponential form as $b^x = m$. If we raise both sides of this equation to the rth power, we get

$$(b^x)^r = m^r \qquad b^{xr} = m^r$$

By the definition of logarithm, we can write the equation $b^{xr} = m^r$ as

$$xr = \log_b m^r$$

Since $x = \log_b m$, we have $r \log_b m = \log_b m^r$.

Power Rule

If $f(x) = x^n$, where n is any real number, then $f'(x) = nx^{n-1}$. Equivalently, $\dfrac{d}{dx}[x^n] = nx^{n-1}$, and if $y = x^n$, then $\dfrac{dy}{dx} = nx^{n-1}$.

Proof

Here we will prove the Power Rule for n being a positive integer. By the definition of derivative, we have $\dfrac{d}{dx}[x^n] = \lim\limits_{h \to 0} \dfrac{(x+h)^n - x^n}{h}$. From the Binomial Theorem in algebra, we know that $(x+h)^n$ can be written as

$$(x+h)^n = x^n + nx^{n-1}h + \frac{n(n-1)}{2}x^{n-2}h^2 + \cdots + nxh^{n-1} + h^n$$

Substituting this expression into the definition of derivative, we get

$$\frac{d}{dx}[x^n] = \lim_{h \to 0} \frac{\left[x^n + nx^{n-1}h + \frac{n(n-1)}{2}x^{n-2}h^2 + \cdots + nxh^{n-1} + h^n \right] - x^n}{h}$$

$$= \lim_{h \to 0} \frac{nx^{n-1}h + \frac{n(n-1)}{2}x^{n-2}h^2 + \cdots + nxh^{n-1} + h^n}{h}$$

$$= \lim_{h \to 0} \left[nx^{n-1} + \frac{n(n-1)}{2}x^{n-2}h + \cdots \underbrace{\left(\begin{array}{c} \text{other terms} \\ \text{with } x \text{ and} \\ h \text{ factors} \end{array} \right)}_{} \cdots + nxh^{n-2} + h^{n-1} \right]$$

Since every term in the brackets except the first contains h as a factor, every term except the first approaches zero as $h \to 0$, and we get

$$\frac{d}{dx}[x^n] = nx^{n-1}$$

Constant Multiple Rule

If $f(x) = k \cdot g(x)$, where k is any real number, then $f'(x) = k \cdot g'(x)$, assuming that g is differentiable. Equivalently, $\frac{d}{dx}[k \cdot g(x)] = k \cdot g'(x)$.

Proof

Recall that the definition of the derivative of a function f is $\frac{d}{dx}[f(x)] = \lim_{h \to 0} \frac{f(x+h) - f(x)}{h}$. Using this definition for $f(x) = k \cdot g(x)$, we get

$$\frac{d}{dx}[k \cdot g(x)] = \lim_{h \to 0} \frac{k \cdot g(x+h) - k \cdot g(x)}{h} \qquad \text{Using the definition of derivative}$$

$$= \lim_{h \to 0} \frac{k[g(x+h) - g(x)]}{h} \qquad \text{Factor } k \text{ from the numerator}$$

$$= k \cdot \lim_{h \to 0} \frac{g(x+h) - g(x)}{h} \qquad \text{From Limit Theorem 3 in Section 2.1}$$

$$= k \cdot g'(x) \qquad \text{By definition, } \lim_{h \to 0} \frac{g(x+h) - g(x)}{h} = g'(x)$$

Sum Rule

If $h(x) = f(x) + g(x)$, where f and g are differentiable functions, then $h'(x) = f'(x) + g'(x)$. Equivalently, $\frac{d}{dx}[f(x) + g(x)] = \frac{d}{dx}[f(x)] + \frac{d}{dx}[g(x)] = f'(x) + g'(x)$.

Proof

We want to show that $\frac{d}{dx}[f(x) + g(x)] = f'(x) + g'(x)$, so, by the definition of derivative,

$$\frac{d}{dx}[f(x) + g(x)]$$

$$= \lim_{h \to 0} \frac{[f(x+h) + g(x+h)] - [f(x) + g(x)]}{h} \qquad \text{Using the definition of derivative}$$

$$= \lim_{h \to 0} \frac{f(x+h) + g(x+h) - f(x) - g(x)}{h} \qquad \text{Removing the parentheses}$$

$$= \lim_{h \to 0} \frac{f(x+h) - f(x) + g(x+h) - g(x)}{h} \qquad \text{Interchanging the two middle terms in the numerator}$$

$$= \lim_{h \to 0} \left[\frac{f(x+h) - f(x)}{h} + \frac{g(x+h) - g(x)}{h} \right] \qquad \text{Rewriting the difference quotient as two fractions}$$

$$= \lim_{h \to 0} \frac{f(x+h) - f(x)}{h} + \lim_{h \to 0} \frac{g(x+h) - g(x)}{h} \qquad \text{Limit Theorem 4 in Section 2.1}$$

$$= f'(x) + g'(x) \qquad \text{By the definition of derivative}$$

Difference Rule

If $h(x) = f(x) - g(x)$, where f and g are differentiable functions, then $h'(x) = f'(x) - g'(x)$. Equivalently, $\frac{d}{dx}[f(x) - g(x)] = \frac{d}{dx}[f(x)] - \frac{d}{dx}[g(x)] = f'(x) - g'(x)$.

Proof

We want to show that $\frac{d}{dx}[f(x) - g(x)] = f'(x) - g'(x)$, so, by the definition of derivative,

$$\frac{d}{dx}[f(x) - g(x)]$$

$$= \lim_{h \to 0} \frac{[f(x + h) - g(x + h)] - [f(x) - g(x)]}{h} \qquad \text{Using the definition of derivative}$$

$$= \lim_{h \to 0} \frac{f(x + h) - g(x + h) - f(x) + g(x)}{h} \qquad \text{Removing the parentheses}$$

$$= \lim_{h \to 0} \frac{f(x + h) - f(x) - g(x + h) + g(x)}{h} \qquad \begin{array}{l}\text{Interchanging the two middle}\\ \text{terms in the numerator}\end{array}$$

$$= \lim_{h \to 0} \frac{[f(x + h) - f(x)] - [g(x + h) - g(x)]}{h} \qquad \begin{array}{l}\text{Factor } -1 \text{ from the last two}\\ \text{terms in the numerator}\end{array}$$

$$= \lim_{h \to 0} \left[\frac{f(x + h) - f(x)}{h} - \frac{g(x + h) - g(x)}{h} \right] \qquad \begin{array}{l}\text{Rewriting the difference}\\ \text{quotient as two fractions}\end{array}$$

$$= \lim_{h \to 0} \frac{f(x + h) - f(x)}{h} - \lim_{h \to 0} \frac{g(x + h) - g(x)}{h} \qquad \text{Limit Theorem 4 in Section 2.1}$$

$$= f'(x) - g'(x) \qquad \text{By the definition of derivative}$$

Product Rule

If $h(x) = f(x) \cdot g(x)$, where f and g are differentiable functions, then

$$h'(x) = f'(x) \cdot g(x) + f(x) \cdot g'(x).$$

Equivalently,

$$\frac{d}{dx}[f(x) \cdot g(x)] = \frac{d}{dx}[f(x)] \cdot g(x) + f(x) \cdot \frac{d}{dx}[g(x)] = f'(x) \cdot g(x) + f(x) \cdot g'(x).$$

Proof

We want to show that $\frac{d}{dx}[f(x) \cdot g(x)] = f'(x) \cdot g(x) + f(x) \cdot g'(x)$, so, by the definition of derivative,

$$\frac{d}{dx}[f(x) \cdot g(x)] = \lim_{h \to 0} \frac{f(x + h) \cdot g(x + h) - f(x) \cdot g(x)}{h}$$

Now we subtract and add the term $f(x) \cdot g(x + h)$ in the numerator to get

$$= \lim_{h \to 0} \frac{f(x + h) \cdot g(x + h) - f(x) \cdot g(x + h) + f(x) \cdot g(x + h) - f(x) \cdot g(x)}{h}$$

Factoring $g(x + h)$ from the first two terms and $f(x)$ from the last two terms gives

$$= \lim_{h \to 0} \frac{[f(x + h) - f(x)] \cdot g(x + h) + f(x) \cdot [g(x + h) - g(x)]}{h}$$

Writing the difference quotient as two fractions yields

$$= \lim_{h \to 0} \left[\frac{f(x+h) - f(x)}{h} \cdot g(x+h) \right] + \lim_{h \to 0} \left[f(x) \cdot \frac{g(x+h) - g(x)}{h} \right]$$

$$= f'(x) \cdot g(x) + f(x) \cdot g'(x) \qquad \blacksquare$$

Quotient Rule

If $h(x) = \frac{f(x)}{g(x)}$, where f and g are differentiable functions, then

$$h'(x) = \frac{f'(x) \cdot g(x) - f(x) \cdot g'(x)}{[g(x)]^2} \quad \text{where } g(x) \neq 0.$$

Equivalently,

$$\frac{d}{dx} \left[\frac{f(x)}{g(x)} \right] = \frac{\frac{d}{dx}[f(x)] \cdot g(x) - f(x) \cdot \frac{d}{dx}[g(x)]}{[g(x)]^2}.$$

Proof

We want to show that $\frac{d}{dx} \left[\frac{f(x)}{g(x)} \right] = \frac{f'(x) \cdot g(x) - f(x) \cdot g'(x)}{[g(x)]^2}$, so, by the definition of derivative,

$$\frac{d}{dx} \left[\frac{f(x)}{g(x)} \right] = \lim_{h \to 0} \frac{\dfrac{f(x+h)}{g(x+h)} - \dfrac{f(x)}{g(x)}}{h}$$

Multiplying the numerator and denominator by $g(x+h) \cdot g(x)$ gives

$$= \lim_{h \to 0} \frac{f(x+h) \cdot g(x) - f(x) \cdot g(x+h)}{h \cdot g(x+h) \cdot g(x)}$$

Subtracting and adding the term $f(x) \cdot g(x)$ in the numerator yields

$$= \lim_{h \to 0} \frac{f(x+h) \cdot g(x) - f(x) \cdot g(x) + f(x) \cdot g(x) - f(x) \cdot g(x+h)}{h \cdot g(x+h) \cdot g(x)}$$

$$= \lim_{h \to 0} \frac{[f(x+h) - f(x)]g(x) - f(x)[-g(x) + g(x+h)]}{h \cdot g(x+h) \cdot g(x)}$$

If we write the numerator in the form of two difference quotients and take the limit, we get

$$= \lim_{h \to 0} \left[\frac{\dfrac{f(x+h) - f(x)}{h} \cdot g(x) - f(x) \cdot \dfrac{g(x+h) - g(x)}{h}}{g(x+h) \cdot g(x)} \right]$$

$$= \frac{f'(x) \cdot g(x) - f(x) \cdot g'(x)}{[g(x)]^2} \qquad \blacksquare$$

Differentiability Implies Continuity

If f is differentiable at $x = c$, then f is continuous at $x = c$.

Proof

Let f be differentiable at $x = c$. So we know that $f'(c)$ exists, and we need to show that $\lim_{x \to c} f(x) = f(c)$ or, equivalently, $\lim_{x \to c} [f(x) - f(c)] = 0$. Assuming $x \neq c$, we begin by multiplying and dividing the expression in the limit argument by $(x - c)$ to get

$$\lim_{x \to c}[f(x) - f(c)] = \lim_{x \to c}\left[(x - c) \cdot \frac{f(x) - f(c)}{x - c}\right]$$

By Limit Theorem 5 in Section 2.1, we have

$$\lim_{x \to c}[f(x) - f(c)] = \lim_{x \to c}(x - c) \cdot \lim_{x \to c}\frac{f(x) - f(c)}{x - c}$$

$$= 0 \cdot f'(c) = 0$$

Hence, $\lim_{x \to c} f(x) = f(c)$ and f is continuous at $x = c$.

Derivative of ln x

If $f(x) = \ln x$, then $f'(x) = \dfrac{1}{x}$.

Proof

Let $f(x) = \ln x$. Then, by the definition of derivative, we have

$$f'(x) = \lim_{h \to 0} \frac{\ln(x + h) - \ln x}{h} \qquad \text{Using the definition of derivative}$$

$$= \lim_{h \to 0} \frac{\ln\left(\dfrac{x + h}{x}\right)}{h} \qquad \text{Logarithm Property 4 in Appendix A}$$

$$= \lim_{h \to 0} \frac{\ln\left(1 + \dfrac{h}{x}\right)}{h} \qquad \text{Rewriting } \tfrac{x+h}{x} \text{ with two terms}$$

Now let $k = \dfrac{x}{h}$. Then, as $h \to 0$, $k \to \infty$ and $h = \dfrac{x}{k}$ and $\dfrac{1}{k} = \dfrac{h}{x}$. So, rewriting the limit in terms of k, we get

$$= \lim_{k \to \infty} \frac{\ln\left(1 + \dfrac{1}{k}\right)}{\dfrac{x}{k}} \qquad \text{Rewriting in terms of } k$$

$$= \lim_{k \to \infty} \frac{1}{x}\left[k \cdot \ln\left(1 + \dfrac{1}{k}\right)\right] \qquad \text{Rewriting without the complex fraction}$$

$$= \lim_{k \to \infty} \frac{1}{x}\left[\ln\left(1 + \dfrac{1}{k}\right)^k\right] \qquad \text{Logarithm Property 5 in Appendix A}$$

Since we know that $\lim\limits_{k \to \infty}\left(1 + \dfrac{1}{k}\right)^k = e$ (from the definition of the number e in Section 1.6), when we take the limit as $k \to \infty$, we get

$$= \frac{1}{x} \cdot \ln e = \frac{1}{x} \qquad \text{Logarithm Property 2 in Appendix A}$$

Fundamental Theorem of Calculus

If f is a continuous function defined on a closed interval $[a, b]$ and F is an anti-derivative of f, then $\int_a^b f(x)\, dx = F(b) - F(a)$.

Proof

Let f be a continuous function with $f(x) > 0$ for all x in $[a, b]$. Now we define an area function A, which represents the area under the graph of f from a to x. See Figure C.1.

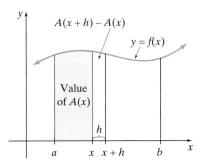

Figure C.1

We need to first show that $\left(\begin{smallmatrix}\text{Area under the graph of } f \\ \text{on } [a, b]\end{smallmatrix}\right) = A(b) - A(a)$. Since f is continuous on $[a, b]$, we know that $\int_a^x f(t)\, dt$ exists, and so we define $A(x) = \int_a^x f(t)\, dt$. We see that $A(a) = \int_a^a f(t)\, dt = 0$, and so $A(b) = A(b) - 0 = A(b) - A(a)$. So we have $\left(\begin{smallmatrix}\text{Area under the graph of } f \\ \text{on } [a, b]\end{smallmatrix}\right) = A(b) - A(a)$. Now we need to show that A is an antiderivative of f. By the definition of derivative, we have $A'(x) = \lim\limits_{h \to 0} \frac{A(x+h) - A(x)}{h}$. Analyzing this limit, we see the following:

- $A(x + h)$ is the area under the graph of f between a and $x + h$.
- $A(x)$ is the area under the graph of f between a and x.
- $A(x + h) - A(x)$ is the area between x and $x + h$.

For a small value of h, x and $x + h$ are values close to one another, and since f is continuous, $f(x + h)$ is close in value to $f(x)$. In other words,

$$A(x + h) - A(x) \approx \left(\begin{smallmatrix}\text{Area of rectangle with height} \\ f(x) \text{ and width } h\end{smallmatrix}\right) = f(x) \cdot h$$

So, for a small value of h, $\frac{A(x+h) - A(x)}{h} \approx \frac{f(x) \cdot h}{h} = f(x)$. Thus, as $h \to 0$ we see that $A'(x) = \lim\limits_{h \to 0} \frac{A(x+h) - A(x)}{h} = f(x)$. Hence, A is an antiderivative of f. ∎

Derivation of the Least-squares Formulas

Given a collection of data points $(x_1, y_1), (x_2, y_2), (x_3, y_3), \dots, (x_n, y_n)$, the linear regression model for the data has the form $y = mx + b$, where

$$m = \frac{n \sum xy - \sum x \cdot \sum y}{n \sum x^2 - \left(\sum x\right)^2} \quad \text{and} \quad b = \frac{\sum y \cdot \sum x^2 - \sum x \cdot \sum xy}{n \sum x - \left(\sum x\right)^2}$$

Proof

Let $(x_1, y_1), (x_2, y_2), (x_3, y_3), \dots, (x_n, y_n)$ be a set of data points. To determine a regression line, we need to find values m and b so that the sum of the squares of the residuals $d = y - (mx + b)$ is a minimum. If we write m and b in the form of a function of two independent variables, we see that we need to minimize

$$\begin{aligned} f(m, b) = \sum d^2 &= \sum [y - (mx + b)]^2 \\ &= (y_1 - mx_1 - b)^2 + (y_2 - mx_2 - b)^2 (y_3 - mx_3 - b)^2 + \cdots \\ &\quad + (y_n - mx_n - b)^2 \end{aligned}$$

The partial derivative of f with respect to m is

$$\begin{aligned} \frac{\partial f}{\partial m} &= 2(y_1 - mx_1 - b)(-x_1) + 2(y_2 - mx_2 - b)(-x_2) \\ &\quad + 2(y_3 - mx_3 - b)(-x_1) + \cdots + 2(y_n - mx_n - b)(-x_n) \\ &= 2[(-x_1y_1 - x_2y_2 - x_3y_3 - \cdots - x_ny_n) + m(x_1^2 + x_2^2 + x_3^2 + \cdots + x_n^2) \\ &\quad + b(x_1 + x_2 + x_3 + \cdots + x_n] \\ &= 2\left[-\sum xy + m\sum x^2 + b\sum x\right] \end{aligned}$$

The partial derivative of f with respect to b is

$$\begin{aligned} \frac{\partial f}{\partial b} &= 2(y_1 - mx_1 - b)(-1) + 2(y_2 - mx_2 - b)(-1) \\ &\quad + 2(y_3 - mx_3 - b)(-1) + \cdots + 2(y_n - mx_n - b)(-1) \\ &= 2[(-y_1 - y_2 - y_3 - \cdots - y_n) + m(x_1 + x_2 + x_3 + \cdots + x_n) \\ &\quad + b(1 + 1 + 1 + \cdots + 1) \\ &= 2\left[-\sum y + m\sum x + bn\right] \end{aligned}$$

Now, if we set $\frac{\partial f}{\partial m} = 0$ and $\frac{\partial f}{\partial b} = 0$ and solve the system of equations, we get

$$\begin{cases} -\sum xy + m\sum x^2 + b\sum x = 0 \\ -\sum y + m\sum x + bn = 0 \end{cases} \quad \text{or} \quad \begin{aligned} m\sum x^2 + b\sum x &= \sum xy \quad &\text{(I)} \\ m\sum x + bn &= \sum y \quad &\text{(II)} \end{aligned}$$

To solve for m, we multiply equation (I) by n and equation (II) by $-\sum x$ and then add the equations to get

$$nm\sum x^2 - m\left(\sum x\right)^2 = n\sum xy - \sum x\sum y$$

$$m\left[n\sum x^2 - \left(\sum x\right)^2\right] = n\sum xy - \sum x\sum y$$

$$m = \frac{n\sum xy - \sum x\sum y}{n\sum x^2 - \left(\sum x\right)^2}$$

To solve for b, we multiply the equation (I) by $-\sum x$ and equation (II) by $\sum x^2$ and then add the equations to get

$$bn\sum x - b\left(\sum x\right)^2 = \sum x^2\sum y - \sum x\sum xy$$

$$b\left[n\sum x - \left(\sum x\right)^2\right] = \sum x^2\sum y - \sum x\sum xy$$

$$b = \frac{\sum y \cdot \sum x^2 - \sum x \cdot \sum xy}{n\sum x - \left(\sum x\right)^2}$$

EXAMPLE 1

Use the least-squares formulas to determine a linear regression function of the form $y = mx + b$ for the following sample of data. Use the regression capabilities of your calculator to verify your answer.

x	19	23	21	15	16	18
y	55	7	20	123	88	76

SOLUTION

When we compute the necessary sums, we get

$$n = 6$$
$$\sum x = 19 + 23 + \cdots + 18 = 112$$
$$\sum y = 55 + 7 + \cdots + 76 = 369$$
$$\sum x^2 = 19^2 + 23^2 + \cdots + 18^2 = 2136$$
$$\sum xy = 19 \cdot 55 + 23 \cdot 7 + \cdots + 18 \cdot 76 = 6247$$

Evaluating the derived formulas for m and b gives

$$m = \frac{6(6247) - (112)(369)}{6(2136) - (112)^2} \approx -14.14,$$

$$b = \frac{(369)(2136) - (112)(6247)}{6(112) - (112)^2} \approx 325.44$$

So the linear regression model for the data is $y = -14.14x + 325.44$. Entering the data in the calculator and using the LINREG command gives the output shown in Figure C.2.

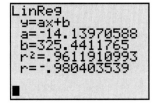

Figure C.2

APPENDIX D   PHOTO AND ILLUSTRATION CREDITS

Answers

Section 1.1 Exercises

1. The ordered pairs are $(-2, 3)$, $(1, 5)$, and $(4, 4)$.
The scatterplot of the points is:

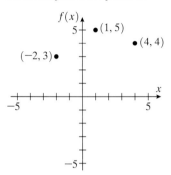

3. The ordered pairs are $(-4, -5)$, $(-3.5, -6)$, $(-2, -8)$, and $(2, -11)$. The scatterplot is:

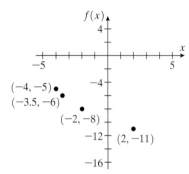

5.

x	y
2	6
4	5
6	4
8	2
10	0

The ordered pairs are $(2, 6)$, $(4, 5)$, $(6, 4)$, $(8, 2)$, and $(10, 0)$.

7.

x	y
10	60
20	50
30	40
40	20
50	10

The ordered pairs are $(10, 60)$, $(20, 50)$, $(30, 40)$, $(40, 20)$, and $(50, 10)$.

9.

x	y
-3	-30
-2	-10
-1	0
0	10
2	10

The ordered pairs are $(-3, -30)$, $(-2, -10)$, $(-1, 0)$, $(0, 10)$, and $(2, 10)$.

11.

Years since 1980	Public Expenditures (in billion of dollars)
0	41
3	44
4	46
5	49
8	51

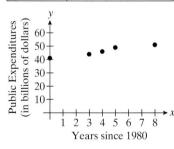

13.

Time	Speed (in miles per hour)
0	0
1	40
2	120
3	160
4	180

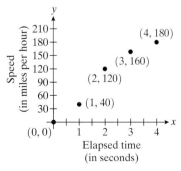

15. Function   **17.** Not a function   **19.** Function   **21.** Not a function   **23.** Function   **25.** $(2, 3)$ **27.** $(1, 1)$ **29.** $(3, 9)$
31. $(0, 0)$ **33.** $\left(\frac{1}{2}, \frac{1}{4}\right)$ **35.** $(-0.25, 0.0625)$ **37. (a)** t **(b)** $y = f(t)$ **(c)** 10   **(d)** 20   **(e)** $[0, \infty)$ **(f)** $[0, \infty)$

39.

Interval Notation	Inequality Notation	Number Line
$[-\frac{1}{2}, 5)$	$-\frac{1}{2} \le x < 5$	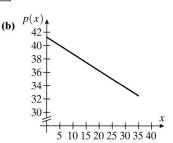
$[-2, \infty)$	$x \ge -2$	
$(-\infty, -3)$	$x < -3$	
$(-1, 10)$	$-1 < x < 10$	
$(-\infty, 3) \cup (3, \infty)$	$x < 3$ or $x > 3$	

41. $(-\infty, \infty)$ **43.** $(-\infty, \infty)$
45. $(-\infty, 5) \cup (5, \infty)$ **47.** $(-\infty, 6]$
49. $(-\infty, \infty)$ **51.** $(-\infty, 0) \cup (0, 2) \cup (2, \infty)$
53. The domain is $(-\infty, \infty)$.; The range is $[\frac{1}{2}, \infty)$.
55. The domain is $(-\infty, \infty)$.; The range is $(-0.43, 0.1)$.

57. (a)

x	0	5	10	15	20	25	30	35
$p(x)$	41.25	40	38.75	37.5	36.25	35	33.75	32.50

(b)

(c) $[32.5, 41.25]$

(d) If 22 units of Teddy Bear Designer Lingerie are sold per day, the price will be $35.75 per unit.

59. $36.75   **61.** The domain is $(-\infty, \infty)$.; The range is $(-\infty, \infty)$.   **63.** The domain is $(-5, \infty)$.; The range is $(-\infty, 7]$.
65. The domain is $(-50, 100]$.; The range is $[0, 10)$.

Section 1.2 Exercises

1. -5 **3.** $-\frac{5}{6}$ **5.** 1.1 **7.** 1 **9.** $-\frac{5}{3}$ **11.** -4 **13.** $\frac{17}{4}$
15. $y - 6 = -3(x - 2)$ **17.** $y - 5 = 4(x - 3)$ **19.** $x = 0$

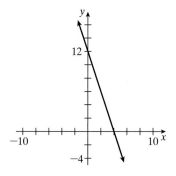 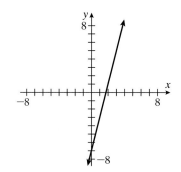 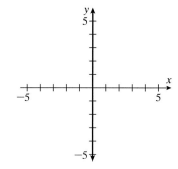

21. $x = -2$

23. $y = 3(x - 8) + 4$ **25.** $y = 5(x - 0) - 13$ **27.** $y = 0.3(x + 4)$
29. $y = -\frac{1}{4}(x + 1) + 2$ **31.** $y = \frac{1}{8}(x + 12)$ **33.** $y = -1.175(x - 0) + 6$
35. x-intercept: $(1.25, 0)$; y-intercept: $(0, 5)$ **37.** x-intercept: $(400, 0)$; y-intercept: $(0, 120)$
39. x-intercept: $(-44.9, 0)$; y-intercept: $(0, 8.98)$ **41. (a)** $f(x) = -1250x + 25,000$
(b) Twelve years   **43. (a)** $f(x) = -6000x + 90,000$ **(b)** 7.5 years
45. The slope of the line is -4.5; decreasing   **47.** Slope of the line is $+4$; increasing
49. The slope of the line is $+4$; increasing   **51.** Slope $= \frac{1}{3}$; increasing
53. Slope $= 14$; increasing   **55.** $C(x) = 147.75x + 2700$ **57.** $C(x) = 575x + 1250$
59. $C(x) = 80x + 650$

61. (a) The fixed cost is $550 and the variable costs are $1.25 per bottle.   **(b)** $C(70) = 637.5$; It costs $637.50 to produce 70 bottles.
(c) 115 bottles   **(d)** $1.25 per bottle   **(e)** $1.25

63.

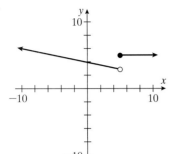

65.

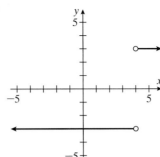

67.

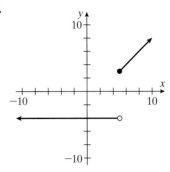

69.

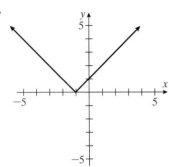

$$f(x) = |x + 1|$$
$$= \begin{cases} -x - 1 & x < -1 \\ x + 1 & x \geq -1 \end{cases}$$

71.

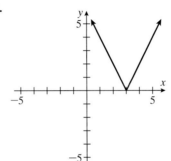

$$g(x) = |6 - 2x|$$
$$= \begin{cases} 6 - 2x & x < 3 \\ -6 + 2x & x \geq 3 \end{cases}$$

73.

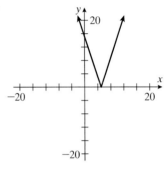

$$g(x) = |3x - 15|$$
$$= \begin{cases} -3x + 15 & x < 5 \\ 3x - 15 & x \geq 5 \end{cases}$$

75. (a) The fixed costs are \$20. The variable costs are \$0.15 per mile.   **(b)** $C(x) = 0.15x + 20$ **(c)** $(0, \infty)$

(d) $C(120) = 38$; It cost \$38 to rent a truck and drive 120 miles.   **(e)** \$0.15; This is called the marginal cost.

77. (a) Since the fixed cost is equal to $C(0)$, then the fixed cost is \$50.   **(b)** It costs \$15 per hour for each additional hour of labor.
This is called the marginal cost.   **(c)** $C(x) = 15x + 50$ **(d)** \$117.50   **79. (a)** Spending is increasing at a rate of \$0.19 billion per year.

(b) \$2.24 billion   **(c)** \$3.19 billion   **(d)** \$0.95 billion more was spent in 1990 than 1985.

Section 1.3 Exercises

1. (a) $(-3, -4)$ **(b)** Increasing on $(-3, \infty)$ and decreasing on $(-\infty, -3)$.   **3. (a)** $(3, 15)$

(b) Increasing on $(-\infty, 3)$ and decreasing on $(3, \infty)$.   **5. (a)** $(-0.6, -4.8)$ **(b)** Increasing on $(-0.6, \infty)$ and decreasing on $(-\infty, -0.6)$.

7. (a) $\left(-\frac{25}{14}, -\frac{209}{280}\right) \approx (-1.79, -0.75)$ **(b)** Increasing on $\left(-\frac{25}{14}, \infty\right)$ and decreasing on $\left(-\infty, -\frac{25}{14}\right)$.

9. (a) $(-1, 1.4)$ **(b)** Increasing on $(-\infty, -1)$ and decreasing on $(-1, \infty)$.   **11.** $x = 4$ or $x = -4$ **13.** $x = 0$ or $x = -2$

15. $x = 2$ or $x = 3$ **17.** $x = \frac{10}{3}$ or $x = -2.5$ **19.** $x = \dfrac{1 + \sqrt{5}}{2}$ and $x = \dfrac{1 - \sqrt{5}}{2}$ **21.** $x = \dfrac{7 + \sqrt{5}}{22}$ and $x = \dfrac{7 - \sqrt{5}}{22}$

23. No real roots   **25.** $x = -4$ and $x = 6$ **27.** $x = 1.5$ **29.** No real zeros   **31. (a)** -5 **(b)** -5 **(c)** -5

33. (a) 5   **(b)** 3   **(c)** 7   **35. (a)** 1   **(b)** 0   **(c)** 2   **37.** 101; This means that during the five-year period, the number of parking

tickets issued increased at an average rate of 101 $\dfrac{\text{tickets}}{\text{year}}$.

39. (a) $s(3) = 96$ feet; This is the height of the rock 3 seconds after it was thrown.

(b) 16 feet per second; The rock traveled an average speed of 16 feet per second over the time interval $[1, 3]$.

41. (a) $g(3) = 2033$; This means that after 3 hours, there were 2033 bacteria in the colony.

(b) 17 bacteria per hour; The bacteria were growing at an average rate of 17 bacteria per hour over the time interval $[3, 6]$.

43. (a) $f(2) = 374.27$; This means that in 1988, there were 374.27 aggravated assaults per 100,000 people.

(b) The greatest assault rate occurred sometime during the middle of 1992.   **(c)** 0.31 aggravated assaults per 100,000 people per year

45. (a) $f(6) = 3326.73$; This means that in 1980, 3326.73 calories were consumed each day per person in the United States.

(b)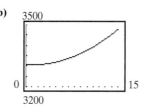

(c) 1985   **(d)** 8.75 calories per day per year; This is the average calories-per-day increase per year between 1975 and 1985.

Section 1.4 Exercises

1. (a) $(f + g)(x) = 10x + 2$; The domain is $(-\infty, \infty)$.   **(b)** $(f - g)(x) = 2x + 4$; The domain is $(-\infty, \infty)$.

(c) $(f \cdot g)(x) = 24x^2 + 6x - 3$; The domain is $(-\infty, \infty)$.   **(d)** $\left(\dfrac{f}{g}\right)(x) = \dfrac{6x + 3}{4x - 1}$; The domain is $(-\infty, 0.25) \cup (0.25, \infty)$.

3. (a) $(f + g)(x) = x^2 + \sqrt{x + 5} + 5$; The domain is $[-5, \infty)$.   **(b)** $(f - g)(x) = -x^2 + \sqrt{x + 5} - 5$; The domain is $[-5, \infty)$.

(c) $(f \cdot g)(x) = x^2\sqrt{x + 5} + 5\sqrt{x + 5}$; The domain is $[-5, \infty)$.   **(d)** $\left(\dfrac{f}{g}\right)(x) = \dfrac{\sqrt{x + 5}}{x^2 + 5}$; The domain is $[-5, \infty)$.

5. 1   **7.** 336   **9.** -11 **11.** $\frac{15}{13}$ **13. (a)** -245; The Handi-Neighbor Hardware Store makes $245 less than the U-Do-It Store when they both sell 20 hammers.   **(b)** 635; Both stores will make $635 combined if each sells 20 hammers.

15. (a) 57.53%

(b) $\left(\dfrac{x + 35}{4x + 45}\right) \cdot 100\%$

17. (a)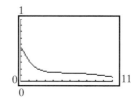

(b) $[1, 3]$

19. (a)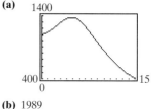

(b) 1989

21. $R(x) = 87.1x$

23. $R(x) = -0.19x^2$

25. $R(x) = \dfrac{-x^2}{2000} + 3x$

27. $R(x) = -0.12x^2 + 30x$

29. (a)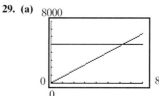

(b) $(625, 5000)$

(c) $5000

31. (a)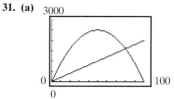

(b) $(0.05, 5.00)$ and $(79.95, 1603.00)$

(c) $5.00 or $1603.00

33. $P(x) = -37x^2 + 1052x - 4300$; $(14.22, 3177.73)$; maximum profit of $\approx$ $3177.73 at $x \approx 14.22$

35. $P(x) = -2.1x^2 + 420x - 9500$; $(100, 11{,}500)$; maximum profit of $11,500 at $x = 100$ **37. (a)** $P(x) = -x^2 + 8.8x - 10.15$

(b) $x \approx 1.37$, $x \approx 7.43$ **(c)** $(4.4, 9.21)$; Maximum profit of $9.21 is attained when $x = 4.4$.   **39. (a)** $P(x) = -10x^2 + 540x - 6650$

(b) $x = 19$, $x = 35$ **(c)** $(27, 640)$; Maximum profit of $640 is attained when $x = 27$.   **41. (a)** $P(x) = -x^2 + 25x - 60$

(b) $x \approx 2.69$, $x \approx 22.31$ **(c)** $(12.5, 96.25)$; Maximum profit of $96.25 is attained when $x = 12.5$.   **43.** Polynomial function; linear

45. Not a polynomial function   **47.** Polynomial function; cubic   **49.** As $x \to -\infty$, $f(x) \to \infty$; as $x \to \infty$, $f(x) \to -\infty$.

51. As $x \to -\infty$, $g(x) \to -\infty$; as $x \to \infty$, $g(x) \to \infty$.   **53.** As $x \to -\infty$, $g(x) \to \infty$; as $x \to \infty$, $g(x) \to \infty$.

55. As $x \to -\infty$, $f(x) \to -\infty$; as $x \to \infty$, $f(x) \to \infty$.   **57.** As $x \to -\infty$, $g(x) \to \infty$; as $x \to \infty$, $g(x) \to \infty$.   **59. (a)** 44; The ball is 44 feet above the ground after 2 seconds.   **(b)** -10; The average rate of change of height in the first 2.5 seconds is -10 feet per second.

61. (a) 2664.38; The sale of 250 Boomer headphones generates $2664.38 in revenue.   **(b)** 10.4521; During sales in the range of 100 to 300 Boomer headphones, the revenue increased by an average of $10.45 per Boomer headphone.   **63. (a)** 19.95; In 1982, 19.95 percent of people taking the SAT intended to study business.   **(b)** 0.93; Over the period 1975 to 1985, the percent of people taking the SAT who intended to study business increased an average of 0.93% per year.   **(c)** -0.87; Over the period 1985 to 1995, the percent of people taking the SAT who intended to study business decreased an average of 0.87% per year.   **(d)** 1975 to 1985 was a period of increasing interest, on average, in business study; 1985 to 1995 was a period of decreasing interest, on average, in business study.

65. (a) As $x \to -\infty$, $f(x) \to \infty$; as $x \to \infty$, $f(x) \to -\infty$.   **(b)** No peaks or valleys   **(c)** f decreases on $(-\infty, \infty)$.

67. (a) As $x \to -\infty$, $y \to -\infty$; as $x \to \infty$, $y \to -\infty$.   **(b)** "Peak" at $(-5, 46)$ **(c)** Decreasing on $(-5, \infty)$ and increasing on $(-\infty, -5)$.   **69. (a)** As $x \to -\infty$, $g(x) \to \infty$ and as $x \to \infty$, $g(x) \to -\infty$.   **(b)** "Peak" at $(0.26, -1.47)$ and a "valley" at $(-2.59, -24.57)$.   **(c)** Decreasing on $(-\infty, -2.59) \cup (0.26, \infty)$ and increasing on $(-2.59, 0.26)$.   **71. (a)** As $x \to -\infty$, $f(x) \to -\infty$ and as $x \to \infty$, $f(x) \to -\infty$ **(b)** "Peak" at $(1.26, 19.56)$ **(c)** Increasing on $(-\infty, 1.26)$ and decreasing on $(1.26, \infty)$.

Section 1.5 Exercises

1. (a) $(-\infty, 0) \cup (0, \infty)$ **(b)** Hole at $x = 0$ **3. (a)** $(-\infty, 2) \cup (2, \infty)$ **(b)** Vertical asymptote at $x = 2$

5. (a) $(-\infty, -2) \cup (-2, \infty)$ **(b)** Hole at $x = -2$ **7. (a)** $(-\infty, -3) \cup (-3, -1) \cup (-1, \infty)$ **(b)** Hole at $x = -1$ and a vertical asymptote at $x = -3$ **9. (a)** $(-\infty, -1.5) \cup (-1.5, 2) \cup (2, \infty)$ **(b)** Hole at $x = -1.5$ and a vertical asymptote at $x = 2$

11. No horizontal asymptote   **13.** $y = 3$ **15.** $y = 3$ **17.** $y = 1$ **19.** $y = 2$ **21.** y-intercept is $(0, 0)$; x-intercept is $(0, 0)$

23. y-intercept is $(0, -\frac{3}{4})$; x-intercept is $(0.5, 0)$ **25.** y-intercept is $(0, -0.5)$; x-intercept is $(-2, 0)$

27. y-intercept is $(0, -\frac{2}{9})$; x-intercept is $(2, 0)$ **29.** y-intercept is $(0, \frac{4}{3})$; x-intercepts are $(-2, 0)$ or $(2, 0)$

31. y-intercept is $(0, 5)$; no x-intercepts   **33.** y-intercept is $(0, -\frac{2}{3})$; x-intercept is $\approx (0.632, 0)$

35. y-intercept is $(0, 0.5)$; x-intercept is $\approx (-0.657, 0)$

37.

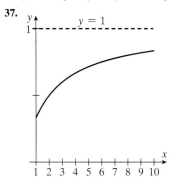

39.

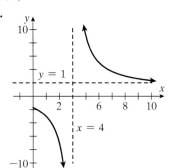

41.

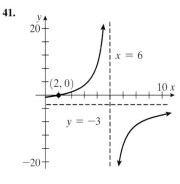

43. (a) $f(85) = 170$; It will cost \$170 million to remove 85% of the pollutants.

(b)

x	5	50	70	90	95
$f(x)$	≈ 1.6	30	70	270	570

(c) \$29,400,000,000   **(d)** $f(100)$ is undefined.

45. (a) $f(10) = 18.2$; After 10 days, the person had about 18 out of 20 items remembered.

(b) $f(100) = 19.82$; After 100 days, the person had about 20 out of 20 items remembered.

47. (a) $S(10) \approx 59.44$; If the company spends \$10,000 on advertising, the income from the sales will be about \$5,944,000.

(b)

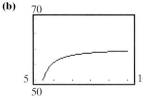

(c) The function has a horizontal asymptote of $y = \frac{120}{2}$ or $y = 60$. This means that the most income the company can generate by advertising is \$6,000,000.

49. $f(x) = (2x + 3)^{1/2}$; $[-1.5, \infty)$ **51.** $f(x) = (5x - 8)^{1/3}$; $(-\infty, \infty)$

53. $g(x) = (6x + 1)^{5/4}$; $[-\frac{1}{6}, \infty)$ **55.** $f(x) = 3(7x - 2)^{-1/2}$; $(\frac{2}{7}, \infty)$

57. $f(x) = \sqrt{8x - 9}$; $[1.125, \infty)$ **59.** $f(x) = \sqrt[3]{7x + 9}$; $(-\infty, \infty)$

61. $f(x) = \sqrt{(4x - 5)^3}$; $[1.25, \infty)$ **63.** $g(x) = \dfrac{1}{\sqrt[3]{7x - 2}}$; $(-\infty, \frac{2}{7}) \cup (\frac{2}{7}, \infty)$ **65. (a)** $g(70) \approx 11.71$; If the tower is 70 feet tall, an observer can see about 11.71 miles into the forest.   **(b)** ≈ 429 feet high   **67. (a)** $g(300) \approx 188$; There are about 188 plant species in 300 square miles of rain forest.   **(b)** 0.0599; From 1728 square feet to 2197 square feet, the average increase in plant species is about 0.0599 species per square foot.   **69. (a)** $f(3) \approx 16,787$; In 1992, there were about 16,787 grocery stores in the United States.

(b) Between 1992 and 1995, the number of grocery stores was decreasing at an average rate of 107.55 stores per year.

71. (a) $f(11) \approx 25.04$; In 1990, there were about 25,040,000 retired workers receiving Social Security benefits.

(b)

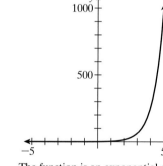

The number of retired workers receiving Social Security benefits is increasing.

(c) 0.623; Between 1981 and 1986, the number of beneficiaries was growing at an average rate of 623,000 people per year.

Section 1.6 Exercises

1. (a)

x	-5	-4	-3	-2	-1	0
$\left(\frac{1}{3}\right)^x$	243	81	27	9	3	1

x	1	2	3	4	5
$\left(\frac{1}{3}\right)^x$	$\frac{1}{3}$	$\frac{1}{9}$	$\frac{1}{27}$	$\frac{1}{81}$	$\frac{1}{243}$

(b)

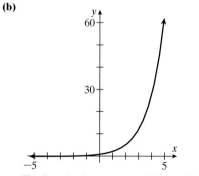

The function is an exponential decay function.

3. (a)

x	-5	-4	-3	-2	-1	0
4^x	$\frac{1}{1024}$	$\frac{1}{256}$	$\frac{1}{64}$	$\frac{1}{16}$	$\frac{1}{4}$	1

x	1	2	3	4	5
4^x	4	16	64	256	1024

(b)

The function is an exponential growth function.

5. (a) All values are rounded to the nearest hundredth or to two significant digits.

x	-5	-4	-3	-2	-1	0
$(2.3)^x$	0.016	0.036	0.082	0.19	0.43	1

x	1	2	3	4	5
$(2.3)^x$	2.3	5.29	12.17	27.98	64.36

(b)

The function is an exponential growth function.

7. (a) All values are rounded to the nearest hundredth.

x	-5	-4	-3	-2	-1	0
$(0.7)^x$	5.95	4.16	2.92	2.04	1.43	1

x	1	2	3	4	5
$(0.7)^x$	0.7	0.49	0.34	0.24	0.17

(b)

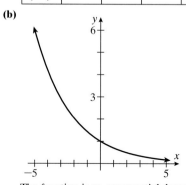

The function is an exponential decay function.

9. (a) All values are rounded to the nearest hundredth
or to two significant digits.

x	-5	-4	-3	-2
e^{2x}	0.000045	0.00034	0.0025	0.018

x	-1	0	1	2
e^{2x}	0.14	1	7.39	54.60

x	3	4	5
e^{2x}	403.43	2980.96	22,026.47

(b)

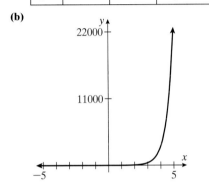

The function is an exponential growth function.

11. (a) All values are rounded to the nearest hundredth.

x	-5	-4	-3	-2	-1	0
$e^{0.3x}$	0.22	0.30	0.41	0.55	0.74	1

x	1	2	3	4	5
$e^{0.3x}$	1.35	1.82	2.46	3.32	4.48

(b)

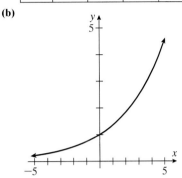

The function is an exponential growth function.

13. (a) All values are rounded to the nearest hundredth
or two significant digits.

x	-5	-4	-3	-2
$e^{-1.6x}$	2980.96	601.85	121.51	24.53

x	-1	0	1	2
$e^{-1.6x}$	4.95	1	0.20	0.041

x	3	4	5
$e^{-1.6x}$	0.0082	0.0017	0.00034

(b)

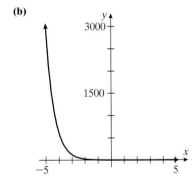

The function is an exponential decay function.

15. (a) $f(7) \approx 41.94$; On the seventh day, the wound is about 42 cm^2.

(b) On the third day the wound is approximately half its original size.

(c) $m_{sec} \approx -18.06$; Between the first and eighth day, the wound was shrinking at an average rate of 18.06 square centimeters per day.

17. (a) $g(5) = 4237$; By the fifth day, 4237 students had heard the rumor.

(b) $m_{sec} = 779.15$; Between the first and fifth day, the rumor was spreading at an average rate of 779.15 students per day.

19. (a) $f(2) \approx 17.04$; In 1992, about 17.04% of the waste
generated in the United States was yard waste.

(b)

20

1 ⌐ . . . . . ⌐ 6
10

Exponential decay

(c) $m_{sec} \approx -0.90$; Between 1991 and 1996, the percentage
of waste that was due to yard waste was shrinking at an
average rate of 0.90% per year.

21. $120   **23.** $3573.05   **25.** $2564.44   **27. (a)** $3173.75   **(b)** $3202.06   **(c)** $3216.87

29. (a) $37,534.15   **(b)** $38,464.55   **(c)** $38,533.00   **31.** $A(5) \approx 7227.37$; The interest made will be $1727.37.

33. $A(5) = 14,353.51$; The interest made will be $4353.51.

35. (a) $A(t) = 4000(1 + \frac{0.0575}{12})^{12t}$

(b)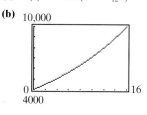

(c) 10,000

Twelve years

37. (a) $f(5) \approx 3789831$; After five minutes, there are approximately 3,789,831 bacteria in the colony.

(b) $m_{sec} \approx 1,076,901.90$; In the first ten minutes, the colony was growing at an average rate of about 1,076,901.90 bacteria per minute.

39. (a) $S(5) \approx 16,321$; After five years on the market, approximately 16,321 Flashfast Cigarette Lighters had been sold.

(b) $m_{sec} \approx 1264.24$; During the first five years, sales were increasing at an average rate of about 1264.24 lighters per year.

41. (a) $f(12) \approx 824.50$; In 1991, there were approximately 824.5 thousand W.W.II veterans receiving compensation for service-connected disabilities.

(c) $m_{sec} \approx -38.31$; Between 1980 and 1990, the number W.W.II veterans receiving compensation was decreasing at an average rate of about 38,310 veterans per year.

(b)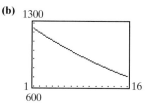

43. (a) The function is an exponential growth model.

(b) $f(21) \approx 218.96$; In 1970, the Federal Government outlays were about $218.96 billion.

(d) $m_{sec} \approx 66.76$; Between 1980 and 1990, the Federal Government outlays were increasing at an average rate of about $66.76 billion per year.

(c)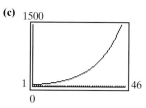

45. $A(7) \approx 1469.61$; After seven years, there will be $1469.61 in the account.   **47.** $A(7) \approx 30,226.90$; After seven years, there will be $30,226.90 in the account.   **49.** $A(7) \approx 14,103.76$; After seven years, there will be $14,103.76 in the account.

51. (a) $f(3) \approx 0.2326$; In 2000, about 23.26% of the households in the United States will own a DVD player.   **(b)** $m_{sec} \approx 0.0547$; Between the years 2000 and 2002, the number of households that own a DVD player will be increasing by an average rate of 5.47% per year.

53. (a) $f(0) \approx 5$; When the influenza initially first broke out, 5 students were infected.   **(b)** $m_{sec} \approx 13$; Between the 3rd and 7th day after the outbreak, the number of infected students was increasing by an average rate of about 13 students per day.

55. (a) The initial size of the seed colony was 8 lizards.   **(b)** After 8 years, there are 723 lizards in the colony.

(c) $m_{sec} \approx 89$; In the first eight years, the population grew at an average rate of 89 lizards per year.

Section 1.7 Exercises

1. $f(g(3)) = 54; f(g(-2)) = -36$ **3.** $f(g(0)) = 2; g(f(0)) = 1$ **5.** $g(f(x)) = 24x - 6; f(g(x)) = 24x + 5$

7. $g(f(x)) = 25x^2 - 15x + 4; f(g(x)) = 5x^2 + 15x + 17$ **9.** $g(f(x)) = \frac{1}{x^3}; f(g(x)) = \frac{1}{x^3}$

11. $g(f(x)) = x + 3; f(g(x)) = \sqrt{x^2 + 3}$ **13.** Let $g(x) = x + 3$ and $f(x) = x^3$.   **15.** Let $g(x) = \frac{1}{x + 3}$ and $f(x) = x^2$.

17. Let $g(x) = x - 2$ and $f(x) = \sqrt[4]{x}$.   **19.** Let $g(x) = \sqrt{x}$ and $f(x) = 2 - 3x$.   **21. (a)** $r(t) = 1.3t$ **(b)** $V(r(t)) = \frac{8.788}{3}\pi t^3$; $V(t)$ is the volume of the sphere as a function of time.   **(c)** $V(6) \approx 1987.80$; After 6 seconds, the volume of the sphere is 1987.80 in^3.

23. $f(g(x)) = f(\frac{1}{8}x) = 8(\frac{1}{8}x) = x$ **25.** $f(g(x)) = f\left(\frac{x + 10}{7}\right) = 7\left(\frac{x + 10}{7}\right) - 10 = x + 10 - 10 = x$

27. $f(g(x)) = f(\sqrt[3]{x - 6}) = (\sqrt[3]{x - 6})^3 + 6 = (x - 6) + 6 = x$ **29.** $f(g(x)) = f(x^2 + 1) = \sqrt{(x^2 + 1) - 1} = \sqrt{x^2} = |x| = x$

since $x \geq 0$ **31.** $5 = \log_2 (32)$ **33.** $-3 = \log_2 \left(\frac{1}{8}\right)$ **35.** $2 = \log_{1/2} \left(\frac{1}{4}\right)$ **37.** $\frac{3}{4} = \log_{81} (27)$ **39.** $5 = \log (100{,}000)$
41. $1 = \ln (e)$ **43.** $\log_2 (3) - \log_2 (5)$ **45.** $\log (8) + \log (20)$ **47.** $\frac{1}{2} \ln (26)$ **49.** $\log_3 (4) - 1.5$ **51.** 1.63 **53.** 2.67 **55.** 3.43
57. -2.95 **59. (a)** $f(4) \approx 8.89$; In 1996, 8.89% of the unemployed workers in the United States labor force were Hispanics.
(b) $x \approx 2$; $x \approx 2$ corresponds to 1994, hence after 1994, the number of unemployed workers in the United States labor force that were
Hispanics dropped below 10%.

61. (a)

x	$f(x)$
1	2.05
3	3.48
5	4.14
7	4.58
9	4.91

(b) $f(6) \approx 4.38$; In 1990, about $4.38 billion were spent on admission to spectator sports.
(c) Between 1985 and 1993, the personal expenditure for admissions to spectators sports was increasing at an average rate of $0.3575 billion per year.
(d) Between 1985 and 1987, the personal expenditure for admissions to spectators sports was increasing at an average rate of $0.715 billion per year.

63. (a) $f(10) \approx 23.05$; In 1989, the Federal Government spent about 23.05 billion of constant 1993–1994 dollars on higher education.
(b) $m_{sec} \approx 0.23$; Between 1989 and 1994, the amount of money the Federal Government spent on higher education was increasing at an
average rate of about 0.23 billion of constant 1993–1994 dollars per year.

Section 1.8 Exercises

1. Quadratic Model   **3.** Cubic Model   **5.** A Logarithmic Model first and then a Power Model as a second choice.
7. Exponential Model   **9.** Logistic Model   **11.** $y = 0.463x + 18.888$ **13.** $y = 15{,}557.133 + 2259.102 \ln (x)$
15. (a) $f(10) = 37.52$; In the tenth year, 37.52% of births were to unwed mothers.   **(b)** $f(20) = 128.82$; In the twentieth year, 128.82%
of births were to unwed mothers.   **(c)** The answer in part b) does not make sense since the percentage cannot be larger than 100%.
This would say that there were more births by unwed mothers than the total number of births, which is impossible.

Chapter 1 Review Exercises

1.

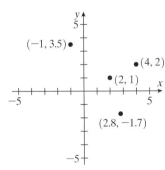

3. r is the independent variable, A is the dependent variable.   **5.** Function   **7.** $(3, 5)$
9. (a) x is the independent variable.   **(b)** y is the dependent variable.   **(c)** 2   **(d)** 2
(e) $[1, 4]$ **(f)** $[1, 5]$ **11.** $[2, \infty)$ **13.**

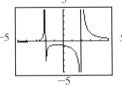

Domain: $(-\infty, -2) \cup (-2, 2) \cup (2, \infty)$
Range: $(-\infty, -0.655) \cup (-0.095, \infty)$

15. The domain is $(-2, \infty)$.; The range is $[-3, \infty)$.   **17.** $\frac{2}{3}$
19. $y - 5 = \frac{5}{4}(x - 0)$ **21.** $y = \frac{3}{4}x - 7$
23. (a) $C(x) = -40x + 800$ **(b)** In five years, the bike will be worth $600.
25. $C(x) = 75x + 2600$ **27.**

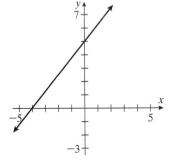

29. (a) Assuming that the cost function is linear, the y-coordinate of the y-intercept will correspond to the fixed costs. Since $(0, 2500)$ is the y-intercept, then the fixed costs are $2500. **(b)** The marginal cost, the cost to produce one additional ceiling fan, is $60 per fan.
(c) $C(x) = 60x + 2500$ **(d)** It will cost $7540 to produce 84 ceiling fans. **31. (a)** $\left(-\frac{3}{2}, 16.5\right)$ **(b)** $(-\infty, 16.5]$
33. (a) $(1.6, -5.24)$ **(b)** The function is increasing on $(-\infty, 1.6)$ and decreasing on $(1.6, \infty)$. **35.** $x = 2.5$ or $x = -4$

37.

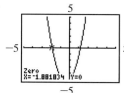 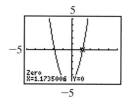

Roots are $x \approx -1.88$ and $x \approx 1.17$

39. (a) $m_{sec} = -1$
(b) $m_{sec} = 1$
(c) $m_{sec} = 5$

41. (a) $f(4) = 82.7$; In 1994, the average American spent 82.7 hours per year watching premium cable television programming.

(b)

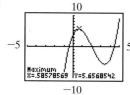

(c) $[82.57, 103.4]$

(d)

x	$f(x)$
0	88.5
1	85.58
2	83.64
3	82.68
4	82.7
5	83.7
6	85.68
7	88.64
8	92.58
9	97.5
10	103.4

(e) In 1992 through 1995 ($x = 2, 3, 4,$ and 5), the average American spent less than 85 hours watching premium cable television programming.

43. (a) $(f + g)(x) = \sqrt{x + 2} + x - 2$; The domain is $[-2, \infty)$.
(b) $(f - g)(x) = \sqrt{x + 2} - x + 2$; The domain is $[-2, \infty)$. **(c)** $(f \cdot g)(x) = (\sqrt{x + 2})(x - 2)$; The domain is $[-2, \infty)$.
(d) $\left(\dfrac{f}{g}\right)(x) = \dfrac{\sqrt{x + 2}}{x - 2}$; The domain is $[-2, 2) \cup (2, \infty)$. **45.** $(f - g)(4) = -29$ **47.** $\left(\dfrac{f}{g}\right)(3) = 0.16$ **49.** $R(x) = 145x - 0.15x^2$

51. $P(x) = -0.08x^2 + 16x - 400$; Vertex: $(100, 400)$; Since a is negative, P has a maximum profit of $400 when $x = 100$ units are produced.
53. (a) Polynomial; quadratic **(b)** Polynomial; constant **(c)** Polynomial; linear **(d)** Polynomial; quartic
55. As $x \to -\infty, g(x) \to \infty$ and as $x \to \infty, g(x) \to \infty$.
57. (a) As $x \to -\infty, f(x) \to -\infty$ and as $x \to \infty, f(x) \to \infty$. **(c)** $f(x)$ is decreasing on $(0.59, 3.41)$ and increasing on $(-\infty, 0.59) \cup (3.41, \infty)$.

(b)

Thus, we have a "peak" at $(0.59, 5.66)$ and a "valley" at $(3.41, -5.66)$.

59. (a) $(-\infty, -1.5) \cup (-1.5, \infty)$ **(b)** Vertical asymptote at $x = -1.5$ **(c)** No horizontal asymptote
61. (a) $(-\infty, -3) \cup (-3, 3) \cup (3, \infty)$ **(b)** Vertical asymptote at $x = -3$; "hole" at $x = 3$ **(c)** Horizontal asymptote of $y = 2$
63. y-intercept is $\left(0, -\frac{16}{9}\right)$; x-intercept is $(4, 0)$ **65.** y-intercept is $(0, 2)$; no x-intercepts

67.

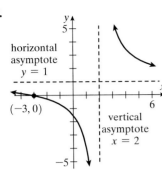

horizontal asymptote $y = 1$

$(-3, 0)$

vertical asymptote $x = 2$

69. $f(x) = (x - 2)^{3/4}$; The domain is $[2, \infty)$.

71. (a) $f(64) = 2$; It will take 2 seconds for an object to fall 64 feet.

(b) It will take 3 seconds for an object to fall 144 feet.

73. (a) $f(12) \approx 123.65$; In 1987, about \$123.65 billion was spent on research and development in the United States.

(b) 200

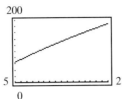

5   21

0

(c) $m_{sec} \approx 8.01$; Between 1983 and 1987, the amount spent on research and development was increasing at an average rate of about \$8.01 billion per year.

75. (a)

x	$f(x)$	x	$f(x)$
-5	≈ 0.0155	1	2.3
-4	≈ 0.0357	2	5.29
-3	≈ 0.0822	3	≈ 12.17
-2	≈ 0.189	4	≈ 27.98
-1	≈ 0.435	5	≈ 64.36
0	1		

(b)

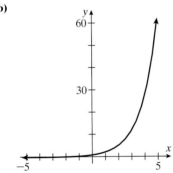

(c) This is an exponential growth function.

77. (a)

x	$f(x)$	x	$f(x)$
-5	$\approx 729,416$	1	≈ 0.0672
-4	$\approx 49,021$	2	≈ 0.0045
-3	≈ 3294.5	3	$\approx 3 \times 10^{-4}$
-2	≈ 221.41	4	$\approx 2 \times 10^{-5}$
-1	≈ 14.88	5	$\approx 1 \times 10^{-6}$
0	≈ 1		

(b)

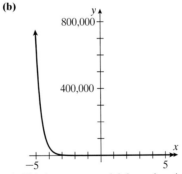

(c) This is an exponential decay function.

79. After seven years, \$2461.60 will be in the account.   **81. (a)** $A(t) = 8000(1.016875)^{4t}$; This is the amount in the account as a function of time.   **(b)** $A(6) \approx 11,953.96$; After six years, \$11,953.96 will be in the account.   **(c)** \$3953.96

83. (a) $f(4) \approx 75,105$; In 1984, there were about 75,105 people in Granaco City.   **(b)** $m_{sec} \approx 3322.13$; Between 1985 and 1988, the population was growing at an average rate of about 3322 people per year.   **(c)** In 1994, the population exceeded 110,000 people.

85. (a) Effective rate ≈ 0.0852 or 8.52%; So, investing in an account that pays 8.2% compounded monthly is equivalent to investing in an account that pays 8.52% compounded annually.   **(b)** Effective rate ≈ 0.0693 or 6.93%; So, investing in an account that pays 6.7% compounded continuously is equivalent to investing in an account that pays 6.93% compounded annually.   **87.** $f(g(4)) = 0$

89. (a) Their domains are all real numbers or $(-\infty, \infty)$.   **(b)** $f(g(x)) = 5x^2 + 3$ **(c)** $g(f(x)) = 25x^2 + 30x + 9$

91. Let $f(x) = x^5$ and $g(x) = x^2 + 2$.   **93.** Let $f(x) = \sqrt{x}$ and $g(x) = 6 - x$.

95.

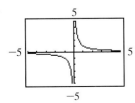

The function is one-to-one.

97. (a) $f(g(x)) = f(x^3 - 5) = \sqrt[3]{(x^3 - 5) + 5} = \sqrt[3]{x^3} = x$ **99.** $-4 = \log(0.0001)$

(b)

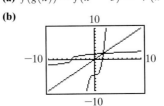

101. $2 \log_2(5) - \log_2(7)$

103. 2.52

105. A Logarithmic Model first and then a Power Model as a second choice.   **107.** $y = 2.9x + 234.867$ **109.** $y = (2.862 \times 10^{-9})x^{5.452}$

CHAPTER 2   Limits, Instantaneous Rate of Change, and the Derivative

Section 2.1 Exercises

1. (a) -1 (b) -1 (c) -1 **3.** (a) 48   (b) 48   (c) 48   **5.** (a) 2   (b) 2   (c) 2   **7.** (a) 12   (b) 12   (c) 12
9. (a) 1   (b) 1   (c) 1   **11.** -6 **13.** 3   **15.** 4   **17.** $\frac{1}{4}$ **19.** -5 **21.** 13   **23.** 6   **25.** 0   **27.** 0   **29.** -2 **31.** $\sqrt{3}$ **33.** 0
35. 0   **37.** 1125   **39.** 5   **41.** 1   **43.** $\frac{1}{4}$ **45.** (a) 1   (b) 1   (c) 1   (d) 2   (e) Does not exist   (f) 2   (g) 3   **47.** $2x$
49. $3x$ **51.** $4x^2$ **53.** $-3x^2$ **55.** (a) $3h + 2$ (b) 3   **57.** (a) $h^2 + 2h + 3$ (b) 2   **59.** (a) $h^2 + 5h + 3$ (b) 5

61. (a) $\dfrac{1}{h + 1}$ (b) -1

63. (a)

2500

0
0   20

(b) $N(10) = 1948$; When $10,000 is spent on advertising, 194,800 items are sold.

(c) $\lim\limits_{x \to 10} N(x) = 1948$; As the level of advertising expenditures approaches $10,000, the number of items sold approaches 194,800.

65. (a)

20000

0
0   25

(b) $AC(10) = 2257.35$; The average cost of producing 10 units of a product is $2257.35 per unit.

(c) $\lim\limits_{x \to 10} AC(x) = 2257.35$; As the level of production approaches 10 units, the average cost approaches $2257.35 per unit.

67. (a) $c(100) = 30$; Rental charges for driving 100 miles will be $30.   (b) $\lim\limits_{m \to 100} c(m) = 30$
(c) $c(200) = 60$; Rental charges for driving 200 miles will be $60.   (d) $\lim\limits_{m \to 200} c(m)$ does not exist.
(e) The charges for $m \geq 200$ were incurred during a second day of driving.

Section 2.2 Exercises

1. (a) $-\infty$ (b) ∞ (c) Does not exist   **3.** (a) ∞ (b) ∞ (c) ∞ **5.** (a) $-\frac{1}{5}$ (b) $-\frac{1}{5}$ (c) $-\frac{1}{5}$
7. (a) $-\infty$ (b) ∞ (c) Does not exist   **9.** Does not exist   **11.** $-\frac{1}{5}$ **13.** $\frac{1}{6}$ **15.** Does not exist
17. (a) Exponential growth   (b) $\lim\limits_{x \to \infty} f(x) = \infty$; $\lim\limits_{x \to -\infty} f(x) = 0$ **19.** (a) Exponential growth   (b) $\lim\limits_{x \to \infty} f(x) = \infty$; $\lim\limits_{x \to -\infty} f(x) = 0$
21. (a) Exponential decay   (b) $\lim\limits_{x \to \infty} f(x) = 0$; $\lim\limits_{x \to -\infty} f(x) = \infty$ **23.** (a) Exponential decay   (b) $\lim\limits_{x \to \infty} f(x) = 0$; $\lim\limits_{x \to -\infty} f(x) = \infty$
25. 2   **27.** $\frac{3}{2}$ **29.** 0   **31.** 0   **33.** $y = 2$ **35.** $y = 0$ **37.** $-\infty$ **39.** Does not exist   **41.** 1   **43.** 0

45. (a)

900

0
0   100

$0 \leq x < 100$

(b) $C(40) = 62$; The cost of removing 40% of the pollutants is $62,000.

(c) $\lim\limits_{x \to 100^-} C(x) = \infty$; As the amount of pollutants removed approaches 100%, the cost grows without bound.

47. (a) 3200; As the amount of money spent on advertising grows without bound, the number of items sold approaches 320,000.

(b)

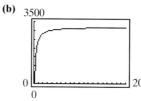

3500

0
0   20

If $1.5 \leq x \leq 3.75$ then $2700 \leq N(x) \leq 3000$.

49. 4.85; As the number of units produced increases without bound, the average cost of production approaches $4.85 per item.

51. (a) 2.5; This means that initially, 2.5 milliliters of medication is administered.

(b) 0; As the amount of time since the medication is administered increases without bound, the amount present in the body approaches 0.

(c)

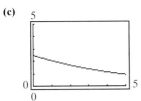

(d) $t \approx 4.58$

53. (a)

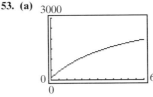

(b) 758 bass

(c) 3500; As time increases without bound, the bass population approaches 3500.

55. (a)

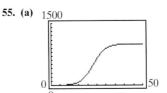

(b) 2 people

(c) 1000; As time increases without bound, the number of people affected by influenza approaches 1000, the entire population.

Section 2.3 Exercises

1. $\dfrac{f(1) - f(-2)}{1 - (-2)} = -1$ **3.** $\dfrac{g(9) - g(1)}{9 - 1} = \dfrac{3}{4}$ **5.** $\dfrac{f(20) - f(0)}{20 - 0} \approx 5.63$ **7.** $\dfrac{f(15) - f(0)}{15 - 0} \approx -0.13$ **9.** $\dfrac{f(10) - f(1)}{10 - 1} \approx 0.26$

11. (a) Positive: A, B and D, E; negative: B, C; zero: C, D **(b)** Positive: B, C; negative: C, D; zero: A, B **13.** 3 **15.** 7 **17.** 99

19. $\frac{3}{4}$ **21.** ≈ 0.63 **23.** $m_{tan} = 7$ **25.** 2 **27.** -6 **29.** -6 **31.** $\frac{1}{2}$ **33.** -13.6

35. (a) $y = -6(x + 2) + 8$
$\quad\quad = -6x - 4$

37. (a) $y = -6(x - 4) - 8$
$\quad\quad = -6x + 16$

39. (a) $y = -13.6(x - 7) - 38.5$
$\quad\quad = -13.6x + 56.7$

(b)

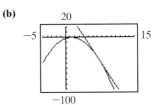

41. $f'(x) = 2$ **43.** $g'(x) = -2$ **45.** $f'(x) = 2x$ **47.** $h'(x) = 2x - 2$ **49.** $g'(x) = 4.2x + 3.2$ **51.** $f'(x) = -4x + 3$

53. $f'(x) = -4.6x + 3.1$ **55.** $g'(x) = 3x^2 + 2x$ **57.** $x = -1: m_{tan} = 2; x = 2: m_{tan} = 2$ **59.** $x = -3: m_{tan} = -8; x = 2: m_{tan} = 2$

61. $x = 0: m_{tan} = 3.2; x = 2: m_{tan} = 11.6$ **63.** $x = 3: m_{tan} = -9; x = 6: m_{tan} = -21$ **65.** $x = -1: m_{tan} = 1; x = 1: m_{tan} = 5$

67. (a) $P'(q) = 1000 - 4q$ **(b)** $P'(200) = 200; P'(300) = -200$; When 200 modems are produced and sold, the profit is increasing at a rate of $200 per modem.; When 300 modems are produced and sold, the profit is decreasing at a rate of $200 per modem.

69. (a) $R(50) = 7500; R(150) = 7500$ **(b)** $R'(q) = 200 - 2q$ **(c)** $R'(50) = 100; R'(150) = -100$; Revenue is increasing at a rate of $100 per set when 50 sets are sold; revenue is decreasing at a rate of $100 per set when 150 sets are sold.

71. (a) $C'(x) = 2x + 15$ **(b)** $C'(40) = 95; C'(100) = 215$; The weekly cost is increasing at a rate of $95 per refrigerator when 40 refrigerators are produced; the weekly cost is increasing at a rate of $215 per refrigerator when 100 refrigerators are produced.

73. (a) $f'(x) = 4.78x + 47.05$ **(b)** $f'(3) = 61.39; f'(4) = 66.17; f'(5) = 70.95$; Imports are increasing at a rate of $61.39 billion per year in 1993, by $66.17 billion per year in 1994, and by $70.95 billion per year in 1995. **(c)** Increasing

75. (a) $f'(x) = -0.82x + 13.53$ **(b)** $f'(6) = 8.61; f'(16) = 0.41; f'(26) = -7.79$; Water consumption was increasing at a rate of 8.61 gallons per day per capita per year in 1965, and was increasing at a rate of 0.41 gallons per day per capita per year in 1975; in 1985, water consumption was decreasing at a rate of 7.79 gallons per day per capita per year. **(c)** Increasing on $1 \leq x \leq 16$, corresponding to the period 1960 to 1975; decreasing on $17 \leq x \leq 31$, corresponding to the period 1976 to 1990.

77. (a) $A(7) \approx 3107.97$; After 7 years, there will be $3107.97 in the account. **(b)** $A'(7) \approx 195.72$; After 7 years, the account was increasing by $195.72 per year.

79. (a) $N(15) = 1458; N'(15) = 2.8$; When \$15,000 is spent on advertising, sales will be 145,800 items, and will increase at a rate of 280 items per thousand advertising dollars. **(b)** $N(25) = 1474.8; N'(25) = 1.008$; When \$25,000 is spent on advertising, sales will be 147,480 items, and will increase at a rate of 101 items per thousand advertising dollars. **(c)** $N'(x) \to 0$; decreasing

(d)

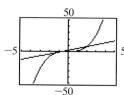

; The result is reasonable. **81.** Tangent line t_1 is associated with the faster growth rate since its slope is larger than the slope of tangent line t_2.

Section 2.4 Exercises

1. $f'(x) = 0$ **3.** $f'(t) = 0$ **5.** $f'(x) = 6x^5$ **7.** $y' = -12x^3$ **9.** $g'(x) = \dfrac{4}{3\sqrt[3]{x}}$ **11.** $f'(x) = \dfrac{1}{\sqrt[3]{x^4}}$ **13.** $h'(x) = \dfrac{8x^3}{3}$

15. $y' = -\dfrac{2\sqrt[3]{x^2}}{3}$ **17.** $f'(x) = 6x^2 + 8x - 7$ **19.** $g'(x) = 6x - 2$ **21.** $y' = -10x - 6$ **23.** $y' = -15x^2 + 7$

25. $f'(x) = \frac{3}{2}x^2 + \frac{6}{5}x - \frac{2}{3}$ **27.** $g'(x) = 2.62x + 2.05$ **29.** $d'(x) = -0.4x + 10.5x^2 - 1.6x^3$ **31.** $y' = 3.45x^2 - 4.6x + 2.53$

33. $f'(x) = \dfrac{3}{2\sqrt{x}} + \dfrac{1}{2} - 10x$ **35.** $y' = \dfrac{1}{3\sqrt[3]{x^2}} + 2x - 9x^2$ **37.** $g'(x) = \dfrac{2}{3\sqrt[3]{x}} + \dfrac{2}{\sqrt{x^3}}$ **39.** $f'(x) = 3.1725x^{0.35}$

41. $AC'(x) = -\dfrac{10}{x^3}$ **43.** $y'(x) = 4x - \dfrac{1}{x^2}$ **45.** $f'(x) = -\dfrac{9}{2\sqrt{x^5}} - \dfrac{2}{\sqrt{x}}$ **47.** $y' = -\dfrac{1.175}{\sqrt{x^3}} + \dfrac{23}{15\sqrt[3]{x^5}}$ **49. (a)** $(-\infty, \infty)$

(b) $f(x) = \dfrac{x^3}{4} - \dfrac{3x^2}{4} + \dfrac{1}{3}$ **(c)** $f'(x) = \dfrac{3x^2}{4} - \dfrac{3x}{2}$ **(d)** $(-\infty, \infty)$ **51. (a)** $(-\infty, 0) \cup (0, \infty)$ **(b)** $f(x) = 2x^2 + 3x - 1 + 3x^{-1}$

(c) $f'(x) = 4x + 3 - \dfrac{3}{x^2}$ **(d)** $(-\infty, 0) \cup (0, \infty)$ **53. (a)** $(-\infty, 0) \cup (0, \infty)$ **(b)** $y = 7x^2 - 50 + x^{-1}$ **(c)** $y' = 14x - \dfrac{1}{x^2}$

(d) $(-\infty, 0) \cup (0, \infty)$ **55. (a)** $(0, \infty)$ **(b)** $h(x) = 2x^{5/2} - 7x^{3/2} + \frac{3}{2}x^{-1/2}$ **(c)** $h'(x) = 5x^{3/2} - \dfrac{21}{2}x^{1/2} - \dfrac{3}{4x^{3/2}}$ **(d)** $(0, \infty)$

57. (a) $f'(x) = 3x^2$ **59. (a)** $y' = -\dfrac{1}{x^2}$ **61. (a)** $f'(x) = \dfrac{2}{3}x^{-1/3} = \dfrac{2}{3\sqrt[3]{x}}$

(b) $f'(-1) = 3$ **(b)** $y'(3) = -\frac{1}{9}$ **(b)** $f'(8) = \frac{1}{3}$

(c) $y = 3x + 2$ **(c)** $y = -\frac{1}{9}x + \frac{2}{3}$ **(c)** $y = \frac{1}{3}x + \frac{4}{3}$

(d)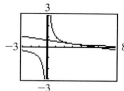

63. (a) $N'(x) = \dfrac{520}{x^2}$ **(b)** $N(10) = 2448; N'(10) = 5.2$; When \$10,000 is spent on advertising, 244,800 CDs are sold and sales are increasing at a rate of 520 CD's per thousand dollars.

65. $C(300) \approx 6334.64; C'(300) \approx 11.58$; When 300 coats are made per week, the total cost is \$6334.64 and the costs are increasing at a rate of \$11.58 per coat.

67. $SD(1) = 93.21; SD'(1) = -186.42$; One mile downwind, the amount of sulphur dioxide is 93.21 ppm and decreasing at a rate of 186.42 ppm per mile.

69. (a) $SD(3) = 22.35; SD(7) = 21.95$; In 1987, the amount of sulphur dioxide released into the atmosphere by the United States was 22,350,000 tons; in 1991, the figure was 21,950,000 tons. **(b)** $SD'(3) = -0.14; SD'(7) = -0.38$; In 1987, the amount of sulphur dioxide released into the atmosphere by the United States was decreasing at a rate of 0.14 million tons per year; in 1991, it was decreasing by 0.38 million tons per year.

71. (a) $\dfrac{N(4) - N(1)}{4 - 1} = 62$; Between 1 and 4 minutes, the average rate of change in the size of bacteria is 62 milligrams per minute.

(b) $N'(4) = 120$; After 4 minutes, the size of bacteria was increasing at a rate of 120 milligrams per minute.

73. (a) $v(t) = s'(t) = -32t$ **(b)** $s(1) = 134$; $s(3) = 6$; After 1 second, the egg is 134 feet above ground; after 3 seconds, it is 6 feet

above ground.   **(c)** $\dfrac{s(3) - s(1)}{3 - 1} = -64$; -64 feet per second   **(d)** $v(1) = -32$; $v(3) = -96$; After 1 second, the instantaneous velocity

is -32 feet per second; after 3 seconds, it is -96 feet per second.   **(e)** $t \approx 3.1$

75. (a) $f'(x) = -0.06x^2 + 1.14x - 4.25$ **(b)** $f'(12) = 0.79$; $f'(15) = -0.65$; The number of tuberculosis cases was increasing at a rate

of 790 per year in 1991 and was decreasing at a rate of 650 per year in 1994.

77. (a) $C(11) = 3405.99$; $C(16) = 3583.04$; In 1985, the number of calories consumed per day per capita was 3405.99; in 1990, it was

3583.04.   **(b)** $C'(11) = 27.16$; $C'(16) = 43.66$; The number of calories consumed per day per capita was increasing by 27.16 calories per

day per capita per year in 1985, and was increasing by 43.66 calories per day per capita per year in 1990.

79. (a) $S(23) = 244.30$; After 23 weeks, weekly sales were $244,300.   **(b)** $S'(23) = -3.26$; After 23 weeks, weekly sales were decreasing

at a rate of $3260 per week.   **(c)** $244,300; decreasing; $3260 per week.

81. (a) $f(4) = 3432.81$; In 1993, there were 3,432,810 members of the Girl Scouts of America.   **(b)** $f'(t) = 27.75t^2 - 247.62t + 484.8$

(c) $f'(4) = -61.68$; In 1993, membership in the Girl Scouts of America was decreasing at the rate of 61,680 members per year.

83. Boy Scouts of America; the instantaneous rate of change for the Boy Scouts was positive and for the Girl Scouts was negative.

Section 2.5 Exercises

1. $f'(x) = 2x(3x + 1)$ **3.** $f'(x) = x^2(15x^2 + 8x - 15)$ **5.** $y' = 3x^3(12x^2 - 45x + 4)$ **7.** $y' = -5x(15x^3 + 15x - 14)$

9. $f'(x) = 12x + 5$ **11.** $y' = 60x^3 + 57x^2 + 12x + 5$ **13.** $g'(x) = 24x^3 + 33x^2 - 58x + 19$ **15.** $y' = 24x + 9\sqrt{x} - \dfrac{4}{\sqrt{x}} - 25$

17. $f'(x) = \frac{172}{5}x^{28/15} - \frac{160}{3}x^{5/3} + \frac{66}{5}x^{6/5} - 20x - 18x^{1/5} + 25$ **19.** $f'(x) = -\dfrac{5}{\sqrt{x}} + \dfrac{15}{2x^{7/2}} - \dfrac{15}{x^4} + 6$

21. $h'(x) = -\dfrac{2}{x^3} - \dfrac{4}{3x^{7/3}} - \dfrac{2}{x^2} - \dfrac{1}{3x^{4/3}}$ **23.** $f'(x) = -\dfrac{1}{(x+1)^2}$ **25.** $y' = \dfrac{10}{(2x+1)^2}$ **27.** $f'(x) = \dfrac{34x^2 + 2x - 13}{(5x^2 + 3x + 2)^2}$

29. $f'(x) = \dfrac{-9x^{1/2} - \frac{3}{2}x^{-1/2} + 30}{(6x - 1)^2}$ **31.** $y' = \dfrac{2(x^3 - 3x^2 + 3x + 2)}{(x-1)^2}$ **33.** $f'(x) = \dfrac{25x^6 - 120x^5 + 45x^4 - 240x^3 + 2x^2 - 16x - 6}{(x-4)^2}$

35. $g'(x) = 6x^2 + 2x - \frac{3}{2}$

37. (a) $f'(x) = 2x(2x^2 - 5)$

(b) $y = -6x + 2$

(c)

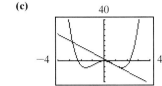

(d)

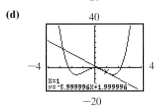

39. (a) $f'(x) = x(5x^3 + 3x + 2)$

(b) $y = 10x - 6$

(c)

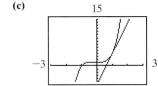

(d)

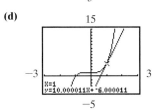

41. (a) $y' = -\dfrac{3}{(x-1)^2}$

(b) $y = -3x + 10$

(c)

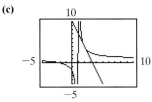

(d)

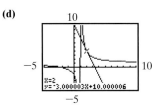

43. (a) $g'(x) = \dfrac{-6(x^2 - 3x + 1)}{(-2x + 3)^2}$ **(c)**   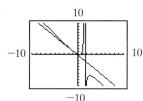   **(d)**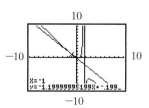

(b) $y = -\frac{6}{5}x - \frac{1}{5}$

45. $f'(x) = 3x^2 + 8x - 7$ **47.** $f'(x) = 24x^3 + 42x^2 - 2x - 21$ **49.** $y' = 21x^{5/2} - \dfrac{15}{2}x^{3/2} + 6\sqrt{x} - \dfrac{1}{\sqrt{x}}$

51. $q(3) = 85.5; q'(3) = 27;$ Three months after it hit the market, monthly sales of the new computer were 8550 units and sales were increasing at the rate of 2700 units per month.   **53. (a)** $R(t) = q(t) \cdot p(t) = (30t - 0.5t^2)(2200 - 34t^2)$ in hundreds of dollars.

(b) $R'(t) = (30 - t)(2200 - 34t^2) + (30t - 0.5t^2)(-68t)$ **(c)** $R(3) = 161{,}937; R'(3) = 33{,}696;$ Three months after it hit the market, monthly revenues for the new computer were \$16,193,700 and increasing at the rate of \$3,369,600 per month.

55. $p(6) = 976; p'(6) = -408;$ Six months after it hit the market, the retail price of the new computer was \$976 and it was decreasing at the rate of \$408 per month.

57. (a) $R'(t) = (30 - t)(2200 - 34t^2) + (30t - 0.5t^2)(-68t)$ **(b)**   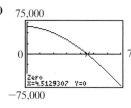   **(c)**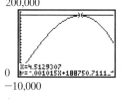

$\quad\quad\quad = (34t^3 - 1020t^2 - 2200t + 66{,}000) + (34t^3 - 2040t^2)$

$\quad\quad\quad = 68t^3 - 3060t^2 - 2200t + 66{,}000$

$x = 4.5129307$ $\dfrac{dy}{dx} = -0.001015$

(d) Maximized; To the left of $t = 4.5129307$ the values of $R'(t)$ are positive so the slope of $R(t)$ is positive and the function is increasing. To the right of $t = 4.5129307$ the values of $R'(t)$ are negative so the slope of $R(t)$ is negative and the function is decreasing. So, $R(4.5129307) \approx 188{,}751$ must be a maximum.

59. $p(3) = 211; p'(3) = -6;$ Three months after being introduced, the retail price of the CD-ROM drive is \$211 and the price is decreasing at the rate of \$6 per month.   **61. (a)** $\dfrac{f(3)}{g(3)} \approx 35.79;$ The average weekly hours in 1992 was 35.79.

(b) $h(t)$ gives the average weekly hours.   **(c)** $h(3) \approx 35.79; h'(3) \approx 0.03;$ In 1992, the average weekly hours was 35.79 and was increasing at the rate of about 0.03 $\dfrac{\text{hours per week}}{\text{year}}$.   **63.** $C(50) = 50; C'(50) = 2;$ The cost of removing 50% of the city's pollutants is \$50,000 and is increasing at the rate of \$2000 per %.   **65.** $P(5) \approx 818.18; P'(5) \approx 112.40;$ Five months after stocking the lake, the bass population is 818 and is increasing at the rate of 112 bass per month.   **67. (a)** $SD(5) = 22.27; CP(5) = 245.76;$ In 1989, the amount of sulphur dioxide was 22,270,000 tons and the civilian population was 245,760,000 persons.

(b) $\dfrac{SD(5)}{CP(5)} \approx 0.09;$ In 1989, the amount of sulphur dioxide per capita was about 0.09 tons.

(c) $H'(t) = \dfrac{-0.1028t^3 - 13.2807t^2 + 125.771t - 346.139}{(2.57t + 232.91)^2}$ **(d)** $H'(5) \approx -0.001;$ In 1989, the per capita emission of sulphur dioxide was decreasing at the rate of about 0.001 tons per year.   **(e)**   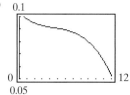   ; decreasing

Section 2.6 Exercises

1. Yes   **3.** No; $\lim\limits_{x\to 3} f(x)$ does not exist.   **5.** Yes; $f(1)$ is defined, $\lim\limits_{x\to 1} f(x)$ exists, and $\lim\limits_{x\to 1} f(x) = f(1)$

7. Yes; $f(3)$ is defined, $\lim\limits_{x\to 3} f(x)$ exists, and $\lim\limits_{x\to 3} f(x) = f(3)$ **9.** g continuous at $x = -2$ **11.** h continuous at $x = 3$

13. g not continuous at $x = 3$ **15.** f not continuous at $x = 5$ **17.** f not continuous at $x = 1$ **19.** $(-\infty, \infty)$ **21.** $(-\infty, -5) \cup (-5, \infty)$

23. $(-\infty, -1) \cup (-1, 3) \cup (3, \infty)$ **25.** $(-\infty, \infty)$ **27.** $(-3, \infty)$ **29.** $(-\infty, \infty)$ **31.** $[-\frac{3}{2}, \infty)$

33. $x = -2$ function not continuous   **35.** $x = -2$ function has corner   **37.** $x = 2$ function has sharp turn

$\quad\ x = 2$ function not continuous $\qquad x = 1$ function has corner   **39.** No values where derivative does not exist

41. (a)

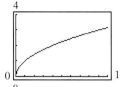

(b) $f'(x) = \dfrac{1}{2\sqrt{x}}$ **(c)** $(-\infty, 0]$

(d) Vertical tangent at $x = 0$.

43. (a)

(b) $f'(x) = \dfrac{2}{3x^{1/3}}$

(c) 0 **(d)** Sharp turn

45. (a)

(b) $f'(x) = \dfrac{x^2 + 2x + 1}{(x + 1)^2} = 1, x \neq -1$

(c) -1 **(d)** Discontinuous

47. (a)

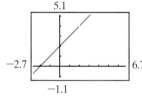

(b) $f'(x) = \dfrac{x^2 - 4x + 4}{(x - 2)^2} = 1, x \neq 2$

(c) 2

(d) Discontinuous

49. The charges for $m \geq 200$ were incurred during a second day of driving.

51. (a) $C(x) = \begin{cases} 1.50x, & 0 < x < 2 \\ 1.00x, & 2 \leq x \leq 6 \end{cases}$

(b)

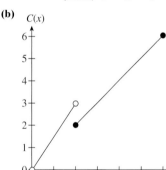

(c) $C(1.5) = 2.25$; 1.5 pounds of hamburger costs \$2.25.

(d) $C(2) = 2$; 2 pounds of hamburger costs \$2.00.

(e) No; $\lim\limits_{x\to 2^-} C(x) \neq \lim\limits_{x\to 2^+} C(x)$

(f) No

Chapter 2 Review Exercises

1. (a) 1   **(b)** 1   **(c)** 1   **3.** 23.591   **5.** 36   **7.** 8   **9.** Does not exist   **11. (a)** 2   **(b)** 0   **(c)** $-\frac{3}{2}$ **(d)** Does not exist

(e) 1   **(f)** 1   **13.** $2x^2 - 9$ **15. (a)** $3(2 + h)^2$ **(b)** 12   **17. (a)** $C(400) = 38,000$; When 400 teddy bears are made, the costs are

\$38,000.   **(b)** $\lim\limits_{x\to 100} C(x) = 37,000$; As the number of teddy bears produced approaches 100, the costs approach \$37,000.

(c) $AC(25) = 1460$; When 25 teddy bears are made, the average cost per bear is \$1460.

19. (a)

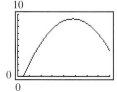

(b) $R(20) = 8.96$; $R(25) = 9.26$; $R(30) = 8.56$ **(c)** 0   **(d)** 24 tablespoons

21. (a) ∞ **(b)** ∞ **(c)** ∞ **23. (a)** $(-\infty, 0) \cup (0, \infty)$ **(b)** Does not exist

25. (a) Exponential growth   **(b)** ∞ **(c)** 0   **27. (a)** Exponential growth   **(b)** ∞ **(c)** 0

29. -2 **31.** $-\infty$ **33.** $y = 0$ **35.** No; $\lim\limits_{x\to 0} f(x)$ does not exist

37. $g(x)$ is continuous at $x = 9$ since $g(9)$ defined, $\lim\limits_{x\to 9} g(x)$ exists, and $\lim\limits_{x\to 9} g(x) = g(9)$; $g(x)$ is not continuous at $x = -3$ since $g(-3)$ is not

defined.   **39.** $(-\infty, -3) \cup (-3, 5) \cup (5, \infty)$

41. (a)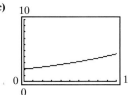

100

0 —— 100
0

(b) 0   **(c)** ∞

43. (a) 2.4; On April 1, the mosquito population is 2,400,000.

(b) ∞; As time increases without bound, so does the population.

(c) 10

0 —— 100
0

(d) 92 days

45. (a) 30

0 —— 10
0

(b) $21,875   **(c)** 25; As time increases without bound, the annual profit approaches $25,000.

47. $\dfrac{g(9) - g(4)}{9 - 4} = \dfrac{4}{5}$ **49.** $\dfrac{f(7.9) - f(6.2)}{7.9 - 6.2} = -\dfrac{50}{413} \approx -0.1211$ **51.** 2.5   **53.** ≈ 0.4610

55. $f'(7) = 12$ **57.** $f'(3) = -\frac{2}{27}$ **59.** $f'(x) = 3$ **61.** $h'(x) = 6.8x + 1.9$ **63.** $m_{\text{tan}} = 6$

65. $m_{\text{tan}} = -\frac{1}{49}$

67. (a) $A(6) = 4894.40$; After 6 years, there is $4894.40 in the account.   **(b)** $A'(6) = 399.29$; After 6 years, the account is increasing at the rate of $399.29 per year.   **69. (a)** $P'(4) = 5$ **(b)** $y = 5t + 40$ **(c)** $m_{\text{tan}} = 5$ means the fish population is growing at the rate of 5 fish per year after 4 years.   **71.** $g'(x) = \dfrac{3}{5x^{4/5}}$ **73.** $f'(x) = -\frac{14}{15}x^{1/6}$ **75.** $h'(x) = -27x^2 + 1$ **77.** $g'(x) = 12.46x + 1.98$

79. $f'(x) = \dfrac{1}{\sqrt{x}} + 14x - \dfrac{3}{2}x^2$ **81.** $h'(x) = 7.8832x^{2.79}$ **83. (a)** $(-\infty, 0) \cup (0, \infty)$ **(b)** $f(x) = 6x^2 + 25x - \dfrac{9}{x} + \dfrac{11}{x^2}$

(c) $f'(x) = 12x + 25 + \dfrac{9}{x^2} - \dfrac{22}{x^3}$ **(d)** $(-\infty, 0) \cup (0, \infty)$ **85. (a)** $f'(x) = \dfrac{3}{4}x^{-1/4} = \dfrac{3}{4x^{1/4}}$ **(b)** $m_{\text{tan}} = \frac{3}{8}$ **(c)** $y = \frac{3}{8}x + 2$

(d) 16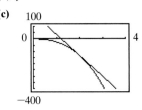

0 —— 32
0

87. $h'(4) = 1600$; After 4 seconds, the rate of change of altitude (i.e., the velocity) is 1600 feet per second.

89. (a) $S(t) = 1.172(2)^{t/1.5}$, where t is the number of years after 1985.

(b) $S'(13) \approx 220.084$; In 1998, clock speed is increasing by 220.084 MHz per year.

91. $g'(x) = -3x(30x^3 - 8x^2 + 27x + 20)$ **93.** $f'(x) = 10x + \dfrac{9}{2}\sqrt{x} - 14x^{2/5} - \dfrac{27}{5x^{1/10}} - \dfrac{27}{2\sqrt{x}} - 45$

95. $h'(x) = -\dfrac{5}{(x-1)^2}$ **97.** $g'(x) = \dfrac{-9x + 12\sqrt{x} + 3}{2\sqrt{x}(3x+1)^2}$

99. (a) $f'(x) = -3x(9x + 10)$

(b) $y = -168x + 204$

(c) 100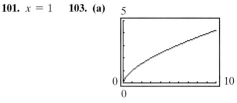

0 —— 4

−400

(d) 100

0 —— 4

X=2
Y=-168.000009X+204.0000_

−400

101. $x = 1$ **103. (a)** 5

0 —— 10
0

(b) $f'(x) = \dfrac{5}{8}x^{-3/8} = \dfrac{5}{8x^{3/8}}$

(c) $(-\infty, 0]$

(d) Vertical tangent at $x = 0$.

105. $f'(x) = \dfrac{33}{2}x^{7/4} + \dfrac{175}{4}x^{3/4} - \dfrac{27}{4x^{1/4}}$

CHAPTER 3   Applications of the Derivative

Section 3.1 Exercises

1. $dy = 6\,dx$ **3.** $dy = (-6x + 2)dx$ **5.** $dy = \dfrac{-5}{(x-1)^2}\,dx$ **7.** $dy = \dfrac{1}{(x+1)^2}\,dx$ **9.** $dy = \left(\dfrac{1}{2\sqrt{x}} - \dfrac{2}{x^2}\right)dx$

11. $dy = \left(-\dfrac{1}{2\sqrt{x^3}} + \dfrac{2}{3\sqrt[3]{x}}\right)dx$ **13.** $dy = \dfrac{-4x}{(x^2-1)^2}\,dx$ **15.** $dy = \left(5.1x^{0.7} + \dfrac{5.6}{x^{0.2}}\right)dx$ **17.** $\triangle y = 0.21; dy = (2x - 2)dx = 0.2$

19. $\triangle y = -398; dy = (5 - 4x)dx = -390$ **21.** $\triangle y \approx 1.58; dy = \dfrac{270}{x^2}\,dx \approx 1.67$ **23.** $\triangle y \approx 0.0349; dy = \dfrac{1}{2\sqrt{x}}\,dx \approx .0354$

25. $\triangle y \approx -0.0802; dy = \dfrac{-4x}{(x^2-1)^2}\,dx \approx -0.0889$ **27.** $\triangle y = 1.4606; dy = (24x^3 - 12x^2)dx = 1.2$

29. Linear approximation: $\frac{51}{10} = 5.1000$; Calculator value: 5.0990   **31.** Linear approximation: $\frac{80}{27} \approx 2.9630$; Calculator value: 2.9625

33. Linear approximation: 1.9938; Calculator value: 1.9937   **35.** Linear approximation: $\frac{97}{343} \approx 0.2828$; Calculator value: 0.2828

37. $dA = 2\pi r\,dr = 2\pi \approx 6.28$ **39.** $dy = \dfrac{55}{\sqrt{x}}\,dx \approx \$13.91$ **41.** $dy = (0.66x^2 - 4.7x + 14.32)dx = \$1,401,960$

43. $dy = (120 - 4.8x)dx = 7200$ volleyballs   **45.** $\triangle y = 6960$ volleyballs, $dy = 7200$ volleyballs   **47. (a)** $y = 0.21(x - 2) + 65.2$

(b) When $x = 3$, the y-value on the tangent line is $y = 65.41$. This means that if the percentage of people living alone in Florida continued at the 1992 rate, the percentage of people living alone in Florida in 1993 would be 65.41%. Using the model, $f(3) = 65.49\%$, or a difference of only 0.08%.

Section 3.2 Exercises

1. $MC(x) = 23$ **(b)** $MC(10) = 23$. The approximate cost of producing the next, or 11th, unit is \$23.   **(c)** $C(11) - C(10) = \$23$

3. (a) $MC(x) = x + 12.7$ **(b)** $MC(11) = 23.7$. The approximate cost of producing the next, or 12th, unit is \$23.70.

(c) $C(12) - C(11) = \$24.20$ **5. (a)** $MC(x) = 0.6x^2 - 6x + 50$ **(b)** $MC(30) = 410$. The approximate cost of producing the next, or 31st, unit is \$410.   **(c)** $C(31) - C(30) = \$425.20$ **7. (a)** $R(x) = 6x; P(x) = x - 500$ **(b)** $MP(x) = 1; MC(x) = 5$

9. (a) $R(x) = -\dfrac{x^2}{20} + 15x; P(x) = -\dfrac{3x^2}{50} + 8x - 1000$ **(b)** $MP(x) = -\dfrac{3x}{25} + 8; MC(x) = \dfrac{x}{50} + 7$

11. (a) $R(x) = -0.005x^2 + 7x; P(x) = 0.001x^3 - 0.005x^2 + 3x - 100$ **(b)** $MP(x) = 0.003x^2 - 0.01x + 3; MC(x) = -0.003x^2 + 4$

13. $MC(x) = 40 - 0.002x; MC(200) = 39.6$. This means that the next, or 201st, tire to be produced is about \$39.60.

15. (a) $AC(x) = \dfrac{100 + 40x - 0.001x^2}{x}; AC(200) = 40.3$. At a production level of 200 tires per day, it costs the Country Day Company, on average, \$40.30 to produce each lawn tractor tire.   **(b)** $MAC(x) = -0.001 - \dfrac{100}{x^2}; MAC(200) = -0.0035$. When 200 lawn tractor tires have been produced, the average cost per tire decreases by 0.35¢ for each additional tire produced.

17. $MP(x) = 5 + \dfrac{1}{2\sqrt{x}}; MP(55) = 5 + \dfrac{\sqrt{55}}{110} \approx 5.067$. At a sales level of 55 magazine subscriptions, there is about a \$5.07 profit in selling the 56th magazine subscription.   **19. (a)** $AP(x) = 5 + \dfrac{1}{\sqrt{x}}; AP(55) = 5 + \dfrac{\sqrt{55}}{55} \approx 5.13$. At 55 magazine subscriptions, the profit from selling each magazine is approximately \$5.13.   **(b)** $MAP(x) = -\dfrac{1}{2x^{3/2}}; MAP(55) = -\dfrac{\sqrt{55}}{6050} \approx -.00123$. Once 55 magazine subscriptions have been sold, the average profit per subscription decreases by 0.12¢.   **21. (a)** $C(q) = 12q + 1200; MC(q) = 12$

(b) $MC(100) = 12; MC(150) = 12$. This means that producing the 101st or 151st pocket pager has an approximate cost of \$12.

(c) Since $MC(q) = 12$, producing the 101st pocket pager or 151st pocket pager has exactly the same additional cost—\$12.

23. (a) $P(x) = -\dfrac{x^2}{95} + \dfrac{39x}{2} - 5500$ **(b)** $P(500) = \dfrac{30,750}{19} \approx 1618.42$. The New Joy Company makes a total profit of \$1618.42 when it sells 500 items.   **(c)** $MP(500) = \frac{341}{38} \approx 8.97$. At a production level of 500 units, the New Joy Company would realize about a \$8.97 profit for selling the 501st unit.   **25. (a)** $R(x) = -\frac{x^2}{30} + 300x; C(x) = 30x + 150{,}000$ **(b)** $P(x) = -\frac{x^2}{30} + 270x - 150{,}000$; Smallest

production level with profit: 601; Largest production level with profit: 7499   **(c)** $P'(1000) = MP(1000) = \frac{610}{3} \approx 203.33$. This means that

at a production level of 1000 hand held computer devices, there is about $203.33 profit for selling the 1001st hand held computer device.

27. $C(2001) \approx 57{,}070 + 30(1) = 57{,}100$

Section 3.3 Exercises

1. (a) 2   **(b)** $\varepsilon_x = \pm\frac{2}{12} = \pm\frac{1}{6}$ **(c)** 16.67%   **3. (a)** 150   **(b)** $\varepsilon_x = \pm\frac{150}{4150} = \pm\frac{3}{83}$ **(c)** 3.61%   **5. (a)** 1500

(b) $\varepsilon_x = \pm\frac{1500}{43{,}500} = \pm\frac{1}{29}$ **(c)** 3.45%   **7.** $\varepsilon_y = \pm0.3$ **9.** $\varepsilon_y \approx \pm0.1125$ **11.** $\varepsilon_y \approx \pm0.0125$ **13.** $\varepsilon_y \approx \pm0.035$ **15.** $\varepsilon_y = \pm0.57$

17. The predicted percentage of votes could be 3% more or 3% less for either candidate. Thus, the margin of error does not rule out the

possibility that the first candidate garners only 49.5% of the votes while the second candidate wins with 50.5% of the votes.

19. 25%   **21.** 8.3%   **23. (a)** 12.5%   **(b)** $R(5) \approx 2236; MR(5) \approx 671$. This means that the additional revenue generated by selling the

next 1000 bolts after 5000 bolts are sold is approximately $671.   **(c)** $R(6) \approx 2236 + 671 = \$2907$. (by linear approximation)

(d) $R(6) \approx 2939$. Relative error of approximation is ±0.011 or $\pm1.1\%$ **25.** $Rel(x) = \frac{2}{x}, Rel(5) = \frac{2}{5}$

27. $Rel(x) = \frac{1}{x - 100}, Rel(125) = \frac{1}{25}$ **29.** $Rel(x) = \frac{1}{2x}, Rel(8) = \frac{1}{16}$ **31.** $Rel(x) = \frac{6x + 5}{3x^2 + 5x}, Rel(2) = \frac{17}{22}$

33. $Rel(25) \approx 0.049$. This means that the poverty threshold in the United States in 1985 was increasing at approximately 4.9% per year.

Chapter 3 Review Exercises

1. $dy = 4\,dx$ **3.** $dy = -\frac{8}{(x - 5)^2}\,dx$ **5.** $dy = \left(\frac{1}{2\sqrt{x}} + \frac{12}{x^5}\right)dx$ **7.** $dy = \left(4x^3 - 2 + \frac{2}{3\sqrt[3]{x}}\right)dx$ **9.** $dy = (20x^4 - 21)dx$

11. $dy = \left(3.4x^{0.7} - \frac{4}{x^{0.2}}\right)dx$ **13.** $\triangle y = 2.34; dy = 2.3$ **15.** $\triangle y = 0.1; dy \approx 0.1051$ **17.** $\triangle y \approx -0.340; dy = -0.4$

19. Linear approximation: 8.0625; Calculator: 8.0623   **21.** Linear approximation: 2.0094; Calculator: 2.0093

23. (a) $dy = (90 - 5.4x)dx$ **(b)** An advertising increase of $400 to $500 would increase sales by approximately 6840 T-shirts.

25. For exercise 23, $\triangle y = 6570$ T-shirts, 270 fewer than the approximation. For exercise 24, $\triangle y = 6593$ T-shirts, 187 fewer than the

approximation.   **27. (a)** $MC(x) = 18$ **(b)** $MC(12) = 18$. It costs approximately $18 to produce the 13th unit.

(c) $C(13) - C(12) = 18$. The answers are exactly the same.   **29. (a)** $MC(x) = 26.7$ **(b)** $MC(8) = 26.7$. It costs $26.70 to produce

the 9th unit.   **(c)** $C(9) - C(8) = 26.7$. The answers are exactly the same.   **31. (a)** $MC(x) = \frac{1}{2}x + 12$

(b) $MC(31) = \frac{55}{2} = 27.5$. It costs $27.50 to produce the 32nd unit.   **(c)** $C(32) - C(31) = 27.75$. The answers vary by $0.25.

33. (a) $R(x) = 11x$ **(b)** $P(x) = 4x - 250$ **(c)** $MP(x) = 4$ **(d)** $MC(x) = 7; MR(x) = 11$

(e) $MR(x) - MC(x) = 4; MP(x) = MR(x) - MC(x)$ exactly.   **35. (a)** $R(x) = -\frac{1}{15}x^2 + 50x$ **(b)** $P(x) = -\frac{1}{6}x^2 + 47x - 850$

(c) $MP(x) = -\frac{1}{3}x + 47$ **(d)** $MC(x) = \frac{1}{5}x + 3; MR(x) = -\frac{2}{15}x + 50$ **(e)** $MR(x) - MC(x) = -\frac{1}{3}x + 47$.

$MP(x) = MR(x) - MC(x)$ exactly.   **37. (a)** $R(x) = -0.005x^2 + 10x$ **(b)** $P(x) = 0.001x^3 - 0.005x^2 + 2x - 100$

(c) $MP(x) = 0.003x^2 - 0.01x + 2$ **(d)** $MC(x) = -0.003x^2 + 8; MR(x) = -0.01x + 10$

(e) $MR(x) - MC(x) = 0.003x^2 - 0.01x + 2. MP(x) = MR(x) - MC(x)$ exactly.   **39. (a)** $C(q) = 85q + 5600$ **(b)** $MC(q) = 85$

41. (a) $MAC(q) = -\frac{5600}{q^2}$ **(b)** $MAC(250) = -0.0896$. This means that when 250 bicycles have been produced the average cost per

bicycle decreases by about $0.09 for each additional bicycle produced.   **43. (a)** $P(q) = -0.02q^2 + 65q - 5600$

(b) At least 1500 bicycles should be manufactured to maximize profit.   **45. (a)** $MC(250) = 125; MC(500) = 125$; At production levels

of 250 and 500 bicycles, it costs about $125 to produce one additional bicycle.   **(b)** They are equal because the marginal cost function is a

constant function.   **47. (a)** $R(q) = -0.022q^2 + 180q$ **(b)** $MR(q) = -0.044q + 180$ **(c)** $MR(500) = 158$. This means that the

revenue gained from producing and selling the 501st bicycle is about $158.

49. (a) $P(x) = R(x) - C(x) = (6x) - (0.0002x^2 + 2x + 1250) = -0.0002x^2 + 4x - 1250$

(b) $P(3000) = 8950$. If the Coloraura Company produces and sells 3000 water color sets, it will make a total profit of \$8950.

(c) $MP(x) = -0.0004x + 4$ **(d)** $MP(3000) = 2.8$. The approximate profit by producing the 3001st water color set is \$2.80.

(e) $P(3001) \approx P(3000) + MP(3000) = \$8952.80$ **(f)** The error is 0.0002 or 0.000002%.

51. $MAC(x) = \dfrac{d}{dx}\left[\dfrac{C(x)}{x}\right] = \dfrac{C'(x) \cdot x - C(x) \cdot 1}{x^2} = \dfrac{C'(x) - \dfrac{C(x)}{x}}{x} = \dfrac{MC(x) - AC(x)}{x}$

53. (a) $C(x) = 150x + 10{,}000; R(x) = -\dfrac{x^2}{75} + 250x$

(b)   1,200,000

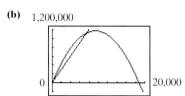

0   20,000

−200,000

Graphically, the break even point
is approximately 102 bookshelves.

(c) $P(x) = -\dfrac{x^2}{75} + 100x - 10{,}000$

(d) Smallest: 102 bookshelves. Largest: 7398 bookshelves.

(e) $MP(x) = -\frac{2}{75}x + 100$

(f) $MP(2500) = \frac{100}{3} \approx 33.33$. At a production level of 2500, there is a \$33.33 gain in profit in selling the 2501st bookshelf.

55. $C(201) \approx C(200) + MC(200) = 30{,}830 + 80 = \$30{,}910$ **57.** $R(101) \approx R(100) + MR(100) = 38{,}000 + 360 = \$38{,}360$

59. $R(301) \approx R(300) + MR(300) = 102{,}000 + 280 = \$102{,}280$ **61. (a)** $MAP(x) = -6000x - 50 + \dfrac{123{,}450}{x^2}$

(b) $MAP(1.5) \approx 45{,}816.67$. At a production level of 1500 printers, the average profit for the next 1000 printers is about \$45,816.67.

63. $\varepsilon_y = \pm 0.05$ **65.** $\varepsilon_y = \pm 0.152$ **67.** $\varepsilon_y = \pm 0.1125$ **69. (a)** 215   **(b)** $\varepsilon_x \approx \pm 0.163$ **(c)** $\pm 16.3\%$ **71. (a)** 250   **(b)** $\varepsilon_x = \pm 0.016$

(c) $\pm 1.6\%$ **73. (a)** 94,250   **(b)** $\varepsilon_x \approx \pm 0.00106$ **(c)** $\pm 0.1\%$ **75.** 54% to 46%   **77. (a)** 5.1%   **(b)** $C(88) = 5{,}374{,}400$.

$MC(88) = 91{,}600$. It will cost the company \$91,600 to produce an additional 1000 modems after 88,000 have already been made.

(c) $C(89) \approx C(88) + MC(88) = 5{,}466{,}000$. The total cost of producing 89,000 modems is about \$5,466,000.

(d) $C(89) = 5{,}466{,}350$. Relative error is about -0.0064%.   **79.** Highest: 80%. Lowest: 76%   **81.** $Rel(x) = \dfrac{21x^2 - 2}{7x^3 - 2x}, Rel(5) \approx 0.605$

83. $Rel(x) = -\dfrac{x - 8}{x^2 - 4x}, Rel(8) = 0$ **85.** $Rel(x) = \dfrac{1}{2x}, Rel(2) = \dfrac{1}{4}$ **87.** $Rel(x) = \dfrac{7}{-7x + x^2}, Rel(2) = \dfrac{-7}{10}$

89. (a) $f(18) = 1288.20$. This means that in 1988, there were about 1,288,200 United States households with a male householder and related children under 18.   **(b)** $f'(x) = 3.04x + 19.9, f'(18) = 74.62$. This means that if the number of United States households with a male householder and related children under 18 continued at 1988 rates, such households would increase at a rate of 74,620 per year.

(c) $Rel(18) \approx 0.0579$. This means that the number of such households was growing at a rate of about 5.79% in 1988.

CHAPTER 4   Additional Differentiation Techniques

Section 4.1 Exercises

1. $2(x + 1)$ **3.** $3(x - 5)^2$ **5.** $-2(2 - x)$ **7.** $6(2x + 4)^2$ **9.** $-10(5 - 2x)^4$ **11.** $30x(3x^2 + 7)^4$

13. $2(x^3 - 2x^2 + x)(3x^2 - 4x + 1)$ **15.** $27x^2(x^3 - 4)^2$ **17.** $50(5x^2 - 3x - 1)^9(10x - 3)$ **19.** $55(4x^2 - x - 4)^{54}(8x - 1)$

21. $(2x - 4)^{-1/2}$ **23.** $\frac{1}{3}(x^2 + 2x)^{-2/3}(2x + 2)$ **25.** $-10(5x - 2)^{-3}$ **27.** $-(x^2 + 2x + 4)^{-3/2}(x + 1)$ **29.** $-\frac{1}{4}(3x^3 - x)^{-5/4}(9x^2 - 1)$

31. $-\dfrac{3}{(3x + 4)^2}$ **33.** $\dfrac{-10}{(x - 2)^3}$ **35.** $\dfrac{-2(2x + 2)}{(x^2 + 2x + 3)^2}$ **37.** $y = 6x - 5$ **39.** $y = -4x + 5$ **41.** $y = 256x - 496$ **43.** $x = 2$

45. $\dfrac{x}{\sqrt{x^2 + 5}}$ **47.** $\frac{2}{3}(2x - 1)^{-2/3}$ or $\dfrac{2}{3(\sqrt[3]{2x - 1})^2}$ **49.** $-5(2x - 8)^{-3/2}$ or $-\dfrac{5}{(\sqrt{2x - 8})^3}$ **51.** $-\frac{64}{3}(5x^2 - 6x + 3)^{-4/3}(10x - 6)$

53. $4(x - 4)^2(x - 1)$ **55.** $(x^2 + 3x)^{1/2} + \dfrac{x(2x + 3)}{2(x^2 + 3x)^{1/2}}$ **57.** $\dfrac{3x^3 - 24x^2}{(3x - 8)^3}$ **59.** $(x + 3)^2(2x - 1)(10x + 9)$

61. $\dfrac{1}{2\sqrt{\dfrac{x+3}{x-3}}} \cdot \dfrac{-6}{(x-3)^2}$ **63.** $0.23(4x^2 + 5x + 6)^{-0.77}(8x + 5)$ **65.** $1.03(x+3)^{0.03}$ **67.** $1.7568(x+1)^{0.22}$

69. The number of persons expected to survive to a given age (near age 70) is decreasing at a rate of about $36.5\ \dfrac{\text{persons}}{\text{year of age}}$.

71. Ten years after the beginning of the study, enrollment is increasing at a rate of $192.09\ \dfrac{\text{students}}{\text{year}}$.

73. (a) $4.5(3x + 6)^{0.5}$ **(b)** With production at a level of 500, the cost of producing the next 100 is about \$20,622.

75. (a) $\dfrac{3}{2\sqrt{3x+5}}$ **(b)** With production at a level of $15\ \dfrac{\text{cameras}}{\text{shift}}$, the cost of producing the 16th camera is about \$21.20.

77. (a) $-\dfrac{25}{\sqrt{22,500 - 50x}}$ **(b)** At a production level of 2500 dolls, the price is decreasing at a rate of about $0.171\ \dfrac{\text{dollars}}{\text{hundred dolls}}$.

79. (a) $\dfrac{-125}{(\sqrt{2x+5})^3}$ **(b)** At a production level of 2000 units, the price is decreasing at a rate of about $0.414\ \dfrac{\text{dollars}}{\text{hundred units}}$.   **(c)** $\dfrac{125x}{\sqrt{2x+5}}$
(d) $\dfrac{125x + 625}{(\sqrt{2x+5})^3}$ **(e)** When the production level is 2000, an increase of 100 units produced and sold increases revenue by about \$10.35.

81. (a) It will take the subject 72 minutes to learn a list of 12 items.   **(b)** $2\sqrt{x-3} + \dfrac{x}{\sqrt{x-3}}$ **(c)** When the list has 12 items, the time to learn additional items is increasing at a rate of $10\ \dfrac{\text{min}}{\text{item}}$.   **83. (a)** There will be about 2040 persons surviving 75 years.   **(b)** $-\dfrac{200}{\sqrt{101 - t}}$
(c) The number of persons expected to survive to a given age (near age 75), is decreasing at a rate of about $39.2\ \dfrac{\text{persons}}{\text{year of age}}$.

85. (a) There are about 17,864 bacteria present 8 hours after the first observation.   **(b)** $73.5(10 + 0.5t)^{1.1}$
(c) 8 hours after the 1st observation, the number of bacteria present is increasing at a rate of about $1340\ \dfrac{\text{bacteria}}{\text{hour}}$.

Section 4.2 Exercises

1. $\dfrac{5}{x}$ **3.** $\dfrac{6}{x}$ **5.** $12x^2 \ln x + 4x^2$ **7.** $\dfrac{15x^4 \ln x - 3x^4}{(\ln x)^2}$ **9.** $-\dfrac{12}{x}$ **11.** $\dfrac{1}{x+7}$ **13.** $\dfrac{2}{2x-5}$ **15.** $\dfrac{2x}{x^2+3}$ **17.** $\dfrac{1}{2x+5}$

19. $\dfrac{6(\ln x)^5}{x}$ **21.** $\dfrac{1}{2\sqrt{x}}\ln\sqrt{x} + \dfrac{1}{2\sqrt{x}}$ **23.** $\dfrac{2x+2}{\ln(x+5)} - \dfrac{x^2 + 2x + 3}{[\ln(x+5)]^2(x+5)}$ **25.** $\dfrac{1}{x \ln 10}$ **27.** $\dfrac{6}{x \ln 3}$ **29.** $2x \log_9 x + \dfrac{x}{\ln 9}$

31. $\dfrac{5}{(5x+3)\ln 2}$ **33.** $\dfrac{1}{\ln 10(x+3)} - \dfrac{2x}{\ln 10(x^2+1)}$ **35.** $y = \dfrac{x}{2} + \ln 2 - 1$ **37.** $y = x - 1$ **39.** $y = 4x - 4$

41. $y = \dfrac{6}{e}x - 5$ **43. (a)** $\dfrac{12}{t}$ **(b)** Twelve hours after the start of the experiment there are about 780 bacteria and the number is growing at a rate of about $1\ \dfrac{\text{bacterium}}{\text{hour}}$.   **45. (a)** $\dfrac{5}{x \ln 2}$ **(b)** The rate of growth after two years is about $3.61\ \dfrac{\text{thousands of prescriptions}}{\text{year}}$ while the growth rate after ten years is $0.72\ \dfrac{\text{thousands of prescriptions}}{\text{year}}$.   **47. (a)** $\dfrac{1.75}{x}$ **(b)** In 1972 the life expectancy of African American females is about 70.33 years and is increasing at a rate of about $0.58\ \dfrac{\text{years}}{\text{birth year}}$.   **(c)** $y = 0.5833x + 68.58$; In 1984 the life expectancy of African American females is 73.15 years according to the model and 77.33 years according to the tangent line approximation.

49. (a)

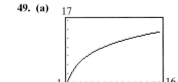

(b) $\dfrac{2}{x}$

(c) In 1984, the consumption growth rate was $0.4\ \dfrac{\text{gallons/capita}}{\text{year}}$ while in 1989 it was $0.2\ \dfrac{\text{gallons/capita}}{\text{year}}$.

Section 4.3 Exercises

1. $7e^x$ **3.** $8 + 2e^x + 2xe^x$ **5.** $\dfrac{10e^x}{(5 - e^x)^2}$ **7.** $8xe^x + 4x^2e^x$ **9.** $-\dfrac{e^x}{2\sqrt{12 - e^x}}$ **11.** $\dfrac{e^x(x^3 - 1) - 3x^2(e^x - 10)}{(x^3 - 1)^2}$ **13.** $\dfrac{-2e^x}{(e^x - 1)^2}$

15. $2e^x + 2xe^x - 1$ **17.** $2e^{2x-1}$ **19.** $\dfrac{e^{\sqrt{x}}}{2\sqrt{x}}$ **21.** 1   **23.** $5e^{2x} + 10xe^{2x}$ **25.** 0   **27.** $\dfrac{2x - e^{-x}}{x^2 + e^{-x}}$ **29.** $10^x \ln 10$ **31.** $-3^{-x} \ln 3$

33. $3x^2 \cdot 0.3^x + x^3 \cdot 0.3^x \ln(0.3)$ **35.** $10^{x+3} \ln 10$ **37.** $\dfrac{-9^{1/x}}{x^2} \ln 9$ **39.** $e^x + xe^x - 2 \cdot 5^{2x} \ln 5$ **41.** $\dfrac{1}{x} 5^{x^2} + 2(\ln 5x) 5^{x^2} x \ln 5$

43. (a) $0.42e^{2t}$ **(b)** Three days after exposure the tumor diameter is growing at a rate of about $0.765 \dfrac{\text{millimeters}}{\text{day}}$.

45. (a) $12(0.8)^t \ln 0.8$ **(b)** One year after the factory opened the population was decreasing at a rate of about 214 per year compared to a rate of about 45 per year eight years after opening.   **47. (a)** Growth model because $0.19 > 0$ and $e > 1$.   **(b)** $50.179e^{0.19x}$

(c) In 1985 the average salary grew at a rate of about $88,730 \dfrac{\text{dollars}}{\text{year}}$ while in 1991 it grew at a rate of about $277,438 \dfrac{\text{dollars}}{\text{year}}$.

49. (a)

(c) The amount accumulated after five years is $2768.06 and is growing at a rate of $179.92/year.

(b) $130e^{0.065t}$

51. (a) Growth model because $1 > 0$ and $1.28 > 1$.   **(b)** $28.69 \cdot 1.28^x \ln 1.28$. The amount invested grew at a rate of about $9.07 \dfrac{\text{billions of dollars}}{\text{year}}$ in 1985 and $107.03 \dfrac{\text{billions of dollars}}{\text{year}}$ in 1995.

(c) $g(x) = 28.69e^{0.25x}$

Section 4.4 Exercises

1. -2 **3.** $-\dfrac{1}{6y}$ **5.** $\dfrac{-x}{y}$ **7.** $\dfrac{4x^3 - 15x^2}{3y^2}$ **9.** $\left(\dfrac{y}{x}\right)^{3/4}$ **11.** $-\dfrac{2x + y}{x}$ **13.** $\dfrac{x^2 + 4y}{y^2 - 4x}$ **15.** $\dfrac{6x - y}{x + y}$ **17.** $\dfrac{-y}{x(\ln x + 1)}$

19. $\dfrac{e^y + 2x}{-xe^y + 2y}$ **21.** $\dfrac{3x^2}{5^y \ln 5}$ **23.** $\dfrac{1}{10^{y-2} \ln 10}$ **25.** $6x(\sqrt[3]{y})^2$ **27.** $-\dfrac{1}{2}$ **29.** $\dfrac{4x^3 - y}{x}$ **31.** $\dfrac{y - 2xy^2}{x}$ **33.** $y = -\dfrac{3}{2}x + \dfrac{13}{2}$

35. $y = 2$ **37.** $y = 2x - 1$ **39.** $y = 2x - 2$ **41. (a)** $-2x$ **(b)** At the demand level of 1100 frames the price is decreasing at a rate of $22 per hundred frames.   **43. (a)** $-\dfrac{p + 2}{x}$ **(b)** At the demand level of 2000 tents the price is decreasing at a rate of $2.50 per hundred tents.   **45.** $2\dfrac{dx}{dt} + 3\dfrac{dy}{dt} = 0$ **47.** $2x\dfrac{dx}{dt} - 3\dfrac{dy}{dt} = 0$ **49.** $2x\dfrac{dx}{dt} + 2y\dfrac{dy}{dt} = 5\dfrac{dx}{dt}$

51. $5\dfrac{dx}{dt} y + 5x\dfrac{dy}{dt} + 4y^3\dfrac{dy}{dt} = \dfrac{dx}{dt}$ **53.** $-\dfrac{2}{7}$ **55.** -4 **57.** $\dfrac{3}{2}$ **59.** $\dfrac{3}{\pi} \dfrac{\text{inches}}{\text{minute}}$ **61.** $\dfrac{5}{3} \dfrac{\text{inches}}{\text{minute}}$ **63.** $0.32\pi \dfrac{\text{inches}^2}{\text{minute}}$

65. $\dfrac{32}{3} \dfrac{\text{feet}}{\text{second}}$ **67. (a)** $\dfrac{dR}{dt} = 250\dfrac{dx}{dt} - \dfrac{4}{5}x\dfrac{dx}{dt}$ **(b)** $34,000 \dfrac{\text{dollars}}{\text{month}}$ **69. (a)** $2x\dfrac{dx}{dt} + 2y\dfrac{dy}{dt} = 0$ **(b)** $-6.87 \dfrac{\text{feet}}{\text{second}}$

71. (a) $\dfrac{dy}{dt} = -\dfrac{2500}{(1 + x)^2}\dfrac{dx}{dt}$ **(b)** $-160 \dfrac{\text{bass}}{\text{year}}$ **73.** $620 \dfrac{\text{dollars}}{\text{week}}$

Section 4.5 Exercises

1. $\frac{11}{10}$ **3.** $\frac{16}{15}$ **5.** $\frac{13}{19}$ **7. (a)** One thousand units will be sold at a price of 5, while 800 units will be sold at a price of 6.

(b) $E_a = 1$: The relative changes in price and quantity sold are equal.   **9.** Since $E_a = 1.9125 > 1$, the demand is elastic.

11. Since $E_a \approx 2.01 > 1$, the demand is elastic.   **13.** $x = \dfrac{-p}{100} + 6$ **15.** $x = \sqrt{\dfrac{300}{p - 10}}$ **17.** $x = -25p + 300$

19. $x = \sqrt{300 - p^2}$ **21.** $x = -10\ln\left(\dfrac{p}{100}\right)$ **23.** 0.294, inelastic   **25.** 0.941, inelastic   **27.** 1.0, unitary   **29.** $\frac{3}{4}$, inelastic

31. 2.0, elastic   **33.** 4.0, elastic   **35.** 0.0350, inelastic   **37.** 2.0, elastic   **39. (a)** $E(4) = \frac{1}{3}$, inelastic   **(b)** Raised

41. (a) $E(15) = 3$, elastic   **(b)** Lowered   **43. (a)** 0.190, raise   **(b)** $12.50   **45. (a)** 0.58, raise   **(b)** $1.83
47. (a) 0.511, raise   **(b)** $116.71

Chapter 4 Review Exercises

1. $3(x + 2)^2$ **3.** $-3(8 - x)^2$ **5.** $8(2x + 5)^3$ **7.** $6(x^2 - 5x + 3)(2x - 5)$ **9.** $\frac{1}{3}(2x^2 - 5x + 7)^{-2/3}(4x - 5)$ **11.** $-\dfrac{6}{(2x + 9)^2}$

13. $-\dfrac{18}{(3x + 5)^4}$ **15.** $y = 400x + 368$ **17.** $y = 32x - 80$ **19.** $\dfrac{7}{2\sqrt{7x - 12}}$ **21.** $\dfrac{-6}{(\sqrt{4x + 5})^3}$ **23.** $(x^2 + 5)^3 + 6x^2(x^2 + 5)^2$

25. $\dfrac{3}{\sqrt{5x - 6}} - \dfrac{5}{2}\dfrac{3x - 7}{(\sqrt{5x - 6})^3}$ **27.** $0.67(3x^2 - x + 1)^{-0.33}(6x - 1)$ **29.** $-0.7(x^3 - x^2 + 5x + 1)^{-1.7}(3x^2 - 2x + 5)$

31. (a) $15(10x - 8)^{0.5}$ **(b)** With production at a level of 700 windows, the cost is increasing at a rate of about 11,811 $\dfrac{\text{dollars}}{\text{hundred windows}}$.

(c) $\dfrac{(10x - 8)^{1.5} + 480}{x}$ **(d)** $15\dfrac{(10x - 8)^{0.5}}{x} - \dfrac{(10x - 8)^{1.5} + 480}{x^2}$ **(e)** With production at a level of 700 windows, the average cost is

decreasing at a rate of about 289 $\dfrac{\text{dollars}}{\text{hundred windows}}$.   **33.** $\dfrac{5}{x}$ **35.** $\dfrac{6x}{3x^2 - 5}$ **37.** $\dfrac{2x + 5}{\ln(x + 4)} - \dfrac{x^2 + 5x - 2}{[\ln(x + 4)]^2(x + 4)}$

39. $12x^2 \log_2 x + \dfrac{4x^2}{\ln 2}$ **41.** $\dfrac{1}{\ln 4}\left(\dfrac{2x}{x^2 + 5} - \dfrac{2}{2x - 3}\right)$ **43.** $\dfrac{1}{2\sqrt{(e^x - 5)}}e^x$ **45.** $2e^{2x-13}$ **47.** $2\dfrac{e^{2x}}{\ln(x - 4)} - \dfrac{e^{2x}}{(\ln(x - 4))^2(x - 4)}$

49. (a) $y = 4e^{-4}x + 12$ **51. (a)** $y = 2x$ **53.** $-5^{-x}\ln 5$

(b)   **(b)**   **55.** $\frac{1}{3}\sqrt[3]{10^x}\ln 10$

57. $\dfrac{0.4^{\sqrt{x}}}{2\sqrt{x}}\ln(0.4)$

59. $\dfrac{1}{x\ln 7} - \dfrac{1}{x\ln 3}$

61. (a) Growth   **(c)** $2149.6 \cdot (1.036)^x \ln 1.036$ **(d)** In 1976 the number of stations was growing at a rate of about

97 $\dfrac{\text{stations}}{\text{year}}$ while in 1981 it was growing at a rate of about 116 $\dfrac{\text{stations}}{\text{year}}$.

(b) 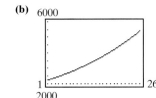   **63.** $\frac{3}{5}$ **65.** $\dfrac{2x - 30x^5}{3y^2}$ **67.** $\dfrac{2xy}{3y^2 - x^2}$ **69.** $\dfrac{y\ln y}{y - x}$ **71.** $-\frac{2}{3}$

73. $\dfrac{2 - 3x^2y}{x^3 - 1}$ **75.** $y = \frac{5}{4}x - \frac{9}{4}$ **77.** $y = \frac{4}{3}x - \frac{1}{3}$

79. (a) $-\dfrac{p + 10}{x}$ **(b)** At a demand level of 15,000 lamps and a price of $30, the price is decreasing at a rate of 2.67 $\dfrac{\text{dollars}}{\text{thousand lamps}}$

81. $3x^2\dfrac{dx}{dt} + 3y^2\dfrac{dy}{dt} = 6\dfrac{dx}{dt}$ **83.** $\dfrac{dx}{dt}y + x\dfrac{dy}{dt} - 3y^2\dfrac{dy}{dt} = 2\dfrac{dx}{dt}$ **85.** 20   **87.** -11 **89. (a)** $\dfrac{dS}{dt} = 8\pi r\dfrac{dr}{dt}$ **(b)** 140.7 $\dfrac{\text{centimeters}^2}{\text{second}}$

91. 1.25   **93.** 1.60   **95.** Since $E_a = 1.38 > 1$, the demand is elastic.   **97.** $\dfrac{12}{p - 5}$ **99.** $-50.0\ln\dfrac{p}{800}$

101. (a) $E(p) = \dfrac{2p^2}{400 - p^2}$ **(b)** $E(12) = 1.125$, elastic   **103. (a)** $E(p) = \dfrac{3p}{(120 - 3p)\ln(120 - 3p)}$ **(b)** $E(5) = 0.0307$, inelastic

105. (a) $E(p) = \dfrac{5p}{(150 - 5p)\ln(150 - 5p)}$; $E(25) = 1.55$, lower   **(b)** $p = 23.34$

CHAPTER 5   Further Applications of the Derivative

Section 5.1 Exercises

1. Increasing: $(-1, 2)$; Decreasing: $(-\infty, -1) \cup (2, \infty)$ **3.** Increasing: $(-\infty, -1)$; Constant: $[-1, 1]$; Decreasing: $(1, \infty)$

5. Increasing: $(-2, 0) \cup (1, \infty)$; Decreasing: $(-\infty, -2) \cup (0, 1)$ **7. (a)** $(-\infty, -2), (0, 2)$ **(b)** $(-2, 0), (2, \infty)$

(c) Zero: $x = \{-2, 0, 2\}$ **9. (a)** $(-\infty, -2), (0, 1), (2, \infty)$ **(b)** $(-2, 0)$ **(c)** Zero: $x = 0$; Undefined: $x = \{-2, 2\}$

11. (a) $(-\infty, 2)$ and $(2, \infty)$ **(b)** None   **(c)** Undefined: $x = 2$ **13.** None   **15.** $x = \frac{5}{2}$ **17.** $x = \{-5, 3\}$ **19.** $x = -\frac{1}{2}$

21. None   **23.** $x = -1$ **25.** $x = -1$ **27.** None

29. (a) Increasing: $(\frac{5}{2}, \infty)$

Decreasing: $(-\infty, \frac{5}{2})$

(b) Relative minimum at $x = \frac{5}{2}$

(c)

31. (a) Increasing: $(-\infty, -5), (3, \infty)$

Decreasing: $(-5, 3)$

(b) Relative maximum at $x = -5$

Relative minimum at $x = 3$

(c)

33. (a) Increasing: $(-\frac{1}{2}, \infty)$

(b) None

(c)

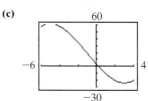

35. (a) Increasing: $(-\infty, \infty)$

(b) None

(c)

37. (a) Increasing: $(-\infty, -1) \cup (-1, \infty)$

(b) None

(c)

39. Relative maximum at $x = \frac{5}{2}$

41. Relative maximum at $x = -1$,

Relative minimum at $x = 0$

43. Relative maximum at $x = -3$,

Relative minimum at $x = 1$

45.

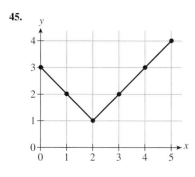

47.

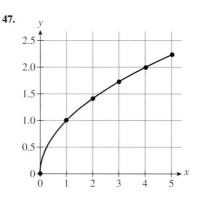

49. (a)

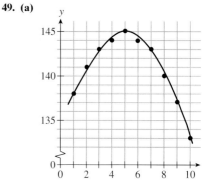

(b) $\dfrac{\text{visits per week}}{\text{years elapsed}}$

51. (a) Increasing: $(0, 500)$; Decreasing: $(500, 1000)$ **(b)** Relative maximum $= (500, 250{,}000)$. When 500 golf clubs are produced and sold, a maximum profit of \$250,000 is realized.   **53. (a)** Increasing: $[1, 16.5)$; Decreasing: $(16.5, 31]$

(b) Relative maximum $= (16.5, 442.31)$. United States water consumption was at its maximum of 442.31 $\dfrac{\text{gallons/day}}{\text{capita}}$ between

1975 and 1976.   **55.** $g'(t) = 1.38t - 9.42$, a linear function. Over the entire interval $[1, 6]$, $g'(t) < 0$. Therefore, United States total petroleum production was decreasing from 1990 to 1995.   **57. (a)** \$6.67   **(b)** $R(x) = x \cdot p(x) = 15.22xe^{-0.015x}$

(c) Increasing: $(0, 67)$; Decreasing: $(67, \infty)$ **(d)** Relative maximum $= (67, 373.27)$. This means that the maximum revenue of $373.29 is obtained when 67 pizzas are sold. **59. (a)** $p'(t) = \dfrac{-230(t - 3)}{(t + 3)^3}$; Critical value: $t = 3$ **(b)** Increasing: $[0, 3)$; Decreasing: $(3, 20]$

(c) Relative maximum $= (3, 19.17)$. The concentration of medicine is highest at 19.17% exactly 3 hours after administration.

61. (a) $AC(x) = \dfrac{50}{x} + 1 + \dfrac{x}{40}$; Critical value: $x \approx 44.72$ **(b)** Increasing: $(44.72, 100]$; Decreasing: $(0, 44.72)$

(c) Relative minimum $= (44.72, 3.24)$. This means that average production costs are lowest when 45 coffee mugs are made, averaging $3.24 per mug. **63. (a)** $P'(x) = R'(x) - C'(x)$ so $P'(x) = 0$ when $R'(x) - C'(x) = 0$; that is, when $R'(x) = C'(x)$.

(b) When $R'(x) > C'(x)$, then $R'(x) - C'(x) = P'(x) > 0$ so $P(x)$ is increasing.

(c) When $R'(x) < C'(x)$, then $R'(x) - C'(x) = P'(x) < 0$ so $P(x)$ is decreasing.

Section 5.2 Exercises

1. $f'(x) = -20x^4 - 18x^2 + 7$; $f''(x) = -80x^3 - 36x$; $f'''(x) = -240x^2 - 36$ **3.** $y' = 21x^2 - 6x + 4$; $y'' = 42x - 6$; $y''' = 42$

5. $f'(x) = e^x$; $f''(x) = e^x$; $f'''(x) = e^x$ **7.** $f'(x) = \dfrac{1}{2\sqrt{x}}$; $f''(x) = -\dfrac{1}{4\sqrt{x^3}}$; $f'''(x) = \dfrac{3}{8\sqrt{x^5}}$ **9.** $f'(x) = \dfrac{1}{x}$; $f''(x) = -\dfrac{1}{x^2}$; $f'''(x) = \dfrac{2}{x^3}$

11. (a) Concave up: $(1, \infty)$; Concave down: $(-\infty, 1)$ **(b)** f' increasing: $(1, \infty)$; f' decreasing: $(-\infty, 1)$

13. (a) Concave up: $(-\infty, \infty)$ **(b)** f' increasing: $(-\infty, \infty)$ **15. (a)** Concave up: $(-\infty, -1) \cup (1, \infty)$; Concave down: $(-1, 1)$

(b) f' increasing: $(-\infty, -1) \cup (1, \infty)$; f' decreasing: $(-1, 1)$ **17.** Concave up: $(-2, \infty)$; Concave down: $(-\infty, -2)$; Inflection point: $(-2, -25)$

19. Concave up: $\left(-\infty, \frac{1}{6}\right)$; Concave down: $\left(\frac{1}{6}, \infty\right)$; Inflection point: $\left(\frac{1}{6}, \frac{-134}{9}\right)$ **21.** Concave up: $(-\infty, \infty)$; No inflection point(s)

23. Concave down: $(-\infty, \infty)$; No inflection point(s) **25.** Concave down: $(-\infty, 0)$; Concave up: $(0, \infty)$; No inflection point(s)

27. Concave down: $(-\infty, 0) \cup (0, \infty)$; No inflection point(s) **29.** Concave up: $(-\infty, \infty)$; No inflection point(s)

31. Concave up: $(-\infty, \infty)$; No inflection point(s) **33.** Concave down: $(-1, \infty)$; No inflection point(s)

35. (a) Increasing: $\left(-\frac{3}{4}, \infty\right)$; Decreasing: $\left(-\infty, -\frac{3}{4}\right)$ **(b)** Relative minimum at $x = -\frac{3}{4}$ **(c)** Concave up: $(-\infty, \infty)$

(d) No inflection points **37. (a)** Increasing: $(-\infty, -5) \cup (1, \infty)$; Decreasing: $(-5, 1)$

(b) Relative maximum at $x = -5$, relative minimum at $x = 1$ **(c)** Concave down: $(-\infty, -2)$; Concave up: $(-2, \infty)$ **(d)** $(-2, 41)$

39. (a) Increasing: $\left(\dfrac{5 - \sqrt{43}}{9}, \dfrac{5 + \sqrt{43}}{9}\right)$; Decreasing: $\left(-\infty, \dfrac{5 - \sqrt{43}}{9}\right) \cup \left(\dfrac{5 + \sqrt{43}}{9}, \infty\right)$

(b) Relative minimum at $x = \dfrac{5 - \sqrt{43}}{9}$, Relative maximum at $x = \dfrac{5 + \sqrt{43}}{9}$ **(c)** Concave down: $\left(\frac{5}{9}, \infty\right)$; Concave up: $\left(-\infty, \frac{5}{9}\right)$

(d) $x = \left(\frac{5}{9}, \frac{-1667}{243}\right)$ **41. (a)** Increasing: $(-\infty, -\sqrt{5}) \cup (\sqrt{5}, \infty)$; Decreasing: $(-\sqrt{5}, 0) \cup (0, \sqrt{5})$

(b) Relative maximum at $x = -\sqrt{5}$, Relative minimum at $x = \sqrt{5}$ **(c)** Concave up: $(0, \infty)$; Concave down: $(-\infty, 0)$ **(d)** None

43. (a) Increasing: $(1, \infty)$; Decreasing: $(-\infty, 1)$ **(b)** Relative minimum at $x = 1$ **(c)** Concave down: $(-\infty, 1) \cup (1, \infty)$ **(d)** None

45. (a) Increasing: $(-\infty, \infty)$ **(b)** None **(c)** Concave up: $(-\infty, \infty)$ **(d)** None

47. (a) Increasing: $(-1, \infty)$ **(b)** None **(c)** Concave down: $(-1, \infty)$ **(d)** None

49.

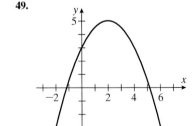

Graphs may vary.

51.

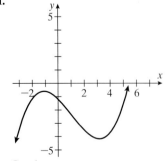

Graphs may vary.

53. If $a > 0$, then $f''(x) = 2a > 0$ for all x, so f is concave up on $(-\infty, \infty)$. A similar argument shows that $a < 0$ implies f is concave down on $(-\infty, \infty)$.

55. (a) $C(5) = 50$. It costs $5000 to produce 5 units.; $C'(5) = 28$. It costs approximately $2800 to produce the 6th unit.

(b) Decreasing: $(0, 2)$; Increasing: $(2, \infty)$; Relative minimum: $(2, 20)$

(c) Inflection point for $C(x)$ is $(2, 20)$.

57. (a) $P(35) = 393{,}812.50$. The profit from selling 3500 refrigerators is \$393,812.50.; $P'(35) = 21{,}197.50$. The profit gained from selling the next 100 refrigerators is about \$21,197.50.   **(b)** Increasing: $(0, 64.49)$; Decreasing: $(64.49, \infty)$; Marginal profit is maximized at 6449 refrigerators.   **59.** The marginal profit is increasing where $\dfrac{d}{dx} MP(x) > 0$. Since $MP(x) = P'(x)$, $\dfrac{d}{dx} MP(x) = \dfrac{d}{dx} P'(x) = P''(x) > 0$. Hence P is concave up. A similar argument shows that P is concave down where the marginal profit is decreasing.

61. $x = 15$. The point at which the rate of change of total sales will no longer increase as a result of advertising is \$15,000,000.; This is the same point at which $TS(x)$ changes from concave up to concave down.   **63. (a)** $b(5) \approx 299$. The total number of births to women 35 to 39 years old in the United States in 1989 was about 299,000; $b'(5) \approx 14.953$. In 1989 the total number of births to women 35 to 39 years old in the United States was increasing at a rate of about $14{,}953 \dfrac{\text{births}}{\text{year}}$.   **(b)** On $[1, 10]$, $b'(x) = 50x^{-3/4} > 0$ so $b(x)$ is increasing on $[1, 10]$.

(c) $b''(x) = -\dfrac{75}{2x^{(7/4)}}$. Over the interval $[1, 10]$, $b''(x) < 0$. Therefore, $b(x)$ is concave down during the entire interval.

(d) Increasing, increasing, slower   **65. (a)** $f(50) \approx 1.00$. At 50 mm-Hg, blood flow is about 1.00 mL/min.; $f'(50) \approx 0.026$. When arterial pressure is 50 mm-Hg, blood flow is increasing at a rate of about $0.026 \dfrac{\text{mL/min}}{\text{mm} - \text{Hg}}$.   **(b)** $f''(x) = 0.278(1.026)^x [\ln(1.026)]^2$ which is always greater than zero over the interval $[20, 120]$. This means that f is concave up over the entire interval, which implies that the rate of blood flow is increasing over the entire interval.   **67. (a)** $CI(20) \approx 3.61$. This means that a 20-year-old has a cardiac index of about $\dfrac{3.61 \text{ L/min}}{m^2}$.;

$CI'(20) \approx -0.045$. At the age of 20, one's cardiac index is decreasing at a rate of approximately $\dfrac{\frac{\text{L/min}}{m^2}}{\text{year}}$.   **(b)** Decreasing: $[10, 80]$

(c) $CI''(x) = \dfrac{2.38875}{x^{(9/4)}}$, which is positive over the interval $[10, 80]$ and therefore $CI(x)$ is concave up. This means that CI is decreasing progressively more slowly over the entire interval.   **69. (a)** $f'(x) = 20.56x + 113.98$, which is > 0 over the interval $[1, 13]$. Therefore, the federal debt increased from 1980 to 1992.   **(b)** $f''(x) = 20.56 > 0$ over the interval $[1, 13]$. Since $f''(x)$ is the first derivative of $f'(x)$, which represents the rate of change of the federal debt, we know that $f''(x) > 0$ implies that the rate of change of the federal debt was increasing.

Section 5.3 Exercises

1. Positive: $(-5, -1) \cup (3, 7)$; Negative: $(-\infty, -5) \cup (-1, 3) \cup (7, \infty)$ **3.** Positive: $(-\infty, -3) \cup (1, 5)$; Negative: $(-3, 1) \cup (5, \infty)$

5. Concave up: $(-1, 4)$; Concave down: $(-\infty, -1) \cup (4, \infty)$ **7.** Positive: $(-\infty, -5) \cup (-\frac{5}{3}, \infty)$; Negative: $(-5, -\frac{5}{3})$

9. Positive: $(-3, 1) \cup (5, \infty)$; Negative: $(-\infty, -3) \cup (1, 5)$ **11.** Concave up: $(-\infty, -1) \cup (3, \infty)$; Concave down: $(-1, 3)$

13. Relative minimum at $x = \frac{1}{3}$ **15.** Relative maximum at $x = \dfrac{2 - \sqrt{43}}{3}$, Relative minimum at $x = \dfrac{2 + \sqrt{43}}{3}$

17. Relative maximum at $x = -3$, Relative minimum at $x = -2$ **19.** Relative maximum at $x = -2$, Relative minimum at $x = 1$

21. Relative minimum at $x \approx -2.25$ and $x \approx 1.57$, Relative maximum at $x \approx -0.07$

23. Relative minimum at $x = -2$ and $x = 2$, Relative maximum at $x = 0$ **25.** Relative maximum at $x = 0$

27. Relative minimum at $x = 0$

29.

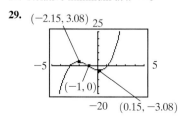

$(-2.15, 3.08)$ $(−1, 0)$ $(0.15, −3.08)$

31.

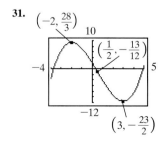

$\left(-2, \frac{28}{3}\right)$ $\left(\frac{1}{2}, -\frac{13}{12}\right)$ $\left(3, -\frac{23}{2}\right)$

33.

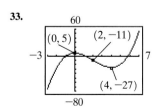

$(0, 5)$ $(2, -11)$ $(4, -27)$

35.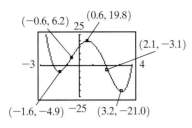

37.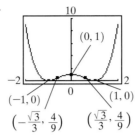

39.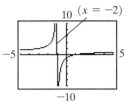

Horizontal asymptote: $y = 1$

41.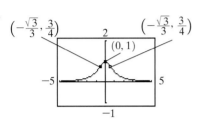

43.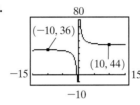

45.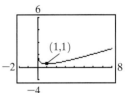

47.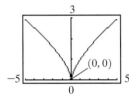

49. Answers will vary. Samples:

(a)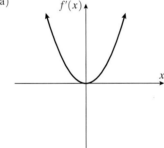

(b)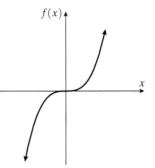

51. Answers will vary. Samples: (a)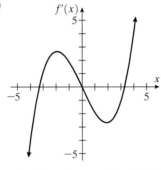

(b)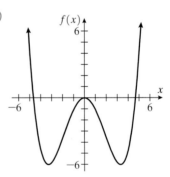

53. (a) $R'(x) = 216 - 0.24x^2$; Increasing: $[0, 30)$; Decreasing: $(30, 50]$ (b) Concave down: $(0, 50)$

55. (a) $AP(x) = -.023x^2 + 4.45x - 15 - \dfrac{2}{x}$ **57.** $AP(x) = P'(x)$ **59.** (a) $C'(x) = 3 + \dfrac{2x}{15}$ (b)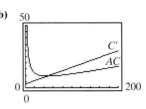

(b)

(c) $(47.4, 9.32)$

(d) Yes

61. (a) $AC(t) = 100t + 400 + \dfrac{11,000}{t}$ (b) 10.5   (c) \$2498

63. **(a)** $f'(x) = \dfrac{3800}{x}$. Since $f'(x) > 0$ over the interval $[40, 90]$. f is increasing.

(b) $f''(x) = -\dfrac{3800}{x^2}$. Since $f''(x) < 0$ over the interval $[40, 90]$, f is concave down.

65. **(a)** $f(20) \approx 62.8$. At winds of 20 mph, 62.8% of total heat loss occurs by convection.; $f'(20) \approx 0.79$. At winds of 20 mph, the percent of heat loss by convection is increasing by about 0.79% per mph.

(b) $f'(x) = \dfrac{16.694}{x + 1}$. Since $f'(x) > 0$ for the interval $[0, 60]$, f is increasing.

(c) $f''(x) = -\dfrac{16.694}{(x + 1)^2}$. Since $f''(x) < 0$ for the interval $[0, 60]$, f is concave down.

(d)

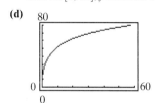

67. **(a)** $f(4) = 1380.51$. In 1992, 1,380,510,000 metric tons of carbon content in CO_2 emissions were released in the atmosphere.; $f'(4) = 21.16$. In 1992, the carbon content in CO_2 emissions was increasing by 21,160,000 metric tons per year.

(b) Increasing: $(2.5, 7]$; Decreasing: $[1, 2.5)$;

Relative minimum: $t = 2.5$

(c) Concave up: $[0, 5)$; Concave down: $(5, 7]$

(d)

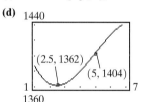

Section 5.4 Exercises

1. Absolute minimum: $(1, -8)$; Absolute maximum: $(-2, 1)$ **3.** Absolute minimum: $(2, -4)$; Absolute maximum: $\left(\dfrac{2 - \sqrt{19}}{3}, 8.21\right)$

5. Absolute minimum: $(-1, -4)$ and $(2, -4)$; Absolute maximum: $(0, 0)$ and $(3, 0)$

7. Absolute minimum: $(2.89, -60.42)$; Absolute maximum: $(-0.34, 25.68)$ **9.** Absolute minimum: $(0.75, 4.89)$; Absolute maximum: $(-2, 29)$

11. Absolute minimum: $\left(\dfrac{\sqrt{2}}{2}, 0\right)$; Absolute maximum: $(2, 2401)$ **13.** Absolute minimum: $(-1, -1)$; Absolute maximum: $(1, 1)$

15. Absolute minimum: $(1, -1)$; Absolute maximum: $(0, -\frac{1}{2})$ **17.** Absolute minimum: $(4, \frac{1}{17})$; Absolute maximum: $(0, 1)$

19. \$46 per day **21.** **(a)** 7 days **(b)** $\dfrac{15 \text{ harmful bacteria}}{\text{cm}^3}$ **23.** Highest: 1986—7.125%; Lowest: 1995—5.073%

25. Absolute maximum: 8620, in year 1974; Absolute minimum: 6809, in year 1995

27. Absolute maximum: 10,434, in year 1960; Absolute minimum: 6856, in year 1995

29. **(a)** The number of Catholic elementary *and* secondary schools in the United States from 1960 to 1995.

(b) Absolute maximum: 12,799, in year 1960; Absolute minimum: 8072, in year 1995

31. Absolute minimums: $(1.6, 1.52)$ and $(6.1, 1.52)$. Natural gas was cheapest in 1990 and 1995.; Absolute maximum: $(3.8, 2.03)$. Natural gas was most expensive in 1992. **33.** $100' \times 100'$ **35.** **(a)** 8654 magazines **(b)** \$1.88/issue **37.** **(a)** 2.24 hours

(b) $\dfrac{0.63 \text{ mg}}{\text{cm}^3}$ **39.** **(a)** $R(x) = 0.14x^3 - 15.52x^2 + 561.21x$ **(b)** Maximum revenue is \$7211.05 **(c)** Demand is 55 units/day

(d) Price is \$131.11/unit **41.** **(a)** Relative maximum: $(6.5, 76.3)$. In the year 1975, total energy consumption in the United States was 76.3 quadrillion Btu.; Relative minimum: $(11.6, 75.0)$. In the year 1980, total energy consumption in the United States was 75 quadrillion Btu.

(b) Absolute minimum: $(1, 66.1)$. In the year 1970, energy consumption was at its lowest. Absolute maximum: $(25, 106.4)$. Energy consumption was at its maximum in 1994. **43.** Absolute minimum: $(1, 344)$. Water usage was least in 1960.; Absolute maximum: $(16.5, 442)$. Water usage was greatest in 1975. **45.** Absolute minimum: $(10, 1.52)$. When blood volume is 10% cells, it takes 1.52 times as much pressure to force bood through a tube as to force water through a tube.; Absolute maximum: $(80, 9.64)$. When blood volume is 80% cells, it takes 9.64 times as much pressure to force blood through a tube as to force water through a tube. **47.** **(a)** $C(x) = 1000 + 60x$

(b) $R(x) = -\frac{5}{7}x^2 + \frac{1200}{7}x$ **(c)** $x = 78$ maximizes profit **49.** **(a)** $R(x) = -\frac{17}{18,000}x^2 + \frac{77}{6}x$ **(b)** Revenue is maximized at $x = 6794$

Section 5.5 Exercises

1. $(\sqrt{3}, 4\sqrt{3})$ **3.** $\left(\dfrac{\sqrt{2}}{2}, 4\sqrt{2} - 3\right) \approx (0.71, 2.66)$ **5.** $(\sqrt[4]{2}, 6\sqrt{2} - 2) \approx (1.19, 6.49)$ **7.** $\left(\sqrt[4]{\dfrac{2}{3}}, 2\sqrt{6} - 1\right) \approx (0.90, 3.90)$

9. $\left(\dfrac{\sqrt{6}}{3}, -2\sqrt{6}\right) \approx (0.82, -4.90)$ **11.** $\left(\dfrac{\sqrt{6}}{2}, -2\sqrt{6}\right) \approx (1.22, -4.90)$ **13.** $\left(\sqrt[4]{\dfrac{2}{3}}, 2 - 2\sqrt{6}\right) \approx (0.90, -2.90)$

15. $\left(\dfrac{\sqrt{10}}{2}, 3 - 2\sqrt{10}\right) \approx (1.58, -3.32)$

17. $x = 200$ ft, $y = 66.67$ ft, area $= 13{,}333.3$ ft^2 **19.** $x \approx 29.9$ ft, $y \approx 50.2$ ft   **21.** $\frac{4}{3} \times \frac{40}{3} \times \frac{10}{3}$ **23.** $20 \times 20 \times 15$

25. (a) $AC(x) = \dfrac{5}{x} + 0.5 + 0.003x$ **(b)** $x \approx 41$ **27.** $AC(t) = 100t + 600 + \dfrac{15{,}000}{t}, t \approx 12.25$ **29.** \$102.50 per room

31. 60 trees per acre   **33.** 35 trees per acre   **35.** Optimal population: 12,500 deer; Sustainable harvest: 4688 deer

37. Optimal population: 60,000 pheasants; Sustainable harvest: 36,000 pheasants   **39.** To minimize inventory costs, each order should have 25 desks, placed 8 times per year.   **41.** To minimize inventory costs, each order should have 240 carrying cases, placed semi-annually.

43. To minimize inventory costs, each order should have 160 microwave ovens, placed 15 times per year.

45. To minimize inventory costs, each order should have 15 (or 20) cars, placed 20 (or 15) times per year.

47. To minimize costs, the publisher should print 2500 books per run, twice a year.

49. To minimize costs, the toy manufacturer should make 5000 toys per run, 10 times per year.

Chapter 5 Review

1. (a) $(-\infty, -2) \cup (3, \infty)$ **(b)** $(-2, 3)$ **(c)** $x = \{-2, 3\}$ **3. (a)** $(-1, 1)$ **(b)** $(1, \infty)$ **(c)** $(-\infty, -1]$ and $x = 1$

5. (a) $(-\infty, -3) \cup (1, \infty)$ **(b)** $(-3, 1)$ **(c)** $x = \{-3, 1\}$ **7. (a)** $(-4, -1) \cup (3, \infty)$ **(b)** $(-\infty, -4) \cup (-1, 3)$ **(c)** $x = \{-4, -1, 3\}$

9. (a) No critical values
  (b) Increasing: $(-\infty, \infty)$
  (c) No relative extrema
  (d)

11. (a) $x = \{-4, 2\}$
  (b) Increasing: $(-\infty, -4) \cup (2, \infty)$
  Decreasing: $(-4, 2)$
  (c) Relative maximum: $(-4, 100)$
  Relative minimum: $(2, -8)$
  (d)

13. (a) $x = \frac{2}{3}$
  (b) Increasing: $(\frac{2}{3}, \infty)$
  Decreasing: $(-\infty, \frac{2}{3})$
  (c) Relative minimum: $(\frac{2}{3}, 0)$
  (d)

15. Relative maximum at $x = \frac{8}{3}$ **17.** Relative maximum at $x = -3$, Relative minimum at $x = 3$

19.

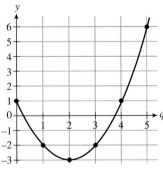

21. (a) $p(30) \approx 1.83$ **(b)** $R(x) = 4.5xe^{-0.03x}$
  (c) Increasing: $(0, 33.33)$; Decreasing: $(33.33, \infty)$
  (d) Relative maximum $\approx (33.33, 55.18)$; If 33,330 bottles of dog shampoo are sold, \$55,180 will be generated in revenue.
  (e)

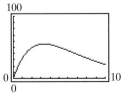

23. (a) $AC(x) = \dfrac{800}{x} + 15 + \dfrac{x}{18}$ **(b)** $x = 120$ **(c)** Increasing: $(120, 500)$; Decreasing: $(0, 120)$

(d) Relative minimum $\approx (120, 28.33)$. At 120 telephones per day, the average cost is minimized—\$28.33 per phone.

25. $f'(x) = 3x^2 - 14x; f''(x) = 6x - 14; f'''(x) = 6$ **27.** $y' = \dfrac{1}{3(x-5)^{2/3}}; y'' = -\dfrac{2}{9(x-5)^{5/3}}; y''' = \dfrac{10}{27(x-5)^{8/3}}$

29. (a) Concave down: $(-\infty, \frac{1}{3})$; Concave up: $(\frac{1}{3}, \infty)$ **(b)** Inflection point: $(\frac{1}{3}, -\frac{146}{27})$

31. (a) Concave up: $(0, \infty)$ **(b)** No inflection point(s)   **33. (a)** Concave up: $(-\infty, 0)$; Concave down: $(0, \infty)$

(b) No inflection point(s)   **35. (a)** Concave up: $(-\infty, \infty)$ **(b)** No inflection points

37. (a) Increasing: $(-\infty, -3) \cup (2, \infty)$; Decreasing: $(-3, 2)$ **(b)** Relative maximum: $(-3, 81)$; Relative minimum: $(2, -44)$

(c) Concave up: $(-\frac{1}{2}, \infty)$; Concave down: $(-\infty, -\frac{1}{2})$ **(d)** Inflection point: $(-\frac{1}{2}, \frac{37}{2})$

39. (a) Decreasing: $(-\infty, 2)$ **(b)** No relative extrema   **(c)** Concave down: $(-\infty, 2)$ **(d)** No inflection point(s)

41. (a) $P(4) = 155$. When 400 cookies are baked and sold, a \$155 profit is made.; $P'(4) = 68$. When 400 cookies are sold, the profit for selling the next 100 cookies is about \$68.   **(b)** Increasing $\approx (0, 9.02)$; Decreasing $\approx (9.02, \infty)$; Relative maximum $\approx (9.02, 357.51)$. If 902 cookies are baked and sold, profits will be maximized at \$357.51.   **(c)** Increasing: $(0, 2)$; Decreasing: $(2, \infty)$. The concavity of P changes a $x = 2$.   **(d)** $(2, 11)$

43. (a)

1500

0   100
0

(b) $p(84) \approx 1369.33$. Eighty-four years after first being introduced, there are about 1369 goats in the region. $p'(84) \approx 2.7$. Eighty-four years after first being introduced, the number of goats is increasing at a rate of about 2.7 $\dfrac{\text{goats}}{\text{year}}$.   **(c)** $p''(t) = \dfrac{487.62(43e^{-.18x} - e^{-.09x})}{(1 + 43e^{-.09x})^3}$

(d) Concave up: $(0, 41.79)$; Concave down: $(41.79, 100)$

(e) about 42 years after they were first introduced

45. $f > 0: (1, 5); f < 0: (-5, 1)$ **47.** $f' > 0: (-5, -3) \cup (-1, 3); f' < 0: (-3, -1) \cup (3, 5)$

49. Concave up: $(-2, 1)$; Concave down: $(-5, -2) \cup (1, 5)$ **51.** Inflection points: $(-2, -3)$ and $(1, 0)$

53. Relative maximum: $(-\frac{2}{3}, \frac{76}{27})$; Relative minimum: $(4, -48)$ **55.** Relative maximum: $(0, \frac{1}{9})$

57.

$(-1, \frac{21}{2})$ 150 $(\frac{1}{4}, \frac{43}{16})$

-5 5

-150 $(\frac{3}{2}, -\frac{41}{8})$

59.

$5\,(0, \frac{1}{4})$

-5 5

$(x = -2)$ -5 $(x = 2)$

61.

100

-5 5

-10

$(1.38, -1.54)$

63. Answers will vary. Sample: **(a)**

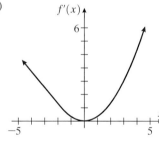

(b)

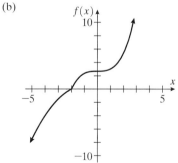

65. (a) $AC(x) = 15 + \dfrac{200}{x}$ **(b)** $AC(x)$ is always decreasing, so it has no minimum.   **(c)** $AP(x) = 70 - 0.12x^2 - \dfrac{200}{x}$ **(d)** $x \approx 9$

67. (a) $f(9) \approx 26$. In 1989, there were 26 African-American representatives. $f'(9) \approx 1.5$. In 1989 the number of African-American representatives was increasing at a rate of about 1.5 $\dfrac{\text{members}}{\text{year}}$.   **(b)** $f'(t) = 0.9106e^{0.058t}$. Over the interval $[1, 15]$, $f'(t)$ is always > 0. Therefore f, which represents the number of African-American representatives, is always increasing.

(c) $f''(t) \approx 0.053e^{0.058t}; f''(t) > 0$ over the interval $[1, 15]$. Therefore, f is concave up and is increasing at a faster rate.

(d)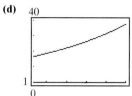

69. Absolute maximum: $(5, 320)$; Absolute minimum: $(-5, -20)$

71. Absolute maximum: $(5, \frac{1310}{3})$; Absolute minimum: $(-5, -\frac{1310}{3})$

73. Absolute maximum: $(10, \frac{52}{5})$; Absolute minimum: $(2, 4)$

75. Absolute maximum: $(19.75, 304.81)$. In 1991, \$304.81 billion dollars were spent on defense.; Absolute minimum: $(1, 70.85)$. In 1973, \$70.85 billion dollars were spent on defense.

77. 40 ft $\times$ 30 ft **79. (a)** $R(x) = -0.002x^3 + 1.5x^2 - 1.7x$ **(b)** \$124,150 per week **(c)** 500 units per week **(d)** \$248.30 per unit

81. $(\sqrt{2}, 6\sqrt{2}) \approx (1.41, 8.49)$ **83.** $\left(\frac{\sqrt{6}}{2}, 10\right) \approx (1.22, 10)$ **85.** $\left(\frac{\sqrt{3}}{2}, -4\sqrt{3}\right) \approx (0.87, -6.93)$ **87.** $(\sqrt{3}, -1) \approx (1.73, -1)$

89. ≈ 10 ft^2 **91.** $18 \times 9 \times 3$ inches **93.** 36 trees

CHAPTER 6 Integral Calculus

Section 6.1 Exercises

1. Yes **3.** No **5.** Yes **7.** No **9.** Yes **11.** $\frac{x^5}{5} + C$ **13.** $\frac{x^{3.31}}{3.31} + C$ **15.** $-\frac{1}{2t^2} + C$ **17.** $\frac{4}{9}x^{9/4} + C$ **19.** $\frac{3}{2}x^{2/3} + C$

21. $\frac{2}{35}x^7 + C$ **23.** $x^2 + 3x + C$ **25.** $\frac{1}{3}x^2 + 4x + C$ **27.** $t^3 + t^2 + 10t + C$ **29.** $x - \frac{2}{3}x^3 + \frac{3}{4}x^4 + C$

31. $2.07x^3 + 0.015x^2 - 4.01x + C$ **33.** $-\frac{1}{x} + \frac{3}{2x^2} + C$ **35.** $3x + \frac{4}{3}(\sqrt{x})^3 + C$ **37.** $-\frac{3}{2x^2} + \frac{4}{5}x^{5/2} - 4x + C$ **39.** $\frac{1}{6}t^3 - \frac{1}{3}t + C$

41. $-\frac{1}{20z^2} - \frac{2}{z} + \frac{z^4}{4} + C$ **43.** $\frac{200}{113}t^{1.13} + 5t + C$ **45.** $f(x) = -2x + 4$ **47.** $f(x) = \frac{5}{2}x^2$ **49.** $f(x) = x^2 - 3x + 4$

51. $f(x) = 500t - \frac{t^2}{40} + 40$ **53.** $S(x) = -2x^{-1} - \frac{3}{2}x^{-2} - x + \frac{13}{2}$ **55.** $y = 3(\sqrt[3]{t})^5 + 3(\sqrt[3]{t})^2 + 1$ **57.** $R(t) = -\frac{t^{-2}}{2} - \frac{t^2}{2} + 5$

59. (a) $P(q) = 40q - 0.025q^2$ **(b)** \$7000 **61. (a)** $R(x) = 0.000015x^3 - 0.015x^2 + 3.75x$ **(b)** $p(x) = 0.000015x^2 - 0.015x + 3.75$

(c) \$2.40 **63. (a)** $AC(x) = \frac{100}{x} + 1.50$ **(b)** $C(x) = 100 + 1.50x$ **(c)** The cost of producing 100 banners is \$250.

65. (a)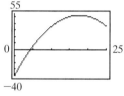

Arrests were decreasing at an increasing rate from 1970 until about 1974 when they began increasing at an increasing rate until about 1988 when they began increasing at a decreasing rate.

(b) $a(t) = -0.1t^3 + 5.28t^2 - 40.31t + 515.76$

(c)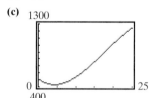

(d) 1,071,630 arrests

Section 6.2 Exercises

1. 0.5 **3.** 0.5 **5.** 0.5 **7.** 0.5 **9.** $\frac{2}{7}$ **11. (a)** 10.0 **(b)** 10.0 **(c)** 0.0 **13. (a)** 20.25 **(b)** 24.75 **(c)** 4.5

15. (a) 13.0 **(b)** 15.0 **(c)** 2.0 **17. (a)** 28.0 **(b)** 60.0 **(c)** 32.0 **19. (a)** 9.625 **(b)** 8.125 **(c)** 1.5

21.

N	LEFTSUM	RTSUM	|LEFTSUM−RTSUM|
10	9.24	6.84	2.4
100	8.1204	7.8804	0.24
1000	8.012004	7.988004	0.024

23.

N	LEFTSUM	RTSUM	|LEFTSUM−RTSUM|
10	36.11837	38.51837	2.4
100	37.21318	37.45318	0.24
1000	37.32133	37.34533	0.024

25.

N	LEFTSUM	RTSUM	|LEFTSUM–RTSUM|
10	32.895	39.195	6.3
100	35.68545	36.31545	0.63
1000	35.9685	36.0315	0.063

27.

N	LEFTSUM	RTSUM	|LEFTSUM–RTSUM|
10	30.24	51.04	20.8
100	38.9664	41.0464	2.08
1000	39.89606	40.10406	0.208

29.

N	LEFTSUM	RTSUM	|LEFTSUM–RTSUM|
10	33.25	39	5.75
100	35.79625	36.37125	0.575
1000	36.05459	36.11209	0.0575

31. (a) $\int_3^9 10 \, dx$ **(b)** Left Sum $= 60$; Right Sum $= 60$

33. (a) $\int_0^{10} \frac{1}{2} x \, dx$ **(b)** Left Sum $= 99$; Right Sum $= 101$

35. (a) $\int_1^4 \frac{1}{x^2} \, dx$ **(b)** Left Sum ≈ 0.7642101; Right Sum ≈ 0.7360851

37. (a) $\int_0^3 (9 - x^2) \, dx$ **(b)** Left Sum $= 18.13455$; Right Sum $= 17.86455$

39. (a) $\int_{-1}^2 (x^2 + 2) \, dx$ **(b)** Left Sum $= 8.95545$; Right Sum $= 9.04545$

41. (a) $\int_1^2 5 \, dx + \int_2^4 (x^2 + 1) \, dx$ **(b)** Left Sum $= 25.5468$; Right Sum $= 25.7868$

43. $y = x^2$

45. $y = \dfrac{1}{x^2}$

47. $y = \sqrt[3]{x}$

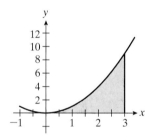

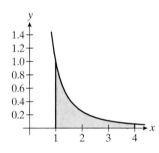

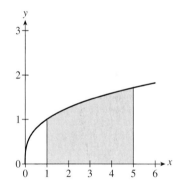

49. $y = \dfrac{2}{\sqrt{x}}$

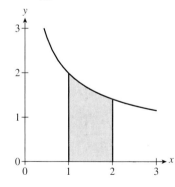

Section 6.3 Exercises

1. $\frac{45}{2}$ **3.** 10   **5.** 14   **7.** $\frac{128}{3}$ **9.** 9   **11.** 8   **13.** $\frac{112}{3}$ **15.** $\frac{165}{4}$ **17.** $\frac{433}{12}$ **19.** $-\frac{85}{2}$ **21.** $-\frac{8}{5}$ **23.** $\frac{20}{3}$

25. **(a)** 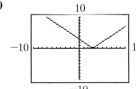   **(b)** $f(x) = \begin{cases} -x + 3, & x < 3 \\ x - 3, & x \geq 3 \end{cases}$ **27.** 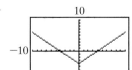   **(b)** $f(x) = \begin{cases} -x - 4, & x < 0 \\ x - 4, & x \geq 0 \end{cases}$

(c) $\frac{29}{2}$

(c) $-4, 4$

(d) Net area $= -12$,

gross area $= 16$

29. $\frac{25}{2}$ **31.** $\frac{14}{3}$ **33.** $\frac{106}{3}$ **35. (a)** -2 **(b)** 20   **(c)** 29   **37. (a)** 4   **(b)** 0   **(c)** $\frac{32}{3}$ **39. (a)** $-3, 3$ **(b)** $-\frac{70}{3}$ **(c)** 30

41. (a) $-1, 2$ **(b)** $\frac{25}{6}$ **(c)** $\frac{79}{6}$ **43. (a)** $-2, 0, 1$ **(b)** $\frac{16}{3}$ **(c)** $\frac{37}{6}$

45. (a) At a production level of 10 $\frac{\text{shoes}}{\text{shift}}$, the cost is increasing at a rate of 42.00 $\frac{\text{dollars}}{\text{shoe}}$.

(b) Total increase in cost of producing 0 to 50 shoes is $1500.

47. (a) The total increase in revenue from producing and selling 20th to 30th satellite dish is $65,500.

(b)

$$R'$$

```
20000
15000
10000
 5000
        15  20  25  30  35   x
```

49. (a) $P(x) = 0.1x^3 + 0.1x^2 - 136.0$

(b) The total increase in profit from producing and selling 10 to 20 machines is $730.

51. (a) In 1976 the rate of imports was 501.12 $\frac{\text{thousands of barrels imported each day}}{\text{year}}$.

(b) 2409.7 thousand of barrels imported each day.

53. (a) In 1989, military expenditures by NATO countries was decreasing at a rate of 10.17 $\frac{\text{billions of dollars}}{\text{year}}$.

(b) The total decrease in military expenditures from 1989 to 1993 was 57.32 billion dollars.

55. (a) In 1990, jet fuel consumption was increasing at a rate of 1903.72 $\frac{\text{millions of gallons}}{\text{year}}$.   **(b)** 11,459.2 millions of gallons

57. (a) $F(x) = 0.09x^4 - 3.42x^3 + 39.26x^2 - 123.18x + 2961$ **(b)** From 1985 to 1995, the total increase in the number of employees was 5.2 thousand.   **59. (a)** 3 $\frac{\text{feet}}{\text{second}}$ **(b)** Between 10 and 20 seconds the object moved 45 feet.

61. (a) 16 $\frac{\text{feet}}{\text{second}}$ **(b)** Between 3 and 5 seconds the object moved 36 feet.

Section 6.4 Exercises

1. $\frac{1}{3}(3x + 4)^3 + C$ **3.** $\frac{(4x^2 + 1)^4}{4} + C$ **5.** $\frac{2}{3}(\sqrt{(t^2 - 1)})^3 + C$ **7.** $\frac{(x^3 - 2x)^2}{2} + C$ **9.** $\frac{3}{2}(\sqrt[3]{x^3 - 5})^2 + C$

11. $-\frac{1}{2(2x^2 + 3)^2} + C$ **13.** $\frac{1}{20}(4x + 7)^5 + C$ **15.** $\frac{1}{12}(t^2 - 3)^6 + C$ **17.** $\frac{2}{3}\sqrt{(x^3 - 5)} + C$ **19.** $-\frac{1}{8(5x^2 - 2)} + C$

21. $\frac{2}{3}(\sqrt{1 - x^{-1}})^3 + C$ **23.** $\frac{1}{11}(t - 7)^{11} + C$ **25.** $-\frac{1}{8(x^2 + 2x + 5)^4} + C$ **27.** $\frac{3}{2}(x^2 + 3x - 10)^2 + C$

29. $-\frac{2}{3}(\sqrt{x + 1})^3 + \frac{2}{5}(\sqrt{x + 1})^5 + C$ **31.** $\frac{(t - 1)^6}{6} + \frac{(t - 1)^7}{7} + C$ **33.** $6\sqrt{x - 1} + 2(\sqrt{x - 1})^3 + C$ **35.** 4   **37.** 0   **39.** 52

41. $2\sqrt{3}$ **43.** $\frac{1}{8}$ **45.** 4   **47.** $\frac{9}{20}$ **49.** $\frac{18}{19}$ **51.** $\frac{1640}{6561}$ **53.** $\frac{52}{3}$ **55.** 0   **57.** $\frac{24}{5}$ **59.** $\frac{1}{2}$ **61.** $\frac{16}{3}\sqrt{2}$

63. (a) When daily production is 75 fobs, it costs about $6 to produce the 76th fob.

(b) The total increase in daily cost of producing from 0 to 75 fobs is $250.   **(c)** $C(x) = 10\sqrt{(x^2 + 10,000)} + 2000$ **(d)** 3000 dollars

65. (a) When monthly production and sales is 70 flags the profit from the 71st flag is about $144.40.

(b) The total increase in profit on the 20th to 100th flags sold each month is 12,000 dollars.   **(c)** $\sqrt{(2x^2 - 400)} - 20$

67. (a) ≈ 1.59 feet   **(b)** $s(t) = \sqrt{(t^2 + 9)} + 3$ **69. (a)** Five years after the town was incorporated homes were being built at a rate

of about $33 \dfrac{\text{homes}}{\text{year}}$.   **(b)** Ninety-four homes were built during the first five years after the town was incorporated.

(c) $h(t) = 10\sqrt{(t^3 + 4)} - 20$ **(d)** There were 94 homes in the town after five years.

Section 6.5 Exercises

1. $2\ln|x| + C$ **3.** $-\frac{2}{3}\ln|t| + C$ **5.** $(1 + e)\ln|x| + C$ **7.** $3\ln 5$ **9.** $\frac{1}{2}(\ln 5 - \ln 2)$ **11.** $\frac{1}{2}\ln|2x + 3| + C$

13. $\frac{1}{2}\ln(t^2 + 1) + C$ **15.** $\frac{1}{2}\ln|x^2 - 4x + 9| + C$ **17.** $\frac{1}{4}\ln(x^4 + 10) + C$ **19.** $\ln|\ln 2x| + C$ **21.** $4\ln(\sqrt{2} + 4) - 4\ln 5$

23. $\frac{2}{3}\ln 7 - \frac{2}{3}\ln 2$ **25. (a)** The sales rate after two days is $20 \dfrac{\text{dresses}}{\text{day}}$.   **(b)** $40\ln 2 \approx 28$ **27. (a)** In 1981, enrollment was increasing

by $15.405 \dfrac{\text{thousands of students}}{\text{year}}$.   **(b)** From 1976 to 1981 foreign enrollment increased by about 165.61 thousand students.

29. (a) In 1996, the number of employees was increasing at a rate of $3.1435 \dfrac{\text{thousands of employees}}{\text{year}}$.   **(b)** $f(x) = 81.356 + 53.44\ln(x + 1)$

31. $2e^x + C$ **33.** $e^x - x + C$ **35.** $4e^t + \frac{1}{2}t^2 - t + C$ **37.** $e - 2$ **39.** $\frac{1}{2}(e^2 - e + 4)$ **41.** $\frac{1}{4}e^{4x+1} + C$ **43.** $e^{x^2+1} + C$

45. $\ln(e^x + 2) + C$ **47.** $3\ln|1 - e^{-x}| + C$ **49.** $2e^{\sqrt{x}} + C$ **51.** $-\frac{1}{4}[e^{-12} - e^{-4}]$ **53.** $2 - e^{-1}$ **55.** $\frac{1}{2}e^{x^2} + C$

57. $\frac{1}{8}(e^{4x} + 4)^2 + C$ **59. (a)** At a production level of 20 faucets the cost to produce the 21st faucet is about $1.56.

(b) \$100.39   **61. (a)** In 1975, infant deaths were decreasing at a rate of about $0.00978 \dfrac{\text{infant deaths per 1000 live births}}{\text{year}}$.

(b) From 1965 to 1975 the total decrease in infant mortality was $1.93 \dfrac{\text{infant deaths}}{\text{1000 live births}}$.   **63. (a)** In 1950 the United States population was

increasing at a rate of about $1.973 \dfrac{\text{millions}}{\text{year}}$.   **(b)** 72.539 million   **65.** $\dfrac{1}{\ln 5}5^x + C$ **67.** $-\dfrac{5}{\ln 2}\left(\dfrac{1}{2}\right)^x + C$ **69.** $-\dfrac{1}{\ln 7}7^{-x} + C$

71. $\dfrac{333{,}000}{\ln 10}$ **73.** $\dfrac{21}{16\ln 2}$ **75. (a)** In 1990 the prison population was growing at a rate of $59{,}603 \dfrac{\text{inmates}}{\text{year}}$.   **(b)** about 244,198 inmates

Section 6.6 Exercises

1. Yes   **3.** No   **5.** Yes   **7.** No   **9.** Yes   **11.** Yes   **13.** Yes   **15.** $y = -2x + C$ **17.** $y = x - \frac{3}{2}x^2 + C$

19. $y = \frac{5}{4}x^4 + \frac{1}{2}x^2 - 2x + C$ **21.** $y = \frac{2}{3}(\sqrt{x})^3 + 2x + C$ **23.** $y = -\dfrac{5}{x} + C$

25.

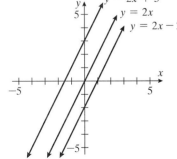

27.

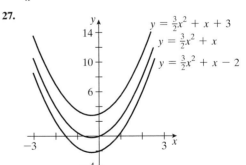

29.

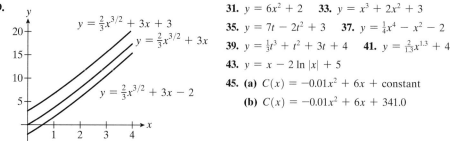

$y = \frac{2}{3}x^{3/2} + 3x + 3$

$y = \frac{2}{3}x^{3/2} + 3x$

$y = \frac{2}{3}x^{3/2} + 3x - 2$

31. $y = 6x^2 + 2$ **33.** $y = x^3 + 2x^2 + 3$

35. $y = 7t - 2t^2 + 3$ **37.** $y = \frac{1}{4}x^4 - x^2 - 2$

39. $y = \frac{1}{3}t^3 + t^2 + 3t + 4$ **41.** $y = \frac{2}{1.3}x^{1.3} + 4$

43. $y = x - 2\ln|x| + 5$

45. (a) $C(x) = -0.01x^2 + 6x + \text{constant}$

  (b) $C(x) = -0.01x^2 + 6x + 341.0$

47. (a) Ten months after its introduction, the percentage of drinkers accepting the new formula is increasing at a rate of $11.1\ \dfrac{\text{percent}}{\text{month}}$.

(b) The percentage of drinkers accepting the new formula after t months is given by $A(t) = 0.5t^2 + 1.1t + C$ where C is determined by initial conditions. **(c)** $A(t) = 0.5t^2 + 1.1t + 5$ **49. (a)** In 1989, deaths from heart disease were decreasing at a rate of $1.7567\ \dfrac{\text{deaths}}{\text{100,000 Americans}}$. **(b)** $D(t) = 336.17t^{-0.06} + .01$ **(c)** $D(15) \approx 285.76$; In 1994, the number of deaths from heart disease was about $285.76\ \dfrac{\text{deaths}}{\text{100,000 Americans}}$. **51.** $y^2 = x^2 + C$ **53.** $y = Ce^{x^3}, y > 0$ **55.** $e^{-y} = C - e^x$ **57.** $y = Ce^{x(x+2)}, y > 0$

59. $y^{1/2} = \frac{1}{3}x^{3/2} + C$ **61.** $y = Ce^{e^x}$ **63.** $y = e^{x^2}$ **65.** $\dfrac{1}{y^3} = -3x^2 + 1$ **67.** $e^y = e^x$ or $y = x$ **69.** $y = 10e^{(x^4-81)/4}$

71. $y = \sqrt{2(1 + x^2)}$

Section 6.7 Exercises

1. (a) $P' = 0.05P, P(0) = 50,000$ **(b)** $P = 50,000e^{0.05t}$ **(c)** 82,436

3. (a) $P' = 0.12P$ **(b)** $P = 5000e^{0.12t}$ **(c)** $P(3) = 7167, P(12) = 21,103, P(24) = 89,071$

5. (a) $N(t) = 39.9e^{0.01728t}$

  (b) 953.25 millions of board feet

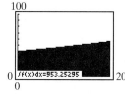

7. (a) $y' = 0.034y, y(0) = 6$

  (b) In 1993 the percentage of wood waste generated in the United States was 6.64 and was growing at a rate of $0.2259\ \dfrac{\text{percent}}{\text{year}}$.

9. (a) $f(t) = 3000e^{(-\ln 2/14.2)t}$ **(b)** 8.31

11. (a) $P' = -0.01822P, P(0) = 920,000$ **(b)** $P = 920,000e^{-0.01822t}$ **(c)** 662,760

13. (a) $y' = -0.09y, y(0) = 6.54$ **(b)** In 1993 the percentage of Caucasians unemployed was 4.99 and was decreasing at a rate of about $0.449\ \dfrac{\text{percent}}{\text{year}}$. **15. (a)** $y' = (10 - y)$ **(b)** $f(0) = 0, L = 10$ **17. (a)** $y' = 2(8 - y)$ **(b)** $f(0) = 0, L = 8$

19. (a) $y' = 0.5(20 - y)$ **(b)** $f(0) = 0, L = 20$ **21. (a)** $y' = 0.1(12 - y)$ **(b)** $f(0) = 0, L = 12$ **23. (a)** $y' = 0.2(1.5 - y)$ **(b)** $f(0) = 0, L = 1.5$ **25. (a)** $y' = 0.1(15.5 - y)$ **(b)** $f(0) = 0, L = 15.5$ **27. (a)** Two weeks after installation about 1354 employees have learned the new system and about $494\ \dfrac{\text{employees}}{\text{week}}$ are learning it. **(b)** about 5.97, or 6, weeks

29. (a) 300 **(b)** $-\ln\frac{15}{13}$

  (c)

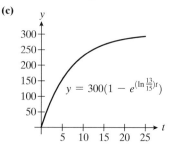

$y = 300(1 - e^{(\ln \frac{13}{15})t})$

  (d) No

31. (a) $L = 75, k \approx 0.14876$

  (b) $y = 75(1 - e^{-0.14876t})$

  (c) $y'(5) = 5.303, y'(25) = 0.27065$

33. (a) $y' = \frac{1}{15}y(15 - y)$ (b) $y(0) = \frac{15}{2}, L = 15$ **35.** (a) $y' = \frac{1}{10}y(10 - y)$ (b) $y(0) = \frac{5}{3}, L = 10$

37. (a) $y' = \frac{1}{30}y(90 - y)$ (b) $y(0) = 45, L = 90$ **39.** (a) $y' = \frac{1}{625}y(2500 - y)$ (b) $y(0) = \frac{1250}{3}, L = 2500$

41. (a) $y' = \frac{3}{10,000}y(4000 - y)$ (b) $y(0) = \frac{4000}{7}, L = 4000$

43. (a) After 8 completed games, 10,019 students can be expected to toss toilet paper.   (b) 6

45. (a) 30,000   (b) 5.7333×10^{-5} (c) $y = \dfrac{30,000}{1 + 3000^{-1.72t}}$

47. (a) $\dfrac{dy}{y(L - y)} = k\, dt$ (b) $\dfrac{d}{dt}\left[\dfrac{1}{L}\ln\left(\dfrac{y}{L - y}\right)\right] = \dfrac{1}{L} \cdot \dfrac{1}{\left(\dfrac{y}{L - y}\right)} \cdot \dfrac{1(L - y) - y(-1)}{(L - y)^2} = \dfrac{1}{y(L - y)}$

(c) $\dfrac{1}{L}\ln\left(\dfrac{y}{L - y}\right) = kt + C \Rightarrow \dfrac{y}{L - y} = e^{kLt+CL} \Rightarrow y = \dfrac{Le^{kLt+CL}}{1 + e^{kLt+CL}} = \dfrac{L}{e^{-kLt-CL} + 1} = \dfrac{L}{1 + ae^{-kLt}}$ where $a = e^{-CL}$

(d) $y' = -\dfrac{L}{(1 + ae^{-kLt})^2} \cdot ae^{-kLt} \cdot (-kL) = k\left(\dfrac{L}{1 + ae^{-kLt}}\right)\left(\dfrac{Lae^{-kLt}}{1 + ae^{-kLt}}\right) = k\left(\dfrac{L}{1 + ae^{-kLt}}\right)\left(L - \dfrac{L}{1 + ae^{-kLt}}\right) = ky(L - y)$;

Since $k > 0$ and $0 < y < L$, we have $ky(L - y) = y' > 0$.

(e) As $t \to \infty$, $ae^{-kLt} \to 0$. Hence $\displaystyle\lim_{t \to \infty}\left(\dfrac{L}{1 + ae^{-kLt}}\right) = \dfrac{L}{1} = L$.

Chapter 6 Review

1. No   **3.** Yes   **5.** $\dfrac{x^8}{8} + C$ **7.** $-\dfrac{1}{4t^4} + C$ **9.** $0.0625x^8 - 3x^2 + C$ **11.** $\dfrac{x^3}{3} + 0.2x^2 - \dfrac{2}{7}x^{7/2} + C$

13. $(-\infty, 2)$: decreasing; $(2, \infty)$: increasing; Concave up with relative minimum at $x = 2$.   **15.** $f(x) = \frac{3}{2}x^2 - 4x + 11$

17. $y = \frac{2}{3}(\sqrt{x})^3 - 23$ **19.** (a) $P(q) = 48q - 0.015q^2 - 3050.0$ (b) \$17,200.00   **21.** $\frac{1}{3}$ **23.** $\frac{1}{8}$ **25.** (a) 50   (b) 75

27. (a) 6.875   (b) 11.375

29.

N	LEFTSUM	RTSUM	|LEFTSUM–RTSUM|
10	705.6	961.6	256
100	816.576	842.176	25.6
1000	828.0538	830.6138	2.56

31.

N	LEFTSUM	RTSUM	|LEFTSUM–RTSUM|
10	211.8298	391.0298	179.2
100	285.5104	303.4304	17.92
1000	293.5047	295.2967	1.792

33. 120   **35.** $\frac{104}{3}$ **37.** 99   **39.** 24   **41.** 15   **43.** (a) $\frac{16}{3}$ (b) $\frac{32}{3}$

45. (a)

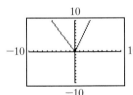

(b) 31

47. (a)

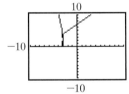

(b) $\frac{331}{6}$

49. (a) x-int: 3
(b) Net area $= 20$, gross area $= 29$

51. (a) x-int: $-1, 2$
(b) Net area $= 18$, gross area $= 27$

53. (a) At a production level of 160 calculators, the cost of producing the 161st is \$20.   (b) The total cost of producing the first 200 calculators is \$3400.   **55.** (a) In 1981, spending was increasing at a rate of $22.774 \dfrac{\text{billions of dollars}}{\text{year}}$.   (b) \$155.78 billion

57. $\frac{1}{4}(5x - 7)^4 + C$ **59.** $\frac{1}{2}(x^2 - 5x + 3)^2 + C$ **61.** $\frac{1}{27}(3x + 6)^9 + C$ **63.** $\frac{1}{2}\ln(x^2 + 4x + 6) + C$

65. $3\left(\dfrac{(t + 4)^4}{4} - \dfrac{4(t + 4)^3}{3}\right) + C$ **67.** $\frac{2}{3}(\sqrt{x + 3})^3 - 6\sqrt{x + 3} + C$ **69.** 2106   **71.** $2\sqrt{1327} - 4$ **73.** 0   **75.** $\frac{22,876}{3}$ **77.** 182

79. (a) 299,215,620 meters   (b) $s = \dfrac{(t^3 - 5t)^4}{4} - 4448$ **81.** $\frac{3}{4}\ln|t| + C$ **83.** $5e^x - \frac{3}{2}x^2 + C$ **85.** $6\ln\frac{5}{3}$ **87.** $\dfrac{24}{\ln 5}$

89. $10\ln(\sqrt{x} + 4) + C$ **91.** $-\frac{1}{4}(5 - e^x)^4 + C$ **93.** $2\ln 5 - 4\ln 2$ **95.** $\dfrac{1365}{64\ln 2}$

97. (a) In 1990 United States billed international calls were increasing at the rate of 7659.9 $\dfrac{\text{millions of calls}}{\text{year}}$. **(b)** From 1985 to 1995 the total increase in United States billed international calls was 518,688 million. **99.** $f(t) = 8.2e^{1.37t}$ **101.** $f(t) = 384e^{-0.87t}$

103. (a) $P' = 0.08P$ **(b)** $P(t) = 300e^{0.08t}$ **105. (a)** $f' = -0.17f, f(0) = 342$

(c)

t	$f(t)$
0	300
2	352
4	413
10	668
20	1486

(d)

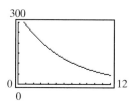

(b)

(c) In 1993 CFCs were released at a rate of about 146.18 $\dfrac{\text{thousands of metric tons}}{\text{year}}$.

(d) In 1993 the decrease in the rate of CFC release was 24.85 $\dfrac{\frac{\text{thousands of metric tons}}{\text{year}}}{\text{year}}$.

CHAPTER 7   Applications of Integral Calculus

Section 7.1 Exercises

1. 12   **3.** 2   **5.** $\frac{14}{3}$ **7.** $\frac{93}{35}$ **9.** $2(e^2 - 1) \approx 12.78$ **11.** Answers will vary. **(a)** 1.59076   **(b)** 0.79538

13. Answers will vary. **(a)** 1.44929   **(b)** 0.72464   **15.** $\approx$ \$6.2 billion   **17.** $\approx$ 22 million   **19.** $\approx$ \$8.98 billion

21. (a) \$4.2 million   **(b)** \$11.4 million   **23.** \$5.10   **25.** $\approx$ 12.7%   **27. (a)** \$135,000   **(b)** $AC = \dfrac{60,000}{x} + 300; AC(500) = 420$

(c) Part (b) is the average cost per diesel engine if 500 engines are built. Part (a) is the average total cost for building 500 engines.

29. \$3484.85   **31.** \$6137.05   **33. (a)** \$3567.35   **(b)** \$26.76   **35. (a)** \$1778.90   **(b)** \$9329.57

37. (a) \$175   **(b)** \$167.71. Between 25 and 75 mugs, total average cost is \$167.71.

39. (a)

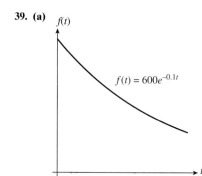

(b) $\displaystyle\int_0^{12} 600e^{-0.1t}dt$ **(c)** 4,193,000 barrels

41. \$6 million

43. $\approx$ \$20,344

45. $\approx$ \$188,235

47. $\approx$ \$679,207

49. $\approx$ \$588,313

51. (a) \$77,400   **(b)** \$80,000   **(c)** \$90,894. Vicki's contributions had longer to earn interest.

53. $\approx$ \$26,963

Section 7.2 Exercises

1. $\displaystyle\int_0^1 [x + 1 - x^2]dx$ **3.** $\displaystyle\int_{-1}^1 [x + 7 - x^2]dx$ **5.** $\displaystyle\int_a^c [g(x) - f(x)]dx + \displaystyle\int_c^b [f(x) - g(x)]dx$ **7.** $\displaystyle\int_0^b [d(x) - k]dx$ **9.** 28

11. $\frac{15}{2}$ **13.** $\frac{59}{6}$ **15.** $\frac{32}{3}$ **17.** $\frac{52}{3}$ **19.** 15   **21.** $\approx$ 30.88   **23.** $\approx$ 23.62   **25.** $\approx$ 2.66   **27.** $\frac{9}{2}$ **29.** 18   **31.** $\frac{1}{2}$ **33.** $\frac{1}{2}$ **35.** \$28,600

37. (a)

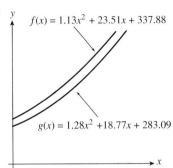

$f(x) = 1.13x^2 + 23.51x + 337.88$

$g(x) = 1.28x^2 + 18.77x + 283.09$

(b) $\int_1^{16} [-0.15x^2 + 4.74x + 54.79]dx$

(c) 1221.45. This means that receipts exceeded expenditures by over $1.2 trillion dollars between 1980 and 1996.

39. (a) $\int_1^2 4[t^{18/5} - t^{5/2}]dt$

(b) ≈ 8.43. Between the first and second minute, the use of the nutrient increases growth by about 8.43 mg.   **(c)** ≈ 8.16 mg

41. Imports exceeded exports by about $77.7 billion between 1989 and 1996.

43. ≈ 24,303 schools

45. (a) ≈ 1209. If the TV personality is used for the ad campaign, sales during the first 20 weeks will be increased by about $1,209,000.

(b) Yes. Sales will increase by $1,209,000, but the personality's salary is only $1 million.   **47.** ≈ $15.7 million

49. (a)

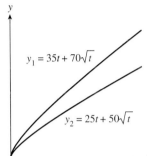

$y_1 = 35t + 70\sqrt{t}$

$y_2 = 25t + 50\sqrt{t}$

(b) ≈ 435. Between ages 4 and 8, the child will learn an additional 435 words by listening to the tapes.

51. Between year 3 and 8, the breakthrough saved Quality Chips about $11.6 million.

53. ≈ $1944.83

55. ≈ $80.94

57. (a) ≈ 97 units   **(b)** ≈ 13 units

59. ≈ 34.73 fewer feet of growth

Section 7.3 Exercises

1. $\int_0^{x_m} [d(x) - d_{mp}]dx$

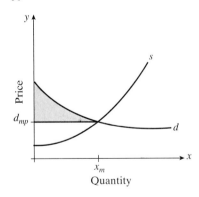

3. $d_{mp} \cdot x_m$

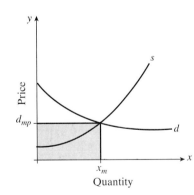

5. $\int_0^{700} [210 - 0.3x]\,dx$

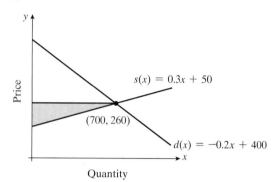

$s(x) = 0.3x + 50$

$(700, 260)$

$d(x) = -0.2x + 400$

Price

Quantity

7. $\int_0^{360} [180 - 0.5x]\,dx$

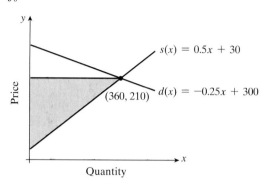

$s(x) = 0.5x + 30$

$(360, 210)$ $d(x) = -0.25x + 300$

Price

Quantity

9. $\int_0^{x_m} [d(x) - d_{mp}]\,dx$

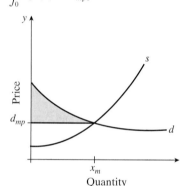

s

d

d_{mp}

Price

x_m

Quantity

11. \$60,000   **13.** $\approx$ \$1666.67   **15.** \$6750   **17.** $\approx$ \$53,333.33   **19.** \$6000

21. $\approx$ \$48,051.10   **23.** $\approx$ \$9937.29   **25.** \$562.50   **27.** \$750   **29.** \$600

31. $\approx$ \$33,333.33   **33.** \$225,000   **35.** $\approx$ \$2919.86   **37.** $\approx$ \$1690.54

39. (a) $(190, 1083)$ **(b)** \$234,650   **(c)** \$102,885

41. (a) $(430, 226)$ **(b)** \$36,980   **(c)** \$46,225

43. (a) $(165, 1654)$ **(b)** \$119,790   **(c)** \$179,685

45. (a) $(35, 255)$ **(b)** $\approx$ \$5716.67   **(c)** $\approx$ \$5716.67

47. (a) $\approx (209, 74)$ **(b)** $\approx$ \$37,112.77   **(c)** $\approx$ \$11,845.42

49. (a) 10   **(b)** \$70   **51.** $\approx$ \$59.13

53. (a) 65 dozen golf balls   **(b)** \$507   **55.** \$415.03

57. (a) $(30, 4)$ **(b)** \$180   **(c)** \$45   **59. (a)** $(25, 5)$ **(b)** \$312.50   **(c)** $\approx$ \$41.67

61. 0.4030   **63.** 0.4143   **65.** 0.4313   **67.** Increasing; inequality in income distribution became greater.

Section 7.4 Exercises

1. $e^{3x}(\frac{2}{3}x - \frac{2}{9}) + C$ **3.** $e^{4x}(\frac{1}{4}x - \frac{1}{16}) + C$ **5.** $e^{-x}(-x - 1) + C$ **7.** $e^{-.03x}(-\frac{100}{3}x - \frac{10,000}{9}) + C$ **9.** $-\dfrac{3}{e^2} + 1 \approx 0.594$

11. $-\dfrac{750}{e^{1/5}} + 625 \approx 10.95$ **13.** $\frac{1}{4}(e^2 + 1) \approx 2.097$ **15.** $x^3(\frac{1}{3}\ln x - \frac{1}{9}) + C$ **17.** $x(\ln|2x| - 1) + C$ **19.** $e^x(x + 2) + C$

21. $e^{-0.1t}(-60t - 600) + C$ **23.** $\dfrac{1250}{9}\left(10 - \dfrac{13}{e^{3/10}}\right) \approx 51.300$ **25.** $e^{-0.8t}(-15t - \frac{375}{2}) + C$ **27.** $(x + 1)^6\left(\dfrac{6x + 13}{42}\right) + C$

29. $(x + 3)^7\left(\dfrac{7x - 19}{56}\right) + C$ **31.** $e^{3x}\left(\dfrac{9x^2 - 6x + 2}{27}\right) + C$ **33.** $e^{-x}(-x^2 - 2x - 2) + C$ **35.** $e^{5x}\left(\dfrac{25x^2 - 10x + 2}{125}\right) + C$

37. $x(\ln x)^2 - 2x\ln x + 2x + C$ **39.** $\dfrac{4x^2(\ln x)^3 - 6x^2(\ln x)^2 + 6x^2\ln x - 3x^2}{8} + C$ **41.** $u = x, dv = e^{-x}dx$ (Parts)

43. $u = \ln x$ (Substitution)   **45.** $u = x + 6, dv = e^x dx$ (Parts)   **47.** $u = 3x, dv = e^{2x}dx$ (Parts)   **49. (a)** $\int_0^{12} 6te^{-0.15t}dt$

(b) $\approx$ 143,243 barrels   **51.** $P(t) = e^{0.2t}(10t - 50) + 50$ **53.** $C(t) = e^{-0.15t}\left(\dfrac{-60t - 400}{9}\right) + \dfrac{418}{9}$ **55.** $\approx$ 12.45 mg

57. $\approx$ \$20.5 million   **59.** $\approx$ \$593,534   **61.** Oil well: $\approx$ \$480,874; Natural gas: $\approx$ \$508,330; The natural gas well is a better investment.

63. (a) $\approx$ 3.98 million. On average, there were 3.98 million men annually enrolled in college in the United States during the 1960s.

(b) $\approx$ 5.21 million. On average, there were 5.21 million men annually enrolled in college in the United States during the 1970s.

(c) $\approx$ 5.76 million. On average, there were 5.76 million men annually enrolled in college in the United States during the 1980s.

65. $\approx$ 38.05. Between 1991 and 1996, paper and paperboard averaged 38.05% of all solid waste in the United States.

67. $\approx$ 2431. United States imports of crude oil averaged 2431 million barrels annually between January 1, 1990 and December 31, 1996.

69. (a) ≈ 22.7. During the 1980s, per capita consumption of cheese averaged 22.7 pounds annually.

(b) ≈ 26.0. During the first half of the 1990s, per capita consumption of cheese averaged 26.0 pounds annually.

71. ≈ 9.3. During the first half of the 1990s, variable costs of operating an automobile averaged 9.3¢ per mile.

Section 7.5 Exercises

1. 22   **3.** ≈ 1.48 **5.** ≈ 0.76 **7.** ≈ 3.83 **9.** $n = 10$: ≈ 6.507; $n = 50$: ≈ 6.404; $n = 100$: ≈ 6.401

11. $n = 10$: ≈ 0.4128; $n = 50$: ≈ 0.4005; $n = 100$: ≈ 0.4001 **13.** $n = 10$: ≈ 0.0690; $n = 50$: ≈ 0.0672; $n = 100$: ≈ 0.0671

15. $n = 10$: ≈ 2.4537; $n = 50$: ≈ 2.4517; $n = 100$: ≈ 2.4517 **17.** $n = 10$: ≈ 0.4091; $n = 50$: ≈ 0.3961; $n = 100$: ≈ 0.3957

19. $n = 10$: ≈ 6.1785; $n = 50$: ≈ 6.1725; $n = 100$: ≈ 6.1723 **21.** $n = 10$: ≈ 4.0511; $n = 50$: ≈ 4.0500; $n = 100$: ≈ 4.0499

23. 9.55   **25.** 100   **27.** $\approx \$3225$ **29.** $\approx \$21,025$ **31.** $\approx \$44,000$ **33. (a)** $\approx 10,349,214$ **(b)** $\approx 23,098,429$

35. (a) $\approx 1,223,140$ thousands of tons   **(b)** $\approx 24,463$ thousands of tons   **37. (a)** ≈ 6478 work stoppages

(b) ≈ 185 work stoppages annually   **39. (a)** ≈ 41.2 million   **(b)** $\approx 1,287,500$ million annually   **41.** $\approx \$12.677$ billion (in 1987 dollars)

43. (a) $\approx 96,538,000$ pregnancies   **(b)** $\approx 6,033,625$ pregnancies annually   **45.** $\approx 66,690$ nursing students annually were pursuing a 2-year degree

Section 7.6 Exercises

1. $\frac{1}{24}$ **3.** Diverges   **5.** $\sqrt{2} \approx 1.41$ **7.** Diverges   **9.** $\frac{1}{e} \approx 0.368$ **11.** $\frac{100}{3e^{3/100}} \approx 32.35$ **13.** Diverges   **15.** Diverges   **17.** $\frac{1}{2}$

19. Diverges   **21.** 5   **23.** $\frac{1}{4}$ **25.** $-\frac{1}{60} \approx -0.0167$ **27.** 1   **29.** 0   **31.** 0   **33.** Diverges   **35.** Diverges   **37.** 600 thousand barrels

39. 16 million ft^3 **41.** ≈ 1166.67 tons   **43.** ≈ 107 ml   **45.** $\approx \$166,667$ **47.** $\approx \$413,793$ **49.** $\approx \$1,076,923$ **51.** $\approx \$218,182$

53. $\approx \$106,667$ **55.** $\approx \$45,454$ **57. (a)** $\approx \$79,263$ **(b)** $\$80,000$ **59. (b)** ≈ 0.1054 **(c)** ≈ 0.4066

61. (b) 0.1875   **(c)** 0.046875   **63. (b)** ≈ 0.2387 **(c)** ≈ 0.2983 **(d)** ≈ 0.4493 **65. (b)** ≈ 0.3023 **(c)** ≈ 0.2109

67. (b) ≈ 0.0670 **(c)** ≈ 0.0607 **(d)** ≈ 0.7788

Chapter 7 Review Exercises

1. 13   **3.** 102   **5.** Answers will vary. **(a)** 11.8624   **(b)** 1.4828   **7. (a)** $\$4900$ **(b)** $AC(x) = \frac{1300}{x} + 18$; $AC(400) = \$21.25$

(c) Part (a) is the average total cost; part (b) is the average cost of one phone.   **9. (a)** $\$2457.29$ **(b)** $\$20.89$ **11.** $\approx \$25,258$

13. $\approx \$18,776$ **15.** $\int_{-\frac{1+\sqrt{17}}{2}}^{\frac{-1+\sqrt{17}}{2}} [4 - x^2 - x]dx$ **17.** 36   **19.** $\frac{125}{6} \approx 20.83$ **21.** $\frac{256}{3} \approx 85.33$ **23.** ≈ 211.43 **25.** $\frac{125}{6} \approx 20.83$

27. $\frac{149}{3} \approx 49.33$ **29.** ≈ 2705. If Lucky is used for the ad campaign, total sales in the first 15 months will increase by 2705 tricycles.

31. $\approx \$795,766$ **33. (a)** 201 units   **(b)** 41 units

35. $\int_0^{60} [-0.4x + 24]dx$

37. $\$1260$

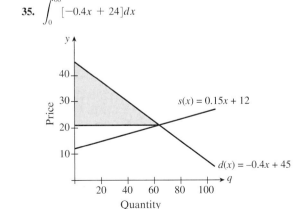

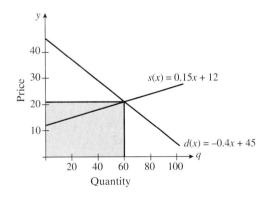

39. $\$4000$ **41.** $\$72$ **43.** $\$40,000$ **45.** $\approx \$1667$ **47. (a)** $(380, 136)$ **(b)** $\$21,660$ **(c)** $\$14,440$ **49. (a)** $(200, 2900)$

(b) $\approx \$213,333$ **(c)** $\$160,000$ **51. (a)** 16 drills   **(b)** $\approx \$42.67$ **53.** $\approx \$2301$ **55. (a)** $(36, 6)$ **(b)** $\$155.52$ **(c)** $\$72$

57. $e^{2x}(2x - 1) + C$ **59.** $x^5\left(\dfrac{5\ln x - 1}{25}\right) + C$ **61.** $e^{-0.2t}(-200 - 40t) + C$ **63.** $e^{-x}(-x^2 - 2x - 2) + C$

65. $x((\ln x)^4 - 4(\ln x)^3 + 12(\ln x)^2 - 24(\ln x) + 24) + C$ **67.** $u = -x^3$ (Substitution) **69.** $u = x, dv = e^{5x}$ (Parts)

71. (a) $\displaystyle\int_0^{12} 8te^{-0.17t}\,dt$ **(b)** $\approx 167{,}395$ barrels **73.** $\approx \$598{,}717$

75. (a) 58,955. The average sales price for a home in the United States between 1976 and 1981 was $58,955.

(b) 95,550. The average sales price for a home in the United States between 1983 and 1987 was $95,550.

(c) 143,405. The average sales price for a home in the United States between 1991 and 1996 was $143,405. **77.** -154

79. ≈ -0.6563 **81.** $n = 10: \approx 74.797; n = 50: \approx 74.798; n = 100: \approx 74.798$

83. $n = 10: \approx 1.679; n = 50: \approx 1.677; n = 100: \approx 1.677$ **85.** $\approx \$12{,}008$ **87.** $\approx \$153{,}500$ **89. (a)** $\approx 6{,}830{,}000$ **(b)** $\approx 683{,}000$

91. $\frac{1}{5}$ **93.** Diverges **95.** Diverges **97.** $\frac{1}{3}$ **99.** Diverges **101.** $\approx 895{,}833$ barrels **103.** $\approx \$88{,}889$

105. (b) ≈ 0.333 **(c)** ≈ 0.556

CHAPTER 8   Calculus of Several Variables

Section 8.1 Exercises

1. 19 **3.** 3 **5.** $\frac{5}{2}$ **7.** $-\frac{13}{5}$ **9.** 0 **11.** Undefined **13.** $2 + e^2$ **15.** $2e$ **17.** All ordered pairs (x, y)

19. All orderd pairs (x, y) such that $x + y \neq 0$ **21.** All ordered pairs (x, y) such that $x - 4y \neq 0$

23.

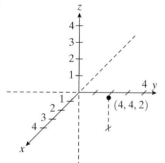

25.

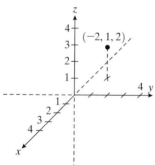

27.

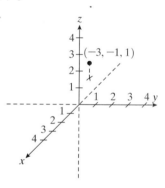

29.

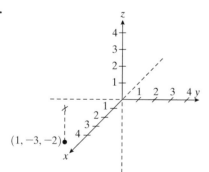

31.

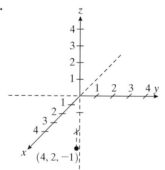

33.

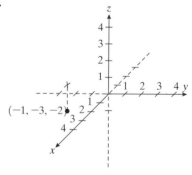

35.

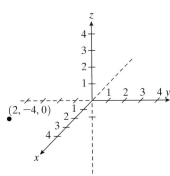

$(2, -4, 0)$

37. Above   **45.**
39. Above
41. Below
43. Below

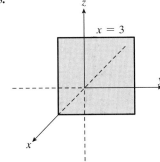

$x = 3$

47.

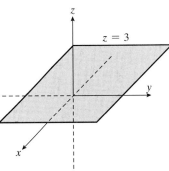

$z = 3$

49.

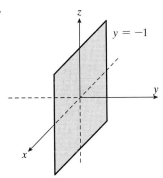

$y = -1$

51. **(a)** $S(5, 3) = 53$; When Seamount spends $5000 each week on newspaper advertising and $3000 each week on radio advertising, weekly sales are $530,000.

(b) $S(3, 5) = 23$; When Seamount spends $3000 each week on newspaper advertising and $5000 each week on radio advertising, weekly sales are $230,000.

53. **(a)** $TS(1, 0.5) = 1.9$; When Tube Town spends $1000 on radio advertising and $500 on television advertising, weekly sales are $19,000.

(b) $TS(0.5, 1) = 3.575$; When Tube Town spends $500 on radio advertising and $1000 on television advertising, weekly sales are $35,750.

55. $R(36, 1) = 36K$, $R(36, 2) = \frac{9}{4}K$

57. A tree with a 30 inch diameter and 12 foot length yields 507 board-feet.   **59.** The monthly cost of producing 50 leaf blowers per month and 30 attachments per month is $2795.00.   **61.** **(a)** $P(x, y) = 85x - 0.8x^2 - 0.25xy + 50.8y + 0.015y^2 - 1000.0$

(b) The monthly profit from sales of 50 leaf blowers per month and 30 attachments per month is $2412.5.

63. **(a)** $C(x, y) = 5x + 4y + 50$ **(b)** $R(x, y) = 12x + 9.5y$ **(c)** $P(x, y) = 7x + 5.5y - 50$ **65.** $f(2000, 20) = 6787.15$;

Two thousand dollars invested at an annual interest rate of 6.125% compounded monthly for 20 years yields $6787.15.

67. $f(0.0725, 25) = 12,251.49$; Two thousand dollars invested at an annual interest rate of 7.25% compounded continuously for 25 years

yields $12,251.49.   **69.** ≈ 7241 **71.** **(a)** A 9 year-old person with a 12 year-old mental age has an IQ of 133.33.

(b) A 12 year-old person with a 9 year-old mental age has an IQ of 75.00.   **(c)** A 12 year-old person with a 12 year-old mental age has

an IQ of 100.00.   **(d)** If $a = m$, then $IQ = 100$. An IQ of 100 represents a person with mental age equal to actual age.

73. **(a)**

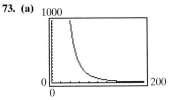

(b) 197   **(c)** 68   **(d)** Answers may vary. Sample answers: (119,40), (81, 129), (40, 1067)

Section 8.2 Exercises

1.

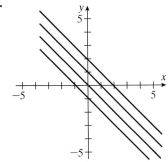

3.

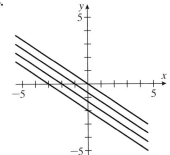

5.

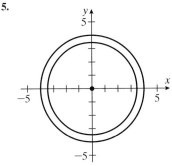

7.

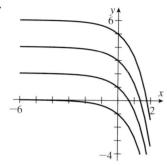

9.

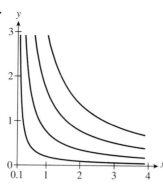

11.

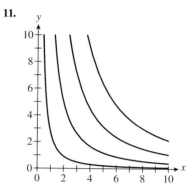

13.

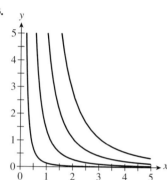

15.

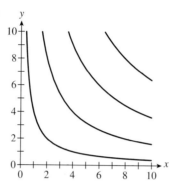

17.

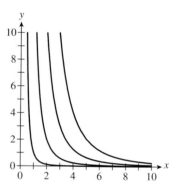

19.

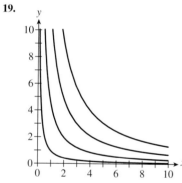

21.

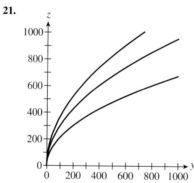

23.

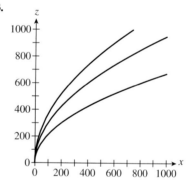

25.

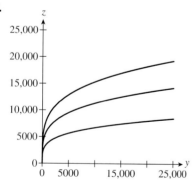

27.

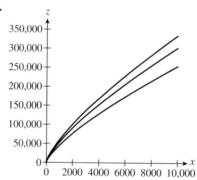

29.

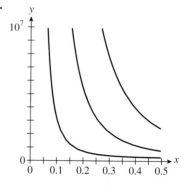

31.

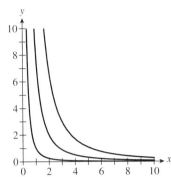

33.

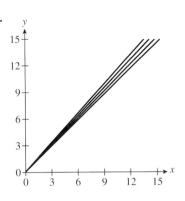

35.

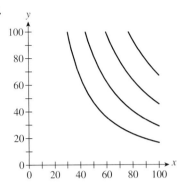

37. (a) Vertical

(b)

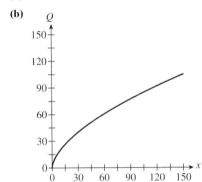

(c) about 137 thousand labor hours

(d) The graph shows the relationship between production and the number of labor hours when capital is held fixed at $18 million.

39. (a) Vertical

(b)

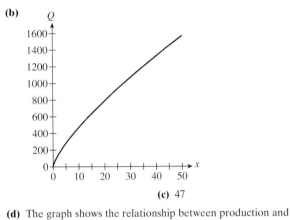

(c) 47

(d) The graph shows the relationship between production and the utilization of capital when labor utilization is held fixed at 100.

41.

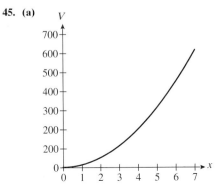

43. (a)

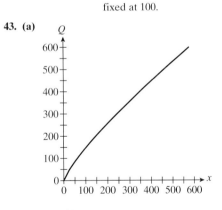

(b) about 528 thousand units of labor

(b) 3   **(c)** about 452

(d) The graph shows the relationship between the volume and the radius of a cylinder with the height held fixed at 4 inches.

45. (a)

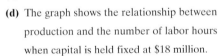

47. (a)

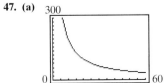

(b) A 15 year-old with a mental age of 20 has an IQ of $133\frac{1}{3}$.

(c) A 16 year-old with a mental age of 20 has an IQ of 125.

Section 8.3 Exercises

1. $f_x(x, y) = 10x, f_y(x, y) = -18y^2$ **3.** $f_x(x, y) = 2y, f_y(x, y) = 2x - 2y$ **5.** $f_x(x, y) = 3x^2y^2, f_y(x, y) = 2x^3y$

7. $f_x(x, y) = 2x + 6xy^3 - y, f_y(x, y) = 9x^2y^2 - 6y^2 - x$ **9.** $f_x(x, y) = 6x^2 + 2, f_y(x, y) = -2y$

11. $f_x(x, y) = 6(2x + 3y)^2, f_y(x, y) = 9(2x + 3y)^2$ **13.** $f_x(x, y) = 5.5815x^{-0.85}y^{0.87}, f_y(x, y) = 32.3727x^{0.15}y^{-0.13}$

15. $f_x(x, y) = e^x \ln y, f_y(x, y) = \dfrac{e^x}{y}$ **17.** $f_x(x, y) = -\dfrac{y}{x^2}, f_y(x, y) = \dfrac{1}{x}$ **19.** $f_x(x, y) = 6x - 6x^2y^4, f_y(x, y) = -8x^3y^3$

21. $f_x(x, y) = 4x(x^2 + y^3), f_y(x, y) = 6y^2(x^2 + y^3)$ **23.** $f_x(x, y) = \dfrac{2x}{x^2 + y^3}, f_y(x, y) = \dfrac{3y^2}{x^2 + y^3}$ **25.** $f_x(x, y) = \dfrac{2xy}{y + x} - \dfrac{x^2y}{(y + x)^2},$

$f_y(x, y) = \dfrac{x^2}{y + x} - \dfrac{x^2y}{(y + x)^2}$ **27.** $f_x(x, y) = 6xy^3 + e^{x+y}, f_y(x, y) = 9x^2y^2 + e^{x+y}$ **29.** $f_x(x, y) = ye^{xy}, f_y(x, y) = xe^{xy} + \dfrac{1}{y}$

31. If we are at the point $(1, 2, 24)$ on the surface and move along the surface in the x-direction parallel to the x-axis, the function is

decreasing at a rate of $20 \dfrac{\text{units of } f}{\text{units of } x}$. **33.** If we are at the point $(2, 3, 38)$ on the surface and move along the surface in the y-direction

parallel to the y-axis, the function is increasing at a rate of $5 \dfrac{\text{units of } f}{\text{units of } y}$. **35.** If we are at the point $(2, 5, \ln 23)$ on the surface and move

along the surface in the x-direction parallel to the x-axis, the function is decreasing at a rate of $\dfrac{1}{23} \dfrac{\text{units of } f}{\text{units of } x}$.

37. If we are at the point $(5, 2, 86.58)$ on the surface and move along the surface in the x-direction parallel to the x-axis, the function is

increasing at a rate of $2.5973 \dfrac{\text{units of } f}{\text{units of } x}$. **39.** If we are at the point $(2, 4, -64 - e^2)$ on the surface and move along the surface in the

x-direction parallel to the x-axis, the function is decreasing at a rate of $-96 - e^{-2} \dfrac{\text{units of } f}{\text{units of } x}$.

41. $f_{xx}(x, y) = -10y, f_{yy}(x, y) = 24y, f_{xy}(x, y) = -10x$ **43.** $f_{xx}(x, y) = 6 - 12xy^2 + 12x, f_{yy}(x, y) = -4x^3, f_{xy}(x, y) = -12x^2y$

45. $f_{xx}(x, y) = -36x, f_{yy}(x, y) = 0, f_{xy}(x, y) = 5$ **47.** $f_{xx}(x, y) = -12xy^2 + 14, f_{yy}(x, y) = 20y^3 - 4x^3, f_{xy}(x, y) = -12x^2y$

49. $f_{xx}(x, y) = y^4e^{xy} - \dfrac{1}{x^2}, f_{yy}(x, y) = 2e^{xy} + 4yxe^{xy} + y^2x^2e^{xy}, f_{xy}(x, y) = 3y^2e^{xy} + y^3xe^{xy}$ **51.** $f_{xx}(x, y) = -1.183x^{-1.35}y^{0.4},$

$f_{yy}(x, y) = -1.248x^{0.65}y^{-1.6}, f_{xy}(x, y) = 1.352x^{-0.35}y^{-0.6}$ **53.** $f_{xx}(x, y) = \dfrac{4y}{x^3}, f_{yy}(x, y) = 0, f_{xy}(x, y) = -\dfrac{2}{x^2}$ **55.** $f_{xx}(x, y) = y^3e^{xy},$

$f_{yy}(x, y) = 2xe^{xy} + yx^2e^{xy}, f_{xy}(x, y) = 2ye^{xy} + y^2xe^{xy}$ **57.** $x = -1, y = -2$ **59. (a)** $S_x(x, y) = 4x, S_y(x, y) = 1$

(b) $S_x(3, 5) = 12$; If \$3000 are spent each week on newspaper advertising and \$5000 are spent each week on radio advertising and the amount spent on radio advertising is kept fixed at \$5000 per week then sales will be increasing at a rate of

\$120,000 $\dfrac{\text{in sales per week}}{\text{thousands of dollars spent on newspaper advertising per week}}$.; $S_y(3, 5) = 1$; If \$3000 are spent each week on newspaper

advertising and \$5000 are spent each week on radio advertising and the amount spent on newspaper advertising is kept fixed at \$3000 then

sales will be increasing at a rate of \$10,000 $\dfrac{\text{in sales per week}}{\text{thousands of dollars spent on radio advertising per week}}$.

61. (a) $TS_x(x, y) = 3x, TS_y(x, y) = 6.4y$ **(b)** $TS_x(1, 0.5) = 3$; If \$1000 are spent each week on radio advertising and \$500 are spent

each week on television advertising and the amount spent on television advertising is kept fixed at \$500 per week then sales will be

increasing at a rate of \$30,000 $\dfrac{\text{in sales per week}}{\text{thousands of dollars spent on radio advertising per week}}$.; $TS_y(1, 0.5) = 3.2$; If \$1000 are spent each week

on radio advertising and \$500 are spent each week on television advertising and the amount spent on radio advertising is kept fixed at

\$1000 per week then sales will be increasing at a rate of \$32,000 $\dfrac{\text{in sales per week}}{\text{thousands of dollars spent on television advertising per week}}$.

63. The rate of change of the total amount accumulated when the point in time is held fixed and the amount of the initial investment is

increased is given by $f_P(P, t) = e^{0.06t}$.; The rate of change of the total amount accumulated when the amount of the initial investment is

held fixed and time is increasing is given by $f_t(P, t) = 0.06Pe^{0.06t} \dfrac{\text{dollars}}{\text{year}}$. **65. (a)** $V_r = \frac{2}{3}\pi rh, V_h = \frac{1}{3}\pi r^2$

(b) $V(3, 5) = 15\pi$; The volume of a cone with radius of 3 and height of 5 is 15π.; $V_r(3, 5) = 10\pi$; For a cone with radius of 3 units and height of 5 units, if the radius increases by 1 unit while the height remains constant at 5 units, the volume will increase by approximately 10π units.; $V_h(3, 5) = 3\pi$; For a cone with radius of 3 units and height of 5 units, if the height increases by 1 unit while the radius remains constant at 3 units, the volume will increase by approximately 3π units.   **67.** $C_w(6, 8.2) \approx 12.2$; For a head with width of 6 inches and length of 8.2 inches, if the width of the head increases by 1 inch while the length remains constant at 8.2 inches, the cephalic index will increase by approximately 12.2 units.; $C_L(6, 8.2) \approx -8.92$; For a head with width of 6 inches and length of 8.2 inches, if the length of the head increases by 1 inch while the width remains constant at 6 inches, the cephalic index will decrease by approximately 8.92 units.

69. (a) $R(x, y) = 350x - 4x^2 + 3xy + 450y - 3y^2$ **(b)** The marginal revenue from an increase in sales of racing bikes is given by $350 - 8x + 3y \dfrac{\text{dollars}}{\text{bikes}}$; The marginal revenue from an increase in sales of mountain bikes is given by $3x + 450 - 6y \dfrac{\text{dollars}}{\text{bikes}}$.

(c) $R(15, 20) = 13{,}050$; With a weekly demand of 15 racing bikes and 20 mountain bikes, the revenue is \$13,050; $R_x(15, 20) = 290$; With a weekly demand of 15 racing bikes and 20 mountain bikes, the marginal revenue from an increase in sales of racing bikes when the demand for mountain bikes is held fixed at 20 is $290 \dfrac{\text{dollars}}{\text{bike}}$.   **71. (a)** Marginal revenue for 2D calculator is $R_x(x, y) = 70 + 0.5y - 0.08x$

(b) Marginal revenue for 3D calculator is $R_y(x, y) = 95 + 0.5x - 0.08y$ **73. (a)** $P(x, y) = 37x + 35.5y - 0.5x^2 - 0.8xy - y^2$

(b) The marginal profit from an increase in sales of two-stroke engines when the sales of four-stroke engines is held fixed is given by $37 - x - 0.8y \dfrac{\text{thousands of dollars}}{\text{thousand two-stroke engine}}$; The marginal profit from an increase in sales of four-stroke engines when the sales of two-stroke engines is held fixed is given by $35.5 - 0.8x - 2y \dfrac{\text{thousands of dollars}}{\text{thousand two-stroke engine}}$.   **(c)** The profit from sales of 18 thousand two-stroke engines and 10 thousand four-stroke engines is 615 thousand dollars.; With sales of 18 thousand two-stroke engines and 10 thousand four-stroke engines, the marginal profit from an increase in sales of two-stroke engines when sales of four-stroke engines is fixed at 10 thousand is $11 \dfrac{\text{thousand dollars}}{\text{thousand engines}}$.   **(d)** With sales of 18 thousand two-stroke engines and 10 thousand four-stroke engines, the marginal profit from an increase in sales of four-stroke engines when sales of two-stroke engines is fixed at 18 thousand is $1.1 \dfrac{\text{thousand dollars}}{\text{thousand engines}}$.   **75. (a)** $f_x(x, y) = 128.8x^{-0.2}y^{0.25}, f_y(x, y) = 40.25x^{0.8}y^{-0.75}$ **(b)** If the company is now using 111 units of labor and 25 units of capital and keeps capital fixed at 25 units, production is increasing at the rate of $112.29 \dfrac{\text{calculators}}{\text{thousands of hours of labor}}$.; If the company is now using 111 units of labor and 25 units of capital and keeps labor fixed at 111 units, production is increasing at the rate of $155.8 \dfrac{\text{calculators}}{\text{capital equipment in millions}}$.   **77. (a)** $f_x(x, y) = 860.598x^{-0.4}y^{0.45}, f_y(x, y) = 645.4485x^{0.6}y^{-0.55}$

(b) If the company is now using 47 units of labor and 8 units of capital and keeps capital fixed at 8 units, production is increasing at the rate of $470.27 \dfrac{\text{engines}}{\text{thousands of hours of labor}}$.; If the company is now using 47 units of labor and 8 units of capital and keeps labor fixed at 47 units, production is increasing at the rate of $2072.1 \dfrac{\text{engines}}{\text{capital equipment in millions}}$.   **79. (a)** $f_x(x, y) = 33x^{-0.45}y^{0.5}, f_y(x, y) = 30x^{0.55}y^{-0.5}$

(b) If the company is now using 220 units of labor and 140 units of capital and keeps capital fixed at 140 units, production is increasing at the rate of $34.474 \dfrac{\text{motorcycles}}{\text{labor in millions of dollars}}$.; If the company is now using 220 units of labor and 140 units of capital and keeps labor fixed at 220 units, production is increasing at the rate of $49.248 \dfrac{\text{motorcycles}}{\text{capital in millions of dollars}}$.   **(c)** Production would increase more with increased spending on capital since the change in productivity with respect to capital is larger than the change in productivity with respect to labor.

Section 8.4 Exercises

1. $(2, -3)$ **3.** $(-1, 2)$ **5.** $(2, -1)$ **7.** $(1, 2)$ **9.** $(1, -4)$ **11.** The point $(3, 2, -10)$ is a relative minimum.
13. The point $(3, -2, -11)$ is a relative minimum.   **15.** The point $(-2, -\frac{2}{3}, \frac{13}{3})$ is a relative maximum.
17. The point $(-2, 4, -11)$ is a relative minimum.   **19.** The point $(3, 3, -7)$ is a relative minimum.
21. The point $(0, 1, 0)$ is a relative maximum and the point $(2, 1, -\frac{16}{3})$ is a saddle point.   **23.** The point $(1, -1, 20)$ is a saddle point and
the point $(1, 3, -44)$ is a relative minimum.   **25.** The point $(0, 0, 0)$ is a saddle point and the point $(1, 1, -1)$ is a relative minimum.
27. The point $(0, 0, 1)$ is a relative maximum.   **29.** The point $(-2, \frac{1}{2}, -18)$ is a relative maximum.   **31. (a)** 3 million on labor, 2 million
on robotics equipment   **(b)** 30 million dollars   **33. (a)** 1500 regular woks, 100 jumbo woks   **(b)** $14,999,800
35. (a) $P(x, y) = -x + 11y - x^2 + xy - 2y^2 - 2$ **(b)** 1.0 million bags of sour cream and chives and 3.0 million bags of barbecue
(c) 14.0 million dollars   **37. (a)** $R(x, y) = 349x - 4x^2 + 3xy + 446y - 3y^2$ **(b)** 88 touring bikes, 118 mountain bikes
(c) 41,744 dollars   **39. (a)** $R(x, y) = 228x - 8x^2 - 2xy + 31y - 0.15y^2$ **(b)** 8 leaf blowers, 50 8-foot attachments
41. 10 inches $\times$ 5 inches $\times$ 2.5 inches

Section 8.5 Exercises

1. $\frac{3}{4}$ **3.** 18   **5.** -16 **7.** $\frac{128}{9}$ **9.** 50   **11.** 108.86   **13.** 5   **15.** 4   **17.** e^4 **19. (a)** $120\,\dfrac{\text{regular models}}{\text{day}}, 60\,\dfrac{\text{joggers' models}}{\text{day}}$

(b) $2160.00   **21. (a)** $80\,\dfrac{\text{pepper spray devices}}{\text{day}}, 40\,\dfrac{\text{siren devices}}{\text{day}}$ **(b)** $960.00   **23. (a)** 39 units by the old line, 65 units by the new line
(b) $4982.00   **25. (a)** 857.1 labor hours per week, 102.9 units of capital investment per week   **(b)** 8268 golf carts per week
27. (a) 60.7 units of labor, 182.1 units of capital   **(b)** For each additional dollar value available to the budget, production increases by
about 7.5492×10^{-5} units.   **(c)** 15.1 units   **29. (a)** 245.3 weekly units of labor, 206.3 weekly units of capital
(b) For each additional dollar value available to the budget, production increases by about 2.3584×10^{-3} units.   **(c)** 117.9 units
31. (a) 360 units of labor, 160 units of capital   **(b)** For each additional dollar value available to the budget, production increases by
about 0.43379 units.   **(c)** 4338 units   **33.** $\frac{625}{2}$ feet by $\frac{625}{4}$ feet   **35.** 50 inches $\times$ 50 inches $\times$ 100 inches

Section 8.6 Exercises

1. $26x^2$ **3.** $26y^2$ **5.** $4x + 6$ **7.** $4 + 6y$ **9.** $\frac{7}{3}x^3 - 3x$ **11.** $\frac{15}{4}y^2 - 3y$

13. (a)

15. (a)

17. (a)

(b) $\displaystyle\int_0^2 \int_0^3 2xy\,dydx = 18$

(c) $\displaystyle\int_0^3 \int_0^2 2xy\,dxdy = 18$

(b) $\displaystyle\int_0^1 \int_0^2 (3x + y)\,dydx = 5$

(c) $\displaystyle\int_0^2 \int_0^1 (3x + y)\,dxdy = 5$

(b) $\displaystyle\int_1^{16} \int_1^4 \sqrt{xy}\,dydx = 196$

(c) $\displaystyle\int_1^4 \int_1^{16} \sqrt{xy}\,dxdy = 196$

19. (a)

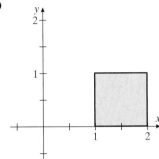

(b) $\int_{1}^{2}\int_{0}^{1} x^2 y^2 \, dy \, dx = \frac{7}{9}$

(c) $\int_{0}^{1}\int_{1}^{2} x^2 y^2 \, dx \, dy = \frac{7}{9}$

21. (a)

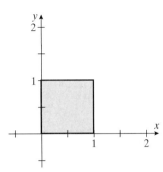

(b) $\int_{0}^{1}\int_{0}^{1} (2 - \frac{1}{2}x^2 + y^2) \, dy \, dx = \frac{13}{6}$

(c) $\int_{0}^{1}\int_{0}^{1} (2 - \frac{1}{2}x^2 + y^2) \, dx \, dy = \frac{13}{6}$

23.

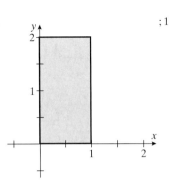

; 1

25.

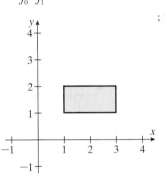

; 16   **27.**

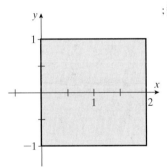

; $\frac{8}{3}e - \frac{8}{3}e^{-1}$ **29.** $\frac{2}{3}$ **31.** $\frac{25}{3}$ **33.** 2   **35.** $\frac{4}{3}e - \frac{4}{3}$

37. $\frac{44}{3}$ **39.** 6   **41.** 3   **43.** $\frac{8}{3}e - \frac{8}{3}$

45. 13918.6   **47.** 11,370   **49.** 36

51. 5959.6 $\frac{\text{people}}{\text{mile}^2}$

53. 148,989

55. $\frac{325}{3}$ parts per million

57. 125.78

Chapter 8 Review

1. $-\frac{6}{11}$ **3.** 103   **5.** All ordered pairs (x, y) such that $x + y \geq 0$

7.

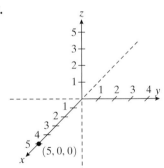

9.

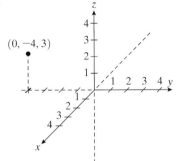

11. Above   **15.**

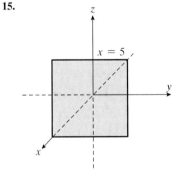

13. Below

17. With sales 4000 $\frac{\text{cups of coffee}}{\text{month}}$ and 3000 $\frac{\text{cups of cappuccino}}{\text{month}}$ there are monthly profits of 800 dollars.

19. (a) $C = 7x + 13y + 5000$ **(b)** $R = 12x + 21y$ **(c)** $P = 5x + 8y - 5000$ **21.** If 4500 dollars is invested at an annual rate of 4.5% compounded continuously for 10 years, the total amount accumulated is 7057.40 dollars.

23.

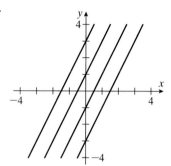

25.

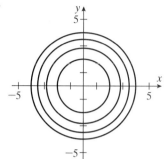

27.

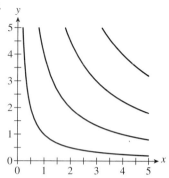

29.

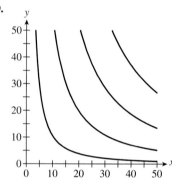

31.

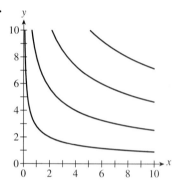

33.

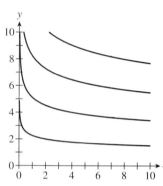

35.

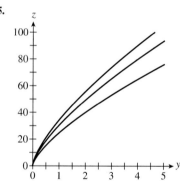

37.

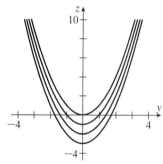

39.

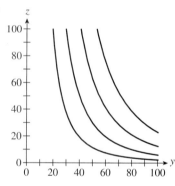

41. (a) Vertical

(b)

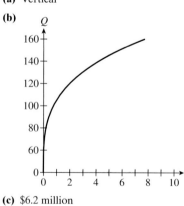

(c) $6.2 million

43. $f_x(x, y) = 5, f_y(x, y) = -6y$

45. $f_x(x, y) = -15(8y^2 - 3x)^4, f_y(x, y) = 80(8y^2 - 3x)^4 y$

47. $f_x(x, y) = -\dfrac{y}{3x^2}, f_y(x, y) = \dfrac{1}{3x}$

49. $f_x(x, y) = e^{x-y} + 15x^2 y, f_y(x, y) = -e^{x-y} + 5x^3$

51. If we are at the point $(3, 5)$ on the surface and move along the surface in the y-direction parallel to the y-axis, the function is changing at a rate of

$$-1750 \, \frac{\text{units of } f}{\text{units of } y}.$$

53. If we are at the point $(3, 2)$ on the surface and move along the surface in the y-direction parallel to the y-axis, the function is changing at a rate of about $9.29 \, \dfrac{\text{units of } f}{\text{units of } y}$.   **55.** $f_{xx}(x, y) = -24x, f_{xy}(x, y) = 10 + 5y^4, f_{yy}(x, y) = 20xy^3$

57. $f_{xx}(x, y) = 2, f_{xy}(x, y) = 4, f_{yy}(x, y) = 8$ **59.** $x = \frac{1}{52}, y = \frac{4}{13}$ **61. (a)** $R = 480.0x - 0.6x^2 + 0.5xy + 720.0y - 0.7y^2$

(b) The marginal revenue from an increase in sales of 15-inch monitors is given by $480.0 - 1.2x + 0.5y \dfrac{\text{dollars}}{\text{monitor}}$.; The marginal revenue

from an increase in sales of 17-inch monitors is given by $0.5x + 720.0 - 1.4y \dfrac{\text{dollars}}{\text{monitor}}$.   **(c)** When 640 15-inch monitors and

850 17-inch monitors have been sold the company receives $137.00 for the 641st 15-inch monitor sold.; When 640 15-inch monitors and
850 17-inch monitors have been sold the company loses $150.00 for the 851st 17-inch monitor sold.

63. (a) $P = -3.5x^2 - 18.5xy + 1.2y^2 + 110.0x + 130y - 600$ **(b)** The marginal profit from an increase in sales of regular VCR's is

given by $-7.0x - 18.5y + 110.0 \dfrac{\text{hundreds of dollars}}{\text{hundreds of VCR's}}$.; The marginal profit from an increase is sales of stereo VCR's is given by

$-18.5x + 2.4y + 130.0 \dfrac{\text{hundreds of dollars}}{\text{hundreds of VCR's}}$.   **(c)** With sales of 8 hundred regular VCR's and 11 hundred stereo VCR's, the marginal

profit from an increase in sales of regular VCR's is $-149.5 \dfrac{\text{hundreds of dollars}}{\text{hundreds of VCR's}}$.; With sales of 8 hundred regular VCR's and

11 hundred stereo VCR's, the marginal profit from an increase in sales of stereo VCR's is $8.40 \dfrac{\text{hundreds of dollars}}{\text{hundreds of VCR's}}$.

(d) The profit from sales of 8 hundred regular VCR's and 11 hundred stereo VCR's is 320 dollars.   **65.** $(-4, 3)$ **67.** $\left(-\frac{4}{3}, \frac{2}{3}\right)$

69. The point $(1, -4, -24)$ is a relative minimum.   **71.** The point $(-1, -1, -1)$ is a saddle point.

73. The point $(2, -4, -8)$ is a relative minimum and the point $(0, 0, 0)$ is a saddle point.   **75.** The point $(0, 2, e^4)$ is a saddle point.

77. (a) $R(x, y) = 12.0x - 0.3x^2 + 0.2xy + 20.0y - 0.2y^2$ **(b)** 44 basic haircuts, 72 deluxe haircuts   **(c)** $984.00

79. (a) 280,000 fine-tip pens, 440,000 large-tip pens   **(b)** $375,600   **81.** 125   **83.** -5 **85.** 22.888   **87. (a)** about 4.8 units of labor,
about 2.0 units of capital   **(b)** about 94,692 telephones   **89.** 10 feet by 30 feet   **91.** $258y^3$ **93.** $16x^2 - 1280$

95. (a)

97. (a)

99. 200

101. $\frac{20}{3}$

103. 195

105. 392.8

107. 104.667 parts per million

(b) $\displaystyle\int_0^5 \int_0^2 (3x - 4xy)\,dy\,dx = -25$

(c) $\displaystyle\int_0^2 \int_0^5 (3x - 4xy)\,dx\,dy = -25$

(b) 208

APPENDIX A   Essentials of Algebra

Appendix A Exercises

1. $-4x^{-6}$ **3.** $1.6x^{1/2}$ **5.** $8x^{-2/3}$ **7.** $\sqrt[7]{x^3}$ **9.** $6.3\sqrt[5]{x^4}$ **11.** $\dfrac{2}{\sqrt[3]{x^2}}$ **13.** $x = 4, y = 6$ **15.** $x = 5, y = \frac{3}{2}$

17. $x = \frac{3}{4}, y = \frac{1}{2}, z = \frac{1}{3}$ **19.** $x = -2, y = 5$ **21.** $\log_b 8 + 3\log_b x - \log_b 125 - 3\log_b y$ **23.** $\log x + 2\log y - \log(y + 4)$

25. $\frac{1}{2}\ln 5 + \frac{3}{2}\ln x - \frac{9}{2}\ln y$ **27.** $x = \frac{26}{3}$ **29.** $x = 4$ **31.** $x = \dfrac{\ln 6}{\ln 3} \approx 1.63$ **33.** $x = \frac{1}{5}$

Index